Digital Communications

Digital Communications

Ian Glover
UNIVERSITY OF BRADFORD

Peter Grant
UNIVERSITY OF EDINBURGH

PRENTICE HALL
London New York Toronto Sydney Tokyo Singapore
Madrid Mexico City Munich Paris

First published 1998 by
Prentice Hall Europe
Campus 400, Maylands Avenue
Hemel Hempstead
Hertfordshire, HP2 7EZ
A division of
Simon & Schuster International Group

Printed and bound in Great Britain by
Hartnolls Ltd, Bodmin, Cornwall

Library of Congress Cataloging-in-Publication Data

Glover, Ian.
    Digital communications / Ian Glover and Peter Grant.
        p.   cm.
    Includes bibliographical references and index.
    ISBN 0-13-565391-6
    1. Digital communications.   I. Grant, Peter M.   II. Title.
TK5103.7.G56   1997
621.382--dc20   96-26711
                                                            CIP

British Library Cataloguing-in-Publication Data

A catalogue record for this book is available
from the British Library

ISBN 0-13-565391-6

1 2 3 4 5   02 01 00 99 98

Contents

2. Part Two   Digital communications principles, 167

5 Sampling, multiplexing and PCM, 169

Preface

Digital communications is a rapidly advancing applications area. The most significant current activities are in the development of mobile communications equipment for personal use, in the expansion of the available bandwidth (and hence information carrying capacity) of the backbone transmission structure through developments in optical fibre, and in the ubiquitous use of networks for data communications.

The aim of this book is fourfold: (1) to present the mathematical theory of signals and systems as required to understand modern digital communications equipment and techniques, (2) to apply and extend these concepts to information transmission links which are robust in the presence of noise and other impairment mechanisms, (3) to show how such transmission links are used in fixed and mobile data communication systems for voice and video transmission, and (4) to introduce the operating principles of modern communications networks formed by the interconnection of many transmission links using a variety of topological structures.

The material is set in an appropriate historical context. Most of the chapters include substantive numerical examples to illustrate the material developed and conclude with problem questions which have been designed to help readers assess their comprehension of this material.

In Chapter 1, we summarise the history of communication systems and introduce some basic concepts such as accessing, modulation, multiplexing, coding and switching, for line and radio transmission. Chapter 1 also includes a review of the advantages of digital communication systems over the older analogue systems which they are now, largely, replacing.

The next 18 chapters are organised in four parts reflecting the four aims referred to above. Specifically Chapters 2 through 4 are devoted to a basic theory of periodic, transient and random signals and the concept of linear transmission systems. Chapters 5 through 13 cover the fundamentals of digital communications and include sampling and multiplexing, baseband line transmission, decision and information theory, and error control coding. This second part also includes a description of the many bandpass modulation schemes used in modern systems, the calculation of received power and associated signal-to-noise ratio for a communications link, and an indication of how the performance of a system can be assessed by simulation, before any actual hardware construction is attempted.

Part 3, Chapters 14 through 16, describes how the principles of digital communications are applied in fixed point-to-point terrestial, and satellite based, microwave systems, in mobile and cellular radio systems, and in video (TV) transmission and storage systems. The fourth part, Chapters 17 through 19, is devoted to communication networks. This starts with a discussion of queuing theory before describing network topology and access techniques, and concludes by examining public networks and ISDN, the internationally agreed standard for the worldwide digital telecommunications network.

To assist the reader, the book includes a list of abbreviations and also a list of notations and conventions used for the mathematical material.

An extensive Reference list and Bibliography is provided at the end of the book, before the Index. All publications referred to in the text are compiled in this list. Each reference is identified in the text by the name(s) of the author(s) and, where necessary, the year of publication in square brackets.

The book is aimed at readers who are completing a graduate level BEng/MEng degree, or starting a postgraduate level MSc degree in Communications, Electronics or Electrical Engineering. It is assumed that these readers will have competence in the mathematical concepts required to handle comfortably the material in Part One.

The book has been compiled from lecture notes associated with final year BEng/MEng/MSc core, and optional, courses in signal theory and digital communications as provided at the Universities of Bradford and Edinburgh in the 1990s. We have deliberately extended our coverage, however, to include some practical aspects of the implementation of digital PCM, SDH, packet speech systems, and the capability of optical and microwave long-haul communication systems. With this balance between theory, applications and systems implementation we hope that this text will be useful both in academia and in the rapidly growing communications industry.

To aid the instructor and the student we provide outline solutions to the majority of the end of chapter problems on the worldwide web at the Edinburgh server address: http://www.ee.ed.ac.uk/~pmg/DIGICOMMS/index.html

Bradford and Edinburgh   Ian Glover and Peter Grant
April 1997

Acknowledgements

Parts of this book have been developed from BEng, MEng and MSc courses provided at the Universities of Edinburgh and Bradford. Three of these courses were first taught by Dr James Dripps at Edinburgh, and Professor Peter Watson and Dr Neil McEwan at Bradford, and we acknowlegde their initial shaping of these courses which is reflected in the book's content and structure. We are grateful to Dr Dripps for having provided draft versions of Chapters 7 and 9 and also for giving us access to material which now forms parts of Chapters 6, 10, 17 and 18. We are grateful to Dr McEwan providing the original versions of sections 2.5.1, 4.3.1, 4.3.2 and 4.3.3 in the form of his teaching notes. Some of the material in Chapters 2, 3, 4, 8 and 11 had its origins in notes taken during lectures delivered at Bradford by Professor Watson and Dr McEwan. We also acknowledge Dr Brian Flynn for assistance with parts of Chapter 19, Dr Angus McLachlan for providing initial thoughts on Chapter 12, Dr Tom Crawford (of Hewlett Packard, Telecomms Division, South Queensferry) for giving us access to further material for Chapter 19 and providing some initial insights into Chapter 6. We are grateful to Dr David Parish of Loughborough University of Technology, for providing an initial draft of Chapter 16, Professor Paddy Farrell (of Victoria University, Manchester) for helpful comments on Chapter 10 and Dr David Cruickshank at Edinburgh for assistance with the problem solutions which are provided on the WWW.

We would like to thank all those colleagues at the Universities of Bradford and Edinburgh who have provided detailed comments on sections of this text. Thanks must also go to the many students who have read and commented on earlier versions of this material, helped to refine the end of chapter problems and particularly Yoo-Sok Saw and Paul Antoszczyszn who generously provided figure material for Chapter 16.

Special thanks are due to Joan Burton, Liz Paterson, Diane Armstrong and Beverley Thomas for their perseverance over several years in typing the many versions of the individual chapters, as they have evolved from initial thoughts into their current form. We also acknowledge Bruce Hassall's generous assistance with the preparation of the final version of the text in the appropriate typefont and text format.

Finally we must thank our respective families, Nandini and Sonia, and Marjory, Lindsay and Jenny for the considerable time that we required to write this book.

Ian Glover and Peter Grant

Abbreviations/symbols

AC   Alternating current (implying sinusoidal signal)
ACF   Autocorrelation function
A/D or ADC   Analogue to digital converter
ADM   Add and drop multiplexor; adaptive delta modulation
ADPCM   Adaptive differential pulse code modulation
AGC   Automatic gain control
ALOHA   (not an abbreviation but Hawaiian for 'hello')
AM   Amplitude modulation
AMI   Alternate mark inversion
AMPS   Advanced mobile phone system (USA)
ANSI   American National Standards Institute
APD   Avalanche photodiode
APK   Amplitude/phase keying
ARPANET   Advanced Research Projects Agency Network
ARQ   Automatic repeat request
ASCII   American Standard Code for Information Interchange
ASK   Amplitude shift keying
ATM   Asynchronous transfer mode

BASK   Binary amplitude shift keying
BCH   Bose–Chaudhuri–Hocquenghem
BER   Bit error ratio/rate
BFSK   Binary frequency shift keying
BMV   Branch metric value
$BO_{i/o}$ Back-off (input/output)
BPSK   Binary phase shift keying
BRZ   Bipolar return to zero
BSS   Broadcast satellite service
BT   British Telecom

CCITT   Comité Consultatif International Télégraphique et Téléphonique
CCIR   Comité Consultatif International des Radiocommunications

CD	Cumulative distribution, compact disc
CDMA	Code division multiple access
CD-ROM	Compact disc read-only memory
CELP	Codebook of excited linear prediction
CEPT	Confederation of European PTT Administrations
CFMSK	Continuous frequency minimum shift keying
CIR	Carrier to interference ratio
CMI	Coded mark inversion
CMOS	Complementary metal oxide silicon (transistor)
CNR	Carrier-to-noise ratio
CODEC	Coder/decoder
CPD	Centre point detection
CP(S)M	Continuous phase (shift) modulation
CRC	Cyclic redundancy check
CRT	Cathode-ray tube
CSDN	Circuit-switched data network
CSMA/CD	Carrier sense multiple access/collision detection
CW	Continuous wave
DA	Demand assigned
DAC	Digital to analogue converter
DBS	Direct broadcast satellite
DC	Direct current
D/C	Downconverter
DCE	Data communication equipment
DCT	Discrete cosine transform
DECT	initially Digital European cordless telecommunications now Digital enhanced cordless telecommunications
DEPSK	Differentially encoded phase shift keying
DFB	Distributed feedback (laser)
DFS	Discrete Fourier series
DFT	Discrete Fourier transform
DM	Delta modulation
DMPSK	Differential M-symbol phase shift keying
DPCM	Differential pulse code modulation
DPSK	Differential phase shift keying
DSB	Double sideband
DSI	Digital speech interpolation
DSMX	Digital system multiplexor
DSP	Digital signal processing
DTE	Data terminal equipment
DTI	Department of Trade and Industry (UK)

ECMA	European Computer Manufacturers Association
EDFA	Erbium doped fibre amplifier
EFTPOS	Electronic funds transfer at point of sale
EIRP	Effective isotropic radiated power
EMI	Electromagnetic interference
ERMES	European Radio Message System
ESD	Energy spectral density
ETSI	European Telecommunications Standards Institute (formerly CEPT)
FDDI	Fibre distributed data interface
FDM	Frequency division multiplex
FDMA	Frequency division multiple access
FECC	Forward error correction coding
FEXT	Far end crosstalk
FFSK	Fast frequency shift keying
FFT	Fast Fourier transform
FH	Frequency hopped
FIFO	First in first out
FILO	First in last out
FIR	Finite impulse response
FIRO	First in random out
FM	Frequency modulation
FPLMTS	Future public land mobile telecommunications system
FS	Fourier series
FSK	Frequency shift keying
FSPL	Free space path loss
FT	Fourier transform
FZ	Fresnel zone
GMSK	Gaussian (filtered) minimum shift keying
GPS	Global positioning system
GSM	Groupe Spéciale Mobile (or Global System for Mobile communications)
HACE	Higher-order automatic cross-connect equipment
HDB	High density bipolar
HDSL	High-speed digital subscriber loop
HDTV	High definition television
HEO	High earth orbit
HF	High frequency
HIHE	Highly inclined highly elliptical (orbit)
HPA	High power amplifier
HSLAN	High speed LAN

I	Inphase (signal component)
I+D	Integrate and dump
IEEE	Institute of Electrical and Electronics Engineers
IF	Intermediate frequency
IFA	Intermediate frequency amplifier
ILD	Injection laser diode
INMARSAT	International Maritime Satellite Consortium
INTELSAT	International Telecommunications Satellite Consortium
IP	Intermodulation product
ISDN	Integrated services digital network
ISI	Inter-symbol interference
ISO	International Standards Organisation
ITU	International Telecommunication Union
JANET	Joint Academic Network
JPEG	Joint Photographic Experts Group
LAN	Local area network
LCFS	Last come first served
LED	Light emitting diode
LEO	Low earth orbit satellite
LIFO	Last in first out
LNA	Low noise amplifier
LO	Local oscillator
LOS	Line of sight
LPC	Linear predictive coding
LPF	Lowpass filter
LW	Long wave
LZW	Lempel-Ziv coding
MAC	Medium access control
MAN	Metropolitan area network
MAP	Manufacturers application protocol; maximum a posteriori criterion
MASK	M-symbol amplitude shift keying
MBC	Model based coding
MCPC	Multiple channels per carrier
MFSK	Multiple frequency shift keying
MMF	Multi-mode fibre
MODEM	Modulator/demodulator
MOS	mean opinion score (for speech quality assessment); metal oxide silicon (transistor)
MPEG	Motion Picture Experts Group
MPSK	M-symbol phase shift keying

MQAM	*M*-symbol quadrature amplitude modulation
MSK	Minimum shift keying
MTBF	Mean time between failure
MW	Medium wave
NA	Not applicable
NASA	National Aeronautics and Space Administration
NATO	North Atlantic Treaty Organisation
NEXT	Near end crosstalk
NPSD	Noise power spectral density
NRZ	Non-return to zero
NSF	National Science Foundation (USA)
NT	Network termination
NTSC	National Television Standards Committee (USA)
OAM&P	Operations, administration, maintenance and provisioning
OFDM	Orthogonal frequency division multiplex
OOK	On–off keying
OQPSK	Offset quadrature phase shift keying
OSI	Open Systems Interconnection
PA	Preassigned
PABX	Private automatic branch exchange
PAD	Packet assembly and disassembly
PAL	Phase alternate line (TV)
PAM	Pulse amplitude modulation
PC	Personal computer
PCM	Pulse code modulation
PCN	Personal communications network
pdf	Probability density function
PDH	Plesiochronous digital hierarchy
PDN	Public data network
PEPL	Plane earth path loss
PIN	Positive–intrinsic–negative (diode)
PLL	Phase locked loop
PM	Phase modulation
PMR	Private mobile radio
PMV	Path metric value
PN	Pseudo-noise
POCSAG	Post Office Code Standards Advisory Group
PON	Passive optical network
POTS	Plain old telephone system
PPM	Pulse position modulation

PRBS	Pseudo-random bit sequence
PRK	Phase reversal keying
PSD	Power spectral density
PSDN	Packet switched data network
PSK	Phase shift keying
PSS	Packet Switched Service
PSTN	Public switched telephone network
PTT	Post, telephone and telegraph
PWM	Pulse width modulation
Q	Quadrature (signal component)
QA	Quasi-analytic
QAM	Quadrature amplitude modulation
QPR	Quadrative partial response
QPSK	Quadrature phase shift keying
RA	Random access
RACE	Research in Advanced Communications in Europe
RF(I)	Radio frequency (interference)
RGB	Red green blue
RMS	Root mean square
RV	Random variable
RX	Receive
RZ	Return to zero
SBC	Sub-band coder
SCPC	Single channel per carrier
SCSI	Small computer system interface
SDH	Synchronous digital hierarchy
SDR	Signal to distortion ratio
SECAM	Système en couleurs à mémoire
SER	Symbol error rate
SERC	Science and Engineering Research Council
SMF	Single mode fibre
SNR	Signal-to-noise ratio
SN_qR	Signal to quantisation noise ratio
SONET	Synchronous optical network
SPE	Synchronous payload envelope
SSB	Single sideband
SSMA	Spread spectrum multiple access
STM	Synchronous transfer mode; synchronous transport module
STR	Symbol timing recovery
STS	Synchronous transfer structure; synchronous transport signal

TA	Terminal adaptor
TACS	Total access communication system (AMPS derivative)
TCM	Trellis coded modulation
TDM	Time division multiplex
TDMA	Time division multiple access
TE	Terminal equipment
TEM	Transverse electromagnetic
TILS	Time invariant linear system
TV	Television
TWSLA	Travelling wave semiconductor laser amplifier
TWT	Travelling wave tube
TX	Transmit
U/C	Upconverter
UHF	Ultra high frequency
UMTS	Universal mobile telecommunication service
VAN	Value added network
VBR	Variable bit rate
VDU	Video display unit
VHF	Very high frequency
VLSI	Very large scale integrated (circuit)
VSAT	Very small aperture satellite terminal
VSB	Vestigial sideband
VT	Virtual tributary
WAN	Wide area network
WDM	Wavelength division multiplex
WRAC	World Radio Administrative Conference
WWW	World Wide Web
XTR	Crosstalk ratio

Principal symbols

a	core radius in optical fibre
a_e	antenna effective area; effective earth radius
A	A-law PCM compander constant; total path attenuation
A_e	effective aperture area of antenna
A_n	nth real Fourier coefficient
$\mathbf{B}$	magnetic flux density
B	signal bandwidth in Hz
B_n	equivalent noise bandwidth; nth imaginary (real valued) Fourier coefficient
$\mathbf{c}$	codeword vector
C	channel capacity; received carrier power level; constant
$C(m\|v)$	conditional cost in interpreting v as symbol m
$\bar{C}$	average cost of decisions
C_i	ith codeword; cost associated with decision error
$C_n; \tilde{C}_n$	nth complex Fourier coefficient
C_∞	channel capacity in infinite channel bandwidth
$^N C_J$	number of combinations of J digits from N-digit block
$\mathbf{d}$	message vector
$\mathbf{D}$	electric flux density
D	deterministic (impulsive) pdf; antenna diameter
D_{ij}	Hamming distance between codewords i and j
$D_{\min}$	minimum Hamming distance
$D(\theta, \phi)$	antenna directivity
$\mathbf{e}$	error vector
e	number of detectable errors in codeword; water vapour partial pressure
$\mathbf{E}$	electric field strength
E	equivocation; energy; field strength
E_b	energy per bit
E_e	filter output error energy arising from DM bit error
$E_k(f)$	energy spectral density with $k = 1, 2$ for 1, 2 sided spectra
E_{si}	normalised symbol i energy
$\mathrm{erf}(x)$	error function
$\mathrm{erfc}(x)$	complementary error function
f	noise factor (ratio)

f	noise factor (ratio)
Δf	peak frequency deviation
f_b	bit rate
f_c	centre frequency
f_χ	filter cut-off frequency
f_H	highest frequency component
f_L	lowest frequency component
f_{LO}	local oscillator frequency
f_o	symbol clock frequency
f_s	sample frequency
f_{3dB}	half power bandwidth
$f(kT_o)$	decision circuit input voltage at sampling instant
$f(t)$	IF transmitted symbol; generalised function
F	noise figure (dB)
F_N	speech formant N frequency component
$F(v)$	cumulative distribution function for v
$F(x)$	nonlinear PCM companding characteristic
$g(t)$	baseband information signal
$g(kT_s)$	sampled version of $g(t)$
$\tilde{g}(kT_s)$	estimate of $g(t)$
$\hat{g}(kT_s)$	prediction of $g(t)$
G	generator matrix
G	amplifier gain
$G_k(f)$	power spectral density with $k = 1, 2$ for $1, 2$ sided spectra
G_l	(negative or fractional) gain of a lossy device
G_p	processing gain
$G_p(f)$	power spectral density of $p(t)$
G_R	receiver antenna gain
G_T	transmitter antenna gain
$G(kT_s)$	adaptive gain of ADM amplifier
$G(\theta, \phi)$	antenna directional gain
G/T	gain to system noise temperature ratio
G.n	G-series (recommendation n) for telephony multiplex
h_B	height of earth's bulge above chord
$h(t)$	impulse response
H	parity check matrix; magnetic field strength
H	entropy
$H_{eff;max}$	effective; maximum entropy
$H(f)$	frequency response

i_m	length of codeword m
I_m	information content of message m
I_n	nth information bit in codeword
$I_o(z)$	modified Bessel function of first kind and order zero
$I(\theta, \phi)$	radiation intensity from an antenna
$\Im$	imaginary part of
k	vector wavenumber
k	Boltzmann's constant; number of information digits in a codeword; earth profile k factor; scalar wave number
l_m	length of codeword m
L	loss in making incorrect decision; link path length; average codeword length
m	mean value; number of hops; modulation index
m_i	ith bit of codeword
M	number of symbols in an alphabet or levels in PCM system
M	Markov (Poisson) pdf
n	number of bits in a codeword; constraint length in convolutional coder; refractive index; mode number; type of semiconductor material
$n(t)$	noise signal
N	average queue length; noise power
N_e	noise power due to DM bit errors
N_0	noise power spectral density
$N(f)$	voltage spectrum of noise
p	type of semiconductor material
$p(v)$	probability density function of variable v
$p(v|m)$	conditional probability density function of v given m
P	power; pressure
P_b	probability of bit error
P_D	probability of detection
P_e	probability of symbol error
P_{em}	probability of error for symbol m
P_{FA}	probability of false alarm
$P_j(t)$	probability of being in state j at time t
P_m	probability of a queue having length m
P_T	transmitted power
P_1	first parity bit in codeword; probability of being in state 1
$P(j, i)$	joint probability of selecting (symbol) i followed by j

$P(m|v)$ a posteriori probability that symbol m was transmitted, given voltage v detected

q quantisation step size
Q quality factor; f_H/B

r   received codeword data vector
r radius
R range; efficiency (rate) of a digital code; redundancy
R' number of errors in codeword
R_b bit rate
$R_{P;L}$ point; line rain rate
R_r radiation resistance
R_s symbol rate
$R_x(\tau)$ correlation function
RC resistor-capacitor time constant τ_c
$\Re$ real part of

s   syndrome codeword vector
s standard deviation
S_n sum of geometric progression
$S_{(peak)}$ average (peak) signal power
S/N signal-to-noise ratio
sgn (x) signum function

t time; number of correctable errors in codeword
t_g guard time
T   transition (coding) matrix
T average delay in a queue; temperature
T_A antenna aperture temperature
T_B brightness temperature
T_D symbol duration
T_e effective noise temperature
T_o symbol period
T_{ph} physical temperature
T_q noise temperature of q
T_s sample period
T_0 reference temperature (290 K)

$v_i(t)$ voltage waveform of symbol i
$v_n(t)$ noise voltage
$v_N(t)$ Nyquist symbol ISI free pulse

v_{th}	receiver decision threshold voltage
$v(t)$	information signal
V	bipolar violation pulses
$V(f)$	voltage spectral density
$V_N(f)$	voltage spectrum of ISI free pulse
V.n	V-series recomendation n for data modems
W	power density
$\mathbf{x}$	vector x
x_{med}	median value of x
$X(f)$	voltage spectrum of $x(t)$
$\|X(f)\|$	amplitude spectrum
X.n	X-series recommendation n for data networks
α	peak to mean power ratio; attenuation factor; normalised excess bandwidth; earth radius
δ	Dirac delta
$\delta(t)$	impulse function
Δ	stepsize
Δf	excess bandwidth
ΔV	voltage difference
ε	error voltage
ε_q	quantisation error
$\overline{\varepsilon_{de}^2}$	mean square (PCM) decoding error
$\overline{\varepsilon_q^2}$	mean square quantisation error
η	efficiency
η_s	spectral efficiency
η_Ω	antenna ohmic efficiency
λ	free space wavelength; arrival rate
μ	μ-law PCM compander constant; service rate
ρ	normalised correlation coefficient; utilisation factor (λ/μ)
σ	Gaussian standard deviation
σ_g	standard deviation of $g(t)$

τ time between successive arrivals; pulse width

τ_c *RC* time constant

χ^2 chi-square distribution

Special functions

$[\,,]$ scalar product
$\langle\,\rangle$ time average
*   convolution operation
a^* complex conjugate of a

$$\Pi(t) = \begin{cases} 1, & |t| < 0.5 \\ \tfrac{1}{2}, & t = 0.5 \\ 0, & |t| > 0.5 \end{cases}$$

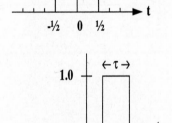

$$\Pi\left(\frac{t - T}{\tau}\right) = \begin{cases} 1, & |t - T| < \tau/2 \\ \tfrac{1}{2}, & |t - T| = \tau/2 \\ 0, & |t - T| > \tau/2 \end{cases}$$

$$\Lambda(t) = \begin{cases} 1 - |t|, & |t| \leq 1.0 \\ 0, & |t| > 1.0 \end{cases}$$

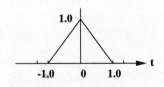

$$\Lambda\left(\frac{t}{\tau}\right) = \begin{cases} 1 - \dfrac{|t|}{\tau}, & |t| \le \tau \\ 0, & |t| > \tau \end{cases}$$

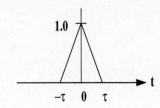

$$\text{sgn}(t) = \begin{cases} 1.0, & t > 0 \\ 0, & t = 0 \\ -1.0, & t < 0 \end{cases}$$

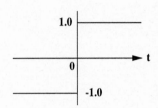

$$u(t) = \begin{cases} 1.0, & t > 0 \\ \tfrac{1}{2}, & t = 0 \\ 0, & t < 0 \end{cases}$$

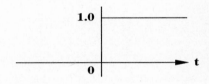

CHAPTER 1

Digital communications overview

1.1 Electronic communications

History, present requirements and future demands

Communication can be defined as the imparting or exchange of information [Hanks]. Telecommunication, which is more narrowly the topic of this book, refers to communication over a distance greater than would normally be possible without artificial aids. In the present day such aids are invariably electrical, electronic or optical and communication takes place by passing signals over wires, through optical fibres or through space using electromagnetic waves.

Modern living demands that we have access to a reliable, economical and efficient means of communication. We use communication systems, particularly the public switched telephone network (PSTN), to contact people all around the world. Telephony is an example of point-to-point communication and normally involves a two-way flow of information. Another type of communication, which (traditionally) involves only one-way information flow, is broadcast radio and television. In these systems information is transmitted from one location but is received at many locations using many independent receivers. This is an example of point-to-multipoint communication.

Communication systems are now very widely applied. Navigation systems, for example, pass signals between a transmitter and a receiver in order to determine the location of a vehicle, or to guide and control its movement. Signalling systems for tracked vehicles, such as trains, are also simple communication systems.

All early forms of communication system (e.g. smoke signals, semaphore, etc.) used digital traffic. The earliest form of electronic communications, telegraphy, was developed in the 1830s, Table 1.1. It was also digital in that the signals, transmitted over wires, were restricted to four types: dots and dashes, representing the Morse coded letters of the alphabet, letter spaces and word spaces. In the 1870s Alexander Graham Bell made analogue communications possible by inventing acoustic transducers to convert speech directly into (analogue) electrical signals.

Table 1.1 *Important events in the history of electronic communications.*

Year	Event	Originator	Information
1837	Line telegraphy perfected	Morse	Digital
1875	Telephone invented	Bell	Analogue
1897	Automatic exchange step by step switch	Strowger	
1901	Wireless telegraphy	Marconi	Digital
1905	Wireless telephony demonstrated	Fessenden	Analogue
1907	First regular radio broadcasts	USA	Analogue
1918	Superheterodyne radio receiver invented	Armstrong	Analogue
1921	First use of land based PMR	Detroit police	Analogue
1928	All electronic television demonstrated	Farnsworth	Analogue
1928	Telegraphy signal transmission theory	Nyquist	Digital
1928	Information transmission	Hartley	Digital
1931	Teletype		Digital
1933	FM demonstrated	Armstrong	Analogue
1934	Radar demonstrated	Kuhnold	
1937	PCM proposed	Reeves	Digital
1939	Commercial TV broadcasting	BBC	Analogue
1943	Matched filtering proposed	North	Digital
1945	Geostationary satellite proposed	Clarke	
1946	ARQ systems developed	Duuren	Digital
1948	Mathematical theory of communications	Shannon	
1955	Terrestrial microwave relay	RCA	Analogue
1960	First laser demonstrated	Maiman	
1962	Satellite communications implemented	TELSTAR 1	Analogue
1963	Geostationary satellite communications	SYNCOM II	Analogue
1966	Optical fibres proposed	Kao & Hockman	
1966	Packet switching		Digital
1970	Medium scale data networks	ARPA/TYMNET	Digital
1970	LANs, WANs and MANs		Digital
1971	The term ISDN coined	CCITT	Digital
1974	Internet concept	Cerf & Kahn	Digital
1978	Cellular radio		Analogue
1978	Navstar GPS launched	Global	Digital
1980	OSI 7 layer reference model adopted	ISO	Digital
1981	HDTV demonstrated	NHK, Japan	Digital
1985	ISDN basic rate access in UK	BT	Digital
1986	SONET/SDH introduced	USA	Digital
1991	GSM cellular system	Europe	Digital
1993	PCN concept launched	Worldwide	Digital
1994	IS-95 CDMA specification	Qualcom	Digital

This led quickly to the development of conventional telephony. Radio communications started around the turn of the century when Marconi patented the first wireless telegraphy system. This was quickly followed by the first demonstration of wireless (or radio) telephony and in 1918 Armstrong invented the superheterodyne radio receiver which is still an important component of much modern day radio receiving equipment. In the 1930s Reeves proposed pulse code modulation (PCM) which laid the foundation for nearly all present-day digital communication systems.

Table 1.1 shows some of the principal events in the development of electronic communications over the last century and a half. The second world war saw rapid, forced, developments in nearly all areas of engineering and technology. Electronics and communications benefited greatly and the new, but associated, discipline of radar became properly established.

In 1945 Arthur C. Clarke wrote his famous article proposing geostationary satellite communications and 1963 saw the launch of the first successful satellite of this type. In 1966 optical fibre communication was proposed by Kao and Hockman and, around the same time, public telegraph and telephone (PTT) operators introduced digital carrier systems.

The first, general purpose, large scale data networks (ARPANET and TYMNET) were developed around 1970, provoking serious commercial interest in packet switching (as an alternative to circuit switching).

The 1970s saw significant improvements in the performance of, and large increases in the volume of traffic carried by, telecommunications systems of all types. Optical fibre losses were dramatically reduced and the capacity of satellites dramatically increased. In the 1980s first analogue, and then digital, cellular radio became an important part of the PSTN. Micro-cellular and personal communications using both terrestrial and satellite based radio technology are now being developed. It seems likely that before long wideband personal communications systems providing voice, data and perhaps even video services, will become possible. Video delivery will require a broadband rather than a narrow (speech) bandwidth connection, Table 1.2.

Increasing demand for traditional services (principally analogue voice communications) has been an important factor in the development of telecommunications technologies. Such developments, combined with more general advances in electronics and computing, have made possible the provision of entirely new (mainly digitally based) communications services. This in turn has stimulated demand still further. Figure 1.1 shows the past and predicted future growth of telecommunications traffic and Figure 1.2 shows the proliferation of services which have been, or are likely to be, offered over the same period.

In telecommunications there are various standards bodies which ensure interoperability of equipment. The International Telecommunications Union (ITU) is an important international communications standards body which only has the power to make recommendations for specifications. Within the ITU are the PTTs (post, telephone and telegraph organisations) from individual nations, e.g. British Telecom and Deutsche Bundespost. In Europe there was, until recently, the Confederation of European PTTs (CEPT), responsible for overseeing the actual implementation of technical standards. CEPT has now been replaced with the European Telecommunications Standards Institute (ETSI) [Temple].

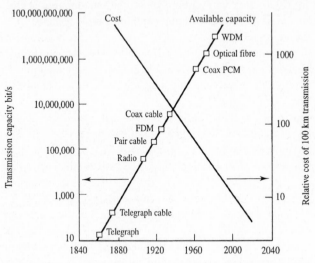

Figure 1.1 *Past and predicted growth of telecommunications traffic (source: Technical demographics, 1995, reproduced with permission of the IEE).*

1.2 Sources and sinks of information

Sources of information can be either natural or man-made. An example of the former might be the air temperature at a given location. An example of the latter might be a set of company accounts. (A third example, speech, falls, in some sense, into both categories.) Digital communication systems represent information, irrespective of its type or origin, by a discrete set of allowed symbols. It is this alphabet of symbols and the device or mechanism which selects them for transmission that, in this context, is usually regarded as the information source. The amount of information conveyed by each symbol, as it is selected and transmitted, is closely related to its selection probability. Symbols likely to be selected more often convey less information than those which are less likely to be selected. Information content (measured in bits) is thus related to symbol rarity.

Sinks of information are, ultimately, people although various types of information storage and display devices (computer disks, magnetic tapes, loudspeakers, VDUs etc.) are usually involved as a penultimate destination.

Transmitters are the devices that impress source information onto an electrical wave (or carrier) appropriate to a particular transmission medium (e.g. optical fibre, cable, free space). Receivers are the devices which extract information from such carriers. They often also reproduce this information in the same form as it was originally generated (e.g. as speech).

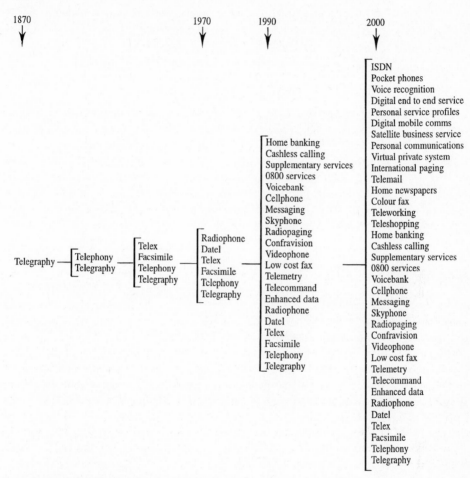

Figure 1.2 *Service proliferation in telecommunications (source: Earnshaw, 1991, reproduced with permission of Peter Peregrinus).*

1.3 Digital communications equipment

An important objective in the design of a communications system is often to minimise equipment cost, complexity and power consumption whilst also minimising the bandwidth occupied by the signal and/or transmission time. (Bandwidth is a measure of how rapidly the information-bearing part of a signal can change and is therefore an important parameter for communication system design. Table 1.2 compares the nominal bandwidth of three common types of information signal.) Efficient use of bandwidth and transmission time ensures that as many subscribers as possible can be accommodated within the constraints of these limited, and therefore valuable, resources.

Table 1.2 *Comparison of nominal bandwidths for several information signals.*

Information signal	Bandwidth
Speech telephony	4 kHz
High quality sound broadcast	15 kHz
TV broadcast (video)	6 MHz

The component parts of a hypothetical digital communications transceiver (transmitter/receiver) are shown in Figure 1.3. Much of the rest of this book is concerned with the operating principles, performance and limitations of a communication system formed by a transmitter/receiver pair linked by a communications channel. Here, however, we give a qualitative overview of such a system, incorporating a brief account of what each block in Figure 1.3 does and why it might be required. (The transceiver in this figure has been chosen to include all the elements commonly encountered in digital communications systems. Not all transceivers will employ all of these elements of course.)

1.3.1 CODECs

At its simplest a transceiver CODEC (coder/decoder) consists of an analogue to digital converter (ADC) in the transmitter, which converts a continuous, analogue, signal into a sequence of code words represented by binary voltage pulses, and a digital to analogue converter (DAC) in the receiver, which converts these voltage pulses back into a continuous, analogue, signal.

The ADC consists of a sampling circuit, a quantiser and a pulse code modulator (Figure 1.3). The sampling circuit provides discrete voltage samples taken, at regular intervals of time, from the analogue signal. The quantiser approximates these voltages by the nearest level from an allowed set of voltage levels. (It is the quantisation process

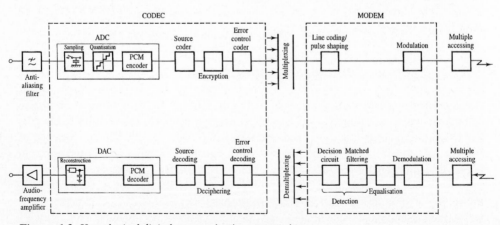

Figure 1.3 *Hypothetical digital communications transceiver.*

which converts the analogue signal to a digital one.) The PCM encoder converts each quantised level to a binary code word, digital ones and zeros each being represented by one of two voltages. An anti-aliasing filter is sometimes included prior to sampling in order to reduce distortion which can occur as a result of the sampling process.

In the receiver's DAC received binary voltage pulses are converted to quantised voltage levels by a PCM decoder which is then smoothed by a low-pass filter to reconstruct (at least a good approximation to) the original, analogue, signal.

Digitisation of analogue signals usually increases the signal's transmission bandwidth but it permits reception at a lower signal-to-noise ratio than would otherwise be the case. This is an example of how one resource (bandwidth) can be traded off against another resource (transmitter power).

CODECs make widespread use of sophisticated digital signal processing techniques to encode efficiently the signal prior to transmission and also to decode the received signals when they are corrupted by noise, distortion and interference. This increases transceiver complexity, but allows higher fidelity, repeatable, almost error-free transmission to be achieved.

1.3.2 Source, security and error control coding

In addition to PCM encoding and decoding a CODEC may have up to three additional functions. Firstly (in the transmitter) it may reduce the number of binary digits (called bits, or sometimes binits) required to convey a given message. This is source coding and can be thought of as effectively removing redundant (i.e. unrequired or surplus) digits. Secondly it may encrypt the source coded digits using a cipher for security. This can yield both privacy (which assures the sender that only those entitled to the information being transmitted can receive it) and authentication (which assures the receiver that the sender is who he/she claims to be). Finally the CODEC may add extra digits to the (possibly source coded and/or encrypted) PCM signal which can be used at the receiver to detect, and possibly correct, errors made during symbol detection. This is error control coding and has the effect of incorporating binary digits at the transmitter which, from an information point of view, are redundant.

In some ways error control coding, which adds redundancy to the bit stream, is the opposite of source coding, which removes redundancy. Both processes may be employed in the same system, however, since the type of redundancy which occurs naturally in the information being transmitted is not necessarily the type best suited to detecting and correcting errors at the receiver.

The source, security and error control, decoding operations in the receiver, Figure 1.3, are the inverse of those in the transmitter.

1.3.3 Multiplexers

In digital communications, multiplexing, to accommodate several simultaneous transmissions usually means, more specifically, time division multiplexing (TDM). Time division multiplexers interleave either PCM code words, or individual PCM binary digits, to allow more than one information link to share the same physical transmission medium (e.g. cable, optical fibre, radio frequency channel). If communication is to occur in real time this implies that the bit rate of the multiplexed signal is at least N times that of each of the N tributary PCM signals and this in turn implies an increased bandwidth requirement.

Demultiplexers split the received composite bit stream back into its component PCM signals.

1.3.4 MODEMs

MODEMs (modulators/demodulators) condition binary pulse streams so that the information they contain can be transmitted over a given physical medium, at a given rate, with an acceptable degree of distortion, in a specified or allocated frequency band. The modulator in the transmitter may change the voltage levels representing individual, or groups of, binary digits. Typically the modulator also shapes, or otherwise filters, the resulting pulses to restrict their bandwidth, and shifts the entire transmission to a convenient, allowed, frequency band. The input to a modulator is thus a baseband digital signal whilst the output is often a bandpass waveform.

The demodulator, in a receiver, reconverts the received waveform into a baseband signal. Equalisation corrects (as far as possible) signal distortion which may have occurred during transmission. Detection converts the demodulated baseband signal into a binary symbol stream. The matched filter, shown as one component of the detector in Figure 1.3, represents one type of signal processing which can be employed, prior to the final digital decision process, in order to improve error performance.

1.3.5 Multiple accessing

Multiple accessing refers to those techniques, and/or rules, which allow more than one transceiver pair to share a common transmission medium (e.g. one optical fibre, one satellite transponder or one piece of coaxial cable). Several different types of multiple accessing are currently in use, each type having its own advantages and disadvantages. The multiple accessing problem is essentially one of efficient and (in some sense) equitable sharing of the limited resource represented by the transmission medium.

1.4 Radio receivers

Many radio receivers (both digital and analogue) incorporate superheterodyning as part of their demodulation process. In these receivers (Figure 1.4) the incoming radio frequency (RF) signal, with carrier frequency f_{RF}, is mixed (i.e. multiplied) with the signal from a local oscillator (LO) of frequency f_{LO}. The sum $(f_{RF} + f_{LO})$ and difference $(f_{RF} - f_{LO})$ frequency products which appear at the mixer output are then filtered to select only the latter which is called the intermediate frequency or IF. The LO frequency is, therefore, always altered or tuned to ensure that the receiver operates with a fixed value of IF (i.e. $f_{RF} - f_{LO}$) irrespective of which RF channel is being received. This allows a considerable effort to be invested in the design of the receiver beyond this point, consisting typically of high gain (fixed frequency) IF amplifiers and high selectivity filters followed by an appropriate IF signal demodulator and/or detector. The superheterodyne receiver can be made more sophisticated by using double frequency conversion in which there are two mixing stages. This enables higher gain and greater selectivity to be achieved in order to increase rejection of unwanted, interfering, signals.

The principal problem with the superheterodyne design is that the receiver is equally sensitive to radio frequency bands centred on $f_{LO} + f_{IF}$ (which is the wanted band) and $f_{LO} - f_{IF}$ (which is an unwanted band). The unwanted 'image' band of frequencies, separated from the wanted RF band by twice the IF frequency, represents a potentially serious source of RF interference and additional noise. A tunable image rejection filter (needing only modest selectivity) can be placed before the mixer in the RF amplifier of Figure 1.4 to attenuate or remove this unwanted band of frequencies.

1.5 Signal transmission

The communications path from transmitter to receiver may use lines or free space. Examples of the former are wire pairs, coaxial cables and optical fibres. The most important use of the latter is radio, although in some situations infrared and optical free space links are also possible (e.g. remote controls for TV, video and hi-fi equipment and also some security systems). Whatever the transmission medium, it is at this point that much of the attenuation, distortion, interference and noise is encountered. Attenuation can be compensated for by introducing amplifiers or signal repeaters at intermediate points along the multiple hop link, Figure 1.5. Distortion may be compensated by equalisers and interference and noise can be minimised by using appropriate predetection signal processing (e.g. matched filters). The nature and severity of transmission medium effects is one of the major influences on the design of transmitters, receivers and repeaters.

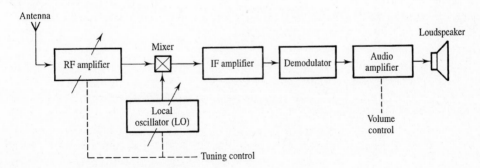

Figure 1.4 *Superheterodyne receiver.*

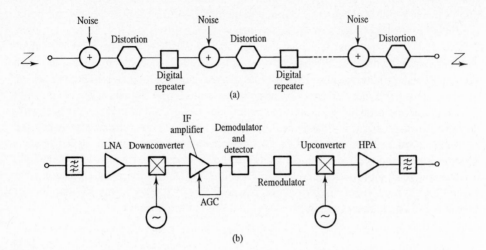

Figure 1.5 *(a) Digital communications (multi-hop) channel; (b) digital repeater (as typically used in terrestrial microwave relay applications).*

1.5.1 Line transmission

The essential advantages of line transmission are:

1. Path loss is usually modest.
2. Signal energy is essentially confined and interference between different systems is seldom severe and often negligible.

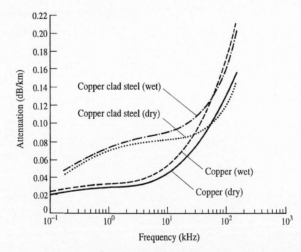

Figure 1.6(a) *Typical attenuation/frequency characteristics for aerial open wire pair lines.*

3. Path characteristics (e.g. attenuation and distortion) are usually stable and relatively easy to compensate for.
4. Capacity is unlimited in that bandwidth can always be reused by laying another line.

The disadvantages of line transmission are:

1. Laying cables in the ground or constructing overhead lines is generally expensive.
2. Extensive wayleaves and planning permission may be needed for underground cables and overhead wires.
3. Broadcasting requires a physical connection to a complex network for each subscriber.
4. Mobile communications services cannot be provided.
5. Networks cannot easily be added to, subtracted from, or otherwise reconfigured.

The degree to which a signal is attenuated by a transmission line depends on the material from which the line is made, its physical construction, and the signal's frequency. Figures 1.6(a) to (e) show some typical attenuation/frequency characteristics for the most common types of line. Open wire has particularly low loss but it is expensive to maintain and susceptible to interference. Loaded cable, Figure 1.6(c) and Table 1.3, is only effective for speech bandwidth signals. Twisted pairs, as used underground, have higher installation cost but lower maintenance costs. (Low loss, circular, waveguides can also be used as a transmission medium but advances in optical fibre technology have, at least for the present, made this technology essentially redundant.) Optical fibre cables have an enormous information carrying capacity with typical bandwidth-distance products of 0.5 GHz-km.

Table 1.3 *Nominal properties of selected transmission lines.*

	Frequency range	Typical attenuation	Typical delay	Repeater spacing
Open wire (overhead line)	0 – 160 kHz	0.03 dB/km @ 1 kHz	3.5 μs/km	40 km
Twisted pairs (multi-pair cables)	0 – 1 MHz	0.7 dB/km @ 1 kHz	5 μs/km	2 km
Twisted pairs (with L loading)	0 – 3.5 kHz	0.2 dB/km @ 1 kHz	50 μs/km	2 km (L spacing)
Coaxial cables	0 – 500 MHz	7 dB/km @ 10 MHz	4 μs/km	1 – 9 km
Optical fibres	1610 – 810 nm	0.2 to 0.5 dB/km	5 μs/km	40 km

Table 1.3 summarises the nominal frequency range of each type of line, their typical attenuations and transmission delays, and typical repeater spacings. The useful bandwidths of the lines, which determine the maximum information transmission rate they can carry, are often, but not always, determined by their attenuation characteristics. Twisted wire pairs, for example, are normally limited to (line coded PCM) data rates of 2 Mbit/s. Coaxial cables, Figure 1.6(d), routinely carry 140 Mbit/s PCM signals but can

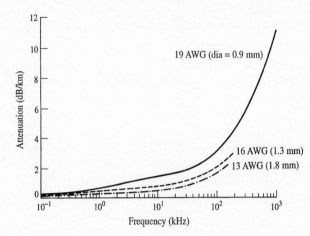

Figure 1.6(b) *Typical characteristics for twisted pair cable transmission lines.*

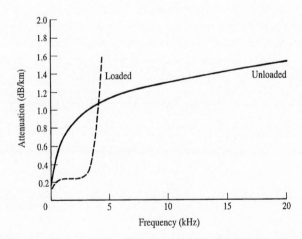

Figure 1.6(c) *Comparison between inductively loaded and unloaded twisted wire pairs.*

handle symbol rates several times greater. Optical fibres have very large bandwidth potential but may be limited to a fraction of this by factors such as the spectral characteristics of optical sources and dispersion effects. Nevertheless, optical fibre PCM bit rates of Gbit/s are possible.

1.5.2 Radio transmission

The advantages of radio transmission are:

1. It is relatively cheap and quick to implement.

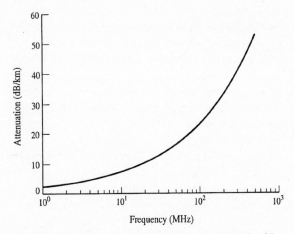

Figure 1.6(d) *Typical attenuation/frequency characteristic for coaxial cable.*

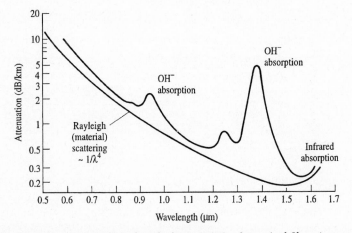

Figure 1.6(e) *Typical attenuation/wavelength characteristics for optical fibres (source: Young, 1994, reproduced with permission of Prentice Hall).*

2. Wayleaves and planning permission are often only needed for the erection of towers to support repeaters and terminal stations.
3. It has an inherent broadcast potential.
4. It has an inherent mobile communications potential.
5. Communications networks can be quickly reconfigured and extra terminals or nodes easily introduced or removed.

The principal disadvantages of radio are:

1. Path loss is generally large due to the tendency of the transmitted signal energy to spread out, most of this energy effectively missing the receive antenna.

2. The spreading of signal energy makes interference between different systems a potentially serious problem.
3. Capacity in a given locality is limited since bandwidth cannot be reused easily.
4. Path characteristics (i.e. attenuation and distortion) tend to vary with time, often in an unpredictable way, making equalisation more difficult and limiting reliability and availability.
5. The time varying nature of the channel can result in anomalous propagation of signals to locations well outside their normal range. This may cause unexpected interference between widely spaced systems.
6. Points 2 and 5 mean that frequency coordination is generally required when planning radio systems. Such coordination is difficult to achieve comprehensively and is expensive.

Table 1.4 *Frequency bands commonly used for radio communication.*

Band	Frequency	Wavelength	Propagation mechanism	Fading process	Noise process	Range	Applications
ELF	30 – 300 Hz	$10^4 - 10^3$ km	Waveguide modes	Diurnal variations due to D-layer	Man-made and atmospherics (lightning discharges)	Worldwide	Submarine
VF	300 – 3000 Hz	1000 – 100 km					
VLF	3 – 30 kHz	100 – 10 km	Surface waves	None			Standards/navigation
LF	30 – 300 kHz	10 – 1 km				1000s km	Maritime mobile, LW b'cast
MF	300 – 3000 kHz	1000 – 100 m	Sky waves	Surface/sky wave intf.		100s km	MW broadcast
HF	3 – 30 MHz	100 – 10 m		Complex ionospherics		4000 km/hop	Amateur
VHF	30 – 300 MHz	10 – 1 m	Line of sight	None	Galactic (synchrotron radiation)	Line of sight	FM broadcast
UHF	300 – 3000 MHz	100 – 10 cm		Ray bending and multipath	Cosmic background		TV broadcast, mobile, LOS
SHF	3 – 30 GHz	10 – 1 cm					Microwave LOS, satellite
EHF	30 – 300 GHz	10 – 1 mm		Rain attenuation	Thermal noise from ground & atmosphere	Short	

The appropriate radio-propagation model for a communication system, the dominant fading and noise processes, and typical system range, all depend on frequency. Table 1.4 shows the electromagnetic spectrum used for radio transmissions and summarises these models and processes. At the lowest frequencies propagation is best modelled by oscillating electromagnetic modes which exist in the cavity between the concentric conducting spheres formed by the earth and its ionosphere. From a few kilohertz up to a few hundred kilohertz vertically polarised radio energy will propagate (by diffraction) around the curved surface of the earth for thousands of kilometers. This is called surface wave propagation and is the mechanism by which long wave radio broadcasts are received.

At slightly higher frequencies in the medium frequency (MF) band some radio energy propagates as a surface wave and some is reflected from the conducting ionosphere as a sky wave. The relative path lengths and phasing of these two signals may result in destructive interference causing fading of the received signal which will vary in severity as the relative strengths and phases of the sky wave and surface wave change. The exact condition of the ionosphere may be critical in this respect, making the quality of signal reception vary, for example, with the time of day or night.

In the high frequency (HF) band the sky wave is usually dominant and ranges of thousands of kilometers are possible, sometimes involving multiple reflections between the ionosphere and ground. At very high frequencies (VHF), and above, signals propagate essentially along line-of-sight paths although reflection, refraction and, at the lower frequencies, diffraction can play an important role in the overall characteristics of the channel. At ultra high frequencies (UHF) currently used both for TV transmissions and cellular radio communications, multipath (i.e. multiple path) propagation caused by reflections from, and diffraction around, buildings and other obstacles in urban areas is the principal cause of signal fading. In the super high frequency (SHF) band (usually called the microwave or centimetric wave band) applications tend to be point-to-point (or fixed-point) communications and the first order fading problem is often due to rain induced attenuation. Extra high frequencies (EHF) and higher are not yet widely used for communication systems, partly due to the significant gaseous background attenuation and large fades which occur in rain. As the electromagnetic spectrum becomes more congested, however, and as the demand for communications becomes yet greater (in terms of both traffic volume and service sophistication) the use of these higher frequency bands will almost certainly become both necessary and economic.

1.6 Switching and networks

Many modern communication systems are concerned exclusively with data traffic. One example is the Internet, over which users can transmit e-mail messages or browse distant information sources and transfer large data and text files. The data networks themselves, and the configuration of computer terminals on a user site, can be organised in many different ways using ring, star or bus connections. In order to ensure interoperability, standards have developed for these topologies and also for the signalling and switching protocols which control the assembly and routing of traffic.

The seven layer ISO model is used throughout data communications networks as the standard hierarchical structure for organising data traffic. The data itself is usually sent as fixed length packets with associated overhead bits which provide addresses, timing or ordering information, and assist in error detection. The physical data interfaces follow evolved standards, e.g. X.25, IEEE 802, which develop progressively to higher data rates as the new high speed (wideband optical) transmission systems are introduced.

With packet data traffic there are inevitable delays while the packets are queued for access to the transmission system. These queues are not serious problems in simple mail networks but, if attempting to transmit speech or video traffic in real time, queue delays, and lost packets due to queue overflow in finite length buffers, can seriously degrade the operation of the communications link.

1.7 Advantages of digital communications

Digital communication systems usually represent an increase in complexity over the equivalent analogue systems. We therefore list here some of the reasons why digital communication has become the preferred option for most new systems and, in many instances, has replaced existing analogue systems.

1. Increased demand for data transmission.
2. Increased scale of integration, sophistication and reliability of digital electronics for signal processing, combined with decreased cost.
3. Facility to source code for data compression.
4. Possibility of channel coding (line, and error control, coding) to minimise the effects of noise and interference.
5. Ease with which bandwidth, power and time can be traded off in order to optimise the use of these limited resources.
6. Standardisation of signals, irrespective of their type, origin, or the services they support, leading to an integrated services digital network (ISDN).

The increase in demand, for voice and data connections, is the principal driving force behind the growth in telecommunications. The traffic, in the backbone network, expressed as equivalent voice circuits, is shown in Figure 1.7. This figure does not merely reflect the explosive growth in mobile communications for the final customer connections but shows world capacity for transmission.

1.8 Summary

The history of electronic communications over the last century and a half has demonstrated an essentially exponential growth in traffic and a continuously increasing demand for greater access to ever more sophisticated services. This trend shows no sign, at present, of changing.

Most modern telecommunications systems are digital and use some form of PCM irrespective of the origin of the information they convey. PCM signals are often coded themselves to improve system performance and/or provide security. Many PCM signals can be combined as a single (time division) multiplex to allow their simultaneous transmission over a single physical medium. Line coding and/or modulation can then be used to match the characteristics of the resulting multiplex to the transmission line or radio channel being used. Multiple accessing techniques allow many transceiver pairs to share a given transmission resource (e.g. cable, fibre, satellite transponder). Switching allows telecommunications networks to be designed which, at reasonable cost, can emulate a fully interconnected set of transceivers.

It is the purpose of this book to describe the operating principles and performance of modern digital communications systems. The description is presented at a systems, rather than a circuit, level and, in view of this, Part One of the book (Chapters 2 to 4) reviews some pertinent mathematical models and properties of signals, noise and

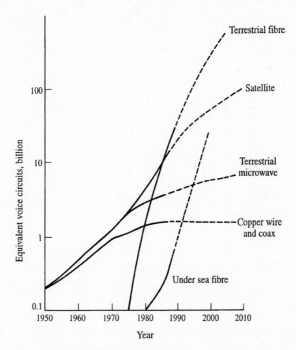

Figure 1.7 *Growth in world transmission capacity (source: Cochrane, 1990, reproduced with permission of British Telecommunications plc.).*

systems. Part Two (Chapters 5 to 13) describes the analogue to digital conversion process, coding, and modulation techniques used to ensure adequate performance of a wide range of digital communications systems (Chapters 5 to 11); Chapter 12 is concerned with physical aspects of noise and the prediction of CNR at the end of a single or multi-hop transmission link; Chapter 13 discusses the computer simulation of communications systems. Part Three (Chapters 14 to 16) discusses modern digital telephony, terrestrial and satellite microwave systems, mobile cellular radio and video coding systems. Part Four (Chapters 17 to 19) describes switching and telecommunications networks including queuing theory and packet data transmission. It also includes a discussion of the current plesiochronous digital telephone network and the evolving synchronous digital hierarchy (SDH) for both telephony and data traffic.

Part One

Signals and systems theory

Signals and systems theory is the body of knowledge related to the definition and description of signals, and the behaviour of systems. In electrical engineering the study of signals is central to telecommunications, whilst the study of systems is probably more closely identified with control. It is obvious, however, that control engineers must be concerned with the signals which form the inputs and outputs of their systems, and conversely, communications engineers must be concerned with the systems which transmit, receive and otherwise process their signals. Nevertheless the closeness of the relationship between signal theory and communications means that the material presented, in Part 1 is biased in favour of signals.

Chapter 2 presents the principal mathematical tools (Fourier series and Fourier transforms) normally used to describe, analyse, and synthesise, waveforms and transient signals. A unifying theme, here, is that of determinism, i.e. the waveforms and signals addressed all allow descriptions which permit their values to be determined, precisely, at any point in time. The choice of Fourier analysis as a technique for splitting complicated signals into their simpler (sinusoidal) component parts leads to the important concepts of spectrum and bandwidth. Much of the communication engineer's effort is directed at conserving spectrum and utilising available signal bandwidth efficiently.

Chapter 2 also introduces ideas of signal orthogonality and correlation which relate to common sense notions of similarity. These concepts are important in communications since only signals which are in some way dissimilar can be assigned different meanings. In digital communications, especially, it is usually a requirement to generate signals which are easily distinguishable.

Chapter 3 deals with random signals (i.e. those which are not deterministic and are thus excluded from Chapter 2). Random signals are important, partly because information cannot be communicated by deterministic signals, and partly because unwanted random signals (constituting noise) always exist in a communications receiver. Such noise has the potential to modify, or obscure, wanted, information bearing signals. Due to their unpredictable nature random signals must be described in terms of their statistical properties. Chapter 3 therefore reviews probability theory and defines the mean, variance, covariance and other statistics, which are used to summarise the behaviour of random signals and noise. The similarity of a signal with a time shifted version of itself determines how rapidly the signal can change with time and provides information about the signal's spectrum. Chapter 3 makes the precise connection between self similarity (or autocorrelation) functions and the Fourier based spectral descriptions presented in Chapter 2.

Chapter 4 is concerned with systems, and in particular the effect that linear systems have on the spectral and autocorrelation properties of signals. The importance, and defining characteristics, of linear systems are discussed and use is made of the Fourier transform to link their equivalent time domain (impulse response) and frequency domain (frequency response) descriptions. The ways in which impulse and frequency responses are used to predict the effect of a system on both deterministic and random signals are thus developed.

The importance of systems to the communications engineer lies in the fact that signals conveying information must be processed many times by subsystems (filters, modulators, amplifiers, equalisers etc.) before they reach their final destination. It is only through a thorough understanding of the modifying effect of these subsystems that one can ensure, in the presence of noise, that signals will remain adequately distinguished to achieve message reception without error.

CHAPTER 2

Periodic and transient signals

2.1 Introduction

Signals and waveforms are central to communications. A *signal* is defined [Hanks] as 'any sign, gesture, token, etc., that serves to communicate information'. It will be shown later that to communicate information such symbols must be in some sense unpredictable or random. The word signal, as applied to electronic communications, therefore implies an electrical quantity (e.g. voltage) possessing some characteristic (e.g. amplitude) which varies unpredictably. A *waveform* is defined as 'the shape of a wave or oscillation obtained by plotting the value of some changing quantity against time'. In electronic communications the term waveform implies an electrical quantity which varies *periodically*, and therefore predictably. Strictly this precludes a waveform from conveying information. However, a waveform can be adapted to convey information by varying one or more of its parameters in sympathy with a signal. Such waveforms are called carriers and typically consist of a sinusoid or pulse train modulated in amplitude, phase or frequency.

Fluctuating voltages and currents can be alternatively classified as either periodic or aperiodic. A periodic signal, if shifted by an appropriate time interval, is unchanged. An aperiodic signal does not possess this property. In this context the term periodic signal is clearly synonymous with waveform. In this chapter our principal concern is with periodic signals and one type of aperiodic signal, i.e. transients. A transient signal is one which has a well defined location in time. This does not necessarily mean it must be zero outside a certain time interval but it does imply that the signal at least tends to zero as time tends to $\pm\infty$. The one sided decaying exponential function is an example of a transient signal which has a well defined start and tends to zero as $t \to \infty$.

If a signal's parameters (amplitude, shape and phase in the case of a periodic signal, amplitude, shape and location in the case of a transient signal) are known, then the signal is said to be deterministic. This means that, in the absence of noise, any future value of the signal can be determined precisely. Signals which are not deterministic must be described using probability theory, as discussed in Chapter 3.

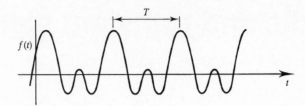

Figure 2.1 *Example of a periodic signal.*

2.2 Periodic signals

A periodic signal is defined as one which has the property:

$$f(t) = f(t \pm nT) \tag{2.1}$$

where n is any integer and T is the repetition period (or simply period) of the signal, Figure 2.1. A consequence of this definition is that periodic signals have no starting time or finishing time, i.e. they are eternal. The normalised power, P, averaged over any T second period, is:

$$P = \frac{1}{T} \int_{t}^{t+T} |f(t)|^2 \, dt \quad (\mathrm{V}^2) \tag{2.2}$$

where the integral is the normalised energy per period. This is clearly a well defined finite quantity. The total energy, E, in a periodic signal, however, is infinite, i.e.:

$$E = \int_{-\infty}^{\infty} |f(t)|^2 \, dt = \infty \quad (\mathrm{V}^2 \, \mathrm{s}) \tag{2.3}$$

For this reason periodic signals (along with some other types of signal) are sometimes called *power* signals. It also means that signals which are strictly periodic are unrealisable. The concept of a strictly periodic signal is, however, both simple and useful. Furthermore it is easy to generate signals which approximate very closely the conceptual ideal.

2.2.1 Sinusoids, cisoids and phasors

An especially simple and useful set of periodic signals is the set of sinusoids. These are generated naturally by projecting a point P, located on the circumference of a rotating disc (with unit radius), onto various planes, Figure 2.2.

If the length of OA in Figure 2.2 is plotted against angular position θ, then the result is the function $\cos \theta$, Figure 2.3(a). If the length of OB is plotted against θ, then the result is $\sin \theta$, Figure 2.3(b). (If the length of $O'C$ on the plane tangent to the disc is plotted against θ then the function $\tan \theta$ results.) If the disc is not of unit radius then the normal (circular) trigonometric ratios are defined by:

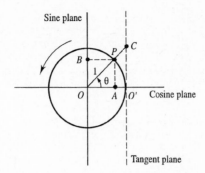

Figure 2.2 *Generation of sinusoids by projection of a radius onto perpendicular planes.*

$$\cos \theta = \frac{OA}{OO'} \qquad\qquad (2.4(a))$$

$$\sin \theta = \frac{OB}{OO'} \qquad\qquad (2.4(b))$$

The angle θ, expressed in degrees or radians, is called the phase of the function and can be related to the time period, T, taken for one revolution, i.e.:

$$\theta = 360 \, \frac{t}{T} \quad \text{degrees} \qquad\qquad (2.5(a))$$

$$\theta = 2\pi \, \frac{t}{T} \quad \text{radians} \qquad\qquad (2.5(b))$$

The angular velocity (or radian frequency), $\omega = d\theta/dt$, of the disc is therefore given by:

$$\omega = \frac{2\pi}{T} \quad \text{rad/s} \qquad\qquad (2.6)$$

and angular position or phase by:

$$\theta = \omega t \quad \text{rad} \qquad\qquad (2.7)$$

$1/T$ is the cyclical frequency of the disc in cycles/s or Hz. The sine and cosine functions plotted against time, t, are shown in Figure 2.4. The functions $\cos \theta$ and $\sin \theta$ are identical in shape but $\cos \theta$ reaches its peak value $T/4$ seconds (i.e. $\pi/2$ radians or 90°) *before* $\sin \theta$. $\cos \theta$ is therefore said to *lead* $\sin \theta$ by $\pi/2$ radians and, conversely, $\sin \theta$ is said to *lag* $\cos \theta$ by $\pi/2$ radians. The relationship between cosine and sine functions can be summarised by:

$$\cos \theta = \sin(\theta + \pi/2) \qquad\qquad (2.8)$$

Notice that the cosine function and sine function have even and odd symmetry respectively about $t = 0$, i.e.:

$$\cos \theta = \cos(-\theta) \qquad\qquad (2.9(a))$$

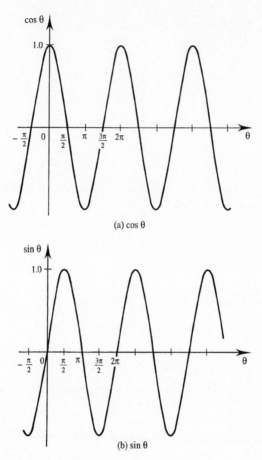

Figure 2.3 *Circular trigonometric functions plotted against phase: (a) cosine function of phase angle; (b) sine function of phase angle.*

$$\sin \theta = -\sin(-\theta) \tag{2.9(b)}$$

A cisoid is a general term which describes a rotating vector in the complex plane. Figure 2.5 shows a cisoid (which makes an angle ϕ with the plane's real axis at time $t = 0$) resolved onto real and imaginary axes. From the definition of the circular trigonometric functions it is clear that the component resolved onto the real axis is:

$$\Re[\text{cisoid}] = \cos(\omega t + \phi) \tag{2.10(a)}$$

and the component resolved onto the imaginary axis is:

$$\Im[\text{cisoid}] = \sin(\omega t + \phi) \tag{2.10(b)}$$

Using Euler's formula (which relates geometrical and algebraic quantities) the real and imaginary components can be expressed together as:

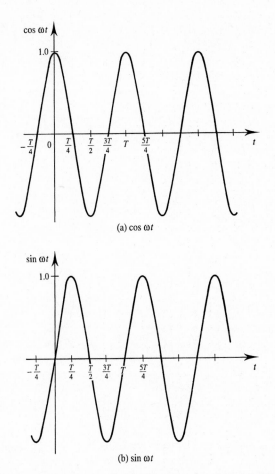

Figure 2.4 *Circular trigonometric functions plotted against time: (a) cosine function of time; (b) sine function of time.*

$$\cos(\omega t + \phi) + j\,\sin(\omega t + \phi) = e^{j(\omega t + \phi)} \tag{2.11}$$

Equation (2.11) is the origin of the term cisoid which is a contraction of $(cos + i \, sin)$us*oid*, where $i = \sqrt{-1}$ replaces j. In three dimensions, with time (or phase) progressing along the axis perpendicular to the complex plane, the cisoid traces out a helical curve, Figure 2.6. For $\phi = 0$ the projection of this helix onto the imaginary/time plane is a sine wave and its projection onto the real/time plane is a cosine wave.

There is a satisfying symmetry relating real sinusoids and complex cisoids in that two, quadrature, sinusoids are required to generate a single cisoid and two, counter rotating, cisoids are required to generate a single sinusoid. If the cisoids are a conjugate pair then the resulting sinusoid is purely real, Figure 2.7.

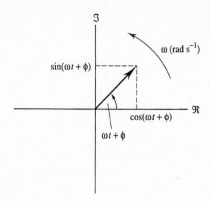

Figure 2.5 *Rotating vector or cisoid.*

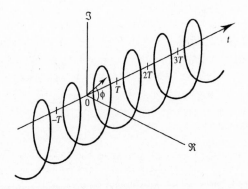

Figure 2.6 *Sketch of cisoid with time progressing perpenducular to the complex plane.*

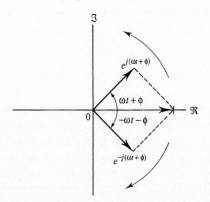

Figure 2.7 *Synthesis of real sinusoid wave from two counter-rotating, conjugate, cisoids.*

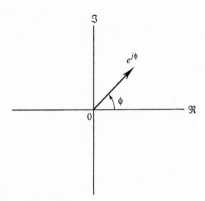

Figure 2.8 *Phasor corresponding to $e^{j(\omega t + \phi)}$.*

Phasors are cisoids which have had their time dependence suppressed. The phasor corresponding to $e^{j(\omega t + \phi)}$ is therefore $e^{j\phi}$ and corresponds to an instantaneous picture of the cisoid at the time $t = 0$, Figure 2.8. Another interpretation of phasors is that they represent a cisoid drawn in a plane which is itself rotating at the same angular frequency as the cisoid. The phasor is therefore stationary with respect to the complex plane. The close relationship between cisoids and phasors is such that a distinction between them is rarely made in practice, the term phasor often being used to describe both.

2.2.2 Fourier series

Almost any periodic signal of practical interest can be approximated by adding together sinusoids with the correct frequencies, amplitudes and phases. An example of a saw-tooth waveform approximated by a sum of sinusoids is shown in Figure 2.9. In general the error between the synthesised approximation and the actual waveform can be made as small as desired by including enough sinusoids in the sum. (This is not true at points of discontinuity, however: see section 2.2.4.) Only one sinusoid at each integer multiple of the fundamental frequency is required in the sum, providing that its amplitude and phase can be chosen freely. The fundamental frequency, f_1, is the reciprocal of the waveform's period, T, i.e.:

$$f_1 = 1/T \tag{2.12}$$

The sinusoid with frequency $f_n = nf_1$ is called the nth harmonic of the fundamental. If the waveform being approximated has a non-zero mean value then, in addition to the set of sinusoids, a 0 Hz, constant, or DC term must be included in the sum. In general, then, the sinusoidal sum, which is called a Fourier series, is given by:

$$v(t) = C_0 + C_1 \cos(\omega_1 t + \phi_1) + C_2 \cos(\omega_2 t + \phi_2) + \cdots \tag{2.13}$$

where C_0 (V) is the DC term, $\omega_1 = 2\pi/T$ (rad/s) is the fundamental frequency and $\omega_2 = 2(2\pi/T)$ (rad/s) is the second harmonic frequency, etc. The series may be truncated

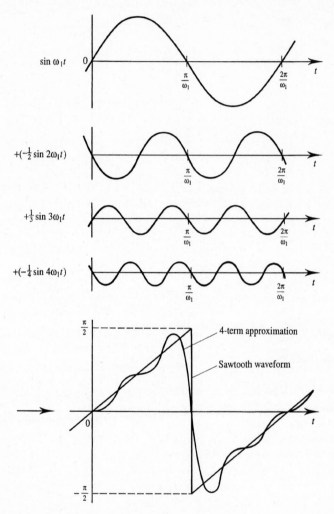

Figure 2.9 *Synthesis of sawtooth waveform by addition of harmonically related sinusoids.*

after a finite number of terms or may extend indefinitely.

Trigonometric forms

The trigonometric form of the Fourier series, expressed by equation (2.13), can be written more compactly as:

$$v(t) = C_0 + \sum_{n=1}^{\infty} C_n \cos(\omega_n t + \phi_n) \qquad (2.14(a))$$

This is the *cosine* form of the series since each term is written as a cosine function (with an explicit phase angle, ϕ). Since each term in the periodic signal is a harmonic of the fundamental, equation (2.14(a)) can be rewritten as:

$$v(t) = C_0 + \sum_{n=1}^{\infty} C_n \cos(n\omega_1 t + \phi_n) \qquad (2.14(b))$$

A slightly different trigonometric form can be created by resolving each sinusoid into cosine and sine components (each with zero phase angle). This gives the *cosine – sine* form of the trigonometric Fourier series, i.e.:

$$v(t) = C_0 + \sum_{n=1}^{\infty} (A_n \cos \omega_n t - B_n \sin \omega_n t) \qquad (2.14(c))$$

Notice that the series is still specified by two real numbers per harmonic but in this case the numbers are cosine and sine amplitudes (or inphase and quadrature amplitudes) rather than amplitude and phase. (The use of a minus sign in equation (2.14(c)) may seem eccentric but its advantage will become clear later.)

If the amplitude, C_n, of the cosine Fourier series is plotted against frequency, $f_n = \omega_n/2\pi$ (Hz), the result is called a discrete, or line, amplitude spectrum, Figure 2.10(a). Similarly, if ϕ_n is plotted against f_n the result is a discrete phase spectrum, Figure 2.10(b). Notice that, for obvious reasons, the phase of the DC (0 Hz) component is not defined. Notice also that the height of the lines in the amplitude spectrum of Figure 2.10(a) represents the peak values of the sinusoidal components. It is possible, of course, to define an RMS amplitude spectrum which would be the same as the peak amplitude spectrum except that each line would be smaller by a factor of $1/\sqrt{2}$.

If the sinusoids of a cosine series are displayed in three dimensions, plotted against time and frequency, Figure 2.11, then the amplitude spectrum corresponds to a projection onto the amplitude-frequency plane. This gives a picture of the 'frequency content' of a signal.

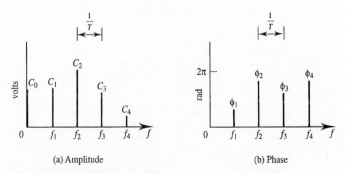

(a) Amplitude (b) Phase

Figure 2.10 *Discrete spectrum of a periodic signal.*

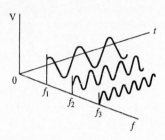

Figure 2.11 *Component sinusoids of a Fourier series plotted against time and frequency.*

Calculation of Fourier coefficients

Since C_0 is the DC, or average, value of the waveform being approximated it is clear that it can be calculated using:

$$C_0 = \frac{1}{T} \int_t^{t+T} v(t)\, dt \tag{2.15}$$

In practice it is easier to calculate the A_n and B_n coefficients associated with the cosine-sine form of the Fourier series than to find the C_n and ϕ_n values of the cosine form. (C_n and ϕ_n can be easily calculated from A_n and B_n as will be shown later.) The essential task in calculating the value of A_1, for example, is to find out how much of the inphase fundamental component, $\cos \omega_1 t$, is contained in $v(t)$. In other words the similarity between $\cos \omega_1 t$ and $v(t)$ must be established. One way of quantifying this similarity is to find their mean product, i.e. $\langle v(t) \cos \omega_1 t \rangle$ where $\langle \ \rangle$ signifies a time average. If $v(t)$ tends to be positive when $\cos \omega_1 t$ is positive and negative when $\cos \omega_1 t$ is negative then $\langle v(t) \cos \omega_1 t \rangle$ will tend to be large and positive indicating a large degree of similarity, Figure 2.12. This would suggest that $v(t)$ contained a large $\cos \omega_1 t$ component. If, conversely, $v(t)$ tends to be negative when $\cos \omega_1 t$ is positive and vice versa then $\langle v(t) \cos \omega_1 t \rangle$ will tend to be large and negative. This would indicate extreme dissimilarity and the conclusion would be that $v(t)$ contained a large $-\cos \omega_1 t$

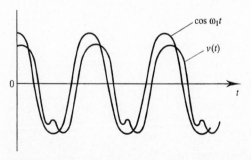

Figure 2.12 *Similar waveforms: $v(t)$ and $\cos \omega_1 t$.*

component. If there was little correlation between the polarity of $v(t)$ and $\cos \omega_1 t$ then $\langle v(t) \cos \omega_1 t \rangle$ would be close to zero and the conclusion would be that $v(t)$ contained almost no $\cos \omega_1 t$ component.

The normal way to find an average value is given by equation (2.15). Therefore:

$$\langle v(t) \cos \omega_1 t \rangle = \frac{1}{T} \int_{t}^{t+T} v(t) \cos \omega_1 t \; dt \tag{2.16}$$

To find the Fourier coefficient, A_1, however, the actual equation used is:

$$A_1 = \frac{2}{T} \int_{t}^{t+T} v(t) \cos \omega_1 t \; dt \tag{2.17}$$

This is because if $v(t)$ was *exactly* like $\cos \omega_1 t$ (i.e. $v(t) = \cos \omega_1 t$) then A_1 should be 1.0. Unfortunately:

$$\langle \cos^2 \omega_1 t \rangle = \langle \tfrac{1}{2}(1 + \cos 2\omega_1 t) \rangle = \tfrac{1}{2} \tag{2.18}$$

The factor of two in equation (2.17) is necessary to make $A_1 = 1$. The general formulae for calculating the cosine-sine Fourier coefficients are therefore:

$$A_n = \frac{2}{T} \int_{t}^{t+T} v(t) \cos \omega_n t \; dt \tag{2.19(a)}$$

$$B_n = -\frac{2}{T} \int_{t}^{t+T} v(t) \sin \omega_n t \; dt \tag{2.19(b)}$$

(B_n quantifies the similarity between $v(t)$ and $-\sin \omega_n t$.) If the cosine series is required the values of C_n and ϕ_n are found easily using simple trigonometry, Figure 2.13, i.e.:

$$C_n = \sqrt{(A_n^2 + B_n^2)} \tag{2.20(a)}$$

$$\phi_n = \tan^{-1}(B_n/A_n) \tag{2.20(b)}$$

A satisfying engineering interpretation of equations (2.19(a)) and (b) is that of 'filtering integrals'. If $v(t)$ is made up of many harmonically related sinusoids the average product of each of these sinusoids with $\cos \omega_n t$ or $\sin \omega_n t$ is only non-zero for that sinusoid which has the same frequency as $\cos \omega_n t$ or $\sin \omega_n t$. This *orthogonality* property is summarised mathematically by:

$$\frac{2}{T} \int_{t}^{t+T} \cos \omega_m t \cos \omega_n t \; dt = \begin{cases} 1, & m = n \\ 0, & m \neq n \end{cases} \tag{2.21(a)}$$

$$\frac{2}{T} \int_{t}^{t+T} \sin \omega_m t \sin \omega_n t \; dt = \begin{cases} 1, & m = n \\ 0, & m \neq n \end{cases} \tag{2.21(b)}$$

$$\frac{2}{T} \int_{t}^{t+T} \cos \omega_m t \sin \omega_n t \; dt = 0 \tag{2.21(c)}$$

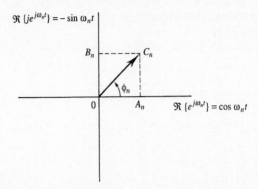

Figure 2.13 *Relationship between amplitude (C_n) of Fourier coefficients and inphase and quadrature components (A_n and B_n).*

These properties and their geometrical interpretation will be discussed further, in a more general context, in section 2.5.

EXAMPLE 2.1
Find the first two Fourier coefficients of a unipolar rectangular pulse train with amplitude 3 V, period 10 ms, duty cycle 20% and pulse leading edge at time $t = 0$. The pulse train $v(t)$ is shown in Figure 2.14.

The DC term is given by:

$$C_0 = \frac{1}{T} \int_{t}^{t+T} v(t)\ dt$$

$$= \frac{1}{0.01} \int_{0}^{0.002} 3\ dt$$

$$= 100\ [3\ t]_0^{0.002} = 0.6\ \text{(V)}$$

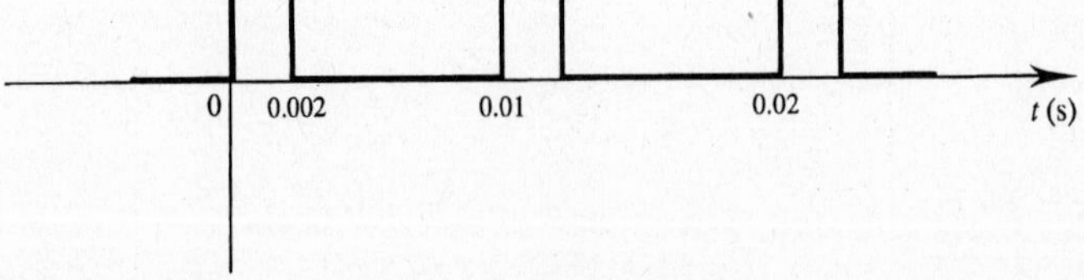

Figure 2.14 *Periodic rectangular pulse train for Example 2.1.*

The inphase coefficients are given by:

$$A_1 = \frac{2}{T} \int_t^{t+T} v(t) \cos 2\pi f_1 t \; dt$$

$$= \frac{2}{0.01} \int_0^{0.002} 3 \cos\left(2\pi \frac{1}{0.01} t\right) dt$$

$$= 600 \left[\frac{\sin(2\pi \; 100 \; t)}{2\pi \; 100} \right]_0^{0.002} = 0.9082 \quad (\text{V})$$

$$A_2 = \frac{2}{T} \int_t^{t+T} v(t) \cos 2\pi f_2 t \; dt$$

$$= \frac{2}{0.01} \int_0^{0.002} 3 \cos\left(2\pi \frac{2}{0.01} t\right) dt$$

$$= 600 \left[\frac{\sin(2\pi \; 200 \; t)}{2\pi \; 200} \right]_0^{0.002} = 0.2806 \quad (\text{V})$$

The quadrature coefficients are given by:

$$B_1 = -\frac{2}{T} \int_t^{t+T} v(t) \sin 2\pi f_1 t \; dt$$

$$= -\frac{2}{0.01} \int_0^{0.002} 3 \sin\left(2\pi \frac{1}{0.01} t\right) dt$$

$$= -600 \left[\frac{-\cos(2\pi \; 100 \; t)}{2\pi \; 100} \right]_0^{0.002} = \frac{3}{\pi} [0.3090 - 1] = -0.6599 \quad (\text{V})$$

$$B_2 = -\frac{2}{T} \int_t^{t+T} v(t) \sin 2\pi f_2 t \; dt$$

$$= -\frac{2}{0.01} \int_0^{0.002} 3 \sin\left(2\pi \frac{2}{0.01} t\right) dt$$

$$= -600 \left[\frac{-\cos(2\pi \; 200 \; t)}{2\pi \; 200} \right]_0^{0.002} = \frac{3}{2\pi} [-0.8090 - 1] = -0.8637 \quad (\text{V})$$

The Fourier coefficient amplitudes are given in equation (2.20(a)) by:

$$C_n = \sqrt{(A_n^2 + B_n^2)}$$

i.e.:

$$C_0 = 0.6 \text{ V}$$

$$C_1 = \sqrt{(0.9082^2 + 0.6599^2)} = 1.1226 \text{ V}$$

$$C_2 = \sqrt{(0.2806^2 + 0.8637^2)} = 0.9081 \text{ V}$$

and the Fourier coefficient phases are given in equation (2.20(b)) by:

$$\phi_n = \tan^{-1}\left(\frac{B_n}{A_n}\right)$$

i.e.:

$$\phi_1 = \tan^{-1}\left(\frac{-0.6599}{0.9082}\right) = -0.6284 \text{ rad or } -36.0°$$

$$\phi_2 = \tan^{-1}\left(\frac{-0.8637}{0.2806}\right) = -1.257 \text{ rad or } -72.0°$$

Note that moving the pulse train to the right or left will change the phase spectrum but not the amplitude spectrum. For example, if the pulse train is moved 0.001 s to the left (such that it has even symmetry about $t = 0$) then the Fourier series will contain cosine waves only and the phase spectrum will be restricted to values of 0° and 180°.

Figure 2.9 shows the decomposition of a sawtooth wave into terms up to the fourth harmonic and also includes the wave reconstructed from these components.

Exponential form

As an alternative to calculating the A_n and B_n coefficients of the cosine-sine Fourier series separately they can be calculated together using:

$$A_n + jB_n = \frac{2}{T} \int_t^{t+T} v(t)(\cos \omega_n t - j \sin \omega_n t) \ dt \tag{2.22}$$

This corresponds to synthesising the function $v(t)$ from the real part of a set of harmonically related cisoids, i.e.:

$$v(t) = C_0 + \sum_{n=1}^{\infty} (A_n \cos \omega_n t - B_n \sin \omega_n t)$$

$$= C_0 + \Re\left\{\sum_{n=1}^{\infty} \tilde{C}_n e^{j\omega_n t}\right\} \tag{2.23}$$

where $\tilde{C}_n = A_n + jB_n$. The tilde (˜) indicates that $\tilde{C}_n$ is generally complex.

   Having a separate DC term, C_0, in equation (2.23) and being required to take the real part of the other terms is, at best, a little inelegant. This can be overcome, however, by using a pair of counter-rotating, conjugate cisoids to represent each real sinusoid in the series, i.e.:

$$v(t) = \sum_{n=-\infty}^{\infty} \tilde{C}'_n \, e^{j\omega_n t} \tag{2.24(a)}$$

where:

$$\tilde{C}'_n = \begin{cases} \tilde{C}_n/2 & \text{for } n > 0 \\ C_0 & \text{for } n = 0 \\ \tilde{C}^*_{-n}/2 & \text{for } n < 0 \end{cases} \tag{2.24(b)}$$

Thus, for example, the pair of cisoids corresponding to $n = \pm 3$ (i.e. the third harmonic cisoids) may look like those shown in Figure 2.15. Notice that the magnitude, $|\tilde{C}'_n|$, of each cisoid in the formulation of equations (2.24) is half that, $|\tilde{C}_n|$, of the corresponding cisoids in equation (2.23) or the corresponding sinusoids in equations (2.14(a)) and (b). Thus the formula for the calculation of Fourier (exponential) coefficients gives results only half as large as that for the trigonometric series, i.e.:

$$\tilde{C}'_n = \frac{1}{T} \int_{t}^{t+T} v(t) \, e^{-j\omega_n t} \, dt \tag{2.25}$$

$e^{-j\omega_n t}$, here, filters out that part of $v(t)$ which is identical to $e^{j\omega_n t}$ (since $(1/T)\int_0^T e^{j\omega_n t} e^{-j\omega_n t} \, dt = 1$). When n is positive, the positively rotating (i.e. anticlockwise) cisoids are obtained and when n is negative (remembering that $\omega_n = n\omega_1$) the negatively rotating (i.e. clockwise) cisoids are obtained. When $n = 0$ equation (2.25) gives the DC term.

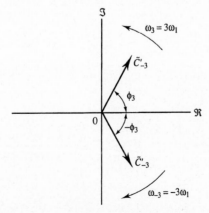

Figure 2.15 *Pair of counter-rotating, conjugate, cisoids (drawn for t = 0) corresponding to the (real) third harmonic of a periodic signal.*

Double sided spectra

If the amplitudes $|\tilde{C}'_n|$ of the cisoids found using equation (2.25) are plotted against frequency f_n the result is called a double (or two) sided (voltage) spectrum. Such a spectrum is shown in Figure 2.16. If $v(t)$ is purely real, i.e. it represents a signal that can exist in a practical single channel system, then each positively rotating cisoid is matched by a conjugate, negatively rotating, cisoid which cancels the imaginary part to zero. The double sided amplitude spectrum thus has *even* symmetry about 0 Hz, Figure 2.16(a). The single sided spectrum (positive frequencies only) representing the amplitudes of the real sinusoids in the trigonometric Fourier series can be found from the double sided spectrum by folding over the negative frequencies of the latter and adding them to the positive frequencies.

If the phase angles of the cisoids are plotted against f_n the result is the double sided phase spectrum which, due to the conjugate pairing of cisoids, will have *odd* symmetry for real waveforms, Figure 2.16(b).

The even and odd symmetries of the amplitude and phase spectra of real signals are summarised by:

$$|\tilde{C}'_n| = |\tilde{C}'_{-n}|$$

(2.26(a))

and:

$$\arg(\tilde{C}'_n) = -\arg(\tilde{C}'_{-n})$$

(2.26(b))

Calculation of coefficients for waveforms with symmetry

For a waveform $v(t)$ with certain symmetry properties, the calculation of some, or all, of the Fourier coefficients is simplified. These symmetries and the corresponding simplifications for the calculation of C_0, A_n and B_n are shown in Table 2.1.

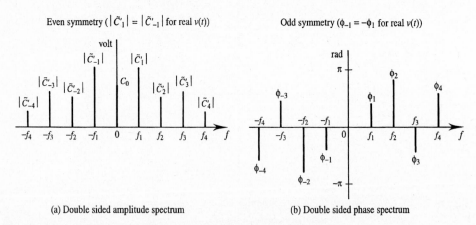

(a) Double sided amplitude spectrum   (b) Double sided phase spectrum

Figure 2.16 *Double sided voltage spectrum of a real periodic signal.*

Table 2.1   *Fourier series formulae for waveforms with symmetry.*

Type of Symmetry	Definition	Example $f(t)$	C_o	A_n	B_n	Non-zero terms
Zero mean	$\int_0^T f(t)\, dt = 0$		$C_o = 0$	$A_n = \dfrac{2}{T}\int_{-\frac{T}{2}}^{\frac{T}{2}} f(t)\cos\omega_n t\, dt$	$B_n = \dfrac{2}{T}\int_{-\frac{T}{2}}^{\frac{T}{2}} f(t)\sin\omega_n t\, dt$	A_n and B_n
Even	$f(t) = f(-t)$		$C_o = \dfrac{1}{T}\int_0^T f(t)\, dt$	$A_n = \dfrac{4}{T}\int_0^{\frac{T}{2}} f(t)\cos\omega_n t\, dt$	$B_n = 0$	A_n and C_o
Odd	$f(t) = -f(-t)$		$C_o = 0$	$A_n = 0$	$B_n = \dfrac{4}{T}\int_0^{\frac{T}{2}} f(t)\sin\omega_n t\, dt$	B_n
Half-wave	$f(t) = -f\left(t + \dfrac{T}{2}\right)$		$C_o = 0$	$A_n = \dfrac{4}{T}\int_0^{\frac{T}{2}} f(t)\cos\omega_n t\, dt$	$B_n = \dfrac{4}{T}\int_0^{\frac{T}{2}} f(t)\sin\omega_n t\, dt$	A_n and B_n, odd harmonics (n odd) only

Discrete power spectra and Parseval's theorem

If the trigonometric coefficients are divided by $\sqrt{2}$ and plotted against frequency the result is a one sided, RMS amplitude spectrum. If each RMS amplitude is then squared, the height of each spectral line will represent the normalised power (or power dissipated in $1\ \Omega$) associated with that frequency, i.e.:

$$\left\{\frac{|\tilde{C}_n|}{\sqrt{2}}\right\}^2 = \frac{|\tilde{C}_n|^2}{2} = P_{n1}\ (V^2) \tag{2.27}$$

Such a one sided power spectrum is illustrated in Figure 2.17(a). A double sided version of the power spectrum can be defined by associating half the power in each line, $P_{n2} = |\tilde{C}_n|^2/4$, with negative frequencies, Figure 2.17(b). The double sided power spectrum is therefore obtained from the double sided amplitude spectrum by squaring each cisoid amplitude, i.e. $P_{n2} = |\tilde{C}'_n|^2$. That the total power in an entire line spectrum is the sum of the powers in each individual line might seem an intuitively obvious statement. This is true, however, only because of the orthogonal nature of the individual sinusoids making up a periodic waveform. Obvious or not, this statement, which applies to any periodic signal, is known as Parseval's theorem. It can be stated in several forms, one of the most useful being:

$$\text{Total power, } P = \frac{1}{T}\int_t^{t+T}|v(t)|^2\,dt = \sum_{n=-\infty}^{\infty}|\tilde{C}'_n|^2 \tag{2.28}$$

The proof of Parseval's theorem is straightforward and is given below:

$$P = \frac{1}{T}\int_t^{t+T} v(t)\,v^*(t)\,dt \tag{2.29(a)}$$

$$v^*(t) = \left[\sum_{n=-\infty}^{\infty}\tilde{C}'_n\,e^{j\omega_n t}\right]^*$$

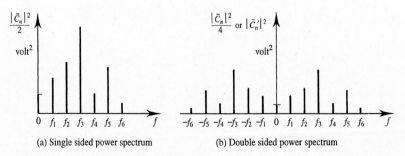

(a) Single sided power spectrum   (b) Double sided power spectrum

Figure 2.17 *Power spectra of a periodic signal.*

$$= \sum_{n=-\infty}^{\infty} (\tilde{C}_n^{\prime *} e^{-j\omega_n t}) \qquad (2.29(\text{b}))$$

Therefore:

$$P = \frac{1}{T} \int_t^{t+T} v(t) \sum_{n=-\infty}^{\infty} (\tilde{C}_n^{\prime *} e^{-j\omega_n t})\, dt$$

$$= \sum_{n=-\infty}^{\infty} \left[\frac{1}{T} \int_t^{t+T} v(t)\, e^{-j\omega_n t}\, dt\; \tilde{C}_n^{\prime *} \right]$$

$$= \sum_{n=-\infty}^{\infty} \tilde{C}_n^{\prime}\, \tilde{C}_n^{\prime *} \qquad (2.29(\text{c}))$$

EXAMPLE 2.2

Find the 0 Hz and first two harmonic terms in the double sided spectrum of the half-wave rectified sinusoid shown in Figure 2.18(a) and sketch the resulting amplitude and phase spectra. What is the total power in these components?

$$\tilde{C}_0^{\prime} = \frac{1}{T} \int_t^{t+T} v(t)\, e^{-j2\pi\, 0 t}\, dt$$

$$= \frac{1}{0.02} \int_{-0.0025}^{0.0075} \sin\left(2\pi\, \frac{1}{0.02}\, t + 2\pi\, \frac{0.0025}{0.02} \right) dt$$

$$= 50 \int_{-0.0025}^{0.0075} \sin(100\pi t + 0.25\pi)\, dt$$

$$= 50 \int_{-0.0025}^{0.0075} \sin(100\pi t) \cos(0.25\pi) + \cos(100\pi t) \sin(0.25\pi)\, dt$$

$$= 50\, \frac{1}{\sqrt{2}} \left[\frac{-\cos 100\pi t}{100\pi} \right]_{-0.0025}^{0.0075} + 50\, \frac{1}{\sqrt{2}} \left[\frac{\sin 100\pi t}{100\pi} \right]_{-0.0025}^{0.0075}$$

$$= \frac{50}{100\pi}\, \frac{1}{\sqrt{2}} \{ -[-0.7071 - 0.7071] + [0.7071 - (-0.7071)] \}$$

$$= \frac{50}{100\pi}\, \frac{1}{\sqrt{2}}\, 4\, (0.7071) = \frac{1}{\pi}\, (= 0.3183)\; (\text{V})$$

$$\tilde{C}_1^{\prime} = \frac{1}{T} \int_t^{t+T} v(t)\, e^{-j2\pi f_1 t}\, dt$$

$$= \frac{1}{0.02} \int_{-0.0025}^{0.0075} \sin\left(2\pi\, \frac{1}{0.02}\, t + \frac{2\pi\, 0.0025}{0.02} \right) e^{-j2\pi \frac{1}{0.02} t}\, dt$$

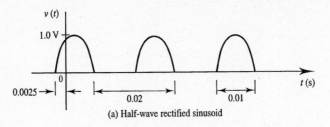

(a) Half-wave rectified sinusoid

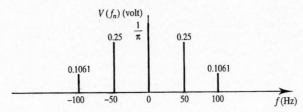

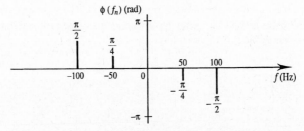

(b) Amplitude and phase spectra of DC and first two harmonics for waveform in Figure 2.18(a)

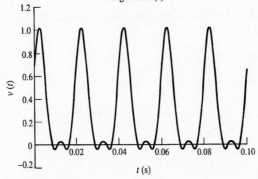

(c) Fourier series approximation to waveform in Example 2.2 (DC plus 2 harmonics)

Figure 2.18 *Waveform, spectra and Fourier series approximation for Example 2.2.*

$$= \frac{50}{\sqrt{2}} \int_{-0.0025}^{0.0075} \sin(100\pi t)\,(\cos 100\pi t - j \sin 100\pi t)\,dt$$

$$+ \frac{50}{\sqrt{2}} \int_{-0.0025}^{0.0075} \cos(100\pi t)\,(\cos 100\pi t - j \sin 100\pi t)\,dt$$

$$= \frac{25}{\sqrt{2}} \left[\int_{-0.0025}^{0.0075} \sin(200\pi t)\,dt + \int_{-0.0025}^{0.0075} 1 + \cos(200\pi t)\,dt \right]$$

$$- j\,\frac{25}{\sqrt{2}} \left[\int_{-0.0025}^{0.0075} (1 - \cos(200\pi t))\,dt + \int_{-0.0025}^{0.0075} \sin(200\pi t)\,dt \right]$$

$$= \frac{25}{\sqrt{2}} \left\{ \left[\frac{-\cos(200\pi t)}{200\pi} \right]_{-0.0025}^{0.0075} + [\,t\,]_{-0.0025}^{0.0075} + \left[\frac{\sin(200\pi t)}{200\pi} \right]_{-0.0025}^{0.0075} \right.$$

$$\left. - j[\,t\,]_{-0.0025}^{0.0075} + j \left[\frac{\sin(200\pi t)}{200\pi} \right]_{-0.0025}^{0.0075} - j \left[\frac{-\cos(200\pi t)}{200\pi} \right]_{-0.0025}^{0.0075} \right\}$$

$$= \frac{25}{\sqrt{2}} \left\{ \left[\frac{0-0}{200\pi} \right] + [0.0075 + 0.0025] + \left[\frac{-1-(-1)}{200\pi} \right] \right\}$$

$$- j\,\frac{25}{\sqrt{2}} \left\{ [0.0075 + 0.0025] - \frac{[-1-(-1)]}{200\pi} + \left[\frac{0-0}{200\pi} \right] \right\}$$

$$= \frac{25}{\sqrt{2}} [0.01 - j0.01]$$

$$= 0.1768 - j0.1768$$

$$= 0.25 \text{ at} -45° \text{ (V)}$$

Since $v(t)$ is real then:

$$\tilde{C}'_{-1} = \tilde{C}'^{*}_{1}$$

$$= 0.1768 + j0.1768$$

$$= 0.25 \text{ at } 45° \text{ (V)}$$

$$\tilde{C}'_2 = \frac{1}{T} \int_{t}^{t+T} v(t)\,e^{-j2\pi f_2 t}\,dt$$

$$= \frac{1}{0.02} \int_{-0.0025}^{0.0075} \sin\left(2\pi\,\frac{1}{0.02}\,t + \frac{2\pi\,0.0025}{0.02} \right) e^{-j2\pi \frac{2}{0.02} t}\,dt$$

$$= 0.1061 \text{ at} - 90° \text{ (V)}$$

Since $v(t)$ is real then:

$$\tilde{C}'_{-2} = \tilde{C}'^{*}_{2} = 0.1061 \text{ at } 90° \text{ (V)}$$

The amplitude and phase spectra are shown in Figure 2.18(b). The sum of the DC term, fundamental and second harmonic is shown in Figure 2.18(c). It is interesting to see that even with so few terms the Fourier series is a recognisable approximation to the half-wave rectified sinusoid. The total power in the DC, fundamental and second harmonic components is given by Parseval's theorem, equation (2.28), i.e.:

$$P = \sum_{n=-2}^{2} |\tilde{C}'_n|^2 \quad (\text{V}^2)$$

$$= 0.1061^2 + 0.25^2 + (1/\pi)^2 + 0.25^2 + 0.1061^2$$

$$= 0.2488 \quad (\text{V}^2)$$

Table 2.2 shows commonly encountered periodic waveforms and their corresponding Fourier series.

2.2.3 Conditions for existence, convergence and Gibb's phenomenon

The question might be asked: how do we know that it is possible to approximate $v(t)$ with a Fourier series and furthermore that adding further terms to the series continues to improve the approximation? Here we give the answer to this question without proof.

A function $v(t)$ has a Fourier series if the following conditions are met:

1. $v(t)$ contains a finite number of maxima and minima per period.
2. $v(t)$ contains a finite number of discontinuities per period.
3. $v(t)$ is absolutely integrable over one period, i.e.:

$$\int_{t}^{t+T} |v(t)| \, dt < \infty$$

The above conditions, called the Dirichlet conditions, are sufficient but not necessary. If a Fourier series does exist it converges (i.e. gets closer to $v(t)$ as more terms are added) at all points except points of discontinuity. Mathematically, this can be stated as follows:

$$\sum_{N} \text{series} \big|_{t_0} \to v(t_0) \text{ as } N \to \infty \text{ for all continuous points } t_0.$$

At points of discontinuity the series converges to the arithmetic mean of the function value on either side of the discontinuity, Figure 2.19(a), i.e.:

Table 2.2 *Fourier series of commonly occurring waveforms.*

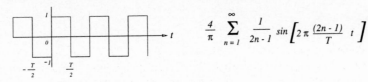

$$\frac{4}{\pi} \sum_{n=1}^{\infty} \frac{1}{2n-1} \sin\left[2\pi \frac{(2n-1)}{T} t\right]$$

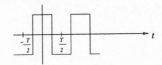

$$\frac{4}{\pi} \sum_{n=1}^{\infty} \frac{(-1)^{n+1}}{2n-1} \cos\left[2\pi \frac{(2n-1)}{T} t\right]$$

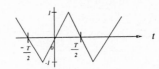

$$\frac{\tau}{T} + \frac{2\tau}{T} \sum_{n=1}^{\infty} sinc\left(\frac{n\tau}{T}\right) \cos\left(2\pi \frac{n}{T} t\right)$$

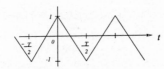

$$\frac{8}{\pi^2} \sum_{n=1}^{\infty} \frac{1}{(2n-1)^2} \cos\left[2\pi\left(\frac{2n-1}{T}\right)t\right]$$

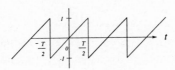

$$\frac{8}{\pi^2} \sum_{n=1}^{\infty} \frac{(-1)^{n+1}}{(2n-1)^2} \sin\left[2\pi\left(\frac{2n-1}{T}\right)t\right]$$

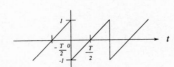

$$\frac{2}{\pi} \sum_{n=1}^{\infty} \frac{(-1)^{n+1}}{n} \sin\left[2\pi\left(\frac{n}{T}\right)t\right]$$

$$-\frac{2}{\pi} \sum_{n=1}^{\infty} \frac{1}{n} \sin\left[2\pi\left(\frac{n}{T}\right)t\right]$$

Table 2.2 ctd. *Fourier series of commonly occurring waveforms.*

$$\frac{2}{\pi} - \frac{4}{\pi} \sum_{n=1}^{\infty} \frac{1}{4n^2-1} \cos\left[2\pi\left(\frac{n}{T}\right)t\right]$$

$$\frac{2}{\pi}\left\{\frac{1}{2} + \frac{\pi}{4} \cos\left[2\pi\left(\frac{1}{T}\right)t\right] - \sum_{n=1}^{\infty} \frac{(-1)^n}{4n^2-1} \cos\left[2\pi\left(\frac{2n}{T}\right)t\right]\right\}$$

$$\frac{1}{T} + \sum_{n=1}^{\infty} \frac{2}{T} \cos\left[2\pi\left(\frac{n}{T}\right)t\right]$$

$$\sum_{N} \text{series}\,\big|_{t_0} \rightarrow \frac{v(t_0^-) + v(t_0^+)}{2} \quad \text{as } N \rightarrow \infty \text{ for all discontinuous points } t_0.$$

At points on either side of a discontinuity the series oscillates with a period T_G given by:

$$T_G = 0.5\, T/N \qquad\qquad (2.30\text{(a)})$$

where T is the period of $v(t)$ and N is the number of terms included in the series. The amplitude, Δ, of the overshoot on either side of the discontinuity is:

$$\Delta = 0.09A \qquad\qquad (2.30\text{(b)})$$

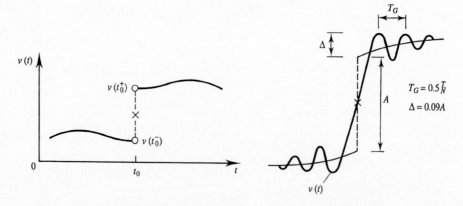

(a) Point of convergence ($\times$) for Fourier series at a discontinuity   (b) Gibb's ears on either side of discontinuity

Figure 2.19 *Overshoot and undershoot of a truncated Fourier series at a point of discontinuity.*

where A is the amplitude of the discontinuity, Figure 2.19(b). The overshoot, Δ, does not decrease as N increases, the resulting spikes sometimes being known as 'Gibb's ears'.

2.2.4 Bandwidth, rates of change, sampling and aliasing

The bandwidth, B, of a signal is defined as the difference (usually in Hz) between two nominal frequencies f_{max} and f_{min}. Loosely speaking f_{max} and f_{min} are, respectively, the frequencies above and below which the spectral components are assumed to be small. It is important to realise that these frequencies are often chosen using some fairly arbitrary rule, e.g. the frequencies at which spectral components have fallen to $1/\sqrt{2}$ of the peak spectral component. It would therefore be wrong to assume always that the frequency components of a signal outside its quoted bandwidth are negligible for all purposes, especially if the precise definition being used for B is vague or unknown.

The $1/\sqrt{2}$ definition of B is a common one and is *usually* implied if no other definition is explicitly given. It is normally called the half power or 3 dB bandwidth since the factor $1/\sqrt{2}$ refers to the voltage spectrum and $20 \log_{10}(1/\sqrt{2}) \approx -3$ dB. The 3 dB bandwidth of a periodic signal is illustrated in Figure 2.20(a). For baseband signals (i.e. signals with

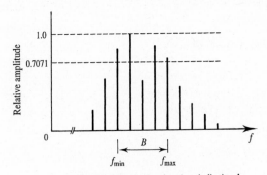

(a) 3 dB bandwidth of a (bandpass) periodic signal

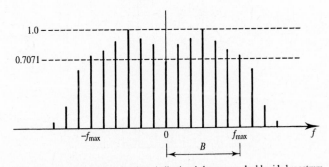

(b) 3 dB bandwidth of a (baseband) periodic signal shown on a double sided spectrum

Figure 2.20 *Definition of 3 dB signal bandwidth.*

significant spectral components all the way down to their fundamental frequency, f_1, or even DC) f_{min} is 0 Hz, *not* $-f_{max}$. This is important to remember when considering two sided spectra. The physical bandwidth is measured using positive frequencies or negative frequencies only, not both, Figure 2.20(b).

In general, if a signal has no significant spectral components above f_H then it cannot change appreciably on a time scale much shorter than about $1/(8 f_H)$. (This corresponds to one eighth of a period of the highest frequency sinusoid present in the signal, Figure 2.21.) A corollary of this is that signals with large rates of change must have high values of f_H. A rectangular pulse stream, for example, contains changes which occur (in principle) infinitely quickly. This implies that it must contain spectral components with infinite frequency. (In practice, of course, such pulse streams are, at best, only approximately rectangular and therefore their spectra can be essentially bandlimited.)

Sampling refers to the process of recording the values of a signal or waveform at (usually) regularly spaced instants of time. A schematic diagram of how this might be achieved is shown in Figure 2.22. It is a surprising fact that if a signal having no spectral components with frequencies above f_H is sampled rapidly enough then the original, continuous, signal can, in principle, be reconstructed from its samples *without error*. The

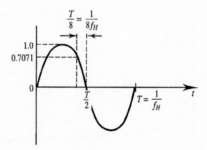

Figure 2.21 *Illustration of minimum time required for appreciable change of signal amplitude.*

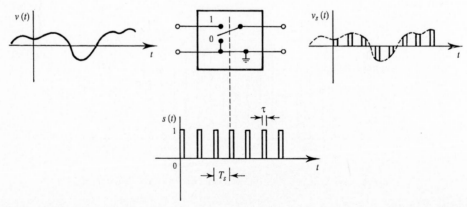

Figure 2.22 *Schematic illustration of sampling.*

minimum sampling rate or frequency, f_s, needed to achieve such ideal reconstruction is related to f_H by:

$$f_s \geq 2f_H \qquad (2.31)$$

Equation (2.31) is called Nyquist's sampling theorem and is of central importance to digital communications. It will be discussed more rigourously in Chapter 5. Here, however, it is sufficient to demonstrate its reasonableness as follows.

Figure 2.23(a) shows a sinusoid which represents the highest frequency spectral component in a certain waveform. The sinusoid is sampled in accordance with equation (2.31), i.e. at a rate higher than twice its frequency. (When $f_s > 2f_H$ the signal is said to be *oversampled*.) Nyquist's theorem essentially says that there is one, and only one, sinusoid which can be drawn through the given sample points. Figure 2.23(b) shows the same sinusoid sampled at a rate $f_s = 2f_H$. (This might be called critical, or Nyquist rate, sampling.) There is still only one frequency of sine wave which can be drawn through

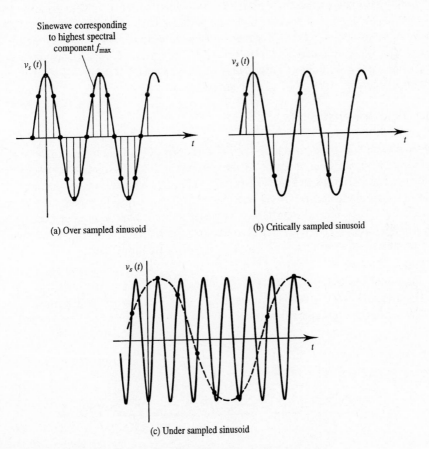

(a) Over sampled sinusoid

(b) Critically sampled sinusoid

(c) Under sampled sinusoid

Figure 2.23 *Demonstration of the sampling theorem and alias frequency.*

the samples. Figure 2.23(c) shows a sinusoid which is now *undersampled* (i.e. $f_s < 2f_H$). The samples could be (and usually are) interpreted as belonging to a sinusoid (shown dotted) of lower frequency than that to which they actually belong. The mistaken identity of the frequency of an undersampled sinusoid is called *aliasing* since the sinusoid inferred from the samples appears under the alias of a new and incorrect frequency. Aliasing is explained more fully later (section 5.3.3) with the aid of frequency domain concepts.

2.3 Transient signals

Signals are said to be transient if they are essentially localised in time. This obviously includes time limited signals which have a well defined start and stop time and which are zero outside the start-stop time interval. Signals with no start time, stop time, or either, are usually also considered to be transient, providing they tend to zero as time tends to $\pm\infty$ and contain finite total energy. Since the power of such signals averaged over all time is zero they are sometimes called *energy* signals. Since transient signals are not periodic they cannot be represented by an ordinary Fourier series. A related but more general technique, namely Fourier transformation, can, however, be used to find a frequency domain, or spectral, description of such signals.

2.3.1 Fourier transforms

The traditional way of approaching Fourier transforms is to treat them as a limiting case of a periodic signal Fourier series as the period, T, tends to infinity. Consider Figure 2.24. The waveform in this figure is periodic and pulsed with interpulse spacing, T_g. The amplitude and phase spectra of $v(t)$ are shown (schematically) in Figure 2.25(a) and (b) respectively. They are discrete (since $v(t)$ is periodic), have even and odd symmetry respectively (since $v(t)$ is real) and have line spacing $1/T$ Hz. If the interpulse spacing is now allowed to grow without limit (i.e. $T_g \to \infty$) then it follows that:

1. Period, $T \to \infty$.
2. Spacing of spectral lines, $1/T \to 0$.
3. The discrete spectrum becomes continuous (as $V(f)$ is defined at all points).
4. The signal becomes aperiodic (since only one pulse is left between $t = -\infty$ and $t = \infty$).

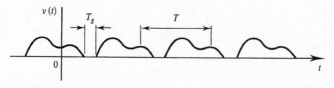

Figure 2.24 *Pulsed, periodic waveform.*

As the spectral lines become infinitesimally closely spaced the discrete quantities in the Fourier series:

$$v(t) = \sum_{n=-\infty}^{\infty} \tilde{C}'_n \, e^{j2\pi f_n t}$$

(2.32)

become continuous, i.e.:

$$f_n \rightarrow f \quad (\text{Hz})$$

$$\tilde{C}'_n(f_n) \rightarrow V(f) \, df \quad (\text{V})$$

$$\sum_{n=-\infty}^{\infty} \rightarrow \int_{-\infty}^{\infty}$$

Since $\tilde{C}'_n(f_n)$, and therefore $V(f) \, df$, have units of V then $V(f)$ has units of V/Hz. $V(f)$ is called a *voltage spectral density*. The resulting 'continuous series', called an *inverse Fourier transform*, is:

$$v(t) = \int_{-\infty}^{\infty} V(f) \, e^{j2\pi ft} \, df$$

(2.33)

The converse formula, equation (2.25), which gives the (complex) Fourier coefficients for a Fourier series, is:

$$\tilde{C}'_n = \frac{1}{T} \int_{t-T/2}^{t+T/2} v(t) \, e^{-j2\pi f_n t} \, dt$$

(2.34)

If this is generalised in the same way as equation (2.32) by letting $T \rightarrow \infty$ then $\tilde{C}'_n \rightarrow 0$ for all n and $v(t)$. This problem can be avoided by calculating $T\tilde{C}'_n$ instead. In this case:

$$f_n \rightarrow f$$

$$T\tilde{C}'_n \rightarrow V(f)$$

$$t \pm \frac{T}{2} \rightarrow \pm\infty$$

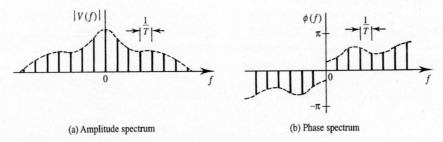

(a) Amplitude spectrum   (b) Phase spectrum

Figure 2.25 *Voltage spectrum of a pulsed, periodic, waveform.*

(Note that $T\tilde{C}'_n$ and $V(f)$ have units of V s or equivalently V/Hz as required.) The *forward* Fourier transform therefore becomes:

$$V(f) = \int_{-\infty}^{\infty} v(t) \, e^{-j2\pi ft} \, dt \qquad (2.35)$$

Equation (2.35) can be interpreted as finding that part of $v(t)$ which is identical to $e^{j2\pi ft}$. This is a cisoid (or rotating phasor) with frequency f and amplitude $V(f) \, df$ (V). For real signals there will be a conjugate cisoid rotating in the opposite sense located at $-f$. This pair of cisoids together constitute a sinusoid of frequency f and amplitude $2V(f) \, df$ (V). A one sided amplitude spectrum can thus be formed by folding the negative frequency components of the two sided spectrum, defined by equation (2.35), onto the positive frequencies and adding.

Sufficient conditions for the existence of a Fourier transform are similar to those for a Fourier series. They are:

1. $v(t)$ contains a finite number of maxima and minima in any finite time interval.
2. $v(t)$ contains a finite number of *finite* discontinuities in any finite time interval.
3. $v(t)$ must be absolutely integrable, i.e.:

$$\int_{-\infty}^{\infty} |v(t)| \, dt < \infty$$

2.3.2 Practical calculation of Fourier transforms

As with Fourier series simplification of practical calculations is possible if certain symmetries are present in the function being transformed. This is best explained by splitting the Fourier transform into cosine and sine transforms as follows:

$$V(f) = \int_{-\infty}^{\infty} v(t) \, e^{-j2\pi ft} \, dt$$

$$= \int_{-\infty}^{\infty} v(t) \cos 2\pi ft \, dt - j \int_{-\infty}^{\infty} v(t) \sin 2\pi ft \, dt \qquad (2.36)$$

The first term in the second line of equation (2.36) is made up of cosine components only. It therefore corresponds to a component of $v(t)$ which has *even* symmetry. Similarly the second term is made up of sine components only and therefore corresponds to an *odd* component of $v(t)$, i.e.:

$$V(f) = V(f)|_{even\ v(t)} + jV(f)|_{odd\ v(t)} \qquad (2.37(a))$$

where:

$$V(f)|_{even\ v(t)} = \int_{-\infty}^{\infty} v(t) \cos 2\pi ft \, dt \qquad (2.37(b))$$

and:

$$V(f)|_{odd\ v(t)} = -\int_{-\infty}^{\infty} v(t) \sin 2\pi ft\ dt \qquad (2.37(c))$$

It follows that if $v(t)$ is purely even (and real) then:

$$V(f) = 2\int_{0}^{\infty} v(t) \cos 2\pi ft\ dt \qquad (2.38(a))$$

Conversely, if $v(t)$ is purely odd (and real) then:

$$V(f) = -2j\int_{0}^{\infty} v(t) \sin 2\pi ft\ dt \qquad (2.38(b))$$

That any function can be split into odd and even parts is easily demonstrated, as follows:

$$v(t) = \frac{v(t) + v(-t)}{2} + \frac{v(t) - v(-t)}{2} \qquad (2.39)$$

The first term on the right hand side of equation (2.39) is, by definition, even and the second term is odd. A summary of symmetry properties relevant to the calculation of Fourier transforms is given in Table 2.3 [after Bracewell].

Table 2.3 *Symmetry properties of Fourier transforms.*

Function	Transform
real and even	real and even
real and odd	imaginary and odd
imaginary and even	imaginary and even
imaginary and odd	real and odd
complex and even	complex and even
complex and odd	complex and odd
real and asymmetrical	complex and Hermitian
imaginary and asymmetrical	complex and antiHermitian
real even plus imaginary odd	real
real odd plus imaginary even	imaginary
even	even
odd	odd

EXAMPLE 2.3
Find and sketch the amplitude and phase spectrum of the transient signal $v(t) = 2e^{-|t|/\tau}$ (V) shown in Figure 2.26(a).

Since $v(t)$ is real and even:

$$V(f) = 2\int_{0}^{\infty} v(t) \cos 2\pi ft\ dt$$

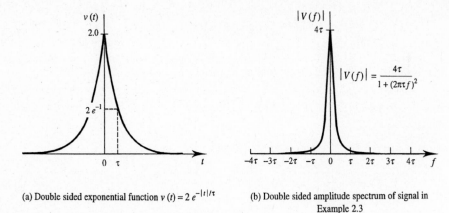

(a) Double sided exponential function $v(t) = 2\,e^{-|t|/\tau}$

(b) Double sided amplitude spectrum of signal in Example 2.3

Figure 2.26 *Double sided exponential function and corresponding amplitude spectrum.*

$$= 4 \int_0^\infty e^{-|t|/\tau}\,\cos 2\pi ft\,dt$$

Using the standard integral:

$$\int e^{ax}\,\cos bx\,dx = \frac{e^{ax}(a\cos bx + b\sin bx)}{a^2 + b^2}$$

$$V(f) = 4\left[\frac{e^{-(1/\tau)t}\,[-1/\tau\cos 2\pi ft + 2\pi f\sin 2\pi ft]}{(1/\tau)^2 + (2\pi f)^2}\right]_0^\infty$$

$$= 4\left[\frac{0 - [-1/\tau + 0]}{(1/\tau)^2 + (2\pi f)^2}\right] = \frac{4\tau}{1 + (2\pi\tau f)^2}\quad\text{(V/Hz)}$$

$|V(f)|$ is sketched in Figure 2.26(b) and since $V(f)$ is everywhere real and positive then $v(t)$ has a null phase spectrum.

2.3.3 Fourier transform pairs

The Fourier transform (for transient functions) and Fourier series (for periodic functions) provide a link between two quite different ways of describing signals. The more familiar description is the conventional time plot such as would be seen on an oscilloscope display. Applying the Fourier transform results in a frequency plot (amplitude and phase). These two descriptions are equivalent in the sense that there is one, and only one, amplitude and phase spectrum pair for each possible time plot. Given a complete time domain description, therefore, the frequency domain description can be obtained exactly and vice versa.

Comprehensive tables of Fourier transform pairs have been compiled by many authors. Table 2.4 lists some common Fourier transform pairs. The notation used here for several of the functions which occur frequently in communications engineering is included at the front of this text following the list of principal symbols. Owing to its central importance in digital communications, the Fourier transform of the rectangular function is derived from first principles below. Later the impulse function is defined as a limiting case of the rectangular pulse. The Fourier transform of the impulse is then shown to be a constant in amplitude and linear in phase.

Table 2.4 *Fourier transform pairs.*

Function	$x(t)$	$X(f)$
Rectangle of unit width	$\Pi(t)$	$\text{sinc}(f)$
Delayed rectangle of width τ	$\Pi\left(\dfrac{t-T}{\tau}\right)$	$\tau\,\text{sinc}(\tau f)e^{-j\omega T}$
Triangle of base width 2τ	$\Lambda\left(\dfrac{t}{\tau}\right)$	$\tau\,\text{sinc}^2(\tau f)$
Gaussian	$e^{-\pi(t/\tau)^2}$	$\tau e^{-\pi(\tau f)^2}$
One sided exponential	$u(t)\,e^{-t/\tau}$	$\dfrac{\tau}{1+j2\pi\tau f}$
Two sided exponential	$e^{-|t|/\tau}$	$\dfrac{2\tau}{1+(2\pi\tau f)^2}$
sinc	$\text{sinc}(2f_\chi t)$	$\dfrac{1}{2f_\chi}\Pi\left(\dfrac{f}{2f_\chi}\right)$
Constant	1	$\delta(f)$
Phasor	$e^{j(\omega_c t+\phi)}$	$e^{j\phi}\delta(f-f_c)$
sine wave	$\sin(\omega_c t+\phi)$	$\dfrac{1}{2j}\left[e^{j\phi}\delta(f-f_c)-e^{-j\phi}\delta(f+f_c)\right]$
cosine wave	$\cos(\omega_c t+\phi)$	$\dfrac{1}{2}\left[e^{j\phi}\delta(f-f_c)+e^{-j\phi}\delta(f+f_c)\right]$
Impulse	$\delta(t-T)$	$e^{-j\omega T}$
Sampling	$\displaystyle\sum_{k=-\infty}^{\infty}\delta(t-kT_s)$	$f_s\displaystyle\sum_{n=-\infty}^{\infty}\delta(f-nf_s)$
Signum	$\text{sgn}(t)$	$\dfrac{1}{j\pi f}$
Heaviside step	$u(t)$	$\dfrac{1}{2}\delta(f)+\dfrac{1}{j2\pi f}$

Fourier transform of a rectangular pulse

The unit rectangular pulse, Figure 2.27(a), is represented here using the notation $\Pi(t)$ and is defined by:

$$\Pi(t) \overset{\Delta}{=} \begin{cases} 1.0, & |t| < \tfrac{1}{2} \\ 0.5, & |t| = \tfrac{1}{2} \\ 0, & |t| > \tfrac{1}{2} \end{cases} \tag{2.40}$$

The voltage spectrum, $V_\Pi(f)$, of this pulse is given by its Fourier transform, i.e.:

$$
\begin{aligned}
V_\Pi(f) &= \int_{-\infty}^{\infty} \Pi(t)\, e^{-j2\pi ft}\, dt \\[2mm]
&= \int_{-\frac{1}{2}}^{\frac{1}{2}} e^{-j2\pi ft}\, dt \\[2mm]
&= \left[\frac{e^{-j2\pi ft}}{-j2\pi f} \right]_{-\frac{1}{2}}^{\frac{1}{2}} = \frac{1}{j2\pi f}\, [e^{j\pi f} - e^{-j\pi f}] \\[2mm]
&= \frac{j2\,\sin(\pi f)}{j2\pi f} = \frac{\sin(\pi f)}{\pi f}
\end{aligned}
\tag{2.41}
$$

The function $\mathrm{sinc}(x)$ is defined by:

$$
\mathrm{sinc}(x) \overset{\Delta}{=} \frac{\sin(\pi x)}{\pi x}
\tag{2.42}
$$

which means that the unit rectangular pulse and unit sinc function form a Fourier transform pair:

$$
\Pi(t) \overset{\text{FT}}{\Leftrightarrow} \mathrm{sinc}(f)
\tag{2.43(a)}
$$

The $\mathrm{sinc}(f)$ function is shown in Figure 2.27(b). Whilst in this case the voltage spectrum can be plotted as a single curve, in general the voltage spectrum of a transient signal is complex and must be plotted either as amplitude and phase spectra or as inphase and quadrature spectra. The amplitude and phase spectra corresponding to Figure 2.27(b) are shown in Figure 2.28(a) and (b).

It is left to the reader to show that the (complex) voltage spectrum of $\Pi[(t - T)/\tau]$ where T is the location of the centre of the pulse and τ is its width is given by:

$$
\Pi\left(\frac{t - T}{\tau} \right) \overset{\text{FT}}{\Leftrightarrow} \tau\, \mathrm{sinc}(\tau f)\, e^{-j2\pi fT}
\tag{2.43(b)}
$$

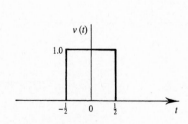

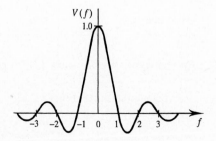

(a) Unit rectangular pulse, $\Pi(t)$

(b) Fourier transform of unit rectangular pulse centred on $t = 0$

Figure 2.27 *Unit rectangular pulse and corresponding Fourier transform.*

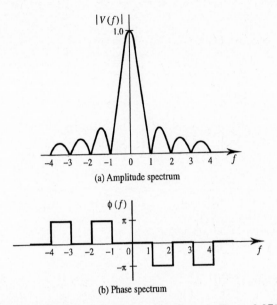

(a) Amplitude spectrum

(b) Phase spectrum

Figure 2.28 *Voltage spectrum of unit rectangular pulse shown in Figure 2.27(a).*

The impulse function and its Fourier transform

Consider a tall, narrow, rectangular voltage pulse of width τ seconds and amplitude $1/\tau$ V occurring at time $t = T$, Figure 2.29. The area under the pulse (sometimes called its strength) is clearly 1.0 V s. The impulse function (also called the Dirac delta function) can be defined as the limit of this rectangular pulse as τ tends to zero, i.e.:

$$\delta(t - T) = \lim_{\tau \to 0} \left(\frac{1}{\tau}\right) \Pi\!\left(\frac{t - T}{\tau}\right) \tag{2.44}$$

This idea is illustrated in Figure 2.30. Whatever the value of τ the strength of the pulse remains unity. Mathematically the impulse might be described by:

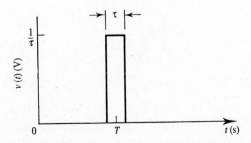

Figure 2.29 *Tall, narrow rectangular pulse of unit strength.*

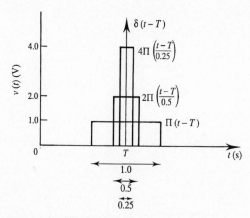

Figure 2.30 *Development of unit strength impulse, $\delta(t - T)$, as a limit of a sequence of unit strength rectangular pulses.*

$$\delta(t - T) = \begin{cases} \infty, & t = T \\ 0, & t \neq T \end{cases} \qquad (2.45(a))$$

$$\int_{-\infty}^{\infty} \delta(t - T) \, dt = \int_{T^-}^{T^+} \delta(t - T) \, dt = 1.0 \qquad (2.45(b))$$

More strictly the impulse is *defined* by its sampling, or *sifting*, property under integration, i.e.:

$$\int_{-\infty}^{\infty} \delta(t - T) \, f(t) \, dt = f(T) \qquad (2.46(a))$$

That equation (2.46(a)) is consistent with equations (2.45) is easily shown as follows:

$$\int_{-\infty}^{\infty} \delta(t - T) \, f(t) \, dt = \int_{T^-}^{T^+} \delta(t - T) \, f(t) \, dt$$

$$= \int_{T^-}^{T^+} \delta(t - T) \, f(T) \, dt$$

$$= f(T) \int_{T^-}^{T^+} \delta(t - T) \, dt = f(T) \qquad (2.46(b))$$

Notice that if we insist that the strength of the impulse has units of Vs, i.e. its amplitude has units of V, then the sampled quantity, $f(T)$, in equations (2.46) would have units of V^2s (or joules in 1 Ω). In view of this the impulse is usually taken to have an amplitude measured in s^{-1} (i.e. to have dimensionless strength). This can be reconciled with an

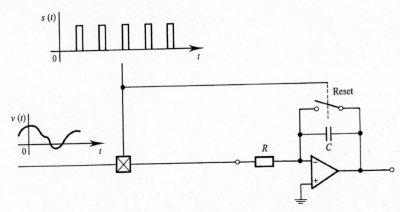

Figure 2.31 *Hypothetical sampling system reconciling physical units of impulse strength (Vs) with units of sampled signal (V).*

equivalent physical implementation of sampling using tall, narrow pulses, a multiplier and integrator by associating dimensions of V^{-1} with the multiplier (required for its output to have units of V) and dimensions of s^{-1} with the integrator (required for its output also to have units of V). Such an implementation, shown in Figure 2.31, is not, of course, used in practical sampling circuits.

As a rectangular pulse gets narrower its Fourier transform (which is a sinc function) gets wider, Figure 2.32. This reciprocal width relationship is a general property of all Fourier transform pairs. Using equation (2.43(b)) it can be seen that as $\tau \to 0$ then $\tau f \to 0$ and $\mathrm{sinc}(\tau f) \to 1.0$. It follows that:

$$\lim_{\tau \to 0} \mathrm{FT}\left\{ \frac{1}{\tau} \Pi\left(\frac{t-T}{\tau} \right) \right\} = e^{-j2\pi fT} \tag{2.47(a)}$$

i.e.:

$$\delta(t-T) \overset{\mathrm{FT}}{\Longleftrightarrow} e^{-j2\pi fT} \tag{2.47(b)}$$

For an impulse occurring at the origin this reduces to:

$$\delta(t) \overset{\mathrm{FT}}{\Longleftrightarrow} 1.0 \tag{2.47(c)}$$

The amplitude spectrum of an impulse function is therefore a constant (measured in V/Hz if $\delta(t)$ has units of V). Such a spectrum is sometimes referred to as white, since all frequencies are present in equal quantities. This is analogous to white light. (From a strict mathematical point of view the impulse function, *as represented here*, does not have a Fourier transform owing to the infinite discontinuity which it contains. The impulse and constant in equation (2.47(c)) can be approximated so closely by tall, thin rectangular pulses and broad sinc pulses, however, that the limiting forms need not be challenged. In any event, if desired, the impulse can be derived as the limiting form of other pulse shapes which contain no discontinuity.)

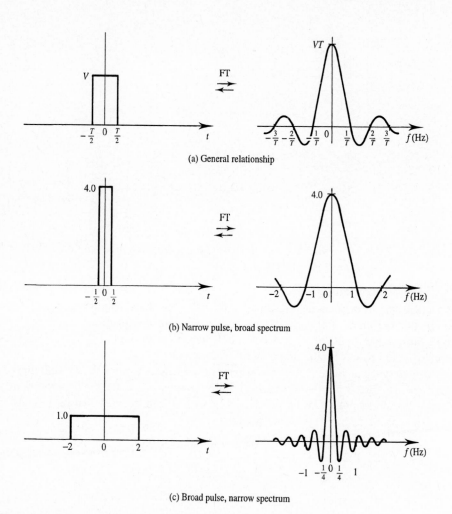

(a) General relationship

(b) Narrow pulse, broad spectrum

(c) Broad pulse, narrow spectrum

Figure 2.32 *Inverse width relationship between Fourier transform pairs.*

2.3.4 Fourier transform theorems and convolution

Since signals can be fully described in either the time or frequency domain it follows that any operation on a signal in one domain has a precisely equivalent operation in the other domain. A list of equivalent operations on $v(t)$ and its transform, $V(f)$, is given in the form of a set of theorems in Table 2.5. Most of the operations in this list (addition, multiplication, differentiation, integration), whether applied to the functions themselves or their arguments, will be familiar. One operation, namely convolution, may be unfamiliar, however, and is therefore described below.

Table 2.5 *Fourier transform theorems.*

Linearity	$av(t) + bw(t)$	$aV(f) + bW(f)$
Time delay	$v(t - T)$	$V(f)e^{-j\omega T}$
Change of scale	$v(at)$	$\dfrac{1}{|a|} V\left(\dfrac{f}{a}\right)$
Time reversal	$v(-t)$	$V(-f)$
Time conjugation	$v^*(t)$	$V^*(-f)$
Frequency conjugation	$v^*(-t)$	$V^*(f)$
Duality	$V(t)$	$v(-f)$
Frequency translation	$v(t)e^{j\omega_c t}$	$V(f - f_c)$
Modulation	$v(t)\cos(\omega_c t + \phi)$	$\tfrac{1}{2}\left[e^{j\phi}V(f - f_c) + e^{-j\phi}V(f + f_c)\right]$
Time differentiation	$\dfrac{d^n}{dt^n} v(t)$	$(j2\pi f)^n\, V(f)$
Integration (1)	$\displaystyle\int_{-\infty}^{t} v(t')dt'$	$(j2\pi f)^{-1}V(f) + \tfrac{1}{2}V(0)\,\delta(f)$
Integration (2)	$\displaystyle\int_{0}^{t} v_e(t')\,dt' + \int_{-\infty}^{t} v_o(t')\,dt'$	$(j2\pi f)^{-1}V(f)$
Convolution	$v(t) * w(t)$	$V(f)W(f)$
Multiplication	$v(t)w(t)$	$V(f) * W(f)$
Frequency differentiation	$t^n v(t)$	$(-j2\pi)^{-n} \dfrac{d^n}{df^n} V(f)$

DC value	$V(0) = \displaystyle\int_{-\infty}^{\infty} v(t)\, dt$
Value at the origin	$v(0) = \displaystyle\int_{-\infty}^{\infty} V(f)\, df$
Integral of a product	$\displaystyle\int_{-\infty}^{\infty} v(t)w^*(t)\, dt = \int_{-\infty}^{\infty} V(f)W^*(f)\, df$

Convolution is normally denoted by * although $\otimes$ is also sometimes used. Applied to two time function $z(t) = f(t) * g(t)$ is defined by:

$$z(t) = \int_{-\infty}^{\infty} f(\tau)\, g(t - \tau)\, d\tau \qquad (2.48(a))$$

Figure 2.33 illustrates time convolution graphically. It can be thought of as a five step process:

1. The arguments of the functions to be convolved are replaced with a dummy variable (in this case τ), Figure 2.33(b).
2. One of the functions (arbitrarily chosen here to be $g(\tau)$) is reversed in its argument (i.e. reflected about $\tau = 0$) giving $g(-\tau)$, Figure 2.33(c).

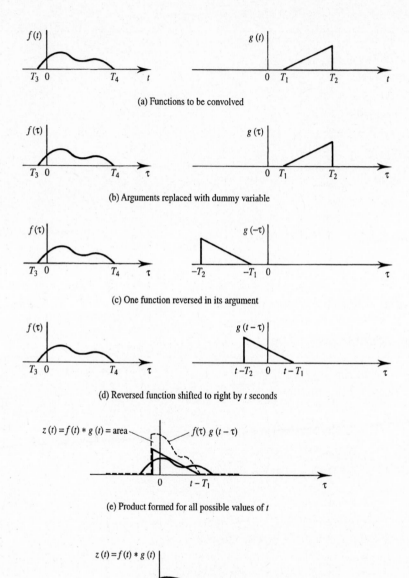

(a) Functions to be convolved

(b) Arguments replaced with dummy variable

(c) One function reversed in its argument

(d) Reversed function shifted to right by t seconds

(e) Product formed for all possible values of t

(f) Area of product found for all possible values of t

Figure 2.33 *Graphical illustration of time convolution.*

3. A variable time shift, t, is introduced into the argument of the reflected function giving $g(t - \tau)$. This is a version of $g(-\tau)$ shifted to the right by t s, Figure 2.33(d).

4. The product function $f(\tau)g(t - \tau)$ is formed for every possible value of t, Figure 2.33(e). (This function changes continuously as t varies and $g(t - \tau)$ slides across $f(\tau)$.)

5. The area under the product function is calculated (by integrating) for every value of t.

There are several important points to note about convolution:

1. The convolution integral, equation (2.48(a)), is sometimes called the superposition integral.

2. It is simply convention which dictates that dummy variables are used so that the result can be expressed as a function of the argument of f and g. There is no reason, in principle, why the alternative definition:

$$z(\tau) = \int_{-\infty}^{\infty} f(t)\, g(\tau - t)\, dt \tag{2.48(b)}$$

should not be used.

3. Convolution is not restricted to the time domain. It can be applied to functions of any variable, for example frequency, f, i.e.:

$$Z(f) = F(f) * G(f) = \int_{-\infty}^{\infty} F(\phi)G(f - \phi)\, d\phi \tag{2.49(a)}$$

or space, x, i.e.:

$$z(x) = f(x) * g(x) = \int_{-\infty}^{\infty} f(\lambda)\, g(x - \lambda)\, d\lambda \tag{2.49(b)}$$

4. The unitary operator for convolution is the impulse function $\delta(t)$ since convolution of $f(t)$ with $\delta(t)$ leaves $f(t)$ unchanged, i.e.:

$$f(t) * \delta(t) = \int_{-\infty}^{\infty} f(\tau)\, \delta(t - \tau)\, d\tau = f(t) \tag{2.50(a)}$$

(This follows directly from the sampling property of $\delta(t)$ under integration.)

5. Convolution in the time domain corresponds to multiplication in the frequency domain and vice versa.

6. Convolution is commutative, associative and distributive, i.e.:

$$f * g = g * f \tag{2.50(b)}$$

$$f * (g * h) = (f * g) * h \tag{2.50(c)}$$

$$f * (g + h) = f * g + f * h \tag{2.50(d)}$$

7. The derivative of a convolution is the derivative of one function convolved with the other, i.e.:

$$\frac{d}{dt}[v(t) * w(t)] \;=\; \frac{dv(t)}{dt} * w(t) \;=\; v(t) * \frac{dw(t)}{dt} \tag{2.50(e)}$$

EXAMPLE 2.4

Convolve the two transient signals, $x(t) = \Pi[(t-1)/2]\sin \pi t$ and $h(t) = 2\,\Pi[(t-2)/2])$, shown in Figure 2.34(a).

For $t < 1$ the picture of the convolution process looks like Figure 2.34(b), i.e. $y(t) = x(t) * h(t) = 0$.

For $1 \le t \le 3$ the picture looks like Figure 2.34(c) and:

$$y(t) = \int_0^{t-1} 2\sin(\pi\tau)\,d\tau$$

$$= 2\left[\frac{-\cos \pi\tau}{\pi}\right]_0^{t-1}$$

$$= \frac{2}{\pi}[1 - \cos \pi(t-1)]$$

For $3 \le t \le 5$ the picture looks like Figure 2.34(d):

$$y(t) = \int_{t-3}^2 2\sin(\pi\tau)\,d\tau$$

$$= 2\left[\frac{-\cos(\pi\tau)}{\pi}\right]_{t-3}^2$$

$$= \frac{2}{\pi}[\cos \pi(t-3) - 1]$$

For $t > 5$ there is no overlapping of the two functions; therefore $y(t) = 0$. Figure 2.34(e) shows a sketch of $y(t)$. Note that the convolved output signal has a duration of 4 time units (i.e. the sum of the durations of the input signals) and, when the two input signals exactly overlap at $t = 3$, the output is 0 as expected.

EXAMPLE 2.5

Convolve the function $\Pi(t - \tfrac{1}{2})$ with itself and show that the Fourier transform of the result is the square of the Fourier transform of $\Pi(t - \tfrac{1}{2})$.

$$z(t) = \Pi(t - \tfrac{1}{2}) * \Pi(t - \tfrac{1}{2}) = \int_{-\infty}^{\infty} \Pi(\tau - \tfrac{1}{2})\,\Pi(t - \tau - \tfrac{1}{2})\,d\tau$$

For $t < 0$, $z(t) = 0$ (by inspection), Figure 2.35(a).

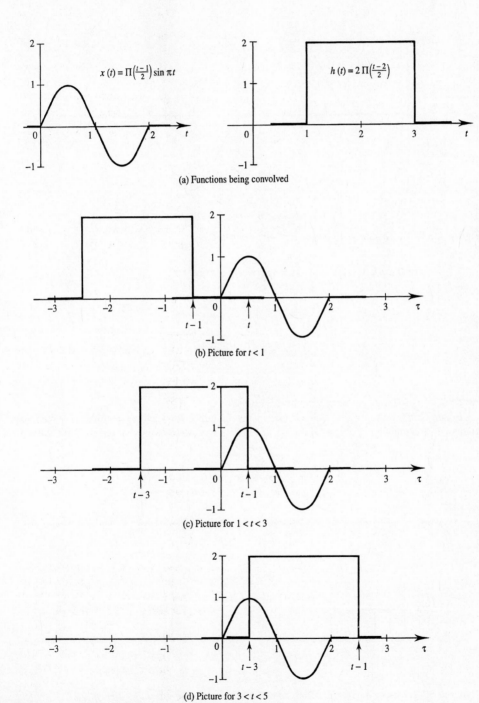

(a) Functions being convolved

(b) Picture for $t < 1$

(c) Picture for $1 < t < 3$

(d) Picture for $3 < t < 5$

Figure 2.34 *Graphical illustration of time convolution.*

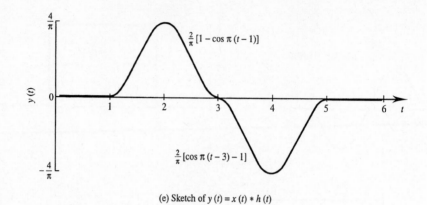

(e) Sketch of $y(t) = x(t) * h(t)$

Figure 2.34-ctd. *Graphical illustration of time convolution.*

For $0 < t < 1$, $z(t) = \displaystyle\int_0^t (1 \times 1)\, d\tau$ (Figure 2.35(b))

$$= [\,\tau\,]_0^t = t$$

For $1 < t < 2$, $z(t) = \displaystyle\int_{t-1}^1 (1 \times 1)\, d\tau$ (Figure2.35(c))

$$= [\,\tau\,]_{t-1}^1 = 2 - t$$

For $t > 2$, $z(t) = 0$ (by inspection), Figure 2.35(d).

Figure 2.35(e) shows a sketch of $z(t)$. This function, for obvious reasons, is called the triangular function which, if centred on $t = 0$, is denoted by $\Lambda(t)$. (Note that the absolute width of

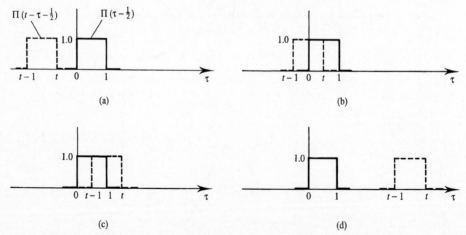

Figure 2.35 *Illustration (a) – (d) of self convolution of a rectangular pulse, Example 2.5.*

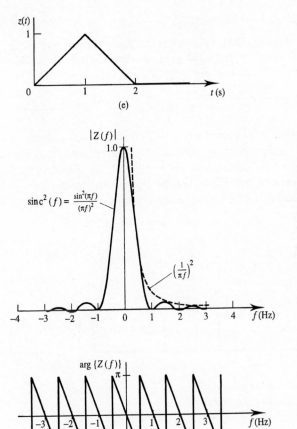

Figure 2.35-ctd. *Convolution result (e) and corresponding amplitude and phase spectra (f).*

$\Pi(t)$ is 1.0 whilst the width of $\Lambda(t)$ is 2.0.) Since the triangular function is centred here on $t = 1$ then we can use the time delay theorem and the tables of Fourier transform pairs to obtain:

$$\Lambda(t - 1) \stackrel{FT}{\Leftrightarrow} \text{sinc}^2(f)\, e^{-j\omega 1}$$

The square of the Fourier transform of $\Pi(t - \frac{1}{2})$, using the time delay theorem and the table of pairs again, is given by:

$$\left[FT\left\{ \Pi\left(t - \frac{1}{2}\right)\right\} \right]^2 = \left[\text{sinc}(f)\, e^{-j\omega/2} \right]^2$$

$$= \text{sinc}^2(f)\, e^{-j\omega}$$

The amplitude and phase spectra of $z(t)$ are given by:

$$|Z(f)| = \text{sinc}^2(f) \quad \text{(V/Hz)}$$

$$\arg(Z(f)) = -2\pi f \quad \text{(rad)}$$

and are shown in Figure 2.35(f). (Notice that the phase spectrum really takes this form of a simple straight line with intercept zero, and gradient -2π (rad/Hz). It is conventional, however, to constrain its plot to the range $[-\pi, \pi]$ or sometimes $[0, 2\pi]$.)

Fourier transforms and Fourier series are, clearly, closely related. In fact by using the impulse function (section 2.3.3) a Fourier transform of a periodic function can be defined. Consider the periodic rectangular pulse stream $\sum_{n=-\infty}^{\infty}\Pi[(t-nT)/\tau]$, shown in Figure 2.36(a). This periodic waveform can be represented by the convolution of a transient signal (corresponding to the single period given by $n = 0$) with the periodic impulse train $\sum_{n=-\infty}^{\infty}\delta(t-nT)$, i.e.:

$$\sum_{n=-\infty}^{\infty} \Pi\left(\frac{t-nT}{\tau}\right) = \Pi\left(\frac{t}{\tau}\right) * \sum_{n=-\infty}^{\infty} \delta(t-nT) \qquad (2.51(a))$$

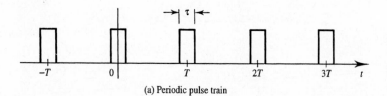

(a) Periodic pulse train

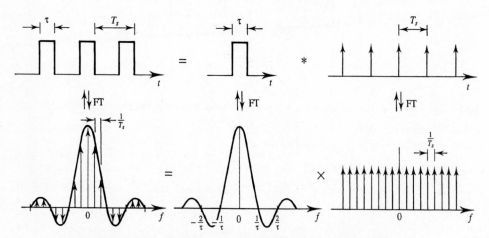

(b) Time and frequency domain representation of a periodic pulse train showing spectral lines arising from periodicity and spectral envelope arising from pulse shape

Figure 2.36 *Time and frequency domain representation of a periodic pulse train.*

(Each impulse in the impulse train reproduces the rectangular pulse in the convolution process.) In the same way that the Fourier transform of a single impulse can be defined (as a limiting case) the Fourier transform of an impulse train is defined (in the limit) as another impulse train. There is the usual relationship between the width (or period) of the impulse train in time and frequency domains, i.e.:

$$\sum_{k=-\infty}^{\infty} \delta(t - kT_s) \overset{FT}{\Leftrightarrow} f_s \sum_{n=-\infty}^{\infty} \delta(f - nf_s) \qquad (2.51(b))$$

where the time domain period, T_s, is the reciprocal of the frequency domain period, f_s. (The subscript s is used because the impulse train in communications engineering is often employed as a 'sampling function', T_s and f_s, in this context, being the sampling period and sampling frequency respectively.) The voltage spectrum of a rectangular pulse train can therefore be obtained by taking the Fourier transform of equation (2.51(a)), Figure 2.36(b), i.e.:

$$FT\left\{ \sum_{n=-\infty}^{\infty} \Pi\left(\frac{t - nT}{\tau} \right) \right\} = FT\left\{ \Pi\left(\frac{t}{\tau} \right) \right\} FT\left\{ \sum_{n=-\infty}^{\infty} \delta(t - nT) \right\}$$

$$= \tau \, \text{sinc}(\tau f) \frac{1}{T} \sum_{n=-\infty}^{\infty} \delta\left(f - \frac{n}{T} \right) \quad \text{(V/Hz)} \quad (2.51(c))$$

This shows that the spectrum is given by a periodic impulse train (i.e. a line spectrum) with impulse (or line) separation of $1/T$ and impulse (or line) strength of $(\tau/T)\text{sinc}(\tau f)$. $((\tau/T)\text{sinc}(\tau f))$ is usually called the spectrum envelope and, although real here, is potentially complex.)

The technique demonstrated here works for any periodic waveform, the separation of spectral lines being given by $1/T$ and the spectral envelope being given by $1/T$ times the Fourier transform of the single period contained in interval $[-T/2, T/2]$.

EXAMPLE 2.6
Sketch the Fourier spectra for a rectangular pulse train comprising pulses of amplitude A V and width 0.05 s with the following pulse repetition periods: (a) ¼ s; (b) ½ s; (c) 1 s.

The spectral envelope is controlled by the rectangular pulses of width 0.05 s. The spectrum is sinc x shaped, Figure 2.32, with the first zeros at $\pm 1/0.05 = \pm 20$ Hz. In all cases the waveform is periodic so the frequency spectrum can be represented by a Fourier series in which the lines, Figure 2.37, are spaced by $1/T_s$ Hz where T_s is the period in seconds.

Thus for (a) the lines occur every 4 Hz and the 0 Hz component, C_0 in equation (2.15), has a magnitude of $A/20 \times 1/T_s = 4A/20 = A/5$. The other components $C_1, C_2, C_3, \cdots, C_n$ follow the sinc x envelope as shown in Figure 2.37(b).

For the case (b) where T_s is ½ s then the spectral lines are now $1/T_s = 2$ Hz apart which is half the spacing of the ¼ s period case in part (a). The envelope of the Fourier spectrum is unaltered but the C_0 term reduces in amplitude owing to the longer period. Thus $C_0 = A/20 \times 2 = A/10$ and the waveform and spectrum are shown in Figure 2.37(b).

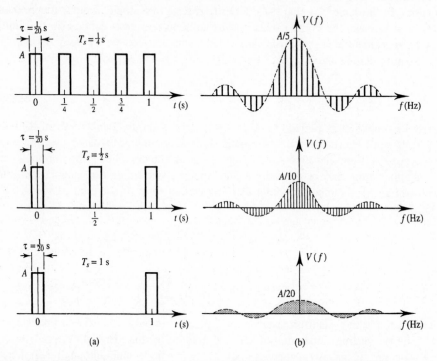

Figure 2.37 *Pulsed waveform and corresponding frequency spectra with specific values for pulse repetition period.*

For (c) the line spacing becomes 1 Hz and the amplitude at 0 Hz drops to $A/20$.

2.4 Power and energy spectra

As an alternative to plotting peak or RMS voltage against frequency the quantity:

$$G_1(f) = |V_{RMS}(f)|^2 \quad (V^2) \tag{2.52(a)}$$

can be plotted for periodic signals. This is a line spectrum the ordinate of which has units of V^2 (or watts in a 1 Ω resistive load) representing *normalised* power. (The subscript 1 indicates that the spectrum is single sided.) If the impedance level, R, is not 1 Ω then the absolute (i.e. non-normalised) power spectrum is given by:

$$G_1(f) = \frac{|V_{RMS}(f)|^2}{R} \quad (W) \tag{2.52(b)}$$

Figure 2.38(a) shows such a power spectrum for a periodic signal. Although each line in Figure 2.38(a) no longer represents a rotating phasor, two sided power spectra are still often defined by associating half the power in each spectral line with a negative frequency, Figure 2.38(b). Notice that this means that the total power in a signal is the sum of the powers in all its spectral lines irrespective of whether a one or two sided spectral representation is being used.

For a transient signal the two sided voltage spectrum $V(f)$ has units of V/Hz and the quantity:

$$E_2(f) = |V(f)|^2 \quad (V^2 \text{ s/Hz}) \tag{2.53(a)}$$

therefore has units of V^2/Hz^2 or V^2 s/Hz. The corresponding non-normalised spectrum is given by:

$$E_2(f) = \frac{|V(f)|^2}{R} \quad (\text{J/Hz}) \tag{2.53(b)}$$

where R is load resistance (in Ω). The quantity $E_2(f)$ now has units of W s/Hz or J/Hz and is therefore called an *energy* spectral density. Like power spectra, energy spectra can be presented as two or one sided, Figure 2.39(a) and (b). Note that energy spectra are always continuous (never discrete) whilst power spectra can be either discrete (as is the case for periodic waveforms) or continuous (as is the case for random signals). Continuous power spectra, i.e. power spectral densities, are discussed in section 3.3.3 and, in the context of linear systems, in section 4.6.1.

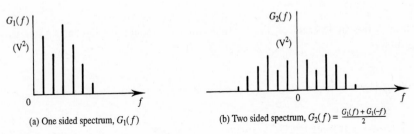

(a) One sided spectrum, $G_1(f)$

(b) Two sided spectrum, $G_2(f) = \frac{G_1(f) + G_1(-f)}{2}$

Figure 2.38 *Power spectra of a periodic signal.*

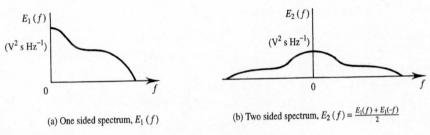

(a) One sided spectrum, $E_1(f)$

(b) Two sided spectrum, $E_2(f) = \frac{E_1(f) + E_1(-f)}{2}$

Figure 2.39 *Energy spectral densities of a transient signal.*

2.5 Generalised orthogonal function expansions

Fourier series and transforms constitute a special case of a more general mathematical technique, namely the orthogonal function expansion. The concept of orthogonal functions is closely connected with that of orthogonal (i.e. perpendicular) vectors. For this reason the important characteristics and properties of vectors are now reviewed.

2.5.1 Review of vectors

Vectors possess magnitude, direction and sense. They can be added, Figure 2.40(a), i.e.:

$$\mathbf{a}, \mathbf{b} \rightarrow \mathbf{a} + \mathbf{b}$$

and multiplied by a scalar, λ, Figure 2.40(b), i.e.:

$$\mathbf{a} \rightarrow \lambda \mathbf{a}$$

Their properties [Spiegel] include commutation, distribution and association, i.e.:

$$\mathbf{a} + \mathbf{b} = \mathbf{b} + \mathbf{a} \tag{2.54(a)}$$

$$\lambda(\mathbf{a} + \mathbf{b}) = \lambda\mathbf{a} + \lambda\mathbf{b} \tag{2.54(b)}$$

$$\lambda(\mu\mathbf{a}) = (\lambda\mu)\mathbf{a} \tag{2.54(c)}$$

A scalar product of two vectors, Figure 2.41, can be defined by:

$$\mathbf{a} \cdot \mathbf{b} = |\mathbf{a}|\,|\mathbf{b}|\cos\theta \tag{2.55}$$

where θ is the angle between the vectors and the modulus $|\ |$ indicates their length or magnitude.

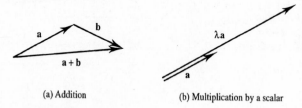

(a) Addition   (b) Multiplication by a scalar

Figure 2.40 *Fundamental vector operations.*

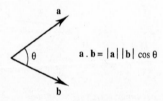

Figure 2.41 *Scalar product of vectors.*

The distance between two vectors, i.e. their difference, is found by reversing the sense of the vector to be subtracted and adding, Figure 2.42:

$$\mathbf{a} - \mathbf{b} = \mathbf{a} + (-\mathbf{b}) \tag{2.56}$$

A vector in three dimensions can be specified as a weighted sum of any three non-coplanar *basis* vectors:

$$\mathbf{a} = a_1\mathbf{e}_1 + a_2\mathbf{e}_2 + a_3\mathbf{e}_3 \tag{2.57}$$

If the vectors are mutually perpendicular, i.e.:

$$\mathbf{e}_1 . \mathbf{e}_2 = \mathbf{e}_2 . \mathbf{e}_3 = \mathbf{e}_3 . \mathbf{e}_1 = 0 \tag{2.58(a)}$$

then the three vectors are said to form an *orthogonal* set. If, in addition, the three vectors have unit length:

$$\mathbf{e}_1 . \mathbf{e}_1 = \mathbf{e}_2 . \mathbf{e}_2 = \mathbf{e}_3 . \mathbf{e}_3 = 1 \tag{2.58(b)}$$

then they are said to form an *orthonormal* set. The above concepts can be extended to vectors with any number of dimensions, N. Such a vector can be represented as the sum of N basis vectors, i.e.:

$$\mathbf{x} = \sum_{i=1}^{N} \lambda_i \mathbf{e}_i \tag{2.59}$$

If $\mathbf{e}_i$ is an orthonormal basis it is easy to find the values of λ_i:

$$\mathbf{x} . \mathbf{e}_j = \sum_{i=1}^{N} \lambda_i \mathbf{e}_i . \mathbf{e}_j = \lambda_j \tag{2.60}$$

(since $\mathbf{e}_i . \mathbf{e}_j = 0$ for $i \neq j$ and $\mathbf{e}_i . \mathbf{e}_j = 1$ for $i = j$). Scalar products of vectors expressed using orthonormal bases are also simple to calculate:

$$\mathbf{x} . \mathbf{y} = \sum_{i=1}^{N} \lambda_i \mathbf{e}_i . \sum_{j=1}^{N} \mu_j \mathbf{e}_j \tag{2.61(a)}$$

i.e.:

$$\mathbf{x} . \mathbf{y} = \sum_{i=1}^{N} \lambda_i \mu_i \tag{2.61(b)}$$

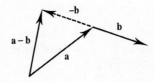

Figure 2.42 *Subtraction of vectors.*

Equation (2.61(b)) is the most general form of Parseval's theorem, equation (2.28). A special case of this theorem is:

$$\mathbf{x} . \mathbf{x} = \sum_{i=1}^{N} \lambda_i^2 \tag{2.61(c)}$$

which has already been discussed in the context of the power contained in a periodic waveform (section 2.2.3).

If the number of basis vectors available to express an N-dimensional vector is limited to M ($M < N$) then, provided the basis is orthonormal, the best approximation (in a least square error sense) is given by:

$$\mathbf{x} \simeq \sum_{i=1}^{M} \lambda_i \mathbf{e}_i = \mathbf{x}_M \tag{2.62}$$

where $\lambda_i = \mathbf{x} . \mathbf{e}_i$ as before. This is easily proved by calculating the squared error:

$$|\mathbf{x} - \mathbf{x}_M|^2 = \left| \mathbf{x} - \sum_{i=1}^{M} \lambda_i \mathbf{e}_i \right|^2$$

$$= |\mathbf{x}|^2 - 2 \sum_{i=1}^{M} \lambda_i (\mathbf{e}_i . \mathbf{x}) + \sum_{i=1}^{M} \lambda_i^2$$

$$= |\mathbf{x}|^2 + \sum_{i=1}^{M} (\lambda_i - \mathbf{e}_i . \mathbf{x})^2 - \sum_{i=1}^{M} (\mathbf{e}_i . \mathbf{x})^2 \tag{2.63}$$

The right hand side of equation (2.63) is clearly minimised by putting $\lambda_i = \mathbf{x} . \mathbf{e}_i$.

From the definition of the scalar product it is apparent that:

$$|\mathbf{x} . \mathbf{y}| \leq |\mathbf{x}| \, |\mathbf{y}| \tag{2.64}$$

This holds for any kind of vector providing that the scalar product is defined to satisfy:

$$\mathbf{x} . \mathbf{x} \geq 0, \text{ for all } \mathbf{x} \tag{2.65(a)}$$

$$\mathbf{x} . \mathbf{x} = 0, \text{ only if } \mathbf{x} = \mathbf{0} \tag{2.65(b)}$$

where $\mathbf{0}$ is a null vector. (Equation (2.64) is a particularly simple form of the Schwartz inequality, see equation 2.71(a).) In order to satisfy equation (2.65(a)) the scalar product of *complex* vectors must be defined by:

$$\mathbf{x} . \mathbf{y} = \sum_{i=1}^{N} \lambda_i^* \mu_i \tag{2.66}$$

2.5.2 Vector interpretation of waveforms

Nyquist's sampling theorem (section 2.2.5) asserts that a *periodic* signal having a highest frequency component located at f_H Hz is *fully* specified by N samples spaced $1/2f_H$ s

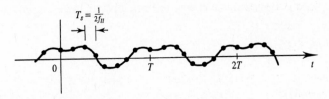

Figure 2.43 *Nyquist sampling of a periodic function (N = 8).*

apart, Figure 2.43. Since all the 'information' in a periodic waveform is contained in this set of independent samples (which repeat indefinitely), each sample value can be regarded as being the length of one vector belonging to an orthogonal basis set. A waveform requiring N samples per period for its specification can therefore be interpreted as an N-dimensional vector. For sampling at the Nyquist rate the intersample spacing is $T_s = 1/2f_H$ and:

$$N = T/T_s = 2f_H T \qquad (2.67(a))$$

where T is the period of the waveform. In a slightly more general form:

$$N = 2BT \qquad (2.67(b))$$

where B is waveform bandwidth. This is called the dimensionality theorem. For transient signals an infinite number of samples would be required to retain all the signal's information.

Functions can be added and scaled in the same way as vectors to produce new functions. A scalar product for certain periodic signals can be defined as a continuous version of equation (2.66) but with a factor $1/T'$ so that the scalar product has dimensions of V^2 and can be interpreted as a *cross-power*, i.e.:

$$[f(t), g(t)] = \frac{1}{T'} \int_0^{T'} f^*(t)\, g(t)\, dt$$

$$= \langle f^*(t)\, g(t) \rangle \quad (V^2) \qquad (2.68(a))$$

where T' is the period of the product $f^*(t)g(t)$. (The notation $[f(t), g(t)]$ is used here to denote the scalar product of functions rather than $\mathbf{f} \cdot \mathbf{g}$ as used for vectors.) More generally, the definition adopted is:

$$[f(t), g(t)] = \lim_{T' \to \infty} \frac{1}{T'} \int_{-T'/2}^{T'/2} f^*(t)\, g(t)\, dt$$

$$= \langle f^*(t)\, g(t) \rangle \quad (V^2) \qquad (2.68(b))$$

since this includes the possibility that $f^*(t)g(t)$ has infinite period. The corresponding definition for transient signals is a *cross-energy*, i.e.:

$$[f(t), g(t)] = \int_{-\infty}^{\infty} f^*(t)\, g(t)\, dt \quad (V^2\, s) \qquad (2.68(c))$$

The scalar products of periodic and transient signals with themselves, which represent average signal power and total signal energy respectively, are therefore:

$$[f(t), f(t)] \equiv \frac{1}{T} \int_0^T |f(t)|^2 \, dt \quad (\text{V}^2) \tag{2.69(a)}$$

$$[f(t), f(t)] \equiv \int_{-\infty}^{\infty} |f(t)|^2 \, dt \quad (\text{V}^2 \, \text{s}) \tag{2.69(b)}$$

Interpreting the signals as vectors, equations (2.69) correspond to finding the vector's square magnitude, i.e.:

$$\mathbf{x} \cdot \mathbf{x} = \sum_{i=1}^{N} \lambda_i^* \lambda_i$$

$$= \sum_{i=1}^{N} |\lambda_i|^2 = |\mathbf{x}|^2 \tag{2.70}$$

Using the definition of a scalar product for complex vectors, equation (2.66), the Schwartz inequality, equation (2.64), becomes:

$$\left| \int_{-\infty}^{\infty} f^*(t) \, g(t) \, dt \right| \leq \left[\int_{-\infty}^{\infty} |f(t)|^2 \, dt \right]^{\frac{1}{2}} \left[\int_{-\infty}^{\infty} |g(t)|^2 \, dt \right]^{\frac{1}{2}} \tag{2.71(a)}$$

where the equality holds if and only if:

$$g(t) = C \, f^*(t) \tag{2.71(b)}$$

Equations (2.71(a)) and (b) can be used to derive the optimum response of predetection filters in digital communications receivers.

2.5.3 Orthogonal and orthonormal signals

Consider a periodic signal, $f(t)$, with only three dimensions[1], i.e. with $N = 2BT = 3$, Figure 2.44. In sample space, Figure 2.45(a), the signal is represented by:

$$\mathbf{f} = \sum_{i=1}^{3} F_i \, \hat{\mathbf{a}}_i \tag{2.72(a)}$$

where $\hat{\mathbf{a}}_i$ represents an orthonormal sample set. The same function $\mathbf{f}$ could, however, be described in a second orthonormal coordinate system, Figure 2.45(b), rotated with respect

---

[1]   In reality the number of dimensions, N, must be even since for a periodic signal the maximum frequency, B Hz, must be an integer multiple, n, of the fundamental frequency $1/T$ Hz. The dimensionality theorem can therefore be written as $N = 2(n/T)T = 2n$. Choosing $N = 2$, however, trivialises this in that the orthogonal functions become phasors whilst choosing $N = 4$ precludes signal vector visualisation in 3-dimensional space.

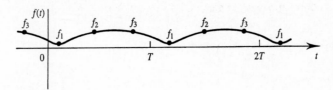

Figure 2.44 *Three dimensional (i.e. 3-sample) function.*

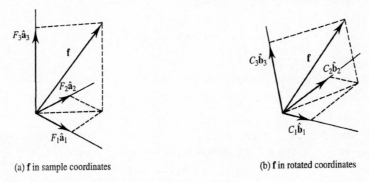

(a) **f** in sample coordinates

(b) **f** in rotated coordinates

Figure 2.45 *Vector interpretation of a 3-dimensional signal.*

to the first, i.e.:

$$\mathbf{f} = \sum_{i=1}^{3} C_i\, \hat{\mathbf{b}}_i \qquad (2.72(\text{b}))$$

Each unit vector $\hat{\mathbf{b}}_j$ itself represents a three sample function:

$$\hat{\mathbf{b}}_j = \sum_{i=1}^{3} (\hat{\mathbf{b}}_j \cdot \hat{\mathbf{a}}_i)\hat{\mathbf{a}}_i \qquad (2.73)$$

This demonstrates the important idea that a function or signal having N dimensions (in this case three) can be expressed in terms of a weighted sum of N other orthonormal, N-dimensional, functions. (Actually these *basis* functions do not have to be orthonormal or even orthogonal providing none can be exactly expressed as a linear sum of the others.)

Generalising the vector notation to make it more appropriate for signals (which may include the case of transient signals where $N = \infty$) **f** is replaced by $f(t)$ and $\mathbf{b}_i$ (which represents an orthogonal but not necessarily orthonormal set) is replaced by $\phi_i(t)$. Equation (2.72(b)) then becomes:

$$f(t) = \sum_{i=1}^{N} C_i\, \phi_i(t) \qquad (2.74)$$

If the basis function set $\phi_i(t)$ is orthonormal over an interval $[a, b]$ then:

$$\int_a^b \phi_i(t)\,\phi_j^*(t)\,dt = \begin{cases} 1, & i = j \\ 0, & i \neq j \end{cases} \tag{2.75}$$

which corresponds to the vector property:

$$\mathbf{b}_i \cdot \mathbf{b}_j = \begin{cases} 1, & i = j \\ 0, & i \neq j \end{cases} \tag{2.76}$$

If the functions are orthogonal but not orthonormal then the upper expression on the right hand side of equation (2.75) does not apply.

EXAMPLE 2.7

Consider the functions shown in Figure 2.46. Do these functions form an orthogonal set over the range $[-1, 1]$? Do these functions form an orthonormal set over the same range?

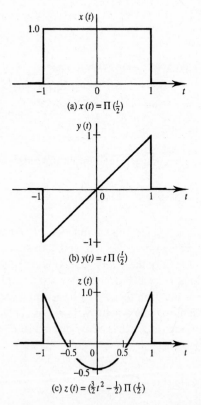

(a) $x(t) = \Pi\left(\frac{t}{2}\right)$

(b) $y(t) = t\,\Pi\left(\frac{t}{2}\right)$

(c) $z(t) = \left(\frac{3}{2}t^2 - \frac{1}{2}\right)\Pi\left(\frac{t}{2}\right)$

Figure 2.46 *Three functions tested for orthogonality and orthonormality in Example 2.7.*

To determine whether the functions form an orthogonal set we use equation (2.75):

$$\int_a^b \phi_i(t)\, \phi_j^*(t)\, dt = 0 \quad (i \neq j)$$

(Since all the functions are real the conjugate symbol is immaterial here.)

First examine $x(t)$ and $y(t)$. Since the product is odd about zero the integral must be zero, i.e. $x(t)$ and $y(t)$ are orthogonal.

Now examine $x(t)$ and $z(t)$.

$$\int_{-1}^1 x(t)\, z(t)\ dt = \int_{-1}^1 \left(\frac{3}{2} t^2 - \frac{1}{2} \right) dt$$

$$= \frac{3}{2} \left[\frac{t^3}{3} \right]_{-1}^1 - \frac{1}{2} [\, t\,]_{-1}^1 = 0$$

i.e. $x(t)$ and $z(t)$ are orthogonal.

Finally we examine $y(t)$ and $z(t)$. Here the product is again odd about zero and the integral is therefore zero by inspection, i.e. $y(t)$ and $z(t)$ are orthogonal.

To establish the normality or otherwise of the functions we test the square integral against 1.0:

$$\int_a^b |\, \phi_i(t)\, |^2\, dt = 1 \quad \text{for normal} \ \ \phi_i(t)$$

The square integral of $x(t)$ is 2.0 by inspection. We need go no further, therefore, since if any function in the set fails this test then the set is not orthonormal.

2.5.4 Evaluation of basis function coefficients

Equation (2.74) can be multiplied by $\phi_j^*(t)$ to give:

$$f(t)\, \phi_j^*(t) = \sum_{i=1}^N C_i\, \phi_i(t)\, \phi_j^*(t) \tag{2.77}$$

Integrating and reversing the order of integration and summation on the right hand side:

$$\int_a^b f(t)\, \phi_j^*(t)\, dt = \sum_{i=1}^N C_i \int_a^b \phi_i(t)\, \phi_j^*(t)\, dt \tag{2.78}$$

Since the integral on the right hand side is zero for all $i \neq j$ equation (2.78) can be rewritten as:

$$\int_a^b f(t)\, \phi_j^*(t)\, dt = C_j \int_a^b |\phi_j(t)|^2\, dt \tag{2.79}$$

and rearranging equation (2.79) gives an explicit formula for C_j, i.e.:

$$C_j = \frac{\int_a^b f(t)\, \phi_j^*(t)\, dt}{\int_a^b |\phi_j(t)|^2\, dt} \qquad (2.80)$$

If the basis functions $\phi_i(t)$ are orthonormal over the range $[a, b]$ then this reduces to:

$$C_j = \int_a^b f(t)\, \phi_j^*(t)\, dt \qquad (2.81)$$

For the special case of $\phi_j(t) = e^{j2\pi f_j t}$, $a = t$ and $b = t + T$, equation (2.81) gives the coefficients for an orthogonal function expansion in terms of a set of cisoids. This, of course, is identical to $T\tilde{C}_n'$ in the Fourier series of equation (2.25).

2.5.5 Error energy and completeness

When a function is *approximated* by a superposition of N basis functions over some range, T, i.e.:

$$f(t) \simeq f_N(t) \qquad (2.82(a))$$

where:

$$f_N(t) = \sum_{i=1}^{N} C_i\, \phi_i(t) \qquad (2.82(b))$$

then the 'error energy', E_e, is given by:

$$E_e = \int_t^{t+T} |f(t) - f_N(t)|^2\, dt \qquad (2.83)$$

(Note that E_e/T is the mean square error.) The basis set $\phi_i(t)$ is said to be *complete* over the interval T, for a given class of signals, if $E_e \to 0$ as $N \to \infty$ for those signals. Calculation of the coefficients in an orthogonal function expansion using equation (2.80) or (2.81) results in a minimum error energy approximation.

EXAMPLE 2.8
The functions shown in Figure 2.47(a) are the first four elements of the orthonormal set of Walsh functions [Beauchamp, Harmuth]. Using these as basis functions find a minimum error energy approximation for the function, $f(t)$, shown in Figure 2.47(b). Sketch the approximation.

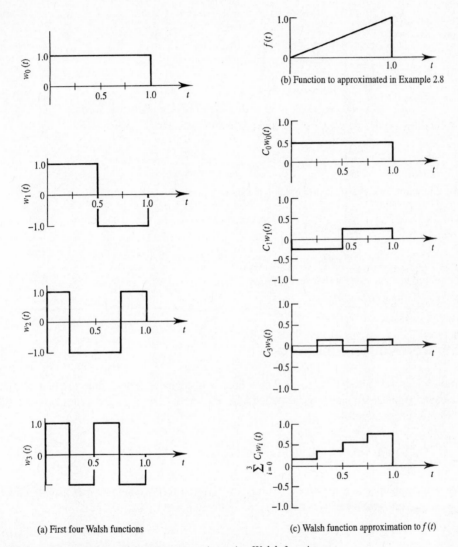

(b) Function to approximated in Example 2.8

(a) First four Walsh functions

(c) Walsh function approximation to $f(t)$

Figure 2.47 *Orthonormal function expansion using Walsh functions.*

Each of the coefficients is found in turn using equation (2.81), i.e.:

$$C_0 = \int_0^1 f(t)\, w_0(t)\, dt = \int_0^1 t\, dt = \left[\frac{t^2}{2}\right]_0^1 = 0.5$$

$$C_1 = \int_0^1 f(t)\, w_1(t)\, dt = \int_0^{0.5} t\, dt + \int_{0.5}^{1.0} -t\, dt$$

$$= \left[\frac{t^2}{2}\right]_0^{0.5} - \left[\frac{t^2}{2}\right]_{0.5}^{1.0} = -0.25$$

$$C_2 = \int_0^1 f(t) \, w_2(t) \, dt = \int_0^{0.25} t \, dt + \int_{0.25}^{0.75} -t \, dt + \int_{0.75}^{1.0} t \, dt$$

$$= \left[\frac{t^2}{2} \right]_0^{0.25} - \left[\frac{t^2}{2} \right]_{0.25}^{0.75} + \left[\frac{t^2}{2} \right]_{0.75}^{1.0} = 0$$

$$C_3 = \int_0^1 f(t) \, w_3(t) \, dt = \int_0^{0.25} t \, dt + \int_{0.25}^{0.5} -t \, dt + \int_{0.5}^{0.75} t \, dt + \int_{0.75}^{1.0} -t \, dt$$

$$= \left[\frac{t^2}{2} \right]_0^{0.25} - \left[\frac{t^2}{2} \right]_{0.25}^{0.5} + \left[\frac{t^2}{2} \right]_{0.5}^{0.75} - \left[\frac{t^2}{2} \right]_{0.75}^{1.0} = -0.125$$

The minimum error energy approximation is therefore given in equation (2.82(b)) by:

$$f_N(t) = \sum_{i=0}^{3} C_i \, w_i(t)$$

$$= 0.5 w_0(t) - 0.25 w_1(t) - 0.125 w_3(t)$$

The approximation $f_N(t)$ is sketched in Figure 2.47(c).

2.6 Correlation functions

Attention is restricted here to *real* functions and signals. The scalar product of two transient signals, $v(t)$ and $w(t)$, defined by equation (2.68(c)), and repeated here for convenience, is therefore:

$$[v(t), w(t)] = \int_{-\infty}^{\infty} v(t) \, w(t) \, dt \tag{2.84(a)}$$

Since this quantity is a measure of similarity between the two signals it is usually called the (*cross*) *correlation* of $v(t)$ and $w(t)$, normally denoted by $R_{vw}(0)$, i.e.:

$$R_{vw}(0) = [v(t), w(t)] \tag{2.84(b)}$$

(Recall that $[v(t), w(t)]$ in section 2.5.2 was called a cross-energy.) More generally a cross correlation *function*, $R_{vw}(\tau)$, can be defined, i.e.:

$$R_{vw}(\tau) = [v(t), w(t - \tau)]$$

$$= \int_{-\infty}^{\infty} v(t) \, w(t - \tau) \, dt \tag{2.85}$$

This is a measure of the similarity between $v(t)$ and a time shifted version of $w(t)$. The value of $R_{vw}(\tau)$ depends not only on the similarity of the signals, however, but also on their magnitude. This magnitude dependence can be removed by normalising both functions such that their associated normalised energies are unity, i.e.:

$$\rho_{vw}(\tau) = \frac{\displaystyle\int_{-\infty}^{\infty} v(t)\, w(t-\tau)\, dt}{\sqrt{\left(\displaystyle\int_{-\infty}^{\infty} |v(t)|^2\, dt\right)} \sqrt{\left(\displaystyle\int_{-\infty}^{\infty} |w(t)|^2\, dt\right)}} \tag{2.86}$$

The normalised cross correlation function, $\rho_{vw}(\tau)$, has the following properties:

1. $-1 \le \rho_{vw}(\tau) \le 1$
2. $\rho_{vw}(\tau) = -1$ if, and only if, $v(t) = -kw(t-\tau)$
3. $\rho_{vw}(\tau) = 1$ if, and only if, $v(t) = kw(t-\tau)$

$\rho_{vw}(\tau) = 0$ indicates that $v(t)$ and $w(t-\tau)$ are orthogonal and hence have no similarity whatsoever. (Later, in Chapter 11, we will use two separate parallel channel carriers to construct a four-phase modulator. By using orthogonal carriers $\sin \omega t$ and $\cos \omega t$ we ensure that there is no interference between the parallel channels, equation (2.21(c)).)

For real *periodic* waveforms, $p(t)$ and $q(t)$, the correlation or scalar product, defined in equation (2.68(b)) and repeated here, is:

$$R_{pq}(0) = [p(t), q(t)]$$

$$= \lim_{T' \to \infty} \frac{1}{T'} \int_{-T'/2}^{T'/2} p(t)\, q(t)\ dt$$

$$= \langle p(t)\, q(t) \rangle \tag{2.87}$$

The generalisation to a cross correlation function is therefore:

$$R_{pq}(\tau) = [p(t), q(t-\tau)]$$

$$= \lim_{T' \to \infty} \frac{1}{T'} \int_{-T'/2}^{T'/2} p(t)\, q(t-\tau)\ dt \tag{2.88}$$

and the normalised cross correlation function, $\rho_{pq}(\tau)$, becomes:

$$\rho_{pq}(\tau) = \frac{\langle p(t)\, q(t-\tau) \rangle}{\sqrt{\langle |p(t)|^2 \rangle}\ \sqrt{\langle |q(t)|^2 \rangle}}$$

$$= \frac{\displaystyle\lim_{T' \to \infty} \left(\frac{1}{T'}\right) \int_{-T'/2}^{T'/2} p(t)\, q(t-\tau)\, dt}{\sqrt{\left(\left(\dfrac{1}{T}\right)\displaystyle\int_{t}^{t+T} |p(t)|^2\, dt\right)} \sqrt{\left(\left(\dfrac{1}{T}\right)\displaystyle\int_{t}^{t+T} |q(t)|^2\, dt\right)}} \tag{2.89}$$

(Notice that the denominator of equation (2.89) is the geometric mean of the normalised powers of $p(t)$ and $q(t)$. For periodic signals $\rho_{pq}(\tau)$ therefore represents $R_{pq}(\tau)$ after $p(t)$ and $q(t)$ have been normalised to an RMS value of 1.0.)

EXAMPLE 2.9
Find the normalised cross correlation function of the sinusoid and the square wave shown in Figure 2.48(a).

Since the periods of the two functions in this example are the same we can perform the averaging in equation (2.89) over the (finite) period of the product, T', i.e. :

$$\rho_{pq}(\tau) = \frac{\left(\dfrac{1}{T'}\right) \displaystyle\int_{-T'/2}^{T'/2} p(t)q(t-\tau)\, dt}{\sqrt{\left[\left(\dfrac{1}{T}\right)\displaystyle\int_{0}^{T} p^2(t)\, dt\right]}\;\sqrt{\left[\left(\dfrac{1}{T}\right)\displaystyle\int_{0}^{T} q^2(t)\, dt\right]}}$$

It is clear that here T' can be taken equal to the period, T, of the sinusoid and square wave. Furthermore the RMS values of the two functions are (by inspection) $1/\sqrt{2}$ and 1.0. The cross correlation function is therefore given by:

$$\rho_{pq}(\tau) = \frac{\dfrac{1}{0.02} \displaystyle\int_{-0.01}^{0.01} \left[\Pi\left(\frac{t-0.005}{0.01}\right) - \Pi\left(\frac{t+0.005}{0.01}\right)\right]\cos[2\pi\,50(t-\tau)]\, dt}{(1/\sqrt{2}) \times 1.0}$$

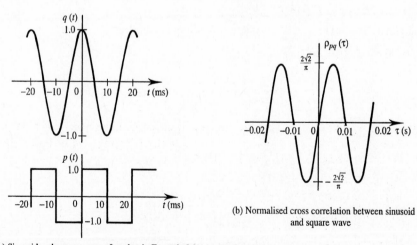

(a) Sinusoid and square wave referred to in Example 2.9

(b) Normalised cross correlation between sinusoid and square wave

Figure 2.48 *Normalised cross correlation of sinusoid and square wave.*

$$= \frac{\sqrt{2}}{0.02} \left\{ \int_0^{0.01} \cos[2\pi\ 50(t - \tau)]\ dt - \int_{-0.01}^0 \cos[2\pi\ 50(t - \tau)]\ dt \right\}$$

Using change of variable $x = t - \tau$:

$$\rho_{pq}(\tau) = \frac{\sqrt{2}}{0.02} \left\{ \int_{-\tau}^{0.01 - \tau} \cos(2\pi\ 50x)\ dx - \int_{-\tau-0.01}^{-\tau} \cos(2\pi\ 50x)\ dx \right\}$$

$$= \frac{\sqrt{2}}{0.02} \left\{ \left[\frac{\sin(2\pi\ 50x)}{2\pi\ 50} \right]_{-\tau}^{0.01 - \tau} - \left[\frac{\sin(2\pi\ 50x)}{2\pi\ 50} \right]_{-\tau-0.01}^{-\tau} \right\}$$

$$= \frac{\sqrt{2}}{0.02\ 2\pi 50} \left\{ \sin[2\pi 50(0.01 - \tau)] + 2\ \sin(2\pi 50\tau) - \sin[2\pi 50(\tau + 0.01)] \right\}$$

$$= \frac{1}{\sqrt{2}\ \pi}\ 4\ \sin(2\pi\ 50\tau) = \frac{4}{\sqrt{2}\ \pi}\ \sin(2\pi\ 50\tau)$$

Figure 2.48(b) shows a sketch of $\rho_{pq}(\tau)$.

If the two signals being correlated are identical then the result is called the *auto*correlation function, $R_{vv}(\tau)$ or $R_v(\tau)$. For real transient signals the autocorrelation function is therefore defined by:

$$R_v(\tau) = \int_{-\infty}^{\infty} v(t)\ v(t - \tau)\ dt \tag{2.90(a)}$$

and for real periodic signals by:

$$R_p(\tau) = \frac{1}{T} \int_t^{t+T} p(t)\ p(t - \tau)\ dt \tag{2.90(b)}$$

Normalised autocorrelation functions can be defined by dividing equations (2.90(a)) and (b) by the energy in $v(t)$ and power in $p(t)$, respectively, dissipated in 1 Ω. There are several properties of (real signal) autocorrelation functions to note:

1. $R_x(\tau)$ is real.
2. $R_x(\tau)$ has even symmetry about $\tau = 0$, i.e.:

$$R_x(\tau) = R_x(-\tau) \tag{2.91}$$

3. $R_x(\tau)$ has a maximum (positive) magnitude at $\tau = 0$, i.e.:

$$|R_x(\tau)| \leq R_x(0), \quad \text{for } any\ \tau \neq 0. \tag{2.92}$$

4a.   If $x(t)$ is periodic and has units of V then $R_x(\tau)$ is also periodic (with the same period as $x(t)$) and has units of V^2 (i.e. normalised power).

4b.   If $x(t)$ is transient and has units of V then $R_x(\tau)$ is also transient with units of V^2 s (i.e. normalised energy).

5a.   The autocorrelation function of a transient signal and its (two sided) energy spectral density are a Fourier transform pair, i.e.:

$$R_v(\tau) \overset{FT}{\Leftrightarrow} E_v(f) \tag{2.93}$$

This theorem is proved as follows:

$$E_v(f) = |V(f)|^2 = V(f)\,V^*(f)$$

$$= FT\left\{v(t) * v^*(-t)\right\} = FT\left\{\int_{-\infty}^{\infty} v(t)\,v^*(-\tau + t)\,dt\right\}$$

$$= FT\,\{R_v(\tau)\} \quad (V^2 \text{ s/Hz}) \tag{2.94(a)}$$

i.e.:

$$E_v(f) = \int_{-\infty}^{\infty} R_v(\tau)\,e^{-j2\pi f\tau}\,d\tau \tag{2.94(b)}$$

(The last line of equation (2.94(a)) is obvious for real $v(t)$ but see also the more general definition of $R_v(\tau)$ given in equation (2.96(b)).)

5b.   The autocorrelation function of a periodic signal and its (two sided) *power* spectral density (represented by a discrete set of impulse functions) are a Fourier transform pair, i.e.:

$$R_p(\tau) \overset{FT}{\Leftrightarrow} G_p(f) \tag{2.95}$$

(Since $R_p(\tau)$ is periodic and $G_p(f)$ consists of a set of discrete impulse functions, $G_p(f)$ could also be interpreted as a power spectrum derived as the Fourier series of $R_p(\tau)$.) Equation (2.95) also applies to stationary random signals which are discussed in Chapter 3.

EXAMPLE 2.10
What is the autocorrelation function, and decorrelation time, of the rectangular pulse shown in Figure 2.49(a)? From a knowledge of its autocorrelation function find the pulse's energy spectral density.

$$R_v(\tau) = \int_{-\infty}^{\infty} 2\,\Pi(t - 1.5)\, 2\,\Pi(t - 1.5 - \tau)\ dt$$

For $\tau = 0$ (see Figure 2.49(b)):

$$R_v(0) = \int_{1}^{2} 2^2\ dt = 4[\ t\ ]_1^2 = 4 \ (V^2 \text{ s})$$

For $0 < \tau < 1$ (see Figure 2.49(c)):

$$R_v(\tau) = \int_{1+\tau}^{2} 2^2 \, dt = 4[\, t \,]_{1+\tau}^{2} = 4(1 - \tau) \ (\mathrm{V}^2 \, \mathrm{s})$$

For $-1 < \tau < 0$ (see Figure 2.49(d)):

$$R_v(\tau) = \int_{1}^{2+\tau} 2^2 \, dt = 4[\, t \,]_{1}^{2+\tau} = 4(1 + \tau) \ (\mathrm{V}^2 \, \mathrm{s})$$

For $|\tau| > 1$ the rectangular pulse and its replica do not overlap; therefore in these regions $R_v(\tau) = 0$. Figure 2.49(e) shows a sketch of $R_v(\tau)$. Notice that the location of this rectangular pulse in time (i.e. at $t = 1.5$) does not affect the location of $R_v(\tau)$ on the time delay axis. Notice also that the symmetry of $R_v(\tau)$ about $\tau = 0$ means that in practice the function need only be found for $\tau > 0$. The decorrelation time, τ_o, of $R_v(\tau)$ is given by:

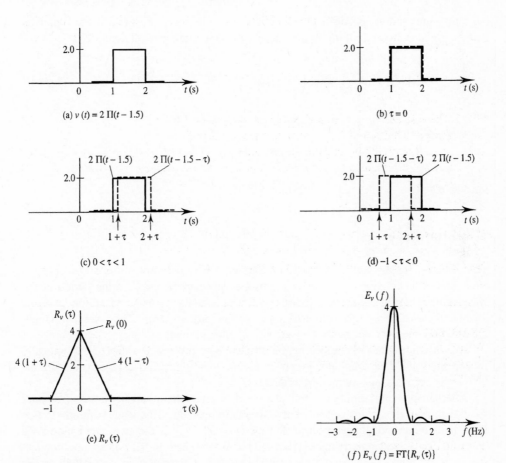

(a) $v\,(t) = 2 \, \Pi(t - 1.5)$

(b) $\tau = 0$

(c) $0 < \tau < 1$

(d) $-1 < \tau < 0$

(e) $R_v(\tau)$

(f) $E_v(f) = \mathrm{FT}\{R_v(\tau)\}$

Figure 2.49 *Autocorrelation (e) of a rectangular pulse (a) – (e) and spectral density (f).*

$$\tau_o = 1 \quad (s)$$

(Other definitions for τ_o, e.g. ½ energy, $1/e$ energy, etc., could be adopted.)

The energy spectral density, $E_v(f)$, is given by:

$$E_v(f) = \text{FT} \{R_v(\tau)\} = \text{FT} \{4 \, \Lambda(t)\}$$

$$= 4 \, \text{sinc}^2(f) \quad (V^2 \, s/Hz) \quad \text{(Using Table 2.4)}$$

$E_v(f)$ is sketched in Figure 2.49(f). Notice that the area under $R_v(\tau)$ is equal to $E_v(0)$ and the area under $E_v(f)$ is equal to $R_v(0)$. This is a good credibility check on the answer to such problems. Also notice that the (first null) bandwidth of the rectangular pulse and the zero crossing definition of decorrelation time are consistent with the rule:

$$B \approx \frac{1}{\tau_o}$$

Note that auto and cross correlation functions are only necessarily real if the functions being correlated are real. If this is not the case then the more general definitions:

$$R_{pq}(\tau) = \lim_{T' \to \infty} \frac{1}{T'} \int_{-T'/2}^{T'/2} p(t) \, q^*(t - \tau) \, dt \qquad (2.96(a))$$

and

$$R_{vw}(\tau) = \int_{-\infty}^{\infty} v(t) \, w^*(t - \tau) dt \qquad (2.96(b))$$

must be adopted.

2.7 Summary

Deterministic signals can be periodic or transient. Periodic waveforms are unchanged when shifted in time by nT seconds where n is any integer and T is the period of the waveform. They have discrete (line) spectra and, being periodic, exist for all time. Transient signals are aperiodic and have continuous spectra. They are essentially localised in time (whether or not they are strictly time limited).

All periodic signals of engineering interest can be expressed as a sum of harmonically related sinusoids. The amplitude spectrum of a periodic signal has units of volts, and the phase spectrum has units of radians or degrees.

Alternatively the amplitudes and phases of a set of harmonically related, counter rotating, conjugate cisoids can be plotted against frequency. This leads naturally to two sided amplitude and phase spectra. For purely real signals the two sided amplitude spectrum has even symmetry about 0 Hz and the phase spectrum has odd symmetry about 0 Hz. If the power associated with each sinusoid in a Fourier series is plotted against frequency the result is a power spectrum with units of V^2 or W. Two sided power spectra can be defined by associating half the total power in each line with a positive frequency and half with a negative frequency. The total power in a waveform is the sum of the powers in each spectral line. This is Parseval's theorem.

Bandwidth refers to the width of the frequency band in a signal's spectrum which contains significant power (or, in the case of transient signals, energy). Many definitions of bandwidth are possible, the most appropriate depending on the application or context. In the absence of a contrary definition, however, the half-power bandwidth is usually assumed. Signals with rapid rates of change have large bandwidth and those with slow rates of change small bandwidth.

The voltage spectrum of a transient signal is continuous and is given by the Fourier transform of the signal. Since the units of such a spectrum are V/Hz it is normally referred to as a voltage spectral density. A complex voltage spectral density can be expressed as an amplitude spectrum and a phase spectrum. The square of the amplitude spectrum has units of V^2 s/Hz and is called an energy spectral density. The total energy in a transient signal is the integral over all frequencies of the energy spectral density.

Fourier transform pairs are uniquely related (i.e. for each time domain signal there is only one, complex, spectrum) and have been extensively tabulated. Theorems allowing the manipulation of existing transform pairs and the calculation of new ones extend the usefulness of such tables. The convolution theorem is especially useful. It specifies the operation in one domain (convolution) which is precisely equivalent to multiplication in the other domain.

Basis functions other than sinusoids and cisoids can be used to expand signals and waveforms. Such generalised expansions are especially useful when the set of basis functions are orthogonal or orthonormal. Signals and waveforms can be interpreted as multidimensional vectors. In this context the concept of orthogonal functions is related to the concept of perpendicular vectors. The orthogonal property of a set of basis functions allows the optimum coefficients of the functions to be calculated independently. Optimum in this context means a minimum error energy approximation.

Correlation is the equivalent operation for signals to the scalar product for vectors and is a measure of signal similarity. The cross correlation function gives the correlation of two functions for all possible time shifts between them. It can be applied, with appropriate differences in its definition, to both transient and periodic functions. The energy and power spectral densities of transient and periodic signals respectively are the Fourier transforms of their autocorrelation functions. Chapter 3 extends correlation concepts to noise and other random signals. In Chapter 8 correlation is identified as an optimum signal processing technique, often employed in digital communications receivers.

2.8 Problems

2.1. Find the DC component and the first two non-zero harmonic terms in the Fourier series of the following periodic waveforms: (a) square wave with period 20 ms and magnitude +2 V from −5 ms to +5 ms and −2 V from +5 ms to +15 ms; (b) sawtooth waveform with a 2 s period and $y = t$ for

$-1 \le t < 1$; and (c) triangular wave with 0.2 s period and $y = 1/3(1 - 10|t|)$ for $-0.1 \le t < 0.1$.

2.2. Find the proportion of the total power contained in the DC and first two harmonics of the waveforms shown in Table 2.2, assuming a 25% duty cycle for the third waveform.

2.3. Use Table 2.2 to find the Fourier series coefficients up to the third harmonic of the waveform with period 2 s, one period of which is formed by connecting the following points with straight lines: $(t, y) = (0,0)$, $(0,0.5)$, $(1,1)$, $(1,0.5)$, $(2,0)$. (Hint: decompose the waveform into a sum of waveforms which you recognise.)

2.4. The spectrum of a square wave, amplitude ± 1.0 V and period 1.0 ms is bandlimited by an ideal filter to 4.0 kHz such that frequencies below 4.0 kHz are passed (undistorted) and frequencies above 4.0 kHz are stopped. What is the normalised power (in V^2), and what is the maximum rate of change (in V/s), of the waveform at the output of the filter? [0. 90 V^2, $16\pi^2 \times 10^3$ V/s]

2.5. How fast must the bandlimited waveform in Problem 2.4 be sampled if (in the absence of noise) it is to be reconstructed from the samples without error? [8.0 kHz]

2.6. Find (without using Table 2.4) the Fourier transforms of the following functions: (a) $\Pi((t - T)/\tau)$; (b) $\Lambda(t/2)$; (c) $3e^{-5|t|}$; (d) $[(e^{-at} - e^{-bt})/(b - a)] \, u(t)$. [Hint: for (c) recall the standard integral $- \int e^{ax} \cos bx \, dx = \dfrac{e^{ax}(a \cos bx + b \sin bx)}{a^2 + b^2}$].

2.7. Find (using Tables 2.4 and 2.5) the amplitude and phase spectra of the following transient signals: (a) a triangular pulse $3(1 - |t - 1|)\Pi((t - 1)/2)$; (b) a 'split phase' rectangular pulse having amplitude $y = 2$ V from $t = 0$ to $t = 1$ and $y = -2$ V from $t = 1$ to $t = 2$ and $y = 0$ elsewhere; (c) a truncated cosine wave $\cos(2\pi 20t)\Pi(t/0.2)$; (d) an exponentially decaying sinusoid $u(t)e^{-5t} \sin(2\pi 20t)$.

2.8. Sketch the following, purely real, frequency spectra and find the time domain signals to which they correspond: (a) $0.1 \, \text{sinc}(3f)$; (b) e^{-f^2}; (c) $\Lambda(f/2) + \Pi(f/4)$; and (d) $\Lambda(f - 10) + \Lambda(f + 10)$.

2.9. Convolve the following pairs of signals: (a) $\Pi(t/T_2)/T_2$ with $\Pi(t/T_1)/T_1$, $(T_2 > T_1)$; (b) $u(t)\exp(-3t)$ with $u(t - 1)$; (c) $\sin(\pi t)\Pi((t - 1)/2)$ with $2\Pi((t - 2)/2)$; and (d) $\delta(t) - 2\delta(t - 1) + \delta(t - 2)$ with $\Pi(t - 0.5) + 2\Pi(t - 1.5)$.

2.10. Find and sketch the energy spectral densities of the following signals: (a) $10\Pi((t - 0.05)/0.1)$; (b) $6 \, e^{-6|t|}$; (c) $\text{sinc}(100t)$; (d) $-\text{sinc}(100t)$. What is the energy contained in signals (a) and (c), and how much energy is contained in signals (b) and (d) below a frequency of 6.0 Hz? [10 V^2s, 0. 01 V^2s, 5. 99 V^2s, 1. 2 × 10^{-3} V^2s}

2.11. Demonstrate the orthogonality, or otherwise, of the function set: $(1/\sqrt{T}) \, \Pi((t - T/2)/T)$; $(\sqrt{2/T}) \cos((\pi/T) \, t) \, \Pi((t - T/2)/T)$; $2\Lambda((t - T/2)/(T/2)) - \Pi((t - T/2)T)$. Do these functions represent an orthonormal set?

2.12. Find the cross-correlation function of the sinewave, $f(t) = \sin(2\pi 50t)$, with a half wave rectified version of itself, $g(t)$.

2.13. Find, and sketch, the autocorrelation function of the 'split phase' rectangular pulse, where $x(t) = -V_0$ for $-T/2 < t < 0$ and $+V_0$ for $0 < t < T/2$.

2.14. What is the autocorrelation of $v(t) = u(t)e^{-t}$? Find the energy spectral density of this signal and the proportion of its energy contained in frequencies above 2.0 Hz. [5.1%]

CHAPTER 3

Random signals and noise

3.1 Introduction

The periodic and transient signals discussed in Chapter 2 are deterministic. This precludes them from conveying information since nothing new can be learned by receiving a signal which is entirely predictable. Unpredictability or randomness is a property which is essential for information bearing signals. (The definition and quantification of randomness is an interesting topic. Here, however, an intuitive and common-sense notion of randomness is all that is required.)

Whilst one type of random signal creates information (i.e. increases knowledge) at a communication receiver another type of random signal destroys it (i.e. decreases knowledge). The latter type of signal is known as noise [Rice]. The distinction between signals and noise is therefore essentially one of their origin (i.e. an information source or elsewhere) and whether reception is intended or not. In this context interference (signals arising from information sources other than the one expected or intended) is a type of noise. From the point of view of describing information signals and noise mathematically, no distinction is necessary. Since signals and noise are random such descriptions must be, at least partly, in terms of probability theory.

3.2 Probability theory

Consider an experiment with three possible, random, outcomes A, B, C. If the experiment is repeated N times and the outcome A occurs L times then the probability of outcome A is defined by:

$$P(A) \stackrel{\Delta}{=} \lim_{N \to \infty} \left\{ \frac{L}{N} \right\} \tag{3.1}$$

Note that the error, ε, for N experimental trials does not tend to zero for large N but actually increases (on average) as $\sqrt{N}$, Figure 3.1. The ratio ε/N does tend to zero, however, for large N, Figure 3.2, i.e.:

$$\frac{\varepsilon}{N} \to 0 \quad \text{as} \quad N \to \infty$$

(Thus on tossing a coin, for example, the result of achieving close to 50% heads is much more likely to be achieved with a large number of samples, e.g. >50 individual tosses or trials.) Such an experiment could be performed N times with one set of (unchanging) apparatus or N times, simultaneously, with N sets of (identical) apparatus. The former is called a temporal experiment whilst the latter is called an ensemble experiment. If, after N trials, the outcome A occurs L times and the outcome B occurs M times, and if A and B are *mutually exclusive* (i.e. they cannot occur together) then the probability that A or B occurs is:

$$P(A \text{ or } B) = \lim_{N \to \infty} \left\{ \frac{L + M}{N} \right\}$$

$$= \lim_{N \to \infty} \left\{ \frac{L}{N} \right\} + \lim_{N \to \infty} \left\{ \frac{M}{N} \right\}$$

$$= P(A) + P(B) \tag{3.2}$$

This is the basic law of additive probabilities which can be used for any number of mutually exclusive events.

Since the outcome of the experimental trials described above is variable and random it is called (unsurprisingly) a random variable. Such random variables can be discrete or continuous. An example of the former would be the score achieved by the throw of a dice. An example of the latter would be the final position of a coin in a game of shove halfpenny. (In the context of digital communications relevant examples might be the voltage of a quantised signal source and the voltage of an, unquantised, noise source.)

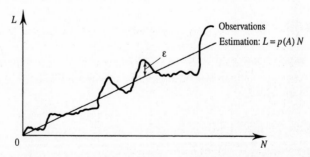

Figure 3.1 *Observations compared with estimation for L outcomes of A after N random trials.*

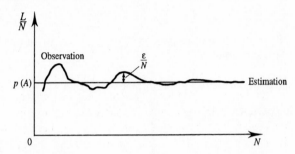

Figure 3.2 *Observation compared with estimation for the fraction L/N of outcomes A after N random trials.*

EXAMPLE 3.1
A dice is thrown once. What is the probability that: (i) the dice shows 3; (ii) the dice shows 6; (iii) the dice shows a number greater than 2; (iv) the dice does not show 5?

(i) Since all numbers between 1 and 6 inclusive are equiprobable, $P(3) = \dfrac{1}{6}$

(ii) As above, $P(6) = \dfrac{1}{6}$

(iii) $P(3) = P(4) = P(5) = P(6) = \dfrac{1}{6}$

$P(3 \text{ or } 4 \text{ or } 5 \text{ or } 6) = P(3) + P(4) + P(5) + P(6) = \dfrac{4}{6} = \dfrac{2}{3}$

(iv) $P(\text{any number but } 5) = 1 - P(5) = 1 - \dfrac{1}{6} = \dfrac{5}{6}$
because total probability, $\sum_{i=1}^{6} P(i)$, must sum to 1.

3.2.1 Conditional probabilities, joint probabilities and Bayes's rule

The probability of event A occurring given that event B is known to have occurred, $P(A|B)$, is called the conditional probability of A on B (or the probability of A conditional on B). The probability of A and B occurring together, $P(A, B)$, is called the joint probability of A and B. Joint and conditional probabilities are related by:

$$P(A, B) = P(B)P(A|B)$$

$$= P(A)P(B|A) \tag{3.3}$$

Rearranging equation (3.3) gives Bayes's rule:

$$P(A|B) = \frac{P(A)P(B|A)}{P(B)} \tag{3.4}$$

EXAMPLE 3.2

Four cards are dealt off the top of a shuffled pack of 52 playing cards. What is the probability that all the cards will be of the same suit?

Given that the first card is a spade the probability that the second card will be a spade is:

$$P(2nd \text{ spade}) = \frac{12}{51} = 0.2353$$

and so on:

$$P(3rd \text{ spade}) = \frac{11}{50} = 0.2200$$

$$P(4th \text{ spade}) = \frac{10}{49} = 0.2041$$

Therefore $P(4 \text{ spades}) = 0.2353 \times 0.2200 \times 0.2041$

$$= 0.01056$$

This is the probability of all the cards being from the same suit.

Notice that intuitively the suit of the first card has been ignored from a probability point of view. A more formal solution to this problem uses Bayes's rule explicitly:

$$P(4 \text{ spades}) = P(\text{spade, 3 spades})$$

$$= P(3 \text{ spades}) \, P(\text{spade} \mid 3 \text{ spades})$$

$$= P(2 \text{ spades}) \, P(\text{spade} \mid 2 \text{ spades}) \, P(\text{spade} \mid 3 \text{ spades})$$

$$= P(\text{spade}) \, P(\text{spade} \mid \text{spade}) \, P(\text{spade} \mid 2 \text{ spades}) \, P(\text{spade} \mid 3 \text{ spades})$$

$$= \frac{13}{52} \times \frac{12}{51} \times \frac{11}{50} \times \frac{10}{49}$$

$$= 0.002641$$

$$P(4 \text{ same suit}) = P(4 \text{ spades or 4 clubs or 4 diamonds or 4 hearts})$$

$$= P(4 \text{ spades}) + P(4 \text{ clubs}) + P(4 \text{ diamonds}) + P(4 \text{ hearts})$$

$$= 0.002641 \times 4 = 0.01056$$

3.2.2 Statistical independence

Events A and B are statistically independent if the occurrence of one does not affect the probability of the other occurring, i.e.:

$$P(A|B) = P(A)$$ (3.5(a))

and:

$$P(B|A) \ = \ P(B) \tag{3.5(b)}$$

It follows that for statistically independent events:

$$P(A, B) \ = \ P(A)P(B) \tag{3.6}$$

EXAMPLE 3.3

Two cards are dealt one at a time, face up, from a shuffled pack of cards. Show that these two events are not statistically independent.

The unconditional probabilities for both events are:

$$P(A) \ = \ P(B) \ = \ \frac{1}{52}$$

The probability of event A is:

$$P(A) \ = \ \frac{1}{52}$$

The probability of event B is:

$$P(B|A) \ = \ \frac{1}{51}$$

The joint probability of the two events is therefore:

$$P(A, B) \ = \ P(A)P(B|A)$$
$$= \ \frac{1}{52} \times \frac{1}{51}$$
$$\neq \ P(A) \ P(B)$$

i.e. the events are not statistically independent.

3.2.3 Discrete probability of errors in a data block

When we consider the problem of performance prediction in a digital coding system we often ask what is the probability of having more than a given number of errors in a fixed length codeword? This is a discrete probability problem. Assume that the probability of single bit (binary digit) error is P_e, that the number of errors is R' and n is the block length, i.e. we require to determine the probability of having more than R' errors in a block of n digits. Now:

$$P(> R' \text{ errors}) = 1 - P(\leq R' \text{ errors}) \tag{3.7(a)}$$

because total probability must sum to 1. We also assume that errors are independent. The above equation may thus be expanded as:

$$P(> R' \text{ errors}) = 1 - [P(0 \text{ error}) + P(1 \text{ error}) + P(2 \text{ errors}) + \cdots + P(R' \text{ errors})] \quad (3.7(b))$$

These probabilities will be calculated individually starting with the probability of no errors. A block representing the codeword is divided into bins labelled 1 to n. Each bin corresponds to one digit in the n digit codeword and it is labelled with the probability of the event in question.

Now considering the general case of j errors in n digits with a probability of error per digit of P_e we can generalise the above equations i.e.:

$$P(j \text{ errors}) = (P_e)^j \, (1 - P_e)^{n-j} \, {}^nC_j \qquad (3.8)$$

where the binomial coefficient ${}^nC_j$ is given by:

$$ {}^nC_j = \frac{n!}{j!(n-j)!} = \binom{n}{j} \qquad (3.9(a))$$

This is the probability of j errors in an n-digit codeword, but what we are interested in is the probability of having more than R' errors. We can write this using equation (3.7(b)) as:

$$P(> R' \text{ errors}) = 1 - \sum_{j=0}^{R'} P(j) \qquad (3.9(b))$$

For large n we have statistical stability in the sense that the number of errors in a given block will tend to the product $P_e \, n$. Furthermore, the fraction of blocks containing a number of errors that deviates significantly from this value will tend to zero, Figure 3.2. The above statistical stability gained from long blocks is very important in the design of effective error correction codewords and is discussed further in Chapter 10.

EXAMPLE 3.4

If the probability of single digit error is 0.01 and hence the probability of correct digit reception is 0.99 calculate the probability of 0 to 2 errors occurring in a ten digit codeword.

If all events (receptions) are independent then the probability of having no errors in the block is $(0.99)^{10} = 0.904382$, Figure 3.3(a).

Now consider the probability of a single error. Initially assume that the error is in the first position. Its probability is 0.01. All other digits are received correctly, so their probabilities are 0.99 and there are 9 of them, Figure 3.3(b). However there are 10 positions where the single error can occur in the 10 digit codeword and therefore the overall probability of a single error occurring is:

$$P(1 \text{ error}) = (0.01)^1 \, (0.99)^9 \, {}^{10}C_1 = 0.091352$$

where ${}^{10}C_1 = 10$, represents the number of combinations of 1 object from 10 objects. We can similarly calculate the probability of 2 errors as:

$$P(2 \text{ errors}) = (0.01)^2 \, (0.99)^8 \, {}^{10}C_2 = 0.00415$$

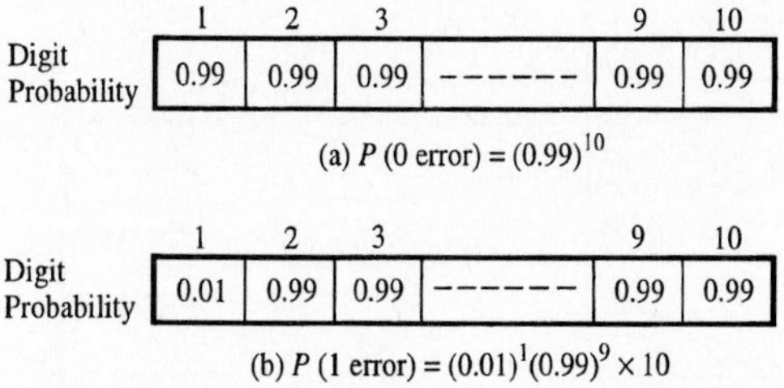

	1	2	3		9	10
Digit Probability	0.99	0.99	0.99	– – – – – –	0.99	0.99

(a) $P\,(0\ \text{error}) = (0.99)^{10}$

	1	2	3		9	10
Digit Probability	0.01	0.99	0.99	– – – – – –	0.99	0.99

(b) $P\,(1\ \text{error}) = (0.01)^{1}(0.99)^{9} \times 10$

Figure 3.3 *Discrete probability of codeword errors with a per digit error probability of 1%.*

Thus the probability of three or more errors, equation (3.7), is $1 - 0.904382 - 0.091352 - 0.00415 = 0.000116$. Notice how the probability of j errors in the data block falls off rapidly with j.

3.2.4 Cumulative distributions and probability density functions

A cumulative distribution (also called a probability distribution) is a curve showing the probability, $P_X(x)$, that the value of the random variable, X, will be less than or equal to some specific value, x, Figure 3.4(a), i.e.:

$$P_X(x) = P(X \le x) \tag{3.10}$$

(Conventionally, upper case letters are used for the name of a random variable and lower case letters for particular values of the random variable. Note, however, that the upper case subscript in the left hand side of equation (3.10) is often omitted.) Some properties of a cumulative distribution (CD) apparent from Figure 3.4(a) are:

1. $0 \le P_X(x) \le 1$ for $-\infty \le x \le \infty$
2. $P_X(-\infty) = 0$
3. $P_X(\infty) = 1$
4. $P_X(x_2) - P_X(x_1) = P(x_1 < X \le x_2)$
5. $\dfrac{dP_X(x)}{dx} \ge 0$

Exceedance curves are also sometimes used to describe the probability behaviour of random variables. These curves are complementary to CDs in that they give the probability that a random variable exceeds a particular value, i.e.:

$$P(X > x) = 1 - P_X(x) \tag{3.11}$$

Probability density functions (pdfs) give the probability that the value of a random variable, X, lies between x and $x + dx$. This probability can be written in terms of a CD as:

$$P_X(x + dx) - P_X(x) = \frac{dP_X(x)}{dx}\, dx \tag{3.12}$$

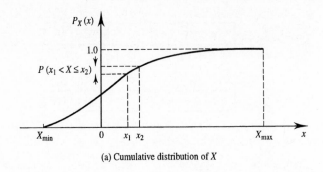

(a) Cumulative distribution of X

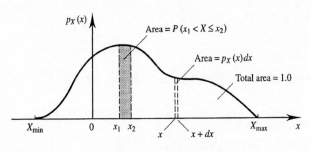

(b) Probability density function of X

Figure 3.4 *Descriptions of a continuous random variable (smooth CD and continuous pdf).*

It is the factor $[dP_X(x)]/dx$, normally denoted by $p_X(x)$, which is defined as the pdf of X. Figure 3.4(b) shows an example. Pdfs and CDs are therefore related by:

$$p_X(x) = \frac{dP_X(x)}{dx} \tag{3.13(a)}$$

$$P_X(x) = \int_{-\infty}^{x} p_X(x')\,dx' \tag{3.13(b)}$$

The important properties of pdfs are:

$$\int_{-\infty}^{\infty} p_X(x)\,dx = 1 \tag{3.14(a)}$$

and

$$\int_{x_1}^{x_2} p_X(x)\,dx = P(x_1 < X \le x_2) \tag{3.14(b)}$$

Both CDs and pdfs can represent continuous, discrete or mixed random variables. Discrete random variables, as typified in Example 3.1, have stepped CDs and purely impulsive pdfs, Figure 3.5. Mixed random variables have CDs which contain discontinuities and pdfs which contain impulses, Figure 3.6.

Pdfs can be measured, in principle, by taking an ensemble of identical random variable generators, sampling them all at one instant and plotting a relative frequency histogram of samples. (The pdf is the limit of this histogram as the size interval shrinks

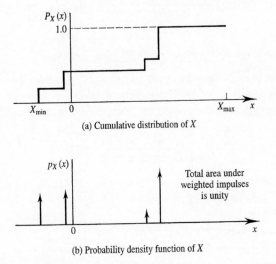

(a) Cumulative distribution of X

(b) Probability density function of X

Figure 3.5 *Probability descriptions of a discrete random variable (stepped CD and discrete pdf).*

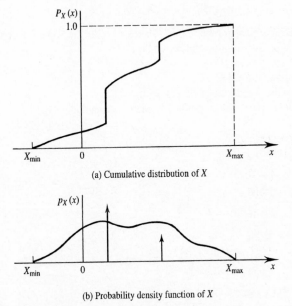

(a) Cumulative distribution of X

(b) Probability density function of X

Figure 3.6 *Probability descriptions of a mixed random variable.*

Figure 3.7 *Determination of pdf for an (ergodic) random process.*

to zero and the number of samples tends to infinity.) Alternatively, at least for ergodic processes (see section 3.3.1), the proportion of time the random variable spends in different value intervals of size Δx could be determined, (see Figure 3.7). In this case:

$$p_X(x)\,\Delta x = \sum_{i=1}^{N} \frac{\Delta t_i}{\text{total observation time}} \tag{3.15}$$

and $p_X(x)$ can be found as the limit of equation (3.15) as $\Delta x \to 0$ (and the observation time and $N \to \infty$). This allows one to calculate the time that the signal $x(t)$ lies between the voltages x and Δx.

EXAMPLE 3.5

A random voltage has a pdf given by:

$$p(V) = k\,u(V+4)\,e^{-3(V+4)} + 0.25\,\delta(V-2)$$

where $u(\ )$ is the Heaviside step function. (i) Sketch the pdf; (ii) find the probability that $V = 2$ V; (iii) find the value of k; and (iv) find and sketch the CD of V.

(i) Figure 3.8(a) shows the pdf of V.
(ii) Area under impulse is 0.25 therefore $P(V = 2) = 0.25$.
(iii) If the area under impulse part of pdf is 0.25 then the area under exponential part of the pdf must be 0.75, i.e.:

$$\int_{-4}^{\infty} k\,e^{-3(V+4)}\,dV = 0.75$$

Use change of variable:

$$V + 4 = x\,, \quad dx/dV = 1$$

when $V = -4$, $x = 0$, and when $V = \infty$, $x = \infty$

Therefore:

$$\int_{0}^{\infty} k\,e^{-3x}\,dx = 0.75$$

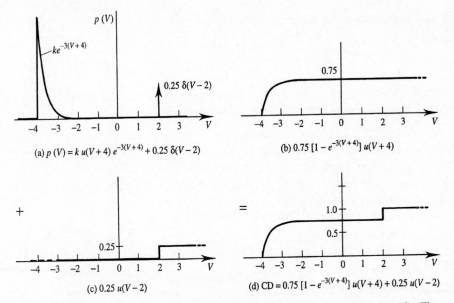

(a) $p(V) = k u(V+4) e^{-3(V+4)} + 0.25 \delta(V-2)$

(b) $0.75 [1 - e^{-3(V+4)}] u(V+4)$

(c) $0.25 u(V-2)$

(d) $CD = 0.75 [1 - e^{-3(V+4)}] u(V+4) + 0.25 u(V-2)$

Figure 3.8 *Pdf of V in Example 3.5, (a) and components (b) and (c) of resulting CD, (d). (The step function is implied in the solution by the lower limit of integration.)*

i.e.:

$$k = \frac{0.75}{[e^{-3x}/-3]_0^\infty}$$

$$= \frac{-3 \times 0.75}{[0-1]} = 2.25$$

(iv)

$$CD = \int_{-\infty}^{V} p(V') \, dV'$$

$$= \int_{-4}^{V} [2.25 e^{-3(V'+4)} + 0.25 \delta(V'-2)] \, dV'$$

Using the same substitution as in part (iii):

$$CD = 2.25 \left[\frac{e^{-3x}}{-3} \right]_0^{V+4} + 0.25 u(V-2)$$

$$= -2.25/3 \times [e^{-3(V+4)} - 1] + 0.25 u(V-2)$$

Figure 3.8 (b) and (c) shows these individual waveforms and (d) shows the combined result.

3.2.5 Moments, percentiles and modes

The first moment of a random variable X is defined by:

$$\bar{X} = \int_{-\infty}^{\infty} x \, p(x) \, dx \qquad (3.16)$$

where $\bar{X}$ denotes an ensemble *mean*. (Sometimes this quantity is called the expected value of X and is written $E[X]$.) The second moment is defined as:

$$\overline{X^2} = \int_{-\infty}^{\infty} x^2 \, p(x) \, dx \qquad (3.17\text{(a)})$$

and represents the mean square of the random variable. (The square root of the second moment is thus the RMS value of X.)

Higher order moments are defined by the general formula:

$$\overline{X^n} = \int_{-\infty}^{\infty} x^n \, p(x) \, dx \qquad (3.17\text{(b)})$$

(The zeroth moment ($n = 0$) is always equal to 1.0 and is therefore not a useful quantity.)

Central moments are the net moments of a random variable taken about its mean. The second central moment is therefore given by:

$$\overline{(X - \bar{X})^2} = \int_{-\infty}^{\infty} (x - \bar{X})^2 \, p(x) \, dx \qquad (3.18\text{(a)})$$

and is usually called the *variance* of the random variable. The square root of the second central moment is the *standard deviation* of the random variable (general symbol s, although for a Gaussian random variable the symbol σ is commonly used). The variance or standard deviation provides a measure of random variable spread, or pdf width. Higher order central moments are defined by the general formula:

$$\overline{(X - \bar{X})^n} = \int_{-\infty}^{\infty} (x - \bar{X})^n \, p(x) \, dx \qquad (3.18\text{(b)})$$

(The zeroth central moment is always 1.0 and the first central moment is always zero. Neither are of any practical use, therefore.) The 3rd and 4th central moments divided by s^3 and s^4 respectively are called *skew* and *kurtosis* [1]. These are a measure of pdf asymmetry and peakiness, the latter being in comparison to a Gaussian function, Figure 3.9. These higher order moments are of current interest for the analysis of non-stationary signals, such as speech, and they are also appropriate for the analysis of non-Gaussian signals.

---

[1] If kurtosis is defined as $[\overline{(X - \bar{X})^4}/s^4] - 3$ then a Gaussian curve will have zero kurtosis.

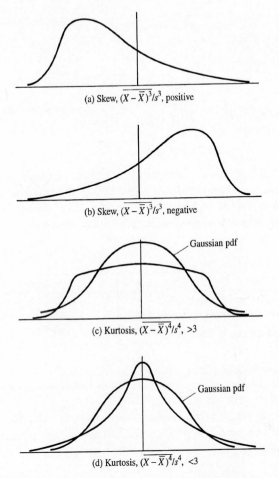

Figure 3.9 *Illustration of skew and kurtosis as descriptors of pdf shape.*

The *p*th *percentile* is the value of X below which p percent of the total area under the pdf lies, Figure 3.10(a), i.e.:

$$\int_{-\infty}^{x} p(x')\, dx' = \frac{p}{100} \tag{3.19}$$

where x is the *p*th percentile. In the special case of $p = 50$ (i.e. the 50th percentile) the corresponding value of x is called the *median* and the pdf is divided into two equal areas, Figure 3.10(b).

The *mode* of a pdf is the value of x for which $p(x)$ is a maximum, Figure 3.11(a). For a pdf with a single mode this can be interpreted as the most likely value of X. In general pdfs can be multimodal, Figure 3.11(b). Moments, percentiles and modes are all examples of *statistics*, i.e. numbers which in some way summarise the behaviour of a

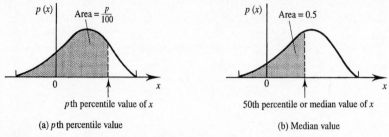

(a) *p*th percentile value

(b) Median value

Figure 3.10 *Illustration of percentiles and median.*

random variable. The difference between probabilistic and statistical models is that statistical models usually give an incomplete description of random variables.

There are some useful electrical interpretations of moments and central moments in the context of random voltages (and currents). These interpretations are summarised below:

Moment	Familiar name	Interpretation
1st	Mean value	DC voltage (or current)
2nd	Mean square value	Total power[1]
2nd central	Variance	AC power[2]

Notes:

(1) This is the total normalised power, i.e. the power dissipated in a 1 Ω load.
(2) This is the power dissipated in a 1 Ω load by the fluctuating (i.e. AC) component of voltage.

The interpretations of 1st and 2nd moments are obvious. The interpretation of 2nd central moment, s^2, becomes clear when the left hand side of equation (3.18(a)) is expanded as shown below, i.e.:

$$s^2 = \langle [x(t) - \langle x(t) \rangle]^2 \rangle$$
$$= \langle x^2(t) - 2x(t)\langle x(t) \rangle + \langle x(t) \rangle^2 \rangle$$
$$= \langle x^2(t) \rangle - \langle x(t) \rangle^2 \qquad (3.20)$$

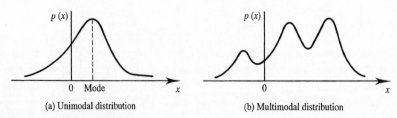

(a) Unimodal distribution

(b) Multimodal distribution

Figure 3.11 *Illustration of modal values.*

(The angular brackets, $\langle \, \rangle$, here indicate time averages, equation (2.16), which for most random variables of engineering interest can be equated to ensemble averages, see section 3.3.1.) Equation (3.20) is an expression of the familiar statistical statement that variance is the mean square minus the square mean. This is clearly also equivalent to total power minus DC power which must be the AC (or fluctuating) power. Figure 3.12 illustrates these electrical interpretations.

Mean value: $\langle x(t) \rangle = \lim\limits_{T \to \infty} \frac{1}{T} \int_{-\frac{T}{2}}^{\frac{T}{2}} x(t)\, dt$

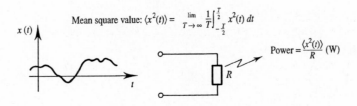

Input   LP filter   Output (as time constant $RC \to \infty$)

(a) Mean value represents DC component of signal.

Mean square value: $\langle x^2(t) \rangle = \lim\limits_{T \to \infty} \frac{1}{T} \int_{-\frac{T}{2}}^{\frac{T}{2}} x^2(t)\, dt$

$\text{Power} = \dfrac{\langle x^2(t) \rangle}{R}\ \text{(W)}$

(b) Mean square value represents power dissipated in 1 Ω

Variance: $\langle [x(t) - \langle x(t) \rangle]^2 \rangle = \langle x^2(t) \rangle - \langle x(t) \rangle^2$

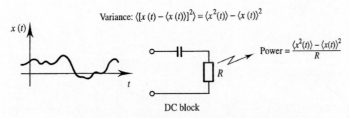

$\text{Power} = \dfrac{\langle x^2(t) \rangle - \langle x(t) \rangle^2}{R}$

DC block

(c) Variance represents AC, or fluctuation, power dissipated in 1 Ω

Figure 3.12 *Engineering interpretations of: (a) mean; (b) mean square; (c) variance.*

EXAMPLE 3.6

A random voltage has a pdf given by:

$$p(V) = u(V) \, 3 \, e^{-3V}$$

Find the DC voltage, the power dissipated in a 1 Ω load and the median value of voltage. What power would be dissipated at the output of an AC coupling capacitor?

Equation (3.16) defines the DC value:

$$\bar{V} = \int_{-\infty}^{\infty} V \, p(V) \, dV = 3 \int_{0}^{\infty} V \, e^{-3V} \, dV$$

Using the standard integral [Dwight, equation 567.9]:

$$\int x^n e^{ax} \, dx = e^{ax} \left[\frac{x^n}{a} - \frac{nx^{n-1}}{a^2} + \frac{n(n-1)x^{n-2}}{a^3} \cdots (-1)^{n-1} \frac{n!x}{a^n} + (-1)^n \frac{n!}{a^{n+1}} \right], \quad n \geq 0$$

$$\bar{V} = 3 \left[e^{-3V} \left(\frac{V}{-3} - \frac{1 \times V^0}{(-3)^2} \right) \right]_0^{\infty}$$

$$= -3 \, [0 - 1/9] = 1/3 \; V$$

The power is given by equation (3.17):

$$\overline{V^2} = \int_{-\infty}^{\infty} V^2 \, p(V) \, dV = \int_{0}^{\infty} V^2 \, 3 \, e^{-3V} \, dV$$

Again using [Dwight, equation 567.9]:

$$\overline{V^2} = 3 \left[e^{-3V} \left(\frac{V^2}{-3} - \frac{2V^1}{(-3)^2} + \frac{2(1) \, V^0}{(-3)^3} \right) \right]_0^{\infty}$$

$$= 3 \left[0 - \left(-\frac{2}{27} \right) \right] = \frac{6}{27} \, (\text{or } 0.2222) \; V^2$$

Median value = 50th percentile, i.e.:

$$\int_{-\infty}^{X_{median}} p(V) \, dV = 0.5$$

Therefore:

$$\int_{0}^{X_{median}} 3 \, e^{-3V} \, dV = 0.5$$

i.e.:

$$3 \left[\frac{e^{-3V}}{-3} \right]_0^{X_{median}} = 0.5$$

$$1 - e^{-3X_{median}} = 0.5$$

$$X_{median} = \frac{\ln(1 - 0.5)}{-3} = 0.2310 \text{ V}$$

The AC coupling capacitor acts as a DC block and the fluctuating or AC power is given by the variance of the random signal as defined in equation (3.20), i.e.:

$$P_{AC} = s^2$$

$$= \langle v^2(t) \rangle - \langle v(t) \rangle^2$$

$$= \overline{V^2} - \overline{V}^2 \quad \text{(signal assumed ergodic, see section 3.3.1)}$$

$$= \frac{6}{27} - \left(\frac{1}{3} \right)^2 = \frac{3}{27} \quad \text{(or 0.1111)} \text{ V}^2$$

3.2.6 Joint and marginal pdfs, correlation and covariance

If X and Y are two random variables a joint probability density function, $p_{X,Y}(x, y)$, can be defined such that $p_{X,Y}(x, y) \, dx \, dy$ is the probability that X lies in the range x to $x + dx$ and Y lies in the range y to $y + dy$. The joint (or bivariate) pdf can be represented by a surface as shown in Figure 3.13(a). For quantitative work, however, it is often more convenient to display $p_{X,Y}(x, y)$ as a contour plot, Figure 3.13(b), and, when investigating bivariate random variables experimentally, sample values of (x, y) can be plotted as a scattergram, Figure 3.13(c). (Contours of constant point density in Figure 3.13(c) correspond, of course, to contours of constant probability density in Figure 3.13(b).)

Just as the total area under the pdf of a single random variable is 1.0, the volume under the surface representing a bivariate random variable is 1.0, i.e.:

$$\int\limits_{X} \int\limits_{Y} p_{X,Y}(x, y) \, dx \, dy = 1.0 \tag{3.21}$$

The probability of finding the bivariate random variable in any particular region, F, of the X,Y plane, Figure 3.13(b), is:

$$P([X, Y] \text{ lies within } F) = \int\limits_{F} \int p_{X,Y}(x, y) \, dx \, dy \tag{3.22}$$

If the joint pdf, $p_{X,Y}(x, y)$, of a bivariate variable is known then the probability that X lies in the range x_1 to x_2 (irrespective of the value of Y) is called a marginal probability of X and is found by integrating over all Y, Figure 3.14. The marginal pdf of X is therefore given by:

$$p_X(x) = \int\limits_{-\infty}^{\infty} p_{X,Y}(x, y) \, dy \tag{3.23(a)}$$

Similarly, the marginal pdf of Y is given by:

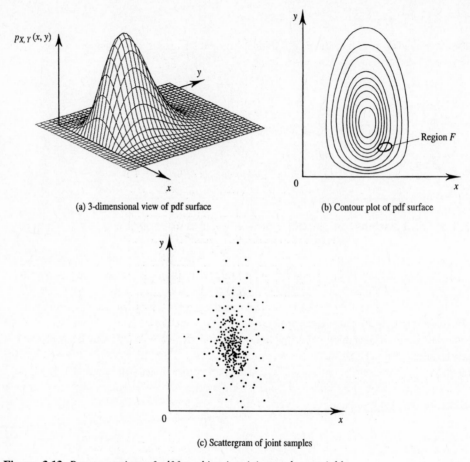

(a) 3-dimensional view of pdf surface

(b) Contour plot of pdf surface

(c) Scattergram of joint samples

Figure 3.13 *Representations of pdf for a bivariate joint random variable.*

$$p_Y(y) = \int\limits_{-\infty}^{\infty} p_{X,Y}(x, y)\, dx \qquad\qquad (3.23(b))$$

EXAMPLE 3.7
Two quantised signals have the following (discrete) joint pdfs:

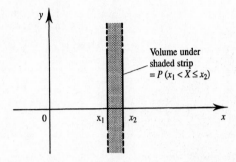

Figure 3.14 *Relationship between a marginal probability and a joint pdf.*

		X			
		1.0	1.5	2.0	2.5
Y	−1.0	0.15	0.08	0.06	0.05
	−0.5	0.10	0.13	0.06	0.05
	0.0	0.04	0.07	0.05	0.05
	0.5	0.01	0.02	0.03	0.05

Find and sketch the marginal pdfs of X and Y.

$$p_X(x) = \int_{-\infty}^{\infty} p_{X,Y}(x, y) \, dy$$

For a discrete joint random variable this becomes column summation:

$$P_X(x) = \sum_y P_{X,Y}(x, y)$$

$$P_X(1.0) = 0.15 + 0.10 + 0.04 + 0.01 = 0.30$$

$$P_X(1.5) = 0.08 + 0.13 + 0.07 + 0.02 = 0.30$$

$$P_X(2.0) = 0.06 + 0.06 + 0.05 + 0.03 = 0.20$$

$$P_X(2.5) = 0.05 + 0.05 + 0.05 + 0.05 = 0.20$$

See Figure 3.15(a) for the marginal pdf of X. Similarly:

$$P_Y(y) = \sum_x P_{X,Y}(x, y)$$

and by row summation:

$$P_Y(-1.0) = 0.15 + 0.08 + 0.06 + 0.05 = 0.34$$

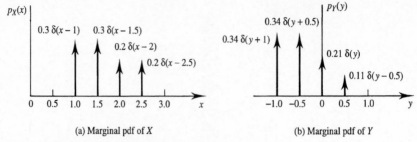

(a) Marginal pdf of X   (b) Marginal pdf of Y

Figure 3.15 *Marginal pdfs in Example 3.7.*

$$P_Y(-0.5) = 0.10 + 0.13 + 0.06 + 0.05 = 0.34$$

$$P_Y(0.0) = 0.04 + 0.07 + 0.05 + 0.05 = 0.21$$

$$P_Y(0.5) = 0.01 + 0.02 + 0.03 + 0.05 = 0.11$$

Figure 3.15(b) shows the marginal pdf of Y.

3.2.7 Joint moments, correlation and covariance

The joint moments of $p_{X,Y}(x,y)$ are defined by:

$$\overline{X^n Y^m} = \int_{-\infty}^{\infty} \int_{-\infty}^{\infty} x^n y^m \, p_{X,Y}(x, y) \, dx \, dy \tag{3.24}$$

In the special case of $n = m = 1$ the joint moment is called the *correlation* of X and Y (see previous correlation definition in section 2.6 for transient and periodic signals). A large positive value of correlation means that when x is high then y, *on average,* will also be high. A large negative value of correlation means that when x is high y, on average, will be low. A small value of correlation means that x gives little information about the magnitude or sign of y. The effect of correlation between two random variables, on a scattergram, is shown in Figure 3.16.

The joint central moments of $p_{X,Y}(x, y)$ are defined by:

$$\overline{(X - \bar{X})^n (Y - \bar{Y})^m} = \int_{-\infty}^{\infty} \int_{-\infty}^{\infty} (x - \bar{X})^n (y - \bar{Y})^m \, p_{X,Y}(x, y) \, dx \, dy \tag{3.25}$$

In this case, when $n = m = 1$, the joint central moment is called the *covariance* of X and Y. This is because the mean values of X and Y have been subtracted before *correlating* the resulting zero mean variables. Covariance therefore refers to the correlation of the *varying* parts of X and Y. If X and Y are already zero mean variables then the correlation and covariance are identical.

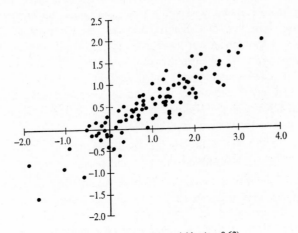

(a) Strongly correlated random variables ($\rho = 0.63$)

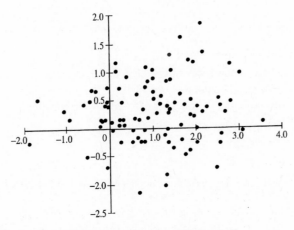

(a) Weakly correlated random variables ($\rho = 0.10$)

Figure 3.16 *Influence of correlation on a scattergram.*

The definitions of joint, and joint central, moments can be extended to multivariate (i.e. more than two) random variables in a straightforward way (e.g. $\overline{X^n Y^m Z^l}$ etc.).

The random variables X and Y are said to be uncorrelated if:

$$\overline{XY} = \bar{X}\,\bar{Y} \tag{3.26}$$

Notice that this implies that the *covariance* (not the correlation) is zero, i.e.:

$$\overline{(X - \bar{X})(Y - \bar{Y})} = 0 \tag{3.27}$$

It also follows that the correlation of X and Y can be zero, only if either $\bar{X}$ or $\bar{Y}$ is zero.

It is intuitively obvious that statistically independent random variables (i.e. random variables arising from physically separate processes) must be uncorrelated. It is not generally the case, however, that uncorrelated random variables must be statistically independent. Indeed the concept of correlation can be applied to deterministic signals such as $\cos \omega t$ and $\sin \omega t$. If concurrent samples are taken from these functions and the correlation is subsequently calculated the result will be zero (due to their orthogonality). It is clearly untrue to say that these are independent processes, however, since one can be derived from the other using a simple delay line. It follows that independence is a stronger statistical condition than uncorrelatedness, i.e.:

$$\text{Independence} \Rightarrow \text{uncorrelatedness}$$
$$\text{Uncorrelatedness} \nRightarrow \text{independence}$$

(An exception to the latter rule is when the random variables are Gaussian. In this special case uncorrelatedness does imply statistical independence: see section 3.2.8.)

The normalised correlation coefficient, ρ, between two random variables (as used in section 2.6, for transient or periodic signals) is the correlation between the corresponding standardised variables, standardisation in this context implying zero mean and unit standard deviation, i.e.:

$$\rho = \overline{\left(\frac{X - \bar{X}}{s_X} \right) \left(\frac{Y - \bar{Y}}{s_Y} \right)} \tag{3.28}$$

where s_X and s_Y are the standard deviations of X and Y respectively. (Since ρ can also be interpreted as the covariance of the random variables with normalised standard deviation then $\rho = 0$ can be viewed as the defining property for uncorrelatedness.)

EXAMPLE 3.8

Find the correlation and covariance of the discrete joint pdf described in Example 3.7.

Correlation is defined in equation (3.24) as:

$$\overline{XY} = \int\limits_{-\infty}^{\infty} \int\limits_{-\infty}^{\infty} xy \, p_{X,Y}(x, y) \, dx \, dy$$

For a discrete joint pdf this becomes:

$$\overline{XY} = \sum_x \sum_y xy \, P_{X,Y}(x, y)$$

$$
\begin{aligned}
= \; & (1.0)\,(-1.0)\,0.15 + (1.0)\,(-0.5)\,0.10 + (1.0)\,(0.0)\,0.04 + (1.0)\,(0.5)\,0.01 \\
+ \; & (1.5)\,(-1.0)\,0.08 + (1.5)\,(-0.5)\,0.13 + (1.5)\,(0.0)\,0.07 + (1.5)\,(0.5)\,0.02 \\
+ \; & (2.0)\,(-1.0)\,0.06 + (2.0)\,(-0.5)\,0.06 + (2.0)\,(0.0)\,0.05 + (2.0)\,(0.5)\,0.03 \\
+ \; & (2.5)\,(-1.0)\,0.05 + (2.5)\,(-0.5)\,0.05 + (2.5)\,(0.0)\,0.05 + (2.5)\,(0.5)\,0.05 \\
= \; & -0.6725
\end{aligned}
$$

Similarly covariance is obtained from equation (3.25) for the discrete pdf as:

$$\overline{(X - \bar{X})(Y - \bar{Y})} = \sum_x \sum_y (x - \bar{X})(y - \bar{Y}) \, P_{X,Y}(x, y)$$

Using the marginal pdfs found in Example 3.7:

$$\bar{X} = \frac{1}{N} \sum_x x \, P_X(x)$$

$$= \frac{1}{4} [(1.0)(0.3) + (1.5)(0.3) + (2.0)(0.2) + (2.5)(0.20)]$$

$$= 0.4125$$

$$\bar{Y} = \frac{1}{N} \sum_y y \, P_Y(y)$$

$$= \frac{1}{4} [(-1.0)(0.34) + (-0.5)(0.34) + (0.0)(0.21) + (0.5)(0.11)]$$

$$= -0.4550$$

and by using each of the 16 discrete probabilities in Example 3.7:

$$
\begin{aligned}
\overline{(X - \bar{X})(Y - \bar{Y})} = \; & (1.0 - 0.4125)(-1.0 + 0.4550)\, 0.15 \\
+ \; & (1.0 - 0.4125)(-0.5 + 0.4550)\, 0.10 \\
+ \; & (1.0 - 0.4125)(0.0 + 0.4550)\, 0.04 \\
+ \; & (1.0 - 0.4125)(0.5 + 0.4550)\, 0.01 \\
+ \; & (1.5 - 0.4125)(-1.0 + 0.4550)\, 0.08 \\
+ \; & (1.5 - 0.4125)(-0.5 + 0.4550)\, 0.13 \\
+ \; & (1.5 - 0.4125)(0.0 + 0.4550)\, 0.07 \\
+ \; & (1.5 - 0.4125)(0.5 + 0.4550)\, 0.02 \\
+ \; & (2.0 - 0.4125)(-0.1 + 0.4550)\, 0.06 \\
& \quad \cdots \qquad \cdots \\
+ \; & (2.5 - 0.4125)(0.5 + 0.4550)\, 0.05 \\
= \; & 0.07825
\end{aligned}
$$

3.2.8 Joint Gaussian random variables

A bivariate random variable is Gaussian if it can be reduced, by a suitable translation and rotation of axes, to the form:

$$p_{X,Y}(x, y) = \frac{1}{2\pi\sigma_X\sigma_Y} e^{-\left[\frac{x^2}{2\sigma_X^2} + \frac{y^2}{2\sigma_Y^2}\right]} \tag{3.29}$$

This idea is illustrated in Figure 3.17. The contours in the x, y plane, given by:

$$\frac{x''^2}{2\sigma_{X''}^2} + \frac{y''^2}{2\sigma_{Y''}^2} = \text{constant} \tag{3.30}$$

are ellipses and the double primed and unprimed coordinate systems are related by:

$$x'' = (x - x_0)\cos\theta + (y - y_0)\sin\theta \tag{3.31(a)}$$

$$y'' = -(x - x_0)\sin\theta + (y - y_0)\cos\theta \tag{3.31(b)}$$

where x_0, y_0 are the necessary translations and θ is the necessary rotation. If $\sigma_{X''} = \sigma_{Y''}$ then the ellipses become circles. The translation of axes removes any DC component in the random variables and the rotation has the effect of reducing the correlation between the random variables to zero. This can be seen by writing the probability density function in the original coordinate system, i.e.:

$$p_{X,Y}(x, y) = \frac{1}{2\pi\sigma_X\sigma_Y\sqrt{(1-\rho^2)}}$$

$$\times e^{-\left[\frac{(x-\bar{X})^2}{2\sigma_X^2(1-\rho^2)} - 2\rho\frac{(x-\bar{X})(y-\bar{Y})}{2\sigma_X\sigma_Y(1-\rho^2)} + \frac{(y-\bar{Y})^2}{2\sigma_Y^2(1-\rho^2)}\right]} \tag{3.32}$$

(for $\bar{X} = \bar{Y} = \rho = 0$ this reduces to equation (3.29)) which can then be written as a product of separate functions in x and y, i.e.:

$$p_{X,Y}(x, y) = \frac{1}{\sqrt{(2\pi)}\sigma_X} e^{-\frac{x^2}{2\sigma_X^2}} \frac{1}{\sqrt{(2\pi)}\sigma_Y} e^{-\frac{y^2}{2\sigma_Y^2}}$$

$$= p_X(x)\, p_Y(y) \tag{3.33}$$

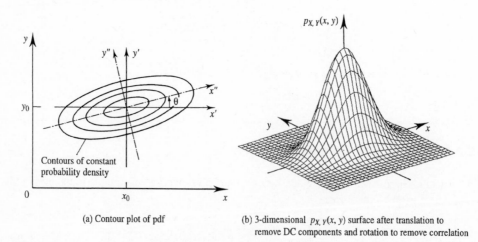

(a) Contour plot of pdf

(b) 3-dimensional $p_{X,Y}(x, y)$ surface after translation to remove DC components and rotation to remove correlation

Figure 3.17 *Joint Gaussian bivariate random variable.*

Equation (3.33) is the necessary and sufficient condition for statistical independence of X and Y. For a multivariate Gaussian function, then, uncorrelatedness does imply independence.

3.2.9 Addition of random variables and the central limit theorem

If two random variables X and Y are added, Figure 3.18, and their joint pdf, $p_{X,Y}(x,y)$ is known, what is the pdf, $p_Z(z)$, of their sum, Z? To answer this question we note the following:

$$Z = X + Y \tag{3.34}$$

Therefore when $Z = z$ (i.e. Z takes on a particular value z) then:

$$Y = z - X \tag{3.35(a)}$$

and when $Z = z + dz$ then:

$$Y = z + dz - X \tag{3.35(b)}$$

Equations (3.35), which both represent straight lines in the X, Y plane, are sketched in Figure 3.19. The probability that Z lies in the range z to $z + dz$ is given by the volume contained under $p_{X,Y}(x,y)$ in the strip between these two lines, i.e.:

$$P(z < Z \le z + dz) = \int_{strip} p_{X,Y}(x, y)\, ds \tag{3.36(a)}$$

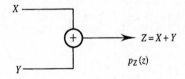

$p_{X,Y}(x, y)$

Figure 3.18 *Addition of random variables.*

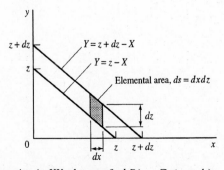

Figure 3.19 *Strip of integration in XY plane to find $P(z < Z \le z + dz)$.*

or:

$$p_Z(z)\, dz \;=\; \int\limits_{strip} p_{X,Y}(x, z-x)\, dx\, dz \tag{3.36(b)}$$

Therefore:

$$p_Z(z) \;=\; \int\limits_{-\infty}^{\infty} p_{X,Y}(x, z-x)\, dx \tag{3.37}$$

If X and Y are statistically independent then:

$$p_{X,Y}(x, z-x) = p_X(x)\, p_Y(z-x) \tag{3.38}$$

and equation (3.37) becomes:

$$p_Z(z) = \int\limits_{-\infty}^{\infty} p_X(x) p_Y(z-x)\, dx \tag{3.39}$$

Equation (3.39) can be recognised as the convolution integral. *The pdf of the sum of independent random variables is therefore the convolution of their individual pdfs.*

The multiple convolution of pdfs which arises when many independent random variables are added has a surprising and important consequence. Since convolution is essentially an integral operation it almost always results in a function which is in some

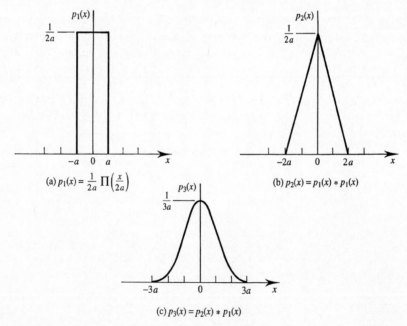

(a) $p_1(x) = \dfrac{1}{2a}\, \Pi\!\left(\dfrac{x}{2a}\right)$

(b) $p_2(x) = p_1(x) * p_1(x)$

(c) $p_3(x) = p_2(x) * p_1(x)$

Figure 3.20 *Multiple self convolution of a rectangular pulse.*

sense smoother (i.e. more gradually varying) than either of the functions being convolved. (This is true provided that the original functions are reasonably smooth which excludes, for instance, the case of impulse functions.) After surprisingly few convolutions this repeated smoothing results in a distribution which approximates a Gaussian function. The approximation gets better as the number of convolutions increases. The tendency for multiple convolutions to give rise to Gaussian functions is called the *central limit theorem* and accounts for the ubiquitous nature of Gaussian noise. It is illustrated for multiple self convolution of a rectangular pulse in Figure 3.20. In the context of statistics the central limit theorem can be stated as follows:

If N statistically independent random variables are added, the sum will have a probability density function which tends to a Gaussian function as N tends to infinity, irrespective of the original random variable pdfs.

A second consequence of the central limit theorem is that the pdf of the *product* of N independent random variables will tend to a log-normal distribution as N tends to infinity, since multiplication of functions corresponds to addition of their logarithms.

If two *Gaussian* random variables are added their sum will also be a Gaussian random variable. In this case the result is exact and holds even if the random variables are correlated. The mean and variance of the sum are given by:

$$\bar{Z} = \bar{X} + \bar{Y} \tag{3.40}$$

and:

$$\sigma_{X \pm Y}^2 = \sigma_X^2 \pm 2\rho\sigma_X\sigma_Y + \sigma_Y^2 \tag{3.41}$$

For uncorrelated (and therefore independent) Gaussian random variables the variances, like the means, are simply added. This is an especially easy case to prove since for independent variables the pdf of the sum is the convolution of two Gaussian functions. This is equivalent to multiplying the Fourier transforms of the original pdfs and then inverse Fourier transforming the result. (The Fourier transform of a pdf is called the *characteristic function* of the random variable.) When a Gaussian pdf is Fourier transformed the result is a Gaussian characteristic function. When Gaussian characteristic functions are multiplied the result remains Gaussian ($e^{-x^2} e^{-x^2} = e^{-2x^2}$). Finally when the Gaussian product is inverse Fourier transformed the result is a Gaussian pdf.

EXAMPLE 3.9
$x(t)$ and $y(t)$ are zero mean Gaussian random currents. When applied individually to 1 Ω resistive loads they dissipate 4.0 W and 1.0 W of power respectively. When both are applied to the load simultaneously the power dissipated is 3.0 W. What is the correlation between X and Y?

Since X and Y have zero mean their variance is equal to their normalised power, i.e.:

$$\sigma_X^2 = 4.0 \quad \text{and} \quad \sigma_Y^2 = 1.0$$

Their standard deviations are therefore:

$$\sigma_X = 2.0 \quad \text{and} \quad \sigma_Y = 1.0$$

Using equation (3.41):

$$\rho = \frac{\sigma_{X \pm Y}^2 - \sigma_X^2 - \sigma_Y^2}{\pm 2 \, \sigma_X \, \sigma_Y}$$

We take the positive sign in the denominator since it is the sum (not difference) which dissipates 3.0 W:

$$\rho = \frac{3.0 - 4.0 - 1.0}{2 \, (2.0) \, (1.0)} = -0.5$$

EXAMPLE 3.10

Two independent random voltages have a uniform pdf given by:

$$p(V) = \begin{cases} 0.5, & |V| \leq 1.0 \\ 0, & |V| > 1.0 \end{cases}$$

Find the pdf of their sum.

$$p(V_1 + V_2) = p_1(V) * p_2(V)$$

Using χ to represent the characteristic function, i.e.-

$$\chi(W) = \text{FT}\{p(V)\}$$

then from the convolution theorem (Table 2.5) and the Fourier transform of a rectangular function (Table 2.4):

$$\chi(W) = \chi_1(W) \, \chi_2(W)$$

$$= 0.5 \, [2 \, \text{sinc}(2W)] \, 0.5 \, [2 \, \text{sinc}(2W)]$$

$$= \text{sinc}^2(2W)$$

The pdf of the sum is then found by inverse Fourier transforming $\chi(W)$, i.e.:

$$p(V_1 + V_2) = 0.5 \, \Lambda \left(\frac{V}{2} \right)$$

$$= \begin{cases} 0.5 \left(1 - \frac{|V|}{2} \right), & |V| \leq 2.0 \\ 0, & |V| > 2.0 \end{cases}$$

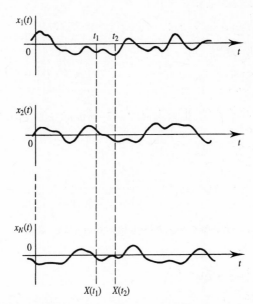

Figure 3.21 *Random process, $X(t)$, as ensemble of sample functions, $x_i(t)$.*

3.3 Random processes

The term random process usually refers to a random variable which is a function of time (or occasionally a function of position) and is strictly defined in terms of an ensemble (i.e. collection) of time functions, Figure 3.21. Such an ensemble of functions may, in principle, be generated using many sets (perhaps an infinite number) of identical sources. The following notation for random processes is adopted here:

1. The random process (i.e. the entire ensemble of functions) is denoted by $X(t)$.
2. $X(t_1)$ or X_1 denotes an ensemble of samples taken at time t_1 and constitutes a random variable.
3. $x_i(t)$ is the ith sample function of the ensemble.

   It is often the case, in practice, that only one sample function can be observed, the other sample functions representing what might have occurred (given the statistical properties of the process) but didn't. It is also the case that each sample function $x_i(t)$ is *usually* a random function of time although this does not have to be so. (For example $X(t)$ may be a set of sinusoids each sample function having random phase.)

   Random processes, like other types of signal, can be classified in a number of different ways. For example, they may be:

1. Continuous or discrete.
2. Analogue or digital (or mixed).
3. Deterministic or non-deterministic.

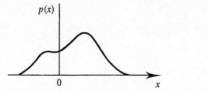

(a) Analogue random process (i.e. continuous pdf)

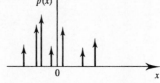

(b) Digital random process (i.e. discrete pdf)

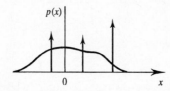

(c) Mixed random process (i.e. continuous plus discrete pdf)

Figure 3.22 *Pdfs of analogue, digital and mixed random processes.*

4. Stationary or non-stationary.
5. Ergodic or non-ergodic.

The first category refers to continuity or discreteness in time or position. (Discrete time signals are also sometimes called a time series.) The second category could be (and sometimes is) referred to as continuous, discrete or mixed which in this context describes the pdf of the process, Figure 3.22. A deterministic random process seems, superficially, to be a contradiction in terms. It describes a process, however, in which each sample function is deterministic. An example of such a process, $x_i(t) = \sin(\omega t + \theta_i)$ where θ_i is a random variable with specified pdf, has already been given. Stationarity and ergodicity are concepts which are central to random processes and they are therefore discussed in some detail below.

3.3.1 Stationarity and ergodicity

Stationarity relates to the time independence of a random process's statistics. There are two definitions:

(a) A random process is said to be stationary in the *strict* (sometimes called narrow) sense if all its pdfs (joint, conditional and marginal) are the same for any value of t, i.e. if none of its statistics change with time.
(b) A random process is said to be stationary in the *loose* (sometimes called wide) sense if its mean value, $\overline{X(t)}$, is independent of time, t, and the correlation, $\overline{X(t_1)X(t_2)}$, depends only on time difference $\tau = t_2 - t_1$.

Ergodicity relates to the equivalence of ensemble and time averages. It implies that each sample function, $x_i(t)$, of the ensemble has the same statistical behaviour as any set of ensemble values, $X(t_j)$, Figure 3.23. Thus for an ergodic process:

$$\langle x_i^n(t) \rangle = \lim_{T \to \infty} \frac{1}{T} \int_{-T/2}^{T/2} x_i^n(t) \, dt$$

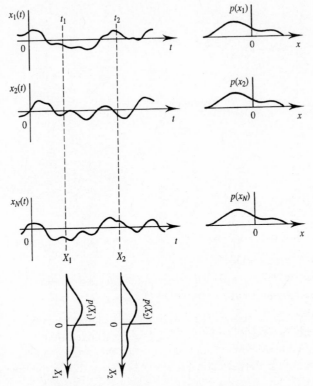

Figure 3.23 *Identity of sample function pdfs, $p(x_i)$, and ensemble random variable pdfs $p(X_i)$, for an ergodic process.*

$$= \int_{-\infty}^{\infty} X_j^n \, p(X_j) \, dx$$

$$= \overline{X^n(t_j)}, \quad \text{for } any \text{ } i \text{ and } j \tag{3.42}$$

It is obvious that an ergodic process must be statistically stationary. The converse is not true, however, i.e. stationary processes need not be ergodic. Ergodicity is therefore a stronger (more restrictive) condition on a random process than stationarity, i.e:

$$\text{Ergodicity} \Rightarrow \text{stationarity}$$
$$\text{Stationarity} \neq\!> \text{ergodicity}$$

3.3.2 Strict and loose sense Gaussian processes

A sample function, $x_i(t)$, is said to belong to a Gaussian random process, $X(t)$, in the *strict* sense if the random variables $X_1 = X(t_1)$, $X_2 = X(t_2)$, $\cdots$, $X_N = X(t_N)$ have an N-dimensional joint Gaussian pdf, Figure 3.24. For an ergodic process the strict sense

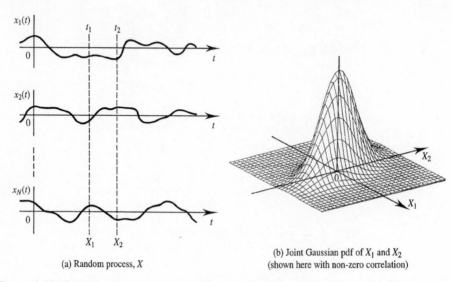

(a) Random process, X

(b) Joint Gaussian pdf of X_1 and X_2
(shown here with non-zero correlation)

Figure 3.24 *Example (drawn for $N = 2$) of the joint Gaussian pdf of random variables (X_1 and X_2) taken from a strict sense Gaussian process.*

Gaussian condition can be defined in terms of a single sample function. In this case if the joint pdf of multiple sets of N-tuple samples, taken with fixed time intervals between the samples of each N-tuple, is N-variate Gaussian then the process is Gaussian in the strict sense. This definition is illustrated in Figures 3.25(a) and (b) for multiple sets of sample pairs (i.e. $N = 2$).

A sample function, $x_i(t)$, is said to belong to a Gaussian random process in the *loose* sense if isolated samples taken from $x_i(t)$ come from a Gaussian pdf, Figure 3.26. The following points can be made about strict and loose sense Gaussian processes:

1. Being a strict sense Gaussian process is a very strong statistical condition, much stronger than being a loose sense Gaussian process. All strict sense Gaussian processes are, therefore, also loose sense Gaussian processes.
2. Examples do exist of processes which are Gaussian in the loose sense but not the strict sense. They are rare in practice, however.
3. A strict sense Gaussian process is the most structureless, random, or unpredictable statistical process possible. It is also one of the most important processes in the context of communications since it describes thermal noise which is present to some degree in all practical systems.
4. A strict sense N-dimensional Gaussian pdf is specified completely by its first and second order moments, i.e. its means, variances and covariances, as all higher moments (Figure 3.9) of the Gaussian pdf are zero.

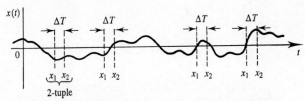

(a) *N*-tuple (*N* = 2) samples with constant sample separation (Δ*T*) taken at random times from a sample function, *x(t)*, of the random process *X(t)*

(b) Joint, *N*-variate (*N* = 2), Gaussian scattergram for *p(x₁, x₂)*

Figure 3.25 *Single sample function definition of ergodic, strict sense Gaussian, random process.*

3.3.3 Autocorrelation and power spectral density

A simple pdf is obviously insufficient to fully describe a random signal (i.e. a sample function from a random process) because it contains no information about the signal's rate of change.

Such information would be available, however, in the joint pdfs, $p(X_1, X_2)$, of random variables, $X(t_1)$ and $X(t_2)$, separated by $\tau = t_2 - t_1$. These joint pdfs are not usually known in full but partial information about them is often available in the form of the correlation, $\overline{X(t_2)X(t_2 - \tau)}$. For ergodic signals the ensemble average taken at any time is equal to the temporal average of any sample function, i.e.:

$$\overline{X(t)\,X(t - \tau)} = \langle x(t)\,x(t - \tau) \rangle$$

$$= \lim_{T \to \infty} \frac{1}{T} \int_{-T/2}^{T/2} x(t)\,x(t - \tau)\,dt$$

$$= R_x(\tau)\ (\text{V}^2) \tag{3.43}$$

The autocorrelation function, $R_x(\tau)$, of a sample function, $x(t)$, taken from a real,

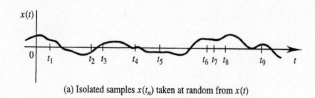

(a) Isolated samples $x(t_n)$ taken at random from $x(t)$

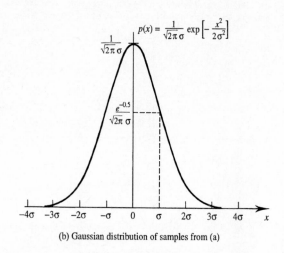

(b) Gaussian distribution of samples from (a)

Figure 3.26 *Definition of a loose sense Gaussian process.*

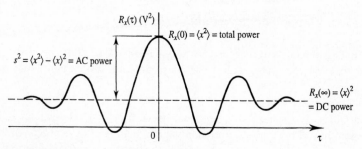

Figure 3.27 *General behaviour of $R_x(\tau)$ for a random process.*

ergodic, random process has the following properties:

1. $R_x(\tau)$ is real.
2. $R_x(\tau)$ has even symmetry (see Figure 3.27), i.e.:

$$R_x(-\tau) = R_x(\tau) \qquad (3.44)$$

3. $R_x(\tau)$ has a maximum (positive) magnitude at $\tau = 0$ which corresponds to the mean square value of (or normalised power in) $x(t)$, i.e.:

$$\langle x^2(t) \rangle = R_x(0) > |R(\tau)|, \quad \text{for } \textit{all } \tau \neq 0 \tag{3.45}$$

4. If $x(t)$ has units of V then $R_x(\tau)$ has units of V^2 (i.e. normalised power).
5. $R_x(\infty)$ is the square mean value of (or normalised DC power in) $x(t)$, i.e.:

$$R_x(\infty) = \langle x(t) \rangle^2 \tag{3.46}$$

6. $R_x(0) - R_x(\infty)$ is the variance, s^2, of $x(t)$, i.e.:

$$R_x(0) - R_x(\infty) = \langle x^2(t) \rangle - \langle x(t) \rangle^2 = s^2 \tag{3.47}$$

7. The autocorrelation function and *two sided* power spectral density of $x(t)$ form a Fourier transform pair, i.e.:

$$R_x(\tau) \overset{FT}{\Leftrightarrow} G_x(f) \tag{3.48}$$

(This is the Wiener-Kintchine theorem [Papoulis] which, although not proved here, can be readily accepted since a similar theorem has been proved in Chapter 2 for transient signals.) Properties of the corresponding power spectral density (most of which are corollaries of the above) include the following:

1. $G_x(f)$ has even symmetry about $f = 0$, i.e.:

$$G_x(-f) = G_x(f) \tag{3.49}$$

2. $G_x(f)$ is real.
3. The area under $G_x(f)$ is the mean square value of (or normalised power in) $x(t)$, i.e.:

$$\int_{-\infty}^{\infty} G_x(f)\, df = \langle x^2(t) \rangle \tag{3.50}$$

4. If $x(t)$ has units of V then $G_x(f)$ has units of V^2/Hz.
5. The area under any impulse in $G_x(f)$ occurring at $f = 0$ is the square mean value of (or normalised DC power in) $x(t)$, i.e.:

$$\int_{0-}^{0+} G_x(f)\, df = \langle x(t) \rangle^2 \tag{3.51}$$

6. The area under $G_x(f)$, excluding any impulse function at $f = 0$, is the variance of $x(t)$ or the normalised power in the fluctuating component of $x(t)$, i.e.:

$$\int_{-\infty}^{0-} G_x(f)\, df + \int_{0+}^{\infty} G_x(f)\, df = \langle x^2(t) \rangle - \langle x(t) \rangle^2 \tag{3.52}$$

7. $G_x(f)$ is positive for all f, i.e.:

$$G_x(f) \geq 0, \quad \text{for } \textit{all } f \tag{3.53}$$

(White noise is a random signal with particularly extreme spectral and autocorrelation properties. It has no self similarity with any time shifted version of itself so its autocorrelation function consists of a single impulse at zero delay, and its power spectral density is flat.)

A normalised autocorrelation function, $\rho_x(\tau)$, can be defined by subtracting any DC value present in $x(t)$, dividing by the resulting RMS value and autocorrelating the result. This is equivalent to:

$$\rho_x(\tau) \; = \; \frac{\langle x(t)x(t-\tau) - \langle x(t)\rangle^2\rangle}{\langle x^2(t)\rangle - \langle x(t)\rangle^2} \tag{3.54}$$

The normalised function, Figure 3.28, is clearly an extension of the normalised correlation coefficient (equation (3.28)) and has the properties:

$$\rho_x(0) \; = \; 1 \tag{3.55(a)}$$

$$\rho_x(\pm\infty) \; = \; 0 \tag{3.55(b)}$$

It can be interpreted as the fraction of $x(t-\tau)$ which is contained in $x(t)$ neglecting DC components. This is easily demonstrated as follows:

Let $f(t)$ be a zero mean stationary random process, i.e.:

$$f(t) \; = \; x(t) - \langle x(t)\rangle \tag{3.56}$$

If the new function:

$$g(t) \; = \; f(t) - \rho f(t-\tau) \tag{3.57}$$

is formed then the value of ρ which minimises $\langle g^2(t)\rangle$ will be the fraction of $f(t-\tau)$ contained in $f(t)$. Expanding $\langle g^2(t)\rangle$:

$$\begin{aligned} \langle g^2(t)\rangle \; &= \; \langle [f(t) - \rho f(t-\tau)]^2\rangle \\ &= \; \langle f^2(t) - 2\rho f(t)f(t-\tau) + \rho^2 f^2(t-\tau)\rangle \\ &= \; \langle f^2(t)\rangle - 2\rho\langle f(t)f(t-\tau)\rangle + \rho^2\langle f^2(t-\tau)\rangle \end{aligned} \tag{3.58}$$

The value of ρ which minimises $\langle g^2(t)\rangle$ is found by solving $d\langle g^2(t)\rangle/d\rho = 0$, i.e.:

$$0 - 2\,R_f(\tau) + 2\rho\,\langle f^2(t-\tau)\rangle \; = \; 0 \tag{3.59}$$

giving:

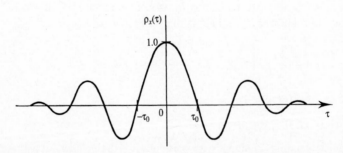

Figure 3.28 *General behaviour of $\rho_x(\tau)$ for a random process. (Shows first null definition of decorrelation time, τ_0.)*

$$\rho = \frac{R_f(\tau)}{\langle f^2(t) \rangle} = \frac{\langle [x(t) - \langle x(t) \rangle][x(t - \tau) - \langle x(t) \rangle] \rangle}{\langle [x(t) - \langle x(t) \rangle]^2 \rangle}$$

$$= \frac{\langle x(t)x(t - \tau) \rangle - \langle x(t) \rangle^2}{\langle x^2(t) \rangle - \langle x(t) \rangle^2} \tag{3.60}$$

EXAMPLE 3.11

Find and sketch the autocorrelation function of the stationary random signal whose power spectral density is shown in Figure 3.29(a).

Using the triangular function, Λ, with f measured in Hz:

$$G_x(f) = 3.0 \left[\Lambda \left(\frac{f - 5000}{1000} \right) + \Lambda \left(\frac{f + 5000}{1000} \right) \right]$$

$$= 3.0 \, \Lambda \left(\frac{f}{1000} \right) * [\delta(f - 5000) + \delta(f + 5000)]$$

$$R_x(\tau) = \text{FT}^{-1} \{ G_x(f) \}$$

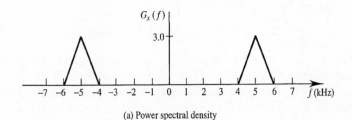

(a) Power spectral density

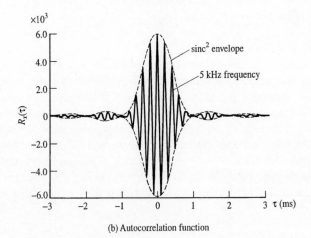

(b) Autocorrelation function

Figure 3.29 *Spectral and temporal characteristics of stationary random signal, Example 3.12.*

$$= 3.0\, \text{FT}^{-1}\left\{ \Lambda\left(\frac{f}{1000} \right) \right\} \text{FT}^{-1}\left\{ \delta(f - 5000) + \delta(f + 5000) \right\}$$

Using Tables 2.4 and 2.5:

$$R_x(\tau) = 3.0 \times 1000\, \text{sinc}^2(1000\tau)\, 2\cos(2\pi\, 5000\tau)$$

Figure 3.29(b) shows a sketch for the solution of Example 3.11.

3.3.4 Signal memory, decorrelation time and white noise

It is physically obvious that practical signals must have a finite memory, i.e. samples taken close enough together must be highly correlated. The decorrelation time, τ_0, of a signal provides a quantitative measure of this memory and is defined as the minimum time shift, τ, required to reduce $\rho_x(\tau)$ to some predetermined, or reference, value, Figure 3.28. The reference value depends on the application and/or preference and can be somewhat arbitrary in the same way as the definition of bandwidth, B (section 2.2.5). Popular choices, however, are $\rho_x(\tau_0) = 1/\sqrt{2}$, 0.5, $1/e$ and 0. Due to the Wiener-Kintchine theorem there is clearly a relationship between B and τ_0, i.e.:

$$B \propto \frac{1}{\tau_0}\ \text{Hz} \tag{3.61}$$

(The constant of proportionality depends on the exact definitions adopted but for reasonably consistent choices is of the order of unity.) Equation (3.61) requires a careful interpretation if the random signal has a passband spectrum (see Example 3.12).

For a random signal or noise with a white power spectral density, equation (3.61) implies that $\tau_0 = 0$, i.e. that the signal has zero memory. In particular the autocorrelation function of white noise will be impulsive, i.e.:

$$R_x(\tau) = C\delta(\tau) \tag{3.62}$$

This means that adjacent samples taken from a white noise process are uncorrelated no matter how closely the samples are spaced. As this is physically impossible it means that white noise, whilst important and useful conceptually, is not practically realisable. (The same conclusion is obvious when considering the total power in a white noise process.)

The common assumption of white, Gaussian, noise processes sometimes gives the impression that Gaussianness and whiteness are connected. This is not true. Noise may be Gaussian or white, or both, or neither. If noise is Gaussian *and* white (and thermal noise, for example, is often modelled in this way) then the fact that adjacent samples from the process are uncorrelated (irrespective of separation) means that they are also independent.

EXAMPLE 3.12

Stating the definition you use, find the decorrelation time of the random signal described in Example 3.11.

Referring to Figure 3.29(b) and using $\rho_x(\tau_0) = 0$, the decorrelation time of $x(t)$ in Example 3.11 is 0.25 cycle of the 5 kHz signal, i.e.:

$$\rho_x(\tau_0) = 0.25 \, T = 0.25 \, \frac{1}{f_c} = \frac{0.25}{5000}$$

$$= 5 \times 10^{-5} \text{ s} \quad \text{(or 50 } \mu\text{s)}$$

(Note that the decorrelation time of the *envelope* of $R_x(\tau)$ in Figure 3.29(b) is 1.0 ms and it is this quantity which is of the order of the reciprocal of the bandwidth.)

3.3.5 Cross correlation of random processes

The cross correlation of functions, taken from two *real* ergodic random processes is:

$$R_{xy}(\tau) = \langle x(t)y(t - \tau) \rangle$$

$$= \lim_{T \to \infty} \frac{1}{T} \int_{-T/2}^{T/2} x(t)y(t - \tau) \, dt \tag{3.63}$$

Some of the properties of this cross correlation function are:

1. $R_{xy}(\tau)$ is real. $\tag{3.64}$

2. $R_{xy}(-\tau) = R_{yx}(\tau)$ $\tag{3.65}$

    (Note that, in general, $R_{xy}(-\tau) \neq R_{xy}(\tau)$.)

3. $[R_x(0)R_y(0)]^{\frac{1}{2}} > |R_{xy}(\tau)|$ for *all* τ. $\tag{3.66}$

    (Note that the maximum value of $R_{xy}(\tau)$ can occur anywhere.)

4. If $x(t)$ and $y(t)$ have units of V the $R_{xy}(\tau)$ has units of V^2 (i.e normalised power) and, for this reason, it is sometimes called a *cross-power*.

5. $[R_x(0) + R_y(0)]/2 > |R_{xy}(\tau)|$,   for *all* τ. $\tag{3.67}$

    (This follows from property 3 since the geometric mean of two real numbers cannot exceed their arithmetic mean.)

6. For *statistically independent* random processes:

    $$R_{xy}(\tau) = R_{yx}(\tau) \tag{3.68(a)}$$

    and if either process has zero mean then:

    $$R_{xy}(\tau) = R_{yx}(\tau) = 0 \quad \text{for *all* } \tau. \tag{3.68(b)}$$

7. The Fourier transform of $R_{xy}(\tau)$ is often called a cross-power spectral density, $G_{xy}(f)$, since its units are V^2/Hz:

$$R_{xy}(\tau) \overset{FT}{\Leftrightarrow} G_{xy}(f) \tag{3.69}$$

If the functions $x(t)$ and $y(t)$ are complex then the cross correlation is defined by:

$$R_{xy}(\tau) = \langle x(t)y^*(t-\tau)\rangle = \langle x^*(t)y(t+\tau)\rangle \tag{3.70}$$

and many of the properties listed above do not apply.

3.4 Summary

Variables are said to be random if their particular value at specified future times cannot be predicted. Information about their probable future values is often available, however, from a probability model. The (unconditional) probability that any of the events, belonging to a subset of mutually exclusive possible events, occurs as the outcome of a random experiment or trial is the sum of the individual probabilities of the events in the subset. The joint probability of a set of statistically independent events is the product of their individual probabilities. A conditional probability is the probability of an event given that some other, specified, event is known to have occurred. Bayes's rule relates joint, conditional and unconditional probabilities.

Cumulative distributions give the probability that a random variable will be less than, or equal to, any particular value. Pdfs are the derivative of the cumulative distribution. The definite integral of a pdf is the probability that the random variable will lie between the integral's limits. Exceedances are the complement of cumulative distributions.

Moments, central moments and modes are statistics of random variables. In general they give partial information about the shape and location of pdfs. Joint pdfs (on definite integration) give the probability that two or more random variables will concurrently take particular values between the specified limits. A marginal pdf is the pdf of one random variable irrespective of the value of any other random variable. The correlation of two random variables is their mean product. The covariance is the mean product of their fluctuating (zero mean) components only, being zero for uncorrelated signals. Statistically independent random variables are always uncorrelated but the converse is not true. The (normalised) correlation coefficient of two random variables is the correlation of their fluctuating components (i.e. covariance) after the standard deviations of both variables have been normalised to 1.0. The pdf of the sum of independent random variables is the convolution of their individual pdfs and, for the sum of many independent random variables, this results in a Gaussian pdf. This is called the central limit theorem. If the random variables are independent then the mean of their sum is the sum of their means and the variance of their sum is the sum of their variances.

Random processes are random variables which change with time (or spatial position). They are defined strictly by an ensemble of functions. Both ensemble and temporal (or spatial) statistics can therefore be defined. A random process is said to be (statistically) stationary in the strict, or narrow, sense if all its statistics are invariant with time (or space). It is said to be stationary in the loose, or wide, sense if its ensemble mean is invariant with time and the correlation between its random variables at different times depends only on time difference. A random process is said to be ergodic if its ensemble and time averages are equal. Random processes which are ergodic are statistically stationary but the converse is not necessarily true.

Gaussian processes are extremely common and important due to the action of the central limit theorem. A process is said to be Gaussian in the strict sense if any pair of (ensemble) random variables has a joint Gaussian pdf. Any sample function of a random process is said to be Gaussian in the loose sense if samples from it are Gaussianly distributed. Not all loose sense Gaussian sample functions belong to strict sense Gaussian processes. Gaussian processes are completely specified by their first and second order moments.

Signal memory is characterised by the signal's autocorrelation function. This function gives the correlation between the signal and a time shifted version of the signal for all possible time shifts. The decorrelation time of a signal is that time shift for which the autocorrelation function has fallen to some prescribed fraction of its peak value. The Wiener-Kintchine theorem identifies the power and energy spectral densities of power and energy signals with the Fourier transform of these signals' autocorrelation functions. The normalised autocorrelation can be interpreted as the fraction of a signal contained within a time shifted version of itself. Signal memory (i.e. decorrelation time) and signal bandwidth are inversely proportional. White noise, with an impulsive autocorrelation function, is physically unrealisable and is memoryless.

Cross correlation relates to the similarity between a pair of different functions, one offset from the other by a time shift. The Fourier transform of a cross correlation function is a cross energy, or power, spectral density depending on whether the function pair represent energy or power signals.

3.5 Problems

3.1. A box contains 30 resistors. 15 of the resistors have nominal values of 1.0 kΩ, 10 have nominal values of 4.7 kΩ and 5 have nominal values of 10 kΩ. 3 resistors are taken at random and connected in series. What is the probability that the 3 resistor combination will have a nominal resistance of: (i) 3 kΩ; (ii) 15.7 kΩ; and (iii) 19.4 kΩ? [0.1121, 0.1847, 0.0554]

3.2. A transceiver manufacturer buys power amplifiers from three different companies (A, B, C). Assembly line workers pick power amplifiers from a rack at random without noticing the supplier. Customer claims, under a one year warranty scheme, show that 8% of all power amplifiers (irrespective of supplier) fail within one year and that 25%, 35% and 40% of all failed power amplifiers were supplied by companies A, B and C respectively. The purchasing department records that power amplifiers have been supplied by companies A, B and C in the proportions 50:40:10 respectively. What is the probability of failure within one year of amplifiers supplied by each company? [0.04, 0.07, 0.32]

3.3. The cumulative distribution function for a continuous random variable, X, has the form:

$$P_X(x) = \begin{cases} 0, & -\infty < x \le -2 \\ a\,(1 + \sin(bx)), & -2 < x \le 2 \\ c, & x > 2 \end{cases}$$

Find: (a) the values of a, b and c that make this a valid CD; (b) the probability that x is negative; and (c) the corresponding probability density function.

3.4. A particular random variable has a cumulative distribution function given by:

$$P_X(x) = \begin{cases} 0, & -\infty < x \le 0 \\ 1 - e^{-x}, & 0 \le x < \infty \end{cases}$$

Find: (a) the probability that $x > 0.5$; (b) the probability that $x \le 0.25$; and (c) the probability that $0.3 < x \le 0.7$. [0.6065, 0.2212, 0.2442]

3.5. The power reflected from an aircraft of complicated shape that is received by a radar can be described by an exponential random variable, w. The pdf of w is:

$$p(w) = \begin{cases} (1/w_o)e^{-w/w_o}, & \text{for } w > 0 \\ 0, & \text{for negative } w \end{cases}$$

where w_o is the average amount of received power. What is the probability that the power received by the radar will be greater than the average received power? [0.368].

3.6. An integrated circuit manufacturer tests the propagation delays of all chips of one particular batch. He discovers that the pdf of the delays is well approximated by a triangular distribution with mean value 8 ns, maximum value 12 ns and minimum value 4 ns. Find: (a) the variance of this distribution; (b) the standard deviation of the distribution; and (c) the percentage of chips which will be rejected if the specification for the device is 10 ns. [2.66, 1.63, 12.5%]

3.7. A bivariate random variable has the joint pdf:

$$p(x, y) = A(x^2 + 2xy) \, \Pi(x)\Pi(y),$$

Find: (a) the value of A which makes this a valid pdf; (b) the correlation of X and Y; (c) the marginal pdfs of X and Y; (d) the mean values of X and Y; and (e) the variances of X and Y. [12, 0.05, 1, 0, 0.15, 0.0833]

3.8. (a) For the zero mean Gaussian pdf: $p_X = (1/\sqrt{2\pi}\sigma)e^{-x^2/(2\sigma^2)}$, prove explicitly that the RMS value $\sqrt{\overline{X^2}}$ is σ.

$$\text{(Hint: } \int_a^b x^2 e^{-x^2} \, dx = -\frac{d}{d\lambda}\left[\int_a^b e^{-\lambda x^2} \, dx\right]_{\lambda = 1}$$

and remember that a complete integral (with limits $\pm\infty$) may be found using the Fourier transform DC value theorem, Table 2.5), and (b) show that, for the Gaussian pdf as defined in part (a), $\overline{X^4} = 3\sigma^4$.

$$\text{(Hint: } \int_0^\infty x^{n-1}e^{-x} \, dx = \Gamma(n), \quad \text{any } n > 0$$

and the gamma function $\Gamma(n)$ has the properties:

$$\Gamma(n + 1) = n\Gamma(n) \quad \text{and} \quad \Gamma(\tfrac{1}{2}) = \sqrt{\pi} \,)$$

3.9. A random signal with uniform pdf, $p_X(x) = \Pi(x)$ is added to a second, independent, random signal with one sided exponential pdf, $p_Y(y) = 3 \, u(y)e^{-3y}$. Find the pdf of the sum.

3.10. $v(t)$, $w(t)$, $x(t)$ and $y(t)$ are independent random signals which have the following pdfs:

$$P_V(v) = \frac{2}{1 + (2\pi v)^2} \qquad P_W(w) = \frac{1}{1 + (\pi w)^2}$$

$$P_X(x) = \frac{2/3}{1 + [(2/3)\pi x]^2} \qquad P_Y(y) = \frac{\frac{1}{2}}{1 + (\frac{1}{2}\pi y)^2}$$

Use characteristic functions to find the pdf of their sum.

3.11. For a tossed dice:

(a) Use convolution to deduce the probabilities of the sum of two thrown dice being 2, 3 etc.

(b) 24 is the largest sum possible on throwing 4 dice. What is the probability of this event from the joint probability of independent events? Check your answer by convolution.

(c) What is the most probable sum for 4 dice? Use convolution to find the probability of this event.

(d) A box containing 100 dice is spilled on the floor. Make as many statements as you can about the sum of the uppermost faces by extending the patterns you see developing in the convolution in parts (a), (b) and (c). [(a) 1/36, 2/36, $\cdots$, 6/36, 5/36, 4/36, $\cdots$, 1/36]; [(b) 7.7 × 10^{-4}]; [(c) 14, 1.13 × 10^{-1}]; [(d) $\Sigma_{min} = 100$, $\Sigma_{max} = 600$, 501 possible Σ's, most likely $\Sigma = 350$, PDF is truncated Gaussian approximation].

3.12. Two, independent, zero mean, Gaussian noise sources (X and Y) each have an RMS output of 1.0 V. A cross-coupling network, is to be used to generate two noise signals (U and V), where $U = (1 - \alpha)X + \alpha Y$ and $V = (1 - \alpha)Y + \alpha X$, with a correlation coefficient of 0.2 between U and V. What must the (voltage) cross-coupling ratio, α, be? [0.8536 or 0.1465]

3.13. A periodic time function, $x(t)$, of period T is defined as a sawtooth waveform with a random 'phase' (i.e. positive gradient zero crossing point), τ, over the period nearest the origin i.e.:

$$x(t) = \frac{2V}{T}(t - \tau), \qquad -T/2 + \tau \le t < T/2 + \tau$$

The pdf of the random variable is:

$$p_T(\tau) = \begin{cases} 1/T, & |\tau| \le T/2 \\ 0, & |\tau| > T/2 \end{cases}$$

Show that the function is ergodic.

3.14. Consider the following time function:

$$X(t) = A\cos(\omega t - \Theta)$$

The phase angle, Θ, is a random variable whose pdf is given as:

$$p_\Theta(\theta) = \frac{1}{2\pi} \qquad \text{for } 0 \le \theta < 2\pi \text{ and zero elsewhere}$$

Find the mean value and variance of θ and of $x(t)$.

3.15. Given that the autocorrelation function of a certain stationary process is:

$$R_{xx}(\tau) = 25 + \frac{4}{1 + 6\tau^2}$$

Find: (a) the mean value, and (b) the variance of the process. [±5, 4]

3.16. $X(t)$ is a deterministic random process defined by:

$$X(t) = \cos(2\pi f t + \Theta) + 0.5$$

where Θ is a uniformly distributed random variable in the range [-π, π], but remains fixed for a given sample waveform of the random process. Calculate $R_{xx}(\tau)$, and identify the source of each term in your answer.

3.17. A stationary random process has an autocorrelation function given by:

$$R(\tau) = \begin{cases} 10(1 - |\tau|/0.05), & |\tau| \le 0.05 \\ 0, & \text{elsewhere} \end{cases}$$

Find: (a) the variance; and (b) the power spectral density of this process. State the relation between bandwidth and decorrelation time of this random process both being defined by the first zero crossing (or touching) point in their respective domains.

3.18. A stationary random process has a power spectral density given by:

$$G(f) = \begin{cases} 5, & 10/2\pi \le |f| \le 20/2\pi \\ 0, & \text{elsewhere} \end{cases}$$

Find: (a) the mean square value; and (b) the autocorrelation function of the process. (If you have access to simple plotting software plot the autocorrelation function.)

3.19. A stationary random process has a bilateral (i.e. double sided) power spectral density given by:

$$G_{xx}(\omega) = \frac{32}{\omega^2 + 16}$$

Find: (a) the average power (on a per-ohm basis) of this random process; and (b) the average power (on a per-ohm basis) of this random process in the range −4 rad/s to 4 rad/s. [4, 2]

3.20. A random variable, $Z(t)$, is defined to be:

$$Z(t) = x(t) + x(t + \tau)$$

$x(t)$ is a sample function from an ergodic process whose autocorrelation function is:

$$R_{xx}(\tau) = e^{-\tau^2}$$

Derive an expression for the autocorrelation of the random process $Z(t)$.

3.21. Show that the autocorrelation function of a non-zero mean random process, $X(t)$, may be written as:

$$R_{xx}(\tau) = R_{x'x'}(\tau) + E[(X(t))^2]$$

where $R_{x'x'}(\tau)$ is the autocorrelation function of a zero-mean random process and E[.] is the expectation operator as defined in equation (3.16).

3.22. The stationary random process $x(t)$ has a power spectral density $G_{xx}(f)$. What is the power spectral density of $y(t) = x(t - T)$.

3.23. Two jointly stationary random processes have sample functions of the form:

$$X(t) = 5\cos(10t + \Theta) \quad \text{and} \quad Y(t) = 20\sin(10t + \Theta)$$

where Θ is a random variable that is uniformly distributed from 0 to 2π. Find the cross correlation function $R_{xy}(\tau)$ of these two processes.

CHAPTER 4

Linear systems

4.1 Introduction

The word system is defined [Hanks] as 'a group or combination of inter-related, inter-dependent, or interacting elements forming a collective entity; ...'. In the context of a digital communications system the interacting elements, for example electronic amplifiers, mixers, detectors etc., are themselves subsystems made up of components such as resistors, capacitors and transistors. An understanding of how systems behave, and are described, is therefore important to the analysis of electronic communications equipment.

This chapter reviews the properties of the most analytically tractable, but also most important, class of system (i.e. linear systems) and applies concepts developed in Chapters 2 and 3 to them. In particular convolution is used to provide a time domain description of the effect of a system on a signal and the convolution theorem is used to link this to the equivalent description in the frequency domain. Towards the end of the chapter the effect of *memoryless non*-linear systems on the pdf of signals and noise is briefly discussed.

4.2 Linear systems

Linear systems constitute one, restricted, class of system. Electronic communications equipment is predominantly composed of interconnected linear subsystems.

4.2.1 Properties of linear systems

Before becoming involved in the mathematical description of linear systems there are two important questions which should be answered:

1. What is a linear system?
2. Why are linear systems so important?

In answering the first question it is almost as important to say what a linear system is not, as to say what it is. Figure 4.1 shows the input/output characteristic of a system which is specified mathematically by the straight line equation:

$$y(t) = mx(t) + C \tag{4.1}$$

This system, perhaps surprisingly, is non-linear (providing $C \neq 0$). A definition of a linear system can be given as follows:

A system is linear if its response to the sum of any two inputs is the sum of its responses to each of the inputs alone.

This property is usually called the principle of *superposition* since responses to component inputs are superposed at the output. In this context linearity and superposition are synonymous. If $x_i(t)$ are inputs to a system and $y_i(t)$ are the corresponding outputs then superposition can be expressed mathematically as:

$$y(t) = \sum_i y_i(t) \tag{4.2(a)}$$

when:

$$x(t) = \sum_i x_i(t) \tag{4.2(b)}$$

Proportionality (also called homogeneity) is a property which follows directly from linearity. It is defined by:

$$y(t) = my_1(t) \tag{4.3(a)}$$

when:

$$x(t) = mx_1(t) \tag{4.3(b)}$$

The system described by Figure 4.1 and equation (4.1) would have this property if $C = 0$ and in this special case is, therefore, linear. For $C \neq 0$, however, the system does not obey proportionality and therefore cannot be linear. (Equations (4.3) represent a necessary and sufficient condition for linearity providing the system is *memoryless* i.e. its instantaneous output depends only on its instantaneous input.) A further property which systems often

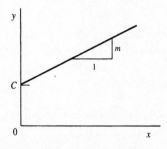

Figure 4.1 *Input/output characteristic of a non-linear system.*

have is *time invariance*. This means that the output of a system does not depend on when the input is applied (except in so far as its location in time). More precisely time invariance can be defined by:

$$y(t) = y_1(t - T) \tag{4.4(a)}$$

when:

$$x(t) = x_1(t - T) \tag{4.4(b)}$$

The majority of communications subsystems obey both equations (4.2) and (4.4) and are therefore called *time invariant linear systems* (TILS).

Time invariant linear systems can be defined using a single formula which also explicitly recognises proportionality, i.e.:

$$\text{If} \quad y_1(t) = S\{x_1(t)\} \quad \text{and} \quad y_2(t) = S\{x_2(t)\} \tag{4.5(a)}$$

$$\text{then} \quad S\{ax_1(t - T) + bx_2(t - T)\} = ay_1(t - T) + by_2(t - T) \tag{4.5(b)}$$

where $S\{\ \}$ represents the functional operation of the system.

4.2.2 Importance of linear systems

The importance of linear systems in engineering cannot be overstated. It is interesting to note, however, that (like periodic signals) linear systems constitute a conceptual ideal that cannot be strictly realised in practice. This is because any device behaves non-linearly if excited by signals of large enough amplitude. An obvious example of this in electronics is the transistor amplifier which saturates when the amplitude of the output approaches the power supply rail voltages (Figure 4.2). Such an amplifier is at least approximately linear, however, over its normal operating range. It is ironic, therefore, that whilst no systems are linear if driven by large enough signals many non-linear systems are at least approximately linear when driven by small enough signals. This is because the transfer characteristic of a non-linear (memoryless) system can normally be represented by a polynomial of the form:

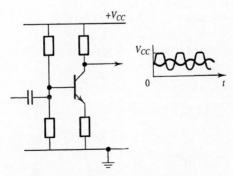

Figure 4.2 *Non-linear behaviour of a simple transistor amplifier.*

$$y(t) = ax(t) + bx^2(t) + cx^3(t) + \cdots \tag{4.5(c)}$$

For small enough input signals (and providing $a \neq 0$) only the first term in equation (4.5(c)) is significant and the system therefore behaves linearly.

An important property of linear systems is that they respond to sinusoidal inputs with sinusoidal outputs of the same frequency (i.e. they conserve the shape of sinusoidal signals).

Other compelling reasons for studying and using linear systems are that:

1. The electric and magnetic properties of free space are linear, i.e.:

$$\mathbf{D} = \varepsilon_o \mathbf{E} \quad (\text{C m}^{-2}) \tag{4.6(a)}$$

$$\mathbf{B} = \mu_o \mathbf{H} \quad (\text{Wb m}^{-2}) \tag{4.6(b)}$$

   (Since free space is memoryless, proportionality is sufficient to imply linearity.)

2. The electric and magnetic properties of many materials are linear over a large range of field strengths, i.e.:

$$\mathbf{D} = \varepsilon_r \varepsilon_o \mathbf{E} \quad (\text{C m}^{-2}) \tag{4.7(a)}$$

$$\mathbf{B} = \mu_r \mu_o \mathbf{H} \quad (\text{Wb m}^{-2}) \tag{4.7(b)}$$

$$\mathbf{J} = \sigma \mathbf{E} \quad (\text{A/m}^{-2}) \tag{4.7(c)}$$

   where ε_r, μ_r and σ are constants. (There are notable exceptions to this, of course, e.g. ferromagnetic materials.)

3. Many general mathematical techniques are available for describing, analysing and synthesising linear systems. This is in contrast to non-linear systems for which few, if any, general techniques exist.

EXAMPLE 4.1

Demonstrate the linearity or otherwise of the systems represented by the diagrams in Figure 4.3.

(a) For input $x_1(t)$ output is $y_1(t) = x_1(t) + f(t)$
    For input $x_2(t)$ output is $y_2(t) = x_2(t) + f(t)$
    For input $x_1(t) + x_2(t)$ output is $x_1(t) + x_2(t) + f(t) \neq y_1(t) + y_2(t)$
    i.e. superposition does not hold and system (a) is, therefore, not linear.

(b) For input $x(t) = x_1(t) + x_2(t)$ the output $y(t)$ is:

$$y(t) = f(t)[x_1(t) + x_2(t)]$$

$$= f(t)x_1(t) + f(t)x_2(t) = y_1(t) + y_2(t)$$

which is the superposition of the outputs due to $x_1(t)$ and $x_2(t)$ alone. System (b) is, therefore, linear.

(c) For input $x(t) = x_1(t) + x_2(t)$ output $y(t)$ is:

$$y(t) = \frac{d}{dt}\left[x_1(t) + x_2(t)\right]$$

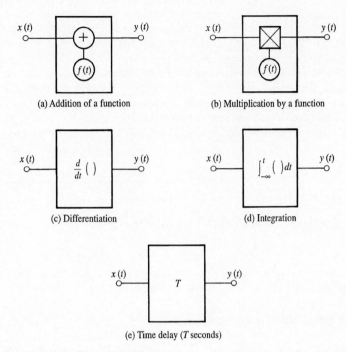

(a) Addition of a function

(b) Multiplication by a function

(c) Differentiation

(d) Integration

(e) Time delay (T seconds)

Figure 4.3 *Systems referred to in Example 4.1.*

$$= \frac{d}{dt} x_1(t) + \frac{d}{dt} x_2(t)$$

$$= y_1(t) + y_2(t)$$

i.e. system (c) is linear.

(d) For input $x(t) = x_1(t) + x_2(t)$ output $y(t)$ is:

$$y(t) = \int_{-\infty}^{t} [x_1(t') + x_2(t')] \, dt'$$

$$= \int_{-\infty}^{t} x_1(t') \, dt' + \int_{-\infty}^{t} x_2(t') \, dt' = y_1(t) + y_2(t)$$

i.e. system (d) is linear.

(e) For input $x(t) = x_1(t) + x_2(t)$ output is:

$$y(t) = x_1(t - T) + x_2(t - T)$$

$$= y_1(t) + y_2(t)$$

i.e. system (e) is linear.

There is an apparent paradox involved in the consideration of the linearity of additive and multiplicative systems. This is that the operation of addition is, by definition, linear, i.e. if two inputs are added the output is the sum of each input by itself. The reason system (a) is not linear is that $f(t)$ is considered to be part of the system, not an input. Conversely multiplication is a non-linear operation in the sense that if two inputs are multiplied then the output is not the sum of the outputs due to each input alone (which would both be zero since each input alone would be multiplied by zero). When $f(t)$ in Figure 4.3(b) is considered to be part of the system, however, superposition holds and the system is therefore linear.

Another point to note is that systems must be either non-linear or time varying (or both) in order to generate frequency components at the output which do not appear at the input. Multiplying by $f(t)$ in Figure 4.3(b) will result in new frequencies at the output providing $f(t)$ is not a constant. This is because in this case we have a time varying linear system. If $f(t) = $ constant then system (b) is a time invariant linear system (in fact a linear amplifier) and no new frequencies are generated.

4.3 Time domain description of linear systems

Just as signals can be described in either the time or frequency domain, so too can systems. In this section time domain descriptions are addressed and the close relationship between linear systems and linear equations is demonstrated.

4.3.1 Linear differential equations

Any system which can be described by a linear differential equation, of the form:

$$a_0 y + a_1 \frac{dy}{dt} + a_2 \frac{d^2 y}{dt^2} + \cdots + a_{N-1} \frac{d^{N-1} y}{dt^{N-1}}$$

$$= b_0 x + b_1 \frac{dx}{dt} + b_2 \frac{d^2 x}{dt^2} + \cdots + b_{M-1} \frac{d^{M-1} x}{dt^{M-1}} \tag{4.8}$$

always obeys the principle of superposition and is therefore linear. If the coefficients a_i and b_i are constants then the system is also time invariant. The response of such a system to an input can be defined in terms of two components. One component, the *free response*, is the output, $y_{free}(t)$, when the input (or forcing function) $x(t) = 0$. (Since $x(t)$ is zero for all t then all the derivatives $d^n x(t)/dt^n$ are also zero.) The free response is therefore the solution of the homogeneous equation:

$$a_0 y + a_1 \frac{dy}{dt} + a_2 \frac{d^2 y}{dt^2} + \cdots a_{N-1} \frac{d^{N-1} y}{dt^{N-1}} = 0 \tag{4.9}$$

subject to the value of the output, and its derivatives, at $t = 0$, i.e.:

$$y(0), \frac{dy}{dt}\Big|_{t=0}, \frac{d^2 y}{dt^2}\Big|_{t=0}, \cdots, \frac{d^{N-1} y}{dt^{N-1}}\Big|_{t=0}$$

These values are called the *initial conditions*. The second component, the *forced response*, is the output, $y_{forced}(t)$, when the input, $x(t)$, is applied but the initial conditions are set to zero, i.e. it is the solution of equation (4.8) when:

$$y(0) = \frac{dy}{dt}\Big|_{t=0} = \frac{d^2y}{dt^2}\Big|_{t=0} = \cdots = \frac{d^{N-1}y}{dt^{N-1}}\Big|_{t=0} = 0 \qquad (4.10)$$

The total response of the system (unsurprisingly, since superposition holds) is the sum of the free and forced responses, i.e.:

$$y(t) = y_{free}(t) + y_{forced}(t) \qquad (4.11)$$

An alternative decomposition of the response of a linear system is in terms of its steady state and transient responses. The steady state response is that component of $y(t)$ which does not decay (i.e. tend to zero) as $t \to \infty$. The transient response is that component of $y(t)$ which does decay as $t \to \infty$, i.e.:

$$y(t) = y_{steady}(t) + y_{transient}(t) \qquad (4.12)$$

4.3.2 Discrete signals and matrix algebra

Consider a linear system with discrete (or sampled) input $x_1, x_2, x_3, \cdots, x_N$ and discrete output $y_1, y_2, y_3, \cdots, y_M$ as shown in Figure 4.4. Each output is then given by a weighted sum of all the inputs [Spiegel]:

$$\begin{bmatrix} y_1 \\ y_2 \\ .. \\ .. \\ y_M \end{bmatrix} = \begin{bmatrix} G_{11} & G_{12} & .... & G_{1N} \\ G_{21} & G_{22} & .... & G_{2N} \\ .. & .. & .. .... & .. \\ .. & .. & .. .... & .. \\ G_{M1} & .. & .... & G_{MN} \end{bmatrix} \begin{bmatrix} x_1 \\ x_2 \\ .. \\ .. \\ x_N \end{bmatrix} \qquad (4.13)$$

i.e.:

$$y_i = \sum_{j=1}^{N} G_{ij}x_j \qquad (4.14)$$

(If the system is a physical system operating in real time then $G_{ij} = 0$ for all values of x_j occurring after y_i.)

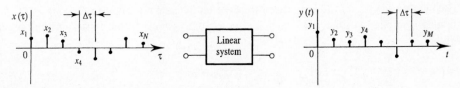

Figure 4.4 *Linear systems with discrete input and output.*

4.3.3 Continuous signals, convolution and impulse response

If the discrete input and output of equation (4.14) are replaced with continuous equivalents, i.e.:

$$y_i \rightarrow y(t)$$

$$x_j \rightarrow x(\tau)$$

(the reason for keeping the input and output variables, τ and t, separate will become clear later) then the discrete summation becomes continuous integration giving:

$$y(t) = \int_0^{N\Delta\tau} G(t, \tau) x(\tau) \, d\tau \qquad (4.15)$$

The limits of integration in equation (4.15) assume that x_1 occurs at $\tau = 0$, and the N input samples are spaced by $\Delta\tau$ seconds. Once again, for physical systems operating in real time, it is obvious that future values of input do not contribute to current, or past, values of output. The upper limit in the integral of equation (4.15) can therefore be replaced by t without altering its value, i.e.:

$$y(t) = \int_0^t G(t, \tau) x(\tau) \, d\tau \qquad (4.16)$$

Furthermore, if input signals are allowed which start at a time arbitrarily distant in the past then:

$$y(t) = \int_{-\infty}^t G(t, \tau) x(\tau) \, d\tau \qquad (4.17)$$

Systems described by equations (4.16) and (4.17) are called *causal* since only past and current input values affect (or cause) outputs. Equations (4.15) to (4.17) are all examples of integral transforms (of $x(\tau)$) in which $G(t, \tau)$ is the transform kernel. Replacing the input to the system described by equation (4.17) with a (unit strength) impulse, $\delta(\tau)$, results in:

$$h(t) = \int_{-\infty}^t G(t, \tau) \delta(\tau) \, d\tau \qquad (4.18)$$

(The symbol $h(t)$ is traditionally used to represent a system's impulse response.) If the impulse is applied at time $\tau = T$ then, assuming the system is time invariant, the output will be:

$$h(t - T) = \int_{-\infty}^t G(t, \tau) \delta(\tau - T) \, d\tau \qquad (4.19)$$

The sampling property of $\delta(\tau - T)$ under integration means that $G(t, T)$ can be interpreted as the response to an impulse applied at time $\tau = T$, i.e.:

$$h(t - T) = G(t, T) \tag{4.20}$$

the surface $G(t, \tau)$ therefore represents the responses for impulses applied at all possible times (Figure 4.5). Replacing T with τ in equation (4.20) (which is a change of notation only), and substituting into equation (4.17) gives:

$$y(t) = \int_{-\infty}^{t} h(t - \tau)x(\tau)\, d\tau \tag{4.21}$$

If non-causal systems are allowed then equation (4.21) is rewritten as:

$$y(t) = \int_{-\infty}^{\infty} h(t - \tau)x(\tau)\, d\tau \tag{4.22}$$

Equations (4.21) and (4.22) can be recognised as convolution, or superposition, integrals. The output of a time invariant linear system is therefore given by the convolution of the system's input with its impulse response, i.e.:

$$y(t) = h(t) * x(t) \tag{4.23}$$

Note that the commutative property of convolution means that equations (4.22) and (4.23) can also be written as:

$$y(t) = x(t) * h(t)$$

$$= \int_{-\infty}^{\infty} h(\tau)x(t - \tau)\, d\tau \tag{4.24}$$

Note also that equations (4.21) to (4.24) are consistent with the definition of an impulse response since in this case:

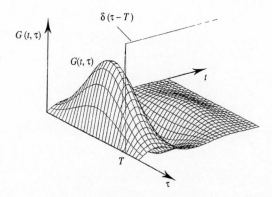

Figure 4.5 *$G(t, \tau)$ for a hypothetical system. Response for an impulse applied at time $\tau = T$ is the curve formed by the intersection of $G(t, \tau)$ with the plane containing $\delta(\tau - T)$.*

$$y(t) \ = \ h(t) * \delta(t) = h(t) \tag{4.25}$$

4.3.4 Physical interpretation of $y(t) = h(t) * x(t)$

The input signal $x(t)$ can be considered to consist of many closely spaced impulses, each impulse having a strength, or weight, equal to the value of $x(t)$ at the time the impulse occurs times the impulse spacing. The output is then simply the sum (i.e. superposition) of the responses to all the weighted impulses. This idea is illustrated schematically in Figure 4.6. It essentially represents a decomposition of $x(t)$ into a set of (orthogonal) impulse functions. Each impulse function is operated on by the system to give a (weighted, time shifted) impulse response and the entire set of impulse responses is then summed to give the (reconstituted) response of the system to the entire input signal. In this context equation (4.22) can be reinterpreted as:

$$y(t) \ = \ \int_{-\infty}^{\infty} h(t - \tau)[x(\tau)\, d\tau] \tag{4.26}$$

where $[x(\tau)d\tau]$ is the weight of the impulse occurring at the input at time τ and $h(t - \tau)$ is the 'fractional' value to which $[x(\tau)d\tau]$ has decayed at the system output by time t (i.e. $t - \tau$ seconds after the impulse occurred at the input). As always, for causal systems, the upper limit in equation (4.26) could be replaced by t corresponding to the condition (see Figure 4.7):

$$h(t - \tau) \ = \ 0, \qquad \text{for } t < \tau \tag{4.27}$$

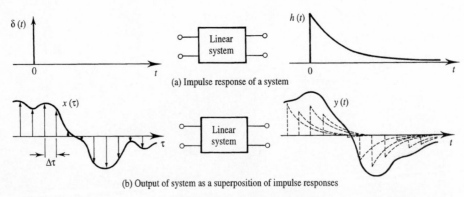

(a) Impulse response of a system

(b) Output of system as a superposition of impulse responses

Figure 4.6 *Decomposition of input into (orthogonal) impulse functions and output formed as a sum of weighted impulse responses.*

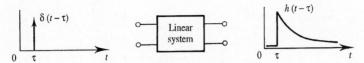

Figure 4.7 *Causal impulse response of a baseband system.*

EXAMPLE 4.2
Find the output of a system having a rectangular impulse response (amplitude A volts, width τ seconds) when driven by an identical rectangular input signal.

Figure 4.8 shows the evolution of the output as $h(t - \tau)$ moves through several different values of t. The result is a triangular function. For a discretely sampled input signal using the standard z-transform notation [Mulgrew and Grant], where $h(n) = x(n) = A + Az^{-1} + Az^{-2} + Az^{-3}$ the output signal is discretely sampled with values $A^2z^{-1} + 2A^2z^{-2} + 3A^2z^{-3} + 4A^2z^{-4} + 3A^2z^{-5}$, etc. Note in this example that, as the impulse response is symmetrical, time reversal of $h(\tau)$ to form $h(-\tau)$ produces a simple shift along the τ-axis.

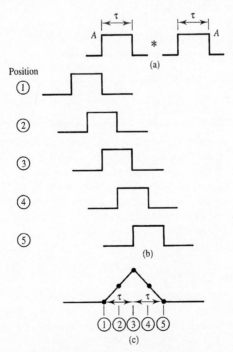

Figure 4.8 *Convolution of two rectangular pulses: (a) pulses; (b) movement of second pulse with respect to first; and (c) values of the convolved output.*

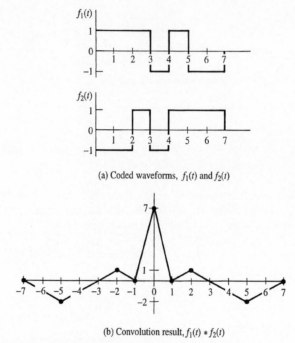

(a) Coded waveforms, $f_1(t)$ and $f_2(t)$

(b) Convolution result, $f_1(t) * f_2(t)$

Figure 4.9 *Convolution of two coded waveforms (Example 4.3).*

EXAMPLE 4.3

Sketch the convolution of the binary coded waveforms $f_1(t)$ and $f_2(t)$ shown in Figure 4.9(a).

This is obtained by time reversing one waveform, e.g. $f_2(t)$, and then sliding it past $f_1(t)$. As each time unit overlaps then, as we are using rectangular pulses, the convolution result is piece-wise linear. The waveforms have been deliberately chosen so that when $f_1(t)$ is time aligned with $f_2(-t)$ then they are identical. The convolution result is shown in Figure 4.9(b). It takes 7 time units to reach the maximum value and another 7 time units to decay again to the final zero value. If $f_2(t)$ is the impulse response of a filter, this represents an example of a matched filter receiver. This type of optimum receiver is discussed in detail later (section 8.3.1).

4.3.5 Step response

Consider the system impulse response shown in Figure 4.10. If the system is driven with a step signal, $u(t)$ (sometimes called the Heaviside step) defined by:

$$u(t) = \begin{cases} 1.0, & t > 0 \\ 0.5, & t = 0 \\ 0, & t < 0 \end{cases} \tag{4.28}$$

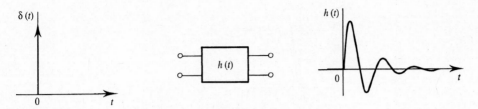

Figure 4.10 *Causal impulse response of a bandpass system.*

then the output of the system (i.e. its step response) is given by:

$$q(t) = \int\limits_{-\infty}^{\infty} h(\tau)u(t-\tau)\,d\tau \qquad (4.29)$$

A graphical interpretation of the integrand in equation (4.29) is shown in Figure 4.11. Since $u(t-\tau) = 0$ for $\tau > t$ and $h(\tau) = 0$ for $\tau < 0$, equation (4.29) can be rewritten as:

$$q(t) = \int\limits_{0}^{t} h(\tau)u(t-\tau)\,d\tau \qquad (4.30)$$

Furthermore, in the region $0 < \tau < t$, $u(t-\tau) = 1.0$, i.e.:

$$q(t) = \int\limits_{0}^{t} h(\tau)\,d\tau \qquad (4.31)$$

The step response is therefore the integral of the impulse response, Figure 4.12. Conversely, of course, the impulse response is the derivative of the step response, i.e.:

$$h(t) = \frac{d}{dt}q(t) \qquad (4.32)$$

Equation (4.32) is particularly useful if the step response of a system is more easily measured than its impulse response.

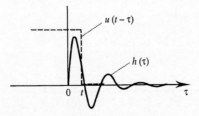

Figure 4.11 *Elements of integrand in equation (4.29).*

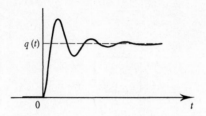

Figure 4.12 *Step response corresponding to impulse response in Figure 4.10.*

EXAMPLE 4.4
Find and sketch the impulse response of the system which has the step response $\Lambda(t-1)$.

The step response is:

$$u(t) \; = \; \Lambda(t-1) \; = \; \begin{cases} t, & 0 \le t \le 1 \\ 2-t, & 1 \le t \le 2 \\ 0, & \text{elsewhere} \end{cases}$$

Therefore the impulse response is:

$$h(t) \; = \; \frac{d}{dt}[\Lambda(t-1)] = \begin{cases} 1, & 0 < t < 1 \\ -1, & 1 < t < 2 \\ 0, & \text{elsewhere} \end{cases}$$

A sketch of $h(t)$ is shown in Figure 4.13.

4.4 Frequency domain description

In the time domain the output of a time invariant linear system is the convolution of its input and its impulse response i.e.:

$$y(t) \; = \; h(t) * x(t) \tag{4.33}$$

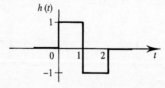

Figure 4.13 *Impulse response of a system with triangular step response (Example 4.4).*

The equivalent frequency domain expression is found by taking the Fourier transform of both sides of equation (4.33) and using the convolution theorem (see Table 2.5):

$$FT\{y(t)\} = FT\{h(t) * x(t)\}$$

$$= FT\{h(t)\} \, FT\{x(t)\} \tag{4.34}$$

$$\text{i.e. } Y(f) = H(f) \, X(f) \tag{4.35}$$

In equation (4.35), $Y(f)$ is the output voltage spectrum, $X(f)$ is the input voltage spectrum and $H(f)$ is the frequency response of the system. All three quantities are generally complex and can be plotted as either amplitude and phase or real and imaginary components. At a particular frequency, f_o, the frequency response is a single complex number giving the voltage gain (or attenuation) and phase shift of a sinusoid of frequency f_o as it passes from system input to output, i.e.:

$$H(f_o) = A(f_o)e^{j\phi(f_o)} \tag{4.36}$$

For a sinusoidal input, $x(t) = \cos 2\pi f_o t$, the output is therefore given by:

$$y(t) = A(f_o) \cos[2\pi f_o t + \phi(f_o)] \tag{4.37}$$

It follows directly from the Fourier transform relationship between $H(f)$ and $h(t)$ that the frequency responses of systems with real impulse responses have *Hermitian* symmetry, i.e.:

$$\Re\{H(f)\} = \Re\{H(-f)\} \tag{4.38(a)}$$

$$\Im\{H(f)\} = -\Im\{H(-f)\} \tag{4.38(b)}$$

where $\Re/\Im$ indicate real/imaginary parts. Equivalently:

$$|H(f)| = |H(-f)| \tag{4.38(c)}$$

$$\phi(f) = -\phi(-f) \tag{4.38(d)}$$

EXAMPLE 4.5

A linear system with the impulse response shown in Figure 4.14(a) is driven by the input signal shown in Figure 4.14(b). Find (i) the voltage spectral density of the input signal, (ii) the frequency response of the system, (iii) the voltage spectral density of the output signal, (iv) the (time domain) output signal.

(i)   The input signal is given by the difference between two rectangular functions:

$$v_{in}(t) = 3.0 \, \Pi\left(\frac{t-1}{2}\right) - 3.0 \, \Pi\left(\frac{t-3}{2}\right)$$

The voltage spectral density of the input is given by the Fourier transform of this:

$$V_{in}(f) = FT\{v_{in}(t)\}$$

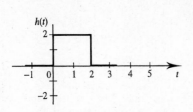

(a) System impulse response

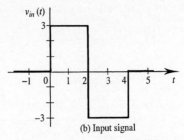

(b) Input signal

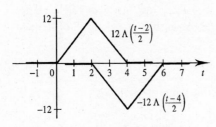

(c) Triangular components of output signal

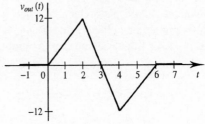

(d) Output signal

Figure 4.14 *Functions for Example 4.5.*

$$= 3.0 \, \text{FT} \left\{ \Pi \left(\frac{t-1}{2} \right) \right\} - 3.0 \, \text{FT} \left\{ \Pi \left(\frac{t-3}{2} \right) \right\}$$

$$= 3.0 \left[2 \, \text{sinc}(2f) \, e^{-j\omega 1} \right] - 3.0 \left[2 \, \text{sinc}(2f) \, e^{-j\omega 3} \right]$$

$$= 6 \, \text{sinc}(2f) \left[e^{j\omega} - e^{-j\omega} \right] e^{-j\omega 2}$$

$$= 6 \, \text{sinc}(2f) \, 2j \sin \omega \, e^{-j2\omega} = 12j \, \text{sinc}(2f) \sin (2\pi f) \, e^{-j4\pi f}$$

$$= 12 \, \frac{\sin^2 (2\pi f)}{2\pi f} \, e^{-j\left(4\pi f - \frac{\pi}{2} \right)}$$

(ii) The frequency response of the system is:

$$H(f) = \text{FT} \{h(t)\}$$

$$= \text{FT} \left\{ 2.0 \, \Pi \left(\frac{t-1}{2} \right) \right\}$$

$$= 2.0 \left[2 \, \text{sinc}(2f) \, e^{-j\omega 1} \right] = 4 \, \text{sinc}(2f) \, e^{-j2\pi f}$$

(iii) The voltage spectral density of the output signal is given by:

$$V_{out} (f) = V_{in}(f) \, H(f)$$

$$= j12 \, \frac{\sin^2 (2\pi f)}{2\pi f} \, e^{-j4\pi f} \, 4 \, \frac{\sin (2\pi f)}{2\pi f} \, e^{-j2\pi f}$$

$$= 12 \frac{\sin^3(2\pi f)}{(\pi f)^2} e^{-j\left(6\pi f - \frac{\pi}{2}\right)} = 12 \frac{\sin^3(2\pi f)}{(\pi f)^2} e^{-j\pi(6f - 0.5)}$$

(iv) The time domain output signal could be found as the inverse Fourier transform of $V_{out}(f)$. It is easier in this example, however, to find the output by convolving the input and impulse response, i.e.:

$$v_{out}(t) = v_{in}(t) * h(t) = \int_{-\infty}^{\infty} v_{in}(\tau) h(t - \tau) \, d\tau$$

Furthermore the problem can be simplified if $v_{in}(t)$ is split into its component parts:

$$v_{out}(t) = \left[3\Pi\left(\frac{t-1}{2}\right) - 3\Pi\left(\frac{t-3}{2}\right) \right] * 2\Pi\left(\frac{t-1}{2}\right)$$

$$= 6\left[\Pi\left(\frac{t-1}{2}\right) * \Pi\left(\frac{t-1}{2}\right) \right] - 6\left[\Pi\left(\frac{t-3}{2}\right) * \Pi\left(\frac{t-1}{2}\right) \right]$$

We know that the result of convolving two rectangular functions of equal width gives a triangular function. Furthermore the peak value of the triangular function is numerically equal to the area under the product of the aligned rectangular functions and occurs at the time shift of the reversed function which gives this alignment. The half width of the triangular function is the same as the width of the rectangular function. Thus:

$$v_{out}(t) = 6\left[2\Lambda\left(\frac{t-2}{2}\right) \right] - 6\left[2\Lambda\left(\frac{t-4}{2}\right) \right]$$

$$= 12\Lambda\left(\frac{t-2}{2}\right) - 12\Lambda\left(\frac{t-4}{2}\right)$$

The two triangular functions making up $v_{out}(t)$ are shown in Figure 4.14(c) and their sum, $v_{out}(t)$, is shown in Figure 4.14(d).

4.5 Causality and the Hilbert transform

All physically realisable systems must be causal, i.e.:

$$h(t) = 0, \qquad \text{for } t < 0 \qquad\qquad\qquad (4.39(a))$$

This is intuitively obvious since physical systems should not respond to inputs before the inputs have been applied. An equivalent way of expressing equation (4.39(a)) is:

$$h(t) = u(t)h(t) \qquad\qquad\qquad (4.39(b))$$

where $u(t)$ is the Heaviside step function. The frequency response of a causal system with real impulse response must therefore satisfy:

$$H(f) = FT\{u(t)\} * FT\{h(t)\}$$

$$= \left[\frac{1}{2} \delta(f) + \frac{1}{j2\pi f} \right] * H(f)$$

$$= \frac{1}{2} H(f) + \left[\frac{1}{j2\pi f} * H(f) \right] \tag{4.40(a)}$$

i.e. $\quad H(f) = \dfrac{1}{j\pi f} * H(f)$ \hfill (4.40(b))

Equation (4.40(b)) is precisely equivalent to equation (4.39(b)).

A necessary and sufficient condition for an amplitude response, $A(f) = |H(f)|$, to be *potentially* causal is:

$$\int_{-\infty}^{\infty} \frac{|\ln A(f)|}{1 + f^2} \, df < \infty \tag{4.41}$$

The expression potentially causal, in this context, means that a system satisfying this criterion will be causal *given that it has a suitable phase response*. Equation (4.41) is called the Paley-Wiener criterion. It has the important implication that a causal system can only have isolated zeros in its amplitude response, i.e. $A(f)$ cannot be zero over a finite band of frequencies.

Returning to the causality condition of equation (4.40(b)), if $H(f)$ is expressed as real and imaginary parts:

$$H_{\Re}(f) + j\, H_{\Im}(f) = \frac{1}{j\pi f} * [H_{\Re}(f) + j\, H_{\Im}(f)]$$

$$= \left[\frac{1}{j\pi f} * H_{\Re}(f) \right] + \left[\frac{1}{\pi f} * H_{\Im}(f) \right] \tag{4.42}$$

and real and imaginary parts are equated, then:

$$H_{\Re}(f) = \frac{1}{\pi f} * H_{\Im}(f) \tag{4.43(a)}$$

$$H_{\Im}(f) = -\frac{1}{\pi f} * H_{\Re}(f) \tag{4.43(b)}$$

The relationship between real and imaginary parts of $H(f)$ in equation (4.43(a)) is called the *inverse* (frequency domain) *Hilbert* transform which can be written explicitly as:

$$H_{\Re}(f) = \frac{1}{\pi} \int_{-\infty}^{\infty} \frac{H_{\Im}(f')}{f - f'} \, df' \tag{4.44}$$

Equation (4.43(b)) is the *forward* Hilbert transform often denoted by:

$$H_{\Im}(f) = \hat{H}_{\Re}(f) \tag{4.45}$$

In the time domain (since the real part of $H(f)$ transforms to the even part of $h(t)$ and the imaginary part of $H(f)$ transforms to the odd part of $h(t)$) the equivalent operations to equations (4.43) are:

$$h_{even}(t) = j \operatorname{sgn}(t) h_{odd}(t) \qquad (4.46)$$

$$h_{odd}(t) = -j \operatorname{sgn}(t) h_{even}(t) \qquad (4.47)$$

Notice that, unlike the Fourier transform, the Hilbert transform does *not* change the domain of the function being transformed. It can therefore be applied either in the frequency domain (as in equation (4.43(b))) or in the time domain. Table 4.1 summarises the frequency and time domain Hilbert transform relationships.

Table 4.1 *Summary of frequency and time domain Hilbert transform relationships.*

$-j \operatorname{sgn}(t) x(t)$	$\overset{\text{FT}}{\rightleftarrows}$	$\hat{X}(f) = \dfrac{-1}{\pi f} * X(f)$
$\downarrow\uparrow$ HT$_f$		HT$_f$ $\uparrow\downarrow$
$x(t)$	$\overset{\text{FT}}{\rightleftarrows}$	$X(f)$
$\uparrow\downarrow$ HT$_t$		HT$_t$ $\downarrow\uparrow$
$x(t) = \dfrac{-1}{\pi t} * x(t)$	$\overset{\text{FT}}{\rightleftarrows}$	$+j \operatorname{sgn}(f) X(f)$

(HT$_t$ *is the time domain Hilbert transform,* HT$_f$ *is the frequency domain Hilbert transform.*)

The time domain Hilbert transform is sometimes called the quadrature filter since it represents an all-pass filter which shifts the phase of positive frequency components by +90° and negative frequency components by −90°. This operation is useful in the representation of bandpass signals and systems as equivalent baseband processes (see section 13.2). It also makes obvious the property that a function and its Hilbert transform are orthogonal.

EXAMPLE 4.6
Establish which of the following systems are causal and which are not: (i) $h(t) = \Lambda(t - 3)$; (ii) $h(t) = e^{-(t-10)^2}$; (iii) $h(t) = u(t) e^{-t}$; (iv) $H(f) = e^{-f^2}$; (v) $H(f) = \Pi(f)$; (vi) $H(f) = \Lambda(f - 3) + \Lambda(f + 3)$; (vii) $H(f) = (1 - jf)/(1 + f^2)$.

(i) $\Lambda(t - 3)$ represents a triangular function which is centred on $t = 3$ and which is zero for $t < 2$ and $t > 4$. It is therefore a causal impulse response.

(ii) $e^{-(t-10)^2}$ represents a Gaussian function centred on $t = 10$. Since it only tends to zero as $t \to \pm\infty$ it represents an acausal impulse response.

(iii) $u(t)e^{-t}$ is causal by definition since the Heaviside factor ensures it is zero for $t < 0$.

(iv) e^{-f^2} represents a Gaussian frequency response. Since Gaussian functions in one domain transform to Gaussian functions in the other domain the impulse response of this system is Gaussian. The system is therefore acausal as in (ii).

(v) $\Pi(f)$ is a strictly bandlimited frequency response. The impulse response cannot therefore be time limited and is thus acausal. (The impulse response is, of course, sinc(t).)

(vi) $\Lambda(f-3) + \Lambda(f+3)$ represents a bandpass triangular amplitude response. It is strictly bandlimited and therefore an acausal system as in (v).

(vii) To test whether $H(f)$ is causal we can find out if $H_\Im(f)$ is the Hilbert transform of $H_\Re(f)$.

$$H_\Im(f) = \frac{-f}{1 + f^2}, \quad H_\Re(f) = \frac{1}{1 + f^2}$$

In the absense of Hilbert transform tables:

$$\hat{H}_\Re(f) = -\frac{1}{\pi f} * \frac{1}{1 + f^2} = -\int_{-\infty}^{\infty} \frac{1}{\pi\phi} \frac{1}{1 + (f - \phi)^2} \, d\phi$$

$$= -\frac{1}{\pi} \int_{-\infty}^{\infty} \frac{1}{\phi \, (\phi^2 - 2f\phi + f^2 + 1)} \, d\phi$$

Using a table of standard integrals (e.g. Dwight, 4th edition, Equation 161.11):

$$\hat{H}_\Re(f) = -\frac{1}{\pi} \left[\frac{1}{2(f^2 + 1)} \ln\left(\frac{\phi^2}{\phi^2 - 2f\phi + f^2 + 1} \right) + \frac{2f}{2(f^2 + 1)} \int \frac{1}{\phi^2 - 2f\phi + f^2 + 1} \, d\phi \right]_{-\infty}^{\infty}$$

The logarithmic factor in the first term in square brackets above tends to zero as $\phi \to \pm\infty$. The integral in the second term is also standard (e.g. Dwight, 4th edition, equation (160.01)) giving:

$$\hat{H}_\Re(f) = -\frac{1}{\pi} \left[\frac{f}{(f^2 + 1)} \frac{2}{\sqrt{4(f^2 + 1) - 4f^2}} \tan^{-1}\left(\frac{(2\phi - 2f)}{\sqrt{4(f^2 + 1) - 4f^2}} \right) \right]_{-\infty}^{\infty}$$

$$= \frac{-1}{\pi} \frac{f}{f^2 + 1} \left[\tan^{-1}(\phi - f) \right]_{-\infty}^{\infty} = -\frac{f}{f^2 + 1}$$

$$= H_\Im(f)$$

Thus $H_\Im(f)$ is the Hilbert transform of $H_\Re(f)$ and $H(f)$ therefore represents a causal system.

4.6 Random signals and linear systems

The effect of a linear system on a deterministic signal is specified completely by:

$$y(t) = h(t) * x(t) \tag{4.48}$$

or, alternatively, by:

$$Y(f) = H(f)X(f) \tag{4.49}$$

Random signals cannot, by definition, be specified as deterministic functions either in the time domain or in the frequency domain. It follows that neither equation (4.48) nor equation (4.49) is particularly useful for information bearing signals or noise. In practice, however, the two properties of such signals which must most commonly be specified are their power spectra and their probability density functions. The effects of linear systems on these signal characteristics are now described.

4.6.1 PSDs and linear systems

The most direct way of deriving the relationship between the power spectral density at the input and output of a linear system is to take the square magnitude of equation (4.49), i.e.:

$$|Y(f)|^2 = |H(f)X(f)|^2$$
$$= |H(f)|^2|X(f)|^2 \tag{4.50}$$

Since $|Y(f)|^2$ and $|X(f)|^2$ are power spectral densities equation (4.50) can be rewritten as:

$$G_y(f) = |H(f)|^2 G_x(f) \quad (\text{V}^2/\text{Hz}) \tag{4.51}$$

If the system input is an energy signal then the power spectral densities, equations (2.52) and (3.49), are replaced by energy spectral densities, equation (2.53):

$$E_y(f) = |H(f)|^2 E_x(f) \quad (\text{V}^2\text{s/Hz}) \tag{4.52}$$

The equivalent time domain description is obtained by taking the inverse Fourier transform of equation (4.50):

$$\text{FT}^{-1}\{Y(f)\,Y^*(f)\} = \text{FT}^{-1}\{H(f)H^*(f)\} * \text{FT}^{-1}\{X(f)X^*(f)\} \tag{4.53}$$

Using the Wiener-Kintchine theorem (or equivalently the conjugation and time reversal Fourier transform theorems, Table 2.5):

$$R_{yy}(\tau) = R_{hh}(\tau) * R_{xx}(\tau) \tag{4.54}$$

where R is the correlation function and the double subscript emphasises auto- or self-correlation. It is almost always the frequency domain description which is the most convenient in practice. As an example of the application of equations (4.51) and (4.52) the noise power spectral density and total noise power at the output of an RC filter are now calculated for the case of white input noise.

EXAMPLE 4.7

Find the output power spectral density for a simple RC filter when it is driven by white noise. What is the total noise power at the filter's output?

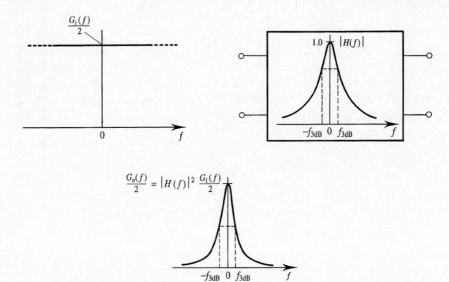

Figure 4.15 *NPSD at output of single pole RC filter driven by white noise. (G(f) is the one sided power spectral density.)*

Figure 4.15 shows the problem schematically. The power spectral density at the filter output is:

$$G_o(f) = |H(f)|^2 \, G_i(f)$$

The frequency response of the filter is given by:

$$H(f) = \frac{1}{1 + j(f/f_{3dB})}$$

where f_{3dB} is the filter −3 dB, or cut-off, frequency. Substituting:

$$G_o(f) = \left| \frac{1}{1 + j(f/f_{3dB})} \right|^2 G_i(f)$$

$$= \frac{1}{1 + (f/f_{3dB})^2} \, G_i(f)$$

Interpreting $G_i(f)$ and $G_o(f)$ as one sided, the total noise power, N, at the filter output is:

$$N = \int_0^\infty G_o(f) \, df$$

$$= \int_0^\infty \frac{1}{1 + (f/f_{3dB})^2} \, G_i(f) \, df$$

Using the change of variable $u = f/f_{3dB}$ and remembering that the input noise is white (i.e. $G_i(f)$ is a constant, Figure 4.15) then:

$$N = G_i \int_0^\infty \frac{1}{1 + u^2} \, f_{3dB} \, du$$

$$= G_i f_{3dB} \int_0^\infty \frac{1}{1 + u^2} \, du = G_i \, f_{3dB} \, [\tan^{-1} u]_0^\infty$$

$$= G_i \, f_{3dB} \, \pi/2 \quad (\text{V}^2)$$

4.6.2 Noise bandwidth

The noise bandwidth, B_N, of a filter is defined as that width which a rectangular frequency response would need to have to pass the same noise power as the filter, given identical white noise at the input to both. This definition is illustrated in Figure 4.16. It can be expressed mathematically as:

$$B_N = \int_0^\infty \frac{|H(f)|^2}{|H(f_p)|^2} \, df \tag{4.55}$$

where f_p is the frequency of peak amplitude response. Notice that noise bandwidth is not equal, in general, to the −3 dB bandwidth. (For the single pole lowpass filter noise bandwidth is larger than the −3 dB bandwidth by a factor $\pi/2$ (see Example 4.7). In this case, for white noise calculations, the use of the 3 dB bandwidth in place of noise bandwidth would therefore lead to a noise power error of 2 dB.)

4.6.3 Pdf of filtered noise

Not only the power spectral density of a random signal is changed when it is filtered but, in general, so is its probability density function. The effect of memoryless systems on the pdf of a signal is discussed in section 4.7. Most communications subsystems, however, have non-zero memory. Unfortunately in this case there is no general, analytical, method of deriving the pdf of the output from the pdf of the input. There is, however, an important exception to this, for which a general result can be derived. This is the pdf of filtered Gaussian noise. Consider Figure 4.17. The output noise, $n_o(t)$, is given by the convolution of the input noise, $n_i(t)$, with the impulse response of the filter or system, i.e.:

$$n_o(t) = \int_{-\infty}^{t} h(t - \tau) \, n_i(\tau) \, d\tau \tag{4.56}$$

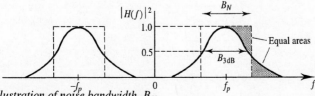

Figure 4.16 *Illustration of noise bandwidth, B_N.*

Notice that the output can be interpreted as a sum of weighted input impulses of strength $n_i(\tau)d\tau$, the weighting factor, $h(t - \tau)$, depending on the time at which the individual input impulses occurred. If $n_i(t)$ is white Gaussian noise then the adjacent impulses are independent Gaussian random variables (since the autocorrelation of $n_i(t)$ is impulsive). The output noise at any instant, e.g. $n_o(t_1)$, is therefore a linear sum of Gaussian random variables and is consequently, itself, a Gaussian random variable. Thus:

Filtered white Gaussian noise is Gaussian

The above result is easily generalised in the following way.

Consider the frequency response in Figure 4.17 to be split into two parts as shown in Figure 4.18. White Gaussian noise at the input of $H_1(f)$ has been shown to result in (non-white) Gaussian noise at the output of $H_2(f)$. Applying the same reasoning, however, the input to $H_2(f)$ is (non-white) Gaussian noise. It follows, by considering the

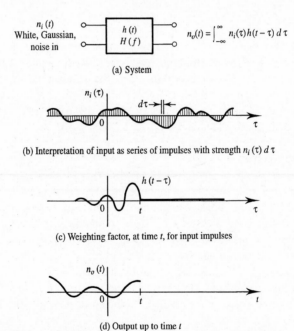

(a) System

(b) Interpretation of input as series of impulses with strength $n_i(\tau)\, d\tau$

(c) Weighting factor, at time t, for input impulses

(d) Output up to time t

Figure 4.17 *Output as linear sum of many independent, Gaussian, random impulses. ($n_i(\tau)$ is white, Gaussian, noise but it can only be drawn as a bandlimited process.)*

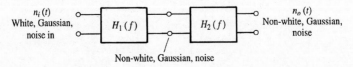

Figure 4.18 *Reinterpretation of $H(f)$ in Figure 4.17 as two cascaded sections.*

input and output of $H_2(f)$ that:

Filtered Gaussian noise is Gaussian

The above result is exact and, to some extent, obvious in that if it were not true then the Gaussian nature of thermal noise, for example, would be obscured by the many filtering processes it is normally subjected to before it is measured.

4.6.4 Spectrum analysers

Spectrum analysers are instruments which are used to characterise signals in the frequency domain. If the signal is periodic the characterisation is partial in the sense that phase information is not usually displayed. If the signal is random, as is the case for noise, the spectral characterisation is essentially complete. Figure 4.19 shows a simplified block diagram of a spectrum analyser and Figure 4.20 shows an alternative conceptual implementation of the same instrument. At a given frequency the display shows either the RMS voltage or mean square voltage which is passed by the filter when it is centred on that frequency. In practice the display y-axis is usually calibrated in dBμ given by $20 \log_{10} (V_{RMS}/10^{-6})$ or dBm given by $10 \log_{10}[(V_{RMS}^2/R_{in})/10^{-3}]$ where dBμ

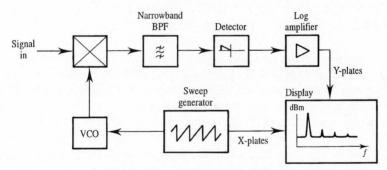

Figure 4.19 *Simplified block diagram of real time, analogue, spectrum analyser.*

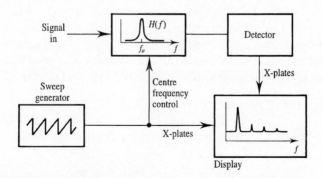

Figure 4.20 *Conceptual model of spectrum analyser.*

indicates dB with respect to a 1 μV and dBm indicates dB with respect to 1 mW. (R_{in}, often 50 Ω, is the input impedance which converts V^2 to W.)

If the signal being measured is periodic, and the filter has a bandwidth which adequately resolves the resulting spectral lines, then the (dBm) display is a faithful representation of the signal's discrete power spectrum. If the signal being measured is a stationary random process, however, then its power spectral density is continuous and the resulting display is the actual power spectral density *correlated* with the filter's squared amplitude response, $|H(f)|^2$ (see section 2.6 and equation (2.85)). If the bandwidth of the filter is narrow compared with the frequency scale over which the signal's power spectral density changes significantly, then the smearing of the spectrum in the correlation process is small and the shape of the spectrum is essentially unchanged. In this case the signal's power spectral density in W/Hz can be found by dividing the displayed spectrum (in watts) by the noise bandwidth, B_N, of the filter (in Hz). On a dB scale this corresponds to:

$$G(f) \ (\text{dB mW Hz}^{-1}) = \text{Display (dBm)} \ - \ 10 \log_{10} B_N \ (\text{dB Hz}) \quad (4.57)$$

4.7 Non-linear systems and transformation of random variables

Non-linear systems are, in general, difficult to analyse. This is principally because superposition no longer applies. As a consequence complicated input signals cannot be decomposed into simple signals (on which the effect of the system is known) and the resulting modified components recombined at the output.

There is one signal characteristic, however, which can often be found at the output of *memoryless* non-linear systems without too much difficulty. This is the signal's probability density function. Mathematically this problem is called a transformation of random variables. An outline of this technique is given below.

Consider a pair of bivariate random variables X, Y and S, T which are related in some deterministic way. Every point in the x, y plane can be mapped into the s, t plane as shown in Figure 4.21. Now consider all the points (x_1, y_1), (x_2, y_2), $\cdots$, in the x, y plane which map into the rectangle centred on s_1, t_1. (There may be none, one, or more than one such point.) Each one of these points (x_n, y_n) has its own small area dA_n in the x, y plane which maps into the rectangle in (s, t). The probability that X, Y lies in any of the areas (x_n, y_n) is equal to the probability that S, T lies in the rectangle at (s_1, t_1), i.e.:

$$\sum_n p_{X,Y}(x_n, y_n) \, dA_n = p_{S,T}(s_1, t_1) \, ds \, dt \quad (4.58)$$

Equation (4.58) can be interpreted as a conservation of probability law.

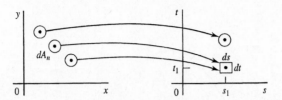

Figure 4.21 *Mapping of random variables using a transformation* $(x, y) \rightarrow (s, t)$.

4.7.1 Rayleigh probability density function

When Gaussian noise (Figure 4.22) is present at the input of an envelope (i.e. amplitude) detector the pdf of the noise at the output is Rayleigh distributed (Figure 4.23). The derivation of this distribution, given below, is a transformation of random variables and uses the conservation of probability law given in equation (4.58).

Let X, Y (quadrature noise components) be independent Gaussian random variables with equal standard deviations, σ, and zero means. Equation (3.33) then simplifies to:

$$p_{X,Y}(x, y) = \frac{1}{\sigma\sqrt{2\pi}} e^{\left(\frac{-x^2}{2\sigma^2} \right)} \frac{1}{\sigma\sqrt{2\pi}} e^{\left(\frac{-y^2}{2\sigma^2} \right)} \tag{4.59}$$

Let R, Θ (noise amplitude and phase) be a new pair of random variables related to X, Y by:

$$r = \sqrt{x^2 + y^2} \tag{4.60(a)}$$

$$\theta = \tan^{-1}(y/x) \tag{4.60(b)}$$

(r, θ can be interpreted as the polar coordinates of the point x, y as shown in Figure 4.24.) The area $d\theta dr$ in the R, Θ plane corresponds to an area $dA = r \, d\theta \, dr$ in the X, Y plane,

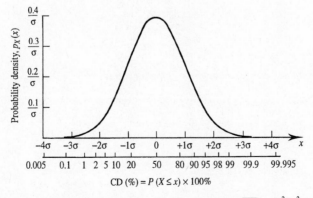

Figure 4.22 *Gaussian probability density function,* $p_X(x) = [1/(\sigma\sqrt{2\pi})]e^{-(x^2/2\sigma^2)}$ *and CD in percent.*

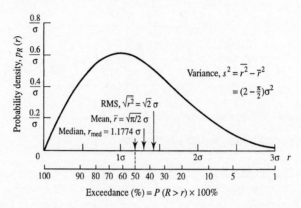

Figure 4.23 *Rayleigh probability density function, $p_R(r) = [r/\sigma^2]e^{-(r^2/2\sigma^2)}$, where σ is the standard deviation of either component in the parent bivariate Gaussian pdf.*

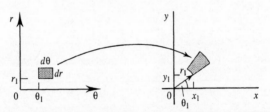

Figure 4.24 *Relationship between (r, θ) and (x, y).*

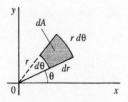

Figure 4.25 *Area in x, y corresponding to rectangle $dr d\theta$ in r, θ.*

Figure 4.25. Conservation of probability requires that:

$$p_{R,\Theta}(r, \theta) \, dr \, d\theta \; = \; p_{X,Y}(x, y) \, r \, d\theta dr \tag{4.61}$$

Therefore:

$$p_{R,\Theta}(r, \theta) \; = \; p_{X,Y}(x, y) \, r$$

$$= \; \frac{r}{\sigma^2 2\pi} \, e^{\left(-\frac{x^2 + y^2}{2\sigma^2}\right)} \; = \; \frac{r}{2\pi\sigma^2} \, e^{\left(\frac{-r^2}{2\sigma^2}\right)} \tag{4.62}$$

Equation (4.62) gives the joint pdf of R and Θ. The (marginal) pdf of R is now given by:

$$p_R(r) = \int\limits_0^{2\pi} p_{R,\Theta}(r,\theta)\, d\theta$$

$$= \frac{r}{2\pi\sigma^2}\, e^{\left(\frac{-r^2}{2\sigma^2}\right)} \int\limits_0^{2\pi} d\theta \qquad (4.63)$$

i.e. $$p_R(r) = \frac{r}{\sigma^2}\, e^{\left(\frac{-r^2}{2\sigma^2}\right)} \qquad (4.64)$$

Equation (4.64) is the Rayleigh probability density function shown in Figure 4.23. Since $p_{R,\Theta}(r,\theta)$ has no θ dependence the marginal probability density function of Θ is uniform, Figure 4.26, i.e.:

$$p_\Theta(\theta) = \frac{1}{2\pi} \qquad (4.65)$$

(Strictly the RHS of equations (4.64) and (4.65) should be multiplied by the Heaviside step function, $u(r)$, and the rectangular function, $\Pi(\theta/2\pi)$, respectively since the probability densities are zero outside these ranges.)

4.7.2 Chi-square distributions

Another transformation of random variables common in electronic communication systems occurs when Gaussian noise is present at the input to a square law device. Let the random variable X representing noise at the input of a square law detector be Gaussianly distributed, i.e.:

$$p_X(x) = \frac{1}{\sigma\sqrt{2\pi}}\, e^{\left(\frac{-x^2}{2\sigma^2}\right)} \qquad (4.66)$$

The detector (Figure 4.27) is characterised by:

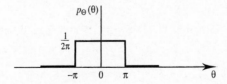

Figure 4.26 *Uniform distribution of Θ.*

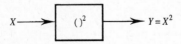

Figure 4.27 *Square-law detector.*

$$Y = X^2 \tag{4.67}$$

which means that two probability areas in X (i.e. $p_X(x)dx$ and $p_X(-x)dx$) both transform to the same probability area $p_Y(y)dy$. This is illustrated in Figure 4.28. Conservation of probability therefore requires that:

$$p_X(x)\, dx + p_X(-x)\, dx = p_Y(y)\, dy \tag{4.68}$$

And by symmetry this means that:

$$2\, p_X(x)\, dx = p_Y(y)\, dy \tag{4.69}$$

Thus:

$$p_Y(y) = 2\, p_X(x)\, \frac{dx}{dy} \tag{4.70}$$

and since $y = x^2$ then:

$$\frac{dy}{dx} = 2x \tag{4.71(a)}$$

and:

$$\frac{dx}{dy} = \frac{1}{2x} \tag{4.71(b)}$$

Therefore:

$$
\begin{aligned}
p_Y(y) &= 2\, p_X(x)\, \frac{1}{2x} \\[2mm]
&= \frac{1}{\sigma\sqrt{2\pi}}\, \frac{1}{x}\, e^{\left(-\frac{x^2}{2\sigma^2}\right)}
\end{aligned}
\tag{4.72}
$$

Using $x = \sqrt{y}$ gives:

$$p_Y(y) = \frac{1}{\sigma\sqrt{2\pi}}\, \frac{1}{\sqrt{y}}\, e^{\left(\frac{-y}{2\sigma^2}\right)}, \quad \text{for } y \geq 0 \tag{4.73}$$

(For $y < 0$, $P_Y(y) = 0$.) Equation (4.73) is, in fact, the special case for $N = 1$ of a more

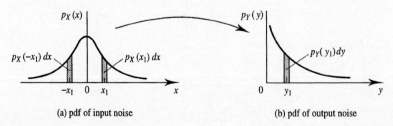

(a) pdf of input noise   (b) pdf of output noise

Figure 4.28 *Mapping of two input points to one output point for a square law detector: (a) pdf of input noise; (b) pdf of output noise.*

general distribution which results from the transformation $Y = \sum_{i=1}^{N} X_i^2$ where X_i are independent Gaussian random variables with equal variance, σ^2, and zero mean. N, here, is the number of degrees of freedom of the distribution. The mean and variance of this generalised chi-square, χ^2, distribution are, respectively:

$$\bar{Y} = N\sigma^2 \qquad\qquad (4.74(\text{a}))$$

and

$$\sigma_Y^2 = 2N\sigma^4 \qquad\qquad (4.74(\text{b}))$$

The pdf for a χ^2 distribution with various degrees of freedom and $\sigma^2 = 1$ is shown in Figure 4.29.

4.8 Summary

Linear systems obey the principle of superposition. Many of the subsystems used in the design of digital communications systems are linear over their normal operating ranges. Linear systems are useful and important because they can be described by linear differential equations.

It is a property of a linear system that its time domain output is given by its time domain input convolved with its impulse response. The impulse response of a linear system is the time derivative of its step response. The output (complex) voltage spectrum of a linear system is the voltage spectrum of its input multiplied by its (complex) frequency response. A system's impulse response and frequency response form a Fourier transform pair.

All physically realisable systems are causal, i.e. their outputs do not anticipate their inputs. The real and imaginary parts of the frequency response of a causal system form a (frequency domain) Hilbert transform pair. The PSD of a random signal at the output of a linear system is given by the PSD at its input multiplied by its squared amplitude response. The autocorrelation of a random signal at the output of a linear system is the

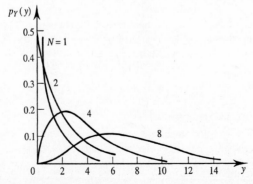

Figure 4.29 *Pdf of a chi-square distribution for several different degress of freedom ($\sigma^2 = 1$).*

convolution of the input signal's autocorrelation with the autocorrelation of the system's impulse response. The noise bandwidth of a system is equal to the width of an ideal rectangular amplitude response which passes the same noise power as the system, for identical white noise at the inputs.

The pdf at the output of memoryless systems, both linear and non-linear, can be found by the method of transformation of random variables. (Linear memoryless systems have a trivial effect since they represent simple scaling factors.) No general, analytical methods are currently known for predicting the pdf at the output of systems with memory. A useful result for the special case of Gaussian noise, however, is that filtered Gaussian noise is Gaussian.

Spectrum analysers are widely used to display the power spectrum of signals. For periodic signals a properly adjusted spectrum analyser will display the signal's discrete line spectrum, usually on a decibel scale of power. For random signals the spectrum analyser displays a good approximation to the signal's power spectral density. In this case care is needed in interpreting the absolute magnitude of the spectrum (in W/Hz) if this information is important. (Often only the shape of the spectrum is required, as the total power in the signal is already known.)

4.9 Problems

4.1. Classify the following systems (input $x(t)$, output $y(t)$, impulse response $h(t)$) as: linear or non-linear, time varying or time invariant, causal or non-causal, memoryless or non-zero memory. (N.B. $u(t)$ is the Heaviside step function.)
(a) $y(t) = 3.7x(t)$, (b) $y(t) = 3.7x(t - 6.2)$, (c) $y(t) = 3.7x(t + 10^{-20})$,
(d) $y(t) = 3.7[x(t - 6.2) + 0.01]$, (e) $y(t) = x(t)\cos(2\pi 50t)$, (f) $y(t) = x^{1.1}(t)e^{-t}$,
(g) $y(t) = \cos(2\pi 50t)[x(t) + x(t - 1)]$, (h) $h(t) = u(t)\cos[2\pi 100(t + 4)]e^{-t}$, (i) $y(t) = x(t)x(t - 2)$,
(j) $y(t) = d/dt\ [x(t + 1)]$, (k) $y(t) = x(t) * u(t)e^{-t}$, (l) $y(t) = x(t/3)$, (m) $y(t) = \int_0^t t'x(t')\ dt'$,
(n) $y(t) = 1/(1 + x(t))$, (o) $y(t) = x(t) + y(t - 1)$, (p) $h(t) = [1 - u(-t)]e^{-t^2}$, (q) $y(t) = \text{sgn}[x(t)]$
4.2. A circuit is described by the linear differential equation:

$$Ry(t) + 2L\frac{dy(t)}{dt} + RLC\frac{d^2y(t)}{dt^2} + L^2C\frac{d^3y(t)}{dt^3} = R\ x(t)$$

where R, L and C are constants, $x(t)$ is the input and $y(t)$ is the output. Find, by taking the Fourier transform of the differential equation, term by term, an expression for the frequency response of the system. What is the amplitude multiplication factor, and phase shift, of a sinusoidal input at the frequency $f = 1/(2\pi\sqrt{LC})$?
4.3. A linear system has the impulse response $h(t) = u(t) - u(t - 2)$. Sketch $h(t)$ and find the system output when its input is $x(t) = \frac{1}{2}\Pi((t - 1)/2) - \frac{1}{2}\Pi((t - 3)/2)$. What is the system's step response and what is its frequency response?
4.4. How might a system with the impulse response given in Problem 4.3 be implemented using integrators, delay lines, invertors (i.e. amplifiers with voltage gain $G_V = -1.0$) and adders?
4.5. The impulse response of a system is given by $h(t) = \Pi((t - 2)/2)$ and the system's input signal is given by $x(t) = (2/3)\ t\ \Pi((t - 1.5)/3)$. Find, and sketch, the system's output.
4.6. The impulse response of a time invariant linear system is $h(t) = u(t)/(1 + t^2)$ where $u(t)$ is the

Heaviside step function. Find, and sketch, the response of this system to a rectangular pulse of unit height and width.

4.7. The amplitude response of a rectangular low pass filter is given by $|H(f)| = \Pi(f/(2f_\chi))$. Find, and sketch, the impulse response of this filter if its phase response is: (a) $\phi(f) = 0$; and (b) $\phi(f) = \text{sgn}(f)\pi/2$.

(N.B. Problems 4.8 and 4.9 presuppose some knowledge of elementary circuit theory.)

4.8. Find the frequency response and impulse response of: (a) an LR; (b) an RL; and (c) a CR filter. (The input is across both components in series and the output is across the second, earthed, component alone. L, R and C denote inductors, resistors and capacitors respectively.)

4.9. An electrical system consists of an RC potential divider (input across series combination of RC, output across C alone) followed by an ideal differentiator described by $y(t) = d/dt\,[x(t)]$. For an impulse applied at the input to the potential divider find and sketch: (i) the response at the potential divider output; and (ii) the response at the differentiator output. Find the frequency response of the entire system and use convolution to calculate the system output when the system input is $x(t) = \Pi((t - T/2)/T)$. Sketch the output if the input pulse width equals the time constant of the potential divider, i.e. $T = RC$.

(N.B. Problem 4.10 presupposes some knowledge of elementary circuit theory and electronics.)

4.10. An ideal operational amplifier is driven at its non-inverting input by a signal via an R_1C potential divider. It is driven at its inverting input by the same signal via a series resistor R_2. Negative feedback is applied using a resistor R_3 connected across the operational amplifier's output and inverting input. Find the impulse response and frequency response of this electronic circuit which is commonly used in signal processing.

4.11. A system has an impulse response $h(t) = u(t)\,e^{-t/\tau}$ and an applied input signal $x(t) = \Pi((t - \tau/2)/\tau)t$. Find the system's output signal.

4.12. A raised cosine filter has the amplitude response:

$$|H(f)| = \begin{cases} \tfrac{1}{2}[1 + \cos(\pi f/2f_\chi)], & |f| \le 2f_\chi \\ 0, & |f| > 2f_\chi \end{cases}$$

Explain (in a few words) why (strictly) this filter is not physically realisable.

4.13. An electrical system consists of 10 cascaded RC filters. (Each filter is a potential divider with input across the series combination of R and C and output across C alone.) If operational amplifier impedance buffers are inserted between all RC filters deduce (without elaborate calculations) the approximate *shape* of the system's amplitude response. (The impedance buffers merely reduce the loading effect of each RC stage on the preceding stage to a negligible level.)

4.14. What is the -3 dB bandwidth of the system with a one sided exponential impulse response $h(t) = u(t)e^{-5t}$? If white Gaussian noise with one sided NPSD of 2.0×10^{-9} V^2/Hz is applied to the input of this system what is the PSD of the noise at the system output? What is the total noise power at the system output? What is the output noise power within the system's -3 dB bandwidth? What is the pdf of the noise at the system output?

4.15. If the noise at the output of the system described in Problem 4.14 is applied (after impedance buffering to avoid loading effects) to a second, identical, system, what will be the total noise power at the (second) system output? What proportion of this total noise power resides in the frequency band below 1.0 Hz? [6.4×10^{-12} V^2, 20%]

4.16. A mobile communications system, consisting of a transmitting mobile and receiving fixed base station, experiences noise at the receiving antenna. Assuming that the noise is spectrally white, and has variance σ^2, calculate the coherence (i.e. autocorrelation) function for the output of a lowpass *RC* filter attached to the antenna output when the spectral density of the transmitted signal is given by:

$$S_{xx}(\omega) = \begin{cases} \frac{1}{2} \left[1 + \cos(\pi\omega/10)\right], & |\omega| < 10 \\ 0, & \text{otherwise} \end{cases}$$

and the square magnitude of the channel frequency response is given by:

$$|H(\omega)|^2 = \frac{A^2}{B^2 + \omega^4}$$

What information does this give you about the system? How might it be measured in practice?

4.17. Find the equivalent noise bandwidth of the finite-time integrator whose impulse response is given by:

$$h(t) = \frac{1}{T} \left[u(t) - u(t - T) \right] \qquad \text{Hint:} \quad \int_0^\infty \frac{\sin^2(ax)}{x^2} \, dx = |a| \frac{\pi}{2}$$

4.18. A linear system has the following impulse response: $h(t) = e^{-5t}$, when $t \geq 0$ and $h(t) = 0$ at other times. The input signal to the above system is a sample function from a random process which has the form:

$$X(t) = M, \qquad -\infty < t < \infty$$

in which M is a random variable that is uniformly distributed from -6 to $+18$. Find: (a) an expression for the output sample function; (b) the mean value of the output; and (c) the variance of the output. [$M/5, 0.2, 1.92$]

4.19. Find the cross-correlation function $R_{xy}(\tau)$ for a single-stage low-pass *RC* filter when the input $x(t)$ has the following autocorrelation function:

$$R_{xx}(\tau) = \frac{\beta N_0}{2} e^{-\beta|\tau|}, \qquad -\infty < \tau < \infty$$

[$0.80 \text{ Hz}, 1.0 \times 10^{-10} \text{ V}^2, 5.0 \times 10^{-11} \text{ V}^2$]

4.20. A random variable x has a pdf: $p_X(x) = u(x)5e^{-5x}$ and a statistically independent random variable y has a pdf: $p_Y(y) = 2u(y)e^{-2y}$. For the random variable $Z = X + Y$ find: (a) $p_Z(0)$; (b) the modal value of z (i.e. that for which $p_Z(z)$ is a maximum); and (c) the probability that $z > 1.0$. [$3.33 \exp(-2z) (1 - \exp(-3z))$, 0, 0.305, 0.22]

4.21. Find the pdf of noise at the output of a full wave rectifier if Gaussian noise with a variance of 1 V^2 is present at its input. (Note the output $y(t)$ of a full wave rectifier is related to its input $x(t)$ by $y(t) = |x(t)|$.)

4.22. A signal with uniform pdf: $p_X(x) = 0.5 \, \Pi((x - 1)/2)$ is processed by a, non-linear, memoryless system with input/output characteristic: $y(t) = 5x(t) + 2$. What is the pdf of the output signal?

4.23. A signal with pdf $p_X(x) = 1/(\pi(1 + x^2))$ is processed by a square law detector (characteristic $Y = X^2$). What is the pdf of processed signal?

Part Two

Digital communications principles

Part Two, by far the largest part of the book, uses the theoretical concepts of Part One to describe and analyse communications links which are robust in the presence of noise and other impairment mechanisms.

Chapter 5 starts with a discussion of sampling and aliasing and demonstrates the practical problems associated with representing an analogue signal in digital (pulse code modulated) form. This highlights the care that must be taken to achieve accurate reconstruction, without distortion, of the original analogue signal in a receiver. Chapter 5 also describes a variety of techniques by which the bandwidth of a PCM signal may be reduced in order to allow the effective use of bandlimited channels. Chapter 6 addresses the fundamentals of binary baseband transmission, covering important aspects of practical decision theory, and describes how the spectral properties of baseband digital signals can be altered by the use of different line coding schemes. Receiver equalisation, employed to overcome transmission channel distortion, is also discussed. Probability theory, is applied to receiver detection and decision processes in Chapter 7, along with a discussion of the Bayes and Neyman-Pearson decision criteria. Chapter 8 investigates optimum pulse shaping at the transmitter, and optimum filtering at the receiver, designed to minimise transmitted signal bandwidth whilst maximising the probability of correct symbol decisions.

Chapter 9 presents the fundamentals of information theory and source coding, introducing the important concepts of entropy and coding efficiency. It is shown how redundancy present in source data may be minimized, using variable length coding schemes to achieve efficient, low bit rate, digital speech, and other, signals. Chapter 10 describes the converse technique, in which transmitted data redundancy is increased to achieve error correction, or detection, in the presence of noise.

Chapter 11 analyses the bandpass binary modulation schemes which employ amplitude, frequency, or phase shift keying. Variants, and hybrid combinations, of these schemes are then examined which are especially spectrally efficient (e.g. QAM), are especially power efficient (e.g. MFSK), or have some other desirable property. Following a detailed discussion of the sources of noise in electronic circuits, Chapter 12 outlines the calculation of received signal power, noise power, and associated signal-to-noise ratio, for simple communications links. Finally Chapter 13 indicates how the performance of a complex communication system can be predicted by simulation before any hardware prototyping is attempted.

Sampling, multiplexing and PCM

5.1 Introduction

Chapter 1 provided an overview of a digital communications system and Chapters 2, 3 and 4 were reviews of some important concepts in the theories of signals, systems and noise. Chapter 5 is the starting point for a more detailed examination of the major functional blocks which make up a complete digital communications system. The discussion of these blocks will primarily be at systems level (i.e. it will concentrate on the inputs, outputs and functional relationships between them) rather than at the implementation level (the design of the electronic circuits which realise these relationships). Referring to Figure 1.3, the principal transmitter subsystems with which this chapter is concerned are the anti-aliasing filter, the sampling circuit, the quantiser, the PCM encoder and the baseband channel multiplexer. In the receiver we are concerned with the demultiplexer, PCM decoder and reconstruction filter. First, however, we give a brief review of pulse modulation techniques. This is mainly because pulse amplitude modulation (in particular) can be identified with the sampling operation preceding quantisation in Figure 1.3 and, as such, constitutes an important part of a digital communications transmitter. In addition, however, pulse modulations (generally) can be used as modulation schemes in their own right for analogue communications systems.

5.2 Pulse modulation

Pulse modulation describes the process whereby the amplitude, width or position of individual pulses in a periodic pulse train are varied (i.e. modulated) in sympathy with the amplitude of a baseband information signal, $g(t)$, Figure 5.1(a) to (d) (adapted from Stremler). Pulse modulation may be an end in itself allowing, for example, many separate information carrying signals to share a single physical channel by interleaving the individual signal pulses as illustrated in Figure 5.2 (adapted from Stremler). Such pulse interleaving is called time division multiplexing (TDM) and is discussed in detail in

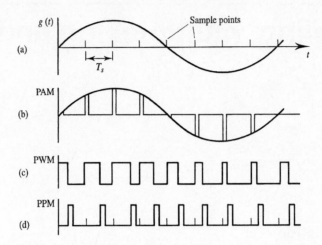

Figure 5.1 *Illustration of pulse amplitude, width and position modulation: (a) input signal.*

section 5.4. Pulse modulation also, however, represents an intermediate stage in the generation of digitally modulated signals. (It is important to realise that pulse modulation is not, in itself, a digital but an analogue technique.) The minimum pulse rate representing each information signal must be twice the highest frequency present in the signal's spectrum. This condition, called the Nyquist sampling criterion, must be satisfied if proper reconstruction of the original continuous signal from the pulses is to be possible. Sampling criteria, and the distortion (aliasing) which is introduced when they are not satisfied, are discussed further in section 5.3.

Since pulse amplitude modulation (PAM) relies on changes in pulse amplitude it requires larger signal-to-noise ratio (SNR) than pulse position modulation (PPM) or pulse width modulation (PWM) [Lathi]. This is essentially because a given amount of additive noise can change the amplitude of a pulse (with rapid rise and fall times) by a greater fraction than the position of its edges (Figure 5.3). PWM is particularly attractive in analogue remote control applications because the reconstructed control signal can easily be obtained by integrating (or averaging) the transmitted PWM signal.

All pulse modulated signals have wider bandwidth than the original information signal since their spectrum is determined solely by the pulse shape and duration. The bandwidth and filtering requirements for pulsed signals are discussed in Chapter 9. If the pulses are short compared with the reciprocal of the information signal bandwidth (or equivalently short with respect to the decorrelation time of the information signal) then the original continuous signal can be reconstructed by lowpass filtering. If the pulse duration is not sufficiently short then equalisation may be necessary after lowpass filtering. The need for, and effect of, equalisation are discussed in the context of sampling in section 5.3.1. Equalisation implementations are described in section 8.5.

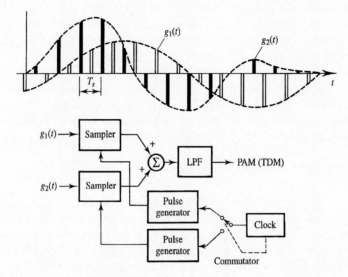

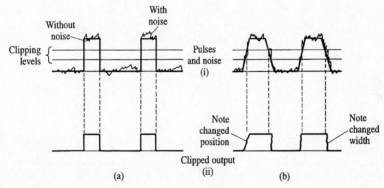

Figure 5.2 *Time division multiplexing of two pulse amplitude modulated signals.*

Figure 5.3 *Effects of noise on pulses: (a) noise induced position and width errors completely absent for ideal pulse; (b) small, noise induced, position and width errors for realistic pulse.*

5.3 Sampling

The process of selecting or recording the ordinate values of a continuous (usually analogue) function at specific (usually equally spaced) values of its abscissa is called sampling. If the function is a signal which varies with time then the samples are sometimes called a time series. This is the most common type of sampling process encountered in electronic communications although spatial sampling of images is also important.

There are obvious similarities between sampling and pulse amplitude modulation. In fact, in many cases, the two processes are indistinguishable, for instance if the pulse duration of the PAM signal is very short. There are, however, two processes both commonly referred to as sampling which should be distinguished. These are flat topped sampling (which is identical to PAM) and natural sampling.

The fundamental property of sampling is that, for a sampling frequency f_s a constant voltage DC input signal and a periodic input signal with a fundamental frequency at integer multiples of f_s both give (to within a multiplicative constant) the same sampled output values. A consequence of this, in the frequency domain, is that the sampled baseband spectrum repeats at f_s and multiples of this sampling frequency, Figure 5.4.

5.3.1 Natural and flat topped sampling

A naturally sampled signal is produced by multiplying the baseband information signal, $g(t)$, by the periodic pulse train, shown previously in Figure 2.36. This is illustrated in Figure 5.4(a), (c) and (e). The important point to note in this case is that the pulse tops follow the variations of the signal being sampled. The spectrum of the information signal is shown in Figure 5.4(b). (The spectrum is, of course, the Fourier transform of $g(t)$ and would normally be complex but is represented here only by its amplitude.) The spectrum

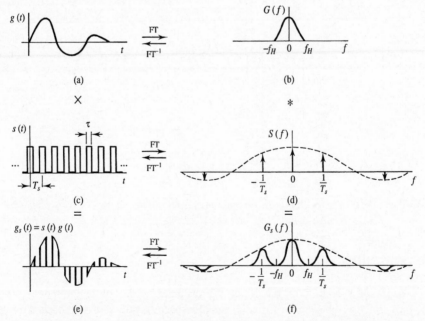

Figure 5.4 *Time and frequency domain illustrations of natural sampling: (a) signal $g(t)$; (b) signal spectrum; (c) sampling function; (d) spectrum of sampling function; (e) sampled signal; (f) spectrum of sampled signal.*

of the periodic pulse train (Figure 5.4(d)) is discrete and consists of a series of equally spaced, weighted, Dirac delta or impulse functions. The spacing between impulses is $1/T_s$ where T_s is the pulse train period and the envelope of weighted impulses is the Fourier transform of a single time domain pulse. (In this case the spectrum has a $\mathrm{sinc}(\tau f)$ shape since the time domain pulses are rectangles of width τ seconds.) Since multiplication in the time domain corresponds to convolution in the frequency domain the spectrum of the sampled signal is found by convolving Figure 5.4(b) with Figure 5.4(d). It is a property of the impulse that, under convolution (section 2.3.4) with a second function, it replicates the second function about the position of the impulse. Each impulse in Figure 5.4 therefore replicates the spectrum of $g(t)$ at a frequency corresponding to its own position. The replicas have the same amplitude weightings as the impulses producing them. The sampled signal spectrum is shown in Figure 5.4(f). It is clear that appropriate lowpass filtering will pass only the baseband spectral version of $g(t)$. It follows that $g(t)$ will appear undistorted (and, in the absence of noise, without error) at the output of the lowpass filter. For obvious reasons such a filter is sometimes called a *reconstruction* filter.

If the pulses produced by the process described above are artificially flattened we have a true PAM signal or flat topped sampling. This can be modelled by assuming that natural sampling proceeds using an impulse train, Figure 5.5(a), (c), (e) (this is sometimes called impulse, or ideal, sampling) and the resulting time series of weighted impulses is convolved with a rectangular pulse, Figure 5.5(e), (g), (i). The resulting spectrum is that of the ideally sampled information signal (Figure 5.5(f)) multiplied with the $\mathrm{sinc}(\tau f)$ spectrum of a single rectangular pulse, Figure 5.5(h), (j). A baseband spectral version of $g(t)$ can be recovered by lowpass filtering but this must then be multiplied by a function which is the inverse of the pulse spectrum ($1/\mathrm{sinc}(\tau f)$, Figure 5.5(k)) if $g(t)$ is to be restored exactly. This process, which is not required for reconstruction of naturally sampled signals, is called equalisation. For proper equalisation the inverse frequency response in Figure 5.5(k) need only exist over the signal bandwidth, $0 \rightarrow f_H$.

5.3.2 Baseband sampling and Nyquist's criterion

The spectral replicas of the information signal in Figures 5.4(f) and 5.5(j) are spaced by $f_s = 1/T_s$ Hz. The baseband spectrum can therefore be recovered by simple lowpass filtering provided that the width of the spectral replicas is less than the spacing, as defined in equation (2.31), i.e.:

$$f_s \geq 2f_H \quad \text{(Hz)} \tag{5.1}$$

where f_H is the highest frequency component in the information signal. If, however, the spacing between spectral replicas is less than their width then they will overlap and reconstruction of $g(t)$ using lowpass filtering will no longer be possible. Equation (5.1) is a succinct statement of Nyquist's sampling criterion. In words this criterion could be stated as follows:

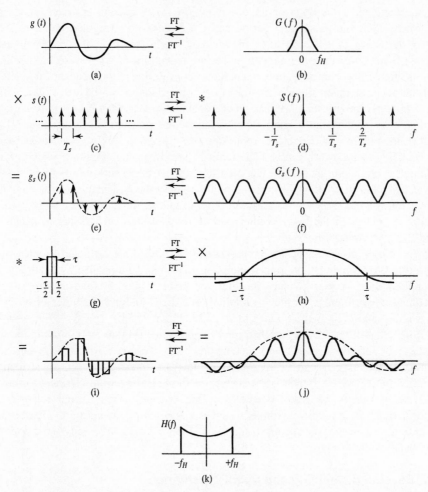

Figure 5.5 *Time and frequency domain illustrations of PAM or flat topped sampling: (a) signal; (b) signal spectrum; (c) sampling function; (d) spectrum of (c); (e) sampled signal; (f) spectrum of (e); (g) finite width sample; (h) spectrum of (g); (i) sampled signal; (j) spectrum of (i); (k) receiver equalising filter to recover g(t).*

A signal having no significant spectral components above a frequency f_H Hz is specified completely by its values at uniform spacings, no more than $1/(2f_H)$ s apart.

Whilst this sampling criterion is valid for any signal it is usually only used in the context of baseband signals. Another less stringent (but more complicated) criterion which can be used for bandpass signals is discussed in section 5.3.5. In the context of sampling a strict distinction between baseband and bandpass signals can be made as follows:

For baseband signals, $B \geq f_L$ (Hz)   (5.2(a))

For bandpass signals, $B < f_L$ (Hz)   (5.2(b))

B in equations (5.2) is the signal's (absolute) bandwidth and f_L is the signal's lowest frequency component. These definitions are illustrated in Figure 5.6.

5.3.3 Aliasing

Figure 5.7 shows the spectrum of an undersampled baseband signal ($f_s < 2f_H$). The baseband spectrum of $g(t)$ clearly cannot be recovered exactly, even with an ideal rectangular lowpass filter. The best achievable, in terms of separating the baseband spectrum from the adjacent replicas, would be to use a rectangular lowpass filter with a cut-off frequency of $f_s/2$. The filtered signal will then be, approximately, that of the original signal $g(t)$ but with the frequencies above $f_s/2$ folded back so that they actually appear below $f_s/2$. (The approximation becomes better as the width of the sampling pulses gets smaller. In the limit of ideal (impulse) sampling the approximation becomes exact.) The spectral components originally representing high frequencies now appear under the alias of lower frequencies. Thus, as in Figure 2.23(c), the sampling represents a high frequency component by a lower frequency sinusoid.

To avoid aliasing a lowpass anti-aliasing filter with a cut-off frequency of $f_s/2$ is often placed immediately before the sampling circuit. Whilst this filter may remove high frequency energy from the information signal the resulting distortion is generally less

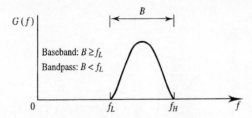

Figure 5.6 *Definitions of baseband and bandpass signals.*

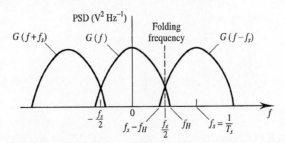

Figure 5.7 *Spectrum of undersampled information signal showing interference between spectral replicas and the folding frequency.*

than that introduced if the same energy is aliased to incorrect frequencies by the sampling process.

5.3.4 Practical sampling, reconstruction and signal to distortion ratio

Two practical points need to be appreciated when choosing the sampling rate in real systems. The first is that f_H must usually be interpreted as the highest frequency component *with significant spectral amplitude*. This is because practical signals start and stop in time and therefore, in principle, have spectra which are not bandlimited in an absolute sense. (For the case of voice signals, Figure 5.8, f_H, in Europe, is usually assumed to be 3.4 kHz.) The second is that whilst it is theoretically possible to reconstruct the original continuous signal from its samples if the sampling rate is exactly twice the highest frequency in its spectrum, in practice it is necessary to sample at a slightly faster rate. This is because ideal, rectangular, anti-aliasing and reconstruction filters (with infinitely steep skirts) are not physically realisable. A practical version of the baseband sampling criterion might therefore be expressed as:

$$f_s \geq 2.2 f_H \quad \text{(Hz)} \tag{5.3}$$

to allow for the transition, or roll-off, into the filter stopband.

A quantitative measure of the distortion introduced by aliasing can be defined as the ratio of unaliased to aliased power in the reconstructed signal. If the reconstruction filter is ideal with rectangular amplitude response then, in the absence of an anti-aliasing filter, the signal to distortion ratio (SDR) is given by:

$$\text{SDR} = \frac{\displaystyle\int_0^{f_s/2} G(f)\, df}{\displaystyle\int_{f_s/2}^{\infty} G(f)\, df} \tag{5.4}$$

where $G(f)$ is the (two sided) power spectral density of the (real) baseband information signal $g(t)$. More generally, if a filter with a frequency response $H(f)$ is used for reconstruction then the integral limits are extended (see Figure 5.7) to give:

$$\text{SDR} \simeq \frac{\displaystyle\int_0^{\infty} G(f)\,|H(f)|^2\, df}{\displaystyle\int_0^{\infty} G(f - f_s)\,|H(f)|^2\, df} \tag{5.5}$$

An approximation sign is used in equation (5.5) for two reasons. Firstly, and most importantly, spectral replicas centred on $2f_s$ Hz and above are assumed to be totally suppressed by $|H(f)|$ in equation (5.5). Secondly, ideal (impulsive) sampling is assumed such that there is no $\text{sinc}(\tau f)$ roll-off in the spectrum of the sampled signal as discussed in section 5.3.1. (In the absence of $1/\text{sinc}(\tau f)$ equalisation then this additional spectral

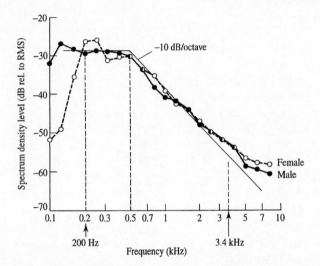

Figure 5.8 *Long-term averaged speech spectra for male and female speakers (source: Furui, 1989, reproduced with permission of Marcel Dekker).*

roll-off should be cascaded with $H(f)$ in equation (5.5) if flat topped sampling is employed.) It should also be appreciated that SDR in equations (5.4) and (5.5) is specifically defined to represent distortion due to aliasing of frequency components. Thus, for example, phase distortion due to the action of a reconstruction filter (with non-linear phase) on the baseband signal's voltage spectrum is not included.

EXAMPLE 5.1
Consider a signal with the power spectral density shown in Figure 5.9(a). Find the alias induced signal to distortion ratio if the signal is sampled at 90% of its Nyquist rate and the reconstruction filter has: (i) an ideal rectangular amplitude response; and (ii) an *RC* lowpass filter response with a 3 dB bandwidth of $f_s/2$.

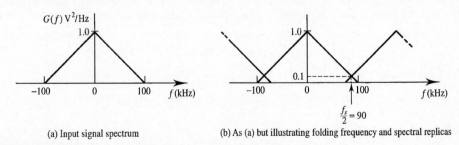

(a) Input signal spectrum   (b) As (a) but illustrating folding frequency and spectral replicas

Figure 5.9 *Signal power spectral density for Example 5.1.*

The sampling rate is given by:

$$f_s = 0.9 \times 2 \, f_H$$

$$= 0.9 \times 2 \times 100 = 180 \; \text{kHz}$$

The folding frequency (Figure 5.9(b)) is thus:

$$\frac{f_s}{2} = 90 \; \text{kHz}$$

(i) For an ideal rectangular reconstruction filter, using equation (5.4):

$$\text{SDR} = \frac{\displaystyle\int_0^{f_s/2} G(f) \, df}{\displaystyle\int_{f_s/2}^{\infty} G(f) \, df} = \frac{\displaystyle\int_0^{10^5} (1 - 10^{-5} f) \, df}{\displaystyle\int_{f_s/2}^{10^5} (1 - 10^{-5} f) \, df}$$

For such a simple $G(f)$, however, we can evaluate the integrals from the area under the triangles in Figure (5.9(b)), i.e.:

$$\text{SDR} = \frac{\frac{1}{2} (100 \times 1) - \frac{1}{2} (10 \times 0.1)}{\frac{1}{2} (10 \times 0.1)}$$

$$= 99 = 20.0 \; \text{dB}$$

(ii) For the *RC* filter:

$$H(f) = \frac{1}{1 + j \, 2\pi \, RC \, f}$$

$$|H(f)|^2 = \frac{1}{1 + (2\pi \, RC)^2 \, f^2}$$

$$f_{3dB} = \frac{f_s}{2} = 90 \; \text{kHz}$$

$$RC = \frac{1}{2\pi \, f_{3dB}} = \frac{1}{2\pi \, 9 \times 10^4} = 1.768 \times 10^{-6} \; \text{s}$$

Reducing the upper limit in the numerator of equation (5.5) to f_H Hz, as this is the maximum signal frequency in the baseband spectrum, and replacing the limits in the denominator with $f_s - f_H$ and $f_s + f_H$ since these are lower and upper frequency limits in the first spectral replica:

$$\text{SDR} = \frac{\displaystyle\int_0^{f_H} (1 - 10^{-5} f) \, [1 + (2\pi \, 1.768 \times 10^{-6})^2 f^2]^{-1} \, df}{\displaystyle\int_{f_s - f_H}^{f_s + f_H} [1 - 10^{-5} |f - f_s|] \, [1 + (2\pi \, 1.768 \times 10^{-6})^2 f^2]^{-1} \, df}$$

Evaluating the numerator:

$$\text{Num} = \int_0^{10^5} \frac{1 - 10^{-5} f}{1 + a^2 f^2} \, df$$

(where $a = 2\pi \, 1.768 \times 10^{-6} = 1.111 \times 10^{-5}$)

$$= \int_0^{10^5} \frac{1}{1 + a^2 f^2} \, df - 10^{-5} \int_0^{10^5} \frac{f}{1 + a^2 f^2} \, df$$

Put $x = af$, $\dfrac{dx}{df} = a$; when $f = 0$, $x = 0$ and when $f = 10^5$, $x = 1.111$

$$\text{Num} = \int_0^{1.111} \frac{1}{1 + x^2} \frac{dx}{a} - 10^{-5} \int_0^{1.111} \frac{1}{a} \frac{x}{1 + x^2} \frac{dx}{a}$$

$$= \frac{1}{a} \left[\tan^{-1} x \right]_0^{1.111} - \frac{10^{-5}}{a^2} \left[\frac{1}{2} \ln(1 + x^2) \right]_0^{1.111}$$

$$= 7.542 \times 10^4 - 3.256 \times 10^4 = 4.286 \times 10^4$$

Evaluating the denominator:

$$\text{Denom} = \int_{f_s - f_H}^{f_s} \frac{1 - 10^{-5} f_s + 10^{-5} f}{1 + a^2 f^2} \, df + \int_{f_s}^{f_s + f_H} \frac{1 + 10^{-5} f_s - 10^{-5} f}{1 + a^2 f^2} \, df$$

$$= (1 - 10^{-5} \times 180 \times 10^3) \int_{80 \times 10^3}^{180 \times 10^3} \frac{1}{1 + a^2 f^2} \, df + 10^{-5} \int_{80 \times 10^3}^{180 \times 10^3} \frac{f}{1 + a^2 f^2} \, df$$

$$+ (1 + 10^{-5} \times 180 \times 10^3) \int_{180 \times 10^3}^{280 \times 10^3} \frac{1}{1 + a^2 f^2} \, df - 10^{-5} \int_{180 \times 10^3}^{280 \times 10^3} \frac{f}{1 + a^2 f^2} \, df$$

Using $x = af$, when $f = 80 \times 10^3$, $x = 0.8888$, when $f = 180 \times 10^3$, $x = 2.000$, and when $f = 280 \times 10^3$, $x = 3.111$

$$\text{Denom} = -0.8 \, \frac{1}{a} \left[\tan^{-1}(x) \right]_{0.8888}^{2.0} + 10^{-5} \frac{1}{a^2} \left[\frac{1}{2} \ln(1 + x^2) \right]_{0.8888}^{2.0}$$

$$+ 2.8 \frac{1}{a} \left[\tan^{-1}(x) \right]_{2.0}^{3.111} - 10^{-5} \frac{1}{a^2} \left[\frac{1}{2} \ln(1 + x^2) \right]_{2.0}^{3.111}$$

$$= -7.201 \times 10^4 \, [1.1071 - 0.7266] + \frac{8.102 \times 10^4}{2} \, [1.6094 - 0.5822]$$

$$+ 2.520 \times 10^5 [1.2598 - 1.1071] - \frac{8.102 \times 10^4}{2} \, [2.3682 - 1.6094]$$

$$= 2.195 \times 10^4$$

$$\text{SDR} = \frac{\text{Num}}{\text{Denom}} = \frac{4.286 \times 10^4}{2.195 \times 10^4} = 1.95 = 2.9 \text{ dB}$$

Comparing the results for (i) and (ii) shows that if significant aliased energy is present a good multipole filter with a steep skirt is essential for signal reconstruction if the SDR is to be kept to tolerable levels. Furthermore the SDR calculated in part (ii) is extremely optimistic since spectral replicas centered on $2f_s$ Hz and above have been ignored. For reconstruction filters with only gentle roll-off equation (5.5) should really be used only as an upper bound on SDR, see Problem 5.3.

5.3.5 Bandpass sampling

In some applications it is desirable to sample a bandpass signal in which the centre frequency is many times the signal bandwidth. Whilst, in principle, it would be possible to sample this signal at twice the highest frequency component in its spectrum and reconstruct the signal from its samples by lowpass filtering it is usually possible to retain all the information needed to reconstruct the original signal whilst sampling at a much lower rate. If advantage is taken of this then there exists one or more frequency *bands* in which the sampling frequency should lie. Thus when sampling a bandpass signal there is generally an upper limit for proper sampling as well as a lower limit. The bandpass sampling criterion can be expressed as follows:

A bandpass signal having no spectral components below f_L Hz or above f_H Hz is specified uniquely by its values at uniform intervals spaced $T_s = 1/f_s$ s apart provided that:

$$2B \left\{ \frac{Q}{n} \right\} \le f_s \le 2B \left\{ \frac{Q-1}{n-1} \right\} \tag{5.6}$$

where $B = f_H - f_L$, $Q = f_H/B$, n is a positive integer and $n \le Q$.

The following comments are made to clarify the use of equation (5.6) and its relationship to the Nyquist baseband sampling criterion.

1. If $Q = f_H/B$ is an integer then $n \le Q$ allows us to choose $n = Q$. In this case $f_s = 2B$ and the correct sampling frequency is exactly twice the signal bandwidth.
2. If $Q = f_H/B$ is not an integer then the lowest allowed sampling rate is given by choosing $n = int(Q)$ (i.e. the next lowest integer from Q). Lower values of n will still allow reconstruction of the original signal but the sampling rate will be unnecessarily high. (Lower values of n may, however, give a wider band of allowed f_s.)
3. If $Q < 2$ (i.e. $f_H < 2B$ or, equivalently, $f_L < B$) then $n \le Q$ means that $n = 1$.
    In this case:

    $$2BQ \le f_s \le \infty \quad \text{(Hz)}$$

    and since $BQ = f_H$ we have:

    $$2f_H \le f_s \le \infty \quad \text{(Hz)}$$

This is a statement of the Nyquist (baseband) sampling criterion.

The validity of the bandpass sampling criterion is most easily demonstrated using convolution (section 2.3.4) for the following special cases. (Convolution results in the sampled signal being replicated at DC and multiples of the sample frequency.)

1. When the spectrum of the bandpass signal $g(t)$ straddles nf_s, i.e. $G(f)$ straddles any of the lines in the spectrum of the sampling signal (Figure 5.10(a)), then convolution results in interference between the positive and negative frequency spectral replicas (Figure 5.10(b)).

2. When the spectrum of $g(t)$ straddles $(n + \frac{1}{2})f_s$, i.e. $G(f)$ straddles any odd integer multiple of $f_s/2$ (Figure 5.10(c)), then similar interference occurs (Figure 10(d)).

3. When the spectrum of $g(t)$ straddles neither nf_s nor $(n + \frac{1}{2})f_s$ (Figure 5.10(e)), then no interference between positive and negative frequency spectral replicas occurs (Figure 5.10(f)) and the baseband (or bandpass) spectrum can be obtained by filtering.

Summarising we have the following conditions for proper sampling:

$$f_H \le n\frac{f_s}{2} \quad \text{(Hz)} \tag{5.7}$$

$$f_L \ge (n-1)\frac{f_s}{2} \quad \text{(Hz)} \tag{5.8}$$

Using $f_L = f_H - B$ we have:

$$\frac{2}{n} f_H \le f_s \le \frac{2}{n-1} (f_H - B) \quad \text{(Hz)} \tag{5.9}$$

Defining $Q = f_H/B$ gives the bandpass sampling criterion of equation (5.6).

EXAMPLE 5.2
The following examples illustrate the use and significance of equation (5.6). First consider a signal with centre frequency 9.5 kHz and bandwidth 1.0 kHz.

The highest and lowest frequency components in this signal are:

$$f_L = 9.0 \text{ kHz} \qquad f_H = 10.0 \text{ kHz}$$

Quotient Q is thus:

$$Q = f_H/B = 10.0/1.0 = 10.0$$

Applying the bandpass sampling criterion of equation (5.6):

$$2 \times 10^3 \left\{ \frac{10}{n} \right\} \le f_s \le 2 \times 10^3 \left\{ \frac{10-1}{n-1} \right\} \quad \text{(Hz)}$$

Since Q is an integer the lowest allowed sampling rate is given by choosing $n = Q = 10$, i.e.:

$$2.0 \le f_s \le 2.0 \text{ (kHz)}$$

The significant point here is that there is zero tolerance in the sampling rate if distortion is to be completely avoided. If n is chosen to be less than its maximum value, e.g. $n = 9$, then:

$$2.222 \le f_s \le 2.250 \text{ (kHz)}$$

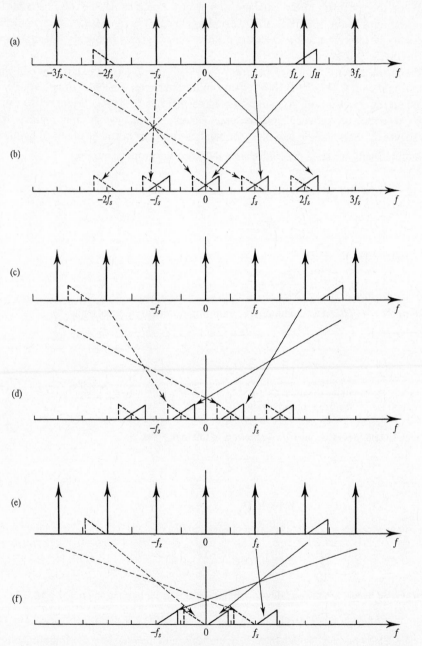

Figure 5.10 *Criteria for correct sampling of bandpass signals: (a) spectrum of signal where $G(f)$*
straddles $2 \times f_s$; (c) spectrum of signal where $G(f)$ straddles $2.5 \times f_s$; (b) & (d)
overlapped spectra after bandpass sampling; (e) bandpass spectrum avoiding the
straddling in (a) & (c); (f) baseband or bandpass spectra, recoverable by filtering.

The sampling rate in this case would be chosen to be 2.236 ± 0.014 kHz. The accuracy required of the sampling clock is therefore ± 0.63%. Now consider a signal with centre frequency 10.0 kHz and bandwidth 1.0 kHz. The quotient $Q = 10.5$ is now no longer an integer. The lowest allowed sampling rate is therefore given by $n = int(Q) = 10.0$. The sampling rate is now bound by:

$$2.100 \leq f_s \leq 2.111 \text{ (kHz)}$$

This gives a required sampling rate of 2.106 ± 0.006 kHz or 2.106 kHz ± 0.26%.

5.4 Analogue pulse multiplexing

In many communications applications different information signals must be transmitted over the same physical channel. The channel might, for example, be a single coaxial cable, an optical fibre, or, in the case of radio, the free space between two antennas. In order for the signals to be received independently (i.e. without cross-talk) they must be sufficiently separated in some sense.

This quality of separateness is usually called orthogonality, section 2.5.3. Orthogonal signals can be received independently of each other whilst non-orthogonal signals cannot. There are many ways in which orthogonality between signals can be provided. An intuitively obvious way, sometimes used in microwave radio communications, is to use two perpendicular polarisations for two independent information channels. Vertical and horizontal, linear, antenna polarisations are usually used but right and left hand circular polarisations may also be used. Such signals are orthogonal in polarisation.

The traditional way of providing orthogonality in analogue telephony and broadcast applications is to transmit different information signals using different carrier frequencies. Such signals (provided their spectra do not overlap) are disjoint in frequency and can be received separately using filters. Tuning the local oscillator in a superheterodyne receiver, Figure 1.4, also allows one signal to be separated from others with disjoint (and therefore orthogonal) frequency spectra. Using different carriers, or frequency bands, to isolate signals from each other in this way is called frequency division multiplexing (FDM).

In FDM telephony, 3.4 kHz bandwidth telephone signals, Figure 5.8, are stacked in frequency at 4 kHz spacings with small frequency guard bands between them to allow separation using practical filters. Figure 5.11 shows how an FDM signal can be generated and Figure 5.12 shows a schematic representation of an FDM signal spectrum generated using single (lower) sideband filters. FDM was the original multiplexing technique for analogue communications and is now experiencing a resurgence in fibre optic systems in which different wavelengths are used for simultaneous transmission of many information signals. In this particular context the term wavelength division multiplexing (WDM) is usually used in preference to FDM.

Orthogonality can be provided in a quite different way for pulse modulated signals. Instead of occupying separate frequency bands (as in FDM) the signals occupy separate time slots. This technique, illustrated in Figure 5.2, is called (analogue) time division

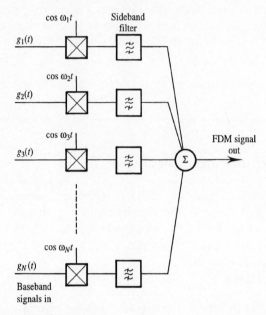

Figure 5.11 *Generation of an FDM signal.*

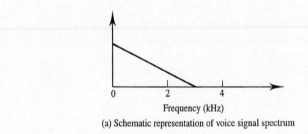

(a) Schematic representation of voice signal spectrum

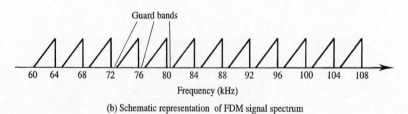

(b) Schematic representation of FDM signal spectrum

Figure 5.12 *FDM example for multiplexed speech channels.*

multiplexing (TDM). In telephony parlance the separate inputs in Figure 5.2 are called tributaries (see Chapter 19).

TDM obviously increases the overall sample rate and therefore the required bandwidth for transmission. It is, however, possible to reduce the bandwidth of the TDM

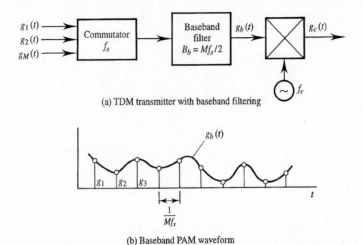

(a) TDM transmitter with baseband filtering

(b) Baseband PAM waveform

Figure 5.13 *Filtered TDM waveform.*

signal dramatically by appropriate filtering since, strictly, it is only necessary that the TDM signal provides the correct amplitude information at the sampling instants. Nyquist's sampling theorem says that a waveform with a bandwidth of $2f_H$ Hz exists which passes through these required points. This is illustrated in Figure 5.13. (At points between the sampling instants the filtered TDM signal is made up of a complicated sum of contributions arising from many pulses. At the sampling points themselves, however, the waveform amplitude is due to a single TDM pulse only.) If a minimum bandwidth TDM signal is formed by filtering then sampling accuracy becomes critical in that samples taken at times other than the correct instant will result in cross-talk between channels.

Cross-talk can also occur between the channels of a TDM signal even if filtering is not explicitly applied. This is because the transmission medium itself may bandlimit the signal. Such bandlimiting effects may often be at least approximated by RC low-pass filtering. In this case the response of the medium results in pulses with exponential rising and falling edges as shown in Figure 5.14. If the guard time between rectangular pulses (i.e. the time between the trailing edge of one pulse and the rising edge of the next) is t_g, and the time constant of the transmission channel is RC, then the amplitude of each pulse will decay to a fraction $e^{-t_g/RC}$ of its peak value by the time the next pulse starts. For the RC characteristic the channel bandwidth is $f_{3dB} = 1/(2\pi RC)$, therefore the cross-talk ratio (XTR) in dB at the pulse trailing edge (i.e. the optimum XTR sampling instant) is:

$$XTR = 20 \log_{10} e^{2\pi f_{3dB}(t_g + \tau)}$$

$$= 54.6 \, f_{3dB} \, (t_g + \tau) \quad (dB) \tag{5.10}$$

Equation (5.10) therefore gives a first order estimate of the required guard time to maintain a desired cross-talk ratio in a bandlimited channel, for a given rectangular pulse width, τ, at the channel input. (If sampling occurs at the centre of the τ s nominal pulse

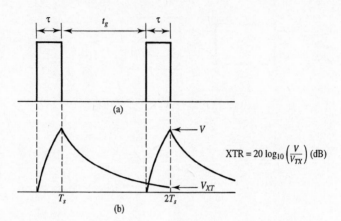

Figure 5.14 *Cross-talk between tributary channels of a TDM signal: (a) signal at channel input;*
(b) signal at channel output.

slot, rather than at its end, then XTR is reduced by approximately 6 dB when $\tau \ll RC$.)

The discussion of TDM in this chapter has been mainly in the context of analogue pulse multiplexing. In Chapter 19 it is shown how (digital) TDM forms the basis of the telephone hierarchy for transmitting multiple simultaneous telephone calls over high speed 2, or 140, Mbit/s data links.

5.5 Quantised pulse amplitude modulation

An information signal which is pulse amplitude modulated becomes discrete (in time) rather than continuous but nevertheless remains analogue in nature since all pulse amplitudes within a specified range are allowed. An alternative way of expressing the analogue property of a PAM signal is to say that the probability density function (pdf) of pulse amplitudes is continuous (Figure 5.15). If a PAM signal is quantised, i.e. each pulse is adjusted in amplitude to coincide with the nearest of a finite set of allowed amplitudes (Figure 5.16) then the resulting signal is no longer analogue but digital and as a consequence has a discrete pdf as illustrated in Figure 5.17 (and previously in Figure 3.4).

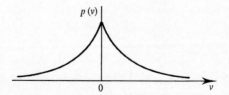

Figure 5.15 *Continuous pdf of typical analogue PAM signal.*

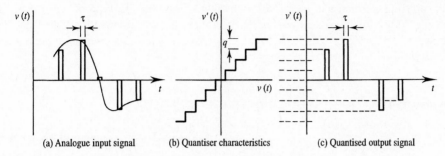

(a) Analogue input signal   (b) Quantiser characteristics   (c) Quantised output signal

Figure 5.16 *Quantisation of a PAM signal.*

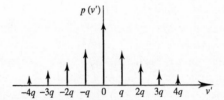

Figure 5.17 *Discrete pdf of quantised PAM signal.*

This digital signal can be represented by a finite set of symbols – the obvious set consisting of one symbol for each quantisation level. Other symbol sets (or alphabets) can be conceived, however, and unique mappings from one set to another established. Probably the simplest and most important alphabet is the binary set consisting of two symbols only, usually denoted by 0 and 1. The rest of this chapter is primarily concerned with the quantisation process and the subsequent efficient coding of quantised levels into binary symbols. In terms of Figure 1.3 the subsystems of principal importance here are the quantiser, pulse code modulation (PCM) encoder and source coder, although digital pulse multiplexing will also be discussed briefly. To some extent the separation of quantiser, PCM encoder and source coder might be misleading since they do not necessarily exist as identifiably separate pieces of hardware in all digital communications systems. For example, quantisation and PCM encoding are implemented together as a binary A/D converter (which may then be followed by a parallel to series converter) in many systems. Similarly some source coders (e.g. delta modulators) effectively *replace* the PCM encoder whilst others take a PCM signal and recode the binary symbols. In some systems (e.g. delta PCM) the source coder precedes the PCM encoder.

Quantising PAM signals is usually a precursor to generating pulse code modulation (PCM) which has some significant advantages over other baseband modulation types. The quantisation process in itself, however, actually degrades the quality of the information signal. This is easy to see since the quantised PAM signal no longer exactly represents the original, continuous analogue, signal but a distorted version of it. Figure 5.18 (which is drawn with PAM pulse width, τ, equal to the sampling period, T_s) shows that the quantised signal can be decomposed into the sum of the analogue signal and the difference between the quantised and the analogue signals. The difference signal is essentially random and can therefore be thought of as a special type of noise process. Like any other signal the power or RMS value of this *quantisation noise* can be calculated or measured. This leads to the concept of a *signal to quantisation noise ratio* (SN_qR).

5.6 Signal to quantisation noise ratio (SN_qR)

To calculate the signal to quantisation noise ratio (SN_qR) of a quantised signal it is convenient to make the following assumptions:

1. Linear quantisation (i.e. equal increments between quantisation levels).
2. Zero mean signal (i.e. symmetrical pdf about 0 V).
3. Uniform signal pdf (i.e. all signal levels equally likely).

The probability density function $p(v)$ of allowed levels is illustrated in the left hand side of Figure 5.18(a). (Narrow rectangles are used to represent the delta functions in the pdf to make interpretation easy.) Each rectangle has an area of $1/M$ where M is the (even) number of quantisation levels. (This is a result of assumption 3.) If the distance between adjacent quantisation levels is q V then the pdf of allowed levels is given by:

$$p(v) = \sum_{k=-M}^{M} (1/M)\, \delta(v - qk/2) \tag{5.11}$$

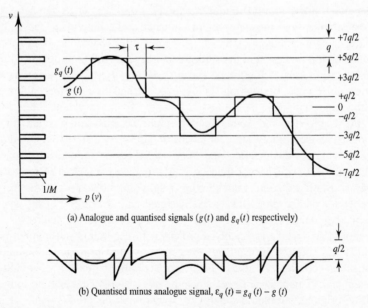

(a) Analogue and quantised signals ($g(t)$ and $g_q(t)$ respectively)

(b) Quantised minus analogue signal, $\varepsilon_q(t) = g_q(t) - g(t)$

Figure 5.18 *Quantisation error interpreted as noise, i.e.* $g_q(t) = g(t) + \varepsilon_q(t)$.

where k takes on odd values only. The mean square signal after quantisation is:

$$\overline{v^2} = \int_{-\infty}^{+\infty} v^2 p(v) \, dv$$

$$= \frac{2}{M} \left[\int_0^{\infty} v^2 \, \delta(v - q/2) \, dv + \int_0^{\infty} v^2 \, \delta(v - 3q/2) \, dv + \cdots \right]$$

$$= \frac{2}{M} \left(\frac{q}{2} \right)^2 [1^2 + 3^2 + 5^2 + \cdots + (M-1)^2]$$

$$= \frac{2}{M} \left(\frac{q}{2} \right)^2 \left[\frac{M(M-1)(M+1)}{6} \right] \tag{5.12}$$

i.e.:

$$\overline{v^2} = \frac{M^2 - 1}{12} q^2 \quad (\text{V}^2) \tag{5.13}$$

Denoting the quantisation error (i.e. the difference between the unquantised and quantised signals) as ε_q, Figure 5.18(b), then it follows from assumption 3 that the pdf of ε_q is uniform:

$$p(\varepsilon_q) = \begin{cases} 1/q, & -q/2 \leq \varepsilon_q < q/2 \\ 0, & \text{elsewhere} \end{cases} \tag{5.14}$$

The mean square quantisation error (or noise) is:

$$\overline{\varepsilon_q^2} = \int_{-q/2}^{+q/2} \varepsilon_q^2 p(\varepsilon_q) \, d\varepsilon_q \tag{5.15}$$

i.e.:

$$\overline{\varepsilon_q^2} = \frac{q^2}{12} \quad (\text{V}^2) \tag{5.16}$$

The signal to quantisation noise ratio is therefore given by:

$$\text{SN}_q\text{R} = \overline{v^2}/\overline{\varepsilon_q^2} = M^2 - 1 \tag{5.17}$$

For large SN$_q$R the approximation SN$_q$R $= M^2$ is often used.

Equation (5.17) represents the average signal to quantisation noise (power) ratio. Since the *peak* signal level is $Mq/2$ V then the peak signal to quantisation noise (power) ratio is:

$$(\text{SN}_q\text{R})_{peak} = \frac{(Mq/2)^2}{\varepsilon_q^2} = 3M^2 \tag{5.18}$$

Expressed in dB the signal to quantisation noise ratios are:

$$SN_qR = 20 \log_{10} M \quad (dB) \tag{5.19}$$

$$(SN_qR)_{peak} = 4.8 + SN_qR \quad (dB) \tag{5.20}$$

5.7 Pulse code modulation

After a PAM signal has been quantised the possibility exits of transmitting not the pulse itself but a number indicating the height of the pulse. Usually (but not necessarily always) the pulse height is transmitted as a binary number. As an example, if the number of allowed quantisation levels were eight then the pulse amplitudes could be represented by the binary numbers from zero (000) to seven (111). The binary digits are normally represented by two voltage levels (e.g. 0 V and 5 V). Each binary number is called a code word and, since each quantised pulse is represented by a code word, the resulting modulation is called pulse code modulation (PCM). Figure 5.19(a) to (e) shows the relationship between an information, PAM, quantised PAM and PCM signal. Figure 5.19 (b) and (c) also illustrates the difference between a pulsed signal with duty cycle (τ/T_s) less than 1.0 and a pulsed signal with duty cycle equal to 1.0. The former are often referred to as return to zero (RZ) signals and the latter as non-return to zero (NRZ) signals.

There is clearly a bandwidth penalty to pay for PCM, if information is to be transmitted in real time, since, in the example given above, three binary pulses are transmitted instead of one quantised PAM pulse. (The penalty here is a factor of three since, for the same pulse duty cycle, each PCM pulse must be one third the duration of the PAM pulse.) The advantage of PCM is that for a given transmitted power the difference between adjacent voltage levels is much greater than for quantised PAM. This means that for a given RMS noise voltage the total voltage (signal plus noise) at the receiver is less likely to be interpreted as representing a level other than that which was transmitted. PCM signals are therefore said to have greater noise immunity than PAM signals.

5.7.1 SN$_q$R for linear PCM

Whilst it is true that PCM signals are more tolerant of noise than the equivalent quantised PAM signals it is also true that both suffer the same degradation due to quantisation noise. For a given number of quantisation levels, M, the number of binary digits required for each PCM code word is $n = \log_2 M$. The PCM peak signal to quantisation noise ratio, $(SN_qR)_{peak}$, is therefore:

$$(SN_qR)_{peak} = 3M^2 = 3(2^n)^2 \tag{5.21}$$

If the ratio of peak to mean signal power, $v_{peak}^2/\overline{v^2}$, is denoted by α then the average signal to quantisation noise ratio is:

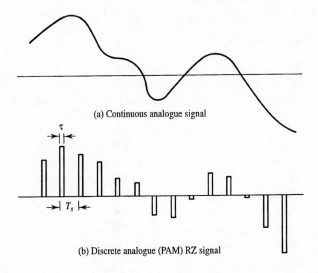

(a) Continuous analogue signal

(b) Discrete analogue (PAM) RZ signal

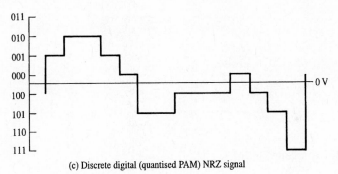

(c) Discrete digital (quantised PAM) NRZ signal

· 001 · 010 · 010 · 001 · 000 · 101 · 101 · 100 · 100 · 100 · 000 · 100 · 101 · 111
(d) Binary coded (quantised) PAM

(e) PCM NRZ signal

Figure 5.19 *Relationship between PAM, quantised PAM and PCM signal.*

$$SN_qR = 3(2^{2n})\,(1/\alpha) \tag{5.22}$$

Expressed in dB this becomes:

$$SN_qR = 4.\,8 + 6n - \alpha_{dB} \tag{5.23}$$

For a sinusoidal signal $\alpha = 2$ (or 3 dB). For a (clipped) Gaussianly distributed random signal (with $v_{peak}/\sigma_g = 4$ where σ_g is the signal's standard deviation, or RMS value, as in section 3.2.5) $\alpha = 16$ (or 12 dB), and for speech $\alpha = 10$ dB. The SN_qR for an *n*-bit PCM voice system can therefore be estimated using the rule of thumb $6(n - 1)$ dB.

EXAMPLE 5.3

A digital communications system is to carry a single voice signal using linearly quantised PCM. What PCM bit rate will be required if an ideal anti-aliasing filter with a cut-off frequency of 3.4 kHz is used at the transmitter and the signal to quantisation noise ratio is to be kept above 50 dB?

From equation (5.23):

$$SN_qR = 4.8 + 6n - \alpha_{dB}$$

For voice signals $\alpha = 10$ dB, i.e.:

$$n = \frac{50 + 10 - 4.8}{6} = 9.2$$

10 bit/sample are therefore required. The sampling rate required is given by Nyquist's rule, $f_s \geq 2f_H$. Taking a practical version of the sampling theorem, equation (5.3), gives:

$$f_s = 2.2 \times 3.4 \text{ kHz} = 7.48 \text{ kHz} \text{ (or k samples/s)}$$

The PCM bit rate (or more strictly binary baud rate) is therefore:

$$R_b = f_s n$$

$$= 7.48 \times 10^3 \times 10 \text{ bit/s}$$

$$= 74.8 \text{ kbit/s}$$

5.7.2 SNR for decoded PCM

If all PCM code words are received and decoded without error then the SNR of the decoded signal is essentially equal to the signal to quantisation noise ratio, SN_qR, as given in equations (5.21) to (5.23). In the presence of channel and/or receiver noise, however, it is possible that one or more symbols in a given code word will be changed sufficiently in amplitude to be interpreted in error. For binary PCM this involves a digital 1 being interpreted as a 0 or a digital 0 being interpreted as a 1. The effect that such an error has on the SNR of the decoded signal depends on which symbol is detected in error. The least significant bit in a binary PCM word will introduce an error in the decoded signal equal to one quantisation level. The most significant bit would introduce an error of many quantisation levels.

The following reasonably simple analysis gives a useful expression for the SNR performance of a PCM system in the presence of noise.

We first assume that the probability of more than one error occurring in a single n-bit PCM code word is negligible. We also assume that all bits in the code word have the same probability (P_e) of being detected in error. Using subscripts $1, 2, \cdots, n$ to denote the significance of PCM code word bits (1 corresponding to the least significant, n corresponding to the most significant) then the possible errors in the decoded signal are:

$$
\begin{aligned}
\varepsilon_1 &= q \\
\varepsilon_2 &= 2q \\
\varepsilon_3 &= 4q \\
&\cdots \\
\varepsilon_n &= 2^{n-1}q
\end{aligned}
\tag{5.24}
$$

The mean square decoding error, $\overline{\varepsilon_{de}^2}$ is the mean square of the possible errors multiplied by the probability of an error occurring in a code word, i.e.:

$$
\begin{aligned}
\overline{\varepsilon_{de}^2} &= nP_e(1/n)[(q)^2 + (2q)^2 + \cdots + (2^{n-1}q)^2] \\
&= P_e(q)^2[4^0 + 4 + 4^2 + 4^3 + \cdots + 4^{(n-1)}]
\end{aligned}
\tag{5.25}
$$

The square bracket is the sum of a geometric progression with the form:

$$
S_n = a + ar + ar^2 + \cdots + ar^{n-1} = \frac{a(r^n - 1)}{r - 1}
\tag{5.26}
$$

where $a = 1$ and $r = 4$. Thus:

$$
\overline{\varepsilon_{de}^2} = P_e q^2 (4^n - 1)/3 \quad (\text{V}^2)
\tag{5.27}
$$

Since the error or noise which results from incorrectly detected bits is statistically independent of the noise which results from the quantisation process we can add them together on a power basis, i.e.:

$$
\text{SNR} = \frac{\overline{v^2}}{\overline{\varepsilon_q^2} + \overline{\varepsilon_{de}^2}}
\tag{5.28}
$$

where $\overline{v^2}$ is the received signal power. Using equation (5.13) for $\overline{v^2}$ and equation (5.16) for $\overline{\varepsilon_q^2}$, and remembering that the number of quantisation levels $M = 2^n$ we have:

$$
\text{SNR} = \frac{M^2 - 1}{1 + 4(M^2 - 1)P_e}
\tag{5.29}
$$

Equation (5.29) allows us to calculate the average SNR of the decoded PCM signal including both quantisation noise and the decoding noise which occurs due to corruption of individual PCM bits by channel or receiver noise. In Chapter 6 expressions are developed which relate P_e to channel SNR. If we denote the channel SNR using the subscript *in* and the decoded PCM SNR using the subscript *out* then for binary, polar, NRZ signalling, using simple centre point decisions (see section 6.2), we have:

$$
\text{SNR}_{out} = \frac{M^2 - 1}{1 + 4(M^2 - 1) \; \tfrac{1}{2} \, \text{erfc} \, (\tfrac{1}{2}\text{SNR}_{in})^{\frac{1}{2}}}
\tag{5.30}
$$

The SNRs in equation (5.30) are linear ratios (not dB values) and the function erfc(x) is the complementary error function (defined later in equation (6.3)). Equation (5.30) is

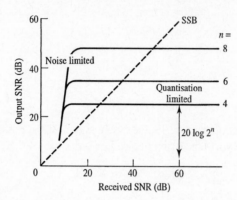

Figure 5.20 *Input/output SNR for PCM.*

sketched for various values of $n = \log_2 M$ in Figure 5.20. The noise immunity advantage of PCM illustrated by this figure is clear. The *x*-axis is the SNR of the received PCM signal. The *y*-axis is the SNR of the reconstructed (decoded) information signal. If the SNR of the received PCM signal is very large then the total noise is dominated by the quantisation process and the output SNR is limited to SN_qR. In practice, however, PCM systems are operated at lower input SNR values near the knee or threshold of the curves in Figure 5.20. The output SNR is then significantly greater than the input SNR. At very low input SNR, when the noise is of comparable amplitude to the PCM pulses, then the interpretation of code words starts to become unreliable. Since even a single error in a PCM code word can change its numerical value by a large amount then the output SNR in this region (i.e. below threshold) decreases very rapidly.

EXAMPLE 5.4
Find the overall SNR for the reconstructed analogue voice signal in Example 5.3 if receiver noise induces an error rate, on average, of one in every 10^6 PCM bits.

From equations (5.29) and (5.17):

$$\text{SNR}_{out} = \frac{SN_qR}{1 + 4\,SN_qR\,P_e}$$

$$SN_qR = 4.8 + 6n - \alpha_{dB} = 4.8 + (6 \times 10) - 10$$

$$= 54.8 \text{ dB (or } 3.020 \times 10^5)$$

$$\text{SNR}_{out} = \frac{3.020 \times 10^5}{1 + 4\,(3.020 \times 10^5)(1 \times 10^{-6})}$$

$$= 1.368 \times 10^5 = 51.4 \text{ dB}$$

The SNR (as a ratio) available with PCM systems increases with the square of the number of quantisation levels while the baud rate, and equivalently the bandwidth, increases with the logarithm of the number of quantisation levels. Thus bandwidth can be exchanged for SNR (as in analogue frequency modulation). Close to threshold PCM is superior to all analogue forms of pulse modulation (and it is also marginally superior to FM) at low SNR. However, all practical PCM systems have a performance which is an order of magnitude below their theoretical optimum.

As PCM signals contain no information in their pulse amplitude they can be regenerated using non-linear processing at each repeater in a long haul system. Such digital, regenerative, repeaters allow accumulated noise to be removed and essentially noiseless signals to be retransmitted to the next repeater in each section of the link. The probability of error does accumulate from hop to hop however. This is discussed in Chapter 6 (section 6.3).

5.7.3 Companded PCM

The expressions for SN_qR derived in section 5.6 (e.g. equations (5.19) and (5.20)) assume that the information signal has a uniform pdf, i.e. that all quantisation levels are used equally. For most signals this is not a valid assumption. If the pdf of the information signal is not uniform but is nevertheless known, and is constant with time, then it is intuitively obvious that to optimise the average SN_qR those quantisation levels used most should introduce least quantisation noise. One way to arrange for this to occur is to adopt non-linear quantisation or, equivalently, companding. Non-linear quantisation is illustrated in Figure 5.21(a). If the information signal pdf has small amplitude for a large fraction of time and large amplitude for a small fraction of time (as is usually the case) then the step size between adjacent quantisation levels is made small for low levels and larger for higher levels. Companding (*comp*ressing-exp*anding*) achieves the same result by compressing the information signal using a non-linear amplitude characteristic (Figure 5.22(a)) prior to linear quantisation and then expanding the reconstructed information signal with the inverse characteristic (Figure 5.22(b)). Ideally the companding characteristic would result in a signal which has a precisely uniform pdf. Whilst this ideal is unlikely to be achieved the compressed signal will have a *more nearly* uniform pdf and therefore better SN_qR than the uncompressed signal.

A rather different problem arises if the information signal has an unknown pdf or if its pdf (measured on some relevant time scale) changes with time. In the case of voice signals, for example, the pdf arising from an individual speaker is usually fairly constant and the *shape* of pdfs arising from different users are usually similar. (A typical voice exceedance curve is shown in Figure 5.23.) The gross signal level, however, can vary widely between speakers, with perhaps a man who habitually shouts whilst using the telephone at one extreme, and a woman who is especially softly spoken at the other extreme. In these cases the companding strategy is normally to maintain, as nearly as possible, a constant SN_qR for all signal levels. (This is quite a different strategy from maximising the average SN_qR.) Since quantisation noise power is proportional to q^2 then RMS quantisation noise voltage is proportional to q. If SN_qR is to be constant for all

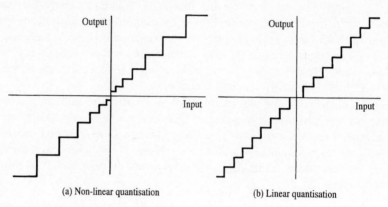

(a) Non-linear quantisation (b) Linear quantisation

Figure 5.21 *Quantisation characteristics.*

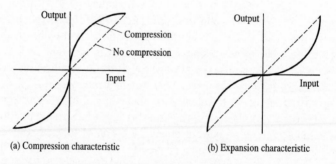

(a) Compression characteristic (b) Expansion characteristic

Figure 5.22 *Typical compression and expansion (compander) characteristics.*

signal levels then q must clearly be proportional to signal level, i.e. v/q must be a constant. If uniform quantisation is used then the signal should be compressed such that increasing the input signal by a given *factor* increases the output signal by a corresponding *additional constant*. Thus equal quantisation increments in the output signal correspond to quantisation increments in the input signal which are equal *fractions* of the input signal. The function which converts multiplicative factors into additional constants is the logarithm and the constant SN_qR compression characteristic is therefore of the form $y = \log x$ (Figure 5.24(a)).

Since information signals can usually take on negative as well as positive values the logarithmic compression characteristic must be reflected to form an odd symmetric function (Figure 5.24(b)). Furthermore the characteristic must obviously be continuous across zero volts and so the two logarithmic functions are modified and joined by a linear section as shown in Figure 5.24(c). The actual compression characteristic used in Europe is the *A*-law defined by:

$$F(x) = \begin{cases} \operatorname{sgn}(x) \dfrac{1 + \ln(A|x|)}{1 + \ln A}, & 1/A < |x| < 1 \\[2mm] \operatorname{sgn}(x) \dfrac{A|x|}{1 + \ln A}, & 0 < |x| < 1/A \end{cases} \qquad (5.31)$$

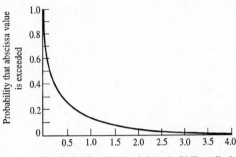

Figure 5.23 *Statistical distribution of single talker speech signal amplitude.*

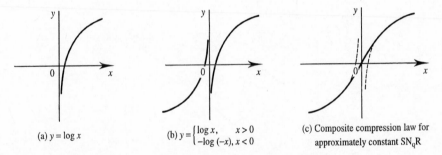

(a) $y = \log x$

(b) $y = \begin{cases} \log x, & x > 0 \\ -\log (-x), & x < 0 \end{cases}$

(c) Composite compression law for approximately constant SN_qR

Figure 5.24 *Development of constant SN_qR compression characteristic.*

where $|x| = |v/v_{peak}|$ is the normalised input signal to the compressor, $F(x)$ is the normalised output signal from the compressor and $\text{sgn}(x)$ is the signum function which is +1 for $x > 0$ and −1 for $x < 0$.

The parameter A in equation (5.31) defines the curvature of the compression characteristic with $A = 1$ giving a linear law (Figure 5.25). The commonly adopted value is $A = 87.6$ which gives a 24 dB improvement in SN_qR over linear PCM for small signals ($|x| < 1/A$) and an (essentially) constant SN_qR of 38 dB for large signals ($|x| > 1/A$) [Dunlop and Smith]. The dynamic range of the logarithmic (constant SN_qR) region of this characteristic is $20 \log_{10}[1/(1/A)] \approx 39$ dB. The overall effect is to allow 11 bit (2048 level) linear PCM, which would be required for adequate voice signal quality, to be reduced to 8 bit (256 level) companded PCM. A 4 kHz voice channel sampled at its Nyquist rate (i.e. 8 kHz) therefore yields a companded PCM bit rate of 64 kbit/s.

The A-law characteristic is normally implemented as a 13-segment piecewise linear approximation to equation (5.31) (in practice 16 segments but with 4 segments near the origin co-linear as illustrated in Figure 5.26). For 8 bit PCM one bit gives polarity, 3 bits indicate which segment the sample lies on and 4 bits provide the location on the segment.

In the USA and Japan a similar logarithmic compression law is used. This is the μ-law given by:

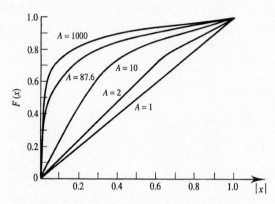

Figure 5.25 *A-law compression characteristic for several values of A.*

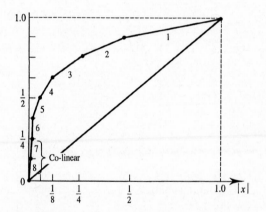

Figure 5.26 *13-segment compression A-law realised by piecewise linear approximation.*

$$F(x) = \text{sgn}(x) \frac{\ln(1 + \mu|x|)}{\ln(1 + \mu)}, \quad 0 \le |x| \le 1 \tag{5.32}$$

The μ-law (with $\mu = 255$) tends to give slightly improved SN_qR for voice signals when compared with the A-law but it has a slightly smaller dynamic range. In practice, like the A-law, the μ-law is usually implemented as a piecewise linear approximation.

A- and μ-law 64 kbit/s companded PCM have been adopted by ITU-T as international toll quality standards (recommendation G.711) for digital coding of voice frequency signals. The sampling rate is 8 kHz and the encoding law uses 8 binary digits per sample. The A- and μ-laws are implemented as 13- and 15-segment piecewise linear curves respectively. For communications between countries using different companding laws (one using A and one using μ) conversion from one to the other is the responsibility of the country using the μ-law.

G.711 PCM communications has a *mean opinion score* (MOS) speech quality, measured by subjective testing, of 4.3 on a scale of 0 to 5. A MOS of 4 allows audible but not annoying degradation, 3 implies that the degradation is slightly annoying, 2 is annoying and 1 is very annoying. If narrow bandwidth is important then other coding techniques are employed (see section 5.8).

5.7.4 PCM multiplexing

Multiplexing of analogue PAM pulses has already been described in section 5.4. If such a multiplexed pulse stream is fed to a PCM encoder the resulting signal is a code word (or *character*) interleaved digital TDM signal. This is illustrated in Figure 5.27(a). In this figure the TDM signal is generated by *first* multiplexing and *then* PCM coding. It is of course possible to generate a character interleaved TDM signal by *first* PCM coding the information signals and *then* interleaving the code words, Figure 5.27(b). There are two reasons for drawing attention to this alternative method of TDM generation.

1. As digital technology has become cheaper and more reliable there has been a tendency for analogue signals to be digitally coded as near their source as possible.
2. Time division multiplexing of PCM signals (rather than PCM encoding of TDM PAM signals) suggests the possibility of interleaving the individual bits of the PCM code

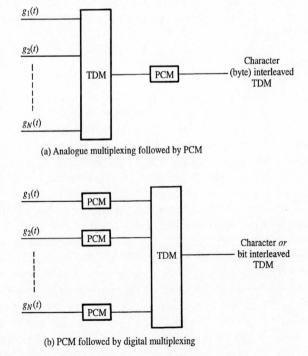

(a) Analogue multiplexing followed by PCM

(b) PCM followed by digital multiplexing

Figure 5.27 *Generation of bit and byte interleaved PCM-TDM signals.*

words rather than the code word (bytes) themselves.

The TDM concept can be extended to higher levels of multiplexing by time division multiplexing two or more TDM signals. A detailed discussion of such multiplexing hierarchies is given in Chapter 19.

5.8 Bandwidth reduction techniques

Bandwidth is a limited, and therefore valuable, resource. This is because all physical transmission lines have characteristics which make them suitable for signalling only over a finite band of frequencies. (The problem of constrained bandwidth is especially severe for the local loop which connects an individual subscriber to their national telephone network.) Bandwidth (or more strictly spectrum) is even limited in the case of radio communications since the transmission properties of the earth's atmosphere are highly variable as function of frequency (see Chapter 14). Furthermore, co-channel (same frequency) radio transmissions are difficult to confine spatially and tend to interfere with each other, even when such systems are widely spaced geographically.

A given installation of transmission lines (wire-pairs, coaxial cables, optical fibres, microwave links and others) therefore represents a finite spectral resource and since adding to this installation (by laying new cables, for instance) is expensive, there is great advantage to be gained in using the existing installation efficiently. This is the real incentive to develop spectrally efficient (i.e. reduced bandwidth) signalling techniques for encoding speech signals.

5.8.1 Delta PCM

One technique to reduce the bandwidth of a PCM signal is to transmit information about the changes between samples instead of sending the sample values themselves. The simplest such system is delta PCM which transmits the difference between adjacent samples as conventional PCM code words. The difference between adjacent samples is, generally, significantly less than the actual sample values which allows the differences to be coded using fewer binary symbols per word than conventional PCM would require. (This reflects the fact that the adjacent samples derived from most, naturally generated, information signals are not usually independent but correlated, see section 2.6.) Block diagrams of a delta PCM transmitter and receiver are shown in Figure 5.28. It can be seen from this figure that the delta PCM transmitter simply represents an adjacent sample differencing operation followed by a conventional (usually reduced word length) pulse code modulator. Similarly the receiver is a pulse code demodulator followed by an adjacent 'sample' summing operation. The reduced number of bits per PCM code word translates directly into a saving of bandwidth. (This saving could, of course, then be traded against signal power and/or transmission time.) The delta PCM system cannot, however, accommodate rapidly varying transient signals as well as a conventional PCM system.

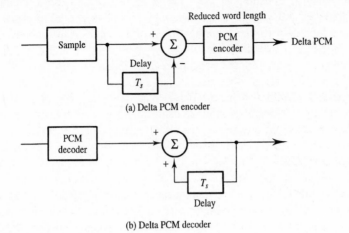

(a) Delta PCM encoder

(b) Delta PCM decoder

Figure 5.28 *Delta PCM transmitter and receiver.*

5.8.2 Differential PCM

The correlation between closely spaced samples of information signals originating from natural sources has already been referred to. An alternative way of expressing this phenomenon is to say that the signal contains redundant information, i.e. the same or similar information resides in two or more samples. An example of this is the redundancy present in a picture or image. To transmit an image over a digital communications link it is normally reduced to a two dimensional array of picture cells (pixels). These pixels are quantised in terms of colour and/or brightness. Nearly all naturally occurring images have very high average correlations, section 2.6, between the characteristics of adjacent pixels. Put crudely, if a given pixel is black then there is a high probability that the adjacent pixels will be at least nearly black. In this context the bandwidth of a random signal, the rate at which it is sampled, the correlation between closely spaced samples and its information redundancy are all closely related. One consequence of redundancy in a signal is that its future values can be predicted (within certain confidence limits) from its current and past values (see Chapter 16).

Differential PCM (DPCM) uses an algorithm to predict an information signal's future value. It then waits until the actual value is available for examination and subsequently transmits a signal which represents a correction to the predicted value. The correction signal is therefore a distillation of the information signal, i.e. it represents the information signal's surprising or unpredictable part. DPCM thus reduces the redundancy in a signal and allows the information contained in it to be transmitted using fewer symbols, less spectrum, shorter time and/or lower signal power. Figure 5.29(a) shows a block diagram of a DPCM transmitter. $g(t)$ is a continuous analogue information signal and $g(kT_s)$ is a sampled version of $g(t)$. k represents the (integer) sample number. $\varepsilon(kT_s)$ is the error between the actual value of $g(kT_s)$ and the value, $\hat{g}(kT_s)$, predicted from previous samples. It is this error which is quantised, to form $\varepsilon_q(kT_s)$, and encoded to give the

DPCM signal which is transmitted. $\tilde{g}(kT_s) = \hat{g}(kT_s) + \varepsilon_q(kT_s)$ is an estimate of $g(kT_s)$ and is the predicted value $\hat{g}(kT_s)$, *corrected* by the addition of the *quantised* error $\varepsilon_q(kT_s)$. From the output of the sampling circuit to the input of the encoder the system constitutes a type of source coder. The DPCM receiver is shown in Figure 5.29(b) and is identical to the predictor loop in the transmitter. The reconstructed signal at the receiver is therefore the estimate, $\tilde{g}(t)$, of the original signal $g(t)$.

The predictor in DPCM systems is often a linear weighted sum of previous samples (i.e. a transversal digital filter) implemented using shift registers [Mulgrew and Grant]. A schematic diagram of such a predictor is shown in Figure 5.30.

5.8.3 Adaptive DPCM

Adaptive DPCM (ADPCM) is a more sophisticated version of DPCM. In this scheme the predictor coefficients (i.e. the weighting factors applied to the shift register elements) are continuously modified (i.e. adapted) to suit the changing signal statistics. (The values of

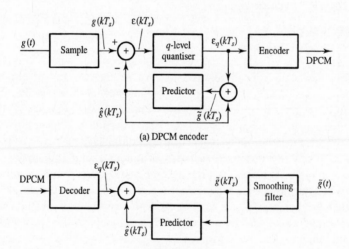

(a) DPCM encoder

(b) DPCM decoder

Figure 5.29 *DPCM transmitter and receiver.*

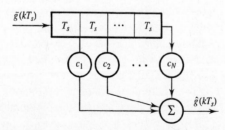

Figure 5.30 *DPCM predictor implemented as a transversal digital filter.*

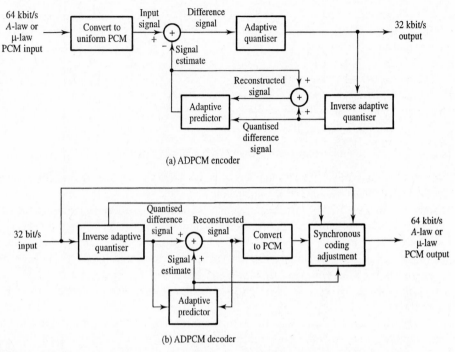

Figure 5.31 *ITU G.721 adaptive differential PCM (ADPCM) transmitter and receiver.*

these coefficients must, of course, also be transmitted over the communications link.) ITU has adopted ADPCM as a reduced bit rate standard. The ITU-T ADPCM encoder takes a 64 kbit/s companded PCM signal (G.711) and converts it to 32 kbit/s ADPCM signal (G.721). The G.721 encoder and decoder are shown in Figure 5.31. The encoder uses 15 level, 4-bit, codewords to transmit the quantised difference between its input and estimated signal. The subjective quality of error free 32 kbit/s ADPCM voice signals is only slightly inferior to 64 kbit/s companded PCM. For probabilities of error greater than 10^{-4} its subjective quality is actually better than 64 kbit/s PCM. 32 kbit/s ADPCM can achieve network quality speech with a MOS of 4.1 at error ratios of 10^{-3} to 10^{-2} for a complexity (measured by a logic gate count) of 10 times simple PCM. It allows an ITU-T 30 + 2 TDM channel (see Chapter 19) to carry twice the number of voice signals which are possible using 64 kbit/s companded PCM. Other specifications are defined by ITU-T, G.726 and G.727, for ADPCM with transmission rates of 16 to 40 kbit/s.

5.8.4 Delta modulation

If the quantiser of a DPCM system is restricted to one bit (i.e. two levels only, $\pm\Delta$) then the resulting scheme is called delta modulation (DM). This can be implemented by replacing the DPCM differencing block and quantiser with a comparator as shown in Figure 5.32(a). An especially simple prediction algorithm assumes that the next sample

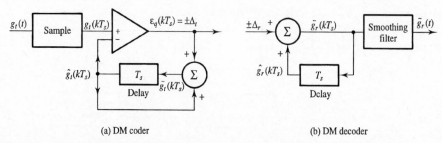

(a) DM coder   (b) DM decoder

Figure 5.32 *DM transmitter (a) and receiver (b) using previous sample (or one tap) prediction.*

value will be the same as the last sample value, i.e. $\hat{g}(kT_s) = \tilde{g}[(k-1)T_s]$. This is called a previous sample predictor and is implemented as a one sample delay. The DM decoder is shown in Figure 5.32(b). Specimen signal waveforms at different points in the DM system are shown in Figure 5.33. Slope overload noise, which occurs when $g(t)$ changes too rapidly for $\tilde{g}(kT_s)$ to follow faithfully, and quantisation noise (also called granular noise) are illustrated in Figure 5.33. There is a potential conflict between the requirements for acceptable quantisation noise and acceptable slope overload noise. To reduce the former the step size Δ should be small and to reduce the latter Δ should be large. One way of keeping both types of noise within acceptable limits is to make Δ small, but sample much faster than the normal minimum (i.e. Nyquist) rate. Typically the DM sampling rate will be many times the Nyquist rate. (This does mean that the bandwidth saving which DM systems potentially offer may be partially or completely eroded.)

If the information signal $g(t)$ remains constant for a significant period of time then $\tilde{g}(kT_s)$ displays a hunting behaviour and the resulting quantisation noise becomes a square wave with a period twice that of the sampling period. When this occurs the quantisation noise is called idling noise. Since the fundamental frequency of this noise is half the sampling frequency, which is usually itself many times the highest frequency in the information signal, then much of the idling noise will be removed by the smoothing filter in the DM receiver.

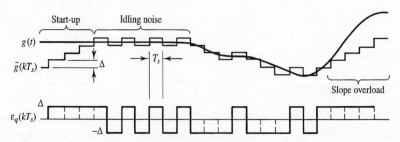

Figure 5.33 *DM signal waveforms illustrating slope overload and quantisation noise.*

A realistic analysis of the SNR performance of DM systems is rather complicated. Here we quote results adapted from [Schwartz, 1990] for a random information signal with Gaussian pdf and white spectrum bandlimited to f_H Hz. In this case the signal to slope overload noise ratio is:

$$\frac{S}{N_{ov}} = \frac{\sigma_g^2}{\sigma_{ov}^2} = 1.2 \left(\frac{1}{2\pi} \frac{\Delta}{4\sigma_g} \frac{f_s}{f_H} \right)^5 e^{1.5 \left(\frac{1}{2\pi} \frac{\Delta}{4\sigma_g} \frac{f_s}{f_H} \right)^2} \tag{5.33}$$

where σ_g is the standard deviation of the (Gaussian) information signal (equal to its RMS value if the DC component is zero), $N_{ov} = \sigma_{ov}^2$ is the variance (i.e. mean square) of the slope overload error, Δ is the DM step size, f_s is the DM sampling rate and f_H is the highest frequency in the baseband information signal (usually the information signal bandwidth). The signal to quantisation noise ratio (neglecting the effect of the smoothing filter) is:

$$SN_qR = \frac{\sigma_g^2}{N_q} = 1.5 \left(\frac{4\sigma_g}{\Delta} \right)^2 \frac{f_s}{f_H} \tag{5.34}$$

For a given peak signal level (assumed here to be $4\sigma_g$) the DM step size Δ can be reduced by a factor of 2 for each factor of 2 increase in f_s without introducing any more slope overload noise. DM SN_qR therefore potentially increases with f_s^3 leading to a $30 \log_{10} 2$ or 9 dB improvement for each octave increase in sampling frequency.

Assuming that the slope overload noise, N_{ov}, and quantisation noise, N_q, are statistically independent then they can be added power-wise to give the total signal to noise ratio, i.e.:

$$SNR = \frac{S}{N_{ov} + N_q} = \frac{1}{(S/N_{ov})^{-1} + (S/N_q)^{-1}} \tag{5.35}$$

Equation (5.35) neglects the effects of channel and/or receiver noise which, if severe, might cause $+\Delta$ and $-\Delta$ symbols to be received in error. Figure 5.34 shows SNR plotted against the normalised step size – sampling frequency product, $(\Delta/4\sigma_g)(f_s/f_H)$, for various values of normalised sampling frequency, f_s/f_H. ($4\sigma_g$ is taken, for practical purposes, to be the maximum amplitude of the Gaussian information signal.) The peaky shape of the curves in Figure 5.34 reflects the fact that slope overload noise dominates for small DM step size and quantisation noise dominates for large step size.

If the SNR is not sufficiently high then the DM receiver will occasionally interpret a received symbol in error (i.e. $+\Delta$ instead of $-\Delta$ or the converse). This is equivalent to the addition of an error of 2Δ (i.e. the difference in the analogue signal represented by $+\Delta$ and $-\Delta$) to the accumulated signal at the DM receiver. The estimated signal $\tilde{g}_r(kT_s)$ at the receiver thereafter follows the variation of the estimated signal at the transmitter $\tilde{g}_t(kT_s)$ but with a constant offset of 2Δ. This situation continues until another error occurs which either cancels the first error or doubles it. The offset is therefore a stepped signal, step transitions occurring at random sample times with an average occurrence rate equal to the BER or P_e times the symbol rate. This is illustrated in Figure 5.35 which shows the result of errors being received in the DM signal and the consequent deviation from the

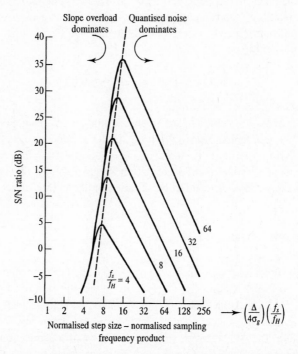

Figure 5.34 *SNR versus normalised step size - sampling frequency product for DM systems
(parameter is normalised sampling frequency, (f_s/f_H)) (source: O'Neal, 1966,
reproduced with permission of ATT Technical Journal).*

ideal output. On average the error (or noise) represented by such a signal increases
without limit as errors accumulate. In practice, however, most of the power in this
'pseudo DC signal' is contained at low frequencies (assuming P_e is small). Provided the
frequency response of the post accumulator (or smoothing) filter goes to zero at 0 Hz then
most of this noise can be removed. A simple way of showing that this is the case is to
consider a post accumulator filter which, close to 0 Hz, has a highpass RC characteristic.
The response of this filter to a step change of 2Δ volts at its input is $2\Delta e^{-t/\tau_c}$ where the
time constant $\tau_c = RC$. The energy, E_e, (dissipated in 1 Ω) at the filter output due to a
single step transition at its input is:

$$E_e = \int_0^\infty [2\Delta \, e^{-t/\tau_c}]^2 \, dt$$

$$= 2\Delta^2 \tau_c \text{ (joules error}^{-1}) \tag{5.36}$$

The average 'noise' power, N_e, in the post filtered signal due to errors is therefore:

$$N_e = \text{BER} \times E_e \quad \text{(W)} \tag{5.37}$$

where BER is the bit (or more strictly symbol) error rate. Using BER $= P_e f_s$, where P_e

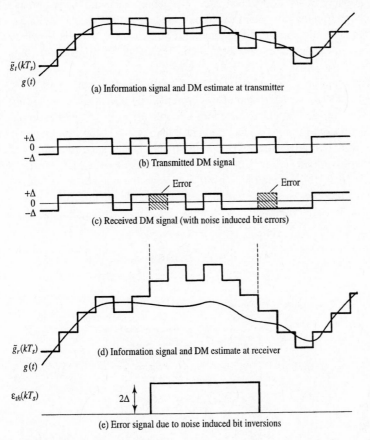

$\tilde{g}_t(kT_s)$
$g(t)$

(a) Information signal and DM estimate at transmitter

$+\Delta$
0
$-\Delta$

(b) Transmitted DM signal

Error   Error

$+\Delta$
0
$-\Delta$

(c) Received DM signal (with noise induced bit errors)

$\tilde{g}_r(kT_s)$
$g(t)$

(d) Information signal and DM estimate at receiver

$\varepsilon_{th}(kT_s)$

2Δ

(e) Error signal due to noise induced bit inversions

Figure 5.35 *Stepped error signal in DM receiver due to thermal noise.*

is the probability of error and f_s is the DM sampling rate, and putting $\tau_c = 1/2\pi f_L$ where f_L is the lowest frequency component with significant amplitude present in the information signal we have:

$$N_e = \frac{\Delta^2}{\pi} \frac{f_s}{f_L} P_e \qquad (5.38)$$

Thus as f_L is increased more of the noise due to bit errors is removed. If an ideal rectangular highpass characteristic is assumed in place of the RC characteristic the result is smaller by a factor of $2/\pi$ (i.e. approximately 2 dB) [Taub and Schilling].

5.8.5 Adaptive delta modulation

In conventional DM the problem of keeping both quantisation noise and slope overload noise acceptably low is solved by oversampling, i.e. keeping the DM step size small and sampling at many times the Nyquist rate. The penalty incurred is the loss of some, or all, of the saving in bandwidth which might be expected with DM. An alternative strategy is to make the DM step size *variable*, making it larger during periods when slope overload noise would otherwise dominate and smaller when quantisation noise might dominate. Such systems are called adaptive DM systems (ADM).

A block diagram of an ADM transmitter is shown in Figure 5.36. The gain block, $G(kT_s)$, controls the variable step size represented by the *constant* amplitude pulses $\pm\Delta$. The step size is varied or adapted according to the history of ε_q. For example, if $\varepsilon_q = +\Delta$ for several adjacent samples it can be inferred that $g(t)$ is rising more rapidly than $\tilde{g}(kT_s)$ is capable of tracking it. Under this condition an ADM system increases the step size to reduce slope overload noise. Conversely if ε_q alternates between $+\Delta$ and $-\Delta$ then the inference is that $g(t)$ is changing slowly and slope overload is not occurring. The step size would therefore be decreased in order to reduce quantisation error. Figure 5.37 illustrates both these conditions. A simple ADM step size adjustment algorithm would be:

$$G(kT_s) = \begin{cases} G[(k-1)T_s]C, & \text{if } \varepsilon_q(kT_s) = \varepsilon_q[(k-1)T_s] \\ G[(k-1)T_s]/C, & \text{if } \varepsilon_q(kT_s) = -\varepsilon_q[(k-1)T_s] \end{cases} \qquad (5.39)$$

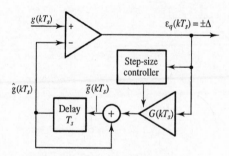

Figure 5.36 *Adaptive delta modulation (ADM) transmitter.*

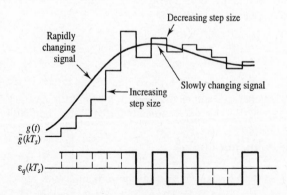

Figure 5.37 *ADM signal waveforms illustrating variable step size.*

where C is a constant > 1.

ADM typically performs with between 8 and 14 dB greater SNR than standard DM. Furthermore, since ADM uses large step sizes for wide variations in signal level and small step sizes for small variations in signal level it has wider dynamic range than standard DM. ADM can therefore operate at much lower bit rates than standard DM, typically 32 kbit/s and exceptionally 16 kbit/s. (16 kbit/s ADM allows reduced quality digital speech to be transmitted directly over radio channels which are allocated 25 kHz channel spacings.)

5.9 Summary

Continuous, analogue, information signals are converted to discrete, analogue, signals by the process of sampling. Pre-sampling anti-aliasing filters are used to limit any resulting distortion to acceptable levels. A naturally sampled signal can be converted to a PAM signal by flattening the pulse tops to give rectangular pulses. An ideal, impulse, sampled signal is similarly converted to a PAM signal by replacing the sample values with rectangular pulses of equivalent amplitude. PWM and PPM have a significant SNR advantage over PAM.

The minimum sampling rate required for an information signal with specified bandwidth is given by Nyquist's sampling theorem, if the signal is baseband, and by the bandpass sampling theorem if the signal is bandpass. In the bandpass case there is one or more allowed sampling rate bands rather than a simple sampling rate minimum. Reconstruction of a continuous information signal from a sampled, or PAM, signal can be achieved using a low-pass filter. In the absence of noise such reconstruction can be essentially error free. Practical filter designs usually mean that modest over-sampling is required if reconstruction is to be ideal.

Pulse modulated signals can be multiplexed together by interleaving the samples of the tributary information signals. The resulting TDM signal has narrower pulses, and therefore greater bandwidth, than is necessary for each of the tributary signals alone. It has the advantage, however, of allowing a single physical channel to carry many real time tributary signals, essentially simultaneously. Guard-times are normally required between the adjacent pulses in a TDM signal to keep cross-talk, generated by the bandlimited channel, to acceptable levels.

Quantisation, the process which converts an analogue signal to a digital signal, results in the addition of quantisation noise, which decreases as the number of quantisation levels increases. Pulse code modulation replaces M quantisation levels with M code words each comprising $n = \log_2 M$ binary digits or bits. A PCM signal then represents each binary digit by a voltage pulse which can have either of two possible amplitudes. The SN_qR of an n-bit, linear, PCM voice signal is approximately $6(n-1)$ dB.

Companding of voice signals prior to PCM encoding increases the SN_qR of the decoded signal by increasing the resolution of the quantiser for small signals at the expense of the resolution for large signals. The ITU-T standard for digitally modulated companded voice signals is 8000 sample/s, 8-bit/sample giving a PCM bit rate of 64 kbit/s. Redundancy in transmitted PCM signals can be reduced by using DPCM techniques and its variants. ADPCM is an ITU standard for reduced bandwidth digital (32 kbit/s) voice transmission which gives received signal quality comparable to standard PCM. ADM, which is a variation on ADPCM, uses one bit quantisation.

5.10 Problems

5.1. (a) Sketch the design of TDM and FDM systems each catering for 12 voice channels of 4 kHz bandwidth.
(b) In the case of TDM, indicate how the information would be transmitted using: (i) PAM; (ii) PPM; (iii) PCM.
(c) Calculate the bandwidths required for the above FDM and TDM systems. You may assume the TDM system uses PPM with 2% resolution. [FDM 48 kHz; TDM 2.4 MHz]

5.2. Two lowpass signals, each band-limited to 4 kHz, are to be multiplexed into a single channel using pulse amplitude modulation. Each signal is impulse-sampled at a rate of 10 kHz. If the time-multiplexed signal waveform is filtered by an ideal lowpass filter (LPF) before transmission:
(a) What is the minimum clock frequency (or baud rate) of the system? [20 kHz]
(b) What is the minimum cut-off frequency of the LPF? [10 kHz]

5.3. Rewrite equation (5.5) such that it takes all spectral replicas into account. Hence find the, aliasing induced, SDR for Example 5.1 (part (ii)) accounting for the spectral replicas centred on f_s and $2f_s$ Hz.

5.4. An analogue bandpass signal has a bandwidth of 40 kHz and a centre frequency of 10 MHz. What is the minimum theoretical sampling rate which will avoid aliasing? What would be the best practical choice for the nominal sampling rate? Would an oscillator with a frequency stability of 1 part in 10^6 be adequate for use as the sampling clock? [80.16 kHz, 80.1603 kHz]

5.5. A rectangular PAM signal, with pulse widths of 0.1 μs, is transmitted over a channel which can be modelled by an RC low-pass filter with a half power bandwidth of 1.0 MHz. If an average XTR of 25 dB or better is to be maintained, estimate the required guard time between pulses. (Assume that pulse sampling at the channel output occurs at the optimum instants.) [0.36 μs]

5.6. Explain why the XTR in Problem 5.5 is degraded if the pulses are sampled (prematurely) at the mid point of the nominal, rectangular pulse, time slots. Quantify this degradation for this particular system. [7.5 dB]

5.7. Twenty-five input signals, each band-limited to 3.3 kHz, are each sampled at an 8 kHz rate then time-multiplexed. Calculate the minimum bandwidth required to transmit this multiplexed signal in the presence of noise if the pulse modulation used is: (a) PAM; (b) quantised PPM with a required level resolution of 5%; or (c) binary PCM with a required level resolution of > 0.5%. (This higher resolution requirement on PCM is normal for speech-type signals because the quantisation noise is quite objectionable.) [(a) 100 kHz, (b) 2 MHz, (c) 800 kHz]

5.8. A hi-fi music signal has a bandwidth of 20 kHz. Calculate the bit rate required to transmit this as a, linearly quantised, PCM signal maintaining a SN_qR of 55 dB. What is the minimum (baseband) bandwidth required for this transmission? (Assume that the signal's peak to mean ratio is 20 dB.) [480 kbit/s, 240 kHz]

5.9. Information is to be transmitted as a linearly quantised, 8-bit, signal over a noisy channel. What probability of error can be tolerated in the detected PCM bit stream if the reconstructed information signal is to have a SNR of 45 dB? [4.1×10^{-6}]

5.10. Show that the peak signal to quantisation noise ratio, $(SN_qR)_{peak}$, for an n-bit, μ-law companded, communications system is given by:

$$(SN_qR)_{peak} = \frac{3 \times 2^{2n}}{[\ln(1 + \mu)]^2}$$

Hence calculate the degraded output SNR in dB for large signal amplitudes in an $n = 8$ bit companded PCM system with typical μ value when compared with a linear PCM system operating at the same channel transmission rate. [10.1 dB]

5.11. An 8-bit A-law companded PCM system is to be designed with piecewise linear approximation as shown in Figure 5.26. 16 segments are employed (with 4 co-linear near the origin) and the segments join at 1/2, 1/4, etc. of the full scale value, as shown in Figure 5.26. Calculate the approximate SNR for full scale and small signal values.
[Full scale 36.3 dB, small signal 71.9 dB]

5.12. A DM communication system must achieve a SNR of 25 dB. The DM step size is 1.0 V. The (zero mean) baseband information signal has a bandwidth of 3.0 kHz and a Gaussian pdf with an RMS value of 0.2 V. Use Figure 5.34 to estimate both the optimum sampling frequency and the gain of the amplifier which is required immediately prior to the modulator input. [96 kHz, 10.5 dB]

CHAPTER 6

Baseband transmission and line coding

6.1 Introduction

This chapter addresses two issues central to baseband signal transmission, digital signal detection and line coding. The detection process described here is restricted to a simple implementation of the decision circuit in Figure 1.3. The analysis of detection processes is important since this leads to the principal objective measure of digital communication systems quality, i.e. the probability of symbols being detected in error, as quantified by probability theory (section 3.2). Appropriate coding of the transmitted symbol pulse shape can minimise the probability of error by ensuring that the spectral characteristics of the digital signal are well matched to the transmission channel. The line code must also permit the receiver to extract accurate PCM bit and word timing signals directly from the received data, for proper detection and interpretation of the digital pulses.

6.2 Baseband centre point detection

The detection of digital signals involves two processes:
1. Reduction of each received voltage pulse (i.e. symbol) to a single numerical value.
2. Comparison of this value with a reference voltage (or, for multisymbol signalling, a set of reference voltages) to determine which symbol was transmitted.

In the case of symbols represented by different voltage levels the simplest way of achieving 1 is to sample the received signal plus noise; 2 is then implemented using one or more comparators. In the case of equiprobable, binary, symbols (zero and one) represented by two voltage levels (e.g. 0 V and 3 V) intuition tells us that a sensible strategy would be to set the reference, V_{ref}, (Figure 6.1(b)) mid-way between the two voltage levels (i.e. at 1.5 V). Decisions would then be made at the centre of each symbol period on the basis of whether the instantaneous voltage (signal plus noise) is above or below this reference. Sampling the instantaneous signal plus noise voltage somewhere near the middle of the symbol period is called *centre point detection*, Figure 6.1(a).

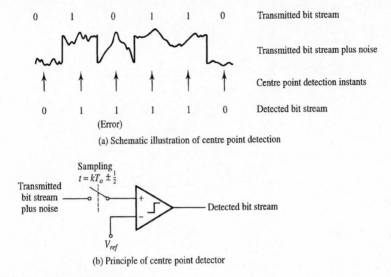

(a) Schematic illustration of centre point detection

(b) Principle of centre point detector

Figure 6.1 *Centre point detection.*

The noise present during detection is often either Gaussian or approximately Gaussian. (This is always the case if thermal noise is dominant, but may also be the case when other sources of noise dominate owing to operation of the central limit theorem.) Since noise with a Gaussian probability density function (pdf) is common and analytically tractable, the bit error rate (BER) of a communications system is often modelled assuming this type of noise alone.

6.2.1 Baseband binary error rates in Gaussian noise

Figure 6.2(a) shows the pdf of a binary information signal which can take on voltage levels V_0 and V_1 only. Figure 6.2(b) shows the probability density function of a zero mean Gaussian noise process, $v_n(t)$, with RMS value σ V. (Since the process has zero mean the RMS value and standard deviation are identical.) Figure 6.2(c) shows the pdf of the sum of the signal and noise voltages. (Whilst Figure 6.2(c) is perhaps no surprise – we can think of the 'quasi-DC' symbol voltages biasing the mean value of the noise to V_0 and V_1 – recall, from Chapter 3, that when independent random variables are added their pdfs are convolved. Figure 6.2(c) is thus the convolution of Figures 6.2(a) and 6.2(b).)

For equiprobable symbols the optimum decision level is set at $(V_0 + V_1)/2$. (This is not the optimum threshold if the symbols are not equiprobable: see section 7.4.4.) Given that the symbol 0 is transmitted (i.e. a voltage level V_0) then the probability, P_{e1}, that the signal plus noise will be above the threshold at the decision instant is given by twice [1] the

---

[1] $p_0(v_n)$ and $p_1(v_n)$, as defined here, each represent a total probability of 0.5. Strictly speaking they are not, therefore, pdfs although their sum is the total signal plus noise pdf irrespective of whether a one or zero is transmitted. The pdf of the signal plus noise conditional on a zero being transmitted is $2p_0(v_n)$.

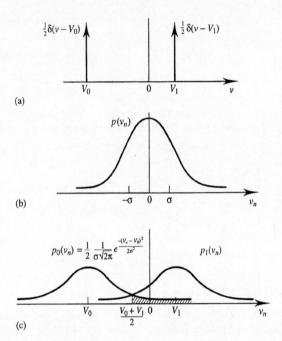

Figure 6.2 *Probability density function of: (a) binary symbol; (b) noise; (c) signal plus noise.*

shaded area under the curve $p_0(v_n)$ in Figure 6.2(c), i.e.:

$$P_{e1} = \int\limits_{(V_0+V_1)/2}^{\infty} \frac{1}{\sigma\sqrt{(2\pi)}} e^{\frac{-(v_n-V_0)^2}{2\sigma^2}} dv_n \tag{6.1}$$

Using the change of variable $x = (v_n - V_0)/\sqrt{2}\sigma$ this becomes:

$$P_{e1} = \frac{1}{\sqrt{\pi}} \int\limits_{(V_1-V_0)/2\sqrt{2}\sigma}^{\infty} e^{-x^2} dx \tag{6.2}$$

The incomplete integral in equation (6.2) cannot be evaluated analytically but can be recast as a complementary error function, $\text{erfc}(x)$, defined by:

$$\text{erfc}(z) \underset{=}{\Delta} \frac{2}{\sqrt{\pi}} \int\limits_{z}^{\infty} e^{-x^2} dx \tag{6.3}$$

Thus equation (6.2) becomes:

$$P_{e1} = \frac{1}{2} \text{erfc} \left(\frac{V_1 - V_0}{2\sigma\sqrt{2}} \right) \tag{6.4}$$

Alternatively, since $\text{erfc}(z)$ and the error function, $\text{erf}(z)$, are related by:

$$\text{erfc}(z) \equiv 1 - \text{erf}(z) \tag{6.5}$$

then P_{e1} can also be written as:

$$P_{e1} = \frac{1}{2}\left[1 - \text{erf}\left(\frac{V_1 - V_0}{2\sigma\sqrt{2}}\right)\right] \tag{6.6}$$

The advantage of using the error (or complementary error) function in the expression for P_{e1} is that this function has been extensively tabulated[2] (see Appendix A).

If the digital symbol one is transmitted (i.e. a voltage level V_1) then the probability, P_{e0}, that the signal plus noise will be below the threshold at the decision instant is:

$$P_{e0} = \int\limits_{-\infty}^{(V_0+V_1)/2} \frac{1}{\sigma\sqrt{(2\pi)}}\, e^{\frac{-(v_n-V_1)^2}{2\sigma^2}}\, dv_n \tag{6.7}$$

It is clear from the symmetry of this problem that P_{e0} is identical to both P_{e1} and the probability of error, P_e, irrespective of whether a one or zero was transmitted. Noting that the probability of error depends on only symbol voltage difference, and not absolute voltage levels, P_e can be rewritten in terms of $\Delta V = V_1 - V_0$, i.e.:

$$P_e = \frac{1}{2}\left[1 - \text{erf}\left(\frac{\Delta V}{2\sigma\sqrt{2}}\right)\right] \tag{6.8}$$

Equation (6.8) is valid for both *unipolar* signalling (i.e. symbols represented by voltages of 0 and ΔV) and *polar signalling* (i.e. symbols represented by voltages of $\pm\Delta V/2$). In fact, it is valid for all pulse levels and shapes providing that ΔV represents the voltage *difference at the sampling instant*. Figure 6.3 shows P_e versus the voltage ratio $\Delta V/\sigma$ expressed both in dB and as a ratio.

Whilst the x-axis of Figure 6.3 is clearly related to signal-to-noise ratio it is not identical to it for all pulse levels and shapes. This is partly because it involves the signal only at the sampling instant (whereas a conventional SNR uses a time averaged signal power) and partly because the use of ΔV neglects any transmitted DC component. For NRZ (non-return to zero, see section 6.4), unipolar, rectangular pulse signalling, Figure 6.4, the normalised peak signal power is $S_{peak} = \Delta V^2$ and the average signal power is $S = \Delta V^2/2$. The normalised Gaussian noise power is $N = \sigma^2$. We therefore have:

$$\frac{\Delta V}{\sigma} = \left(\frac{S}{N}\right)_{peak}^{1/2} = \sqrt{2}\left(\frac{S}{N}\right)^{1/2} \tag{6.9}$$

where $(S/N)_{peak}$ indicates peak signal power divided by average (or expected) noise power. (The peak noise power would in principle be infinite for Gaussian noise.) Substituting for $\Delta V/\sigma$ in equation (6.8) we therefore have:

---

[2] The Q-function, which represents the area under the tail of a (zero mean, unit variance) Gaussian pdf, is defined by $Q(z) = \int_z^\infty (1/\sqrt{2\pi})e^{-(x^2/2)}\, dx$. This function is often used as an alternative to erfc(z) in the formulation of P_e problems and is related to it by $Q(z) = \frac{1}{2}\text{erfc}(z/\sqrt{2})$ or erfc $(z) = 2Q(z\sqrt{2})$.

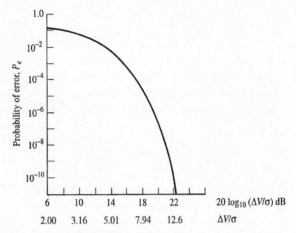

Figure 6.3 *Probability of error versus $\Delta V/\sigma$ (polar NRZ SNR = $20\log_{10}(\Delta V/\sigma) - 6dB$, while unipolar SNR = $20\log_{10}(\Delta V/\sigma) - 3\ dB$).*

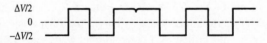

Figure 6.4 *Unipolar rectangular pulse signal.*

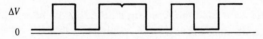

Figure 6.5 *Polar rectangular pulse signal.*

$$P_e = \frac{1}{2}\left[1 - \operatorname{erf}\frac{1}{2\sqrt{2}}\left(\frac{S}{N}\right)^{\!1/2}_{peak}\right] \tag{6.10(a)}$$

or

$$P_e = \frac{1}{2}\left[1 - \operatorname{erf}\frac{1}{2}\left(\frac{S}{N}\right)^{\!1/2}\right] \tag{6.10(b)}$$

For NRZ, polar, rectangular pulse signalling with the same voltage spacing as in the unipolar case, Figure 6.5, the peak and average signal powers are identical, i.e. $S_{peak} = S = (\Delta V/2)^2$ and we can therefore write:

$$\frac{\Delta V}{\sigma} = 2\left(\frac{S}{N}\right)^{\!1/2}_{peak} = 2\left(\frac{S}{N}\right)^{\!1/2} \tag{6.11}$$

Substituting into equation (6.8) we now have:

$$P_e = \frac{1}{2} \left[1 - \text{erf} \frac{1}{\sqrt{2}} \left(\frac{S}{N} \right)^{\frac{1}{2}} \right] \tag{6.12}$$

where a distinction between peak and average SNR no longer exists. Equations (6.10) and (6.12) are specific to a particular signalling format. In contrast equation (6.8) is general and, therefore, more fundamental. For statistically independent, equiprobable binary symbols, such as are being discussed here, each symbol carries one bit of information (see Chapter 9). The probability of symbol error, P_e, in this special case is therefore identical to the probability of bit error, P_b.

Whilst probabilities of symbol and bit error are the quantities usually *calculated*, it is the symbol error rate (SER) or bit error rate (BER) which is usually *measured*. These are simply the number of symbol or bit errors occurring per unit time (usually one second) measured over a convenient (and sometimes specified) period. The symbol error rate is clearly related to the probability of symbol error by:

$$\text{SER} = P_e R_s \tag{6.13}$$

where R_s is the symbol rate in symbol/s or baud. Bit error rate is related to the probability of bit error by:

$$\text{BER} = P_e R_s H = P_e R_b \tag{6.14}$$

where H, the entropy of the source (see Chapter 9), is the average number of information bits carried per symbol and R_b is the information bit rate.

EXAMPLE 6.1
Find the BER of a 100 kbaud, equiprobable, binary, polar, rectangular pulse signalling system assuming ideal centre point decisions, if the measured SNR at the detector input is 12.0 dB.

$$\frac{S}{N} = 10^{\frac{12}{10}} = 15.85$$

Using equation (6.12):

$$P_e = \frac{1}{2} \left[1 - \text{erf} \frac{1}{\sqrt{2}} (15.85)^{\frac{1}{2}} \right]$$

$$= \frac{1}{2} [1 - \text{erf} (2.815)] = 3.45 \times 10^{-5}$$

Using equation (6.14):

$$\text{BER} = P_e R_s H$$

$$= 3.45 \times 10^{-5} \times 100 \times 10^3 \times 1 = 3.45 \text{ bit errors/s}$$

6.2.2 Multilevel baseband signalling

Figure 6.6 shows a schematic diagram of a multilevel (or multisymbol) signal. If the number of, equally spaced, allowed levels is M then the symbol plus noise pdfs look like those in Figure 6.7 (drawn for $M = 4$). The probability of symbol error for the $M - 2$ inner symbols (i.e. any but the symbols represented by the lowest or highest voltage level) is now twice that in the binary case. This is because the symbol can be in error if the signal plus noise voltage is too high *or* too low, i.e.:

$$P_{eM}\big|_{inner\ symbols} = 2P_e \tag{6.15}$$

The symbol error for the two outer levels (i.e. the symbols represented by the lowest and highest voltage levels) is identical to that for the binary case, i.e.:

$$P_{eM}\big|_{outer\ symbols} = P_e \tag{6.16}$$

Once again assuming equiprobable symbols, the average probability of symbol error is:

$$P_{eM} = \frac{M-2}{M}\, 2P_e + \frac{2}{M}\, P_e$$

$$= \frac{2(M-1)}{M}\, P_e \tag{6.17}$$

Substituting for P_e from equation (6.8) we have:

$$P_{eM} = \frac{M-1}{M}\left[1 - \mathrm{erf}\left(\frac{\Delta V}{2\sigma\sqrt{2}}\right)\right] \tag{6.18}$$

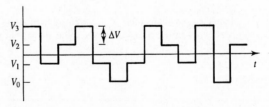

Figure 6.6 *Illustration of waveform for four level baseband signalling.*

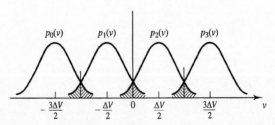

Figure 6.7 *Conditional pdfs for the four level baseband signalling system in Figure 6.6 assuming Gaussian noise ($p_i(v) = p(v|V_i)$).*

where ΔV is the difference between the equally spaced, adjacent, voltage levels, Figure 6.6. (Multilevel signalling is extended from baseband to bandpass systems in Chapter 11 where the different symbols are represented by differences in signal frequency, amplitude or phase.)

EXAMPLE 6.2

A four level, equiprobable, baseband signalling system uses NRZ rectangular pulses. The attenuation between transmitter and receiver is 15 dB and the noise power at the 50 Ω input of an ideal centre point decision detector is 10 μW. Find the average signal power which must be transmitted to maintain a symbol error probability of 10^{-4}.

The standard deviation of the noise voltage is equal to the RMS noise voltage (since the noise has zero mean).

$$\sigma = \sqrt{P\,R}$$

$$= \sqrt{1 \times 10^{-5} \times 50} = 2.236 \times 10^{-2} \quad (V)$$

Rearranging equation (6.18):

$$\Delta V = 2\sigma\sqrt{2}\,\mathrm{erf}^{-1}\!\left[1 - \frac{M\,P_{eM}}{M-1}\right]$$

$$= 2 \times 2.236 \times 10^{-2} \times \sqrt{2}\ \mathrm{erf}^{-1}\!\left[1 - \frac{4 \times 10^{-4}}{4-1}\right]$$

$$= 6.324 \times 10^{-2}\ \mathrm{erf}^{-1}(0.999867) = 0.171 \quad (V)$$

Thus symbol levels are ± 85.5 mV, ± 256 mV in Figures 6.6 and 6.7.

Two of the symbols are represented by received power levels of $0.0855^2/50 = 1.46 \times 10^{-4}$ W and two by received power levels of $0.256^2/50 = 1.31 \times 10^{-3}$ W. Since the symbols are equiprobable the average received power, S_R, must be:

$$S_R = \tfrac{1}{2}(0.146 + 1.31)10^{-3}\ \mathrm{W} = 0.728\ \mathrm{mW}$$

The transmitted power, P_T, is therefore:

$$P_T = S_R \times 10^{\frac{15}{10}}$$

$$= 0.728 \times 31.62 = 23\ \mathrm{mW}$$

6.3 Error accumulation over multiple hops

All signal transmission media (e.g. cables, waveguides, optical fibres) attenuate signals to a greater or lesser extent. This is even the case for the space through which radio waves travel (although for free space, the use of the word attenuate, to describe this effect, might be disputed – see Chapters 12 and 14). For long communication paths attenuation might be so severe that the sensitivity of normal receiving equipment would be inadequate to detect the signal. In such cases the signal is boosted in amplitude at regular intervals along the transmitter–receiver path. The equipment which boosts the signal is called a *repeater* and the path between adjacent repeaters is called a *hop*. (A repeater along with its associated, preceding, hop is called a *section*.) Thus long distance communication is usually achieved using multiple hops (see also later, section 19.4).

Repeaters used on multihop links can be divided into essentially two types. These are *amplifying* repeaters and *regenerative* repeaters. For analogue communications linear amplifiers are required. For digital communications either type of repeater could be used but normally regenerative repeaters are employed.

Figure 6.8 shows a schematic diagram of an *m*-hop link. (*m* hops imply a transmitter, a receiver and *m* −1 repeaters.) If a binary signal with voltage levels $\pm\Delta V/2$ is transmitted then the voltage received at the input of the first amplifying repeater is $\pm\alpha\Delta V/2 + n_1(t)$ where α is the linear voltage attenuation factor (or voltage gain factor < 1) and $n_1(t)$ is the random noise (with standard deviation, or RMS value, σ) added during the first hop. In a well designed system the voltage gain, G_V, of the repeater will be just adequate to compensate for the attenuation over the first hop, i.e. $G_V = 1/\alpha$. At the output of the first repeater the signal is restored to its original level (i.e. $\pm\Delta V/2$) but the noise signal is also amplified to a level $G_V n_1(t)$ and will now have an RMS value of $G_V\sigma$. (Chapter 12 contains a discussion on the origin of such noise.)

Assuming that each hop incurs the same attenuation the signal voltage at the input to the second repeater will again have fallen to $\pm\alpha\Delta V/2$ and the noise voltage from the first hop will have fallen to $n_1(t)$. A similar noise voltage, $n_2(t)$, arising from the second hop will also, however, be added and providing it is statistically independent from $n_1(t)$ will add to it on a power basis. The total noise power (in 1 Ω) will therefore be the sum of the noise variances, i.e. $2\sigma^2$ assuming equal noise is added on each hop. It can therefore be seen that with amplifying repeaters the noise power after *m* hops will be *m* times the noise power after one hop whilst the signal voltage at the receiver will be essentially the same as at the input to the first repeater. The probability of error for an *m*-hop link is therefore given by equation (6.8) with the RMS noise voltage, σ, replaced by $\sigma\sqrt{m}$, i.e.:

$$P_e\big|_{m\ hops} = \frac{1}{2}\left[1 - \mathrm{erf}\left(\frac{\Delta V}{2\sigma\sqrt{2m}}\right)\right] \tag{6.19}$$

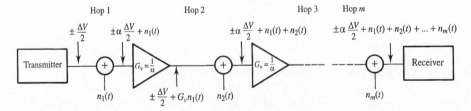

Figure 6.8 *Multihop link utilising linear amplifiers as signal boosters.*

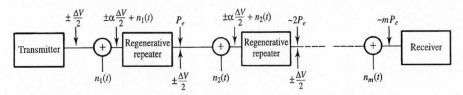

Figure 6.9 *Multihop link utilising regenerative repeaters as signal boosters.*

If the linear amplifiers are replaced with regenerative repeaters, Figure 6.9, the situation changes dramatically. The repeater now uses a decision process to establish whether a digital 0 or 1 is present at its input and a new, and noiseless, pulse is generated for transmission to the next repeater. Noise does not therefore accumulate from repeater to repeater as it does in the case of linear amplifiers. There is, however, an equivalent process in that symbols will be detected in error at each repeater with a probability P_e given by equation (6.8). Providing this probability is small (specifically $mP_e \ll 1$) the probability of any given symbol being detected in error (and therefore inverted) more than once over the m-hops of the link can be neglected. In this case the probability of error (rather than the noise power) accumulates linearly over the hops and after m hops we have:

$$P_e\big|_{m \ hops} = mP_e = \frac{m}{2}\left[1 - \mathrm{erf}\left(\frac{\Delta V}{2\sigma\sqrt{2}} \right) \right] \tag{6.20}$$

where P_e is the one-hop probability given in equation (6.8).

Figure 6.10(a) illustrates the increase in $P_e\big|_{m \ hops}$ as the number of hops increases for both amplifying and regenerative repeaters, clearly showing the benefit of digital regeneration. Figure 6.10(b) shows the typical saving in transmitter power (per repeater) realised using digital regeneration. (The power saving shown is the square of the factor by which ΔV in equation (6.19) must be larger than ΔV in equation (6.20) for a $P_e = 10^{-5}$ after m hops.)

EXAMPLE 6.3
If 15 link sections, each identical to that described in Example 6.1, are cascaded to form a 15-hop link find the probability of bit error when the repeaters are implemented as (i) linear amplifiers and (ii) digital regenerators.

(i) Using equation (6.11):

$$\frac{\Delta V}{\sigma} = 2\left(\frac{S}{N} \right)^{\frac{1}{2}} = 2(15.85)^{\frac{1}{2}} = 7.962$$

Now using equation (6.19):

$$P_e = \frac{1}{2}\left[1 - \mathrm{erf}\left(\frac{7.962}{2\sqrt{2}\sqrt{15}} \right) \right]$$

$$= \tfrac{1}{2}\ [1 - \mathrm{erf}\ (0.727)]\ =\ 0.152$$

(ii) Since $mP_e = 15 \times 3.45 \times 10^{-5} \ll 1.0$ we can use equation (6.20), i.e.:

$$P_e|_{15\ hops}\ \approx\ 15 \times P_e|_{1\ hop}\ =\ 5.175 \times 10^{-4}$$

The advantage of regenerative repeaters in this example is clear.

6.4 Line coding

The discussion of baseband transmission up to now has centred on the BER performance of unipolar, and polar, rectangular pulse representations of the binary symbols zero and one. Binary data can, however, be transmitted using many other pulse types. The choice of a particular pair of pulses to represent the symbols 1 and 0 is called line coding and

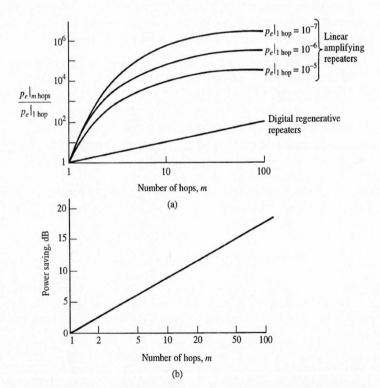

Figure 6.10 *(a) P_e degradation due to multiple hops for different repeater types (source: Stremler, 1990, reproduced with the permission of Addison Wesley) and (b) saving in transmission power using digital regenerative, instead of linear, repeaters for $P_e = 10^{-5}$ (source: Carlson, 1986, reproduced with the permission of McGraw-Hill).*

selection is usually made on the grounds of one or more of the following considerations:

1. Presence or absence of a DC level.
2. Power spectral density – particularly its value at 0 Hz.
3. Spectral occupancy (i.e. bandwidth).
4. BER performance (i.e. relative immunity from noise).
5. Transparency (i.e. the property that any arbitrary symbol, or bit, pattern can be transmitted and received).
6. Ease of clock signal recovery for symbol synchronisation.
7. Presence or absence of inherent error detection properties.

Line coding is usually thought of as the selection, or design, of pulse pairs which retain sharp transitions between voltage levels. Figure 6.11, in which T_o represents the symbol period, shows a variety of pulse types and spectra corresponding to several commonly used line codes. Figure 6.12 shows the transmitted waveforms of these (and a few other) line codes for an example sequence of binary data and Table 6.1 compares some of their important properties.

After line coding the pulses may be filtered or otherwise shaped to further improve their properties, for example their spectral efficiency and/or immunity to intersymbol interference (see Chapter 8). The distinction between line coding and pulse shaping is not always easy to make, and in some cases, it might be argued, artificial. Here, however, we make the distinction if for no other reason than to subdivide our discussion of baseband transmission into manageable parts. The line codes included in Figures 6.11

Table 6.1 *Comparison of line (baseband) code performance.*

	Timing extraction	Error detection	Relative transmitter power (single-point decisions) Average	Peak	First null bandwidth	AC coupled	Trans-parent
Unipolar (NRZ)	Difficult	No	2	4	f_o	No	No
Unipolar (RZ)	Simple	No	1	4	$2f_o$	No	No
Polar (NRZ)	Difficult	No	1	1	f_o	No	No
Polar (RZ)	Rectify	No	½	1	$2f_o$	No	No
Dipolar - OOK	Simple	No	2	4	$2f_o$	Yes	No
Dipolar - split ϕ	Difficult	No	1	1	$2f_o$	Yes	Yes
Bipolar (RZ)	Rectify	Yes	1	4	f_o	Yes	No
Bipolar (NRZ)	Difficult	Yes	2	4	$f_o/2$	Yes	No
HDB3	Rectify	Yes	1	4	f_o	Yes	Yes
CMI	Simple	Yes	See note	See note	$2f_o$	Yes	Yes

Note for CMI transmission at least two samples per symbol are required (e.g. taken 0.25 and 0.75 of the way through the symbol pulse). The difference between these samples results in a detected voltage which is twice the transmitted level for a digital zero, and a voltage of zero for a digital one. The RMS noise, σ, is increased by a factor of $\sqrt{2}$ due to the addition of the independent noise samples. On the (same) equal BER basis as the required transmitter powers given in Table 6.1 this results in a required CMI relative transmitter power (both average and peak) of 2. To compare this fairly with the other line codes, the relative powers required in Table 6.1 must all be reduced by a factor of 0.5 to reflect the factor of 2 improvement in SNR due to double sampling.

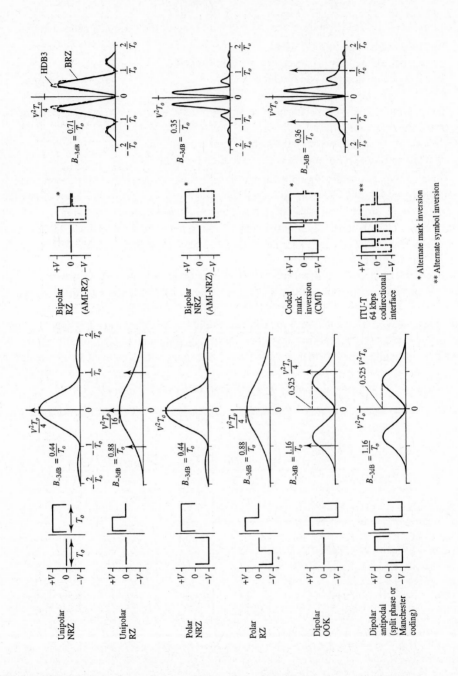

Figure 6.11 *Selection of commonly used line code symbols (0, 1) and associated spectra.*

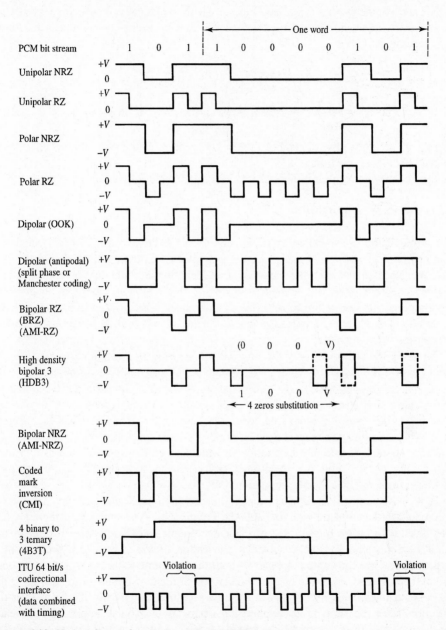

Figure 6.12 *Various line code waveforms for PCM bit sequence.*

and 6.12, and Table 6.1, do not form a comprehensive list, there being many other baseband signalling formats. One notable line code not included here, for example, is the Miller code which has a particularly narrow spectrum centred on $0.4/T_o$ Hz [Stremler, Sklar].

For the simple binary signalling formats (with statistically independent symbols), shown on the left hand side of Figure 6.11, the power spectral densities can be found using [Stremler]:

$$G(f) = p(1-p)\frac{1}{T_o}|F_1(f) - F_2(f)|^2$$

$$+ \frac{1}{T_o^2} \sum_{n=-\infty}^{\infty} |pF_1(nf_o) + (1-p)F_2(nf_o)|^2 \, \delta(f - nf_o) \qquad (6.21)$$

where p is the probability of symbol 1, $f_o = 1/T_o$ represents the symbol rate and $F(f)$ is the symbol voltage spectrum.

The codes which appear in Figures 6.11, 6.12 and Table 6.1 are now described.

6.4.1 Unipolar signalling

Unipolar signalling (also called on-off keying, OOK) refers to a line code in which one binary symbol (denoting a digital zero, for example) is represented by the absence of a pulse (i.e. a space) and the other binary symbol (denoting a digital one) is represented by the presence of a pulse (i.e. a mark). There are two common variations on unipolar signalling, namely non-return to zero (NRZ) and return to zero (RZ). In the former case the duration (τ) of the mark pulse is equal to the duration (T_o) of the symbol slot. In the latter case τ is less than T_o.

Typically RZ pulses fill only the first half of the time slot, returning to zero for the second half. (The mark duty cycle, τ/T_o, would be 50% in this case although other duty cycles can be, and are, used.) The power spectral densities of both NRZ and RZ signals have a $[(\sin x)/x]^2$ shape where $x = \pi\tau f$. RZ signals (assuming a 50% mark duty cycle) have the disadvantage of occupying twice the bandwidth of NRZ signals (see Figure 6.11). They have the advantage, however, of possessing a spectral line at the symbol rate, $f_o = 1/T_o$ Hz (and its odd integer multiples), which can be recovered for use as a symbol timing clock signal, Table 6.1. Non-linear processing must be used to recover a clock waveform from an NRZ signal (see section 6.7).

Both NRZ and RZ unipolar signals have a non-zero average (i.e. DC) level represented in their spectra by a line at 0 Hz, Figure 6.11. Transmission of these signals over links with either transformer or capacitor coupled (AC) repeaters results in the removal of this line and the consequent conversion of the signals to a polar format. Furthermore, since the continuous part of both the RZ and NRZ signal spectrum is non-zero at 0 Hz then AC coupling results in distortion of the transmitted pulse shapes. If the AC coupled lines behave as highpass RC filters (which is typically the case) then the distortion takes the form of an exponential decay of the signal amplitude after each transition. This effect, referred to as signal 'droop', is illustrated in Figure 6.13 for an NRZ signal. Although the long term DC component is zero, after AC coupling 'short term' DC levels accumulate with long strings of ones or zeros. The accumulated DC level is most apparent for the first few symbols after a string represented by a constant voltage. Neither variety of unipolar signal is therefore suitable for transmission over AC coupled lines.

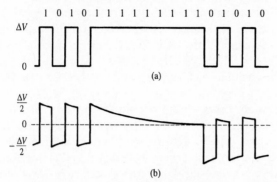

Figure 6.13 *Distortion due to AC coupling of unipolar NRZ signal: (a) input; (b) output.*

Since unipolar voltage levels of 0 or V volts are equivalent (in terms of BER) to polar levels of $\pm V/2$ volts (section 6.2.1) then unipolar signalling requires twice the average, and four times the peak, transmitter power when compared with polar signalling, Table 6.1.

6.4.2 Polar signalling

In polar signalling systems a binary one is represented by a pulse $g_1(t)$ and a binary zero by the opposite (or *antipodal*) pulse $g_0(t) = -g_1(t)$, Figure 6.11. Figure 6.12 compares polar and unipolar signals for a typical data stream. The NRZ and RZ forms of polar signals have identically shaped spectra to the NRZ and RZ forms of unipolar signals except that, due to the opposite polarity of the one and zero symbols, neither contain any spectral lines. Polar signals have the same bandwidth requirements as their equivalent unipolar signals and suffer the same distortion effects (in particular signal droop) if transmitted over AC coupled lines, Table 6.1.

As pointed out in section 6.4.1 polar signalling has a significant power (or alternatively BER) advantage over unipolar signalling. Fundamentally this is because the pulses in a unipolar scheme are only *orthogonal* whilst the pulses in a polar scheme are *antipodal*. Another way of explaining the difference in performance is to observe that the average or DC level transmitted with unipolar signals contains no information and is therefore wasted power.

Polar binary signalling also has the advantage that, providing the symbols are equiprobable, the decision threshold is 0 V. This means that no automatic gain control (AGC) is required in the receiver.

6.4.3 Dipolar signalling

Dipolar signalling is designed to produce a spectral null at 0 Hz. This makes it especially well suited to AC coupled transmission lines. The symbol interval, T_o, is split into positive and negative pulses each of width $T_o/2$ s, Figure 6.11. This makes the total area under either pulse type equal to zero which results in the desirable DC null in the signal's spectrum. Both OOK and antipodal forms of dipolar signalling are possible, the latter being called split phase or Manchester coding (Figure 6.11). A spectral line at the clock frequency ($1/T_o$ Hz) is present in the OOK form but absent in the antipodal form.

Manchester coding is widely used for the distribution of clock signals within VLSI circuits, for magnetic recording and for Ethernet LANs (see Chapter 18).

6.4.4 Bipolar alternate mark inversion signalling

Bipolar signalling (also called alternate mark inversion) uses three voltage levels ($+V$, 0, $-V$) to represent two binary symbols (0 and 1) and is therefore a pseudo-ternary line code. Zeros, as in unipolar signalling, are represented by the absence of a pulse (i.e. 0 V) and ones (or *marks*) are represented alternately by voltage levels of $+V$ and $-V$. Both RZ and NRZ forms of bipolar signalling are possible, Figure 6.11, although the RZ form is more common. Alternating the mark voltage level ensures that the bipolar spectrum has a null at DC and that signal droop on AC coupled lines is avoided. The alternating mark voltage also gives bipolar signalling a single error detection capability and reduces its bandwidth over that required for the equivalent unipolar or polar format (see Figure 6.11).

6.4.5 Pulse synchronisation and HDB*n* coding

Pulse synchronisation is usually required at a repeater or receiver to ensure that the samples, on the basis of which symbol decisions are made, are taken at the correct instants in time. In principle those line codes (such as unipolar RZ and dipolar OOK) which possess a spectral line at $1/T_o$ Hz have an inherent pulse synchronisation capability since all that is required to regenerate a clock signal is for this spectral line to be extracted using a filter or phase locked loop. Other line codes which do not possess a convenient spectral line can often be processed, for example by rectification, in order to generate one (see later Figures 6.26 to 6.29). This is the case for the bipolar RZ (BRZ) line code which is often used in practical PCM systems.

Although rectification of a BRZ signal results in a unipolar RZ signal and therefore a spectral line at $1/T_o$ Hz, in practice there is a problem if long strings of zeros are transmitted. In this case pulse synchronisation might be lost due, for instance, to loss of lock of the pulse timing phase locked loop, Figure 6.30. To prevent this, many BRZ systems use high density bipolar substitution (HDB*n*). Here, when the number of continuous zeros exceeds *n* then they are replaced by a special code. $n = 3$ (HDB3) is the code recommended (G.703) by ITU-T for PCM systems at multiplexed bit rates of 2, 8 and 34 Mbit/s (see section 19.2). In HDB3 a string of four zeros is replaced by either 000*V* or 100*V*. Here *V* is a binary 1 with sign chosen to *violate* the alternating mark rule so that it can be detected as the special sequence representing the all zero code, Figure 6.12. Furthermore, consecutive violation (*V*) pulses alternate in polarity to avoid introducing a DC component. (This is achieved by having the two possibilities 000*V* or 100*V*. The selection depends on the number of digital ones since the last code insertion.)

The HDB spectrum has minor variations compared with the BRZ signal from which it is derived (Figure 6.11). HDB3 is sometimes referred to as B4ZS denoting bipolar signalling with four-zeros substitution.

6.4.6 Coded mark inversion (CMI)

CMI is a polar NRZ code which uses both amplitude levels (each for half the symbol period) to represent a digital 0 and either amplitude level (for the full symbol period) to represent a digital 1. The level used alternates for successive digital ones. CMI is therefore a combination of dipolar signalling (used for digital zeros) and NRZ AMI (used for digital ones). CMI is the code recommended (G.703) by ITU-T for 140 Mbit/s multiplexed PCM (see Chapter 19). The ITU codirectional interface at 64 kbit/s uses a refinement of CMI, Figure 6.11, in which the polarity of consecutive symbols (irrespective of whether they are 1s or 0s) is alternated. Violations of the alternation rule are then used every eighth symbol to denote the last bit of each (8-bit) PCM code word, Figure 6.12.

Table 6.2 *4B3T coding example showing uncoded binary and coded ternary signals.*

Binary input signal	Ternary output signal Running digital sum at end of preceding word equal to	
	−2, −1 or 0	1, 2 or 3
0000	+ 0 −	+ 0 −
0001	− + 0	− + 0
0010	0 − +	0 − +
0011	+ − 0	+ − 0
0100	0 + −	0 + −
0101	− 0 +	− 0 +
0110	0 0 +	0 0 −
0111	0 + 0	0 − 0
1000	+ 0 0	− 0 0
1001	+ + −	− − +
1010	+ − +	− + −
1011	− + +	+ − −
1100	0 + +	0 − −
1101	+ 0 +	− 0 −
1110	+ + 0	− − 0
1111	+ + +	− − −

6.4.7 *nBmT* coding

nBmT is a line code in which *n* binary symbols are mapped into *m* ternary symbols. A coding table for *n* = 4 and *m* = 3 (i.e. 4B3T) is shown as Table 6.2. This code lengthens the transmitted symbols to reduce the signal bandwidth. In Table 6.2 the top six outputs are balanced and hence are fixed. The lower 10 have a polarity imbalance and need occasionally to be inverted to avoid the running digital sum causing DC wander. In the table the left hand column of coded signals or symbols (which sum to a zero or positive value) is used if the preceding running sum is negative. The right hand column, in which the symbols sum to zero or a negative value, is used if the preceding running sum is positive. With three ternary symbols we have 3^3 = 27 possible states and by not transmitting (0, 0, 0) we have 26, comprising the 6 + 10 + 10 unique states in Table 6.2. As 2^4 = 16 this conveniently matches to the 4-bit binary code requirements.

The 4B3T spectrum is further modified from that of HDB3 skewing the energy towards low frequencies and 0 Hz [Flood and Cochrane]. A further development of this concept is 2B1Q where two binary bits are converted into one four-level (quaternary) symbol. This is an example of *M*-ary (compare with *bin*ary, *tern*ary, etc.) signalling. *M*-ary coding is examined further in sections 11.4 and 11.5 in the context of carrier based coded signals.

6.5 Multiplex telephony

PCM is used in conjunction with TDM to realise multichannel digital telephony. The internationally agreed *European* ITU standard provides for the combining of 30 speech channels, together with two subsidiary channels for signal and system monitoring. Each speech channel signal is sampled at 8 kHz and non-linearly quantised (or companded) into 8-bit words (see Chapter 5). The binary symbol rate per speech channel is therefore 64 kbit/s, and for the composite 30 + 2 channel signal multiplex is 32 × 64 kbit/s = 2.048 Mbit/s. For convenience this is often referred to as a 2 Mbit/s signal. A key advantage of the 2 Mbit/s TDM multiplex is that it is readily transmitted over 2 km sections of twisted pair (copper) cables which originally carried only one analogue voice signal. Now, 2 Mbit/s signals are also transmitted over optical fibre circuits. In the USA and Japan the multiplex combines fewer speech channels into a 1.5 Mbit/s signal.

The 2 Mbit/s data transmission system is now examined to highlight the problems associated with transmitting and receiving such signals over a typical twisted pair, wire cable. The principal requirement is that signal fidelity should be sufficiently high to allow the receiver electronics to synchronise to the noisy and distorted incoming data signal and make correct data decisions.

6.6 Digital signal regeneration

The key facet of digital transmission systems is that after one, physically short, section in a link there is usually sufficiently high SNR to detect reliably the received binary data, and (possibly after error correction) regenerate an almost error free data stream, Figure 6.14, for retransmission over the next stage or section. Regeneration allows an increase in overall communications path length with negligible decrease in message quality provided each regenerator operates at an acceptable point on the error rate curve, Figure 6.3. If the path loss on each section is approximately 40 dB then a 3 V transmitted polar signal is received as 30 mV and, allowing for near end crosstalk noise (see later, Figure 6.23) rather than thermal noise, the SNR is typically 18 dB. At this SNR the P_e, found using equation (6.12) and the approximation:

$$\text{erf}(x) \simeq 1 - \left[\frac{e^{-x^2}}{\sqrt{\pi} x} \right], \quad \text{for } x \geq 4 \tag{6.22}$$

is 10^{-15}. (For NRZ polar signalling $(\Delta V/\sigma)$ dB $= (S/N) + 6$ dB, see equation (6.11).) For identical calculations on each section then the P_e on each section will be 10^{-15} and the total error will be the sum of the errors on each section. Thus, on the two section link shown in Figure 6.14, the total error probability is 2×10^{-15}. This was discussed previously in section 6.3 where for an m-section link $(P_e|_{m\ hops})$ was shown to be m times the error on a single hop or regenerative section. (In practice the signal attenuation, and noise introduced, is rarely precisely equal for all sections. Since the slope of the P_e curve in Figure 6.3, in the normal operating region, is very steep it is often the case that the P_e performance of a multi-hop link is dominated by the performance of its worst section.)

Provided the error rate on each section is acceptable then the cumulative or summed rate for the link is low, compared with the error rate when there is no regeneration and the single section loss is very high. The input SNR must typically be 18 dB in a section design, to ensure that there is an extremely low BER on each individual section. Regeneration thus permits long distance transmission with high message quality provided the link is properly sectioned with the appropriate loss on each section. The principal

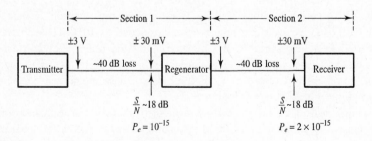

(Near end noise ~3.8 mV_{RMS})

Figure 6.14 *Principle of regenerative repeater in long communications system.*

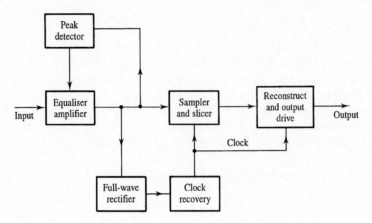

Figure 6.15 *Simplified block diagram of pulse code modulation (PCM) regenerator.*

component of such a digital line system is the digital regenerative repeater or regenerator, Figure 6.15.

6.6.1 PCM line codes

Line coding is used, primarily, to match the spectral characteristics of the digital signal to its channel and to provide guaranteed recovery of timing signals. ITU-T recommends HDB3 for (multiplexed) PCM bit rates of 2, 8 and 34 Mbit/s and CMI for 140 Mbit/s. For the unmultiplexed bit rate (64 kbit/s) ITU-T G.703 has three different line code recommendations, one for each of three PCM equipment interface standards. The code used depends on the type of equipment connected to either side of the PCM interface. Three types of equipment interface are defined, namely codirectional, centralised clock and contradirectional. The interface type depends on the origin of 64 kHz and 8 kHz timing signals and their transmission direction with respect to that of the information or data (Figure 6.16).

For a codirectional interface (Figure 6.16(a)) there is one transmission line for each transmission direction. In order to incorporate the timing signals into the data the 64 kbit/s pulse or bit period is subdivided into four 'quarter-bit' intervals. Binary ones are unipolar NRZ coded as 1100 and binary zeros are coded as 1010. The quarter-bits are then converted to a three level signal by alternating the polarity of consecutive four quarter-bit blocks. (Each quarter-bit has a nominal duration of 3.9 μs.) The alternation in polarity of the blocks is violated every eighth block. These violations represent the last PCM bit of each 8-bit PCM word (Figure 6.12).

For a centralised clock interface (Figure 6.16(b)) there is one transmission line for each transmission direction to carry PCM data and one transmission line from the central clock to the equipment on each side of the interface to carry 64 kHz and 8 kHz timing signals. The data line uses a bipolar code with a 100% duty cycle (i.e. AMI NRZ). The timing signal line code is bipolar with a duty cycle between 50% and 80% (i.e. AMI RZ)

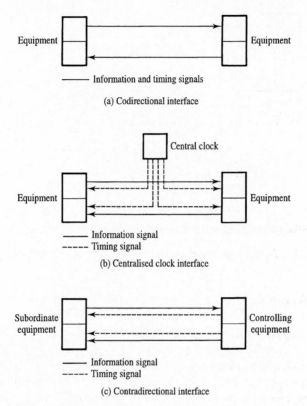

Figure 6.16 *PCM 64 kbit/s interfaces.*

with polarity violations aligned with the last (eighth) bit of each PCM code word.

For a contradirectional interface (Figure 6.16(c)) there are two transmission lines for each transmission direction, one for PCM data and one for the timing signal. The data line code (AMI NRZ) and timing signal (all ones) line code (AMI RZ) are identical to those for a centralised clock interface except that the timing signal must have a 50% duty cycle.

6.6.2 Equalisation

A significant problem in PCM cable, and many other, communication systems is that considerable amplitude and phase distortion may be introduced by the transmission medium. For a 2 Mbit/s PCM system the RZ bipolar pulse has a width of ¼ μs and hence a bandwidth of approximately 2 MHz. When this is compared with typical metallic cable characteristics, Figure 6.17, it can be seen that the received pulse will be heavily distorted and attenuated. A potentially serious consequence of this distortion is that the pulse will be stretched in time as shown in Figure 6.18.

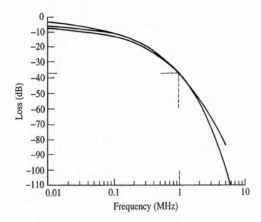

Figure 6.17 *Typical frequency responses for a 2 km length of cable.*

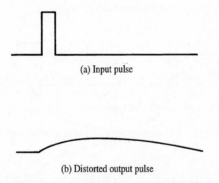

Figure 6.18 *(a) Input and (b) output 2 Mbit/s pulse for a 2 km length of cable.*

Such distortion also results when wideband signals are transmitted over the local loop telephone connection from the exchange to the subscriber. Here the cable bandwidth is only several kHz. In the high-speed digital subscriber loop (HDSL) [Baker, Young *et al.*], however, we can transmit at Mbit/s provided there is still adequate SNR at the receiver. HDSL relies on efficient data coding, such as 2B1Q, i.e. four level signalling (which is an extension of the coding ideas illustrated by Table 6.2) to reduce transmitted signal bandwidth, combined with techniques to equalise (i.e. compensate for) distortion introduced by the restricted bandwidth of the transmission network. HDSL thus achieves high rate transmission on existing networks without requiring replacement of the copper cables by alternative (optical) transmission media.

When we move from considering individual pulses to a data stream, Figure 6.19, the long time domain tails from the individual received symbols cause intersymbol interference (ISI). This is overcome by applying an equalising filter [Mulgrew and Grant] in the receiver which has the inverse frequency response, Figure 6.20, to the raw line or channel characteristic of Figure 6.17. Cascading the effect of the line with the equaliser

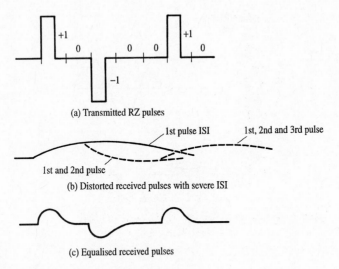

(a) Transmitted RZ pulses

(b) Distorted received pulses with severe ISI

(c) Equalised received pulses

Figure 6.19 *Intersymbol interference in a pulse train.*

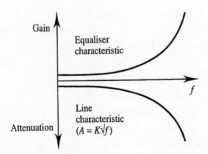

Figure 6.20 *Line equaliser frequency responses.*

provides a flat overall response which reduces the distortion (time stretching) of the pulses as shown in Figure 6.19(c). A detailed discussion of ISI and one aspect of equalisation (Nyquist filtering), which is used to minimise it, is given in Chapter 8.

6.6.3 Eye diagrams

A useful way of quickly assessing the adequacy of a digital communications system is to display its *eye diagram* on an oscilloscope. This is achieved by triggering the oscilloscope with the recovered symbol timing clock signal and displaying the symbol stream using a sweep time sufficient to show 2 or 3 symbol periods. Figure 6.21(a) shows a noiseless, but distorted, binary bit stream and Figure 6.21(b) shows the resulting two level eye pattern when the pulses are overlaid on top of one another. Distortion of the digital signal causes ISI which results in variations of the pulse amplitudes at the sampling instants. The eye opening indicates how much tolerance the system possesses

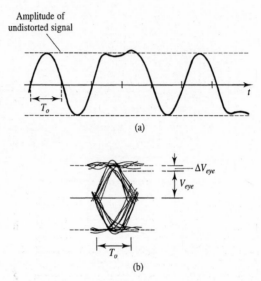

Figure 6.21 *(a) Noiseless but distorted signal; and (b) corresponding eye diagram.*

to noise before incorrect decisions will be made.

It can be seen that there is no ISI free instant in this case which could be chosen for ideal sampling. In addition to partial closing of the eye, distortion also results in a narrowing of the eye in the horizontal (time) dimension. This is because the symbol timing signal is derived from the zero crossings of the distorted bit stream resulting in symbol timing *jitter*. This effectively means that some symbols will be sampled at non-optimum instants. Figure 6.22 shows examples of eye patterns for NRZ binary and multilevel systems. The optimum sampling time clearly occurs at the position of maximum eye opening. Both distortion of the digital signal and additive noise contribute to closing of the eye. The slope of the eye pattern between the maximum eye opening and the eye corner is a measure of sensitivity to timing error.

The trade-off possible between ISI and noise can be estimated from the noise free eye diagram. For example, in Figure 6.21(b) the degradation in eye opening, $\Delta V_{eye}/V_{eye}$, is about 20%. (Notice that ΔV_{eye} is about half the thickness of the eye pattern at the point of maximum opening because distortion results in traces above as well as below the undistorted signal level.) If the signal voltage is increased by a factor $1/(1 - 0.2) = 1.25$ then the eye opening will be restored to the value it would have in the absence of ISI. An increase in SNR of $20 \log_{10} 1.25 = 1.9$ dB would therefore, at least approximately, compensate the BER degradation due to ISI.

In a practical regenerator, the major contribution to intersymbol interference generally arises from the residual amplitude of the pulse at the preceding and succeeding sampling instants and intersymbol interference from other sampling instants can usually be ignored.

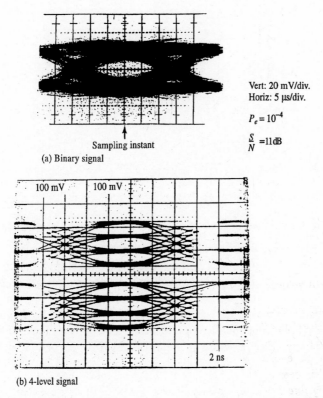

Vert: 20 mV/div.
Horiz: 5 μs/div.

$P_e = 10^{-4}$

$\frac{S}{N} = 11\,dB$

Sampling instant

(a) Binary signal

(b) 4-level signal

Figure 6.22 *(a) Eye diagram for NRZ digital binary signal at $P_e = 10^{-4}$ and SNR of 11 dB (source: Feher, 1983, reproduced with the permission of Prentice Hall); and (b) 4-level signal at 200 Mbit/s (source: Feher, 1981, reproduced with his permission).*

Before regenerator design is examined in more detail we need to discuss one further impairment, i.e. that caused by signal crosstalk.

6.6.4 Crosstalk

Two types of crosstalk arise when bidirectional signals are transmitted across a bundled cable comprising many individual twisted pairs. With approximately 40 dB of insertion loss across the cable a 3 V transmitted pulse is received as a 30 mV signal at the far end. Near end crosstalk (NEXT) results from the capacitive coupling of the 3 V transmitted pulse on an outgoing pair interfering with the 30 mV received pulse on an incoming pair, Figure 6.23.

Far end crosstalk (FEXT), on the other hand, occurs due to coupling of a transmitted pulse on one outgoing pair with a pulse on another outgoing pair. NEXT thus refers to crosstalk between signals travelling in opposite directions and *effectively* takes place near the cable ends whilst FEXT refers to crosstalk between signals travelling in the same

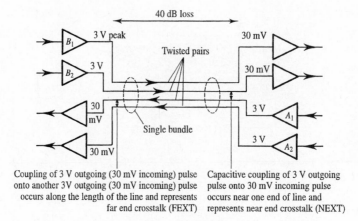

Coupling of 3 V outgoing (30 mV incoming) pulse onto another 3V outgoing (30 mV incoming) pulse occurs along the length of the line and represents far end crosstalk (FEXT)

Capacitive coupling of 3 V outgoing pulse onto 30 mV incoming pulse occurs near one end of line and represents near end crosstalk (NEXT)

Figure 6.23 *Near and far end crosstalk (NEXT and FEXT) due to capacitive cable coupling.*

direction and takes place over the cable's entire length.

Assuming the transmitted pulse spectrum has a $|(\sin x)/x|$ shape and the coupling can be modelled as a high pass RC filter, which introduces a coupling gain of 6 dB/octave, then NEXT has a distorted spectrum as shown in Figure 6.24, in which the spectral magnitude, relative to the normal spectrum, increases with increasing frequency.

Crosstalk and ISI reduction therefore demands a composite equaliser response which differs from Figure 6.20 as shown in Figure 6.25. This has a low frequency portion

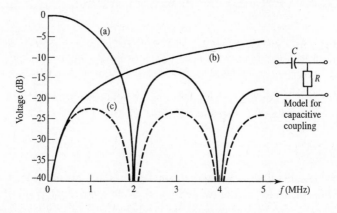

(a) Transmitted spectrum, $V_{dB} = 20 \log_{10} \left| \dfrac{\sin (2\pi T_o f)}{2\pi T_o f} \right|$, $T_o = 0.25$ μs

(b) Coupling loss (reducing at 6 dB/octave for high frequency), $L = 20 \log_{10} \left(\dfrac{2\pi f RC}{\sqrt{1 + (2\pi f RC)^2}} \right)$, $RC = 0.02$ μs

(c) Resultant NEXT spectrum, $V_{dB} + L$

Figure 6.24 *Spectrum of NEXT crosstalk signal.*

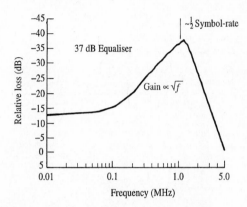

Figure 6.25 *Combined equaliser response for ISI and crosstalk.*

where loss decreases with the square root of frequency, as in Figure 6.20. The high frequency portion of the spectrum, however, is shaped to reduce crosstalk effects. This is a compromise, as it does increase ISI and sensitivity to jitter. The equaliser normally has lowest loss at half the symbol rate (i.e. 1 MHz on a 2 Mbit/s link).

6.7 Symbol timing recovery

After effective equalisation has been implemented it is necessary to derive a receiver clock, from the received signal, in order to time, accurately, the sampling and data recovery process. Symbol timing recovery (STR) is required since most digital systems are self-timed from the received signal to avoid the need for a separate timing channel. It can be achieved by first filtering and then rectifying, or squaring, a bipolar RZ line coded signal, Figure 6.26. Rectifying or squaring removes the alternating pulse format which approximately doubles the received signal's bandwidth. It removes the notch at the symbol rate, $f_o = 1/T_o$, and introduces an f_o clock signal component which can be extracted with a resonant circuit.

As the resonant circuit oscillates at its natural frequency it fills in the gaps left by the zero data bits for which no symbol is transmitted, Figure 6.27. The frequency and phase of this recovered clock must be immune from transmission distortions in, and noise on, the received signal. A problem arises in the plesiochronous multiplex in that extra (justification) bits are added or removed at the individual multiplexers to obtain the correct bit rates on the transmission links (see Chapter 19). This means that symbol timing can become irregular introducing timing jitter. Timing jitter manifests itself as FM modulation on the recovered clock signal, Figure 6.28.

The passive, tuned *LC* circuit, Figure 6.29, has a low Q factor (30 – 100) which does not give good noise suppression. Its wide bandwidth, however, means it is relatively tolerant to small changes in precise timing of the received signals (i.e. jitter). A high Q (1000 to 10000) phase locked loop gives good noise reduction but is no longer so

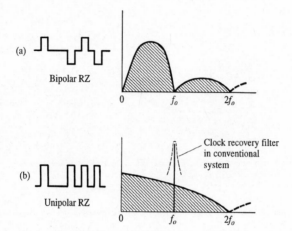

(a) Bipolar RZ

(b) Unipolar RZ

Clock recovery filter in conventional system

Figure 6.26 *Clock recovery by full wave rectification.*

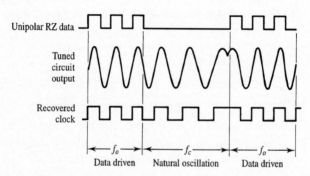

Unipolar RZ data

Tuned circuit output

Recovered clock

f_o f_c f_o

Data driven   Natural oscillation   Data driven

Figure 6.27 *Clock recovery oscillator operation when a string of no symbols (spaces) is received.*

tolerant to jitter effects. In Figure 6.28 we see suppression of the high frequency jitter terms. In both systems an output comparator clips the sinusoid to obtain the recovered clock waveform.

In some systems timing extraction is obtained by detecting the zero crossings of the waveform. The 'filter and square' operation is also used for STR in QPSK receivers, see later Figure 11.24. (Another QPSK STR technique is to delay the received symbol stream by half a symbol period and then form a product with the undelayed signal. This gives a periodic component at the symbol rate which can be extracted using a PLL.)

We may define peak-to-peak jitter (in seconds, or bits) as the maximum peak-to-peak displacement of a bit or symbol with respect to its position in a hypothetical unjittered reference stream. One effect of the jitter is that the regenerated clock edge varies with respect to the correct timing point in the eye diagram, thus increasing the P_e. Although noise can be removed by clipping, the remaining phase modulation still contributes to the jitter.

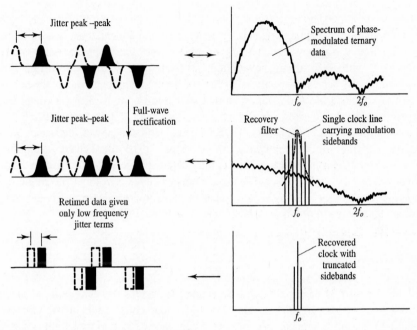

Figure 6.28 *Effect of jitter on clock recovery.*

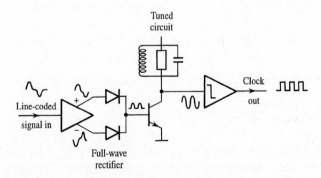

Figure 6.29 *Passive oscillator for clock recovery.*

A significant problem with jitter is that it accumulates over a multihop system. Timing jitter in the incoming data introduces eye jitter, while the clock recovery process in each repeater introduces more clock edge jitter. To minimise this problem data scramblers can be used to stop the accumulation of data dependent jitter. Limits on jitter are specified in the ITU-T G.823/4 recommendations.

6.8 Repeater design

In the receiver a matched filter (see section 8.3) for data bit detection is usually by far the most computationally demanding task. It is typically implemented using a finite impulse response filter [Mulgrew and Grant] to obtain the linear phase requirement. In comparison with this, timing extraction, phase/frequency error determination and correction usually account for only 10 to 20% of the computational load in the receiver.

A more detailed block diagram than that shown in Figure 6.15, for a complete PCM regenerative repeater, is shown in Figure 6.30. Power is fed across the entire PCM link, and each repeater AC couples the HDB3 signal, with the power extracted, from the primary winding of the coupling transformer. (The duplicated sampling and reconstruction stages in Figure 6.30 accommodate separately the positive and negative bipolar pulses.) For an *m* repeater link total supply voltage is *m* times the single repeater requirement, e.g. $m \times 5$ V. Complete repeaters are now available as single integrated circuits which only require transformer connection to the PCM system.

6.9 Summary

The simplest form of baseband digital detection uses *centre point sampling* to reduce the received symbol plus noise to a single voltage, and comparators to test this voltage against appropriate references. A formula giving the single hop probability of error for symbols represented by uniformly spaced voltages in the presence of Gaussian noise has been derived. The probability of error after *m identical* hops is *m* times greater than that after a single hop providing that *regenerative* repeaters are used between hops. The main polar and bipolar signalling techniques have been outlined and those which are preferred for various transmission systems, used in the PCM multiplex hierarchy, discussed. This

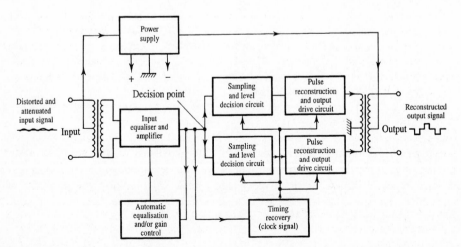

Figure 6.30 *Complete PCM regenerator.*

has involved consideration of the spectral properties of these line codes to ensure that they adequately match the transmission channel frequency response.

This chapter has also addressed some specific problems of data transmission in a metallic wire, twisted pair, cable. Distortion due to the cable frequency response and its compensation via equalisation have been examined. The effects of crosstalk interference and imperfect timing recovery have also been discussed to show their importance in practical data transmission systems. Although we have principally addressed the 2 Mbit/s metallic wire system, these signal degradation mechanisms are present in practically all digital communication systems.

6.10 Problems

6.1. A, baseband, NRZ, rectangular pulse signal is used to transmit data over a single section link. The binary voltage levels adopted for digital 0s and 1s are -2.5 V and $+3.5$ V respectively. The receiver uses a centre point decision process. Find the probability of symbol error in the presence of Gaussian noise with an RMS amplitude of 705 mV. By how much could the transmitter power be reduced if the DC component were reduced to its optimum value? [0.24 dB]

6.2. Find the mean and peak transmitter power required for a single section, 2 km, link which employs, unipolar RZ (50% mark space ratio), rectangular pulse signalling and centre point detection if the probability of bit error is not to exceed 10^{-6}. The specific attenuation of the (perfectly equalised) link transmission line is 1.8 dB/km and the (Gaussian) noise power at the receiver decision circuit input is 1.8 mW. (Assume that the (real) impedance level at transmitter output and decision circuit input are both equal to the characteristic impedance of the transmission line.) [93 mW, 0.37 W]

6.3. A 1.0 Mbaud, baseband, digital signal has 8 allowed voltage levels which are equally spaced between $+3.5$ V and -3.5 V. If this signal is transmitted over a 26 section, regenerative repeater link, and assuming that all the sections of the link are identical, find the RMS noise voltage which could be tolerated at the end of each hop whilst maintaining an overall link symbol error rate of 1 error/s. [91.4 mV]

6.4. What are the two factors which control the power spectral density (PSD) of a line coded communications signal? Using the fact that the Fourier transform of a rectangular pulse has a characteristic $(\sin x)/x$ shape, approximate the PSD and consequent channel bandwidth requirements for a return to zero, on-off keyed, line coded signal incorporating a positive pulse whose width equals one-third the symbol interval. Compare this with a non-return to zero signal, sketching the line coded waveforms and PSDs for both cases.

How do these figures compare with the minimum theoretical transmission bandwidth? What are the practical disadvantages of the on–off keyed waveform?

6.5. A binary transmission scheme with equiprobable symbols uses the absence of a voltage pulse to represent a digital 0 and the presence of the pulse $p(t)$ to represent a digital 1 where $p(t)$ is given by:

$$p(t) = \Pi\{t/(T_o/3)\} - 0.5\,\Pi\{[t - (T_o/3)]/(T_o/3)\} - 0.5\,\Pi\{[t + (T_o/3)]/(T_o/3)\}$$

Sketch the pulse, $p(t)$, and comment on the following aspects of this line code: (a) its DC level; (b) its suitability for transmission using AC coupled repeaters; (c) its (first null) bandwidth; (d) its P_e performance, using ideal centre point decision, compared with unipolar NRZ rectangular pulse signalling; (e) its self clocking properties (referring to equation (6.21) if you wish); and (f) its error detection properties.

6.6. Justify the entries in the last (transparency) column of Table 6.1.

6.7. The power spectral density of a bipolar NRZ signal (Figure 6.11) is given by:

$$G(f) = V^2 T_o \; \text{sinc}^2(T_o f) \; \sin^2(2\pi T_o f)$$

Use this to help verify the power spectral density of the coded mark inversion (CMI) signal given in Figure 6.11. (A plot of your result is probably necessary to verify the PSD.) [Hint: since the correlation of the mark and space parts of the CMI signal is zero for all values of time shift, then the power spectra of these parts can be found separately and summed to get the power spectral density of the CMI signal.]

6.8. Draw a perfectly equalised prototype positive pulse for an AMI signal. Use it to construct the eye diagram for the signal $+1, -1, +1, -1, + \cdots$. Add and label the remaining trajectories for a random signal.

6.9. Explain why a non-linear process is required to recover clock timing from an HDB3 signal.

6.10. For Gaussian variables use the tables of erf(x) supplied in Appendix A to find: (a) The probability that $x \leq 5.5$ if x is a Gaussian random variable with mean $\mu = 3$ and standard deviation $\sigma = 2$; (b) the probability that $x > 5.5$.

(c) Assuming the height of clouds above the ground at some location is a Gaussian random variable x with $\mu = 1830$ metres and $\sigma = 460$ metres, find the probability that clouds will be higher than 2750 metres.

(d) Find the probability that a Gaussian random variable with standard deviation of σ will exceed: (i) σ, (ii) 2σ, and (iii) 3σ. [(a) 0.89, 0.11]; [(b) 2.28×10^{-2}]; [(c) 0.16; 2.3×10^{-2}; 1.3×10^{-3}]

CHAPTER 7

Decision theory

7.1 Introduction

In section 6.2 the detection of baseband binary symbols was described and analysed for the special case of Gaussian noise and equiprobable symbols. The decision reference voltage, against which detected symbols were tested, was set intuitively (and correctly) mid-way between those voltages which would have been detected, if the binary symbols were received without noise. In this chapter a more systematic analysis of symbol decision theory is given, applying the descriptions of random signals and noise developed in Chapter 3. This allows the optimum levels for receiver decision voltages to be found in the more general case of arbitrary noise distributions and unequal symbol probabilities.

For a binary data stream there are two principal types of decision:

(a) Soft (multi-level) decisions.
(b) Hard (2 level) decisions.

Soft decision receivers, Figure 7.1(a), quantise the decision instant signal plus noise voltage using several allowed levels, each represented by a decision word of a few binary bits. Figure 7.1(c) illustrates an 8-level (3-bit) soft decision process which is typical. Each soft decision contains information not only about the most likely transmitted symbol (000 to 011 indicating a likely 0 and 100 to 111 indicating a likely 1) but also information about the confidence or likelihood which can be placed on this decision. The soft decisions are converted to final or hard decisions using an algorithm which inspects a sequence of several PCM words and makes decisions accounting for the confidence levels they represent, in conjunction with the error control decoding rules. The algorithm tends to be heuristic and requires tailoring to the particular application. Although such algorithms can now be implemented using high speed VLSI decoders, these techniques will not be discussed further here.

Hard decisions (Figure 7.1(b)) are more common than soft decisions. The two major decision criteria used in this case are Bayes and Neyman-Pearson. The Bayes decision criterion is used extensively in binary communications while the Neyman-Pearson criterion is used more in radar applications. The principal difference between the two is

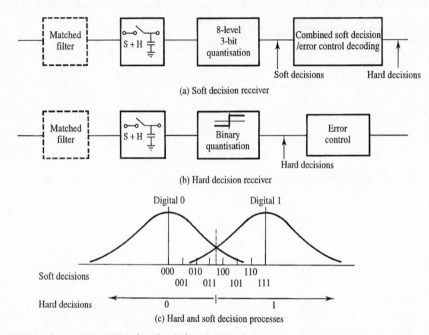

Figure 7.1 *Comparison of hard and soft decision receivers.*

that the Bayes decision rule assumes known a priori source statistics for the occurrence of digital ones and zeros while the Neyman-Pearson criterion makes no such assumption. This is appropriate to radar applications because the a priori probability of the appearance of a target is not normally known.

7.2 A priori, conditional and a posteriori probabilities

There are four probabilities and two probability density functions, see Chapter 3, associated with symbol transmission and reception. These are:

1. $P(0)$ – a priori probability of transmitting the digit 0.
2. $P(1)$ – a priori probability of transmitting the digit 1.
3. $p(v|0)$ – conditional probability density function of detecting voltage v given that a digital 0 was transmitted.
4. $p(v|1)$ – conditional probability density function of detecting voltage v given that a digital 1 was transmitted.
5. $P(0|v)$ – a posteriori probability that a digital 0 was transmitted given that voltage v was detected.
6. $P(1|v)$ – a posteriori probability that a digital 1 was transmitted given that voltage v was detected.

The terms a priori and a posteriori imply reasoning from cause to effect and reasoning from effect to cause respectively. The cause of a communication event, in this context, is the symbol transmitted and the effect is the voltage detected. The a priori probabilities (in communications applications) are usually known in advance. The conditional probabilities are dependent on the data (1 or 0) transmitted. The a posteriori probabilities can only be established after many events (i.e. symbol transmissions and receptions) have been completed. (Note that the lower case notation is used for conditional probability density functions and upper case for a posteriori probabilities. Also, note that the position of the transmitted and detected quantities is interchanged between these cases.)

7.3 Symbol transition matrix

When communications systems operate in the presence of noise (which is always the case) there is the possibility of transmitted symbols being sufficiently corrupted to be interpreted, at the receiver, in error. If the characteristics of the noise are precisely enough known, or many observations of symbol transmissions and receptions are made, then the probability of each transmitted symbol being interpreted at the receiver as any of the other symbols can be found. These *transition* probabilities are usually denoted using the conditional probability notation, $P(i|j)$, where i represents the symbol received and j represents the symbol transmitted. $P(i|j)$ is therefore the probability that a symbol will be interpreted as i given that j was transmitted. The symbol transition probabilities can be arranged as the elements of a matrix to describe the end to end properties of a communications channel. Examples of such matrices are given below.

7.3.1 Binary symmetric channel

In a binary channel four types of communication events can occur. These events are:
  0 transmitted and 0 received
  0 transmitted and 1 received
  1 transmitted and 1 received
  1 transmitted and 0 received
Denoting transmitted symbols with subscript *TX* and received symbols with subscript *RX* (for extra clarity) the binary channel can be represented schematically as shown in Figure 7.2(a). The corresponding symbol transition matrix is:

$$\begin{bmatrix} P(0_{RX}|0_{TX}) & P(0_{RX}|1_{TX}) \\ P(1_{RX}|0_{TX}) & P(1_{RX}|1_{TX}) \end{bmatrix}$$

If the probability, p, of a transmitted 0 being received in error as a 1, is equal to the probability of a transmitted 1 being received as a 0, then the binary channel is said to be symmetric and the transition matrix can be written as:

$$\begin{bmatrix} 1-p & p \\ p & 1-p \end{bmatrix}$$

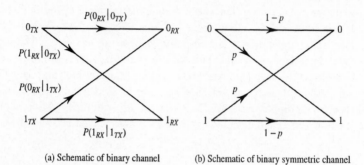

(a) Schematic of binary channel   (b) Schematic of binary symmetric channel

Figure 7.2 *Schematic representations of binary channels.*

Figure 7.2(b) shows the equivalent schematic diagram. The conditional probabilities in the matrix must sum vertically to unity, i.e.:

$$P(0_{RX}|0_{TX}) + P(1_{RX}|0_{TX}) = 1 \tag{7.1}$$

For the binary symmetric case:

$$P(0_{RX}|0_{TX}) = P(1_{RX}|1_{TX}) = 1 - p \tag{7.2}$$

The unconditional probability of receiving a 0 is therefore:

$$P_{0_{RX}} = P(0_{TX})P(0_{RX}|0_{TX}) + P(1_{TX})P(0_{RX}|1_{TX}) \tag{7.3}$$

which can be rewritten for the binary symmetric channel as:

$$P_{0_{RX}} = P(0)(1 - p) + P(1)p \tag{7.4}$$

Similarly:

$$P_{1_{RX}} = P(1)(1 - p) + P(0)p \tag{7.5}$$

Equations (7.4) and (7.5) relate the observed probabilities of received symbols to the a priori probabilities of transmitted symbols and the error probabilities of the channel.

EXAMPLE 7.1 – Multisymbol transmission

For an M symbol communication alphabet the 2×2 transition matrix of the binary channel must be extended to an $M \times M$ matrix. Consider a source with the $M = 6$ symbols, $A\ B\ C\ D\ E\ F$, and a 6-ary receiver which can distinguish these symbols. The transition matrix, **T**, for this system contains 6×6 transition probabilities which are:

$$
\mathbf{T} =
\begin{bmatrix}
\dfrac{1}{2} & 0 & \dfrac{1}{24} & 0 & 0 & \dfrac{1}{8} \\[2mm]
\dfrac{1}{4} & \dfrac{1}{2} & \dfrac{1}{6} & \dfrac{1}{3} & \dfrac{1}{4} & \dfrac{1}{12} \\[2mm]
\dfrac{1}{8} & \dfrac{1}{8} & \dfrac{1}{4} & \dfrac{1}{6} & 0 & \dfrac{1}{12} \\[2mm]
0 & \dfrac{1}{3} & \dfrac{1}{4} & \dfrac{1}{3} & \dfrac{1}{4} & \dfrac{1}{4} \\[2mm]
\dfrac{1}{8} & 0 & \dfrac{1}{6} & \dfrac{1}{12} & \dfrac{1}{2} & \dfrac{1}{8} \\[2mm]
0 & \dfrac{1}{24} & \dfrac{1}{8} & \dfrac{1}{12} & 0 & \dfrac{1}{3}
\end{bmatrix}
$$

(For a transmitted symbol B the probabilities of receiving symbols $A, \cdots, F$ are given by the $M = 6$ entries in the matrix's second column.)

If the a priori transmission probabilities of the six symbols are:

$$P(A) = 0.1$$
$$P(B) = 0.15$$
$$P(C) = 0.4$$
$$P(D) = 0.05$$
$$P(E) = 0.2$$
$$P(F) = 0.1$$

then find the probability of error when (i) only the symbol D is transmitted and (ii) when a random string of symbols is transmitted. Also find, for this latter case, the probability of receiving the symbol C.

The probability of error, P_e, when the symbol D is transmitted is given by the sum of all probabilities in the fourth (i.e. D transmitted) column, excluding the fourth (i.e. D received) element:

$$P_e = \sum_{m=A}^{F} P(m_{RX}|D_{TX}) \quad \text{(excluding the } m = D \text{ term)}$$

$$= 0 + 1/3 + 1/6 + 1/12 + 1/12 = 2/3$$

This is identical to $1 - P(D_{RX}|D_{TX})$, i.e.:

$$P_e = 1 - 1/3 = 2/3$$

(Note that the leading diagonal elements, $P(m|m)$, in the transition matrix are the probabilities of correct symbol reception and $1 - P(m|m)$ are therefore the probabilities of symbol error.)

The probability of error when a (long) random string of symbols $(A, \cdots, F)$ is transmitted is a weighted sum of the probabilities of error for each transmitted symbol where the weighting factors are the a priori probabilities of transmission, i.e.:

$$P_e = \sum_{m=A}^{F} P(m) [1 - P(m_{RX}|m_{TX})]$$

$$= 0.1 \times \frac{1}{2} + 0.15 \times \frac{1}{2} + 0.4 \times \frac{3}{4} + 0.05 \times \frac{2}{3} + 0.2 \times \frac{1}{2} + 0.1 \times \frac{2}{3}$$

$$= 0.625$$

In addition to probabilities of error, the various probabilities of symbol reception can also be found from the transition matrix. The probability of receiving symbol C when symbol D is transmitted is given by inspection of the appropriate matrix element, i.e.:

$$P(C_{RX}|D_{TX}) = 1/6$$

The (unconditional) probability of receiving symbol C when a (long) random string of symbols $(A, \cdots, F)$ is transmitted is the a priori weighted sum of all elements in the third (i.e. C received) row of the transition matrix:

$$P(C_{RX}) = \sum_{m=A}^{F} P(C_{RX}|m_{TX}) \, P(m)$$

$$= 0.1 \times \frac{1}{8} + 0.15 \times \frac{1}{8} + 0.4 \times \frac{1}{4} + 0.05 \times \frac{1}{6} + 0.2 \times 0 + 0.1 \times \frac{1}{12}$$

$$= 0.148$$

7.4 Bayes's decision criterion

This is the most widely applied decision rule in communications systems and, as a consequence, it will be discussed in detail. In essence it operates so as to minimise the average *cost* (in terms of errors or lost information) of making decisions.

7.4.1 Decision costs

In binary transmission there are two ways to lose information:
1. Information is lost when a transmitted digital 1 is received in error as a digital 0.
2. Information is lost when a transmitted digital 0 is received in error as a digital 1.

The cost in the sense of lost information due to mechanisms (1) and (2) is denoted here by C_0 and C_1 respectively. There is no cost (i.e. no information is lost) when correct decisions are made.

7.4.2 Expected conditional decision costs

The expected conditional cost, $C(0|v)$, incurred when a detected voltage v is interpreted by a decision circuit as a digital 0 is given by:

$$C(0|v) = C_0 \, P(1|v) \tag{7.6(a)}$$

where C_0 is the cost if the decision is in error and $P(1|v)$ is the (a posteriori) probability that the decision is in error. By symmetry the corresponding equation can be written:

$$C(1|v) = C_1 \, P(0|v) \tag{7.6(b)}$$

This is the expected conditional cost incurred when v is interpreted as a digital 1.

7.4.3 Optimum decision rule

A rational decision rule to adopt is to interpret each detected voltage, v, as either a 0 or a 1, so as to minimise the expected conditional cost, i.e.:

$$C(1|v) \underset{\underset{0}{>}}{\overset{1}{<}} C(0|v) \tag{7.7}$$

The interpretation of inequality (7.7) is 'if the upper inequality holds then decide 1, if the lower inequality holds then decide 0'. Substituting equations (7.6) into the inequality (7.7) and cross dividing gives:

$$\frac{P(0|v)}{P(1|v)} \underset{\underset{0}{>}}{\overset{1}{<}} \frac{C_0}{C_1} \tag{7.8}$$

If the costs of both types of error are the same, or unknown (in which case the rational assumption must be $C_0 = C_1$), then (7.8) represents a *maximum a posteriori* probability (MAP) decision criterion. Inequality (7.8) is one form of Bayes's decision criterion. It uses a posteriori probabilities, however, which are not usually known. The criterion can be transformed to a more useful form by using Bayes's theorem, equation (3.4), i.e.:

$$P(0|v) = \frac{p(v|0)\ P(0)}{p(v)} \tag{7.9(a)}$$

and:

$$P(1|v) = \frac{p(v|1)\ P(1)}{p(v)} \tag{7.9(b)}$$

Figure 7.3 illustrates the conditional probability density functions for the case of zero mean Gaussian noise. Dividing equation (7.9(a)) by (7.9(b)):

$$\frac{P(0|v)}{P(1|v)} = \frac{p(v|0)\ P(0)}{p(v|1)\ P(1)} \tag{7.10}$$

and substituting equation (7.10) into (7.8):

$$\frac{p(v|0)\ P(0)}{p(v|1)\ P(1)} \underset{\underset{0}{>}}{\overset{1}{<}} \frac{C_0}{C_1} \tag{7.11}$$

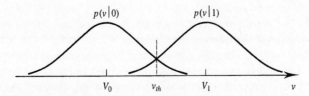

Figure 7.3 *Probability distributions for binary transmissions V_0 or V_1.*

Rearranging the inequality (7.11) gives Bayes's criterion in a form using conditional probability density functions and the a priori source probabilities which are usually known:

$$\frac{p(v|0)}{p(v|1)} \underset{\underset{0}{>}}{\overset{1}{<}} \frac{C_0\, P(1)}{C_1\, P(0)} \tag{7.12}$$

The left hand side of the above inequality is called the *likelihood ratio* (which is a function of v) and the right hand side is a likelihood threshold which will be denoted by L_{th}. If $C_0 = C_1$ and $P(0) = P(1)$ (or, more generally, $C_0 P(1) = C_1 P(0)$), or if neither costs nor a priori probabilities are known (in which case $C_0 = C_1$ and $P(0) = P(1)$ are the most rational assumptions) then $L_{th} = 1$ and equation (7.12) is called a *maximum likelihood* decision criterion. This type of hypothesis testing is used by the receivers described in Chapter 11, e.g. Figure 11.7. Table 7.1 illustrates the relationship between maximum liklihood, MAP and Bayes decision criteria.

Table 7.1 *Comparison of receiver types*

Receiver	A priori prob-abilities known	Decision costs known	Assumptions	Decision criterion
Bayes	Yes	Yes	None	$\dfrac{p(v|0)}{p(v|1)} \underset{\underset{0}{>}}{\overset{1}{<}} \dfrac{C_0\,P(1)}{C_1\,P(0)}$
MAP	Yes	No	$C_0 = C_1$	$\dfrac{p(v|0)}{p(v|1)} \underset{\underset{0}{>}}{\overset{1}{<}} \dfrac{P(1)}{P(0)}$
Max. likelihood	No	No	$C_0 P(1) = C_1 P(0)$	$\dfrac{p(v|0)}{p(v|1)} \underset{\underset{0}{>}}{\overset{1}{<}} 1$

7.4.4 Optimum decision threshold voltage

Bayes's decision criterion represents a general solution to setting the optimum reference or threshold voltage, v_{th}, in a receiver decision circuit. The threshold voltage which minimises the expected conditional cost of each decision is the value of v which satisfies:

$$\frac{p(v|0)}{p(v|1)} = \frac{C_0\, P(1)}{C_1\, P(0)} = L_{th} \tag{7.13}$$

This is illustrated for two conditional pdfs in Figure 7.4. (These pdfs actually correspond to those obtained for envelope detection of OOK rectangular pulses, see Figure 6.11, $p(v|0)$ being a Rayleigh distribution (see section 4.7.1) and $p(v|1)$ being Rician.) If $L_{th} = 1.0$ such as would be the case for statistically independent, equiprobable symbols with equal error costs, then the voltage threshold would occur at the intersection of the two conditional pdfs. (This is exactly the location of the decision threshold selected intuitively for centre point detection of equiprobable, rectangular binary symbols in section 6.2.1.)

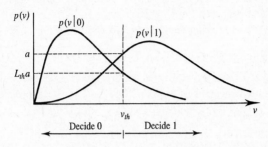

Figure 7.4 *Optimum decision threshold voltage for envelope detected OOK signal.*

In practice, a single threshold, such as that shown in Figure 7.4, is normally adequate against which to test detected voltages. It is possible in principle, however, for the conditional pdfs to intersect at more than one point. This situation is illustrated in Figure 7.5. Assuming the symbols are equiprobable the decision thresholds are then set at the intersection of the condition pdfs. Here, however, this implies more than one threshold, with several decision regions delineated by these thresholds. It is important to appreciate that despite their unconventional appearance the conditional pdfs in equation (7.13) behave in the same way as any other function. Equation (7.13), which can be rewritten in the form:

$$p(v|0) = L_{th} \, p(v|1) \tag{7.14}$$

can therefore be solved using any of the normal techniques including, where necessary, numerical methods.

7.4.5 Average unconditional decision cost

It is interesting to note (and perhaps self evident) that as well as minimising the cost of each decision, Bayes's criterion also minimises the average cost of decisions (i.e. the

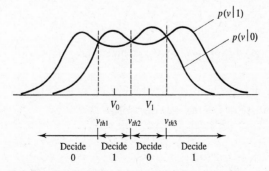

Figure 7.5 *Optimum decision thresholds for equiprobable binary voltages in noise with multimodal pdf.*

costs averaged over all decisions irrespective of type). This leads to the following alternative derivation of Bayes's criterion.

The average cost, $\bar{C}$, of making decisions is given by:

$$\bar{C} = C_1 P(0)P(1_{RX}|0_{TX}) + C_0 P(1)P(0_{RX}|1_{TX}) \tag{7.15}$$

where $P(0)P(1_{RX}|0_{TX})$ is the probability that any given received symbol is a digital 1 which has been detected in error and $P(1)P(0_{RX}|1_{TX})$ is the probability that any given received symbol is a digital 0 also detected in error. From Figure 7.3 it can be seen that:

$$P(1_{RX}|0_{TX}) = \int_{v_{th}}^{\infty} p(v|0) \, dv \tag{7.16(a)}$$

$$P(0_{RX}|1_{TX}) = \int_{-\infty}^{v_{th}} p(v|1) \, dv \tag{7.16(b)}$$

where v_{th} is the receiver decision threshold voltage. Substituting equations (7.16) into equation (7.15) the average cost of making decisions can be written as:

$$\bar{C} = C_1 P(0) \int_{v_{th}}^{\infty} p(v|0) \, dv + C_0 P(1) \int_{-\infty}^{v_{th}} p(v|1) \, dv \tag{7.17}$$

Bayes's decision criterion sets the decision threshold v_{th} so as to minimise $\bar{C}$. To find this optimum threshold $\bar{C}$ must be differentiated with respect to v_{th} and equated to zero. This involves differentiating the integrals in equation (7.17) which is most easily done using [Dwight, equations 69.1 and 69.2], i.e.:

$$\frac{d}{dx} \int_{c}^{x} f(t) \, dt = f(x) \tag{7.18(a)}$$

$$\frac{d}{dx} \int_{x}^{c} f(t) \, dt = -f(x) \tag{7.18(b)}$$

We therefore have:

$$\frac{d\bar{C}}{dv_{th}} = C_1 P(0) [-p(v_{th}|0)] + C_0 P(1) [p(v_{th}|1)] \tag{7.19}$$

Setting equation (7.19) equal to zero for minimum (or maximum) $\bar{C}$:

$$C_0 P(1) \, p(v_{th}|1) = C_1 P(0) \, p(v_{th}|0) \tag{7.20(a)}$$

and rearranging, gives:

$$\frac{P(v_{th}|1)}{P(v_{th}|0)} = \frac{C_1 \, P(0)}{C_0 \, P(1)} \tag{7.20(b)}$$

Solution of equation (7.20) for v_{th} gives the Bayes's criterion (i.e. optimum) decision voltage.

EXAMPLE 7.2 – Binary transmission
Consider a binary transmission system subject to additive Gaussian noise which has a mean value of 1.0 V and a standard deviation of 2.5 V. (Note that, unusually, the standard deviation of the noise is not the same as its RMS value in this example.) A digital 1 is represented by a rectangular pulse with amplitude 4.0 V and a digital 0 by a rectangular pulse with amplitude –4.0 V. The costs (i.e. information lost) due to each type of error are identical (i.e. $C_0 = C_1$) but the a priori probabilities of symbol transmission are different, $P(1)$ being twice $P(0)$, as the symbols are not statistically independent. Find the optimum decision threshold voltage and the resulting probability of symbol error.

Using Bayes's decision criterion the optimum decision threshold voltage is given by solving equation (7.14), i.e.:

$$p(v_{th}|0) = L_{th}\, p(v_{th}|1)$$

where:

$$L_{th} = \frac{C_0}{C_1}\frac{P(1)}{P(0)}$$

Since $C_1 = C_0$ and $P(1) = 2P(0)$ then:

$$L_{th} = 2.0$$

The noise is Gaussian and its mean value adds to the symbol voltages. An equivalent signalling system therefore has symbol voltages of –3 V and 5 V, and noise with zero mean.

The conditional probability density functions are:

$$p(v|0) = \frac{1}{\sigma_0\sqrt{(2\pi)}}\, e^{\frac{-(v-V_0)^2}{2\sigma_0^2}}$$

$$p(v|1) = \frac{1}{\sigma_1\sqrt{(2\pi)}}\, e^{\frac{-(v-V_1)^2}{2\sigma_1^2}}$$

where V_0 and V_1 are the signal voltages representing digital 0s and 1s respectively. Substituting these equations into the above version of equation (7.14) with $\sigma_0 = \sigma_1 = 2.5$, $V_0 = -3$ and $V_1 = 5$:

$$\frac{1}{2.5\sqrt{(2\pi)}}\, e^{\frac{-(v+3)^2}{2(2.5)^2}} = 2.0\,\frac{1}{2.5\sqrt{(2\pi)}}\, e^{\frac{-(v-5)^2}{2(2.5)^2}}$$

Cancelling and taking logs:

$$\frac{-(v+3)^2}{2(2.5)^2} = \ln(2.0) - \frac{(v-5)^2}{2(2.5)^2}$$

$$\frac{-(v^2 + 6v + 9) + (v^2 - 10v + 25)}{2(2.5)^2} = \ln(2.0)$$

$$16 - 16v = 2(2.5)^2 \ln(2.0)$$

or:

$$v = \frac{16 - [2(2.5)^2 \ln (2.0)]}{16} = 0.46 \text{ V}$$

Figure 7.6 illustrates the *un*conditional probability density functions of symbols plus noise and the location of the optimum threshold voltage at $v_{th} = 0.46$ V for this example. (The conditional probability density functions would both enclose unit area.)

The probability of error for each symbol can be found by integrating the error tails of the normalised ($\bar{v} = 0, \sigma = 1$) Gaussian probability density function. The probability of a transmitted 0 being received in error is therefore:

$$P_{e1} = \int_{\frac{v_{th}-V_0}{\sigma_0}}^{\infty} \frac{1}{\sqrt{(2\pi)}} e^{-\frac{v^2}{2}} dv = \int_{1.384}^{\infty} \frac{1}{\sqrt{(2\pi)}} e^{-\frac{v^2}{2}} dv$$

Using the substitution $x = v/\sqrt{2}$:

$$P_{e1} = \frac{1}{\sqrt{\pi}} \int_{1.384/\sqrt{2}}^{\infty} e^{-x^2} dx = 0.5 \, \text{erfc}(1.384/\sqrt{2}) = 0.0831$$

Similarly the probability of a transmitted 1 being received in error is:

$$P_{e0} = \int_{-\infty}^{\frac{v_{th}-V_1}{\sigma_1}} \frac{1}{\sqrt{(2\pi)}} e^{-\frac{v^2}{2}} dv$$

From symmetry P_{e0} can also be written:

$$P_{e0} = \int_{\frac{-(v_{th}-V_1)}{\sigma_1}}^{\infty} \frac{1}{\sqrt{(2\pi)}} e^{-\frac{v^2}{2}} dv = \int_{1.816}^{\infty} \frac{1}{\sqrt{(2\pi)}} e^{-\frac{v^2}{2}} dv$$

Using the same substitution as before this becomes:

$$P_{e0} = 0.5 \, \text{erfc}(1.816/\sqrt{2}) = 0.0347$$

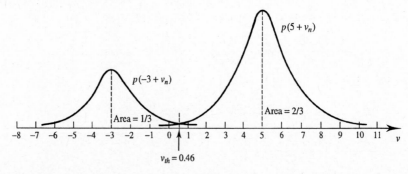

Figure 7.6 *Unconditional symbol plus noise voltage pdf for Example 7.2, $p(V + v_n)$ denotes voltage pdf of symbol voltage V plus noise.*

The overall probability of symbol error is therefore:

$$P_e = P(0)P_{e1} + P(1)P_{e0}$$

$$= \frac{1}{3}(0.0831) + \frac{2}{3}(0.0347) = 0.0508$$

Optimum thresholding, as discussed above, can be applied to centre point detection processes, such as described in Chapter 6, or after predetection signal processing (e.g. matched filtering), as discussed in Chapter 8. The important point to appreciate is that the sampling instant signal plus noise pdfs, used to establish the optimum threshold voltage(s), are those at the decision circuit input irrespective of whether predetection processing has been applied or not.

7.5 Neyman-Pearson decision criterion

The Neyman-Pearson decision criterion requires only a posteriori probabilities. Unlike Bayes's decision rule it does not require a priori source probability information. This criterion is particularly appropriate to pulse detection in Gaussian noise, as occurs in radar applications, where the source statistics (i.e. probabilities of presence, and absence, of a target) are unknown. It also works well when $C_0 \gg C_1$ or, in radar terms, when the information cost of erroneously deciding a target is present is much less than the information cost of erroneously deciding a target is absent. In this context the important probabilities are those for target detection, P_D, and target false alarm, P_{FA}. The conditional probability density functions for these two decisions and a selected decision threshold voltage are shown in Figure 7.7. The two probabilities are:

$$P_D = \int_{v_{th}}^{\infty} p(v|s+n)\, dv \tag{7.21}$$

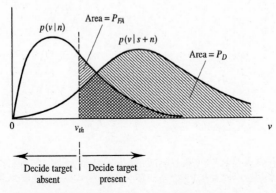

Figure **7.7** *Conditional pdfs and threshold voltage for Neyman-Pearson radar signal detector.*

$$P_{FA} = \int_{v_{th}}^{\infty} p(v|n) \, dv \tag{7.22}$$

where s denotes signal (arising due to reflections from a target) and n denotes noise. The optimum detector consists of a linear filter followed by a threshold detector [Blahut, 1987] which tests the hypothesis as to whether there is noise only present or a signal pulse plus noise. The filter is matched to the pulse (see section 8.3). In the Neyman-Pearson detector the threshold voltage, v_{th}, is chosen to give an acceptable value of P_{FA} and the detection probability then follows the characteristic shown in Figure 7.8. Note here that detection performance is dependent on both the choice of P_{FA} and the ratio of received pulse energy to noise power spectral density, E/N_0.

7.6 Summary

Soft and hard decision processes are both used in digital communications receivers but hard decision processes are simpler to implement and therefore more common. Transition probabilities $P(i|j)$ describe the probability with which transmitted symbol j will be corrupted by noise sufficiently to be interpreted at the receiver as symbol i. These probabilities can be assembled into a matrix to describe the end to end error properties of a communications channel.

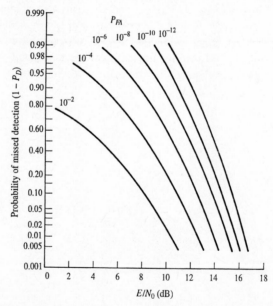

Figure 7.8 *Pulse detection in Gaussian noise using the Neyman-Pearson detection criterion (source: Blahut, 1987, reproduced with the permission of Addison Wesley).*

Bayes's decision criterion is the criterion most often used in digital communications receivers. It is optimum in the sense that it minimises the cost of (i.e. reduces the information lost when) making decisions. Bayes's criterion allows the optimum threshold voltage(s) which delineate(s) decision regions, to be found. Maximum a posteriori probability (MAP) and maximum likelihood detectors are special cases of a Bayes's criterion detector, appropriate when decision costs, or decision costs and a priori probabilities, are unknown.

The Neyman-Pearson decision criterion is normally used in radar applications. It has the advantage over Bayes's criterion that a priori symbol probabilities, whilst known to be very different, need not be quantified. The threshold level of the Neyman-Pearson criterion is set to give acceptable probabilities of false alarm and the probability of detection, given a particular received E/N_0 ratio, follows from this.

7.7 Problems

7.1. Derive the Bayes decision rule from first principles.

7.2. Under what assumptions does the Bayes receiver become a maximum likelihood receiver? Illustrate your answer with the appropriate equations.

7.3. In a baseband binary transmission system a 1 is represented by +5 V and a 0 by −5 V. The channel is subject to additive, zero mean, Gaussian noise with a standard deviation of 2 V. If the a priori probabilities of 1 and 0 are 0.5 and the costs C_0 and C_1 are equal, calculate the optimum position for the decision boundary, and the probability of error for this optimum position. [0 V, 6.23×10^{-3}]

7.4. What degenerate case of Bayes receiver design is being implemented in Problem 7.3? Assume that the decision threshold in Problem 7.3 is incorrectly placed by being moved up from its optimum position by 0.5 V. Calculate the new probability of error associated with this suboptimum threshold. [7.4125×10^{-3}]

7.5. Design a Bayes detector for the a priori probabilities $P(1) = 2/3$, $P(0) = 1/3$ with costs $C_0 = 1$ and $C_1 = 3$. The conditional pdfs for the received variable v are given by:

$$p(v|0) = \frac{1}{\sqrt{(2\pi)}} e^{-v^2/2}$$

$$p(v|1) = \frac{1}{\sqrt{(2\pi)}} e^{-(v-1)^2/2} \qquad [v = 0.905]$$

7.6. A radar system operates in, zero mean, Gaussian noise with variance 5 volts. If the probability of false alarm is to be 10^{-2} calculate the detection threshold. If the expected return from a target at extreme range is 4 volts, what is the probability that this target will be detected? [5.21, 0.2945]

CHAPTER 8

Optimum filtering for transmission and reception

8.1 Introduction

There are two signal filtering techniques which are of basic importance in digital communications. The first is concerned with filtering for transmission in order to minimise signal bandwidth. The second is concerned with filtering at the receiver in order to maximise the SNR at the decision instant (and consequently minimise the probability of symbol error). This chapter examines each of these problems and establishes criteria to be met by those filters providing optimum solutions.

8.2 Pulse shaping for optimum transmissions

Spectral efficiency, η_s, is defined as the rate of information transmission per unit of occupied bandwidth[1], i.e.:

$$\eta_s = R_s H/B \quad \text{(bits/s/Hz)} \tag{8.1}$$

where R_s is the symbol rate, H is entropy, i.e. the average amount of information (measured in bits) conveyed per symbol, and B is occupied bandwidth. (For an alphabet containing M, statistically independent, equiprobable symbols, $H = \log_2 M$ bit/symbol, see Chapter 9.) The same term is also sometimes used for the quantity R_s/B which has units of symbol/s/Hz or baud/Hz. Since spectrum is a limited resource it is often desirable to minimise the bandwidth occupied by a signal of given baud rate. Nyquist's sampling theorem (section 5.3.2) limits the transmission rate of independent samples (or symbols) in a baseband bandwidth B to:

---

[1] In cellular radio applications the term spectral efficiency is also used in a more general sense, incorporating the spatial spectrum 'efficiency'. This quantity is variously ascribed units of voice channels/MHz/km^2, Erlangs/MHz/km^2 or voice channels/cell. (Typically these spectral 'efficiencies' are much greater than unity!) The quantity called spectral efficiency here is then referred to as bandwidth efficiency.

$$R_s \leq 2B \quad \text{(symbol/s)} \tag{8.2}$$

The essential pulse shaping problem is therefore one of how to shape transmitted pulses to allow signalling at, or as close as possible to, the maximum (Nyquist) rate of $2B$ symbol/s.

8.2.1 Intersymbol interference (ISI)

Rectangular pulse signalling, in principle, has a spectral efficiency of 0 bit/s/Hz since each rectangular pulse, strictly speaking, has infinite bandwidth. In practice, of course, rectangular pulses can be transmitted over channels with finite bandwidth if a degree of distortion can be tolerated.

In digital communications it might appear that distortion is unimportant since a receiver must only distinguish between pulses which have been distorted in the same way. This might be thought especially true for OOK signalling in which only the presence or absence of pulses is important. This is not, in fact, the case since, if distortion is severe enough, then pulses may overlap in time. The decision instant voltage might then arise not only from the current symbol pulse but also from one or more preceding pulses. The smearing of one pulse into another is called *intersymbol interference* and is illustrated in Figure 8.1 for the case of rectangular baseband pulses, distorted by an *RC* lowpass channel (as discussed previously in section 5.4).

8.2.2 Bandlimiting of rectangular pulses

The nominal bandwidth of a baseband, unipolar, NRZ signal with baud rate $R_s = 1/T_o$ symbol/s was taken in section 6.4.1 to be $B = 1/T_o$ Hz. (This corresponds to the positive frequency width of the signal's main spectral lobe.) It is instructive to see the effect of limiting a rectangular pulse to this bandwidth before transmission (Figure 8.2). The filtered pulse spectrum, Figure 8.2(f), is then restricted to the main lobe of the rectangular pulse spectrum, Figure 8.2(b). The filtered pulse shape, Figure 8.2(e), is the rectangular pulse convolved with the filter's $\text{sinc}(2Bt)$ impulse response. Figure 8.2(c) shows the rectangular pulse superimposed on the filter's impulse response for two values of time offset, 0 and T_o seconds. At zero offset the convolution integral gives the received pulse peak at $t = 0$. At an offset of $T_o/2$ (not shown) the convolution integral clearly gives a

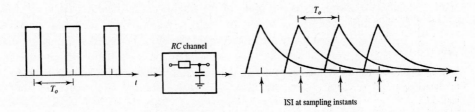

Figure 8.1 *Pulse smearing due to distortion in an RC channel.*

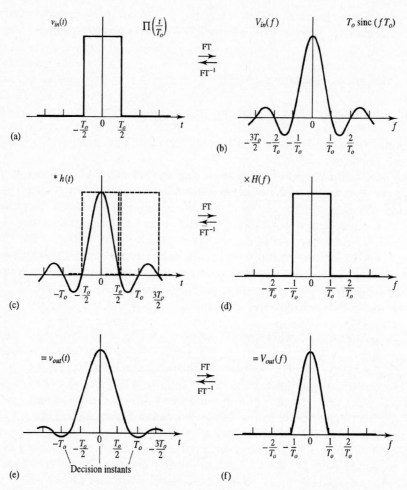

Figure 8.2 *NRZ rectangular pulse distortion due to rectangular frequency response filtering.*

reduced, but still large, positive result. At an offset of T_o the convolution gives a negative value (since the negative first sidelobe of the sinc function is larger than its positive second sidelobe). The zero crossing point of the filtered pulse therefore occurs a little before T_o seconds, Figure 8.2(e). At the centre point sampling instants $(\cdots, -T_o, 0, T_o, 2T_o, \cdots)$ receiver decisions would therefore be based not only on that pulse which should be considered but also, erroneously, on contributions from adjacent pulses. These unwanted contributions have the potential to degrade BER performance.

8.2.3 ISI free signals

The decision instants marked on Figure 8.2(e) illustrate an important point, i.e.:

Only decision instant ISI is relevant to the performance of digital communications systems. ISI occurring at times other than the decision instants does not matter.

If the signal pulses could be persuaded to pass through zero at every decision instant (except, of course, one) then ISI would no longer be a problem. This suggests a definition for an ISI free signal, i.e.:

An ISI free signal is any signal which passes through zero at all but one of the sampling instants.

Denoting the ISI free signal by $v_N(t)$ and the sampling instants by nT_o (where n is an integer and T_o is the symbol period) this definition can be expressed mathematically as:

$$v_N(t) \sum_{n=-\infty}^{\infty} \delta(t - nT_o) = v_N(0)\, \delta(t) \tag{8.3}$$

The important property of ISI free signals, summarised by equation (8.3), is illustrated in Figure 8.3. Such signals suppress all the impulses in a sampling function except one (in this case the one occurring at $t = 0$). A good example of an ISI free signal is the sinc pulse, Figure 8.4. These pulses have a peak at one decision instant and are zero at all other decision instants as required. (They also have the minimum (Nyquist) bandwidth for a given baud rate.) An OOK, multilevel, or analogue PAM, system could, in principle, be implemented using sinc pulse signalling. In practice, however, there are two problems:

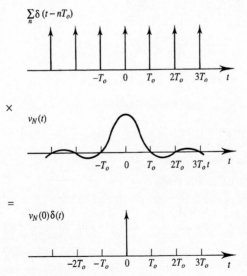

Figure 8.3 *Impulse suppression property of ISI free pulse.*

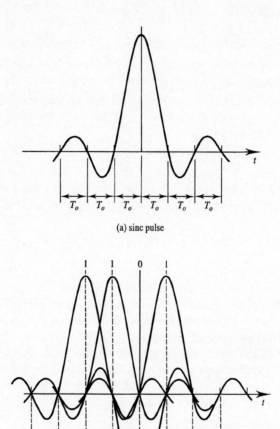

(a) sinc pulse

(b) sinc pulse signalling

Figure 8.4 *ISI free transmission using sinc pulses.*

1. sinc pulses are not physically realisable.
2. sinc pulse sidelobes (and their rates of change at the decision instants) are large and decay only with $1/t$.

The obvious way to generate sinc pulses is by shaping impulses with lowpass rectangular filters. The first problem could be equally well stated, therefore, as 'linear phase lowpass rectangular filters are not realisable'. The second problem means that extremely accurate decision timing would be required at the receiver to keep decision instant ISI to tolerable levels.

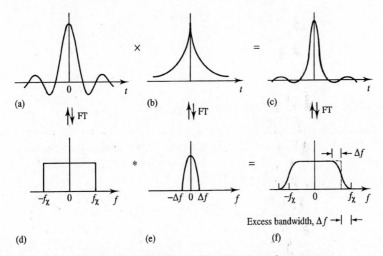

Figure 8.5 *Suppression of sinc pulse sidelobes and its effect on pulse spectrum.*

A more practical signal pulse shape would retain the desirable feature of sinc pulses (i.e. regularly spaced zero crossing points) but have an envelope with much more rapid roll-off. This can be achieved by multiplying the sinc pulse with a rapidly decaying monotonic function, Figure 8.5(a) to (c). In the frequency domain this corresponds to convolving the sinc pulse rectangular spectrum, Figure 8.5(d), with the spectrum of the decaying function, Figure 8.5(e), to obtain the final spectrum, Figure 8.5(f). As long as the decaying function is real and even its spectrum will be real and even which implies that the modified pulse spectrum will have *odd* symmetry about the sinc pulse's cut-off frequency, f_χ, Figure 8.5(f). This suggests an alternative definition for ISI free (baseband) signals, i.e.:

> *An ISI free baseband signal has a voltage spectrum which displays odd symmetry about $1/(2T_o)$ Hz.*

A more general statement, which includes ISI free bandpass signals, can be made by considering frequency translation of the baseband spectrum using the modulation theorem (Figure 8.6), i.e.:

> *An ISI free signal has a voltage spectrum which displays odd symmetry between its centre frequency, f_c, and $f_c \pm 1/T_o$ Hz.*

This property can also be demonstrated by Fourier transforming equation (8.3), replacing the product by a convolution (*), section 2.3.4, to give:

$$V_N(f) * \frac{1}{T_o} \sum_{n=-\infty}^{\infty} \delta\left(f - \frac{n}{T_o}\right) = v_N(0) \qquad (8.4)$$

which, using the replicating action of delta functions under convolution, gives:

$$\frac{1}{T_o} \sum_{n=-\infty}^{\infty} V_N\left(f - \frac{n}{T_o}\right) = v_N(0) \qquad (8.5)$$

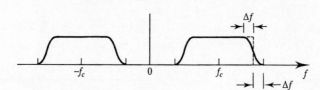

Figure 8.6 *Amplitude spectrum of a bandpass ISI free signal.*

Figure 8.7 is a pictorial interpretation of equation (8.5). The spectrum of the ISI free signal is such that if replicated along the frequency axis with periodic spacing $1/T_o$ Hz the sum of all replicas is a constant. This requires exactly the spectral symmetry described in the definitions given above.

EXAMPLE 8.1
Specify a baseband Nyquist channel which has a piecewise linear amplitude response, an absolute bandwidth of 10 kHz, and is appropriate for a baud rate of 16 kbaud. What is the channel's excess bandwidth?

The cut-off frequency of the parent rectangular frequency response is given by:

$$f_\chi = R_s/2 = 16 \times 10^3/2 = 8 \times 10^3 \text{ Hz}$$

The simplest piecewise linear roll-off therefore starts at $8 - 2 = 6$ kHz, is 6 dB down at $f_\chi = 8$ kHz and is zero ($-\infty$ dB down) at $8 + 2 = 10$ kHz. (The amplitude response, below the start of roll-off, is flat and the phase is linear.) Thus:

$$|H_N(f)| = \begin{cases} 1.0, & |f| < 6000 \text{ (Hz)} \\ 2.5 - 0.25 \times 10^{-3} f, & 6000 \leq |f| \leq 10000 \text{ (Hz)} \\ 0, & |f| > 10000 \text{ (Hz)} \end{cases}$$

The channel's excess bandwidth is 10 kHz -8 kHz $= 2$ kHz.

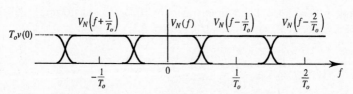

Figure 8.7 *Constant sum property of replicated ISI free signal spectra.*

8.2.4 Nyquist's vestigial symmetry theorem

Nyquist's vestigial symmetry theorem defines a symmetry condition on $H(f)$ which must be satisfied to realise an ISI free baseband impulse response. It can be stated as follows:

If the amplitude response of a lowpass rectangular filter with linear phase and bandwidth f_χ is modified by the addition of a real valued function having odd symmetry about the filter's cut-off frequency, then the resulting impulse response will retain at least those zero crossings present in the original $\text{sinc}(2f_\chi t)$ response, i.e. it will be an ISI free signal.

This 'recipe' for deriving the whole family of Nyquist filters from a lowpass rectangular prototype is illustrated in Figure 8.8. The theorem requires no further justification since it follows directly from the spectral properties of ISI free signals (section 8.2.3). The theorem can be generalised to include filters with ISI free bandpass impulse responses in an obvious way.

8.2.5 Raised cosine filtering

The family of raised cosine filters is an important and popular subset of the family of Nyquist filters. The odd symmetry of their amplitude response is provided using a cosinusoidal, half cycle, roll-off (Figure 8.9). Their (lowpass) amplitude response therefore has the following piecewise form:

$$|H(f)| = \begin{cases} 1.0, & |f| \le (f_\chi - \Delta f) \\ \frac{1}{2}\left\{1 + \sin\left[\frac{\pi}{2}\left(1 - \frac{|f|}{f_\chi}\right)\frac{f_\chi}{\Delta f}\right]\right\}, & (f_\chi - \Delta f) < |f| < (f_\chi + \Delta f) \\ 0, & |f| \ge (f_\chi + \Delta f) \end{cases} \quad (8.6)$$

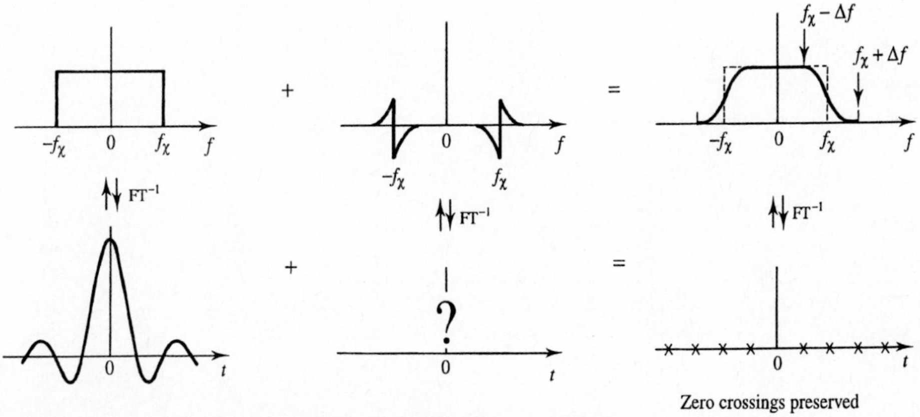

Zero crossings preserved

Figure 8.8 *Illustration of Nyquist's vestigial symmetry theorem.*

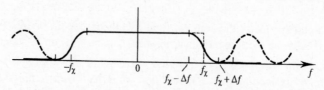

Figure 8.9 *Amplitude response of (linear phase) raised cosine filter ($f_\chi = R_s/2$).*

and their phase response is linear (implying the need for finite impulse response filters [Mulgrew and Grant]). f_χ in equation (8.6) is the cut-off frequency of the prototype rectangular lowpass filter (and the −6 dB frequency of the raised cosine filter). f_χ is related to the symbol period, T_o, by $f_\chi = R_s/2 = 1/(2T_o)$. Δf is the excess (absolute) bandwidth of the filter over the rectangular lowpass prototype. The normalised excess bandwidth, α, given by:

$$\alpha = \frac{\Delta f}{f_\chi} \tag{8.7}$$

is called the roll-off factor and can take any value between 0 and 1. Figure 8.10(a) shows the raised cosine amplitude response for several values of α. When $\alpha = 1$ the characteristic is said to be a *full* raised cosine and in this case the amplitude response simplifies to:

$$|H(f)| = \begin{cases} \frac{1}{2}\left(1 + \cos\left(\dfrac{\pi f}{2f_\chi}\right)\right), & |f| \le 2f_\chi \\ 0, & |f| > 2f_\chi \end{cases}$$

$$= \begin{cases} \cos^2\left(\dfrac{\pi f}{4f_\chi}\right), & |f| \le 2f_\chi \\ 0, & |f| > 2f_\chi \end{cases} \tag{8.8}$$

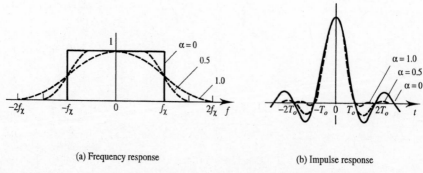

(a) Frequency response   (b) Impulse response

Figure 8.10 *Responses of raised cosine filters with three different roll-off factors.*

(The power spectral density (PSD) of an ISI free signal generated using a full raised cosine filter therefore has a $\cos^4[\pi f/(4f_\chi)]$ shape, Figure 8.11.) The impulse response of a full raised cosine filter (found from the inverse Fourier transform of equation (8.8)) is:

$$h(t) = 2f_\chi \frac{\sin 2\pi f_\chi t}{2\pi f_\chi t} \frac{\cos 2\pi f_\chi t}{1 - (4f_\chi t)^2} \tag{8.9}$$

This is shown in Figure 8.10(b) along with the impulse responses of raised cosine filters with other values of α. The first part of equation (8.9) represents the sinc impulse response of the prototype rectangular filter. The second part modifies this with extra zeros (due to the numerator) and faster decaying envelope ($1/t^3$ in total due to the denominator). The absolute bandwidth of a baseband filter (or channel) with a raised cosine frequency response is:

$$B = \frac{1}{2T_o} (1 + \alpha)$$

$$= \frac{R_s}{2} (1 + \alpha) \quad \text{(Hz)} \tag{8.10}$$

where R_s is the symbol (or baud) rate. For a bandpass raised cosine filter the bandwidth is twice this, i.e.:

$$B = R_s(1 + \alpha) \quad \text{(Hz)} \tag{8.11}$$

(This simply reflects the fact that when baseband signals are converted to bandpass (double sideband) signals by amplitude modulation, their bandwidth doubles.) Impulse signalling over a raised cosine *baseband* channel has a spectral efficiency of 2 symbol/s/Hz when $\alpha = 0$ and 1 symbol/s/Hz when $\alpha = 1$. For binary signalling systems (assuming equiprobable, independent, symbols) this translates to 2 bit/s/Hz and 1 bit/s/Hz respectively (see Chapter 9). For bandpass filters and channels these efficiencies are halved, due to the double sideband spectrum.

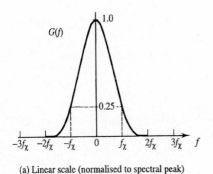

(a) Linear scale (normalised to spectral peak)

(b) Logarithmic scale (dB with respect to spectral peak)

Figure 8.11 *PSD of an ISI free signal generated using a full raised cosine filter.*

EXAMPLE 8.2

What absolute bandwidth is required to transmit an information rate of 8.0 kbit/s using 64-level baseband signalling over a raised cosine channel with a roll-off factor of 40%?

The bit rate, R_b, is given by the number of bits per symbol (H) times the number of symbols per second (R_s), i.e.:

$$R_b = H \times R_s$$

Therefore:

$$R_s = R_b/H$$

$$= \frac{8 \times 10^3}{\log_2 64} = 1.333 \times 10^3 \quad \text{(symbol/s)}$$

$$B = \frac{R_s}{2}(1 + \alpha)$$

$$= \frac{1.333 \times 10^3}{2}(1 + 0.4)$$

$$= 933.1 \quad \text{(Hz)}$$

This illustrates the bandwidth efficiency of a multi-level signal.

8.2.6 Nyquist filtering for rectangular pulses

It is possible to generate ISI free signals by shaping impulses with Nyquist filters. A more usual requirement, however, is to generate such signals by shaping rectangular pulses. The appropriate pulse shaping filter must then have a frequency response:

$$H(f) = \frac{V_N(f)}{\text{sinc}(\tau f)} \tag{8.12}$$

where $V_N(f)$ is the voltage spectrum of an ISI free pulse and $\text{sinc}(\tau f)$ is the frequency response of a hypothetical filter which converts impulses into rectangular pulses with width τ. If $V_N(f)$ is chosen to have a full raised cosine shape ($\alpha = 1$) then:

$$H(f) = \frac{\pi f \tau}{\sin(\pi f \tau)} \cos^2\left(\frac{\pi f}{4 f_\chi}\right) \tag{8.13}$$

Figure 8.12 ($\alpha = 1$) shows the frequency response corresponding to equation (8.13). Responses corresponding to other values of α are also shown.

8.2.7 Duobinary signalling

One of the problems associated with the use of sinc pulses, for ISI free signalling at the Nyquist rate, is the construction of a linear phase, lowpass rectangular filter. Such a filter

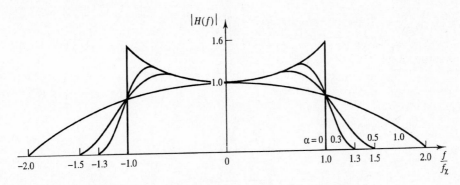

Figure 8.12 *Amplitude response of Nyquist filter for rectangular pulse shaping.*

(or rather an adequate approximation) is required to shape impulses. Duobinary signalling uses not a rectangular filter for pulse shaping but a cosine filter (not to be confused with the raised cosine filter). The amplitude response of a baseband cosine filter, Figure 8.13, is given by:

$$|H(f)| = \begin{cases} \cos \pi f T_o, & |f| \le 1/(2T_o) \\ 0, & |f| \le 1/(2T_o) \end{cases} \qquad (8.14(a))$$

and its (linear) phase response is usually taken to be:

$$\phi(f) = -\frac{\omega T_o}{2} \quad \text{(rad)} \qquad (8.14(b))$$

It has the same absolute bandwidth, $1/(2T_o)$ Hz, as the rectangular filter used for sinc signalling, and duobinary signalling therefore proceeds at the same maximum baud rate. Since its amplitude response has fallen to a low level at the filter band edge the linearity of the phase response in this region is not critical. This makes the cosine filter relatively easy to approximate. The impulse response of the cosine filter is most easily found by expressing its frequency response as a product of cosine, rectangular lowpass and phase factors, i.e.:

$$H(f) = \cos(\pi f T_o) \, \Pi\!\left(\frac{f}{2f_\chi}\right) e^{-j\pi f T_o} \qquad (8.15)$$

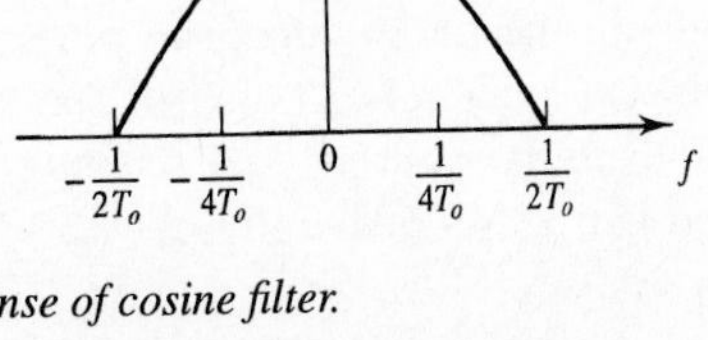

$$|H(f)|$$

Figure 8.13 *Amplitude response of cosine filter.*

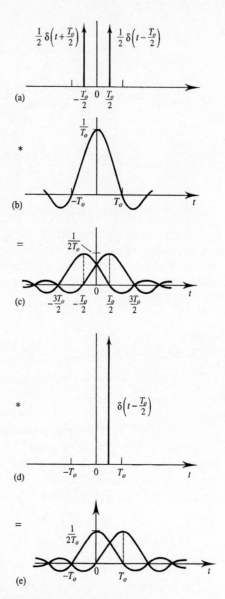

Figure 8.14 *Derivation of cosine filter impulse response.*

where $f_\chi = 1/(2T_o) = f_o/2$, and Π is the rectangular gate function. The product of the first two factors corresponds, in the time domain, to the convolution of a pair of delta functions with a sinc function, Figure 8.14(a) to (c). The third factor simply shifts the resulting pair of sinc functions to the right by $T_o/2$ seconds, Figure 8.14(d) and (e). The impulse response, shown in Figure 8.15, is therefore:

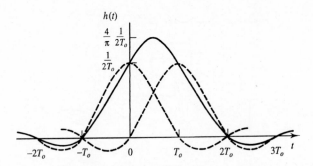

Figure 8.15 *Impulse response (solid curve) for the cosine filter.*

$$h(t) = \frac{1}{2T_o} [\text{sinc } f_o t + \text{sinc } f_o(t - T_o)] \tag{8.16}$$

It can be seen that the more gradual roll-off of the cosine filter (when compared to the rectangular filter) has been obtained at the expense of a significantly lengthened impulse response, which results in severe ISI. Duobinary signalling is important, however, because sampling instant interference occurs between *adjacent* symbols only, Figure 8.16, and is of predictable magnitude.

A useful model of duobinary signalling (which does *not* correspond to its normal implementation) is suggested by equation (8.16). It is clear that this is a superposition of two lowpass rectangular filter impulse responses, one delayed with respect to the other by T_o seconds. The cosine filter could therefore be implemented using a one symbol delay device, an adder and a rectangular filter, Figure 8.17. This implementation is obvious if equation (8.15) is rewritten as:

$$H(f) = \frac{1}{2} (1 + e^{-j2\pi fT_o}) \Pi\left(\frac{f}{2f_\chi}\right) \tag{8.17}$$

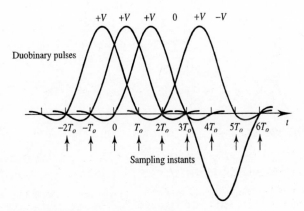

Figure 8.16 *Adjacent symbol ISI for duobinary signalling.*

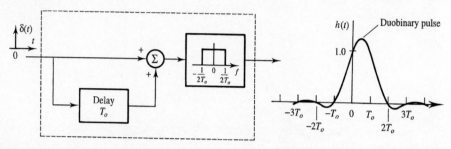

Figure 8.17 *Equivalent model of cosine filter for duobinary signalling.*

Duobinary signalling can therefore be interpreted as adjacent pulse summation followed by rectangular lowpass filtering. Figure 8.18 shows the composite pulses which arise from like and unlike combinations of binary impulse pairs. At the receiver, sampling can take place at the centre of the composite pulses (i.e. at the point of maximum ISI midway between the response peaks of the original binary impulses). This results in *three* possible levels at each decision instant, i.e. $+V$, 0 and $-V$, the level observed depending on whether the binary pulse pair are both positive, both negative or of opposite sign, Figure 8.19. Like bipolar line coding (Chapter 6), duobinary signalling is therefore a form of pseudo-ternary signalling [Lender]. The summing of adjacent pulse pairs at the transmitter can be described explicitly using the notation:

$$z_k = y_k + y_{k-1} \qquad (8.18)$$

where z_k represents the kth (ternary) symbol after duobinary coding and y_k represents the kth (binary) symbol before coding. The decoding process after detection at the receiver is therefore the inverse of equation (8.18), i.e.:

$$\hat{y}_k = \hat{z}_k - \hat{y}_{k-1} \qquad (8.19)$$

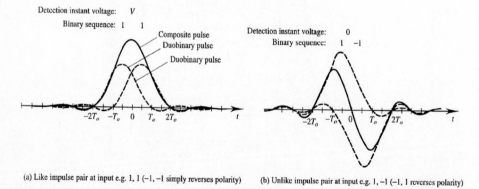

(a) Like impulse pair at input e.g. 1, 1 (−1, −1 simply reverses polarity)   (b) Unlike impulse pair at input e.g. 1, −1 (−1, 1 reverses polarity)

Figure 8.18 *Composite pulses arising from like and unlike combinations of input impulse pairs.*

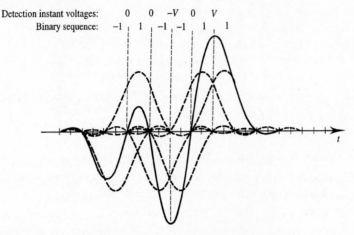

Detection instant voltages: 0   0   −V   0   V

Binary sequence: −1   1   −1   −1   1   1

Figure 8.19 *Duobinary waveform arising from an example binary sequence.*

(where the hats (ˆ) distinguish between detected and transmitted symbols to allow for the possibility of errors). The block diagram of a duobinary receiver is shown in Figure 8.20(a).

The essential advantage of duobinary signalling is that it permits signalling at the Nyquist rate without the need for linear phase, rectangular, lowpass filters (or their equivalent). The disadvantages are:

1. The ternary nature of the signal requires approximately 2.5 dB greater SNR when compared with ideal binary signalling for a given probability of error.

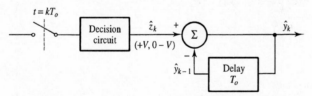

(a) Duobinary receiver (order of decision circuit and decoder could be reversed)

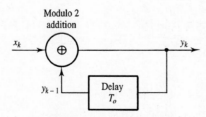

(b) Precoding for duobinary signalling

Figure 8.20 *Duobinary receiver and precoder.*

2. There is no one to one mapping between detected ternary symbols and the original binary digits.
3. The decoding process, $\hat{y}_k = \hat{z}_k - \hat{y}_{k-1}$, results in the propagation of errors.
4. The duobinary power spectral density (which is $\cos^2(\pi T_o f)\,\Pi[f/(2f_\chi)]$, i.e. the square of the cosine filter amplitude response) has a maximum at 0 Hz making it unsuitable for use with AC coupled transmission lines.

Problems 2 and 3 can be solved by adding the following precoding algorithm to the bit stream prior to duobinary pulse shaping:

$$y_k = x_k \oplus y_{k-1} \qquad (8.20)$$

where y_k is the kth precoded bit, x_k is the uncoded bit and $\oplus$ represents modulo 2 addition. Figure 8.20(b) shows the block diagram corresponding to equation (8.20). The effect of precoding plus duobinary coding can be simplified, i.e.:

$$z_k = (x_k \oplus y_{k-1}) + (x_{k-1} \oplus y_{k-2})$$
$$= (x_k \oplus y_{k-1}) + y_{k-1} \qquad (8.21)$$

The truth table for equation (8.21) is shown in Table 8.1. The important property of this table is that $z_k = 1$ when, and only when, $x_k = 1$. The precoded duobinary signal can therefore be decoded on a bit by bit basis (i.e. without the use of feedback loops). One to one mapping, of received ternary symbols to original binary symbols, is thus re-established and error propagation is eliminated.

Table 8.1 *Precoded duobinary truth table.*

x_k	y_{k-1}	$x_k \oplus y_{k-1}$	z_k
0	0	0	0
1	0	1	1
0	1	1	2
1	1	0	1

EXAMPLE 8.3
Find the output data sequence of a duobinary signalling system (a) without precoding and (b) with precoding if the input data sequence is: 1 1 0 0 0 1 0 1 0 0 1 1 1.

(a) Using equation (8.18):

input data, y_k 1100010100111

y_{k-1} ?1100010100111

duobinary data, z_k ?210011110122?

(b) Using equations (8.20) and (8.21) or Table 8.1, and assuming that the initial precoded bit is a digital 1:

$$x_k \qquad 1100010100111$$

$$y_{k-1} \quad (1)011110011101$$

$$z_k \qquad 1122210122111$$

Assuming that, after precoding, the (y_k) binary digits (1,0) are represented by positive and negative impulses, then the duobinary detection algorithm is:

$$\hat{x}_k = \begin{cases} 1, & f(kT_s) = 0 \\ 0, & f(kT_s) = \pm V \end{cases} \qquad (8.22)$$

where $f(kT_s)$ is the received voltage at the appropriate decision instant.

Problem 4 (i.e. the large DC value of the duobinary PSD) can be addressed by replacing the transmitter one bit delay and adder in Figure 8.17 by a two bit $(2T_o)$ delay and subtractor, Figure 8.21(a). This results in *modified duobinary* signalling. The PSD has a null at 0 Hz but is still strictly bandlimited to $1/(2T_o)$ Hz. Figure 8.21(b) is a block diagram of the modified duobinary receiver. The frequency response of the pulse shaping filter, Figure 8.22(a), is:

$$H(f) = \frac{1}{2}(1 - e^{-j4\pi fT_o}) \, \Pi\!\left(\frac{f}{2f_\chi}\right) \qquad (8.23)$$

where, once again, the filter cut-off frequency is $f_\chi = 1/(2T_o)$ Hz. The corresponding

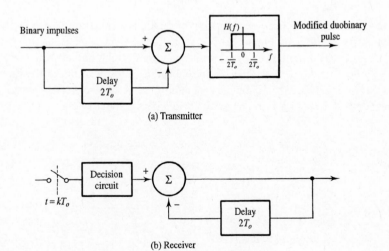

(a) Transmitter

(b) Receiver

Figure 8.21 *Equivalent model for modified duobinary signalling.*

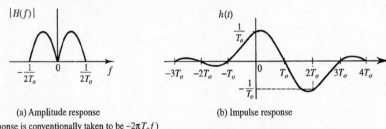

(a) Amplitude response

(b) Impulse response

(Phase response is conventionally taken to be $-2\pi T_o f$)

Figure 8.22 *Characteristics of pulse shaping filter for modified duobinary signalling.*

impulse response is shown in Figure 8.22(b). The amplitude and phase response are:

$$|H(f)| = \begin{cases} |\sin(2\pi f T_o)|, & |f| \le 1/(2T_o) \\ 0, & |f| > 1/(2T_o) \end{cases} \quad (8.24)$$

$$\phi(f) = -\omega T_o, \qquad |f| \le 1/(2T_o) \quad (8.25)$$

Since $(1 - e^{-j4\pi fT_o})$ can be factorised into $(1 - e^{-j2\pi fT_o})(1 + e^{-j2\pi fT_o})$, modified duobinary pulse shaping can be realised by cascading a one bit delay and subtractor (to implement the first factor) with a conventional duobinary pulse shaping filter (to implement the second factor). The block diagram corresponding to this implementation is shown in Figure 8.23. Precoding can be added to avoid modified duobinary error propagation. The appropriate precoding and post-decoding algorithms are illustrated as block diagrams in Figure 8.24.

8.2.8 Partial response signalling

Partial response signalling is a generalisation of duobinary signalling in which the single element transversal filter of Figure 8.17 is replaced with an N-element (tap weighted) filter. This produces a multilevel signal with non-zero correlation between symbols over an $N + 1$ symbol window. Since the ISI introduced as a result of this correlation is of a prescribed form it can be, as in duobinary signalling, effectively cancelled at the receiver. Partial response signalling is also known as correlative coding.

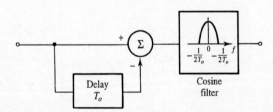

Figure 8.23 *Practical implementation of modified duobinary pulse shaping filter.*

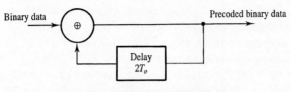

(a) Precoding (⊕ signifies modulo 2 addition)

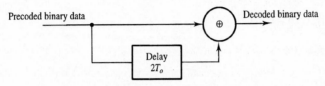

(b) Decoding (⊕ signifies modulo 2 addition)

Figure 8.24 *Precoding and de-precoding for modified duobinary signalling.*

8.3 Pulse filtering for optimum reception

Formulas were derived in sections 6.2.1 and 6.2.2 for the probability of bit error expected when equiprobable, rectangular, baseband symbols are detected, using a centre point decision process in the presence of Gaussian noise. Since this process compares a single sample value of signal plus noise with an appropriate threshold the following question might be asked. *If several samples of the signal plus noise voltage are examined at different time instants within the duration of a single symbol (as illustrated in Figure 8.25) is it not possible to obtain a more reliable (i.e. lower P_e) decision?* The answer to this question is normally yes since, at the very least, majority voting of multiple decisions associated with a given symbol could be employed to reduce the probability of error. Better still, if n samples were examined, an obvious strategy would be to add the samples together and compare the result with n times the appropriate threshold for a single sample. If this idea is extended to its limit (i.e. $n \to \infty$) than the discrete summation of symbol plus noise samples becomes continuous integration of the symbol plus noise

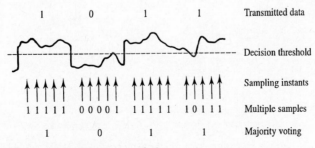

Figure 8.25 *Multiple sampling of single symbols.*

voltage. The post integration decision threshold then becomes $\frac{1}{2} \left(\int_0^{T_o} v_0 \, dt + \int_0^{T_o} v_1 \, dt \right)$ where v_0 and v_1 are the voltage levels representing binary zeros and ones respectively. After each symbol the integrator output would be reset to zero ready for the next symbol. This signal processing technique, Figure 8.26, is a significant improvement on centre point sampling and is called integrate and dump (I+D) detection. It is the optimum detection process for baseband rectangular pulses in that the resulting probability of error is a minimum. It is also easy to implement as shown in Figure 8.27. I+D is a special case of a general and optimum type of detection process, which can be applied to any pulse shape, called *matched filtering*.

8.3.1 Matched filtering

A matched filter can be defined as follows:

> *A filter which immediately precedes the decision circuit in a digital communications receiver is said to be matched to a particular symbol pulse, if it maximises the output SNR at the sampling instant when that pulse is present at the filter input.*

The criteria which relate the characteristics (amplitude and phase response) of a filter to those of the pulse to which it is matched can be derived as follows.

Consider a digital communications system which transmits pulses with shape $v(t)$, Figure 8.28(a). The pulses have a (complex) voltage spectrum $V(f)$, Figure 8.28(b) and a normalised energy spectral density (ESD) $|V(f)|^2$ V²s/Hz, Figure 8.28(d). If the noise power spectral density (NPSD) is white, it can be represented as a constant ESD per pulse period as shown in Figure 8.28(d). If the spectrum is divided into narrow frequency

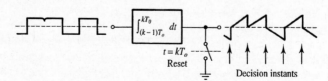

Figure 8.26 *Integrate and dump detection for rectangular pulses.*

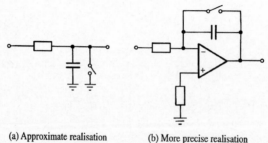

(a) Approximate realisation   (b) More precise realisation

Figure 8.27 *Simple circuit realisations for integrate and dump (I+D) detection.*

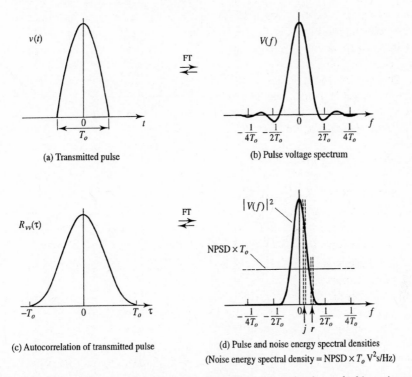

$v(t)$

(a) Transmitted pulse

FT

$V(f)$

$-\dfrac{1}{4T_o}$ $-\dfrac{1}{2T_o}$ 0 $\dfrac{1}{2T_o}$ $\dfrac{1}{4T_o}$ f

(b) Pulse voltage spectrum

$R_{vv}(\tau)$

$-T_o$ 0 T_o τ

(c) Autocorrelation of transmitted pulse

FT

$|V(f)|^2$

NPSD $\times T_o$

$-\dfrac{1}{4T_o}$ $-\dfrac{1}{2T_o}$ 0 $\dfrac{1}{2T_o}$ $\dfrac{1}{4T_o}$ f

j r

(d) Pulse and noise energy spectral densities
(Noise energy spectral density = NPSD $\times T_o$ V^{2}s/Hz)

Figure 8.28 *Relationship between energy spectral densities of signal pulse and white noise to illustrate matched filtering amplitude criterion.*

bands it can be seen from this figure that some bands (such as j) have large SNR and some (such as r) have much smaller SNR. Any band which includes signal energy should clearly make a contribution to the decision process (otherwise signal is being discarded). It is intuitively obvious, however, that those bands with high SNR should be correspondingly more influential in the decision process than those with low SNR. This suggests forming a weighted sum of the individual sub-band signal and noise energies where the weighting is in direct proportion to each band's SNR. Since the NPSD is constant with frequency the SNR is proportional to $|V(f)|^2$. Remembering that the power or energy density passed by a filter is proportional to $|H(f)|^2$, this argument leads to the following statement of the amplitude response required for a matched filter assuming white noise:

The square of the amplitude response of a matched filter has the same shape as the energy spectral density of the pulse to which it is matched.

Now consider the pulse spectrum in Figure 8.28 to be composed of many closely spaced and harmonically related spectral lines (Figure 8.29(c), (d)). The amplitude and phase spectra give the amplitude and phase of each of the cosine waves into which a periodic version of the pulse stream has been decomposed (Figure 8.29(a), (b)). If it can

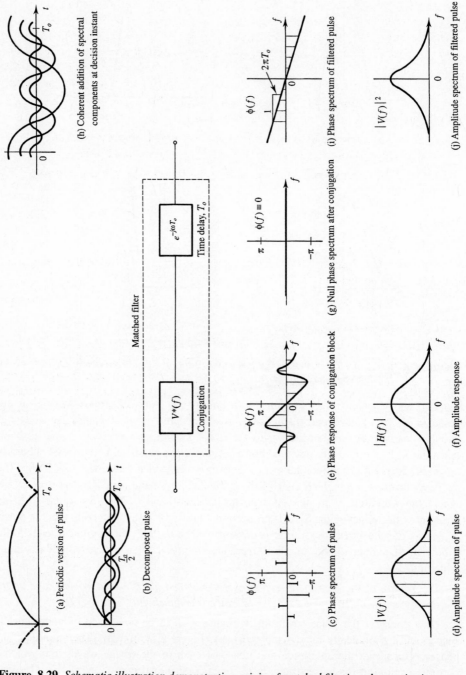

Figure 8.29 *Schematic illustration demonstrating origin of matched filtering phase criterion.*

be arranged for all the cosine waves to reach a peak simultaneously in time then the signal voltage (and therefore the signal power) will be a maximum at that instant.

The filter which achieves this has a phase response which is the opposite (i.e. the negative) of the pulse phase spectrum, Figure 8.29(e). The post-filtered pulse would then have a null phase spectrum, Figure 8.29(g), and the component cosine waves would peak together at time $t = 0, T_o, 2T_o, \cdots$ (Figure 8.29(h)). In practice a linear phase shift, $e^{-j\omega T_o}$, corresponding to a time delay of T_o seconds, must be added (or rather included) in the matched filter's frequency response to make it realisable[1]. This gives us a statement of the phase response required for a matched filter, i.e.:

The phase response of a matched filter is the negative of the phase spectrum of the pulse to which it is matched plus an additional linear phase of $-2\pi fT_o$ rad.

The matched filtering amplitude and phase criteria can be expressed mathematically as:

$$|H(f)|^2 = k^2|V(f)|^2 \qquad (8.26(a))$$

$$\phi(f) = -\phi_v(f) - 2\pi T_o f \quad (\text{rad}) \qquad (8.26(b))$$

where $\phi_v(f)$ is the phase spectrum of the expected pulse and k is a constant. Equations (8.26) can be combined into a single matched filtering criterion, i.e.:

$$H(f) = kV^*(f)\, e^{-j\omega T_o} \qquad (8.27)$$

where the superscript * indicates complex conjugation.

Matched filtering essentially takes advantage of the fact that the pulse or signal frequency components are coherent in nature whilst the corresponding noise components are incoherent. It is therefore possible, using appropriate processing, to add spectral components of the signal voltage-wise whilst the same processing adds noise components only power-wise. The extension of the above arguments to pulses buried in non-white noise is straightforward in which case the matched filtering amplitude response generalises to:

$$|H(f)| = \frac{k|V(f)|}{\sqrt{G_n(f)}} \qquad (8.28)$$

where $G_n(f)$ is noise power spectral density (see Chapter 3). The phase response is identical to that for white noise.

EXAMPLE 8.4

Find the frequency response of the filter which is matched to the triangular pulse $\Lambda(t - 1)$.

---

[1] This is to shift the *(single)* instant of constructive interference, between the (elemental) component sinusoids of the *(single)* aperiodic symbol, from $t = 0$ to $t = T_o$.

The voltage spectrum of the pulse is given by:

$$V(f) = \text{FT} \{\Lambda(t - 1)\}$$

$$= \text{sinc}^2(f)\, e^{-j2\pi f 1}$$

The frequency response of the matched filter is therefore:

$$H(f) = V^*(f)\, e^{-j\omega T_o}$$

$$= \text{sinc}^2(f)\, e^{+j2\pi f}\, e^{-j2\pi f 2}$$

$$= \text{sinc}^2(f)\, e^{-j2\pi f}$$

8.3.2 Correlation detection

We now apply correlation (described in section 2.6) to receiver design. The impulse response of a filter is related to its frequency response by the inverse Fourier transform, i.e.:

$$h(t) = \int_{-\infty}^{\infty} H(f)\, e^{j2\pi ft}\, df \tag{8.29}$$

This equation can therefore be used to transform the matched filtering criterion described by equation (8.27) into the time domain:

$$h(t) = \int_{-\infty}^{\infty} k\, V^*(f)\, e^{j2\pi f(t-T_o)}\, df$$

$$= k \left[\int_{-\infty}^{\infty} V(f)\, e^{j2\pi f(T_o - t)}\, df \right]^* \tag{8.30}$$

i.e.:

$$h(t) = k\, v^*(T_o - t) \tag{8.31}$$

Equation (8.31) is a statement of the matched filtering criterion in the time domain. For a filter matched to a purely real pulse it can be expressed in words as follows:

The impulse response of a matched filter is a time reversed version of the pulse to which it is matched, delayed by a time equal to the duration of the pulse.

Figure 8.30 illustrates equation (8.31) pictorially. The time delay, T_o, is needed to ensure causality (section 4.5) and corresponds to the need for the linear phase factor in equation (8.27).

Equation (8.31) allows the output pulse of a matched filter to be found directly from its input. The output of any time invariant linear filter is its input convolved with its impulse response. The convolution process involves reversing one of the functions

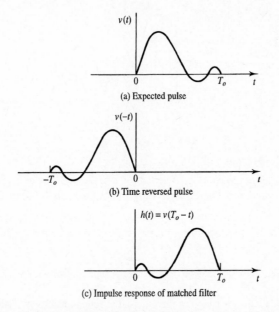

(a) Expected pulse

(b) Time reversed pulse

(c) Impulse response of matched filter

Figure 8.30 *Relationship between expected pulse and impulse response of matched filter.*

(either input or impulse response), sliding the reversed over the non-reversed function and integrating the product. Since the impulse response of the matched filter is a time reversed copy of the expected input, and since convolution requires a further reversal, then the output is given by the integrated sliding product of *either* the input *or* the impulse response with an *un*reversed version of *itself*. This is illustrated in Figure 8.31. The output is thus the *autocorrelation* (section 2.6) of either the input pulse *or* the impulse response. An algebraic proof for a real signal pulse is given below.

Let $v_{in}(t)$, $v_{out}(t)$ and $h(t)$ be the input, output and impulse response of a filter. Then by convolution (section 4.3.4):

$$v_{out}(t) \; = \; v_{in}(t) * h(t) \tag{8.32}$$

If the filter is matched to $v_{in}(t)$ then:

$$v_{out}(t) \; = \; v_{in}(t) * k \, v_{in}(T_o - t)$$

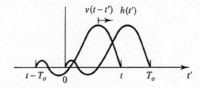

Figure 8.31 *Equivalence of v(t) * h(t) and $R_{hh}(t')$.*

$$= k \int_{-\infty}^{\infty} v_{in}(t') \, v_{in}(T_o + t' - t) \, dt' \qquad (8.33)$$

Putting $T_o - t = \tau$:

$$v_{out}(t) = v_{out}(T_o - \tau) = k \int_{-\infty}^{\infty} v_{in}(t') \, v_{in}(t' + \tau) \, dt'$$

$$= k \, R_{v_{in}v_{in}}(\tau)$$

$$(= k \, R_{hh}(\tau)) \qquad (8.34)$$

Equation (8.34) can be expressed in words as follows:

The output of a filter driven by, and matched to, a real input pulse is, to within a multiplicative constant, k, and a time shift, T_o, the autocorrelation of the input pulse.

EXAMPLE 8.5
What will be the output of a filter matched to rectangular input pulses with width 1.0 ms?

The output pulse is the autocorrelation of the input pulse (equation (8.34)). Thus the output pulse will be triangular with width 2.0 ms.

The correlation property of a matched filter can be realised directly in the time domain. A block diagram of a classical correlator is shown in Figure 8.32. The correlator input pulse, $v_{in}(t)$, is distinguished from the reference pulse by a subscript since the input is strictly the sum of the signal pulse plus noise, i.e.:

$$v_{in}(t) = v(t) + n(t) \qquad (8.35)$$

In digital communications the variable delay, τ, is usually unnecessary (Figure 8.33) since the pulse arrival times are normally known. Furthermore, it is only the peak value of the correlation function, $R_{v_{in}v}(0)$, which is of importance. The correlator output

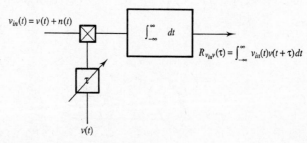

Figure 8.32 *Block diagram of signal correlator.*

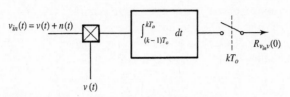

Figure 8.33 *Signal correlator for digital communications receiver.*

reaches this maximum value at the end of the input pulse, i.e. after T_o seconds. This, therefore, represents the correct sampling instant which leads to an optimum (i.e. maximum) decision SNR. (A matched filter response for a baseband coded waveform using the receiver correlation operation was demonstrated previously in Figure 4.9.)

It is interesting to note that an analogous argument to that used in the frequency domain to derive the amplitude response of a matched filter (equation (8.26(a)) can be used to demonstrate the optimum nature of correlation detection. Specifically, if the noise is stationary then its expected amplitude throughout the duration of the signal pulse will be constant. The expected signal to RMS noise *voltage* ratio during pulse reception is therefore proportional to $v(t)$. (The expected signal to noise *power* ratio is proportional to $|v(t)|^2$.) It seems entirely reasonable, then, to weight each instantaneous value of signal plus noise voltage, $v_{in}(t)$, by the corresponding value of $v(t)$ and then to add (i.e. integrate) the result. (This corresponds to weighting each value of signal plus noise *power* by $|v(t)|^2$.)

If the reference signal, $v(t)$, is approximated by n sample values at regularly spaced time instants, $v(\Delta t), v(2\Delta t), \cdots, v((n-1)\Delta t), v(n\Delta t)$, the correlator can be implemented using a shift register, a set of weighting coefficients (the sample values) and an adder (Figure 8.34). This particular implementation has the form of a finite impulse response digital filter [Mulgrew and Grant] and illustrates, clearly, the equivalence of matched filtering and correlation detection.

A single matched filter, or correlation channel, is obviously adequate as a detector in the case of on-off keyed (OOK) systems since the output will be a maximum when a

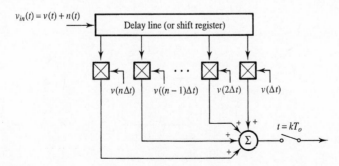

Figure 8.34 *Shift register implementation of matched filter illustrating relationship to correlation.*

pulse is present and essentially zero (ignoring noise) when no pulse is present. For binary systems employing two, non-zero, pulses a possible implementation would include two filters or correlators, one matched to each pulse type as shown in Figure 8.35. (This configuration will be used later for FSK detection, Chapter 11.) If the filter output is denoted by $f(t)$ then the possible sampling instant output voltages are:

$$f(kT_o) = \begin{cases} \displaystyle\int_0^{T_o} v_0^2(t)\,dt - \int_0^{T_o} v_0(t)v_1(t)\,dt, & \text{if symbol 0 is present} \\[3ex] \displaystyle -\int_0^{T_o} v_1^2(t)\,dt + \int_0^{T_o} v_1(t)\,v_0(t)\,dt, & \text{if symbol 1 is present} \end{cases} \tag{8.36}$$

If the signal pulses $v_0(t)$ and $v_1(t)$ are orthogonal but contain equal normalised energy, E_s V²s (i.e. joules of energy dissipated in a 1 Ω load) then the sampling instant voltages will be $\pm E_s$. If the pulses are antipodal (i.e. $v_1(t) = -v_0(t)$) then the sampling instant voltages will be $\pm 2E_s$. The same output voltages can be generated, however, for all (orthogonal, antipodal or other) binary pulse systems using only one filter or correlator by matching to the pulse difference signal, $v_1(t) - v_0(t)$, as shown in Figure 8.36.

For multisymbol signalling the number of channels in the matched filter or correlation receiver can be extended in an obvious way, Figure 8.37. (If antipodal signal pairs are used in an M-ary system only $M/2$ detection channels are needed.)

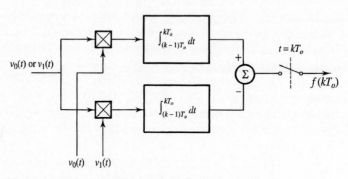

Figure 8.35 *Two channel, binary symbol correlator.*

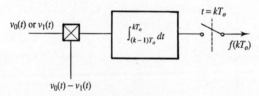

Figure 8.36 *One channel, binary symbol correlator.*

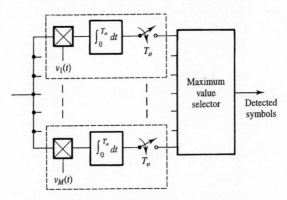

Figure 8.37 *Multichannel correlator for reception of M-ary signals.*

There is a disquieting aspect to equation (8.36) in that it seems dimensionally unsound. $f(kT_o)$ is a voltage yet the RHS of the equation has dimensions of normalised energy (V^2 s). This apparant paradox is resolved by considering the implementation of a correlator in more detail. The upper channel of the receiver in Figure 8.35, for example, contains a multiplier and an integrator. The multiplier must have an associated constant k_m with dimensions of V^{-1} if the output is to be a voltage (which it certainly is). Strictly, then, the output of the multiplier is $k_m v_0^2(t)$ volts when a digital 0 is present at its input. Similarly the integrator has a constant k_I which has dimensions of s^{-1}. (If this integrator is implemented as an operational amplifier, for example, with a resistor, R, in series with its inverting input and a capacitor, C, as its negative feedback element then $k_I = 1/(RC)$ (s^{-1}).) Since these constants affect signal voltages and noise voltages in identical ways they are usually ignored. This is equivalent to arbitrarily assigning to them a numerical value of 1.0 (resulting in an overall 'conversion' constant of 1.0 V/V^2s) and then (for orthogonal symbols) equating the numerical value of voltage at the correlator output with normalised symbol energy at the correlator input.

8.3.3 Decision instant SNR

A clue to the SNR performance of ideal matched filters and correlation detectors comes from equation (8.36). For orthogonal signal pulses the second (cross) terms in these equations are (by definition) zero. This leaves the first terms which represent the normalised energy, E_s, contained in the signal pulses. (The minus signs in equations (8.36) arise due to the subtractor placed after the integrators.) It is important to remember that it is the correlator output *voltage* which is numerically equal (assuming a 'conversion' constant of 1.0 V/V^2s) to the normalised symbol energy E_s, i.e.:

$$f(kT_o) = E_s \ (V) \tag{8.37(a)}$$

The sampling instant normalised signal power at the correlator output is therefore:

$$|f(kT_o)|^2 = E_s^2 \ (V^2) \tag{8.37(b)}$$

Since the noise at the correlator input is a random signal it must properly be described by its autocorrelation function (ACF) or its power spectral density. At the input to the multiplier the ACF of $n(t)$ is:

$$R_{nn}(\tau) = \langle n(t) \, n(t + \tau) \rangle \; (\text{V}^2) \tag{8.38}$$

Assuming that $n(t)$ is white with double sided power spectral density $N_0/2$ (V^2/Hz) then its ACF can be calculated by taking the inverse Fourier transform (Table 2.4) to obtain:

$$R_{nn}(\tau) = \frac{N_0}{2} \, \delta(\tau) \; (\text{V}^2) \tag{8.39}$$

The ACF of the noise after multiplication with $v(t)$ is:

$$R_{xx}(\tau) = \langle x(t) \, x(t + \tau) \rangle$$

$$= \langle n(t)v(t) \, n(t + \tau)v(t + \tau) \rangle \; (\text{V}^2) \tag{8.40}$$

where $x(t) = n(t)v(t)$ and a 'multiplier constant' of 1.0 V/V^2 has been adopted. Since $n(t)$ and $v(t)$ are independent processes equation (8.40) can be rewritten as:

$$R_{xx}(\tau) = \langle n(t) \, n(t + \tau) \rangle \langle v(t) \, v(t + \tau) \rangle$$

$$= \frac{N_0}{2} \, \delta(\tau) \, R_{vv}(\tau) \; (\text{V}^2) \tag{8.41}$$

$\delta(\tau)$ is zero everywhere except at $\tau = 0$, therefore:

$$R_{xx}(\tau) = \frac{N_0}{2} \, \delta(\tau) \, R_{vv}(0) \; \; (\text{V}^2) \tag{8.42}$$

$R_{vv}(0)$ is the mean square value of $v(t)$. Thus:

$$R_{xx}(\tau) = \frac{N_0}{2} \, \delta(\tau) \, \frac{1}{T_o} \int_0^{T_o} v^2(t) \, dt$$

$$= \frac{N_0}{2} \, \delta(\tau) \, \frac{E_s}{T_o} \; (\text{V}^2) \tag{8.43}$$

Using the Wiener-Kintchine theorem (equation (3.48)) the two sided power spectral density of $x(t) = n(t)v(t)$ is the Fourier transform of equation (8.43), i.e.:

$$G_x(f) = \frac{N_0}{2} \, \frac{E_s}{T_o} \; (\text{V}^2/\text{Hz}) \tag{8.44}$$

The impulse response of a device which integrates from 0 to T_o seconds is a rectangle of unit height and T_o seconds duration (i. e. $\Pi[(t - T_o/2)/T_o]$). A good conceptual model of such a device is shown in Figure 8.38. The frequency response of this time windowed integrator (sometimes called a moving average filter) is the Fourier transform of its impulse response, i.e.:

$$H(f) = T_o \, \text{sinc}(T_o f) \, e^{-j\omega T_o/2} \tag{8.45}$$

and its amplitude response is therefore:

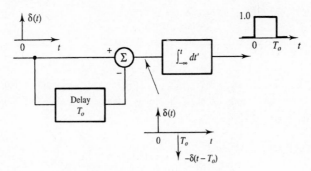

Figure 8.38 *Time windowed integrator (or moving average filter).*

$$|H(f)| = T_o \, |\text{sinc}(T_o f)| \tag{8.46}$$

The NPSD at the integrator output is:

$$G_y(f) = G_x(f) \, |H(f)|^2$$

$$= \frac{N_0}{2} \, E_s T_o \, \text{sinc}^2(T_o f) \tag{8.47}$$

and the total noise power at the correlator output is:

$$N = \frac{N_0}{2} \, E_s T_o \int_{-\infty}^{\infty} \text{sinc}^2(T_o f) \, df$$

$$= \frac{N_0}{2} \, E_s \ (\text{V}^2) \tag{8.48}$$

(The integral of the sinc^2 function can be seen by inspection to be $1/T_o$ using the Fourier transform 'value at the origin' theorem, Table 2.5.) The standard deviation of the noise at the correlator output (or, equivalently, its RMS value since its mean value is zero) is:

$$\sigma = \sqrt{N} = \sqrt{\left(\frac{N_0}{2} \, E_s\right)} \ (\text{V}) \tag{8.49}$$

Equations (8.37(a)) and (8.49) give a decision instant signal to RMS noise voltage ratio of:

$$\frac{f(T_o)}{\sigma} = \sqrt{\left(\frac{2E_s}{N_0}\right)} \tag{8.50}$$

or, alternatively, a decision instant signal to noise power ratio of:

$$\frac{S}{N} = \frac{|f(T_o)|^2}{\sigma^2} = \frac{2E_s}{N_0} \tag{8.51}$$

The important point here is that *the decision instant SNR at the output of a correlation* receiver (or matched filter) depends only on pulse energy and input NPSD. It is independent of pulse shape.

EXAMPLE 8.6

What is the sampling instant signal-to-noise ratio at the output of a filter matched to a triangular pulse of height 10 mV and width 1.0 ms if the noise at the input to the filter is white with a power spectral density of 10 nV^2/Hz?

Energy in the input pulse is given by:

$$E_s = \int_0^{T_o} v^2(t)\, dt = \int_0^{\frac{T_o}{2}} v^2(t)\, dt + \int_{\frac{T_o}{2}}^{T_o} v^2(t)\, dt$$

$$= \int_0^{0.5 \times 10^{-3}} [20\, t]^2\, dt + \int_{0.5 \times 10^{-3}}^{1 \times 10^{-3}} [2 \times 10^{-2} - 20\, t]^2\, dt$$

$$= 400 \left[\frac{t^3}{3} \right]_0^{0.5 \times 10^{-3}} + 4 \times 10^{-4} \left[\frac{t}{1} \right]_{0.5 \times 10^{-3}}^{1 \times 10^{-3}}$$

$$- 80 \times 10^{-2} \left[\frac{t^2}{2} \right]_{0.5 \times 10^{-3}}^{1 \times 10^{-3}} + 400 \left[\frac{t^3}{3} \right]_{0.5 \times 10^{-3}}^{1 \times 10^{-3}}$$

$$= 0.33 \times 10^{-7} \ (\text{V}^2 \text{ s})$$

Using equation (8.51) the sampling instant SNR when a pulse is present at the filter input is:

$$\frac{S}{N} = \frac{2E_s}{N_0} = \frac{2 \times 0.33 \times 10^{-7}}{10 \times 10^{-9}} = 6.67 = 8.2 \text{ dB}$$

8.3.4 BER performance of optimum receivers

A general formula giving the probability of symbol error for an optimum binary receiver (matched filter or correlator) is most easily derived by considering a single channel correlator matched to the binary symbol difference, $v_1(t) - v_0(t)$. When a binary 1 is present at the receiver input the decision instant voltage at the output is given by equation (8.36) as:

$$f(kT_o) = \int_{(k-1)T_o}^{kT_o} v_1(t)\, [v_1(t) - v_0(t)]\, dt$$

$$= E_{s1} - \int_{(k-1)T_o}^{kT_o} v_1(t)v_0(t) \, dt \qquad (8.52)$$

where E_{s1} is the binary 1 symbol energy. When a binary 0 is present at the receiver input the decision instant voltage is:

$$f(kT_o) = -E_{s0} + \int_{(k-1)T_o}^{kT_o} v_1(t)v_0(t) \, dt \qquad (8.53)$$

where E_{s0} is the binary 0 symbol energy. The second term of both equations (8.52) and (8.53) represents the correlation between symbols. Defining the normalised correlation coefficient to be:

$$\rho = \frac{1}{\sqrt{(E_{s0} E_{s1})}} \int_{(k-1)T_o}^{kT_o} v_1(t) \, v_0(t) \, dt \qquad (8.54)$$

then equations (8.52) and (8.53) can be written as:

$$f(kT_o) = \begin{cases} E_{s1} - \rho\sqrt{(E_{s0}E_{s1})}, & \text{for binary 1} \\ -E_{s0} + \rho\sqrt{(E_{s0}E_{s1})}, & \text{for binary 0} \end{cases} \qquad (8.55)$$

(The proper interpretation of equation (8.54) when $v_0(t) = 0$ and therefore $E_{s0} = 0$ (i.e. OOK signalling, section 6.4.1) is $\rho = 0$.) The difference in decision instant voltages representing binary 1 and 0 is:

$$\Delta V = E_{s1} + E_{s0} - 2\rho\sqrt{(E_{s1} E_{s0})} \qquad (8.56)$$

Equation (8.56) also represents the energy, E_s', in the reference pulse of the single channel correlator, i.e.:

$$E_s' = \int_0^{T_o} [v_1(t) - v_0(t)]^2 \, dt = \Delta V \qquad (8.57)$$

The RMS noise voltage at the output of the receiver is given by equation (8.49), i.e.:

$$\sigma = \sqrt{\left(\frac{N_0}{2} E_s'\right)}$$

$$= \left[\frac{N_0}{2} \left(E_{s1} + E_{s0} - 2\rho\sqrt{(E_{s1} E_{s0})}\right)\right]^{1/2} \qquad (8.58)$$

The quantity $\Delta V/\sigma$ is therefore:

$$\frac{\Delta V}{\sigma} = \left[\frac{2}{N_0} \left(E_{s1} + E_{s0} - 2\rho\sqrt{(E_{s1} E_{s0})}\right)\right]^{1/2} \qquad (8.59)$$

For binary symbols of equal energy, E_s, this simplifies to:

$$\frac{\Delta V}{\sigma} = 2\sqrt{\left[\frac{E_s}{N_0}(1-\rho)\right]} \tag{8.60}$$

Equation (8.59) or (8.60) can be substituted into the centre point sampling formula (equation (6.8)) to give the ideal correlator (or matched filter) probability of symbol error. In the (usual) equal symbol energy case this gives:

$$P_e = \frac{1}{2}\left\{1 - \text{erf}\sqrt{\left[\frac{E_s}{2N_0}(1-\rho)\right]}\right\} \tag{8.61}$$

Although strictly speaking equations (8.60) and (8.61) are valid only for equal energy binary symbols they also give the correct probability of error for OOK signalling *providing* that E_s is interpreted as the average energy per symbol (i.e. half the energy of the non-null symbol). For all orthogonal signalling schemes (including OOK) $\rho = 0$. For all antipodal schemes (in which $v_1(t) = -v_0(t)$) $\rho = -1$.

EXAMPLE 8.7

A baseband binary communications system transmits a positive rectangular pulse for digital ones and a negative triangular pulse for digital zeros. If the (absolute) widths, peak pulse voltages, and noise power spectral density at the input of an ideal correlation receiver are all identical to those in Example 8.6 find the probability of bit error.

The energy in the triangular pulse has already been calculated in Example 8.6:

$$E_{s0} = 0.33 \times 10^{-7} \text{ V}^2\text{s}$$

The energy in the rectangular pulse is:

$$E_{s1} = v^2 T_o = (10 \times 10^{-3})^2 \times 1 \times 10^{-3} = 1 \times 10^{-7} \text{ V}^2 \text{ s}$$

Using equation (8.54):

$$\rho = \frac{1}{\sqrt{E_{s0}\,E_{s1}}} \int_0^{T_o} v_1(t)\,v_0(t)\,dt$$

$$= \frac{1}{\sqrt{0.33 \times 10^{-7} \times 1 \times 10^{-7}}} \int_0^{10^{-3}} \left[-10 \times 10^{-3}\,\Pi\left(\frac{t - 0.5 \times 10^{-3}}{10^{-3}}\right)\right.$$

$$\left. \times 10 \times 10^{-3}\,\Lambda\left(\frac{t - 0.5 \times 10^{-3}}{5 \times 10^{-4}}\right)\right] dt$$

$$= -1.74 \times 10^3 \times 2 \left[2 \times 10^3\, t^2/2\right]_0^{0.5 \times 10^{-3}} = -0.87$$

Using equation (8.59):

$$\frac{\Delta V}{\sigma} = \left[\frac{2}{N_0} \left(E_{s1} + E_{s0} - 2\rho \sqrt{(E_{s1} E_{s0})} \right) \right]^{\frac{1}{2}}$$

$$= \left[\frac{2}{10 \times 10^{-9}} (0.33 \times 10^{-7} + 1.0 \times 10^{-7} - 2(-0.87) \sqrt{0.33 \times 10^{-14}}) \right]^{\frac{1}{2}}$$

$$= 6.83$$

Using equation (6.8):

$$P_e = \frac{1}{2} \left[1 - \text{erf} \frac{\Delta V}{2\sqrt{2}\,\sigma} \right]$$

$$= \frac{1}{2} \left[1 - \text{erf} \frac{6.83}{2\sqrt{2}} \right] = 3.2 \times 10^{-4}$$

8.3.5 Comparison of baseband matched filtering and centre point detection

Equation (8.59) can be used to compare the performance of a baseband matched filter receiver with simple centre point detection of rectangular pulses as discussed in Chapter 6. For unipolar NRZ transmission equation (8.59) shows that the detection instant $\Delta V/\sigma$ after matched filtering is related to that for simple centre point detection (CPD) by:

$$\left(\frac{\Delta V}{\sigma} \right)_{MF} = \left(\frac{2 E_{s1}}{N_0} \right)^{\frac{1}{2}}$$

$$= \left(\frac{2 \Delta V^2 T_o}{\sigma^2/B} \right)^{\frac{1}{2}}$$

$$= \sqrt{2}(T_o B)^{\frac{1}{2}} \left(\frac{\Delta V}{\sigma} \right)_{CPD} \tag{8.62}$$

where T_o is the rectangular pulse duration and B is the CPD predetection (rectangular) bandwidth. It may be disturbing to recognise that if rectangular pulse CPD transmission is interpreted literally then B must be infinite to accommodate infinitely fast rise and fall times. However, in this (literal) case $(\Delta V/\sigma)_{CPD}$ is zero due to the infinite noise power implied by a white noise spectrum. In practice, the CPD predetection bandwidth B is limited to a finite value (say 2 or 3 times $1/T_o$) and T_o is interpreted as the symbol period to allow the resulting spreading of the symbol in time. The saving of transmitter power (or allowable increase in noise power spectral density) that matched filtering provides for compared with CPD is therefore:

$$\frac{(\Delta V/\sigma)_{MF}}{(\Delta V/\sigma)_{CPD}} = \sqrt{2}(T_o B)^{\frac{1}{2}}$$

$$= 3.0 + (T_o B)_{dB} \quad \text{(dB)} \tag{8.63}$$

A CPD predetection bandwidth of three times the baud rate $(B = 3/T_o)$, for example, therefore gives a power saving of 7.8 dB. (Although equation (8.63) has been derived for unipolar transmission it is also correct for the polar transmission case.)

For $B = 0.5/T_o$ (the minimum bandwidth consistent with ISI free transmission), then the performance of matched filtering and CPD is the same. This apparent paradox is resolved by appreciating that this minimum bandwidth implies sinc pulse transmission in which case the rectangular predetection CPD filter is itself precisely matched to the expected symbol pulse shape.

8.3.6 Differences between matched filtering and correlation

Although matched filters and correlation detectors give identical detection instant signal and noise voltages at their outputs for identical inputs (and therefore have identical P_e performance) they do not *necessarily* give the same pulse shapes at their outputs. This is because in the case of the correlator (Figure 8.33) the received pulse and reference pulse are aligned in time throughout the pulse duration whereas in the case of the filter (Figure 8.34) the received pulse slides across the reference pulse giving the true ACF (neglecting noise) of the input pulse. Specifically the pulse at the output of the correlator is given by:

$$f(t) = \int_0^t v^2(t') \, dt' \tag{8.64}$$

whilst the pulse at the output of the matched filter (see equation (8.33)) is given by:

$$f(t) = \int_0^t v(t') \, v(t' + T_o - t) \, dt' \tag{8.65}$$

(The lower limits in the integrals of equations (8.64) and (8.65) assume that the pulses start at $t = 0$.) This difference has no influence on P_e providing there are no errors in symbol timing. If, however, decision instants are not perfectly timed then there is the possibility of a discrepancy in matched filter and correlator performance. This is well illustrated by the case of rectangular RF pulse signalling. Figure 8.39 shows the detector output pulses for a matched filter and a correlator. It is clear that, providing the timing instant never occurs after $t = T_o$, the matched filter would suffer greater performance degradation due to symbol timing errors, for this type of pulse, than the correlator.

8.4 Root raised cosine filtering

Nyquist filtering and matched filtering have both been identified as optimum filtering techniques, the former because it results in ISI free signalling in a bandlimited channel and the latter because it results in maximum SNR at the receiver decision instants. Whilst Nyquist filtering was discussed in the context of transmitter filters it is important

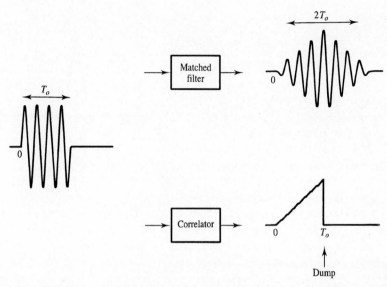

Figure 8.39 *Output pulses of matched filter and correlator for a rectangular RF input pulse.*

to realise that the Nyquist frequency response which gives ISI free detection includes the pulse shaping filter at the transmitter, the frequency response of the transmission medium and any filtering in the receiver prior to the decision circuit, i.e.:

$$H_N(f) = H_T(f) H_{ch}(f) H_R(f) \tag{8.66}$$

where the subscripts denote the Nyquist, transmitter, channel and receiver frequency responses respectively. Assuming that the channel introduces negligible distortion $(H_{ch}(f) = 1)$ then it is clear that the Nyquist frequency response can be split in any convenient way between the transmitter and receiver. It is also clear that if the transmitter and receiver filter are related by:

$$H_T^*(f) = H_R(f) \tag{8.67}$$

then the spectrum of the transmitted pulses (assuming impulses prior to filtering) will be the conjugate of the frequency response of the receiver. Apart from a linear phase factor this is precisely the requirement for matched filtering. It is therefore possible, by judicious splitting of the overall system frequency response, to satisfy both the Nyquist and matched filtering criteria simultaneously. A popular choice for $H_T(f)$ and $H_R(f)$ is the root raised cosine filter (Figure 8.40(a)) derived from equation (8.8), i.e.:

$$H_T(f) = H_R(f)$$

$$= \begin{cases} \sqrt{\cos^2(\pi f/4f_\chi)}, & f \le 2f_\chi \\ 0, & f > 2f_\chi \end{cases} \tag{8.68}$$

where $f_\chi = 1/(2T_o)$. The overall frequency response then has a full raised cosine

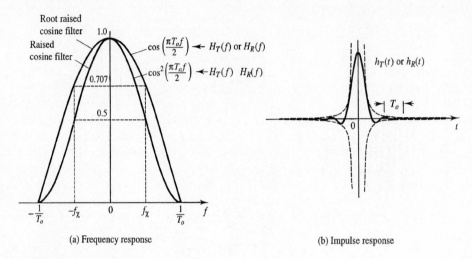

(a) Frequency response

(b) Impulse response

Figure 8.40 *Root raised cosine filter responses.*

characteristic giving ISI free detection. The impulse response of the root raised cosine filter, given by the inverse Fourier transform of equation (8.68), is:

$$h(t) = \frac{8 f_\chi}{\pi} \frac{\cos(4\pi f_\chi t)}{1 - 64 f_\chi^2 t^2} \qquad (8.69)$$

This impulse response, Figure 8.40(b), is, of course, the transmitted pulse shape.

The similarity between the root raised cosine filter and the cosine filter used in duobinary signalling is obvious. The difference is in their bandwidth. The bandwidth B, of the root raised cosine filter is $B = 1/T_o$ Hz. The bandwidth of the (duobinary) cosine filter is $B = 1/(2T_o)$ Hz.

8.5 Equalisation

When a digital signal is transmitted over a realistic channel it can be severely distorted. The communications channel including transmitter filters, multipath effects and receiver filters can be modelled by a finite impulse response (FIR) filter, with the same structure as that shown in Figure 8.34 [Mulgrew and Grant]. The transmitted data can often be effectively modelled as a discrete random binary sequence, $x(kT_o)$, which can take on values of, say, ± 1 V. Gaussian noise samples, $n(kT_o)$, are added to the FIR filter (i.e. channel) output resulting in the received samples, $f(kT_o)$. In the simplest case all the coefficients of the FIR filter would be zero except for one tap which would have weight, h_0. The received signal samples would then be:

$$f(kT_o) = \pm h_0 + n(kT_o) \qquad (8.70)$$

and we could tell what data was being transmitted by simply testing whether $f(kT_o)$ is greater than or less than zero, i.e.:

$$\text{if } f(kT_o) \geq 0 \quad \text{then } x(kT_o) = +1 \quad \text{else } x(kT_o) = -1 \tag{8.71}$$

The received sample $f(kT_o)$ is more often, however, a function of several transmitted bits or symbols as defined by the number of taps with significant weights. This merging of samples represents intersymbol interference, as discussed in section 8.2.

Each channel has a particular frequency response. If we knew this frequency response we could include a filter in the receiver with the 'opposite' or inverse frequency response, as discussed in section 6.6.2. Everywhere the channel had a peak in its frequency response the inverse filter would have a trough and vice versa. The frequency response of the channel in cascade with the inverse filter would ideally, have a, wideband, flat amplitude response and a linear phase response, i.e. as far as the transmitted signal was concerned the cascade of the channel and the inverse filter together would look like a simple delay. Effectively we would have equalised the frequency response of the channel.

Note that the equaliser operation is fundamentally wideband, compared to the matched filter of Figure 8.34. When a signal is corrupted by white noise the matched filter detector possesses a frequency response which is *matched accurately* to the expected signal characteristic. Consequently the matched filter bandwidth equals the signal bandwidth. The equalising filter bandwidth is typically much greater than the signal bandwidth, however, to achieve a commensurate narrower duration output pulse response, than with the matched filter operation.

In practice when we switch on our digital mobile radio, or telephone modem, we have no idea what the frequency response of the channel between the transmitter and the receiver will be. In this type of application the receiver equaliser must, therefore, be adaptive. If such equalisers are to approximate the optimal filter or estimator they require explicit knowledge of the signal environment in the form of correlation functions, power delay profiles, etc. In most situations such functions are unknown and/or time-varying. The equaliser must therefore employ a closed loop (feedback) arrangement in which its frequency response is adapted, or controlled, by a feedback algorithm. This permits it to compensate for time-varying distortions and still achieve performance, close to the optimal estimator function.

Adaptive filters [Mulgrew and Grant] use an adjustable or programmable filter whose impulse response is controlled to pass the desired components of each signal sample and to attenuate the undesired components in order to compensate distortion present in the input signal. This may be achieved by employing a known data sequence or training signal, Figure 8.41. An input data plus noise, sample sequence, $f(kT_o)$, is convolved with a time-varying FIR sequence, $h_i(kT_o)$. The output of the N-tap filter is $\hat{x}(kT_o)$, is given by the discrete convolution operation (see section 13.6):

$$\hat{x}(kT_o) = \sum_{i=0}^{N-1} h_i(kT_o) f((k-i)T_o) \tag{8.72}$$

The filter output, $\hat{x}(kT_o)$, is used as the estimate of the training signal, $x(kT_o)$, and is subtracted from this signal to yield an error signal:

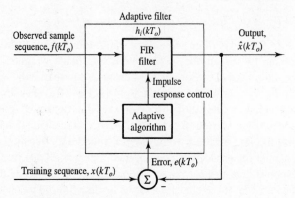

Figure 8.41 *Adaptive filter operation.*

$$e(kT_o) = x(kT_o) - \hat{x}(kT_o) \tag{8.73}$$

The error is then used in conjunction with the input signal, $f(kT_o)$, to determine the next set of filter weight values, $h_i((k+1)T_o)$.

The impulse response of the adaptive filter, $h_i(kT_o)$, is thus progressively altered as more of the observed, and training, sequences become available, such that the output $\hat{x}(kT_o)$ converges to the training sequence $x(kT_o)$ and hence the output of the optimal filter. Adaptive filters again employ a finite impulse response structure, as this is more stable than other filter forms, such as recursive infinite impulse response designs.

Figure 8.42 illustrates how this technique can be applied for practical data communications. Here the input, $f(kT_o)$, is genuine data, $x(kT_o)$, convolved with the communication channel impulse response, plus additive noise. When the transmitter is switched on, however, it sends a training sequence prior to the data. The objective here is to make the output, $\hat{x}(kT_o)$, approximate the training sequence and in so doing 'teach' the adaptive algorithm the required impulse response of the inverse filter.

The adaptive algorithm can then be switched off and genuine data transmitted. On conventional telephone lines the channel response does not change with time once the circuit has been established. Having trained the equaliser the adaptive algorithm can,

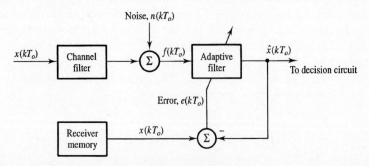

Figure 8.42 *Use of adaptive filter for adaptive equalisation.*

therefore, be disconnected until a new line is dialled up.

In digital mobile radio applications, for example, this is not the case since the channel between the transmitter and receiver will change with time, and also with the position and velocity of the mobile transceiver. Thus the optimum filter is required to track changes in the channel.

One way of tackling this problem is to use what is called 'decision directed' operation, Figure 8.43. If the equaliser is working well then $\hat{x}(kT_o)$ will just be the binary data plus noise. We can remove the noise in a decision circuit which simply tests whether $\hat{x}(kT_o)$ is greater or less than zero. The output of the decision circuit is then identical to the transmitted data. We can use this data to continue training the adaptive filter by changing the switch in Figure 8.43 to its lower position. Here the error is the difference between the binary data and the filter output. This 'decision directed' system will work well provided the receiver continues to make correct decisions about the transmitted data. The decision directed equaliser thus operates effectively in slowly changing channels where the adaptive filter feedback loop can track these changes. If the filter cannot track the signal then the switch in Figure 8.43 must be moved to the upper position and the training sequence retransmitted to initialise the filter, as previously in Figure 8.42. The adaptive filter concept is fundamental to the operation of digital cellular systems which must overcome the channel fading effects described in section 15.2.

A useful way of quickly assessing the performance of a digital communications equaliser is to display its output as an eye diagram on an oscilloscope as described in section 6.6.3.

8.6 Summary

Two types of optimum filtering are important to digital communications. Nyquist filtering constrains the bandwidth of a signal whilst avoiding sampling instant ISI at the

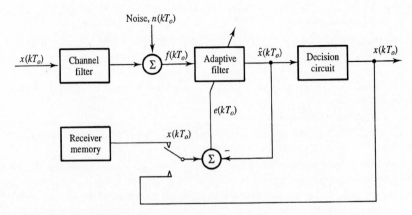

Figure 8.43 *Decision directed adaptive equaliser operation.*

decision circuit input. Matched filtering maximises the sampling instant SNR at the decision circuit input. Both types of filtering are optimum, therefore, in the sense that they minimise the probability of bit error. Nyquist filtering can be implemented entirely at the transmitter if no distortion occurs in the transmission channel or receiver. Implementation is often, however, distributed between transmitter and receiver, distortion in the channel being cancelled by a separate equaliser. The amplitude response of a Nyquist filter has odd symmetry about its −6 dB frequency points and its phase response is linear.

Duobinary signalling uses pulses, transmitted at a baud rate of $1/(2B)$ symbol/s, which suffer from severe, but predictable, ISI. The predictability of the ISI means that it can be cancelled by appropriate signal, or symbol, processing at the receiver, thus effectively allowing ISI free transmission at the maximum theoretical baud rate. Partial response signalling represents a generalisation of the duobinary technique to multilevel signalling. In this case the (predictable) ISI extends across a window of several adjacent symbols.

Matched filtering is implemented at the receiver. The amplitude response of this filter is proportional to the amplitude spectrum of the symbol to which it is matched and its phase response is, to within a linear phase factor, opposite to the phase spectrum of the symbol. Correlation detection is matched filtering implemented in the time domain.

Root raised cosine filters (applied to impulse signalling at the transmitter and the transmitted symbols at the receiver) satisfy both Nyquist and matched filtering criteria, assuming a distortionless, or perfectly equalised, channel. Channels which are time varying may require adaptive equalisation.

8.7 Problems

8.1. A binary information source consists of statistically independent, equiprobable, symbols. If the bandwidth of the baseband channel over which the symbols are to be transmitted is 3.0 kHz what baud rate will be necessary to achieve a spectral efficiency of 2.5 bit/s/Hz? Is ISI free reception at this baud rate possible? What must be the minimum size of the source symbol alphabet to achieve ISI free reception and a spectral efficiency of 16 bit/s/Hz? [7.5 kbaud, no, 256]

8.2. What is the Nyquist filtering criterion expressed in: (a) the time domain; and (b) the frequency domain?

8.3. State Nyquist's vestigial symmetry theorem. Why is this theorem useful in the context of digital communications?

8.4. Which is the more general, the family of Nyquist filters or the family of raised cosine filters? Sketch the amplitude response of: (a) a baseband raised cosine filter with a normalised excess bandwidth of 0.3; and (b) a bandpass full raised cosine filter.

8.5. Given that a Nyquist filter has odd symmetry about its parent rectangular filter's cut-off frequency, demonstrate that the impulse response of the Nyquist filter retains those zeros present in the impulse response of the top hat filter. [Hint: Consider how the odd symmetry of the Nyquist filter's frequency response could be obtained by convolving the rectangular function with an even function.]

8.6 Justify Nyquist's vestigial symmetry theorem, as encapsulated in Figure 8.8, directly (i.e.

without recourse to the argument used in Problem 8.5).

8.7. A baseband binary (2-level) PCM system is used to transmit a single 3.4 kHz voice signal. If the sampling rate for the voice signal is 8 kHz and 256 level quantisation is used, calculate the bandwidth required. Assume that the total system frequency response has a full raised cosine characteristic. [64 kHz]

8.8. An engineer proposes to use sinc pulse signalling for a baseband digital communications system on the grounds that the sinc pulse is a special case of a Nyquist signal. Briefly state whether you support or oppose the proposal and, on what grounds, your case is based.

8.9. A baseband transmission channel has a raised-cosine frequency response with a roll-off factor, $\alpha = 0.4$. The channel has an (absolute) bandwidth of 1200 kHz. An analogue signal is converted to binary PCM with 64-level quantisation before being transmitted over the channel. What is the maximum limit on the bandwidth of the analogue signal? What is the maximum possible spectral efficiency of this system? [143 kHz, 1.43 bit/s/Hz]

8.10. A voice signal is restricted to a bandwidth of 3.0 kHz by an ideal anti-aliasing filter. If the bandlimited signal is then over-sampled by 33% find the ISI free bandwidth required for transmission over a channel, having a Nyquist response with a roll-off factor of 50%, for the following schemes: (a) PAM; (b) 8-level quantisation, binary PCM; and (c) 64-level quantisation, binary PCM. [6 kHz, 18 kHz, 36 kHz]

8.11. A 4-level PCM communications system has a bit rate of 4.8 kbit/s and a raised cosine total system frequency response with a roll-off factor of 0.3. What is the minimum transmission bandwidth required? [N.B. A 4-level PCM system is one in which information signal samples are coded into a 4-level symbol stream rather than the usual binary symbol stream.] [1.56 kHz]

8.12. Demonstrate that duobinary signalling suffers from error propagation whilst precoded duobinary signalling does not.

8.13. (a) What is the principal objective of matched filtering? (b) What is the (white noise) matched filtering criterion expressed in: (i) the time domain; and (ii) the frequency domain?
(c) How is the sampling instant SNR at the output of a matched filter related to the energy of the expected symbol and NPSD at the filter's input?

8.14. Sketch the impulse response of the filter which is matched to the pulse:

$$f(t) = \Pi(t - 0.5) + (2/3)\,\Pi(t - 1.5)$$

What is the output of this filter when the pulse $f(t)$ is present at its input?

8.15. The transmitted pulse shape of an OOK, baseband, communication system is $1000t\Pi([t - 0.5 \times 10^{-3}]/10^{-3})$. What is the impulse response of the predetection filter which maximises the sampling instant, SNR (i.e. the matched filter)? If the noise at the input to the predetection filter is white and Gaussian with a one-sided power spectral density of $2.0 \times 10^{-5} V^2/Hz$, what probability of symbol error would you expect in the absence of intersymbol interference? [2×10^{-4}]

8.16. (a) A polar binary signal consists of +1 or −1 V pulses during the interval $(0, T)$. Additive white Gaussian noise having a two sided NPSD of 10^{-6} W/Hz is added to the signal. If the received signal is detected with a matched filter, determine the maximum bit rate that can be sent with an error probability, P_e, of less than or equal to 10^{-3}. Assume that the impedance level is 50 ohms.
What is the sampling instant SNR in dB at the output of the filter?
Can you say anything about the SNR at the filter input? [2.1 kbit/s, 9.8 dB]

8.17. The time-domain implementation of a matched filter is called a *correlation* detector. In view of the fact that the output of a filter is the *convolution* of its input with its impulse response explain why this terminology is appropriate.

8.18. A binary baseband communications system employs the transmitted pulses $v_0(t) = -\Lambda(t/10^{-3})$ V and $v_1(t) = \Pi(t/10^{-3})$ V to represent digital zeros and ones respectively. What is the ideal, single channel, correlator reference signal for this system? If the loss from transmitter to receiver is 40.0 dB what value of noise power spectral density at the correlator input can be tolerated whilst maintaining a probability of symbol error of 10^{-6}? [5.15×10^{-9} V^2/Hz]

8.19. A digital communications receiver uses root raised cosine filtering at both its transmitter and receiver. Show that the transmitted pulse has the form:

$$v(t) = \frac{4}{\pi} R_s \frac{\cos(2\pi R_s t)}{1 - 16 R_s^2 t^2}$$

where R_s is the baud rate.

CHAPTER 9

Information theory and source coding

9.1 Introduction

There may be a number of reasons for wishing to change the form of a digital signal as supplied by an information source prior to transmission. In the case of English language text, for example, we start with a data source consisting of about 40 distinct symbols (the letters of the alphabet, integers and punctuation). In principle we could transmit such text using a signal alphabet consisting of 40 distinct voltage waveforms. This would constitute an M-ary system where $M = 40$ unique signals. It may be, however, that for one or more of the following reasons this approach is inconvenient, difficult or impossible:

1. The transmission channel may be physically unsuited to carrying such a large number of distinct signals.
2. The relative frequencies with which different source symbols occur will vary widely. This will have the effect of making the transmission inefficient in terms of the time it takes and/or bandwidth it requires.
3. The data may need to be stored and/or processed in some way before transmission. This is most easily achieved using binary electronic devices as the storage and processing elements.

For all these reasons (and perhaps others) sources of digital information are almost always converted as soon as possible into binary form, i.e. each symbol is encoded as a binary word. After appropriate processing the binary words may then be transmitted directly, as either baseband or bandpass signals, or may be recoded into another multi-symbol alphabet. (In the latter case it is unlikely that the transmitted symbols map directly onto the original source symbols.) The body of knowledge, information theory, concerned with the representation of information by symbols gives theoretical bounds on the performance of communication systems and permits assessment of practical system efficiency. The landmarks in information theory were developed by Hartley and Nyquist in the 1920s and are summarised in [Shannon, 1948].

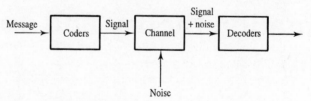

Figure 9.1 *The communications channel model.*

The simplified communications system shown in Figure 9.1, may include several encoders and decoders to implement one or more of the following processes:

- Formatting (which transforms information from its original, or natural, form to a well defined, and standard, digital form, e.g. PCM, see Chapter 6).
- Source coding (which reduces the average number of symbols required to transmit a given message).
- Encryption (which codes messages using a cipher to prevent unauthorised reception or transmission).
- Error control coding (which allows a receiver to detect, and sometimes correct, symbols which are received in error, see Chapter 10).
- Line coding/pulse shaping (which ensures the transmitted symbol waveforms are well suited to the characteristics of the channel, see Chapters 6 and 8).

It is the second of these processes, source coding, which is the principal concern of this chapter.

9.2 Information and entropy

9.2.1 The information measure

The concept of information content is related to predictability or scarcity value. That is, the more predictable or probable a particular message, the less information is conveyed by transmitting that message. For example, the football score Manchester United 7, Bradford Academicals 0, contains little information. The information content of Bradford Academicals 7, Manchester United 0, on the other hand is enormous![1]

In essence highly probable messages contain little information and we can write:

$$P(message) = 1 \text{ carries zero information}$$

$$P(message) = 0 \text{ carries infinite information}$$

The definition of information content for a symbol m should be such that it monotonically decreases with increasing message probability, $P(m)$, and it goes to zero

[1]Readers must be aware that Manchester United is a premier division team while Bradford Academicals is a non-league team and thus the chances are very remote for the second score to occur.

for a probability of unity. Another desirable property is that of additivity. If one were to communicate two (independent) messages in sequence, the total information content should be equal to the sum of the individual information contents of the two messages. We know that the total probability of the composite message is the product of the two individual, independent, probabilities. Therefore, the definition of information must be such that when probabilities are multiplied information is added.

The required properties of an information measure are summarised in Table 9.1.

Table 9.1 *Information measures.*

Message	P (message)	Information content
m_1	$P(m_1)$	I_1
m_2	$P(m_2)$	I_2
$(m_1 + m_2)$	$P(m_1)P(m_2)$	$I_1 + I_2$

The logarithm operation clearly satisfies these requirements. We thus *define* ($\underset{=}{\Delta}$) the information content, I_m, of a message, m, as:

$$I_m \underset{=}{\Delta} \log \frac{1}{P(m)} \equiv -\log P(m) \tag{9.1}$$

This definition satisfies the additivity requirement, the monotonicity requirement, and for $P(m) = 1$, $I_m = 0$. Note that this is true regardless of the base chosen for the logarithm. Base 2 is usually chosen, however, the resulting quantity of information being measured in bits:

$$I_1 + I_2 = -\log_2 P(m_1) - \log_2 P(m_2) = -\log_2 [P(m_1)\ P(m_2)] \quad \text{(bits)} \tag{9.2}$$

Table 9.2 *Word length and symbol probabilities.*

Vocabulary size	No. binary digits	Symbol probability
2	1	1/2
4	2	1/4
8	3	1/8
•	•	•
•	•	•
•	•	•
128	7	1/128

9.2.2 Multisymbol alphabets

Consider, initially, vocabularies of equiprobable message symbols represented by fixed length binary codewords. Thus a vocabulary size of four symbols is represented by the binary digit pairs 00, 01, 10, 11. The binary word length and symbol probabilities for other vocabulary sizes are shown in Table 9.2. The ASCII vocabulary used in teleprinters contains 128 symbols and therefore uses a $\log_2 128 = 7$ digit (fixed length) binary code word to represent each symbol. ASCII symbols are not equiprobable, however, and for this particular code each symbol does not, therefore, have a selection probability of 1/128.

9.2.3 Commonly confused entities

Several information-related quantities are sometimes confused. These are:

- Symbol   A member of a source alphabet. May or may not be binary, e.g. 2 symbol binary, 4 symbol PSK (see Chapter 11), 128 symbol ASCII.
- Baud   Rate of symbol transmission, i.e. 100 baud = 100 symbol/s.
- Bit   Quantity of information carried by a symbol with selection probability $P = 0.5$.
- Bit rate   Rate of information transmission (bit/s). (In the special, but common, case of signalling using independent, equiprobable, binary symbols, the bit rate equals the baud rate.)
- Message   A meaningful sequence of symbols. (Also often used to mean a source symbol.)

9.2.4 Entropy of a binary source

Entropy (H) is defined as the average amount of information conveyed per symbol. For an alphabet of size 2 and assuming that symbols are statistically independent:

$$H \underset{=}{\Delta} \sum_{m=1}^{2} P(m) \log_2 \frac{1}{P(m)} \quad \text{(bit/symbol)} \tag{9.3}$$

For the two symbol alphabet (0, 1) if we let $P(1) = p$ then $P(0) = 1 - p$ and:

$$H = p \log_2 \frac{1}{p} + (1 - p) \log_2 \frac{1}{1 - p} \quad \text{(bit/symbol)} \tag{9.4}$$

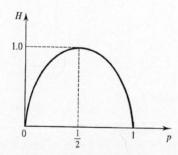

Figure 9.2 *Entropy for binary data transmission versus the selection probability, p, of a digital 1.*

The entropy is maximised when the symbols are equiprobable as shown in Figure 9.2. The entropy definition of equation (9.3) holds for all alphabet sizes. Note that, in the binary case, as either of the two messages becomes more likely, the entropy decreases. When either message has probability 1, the entropy goes to zero. This is reasonable since, at these points, the outcome of the transmission is certain. Thus, if $P(0) = 1$, we know the symbol 0 will be sent repeatedly. If $P(0) = 0$, we know the symbol 1 will be sent repeatedly. In these two cases, no information is conveyed by transmitting the symbols.

For the ASCII alphabet, source entropy would be given by $H = -\log_2 (1/128) = 7$ bit/symbol if all the symbols were both *equiprobable* and *statistically independent*. In practice H is less than this, i.e.:

$$H \underset{=}{\Delta} \sum_{m=1}^{128} P(m) \log_2 \frac{1}{P(m)} \quad < 7 \text{ bit/symbol} \tag{9.5}$$

since the symbols are neither, making the code less than 100% efficient. Entropy thus indicates the *minimum* number of binary digits required per symbol (averaged over a long sequence of symbols).

9.3 Conditional entropy and redundancy

For sources in which each symbol selected is not statistically independent from all previous symbols (i.e. sources with memory) equation (9.3) is insufficiently general to give the entropy correctly. In this case the joint and conditional statistics (section 3.2.1) of symbol sequences must be considered. A source with a memory of one symbol, for example, has an entropy given by:

$$H = \sum_i \sum_j P(j,i) \log_2 \frac{1}{P(j|i)} \quad \text{(bit/symbol)} \tag{9.6}$$

where $P(j,i)$ is the probability of the source selecting i and j and $P(j|i)$ is the probability that the source will select j given that it has previously selected i. Bayes's theorem, equation (3.3), can be used to re-express equation (9.6) as:

$$H = \sum_i P(i) \sum_j P(j|i) \log_2 \frac{1}{P(j|i)} \quad \text{(bit/symbol)} \tag{9.7}$$

(For independent symbols $P(j|i) = P(j)$ and equation (9.7) reduces to equation (9.3).) The effect of having dependency between symbols is to increase the probability of selecting some symbols at the expense of others *given a particular symbol history*. This reduces the average information conveyed by the symbols, which is reflected in a reduced entropy. The difference between the actual entropy of a source and the (maximum) entropy, H_{max}, the source could have if its symbols were independent and equiprobable is called the *redundancy* of the source. For an M symbol alphabet redundancy, R, is therefore given by:

$$R = H_{\max} - H$$

$$= \log_2 (M) - H \quad \text{(bit/symbol)} \tag{9.8}$$

(It is easily shown that the quantity $R/(H^2 + RH)$ relates to the number of symbols per bit of information which are transmitted unnecessarily from an information theory point of view. There may well be good reasons, however, to transmit such redundant symbols, e.g. for error control purposes as discussed in Chapter 10.)

EXAMPLE 9.1

Find the entropy, redundancy and information rate of a four symbol source (A, B, C, D) with a baud rate of 1024 symbol/s and symbol selection probabilities of 0.5, 0.2, 0.2 and 0.1 under the following conditions:

(i)   The source is memoryless (i.e. the symbols are statistically independent).

(ii) The source has a one symbol memory such that no two consecutively selected symbols can be the same. (The long term relative frequencies of the symbols remain unchanged, however.)

(i)

$$H = \sum_{m=1}^{M} p(m) \log_2 \frac{1}{p(m)}$$

$$= 0.5 \log_2 \left(\frac{1}{0.5} \right) + 2 \times 0.2 \log_2 \left(\frac{1}{0.2} \right) + 0.1 \log_2 \left(\frac{1}{0.1} \right)$$

$$= 0.5 \frac{\log_{10} 2}{\log_{10} 2} + 0.4 \frac{\log_{10} 5}{\log_{10} 2} + 0.1 \frac{\log_{10} 10}{\log_{10} 2}$$

$$= 1.761 \quad \text{bit/symbol}$$

$$R = H_{\max} - H$$

$$= \log_2 M - H$$

$$= \log_2 4 - 1.761 = 2.0 - 1.761 = 0.239 \quad \text{bit/symbol}$$

$$R_i = R_s H = 1024 \times 1.761 = 1.803 \times 10^3 \quad \text{bit/s}$$

where R_i is the information rate and R_s is the symbol rate.

(ii) The appropriate formula to apply to find the entropy of a source with one symbol memory is equation (9.7). First, however, we must find the conditional probabilities which the formula contains. If no two consecutive symbols can be the same then:

$$P(A|A) = P(B|B) = P(C|C) = P(D|D) = 0$$

Since the (unconditional) probability of A is unchanged, $P(A) = 0.5$, then every, and only every, alternate symbol must be A, $P(\bar{A}|A) = P(A|\bar{A}) = 1.0$, where $\bar{A}$ represents not A. Furthermore if every alternate symbol is A then no two non-A symbols can occur consecutively, $P(\bar{A}|\bar{A}) = 0$. Writing the above three probabilities explicitly:

$$P(B|A) + P(C|A) + P(D|A) = 1.0$$

$$P(A|B) = P(A|C) = P(A|D) = 1.0$$

$$P(B|C) = P(B|D) = P(C|B) = P(C|D) = P(D|B) = P(D|C) = 0$$

Since the (unconditional) probabilities of B, C and D are to remain unchanged, i.e.:

$$P(B) = P(C) = 0.2 \text{ and } P(D) = 0.1$$

the conditional probability $P(B|A)$ must satisfy:

$$P(B) = P(B|A)P(A) + P(B|B)P(B) + P(B|C)P(C) + P(B|D)P(D)$$

$$= P(B|A)0.5 + 0 + 0 + 0$$

i.e. $P(B|A) = \dfrac{0.2}{0.5} = 0.4$

Similarly:

$$P(C) = P(C|A)P(A) + P(C|B)P(B) + P(C|C)P(C) + P(C|D)P(D)$$

$$= P(C|A)\,0.5 + 0 + 0 + 0$$

i.e. $P(C|A) = \dfrac{0.2}{0.5} = 0.4$

$$P(D) = P(D|A)P(A) + P(D|B)P(B) + P(D|C)P(C) + P(D|D)P(D)$$

$$= P(D|A)\,0.5 + 0 + 0 + 0$$

i.e. $P(D|A) = \dfrac{0.1}{0.5} = 0.2$

We now have numerical values for all the conditional probabilities which can be substituted into equation (9.7):

$$H = \sum_i P(i) \sum_j P(j|i) \log_2 \frac{1}{P(j|i)} \quad \text{(bit/symbol)}$$

$$= P(A)\left[P(A|A) \log_2 \frac{1}{P(A|A)} + P(B|A) \log_2 \frac{1}{P(B|A)} + P(C|A) \log_2 \frac{1}{P(C|A)} + P(D|A) \log_2 \frac{1}{P(D|A)} \right]$$

$$+ P(B)\left[P(A|B) \log_2 \frac{1}{P(A|B)} + P(B|B) \log_2 \frac{1}{P(B|B)} + P(C|B) \log_2 \frac{1}{P(C|B)} + P(D|B) \log_2 \frac{1}{P(D|B)} \right]$$

$$+ P(C)\left[P(A|C) \log_2 \frac{1}{P(A|C)} + P(B|C) \log_2 \frac{1}{P(B|C)} + P(C|C) \log_2 \frac{1}{P(C|C)} + P(D|C) \log_2 \frac{1}{P(D|C)} \right]$$

$$+ P(D)\left[P(A|D) \log_2 \frac{1}{P(A|D)} + P(B|D) \log_2 \frac{1}{P(B|D)} + P(C|D) \log_2 \frac{1}{P(C|D)} + P(D|D) \log_2 \frac{1}{P(D|D)} \right]$$

$$= 0.5\left[0.0 \log_2 (1/0) + 0.4 \log_2 2.5 + 0.4 \log_2 2.5 + 0.2 \log_2 5.0 \right]$$

$$+ 0.2\left[1.0 \log_2 1.0 + 0 + 0 + 0 \right] + 0.2\left[1.0 \log_2 1.0 + 0 + 0 + 0 \right]$$

$$+ \, 0.1 \left[\; 1.0 \log_2 1.0 + 0 + 0 + 0 \; \right]$$

$$= 0.5 \left[\; 0.4 \log_{10} 2.5 + 0.4 \log_{10} 2.5 + 0.2 \log_{10} 5.0 \; \right] / \log_{10} 2 \; = \; 0.761 \; \text{ bit/symbol}$$

$$R \; = \; H_{\max} - H$$

$$= \; \log_2 4 - 0.761 \; = \; 1.239 \; \text{ bit/symbol}$$

(Notice that in this case the symbol *A* carries no information at all since its occurrences are entirely predictable.)

$$R_i \; = \; R_s H = 1024 \times 0.761 = 779 \;\; \text{bit/s}$$

9.4 Information loss due to noise

The information transmitted by a memoryless source (i.e. a source in which the symbols selected are statistically independent) is related only to the source symbol probabilities by equation (9.1), i.e.:

$$I_{TX}(i_{TX}) \; = \; \log_2 \frac{1}{P(i_{TX})} \quad \text{(bits)} \tag{9.9}$$

where $I_{TX}(i_{TX})$ is the information transmitted when the ith source symbol is selected for transmission and $P(i_{TX})$ is the (a priori) probability that the ith symbol will be selected (see section 7.2 for the definition of a priori probability). The use of a subscript *TX* with i, although not really necessary, will help with what can be a confusing notation later in the chapter. For a noiseless channel there is no doubt on detecting a given received voltage v_{RX} corresponding to a given transmitted symbol, $(v_{RX} : i_{TX})$, which source symbol was selected for transmission. The quantity of information, in this case, gained by receiving v_{RX} is therefore identical to that transmitted, i.e.:

$$I_{RX}(v_{RX} : i_{TX}) \; = \; I_{TX}(i_{TX})$$

$$= \; \log_2 \frac{1}{P(i_{TX})} \quad \text{(bits)} \tag{9.10}$$

For a noisy channel, however, there is some uncertainty given a detected voltage, v_{RX}, about which symbol was actually selected at the source. This uncertainty is related to the (a posteriori) probability, $P(i_{TX}|v_{RX})$, that symbol i was transmitted given that voltage v_{RX} was detected (see again section 7.2 for the definition of an a posteriori probability). (For the noiseless channel $P(i_{TX}|v_{RX})$ is unity for one particular source symbol and zero for all others.) Intuition tells us that for a noisy channel the information received should be less than that transmitted by an amount related to the uncertainty in the decision process. In fact the received information is given by:

$$I_{RX}(v_{RX} : i_{TX}) \; = \; \log_2 \frac{P(i_{TX}|v_{RX})}{P(i_{TX})} \quad \text{(bits)} \tag{9.11}$$

Using $P(i_{TX}|v_{RX})$ in equation (9.11) corresponds to a soft decision process (see Figure 7.1). For hard decision processes the detected voltage v_{RX} is converted immediately to a received symbol i_{RX}. In this case equation (9.11) is rewritten as:

$$I_{RX}(i_{RX}) = \log_2 \frac{P(i_{TX}|i_{RX})}{P(i_{TX})} \quad \text{(bits)} \tag{9.12}$$

where $P(i_{TX}|i_{RX})$ can be found from (but is not an element of) the end-to-end symbol transition matrix, $P(i_{RX}|j_{TX})$, described in section 7.3. (Note the interchanged position of *TX* and *RX* quantities.) Care must be taken not to confuse the conditional probabilities here (which relate to channel induced uncertainties, see section 7.2) with those in section 9.3 (which relate to source memory).

The reduction in information content of each symbol due to noise means that the effective entropy, H_{eff}, of the symbols on reception is less than their transmission entropy, H. The effective (or reception) entropy is defined in the normal way, i.e.:

$$H_{eff} = \sum_i P(i_{RX})I_{RX}(i_{RX}) \quad \text{(bit/symbol)} \tag{9.13}$$

where $P(i_{RX})$ is the (unconditional) symbol reception probability. The difference between the transmission entropy and the effective entropy is called the equivocation, E, i.e.:

$$H_{eff} = H - E \quad \text{(bit/symbol)} \tag{9.14}$$

The equivocation essentially accounts for the uncertainty with which the transmitted symbols can be assumed to be the same as the received symbols. The information associated with this uncertainty and, therefore, the definition of equivocation is:

$$E = \sum_{i,j(i\neq j)} P(i_{TX}, j_{RX}) \log_2 \frac{1}{P(i_{TX}|j_{RX})} \quad \text{(bit/symbol)} \tag{9.15}$$

The similarity between equations (9.15) and (9.3) is notable, the essential difference being that $P(i_{TX}|j_{RX})$ relates to the probability of symbol error. Multiplying by the joint probability $P(i_{TX}, j_{RX})$ and summing over all i, j $(i\neq j)$ simply averages the resulting spurious 'information' over all types of error. Using $P(i_{TX}, j_{RX}) = P(j_{RX})P(i_{TX}|j_{RX})$ equation (9.15) can be rewritten as:

$$E = \sum_j P(j_{RX}) \sum_{i(i\neq j)} P(i_{TX}|j_{RX}) \log_2 \frac{1}{P(i_{TX}|j_{RX})} \tag{9.16}$$

where $P(j_{RX})$ is an (unconditional) symbol reception probability. An alternative interpretation of equivocation is as negative information added by noise.

EXAMPLE 9.2
Consider a 3 symbol source A, B, C with the following transition matrix:

$$
\begin{array}{cccc}
 & A_{TX} & B_{TX} & C_{TX} \\
A_{RX} & 0.6 & 0.5 & 0 \\
B_{RX} & 0.2 & 0.5 & 0.333 \\
C_{RX} & 0.2 & 0 & 0.667
\end{array}
$$

(Note that the elements of the transition matrix are the conditional probabilities $P(i_{RX}|j_{TX})$ and that columns and rows are labelled i_{TX} and i_{RX} only for extra clarity.) For the specific a priori (i.e. transmission) probabilities: $P(A) = 0.5$; $P(B) = 0.2$; $P(C) = 0.3$, find: (i) the (unconditional) symbol reception probabilities $P(i_{RX})$ from the conditional and a priori probabilities; (ii) the equivocation; (iii) the source entropy; and (iv) the effective entropy.

(i) The symbol reception probabilities are found as follows:

$$
\begin{aligned}
P(A_{RX}) &= P(A_{TX})\,P(A_{RX}|A_{TX}) + P(B_{TX})\,P(A_{RX}|B_{TX}) + P(C_{TX})\,P(A_{RX}|C_{TX}) \\
&= 0.5 \times 0.6 + 0.2 \times 0.5 + 0.3 \times 0 = 0.4
\end{aligned}
$$

$$
P(B_{RX}) = 0.5 \times 0.2 + 0.2 \times 0.5 + 0.3 \times 0.333 = 0.3
$$

$$
P(C_{RX}) = 0.5 \times 0.2 + 0.2 \times 0 + 0.3 \times 0.667 = 0.3
$$

(ii) To find the equivocation from equation (9.16) we require the probabilities $P(i_{TX}|j_{RX})$ which can be found in turn from $P(i_{RX}|j_{TX})$ and $P(i_{RX})$ using Bayes's rule, equation (3.3):

$$
\begin{aligned}
P(A_{TX}|A_{RX}) &= \frac{P(A_{TX}, A_{RX})}{P(A_{RX})} \\
&= \frac{P(A_{RX}|A_{TX})\,P(A_{TX})}{P(A_{RX})} = \frac{0.6 \times 0.5}{0.4} = 0.75
\end{aligned}
$$

$$
\begin{aligned}
P(B_{TX}|A_{RX}) &= \frac{P(B_{TX}, A_{RX})}{P(A_{RX})} \\
&= \frac{P(A_{RX}|B_{TX})P(B_{TX})}{P(A_{RX})} = \frac{0.5 \times 0.2}{0.4} = 0.25
\end{aligned}
$$

$$
P(C_{TX}|A_{RX}) = \frac{P(A_{RX}|C_{TX})P(C_{TX})}{P(A_{RX})} = \frac{0 \times 0.3}{0.4} = 0
$$

$$
P(A_{TX}|B_{RX}) = \frac{P(B_{RX}|A_{TX})P(A_{TX})}{P(B_{RX})} = \frac{0.2 \times 0.5}{0.3} = 0.333
$$

$$
P(B_{TX}|B_{RX}) = \frac{P(B_{RX}|B_{TX})P(B_{TX})}{P(B_{RX})} = \frac{0.5 \times 0.2}{0.3} = 0.333
$$

$$
P(C_{TX}|B_{RX}) = \frac{P(B_{RX}|C_{TX})\,P(C_{TX})}{P(B_{RX})} = \frac{0.333 \times 0.3}{0.3} = 0.333
$$

$$
P(A_{TX}|C_{RX}) = \frac{P(C_{RX}|A_{TX})P(A_{TX})}{P(C_{RX})} = \frac{0.2 \times 0.5}{0.3} = 0.333
$$

$$
P(B_{TX}|C_{RX}) = \frac{P(C_{RX}|B_{TX})P(B_{TX})}{P(C_{RX})} = \frac{0 \times 0.2}{0.3} = 0
$$

$$
P(C_{TX}|C_{RX}) = \frac{P(C_{RX}|C_{TX})\,P(C_{TX})}{P(C_{RX})} = \frac{0.667 \times 0.3}{0.3} = 0.667
$$

Equation (9.16) now gives the equivocation:

$$E = P(A_{RX}) \left[P(B_{TX}|A_{RX}) \log_2 \frac{1}{P(B_{TX}|A_{RX})} + P(C_{TX}|A_{RX}) \log_2 \frac{1}{P(C_{TX}|A_{RX})} \right]$$

$$+ P(B_{RX}) \left[P(A_{TX}|B_{RX}) \log_2 \frac{1}{P(A_{TX}|B_{RX})} + P(C_{TX}|B_{RX}) \log_2 \frac{1}{P(C_{TX}|B_{RX})} \right]$$

$$+ P(C_{RX}) \left[P(A_{TX}|C_{RX}) \log_2 \frac{1}{P(A_{TX}|C_{RX})} + P(B_{TX}|C_{RX}) \log_2 \frac{1}{P(B_{TX}|C_{RX})} \right]$$

$$= 0.4 \left[0.25 \log_2 \left(\frac{1}{0.25} \right) + 0 \right]$$

$$+ 0.3 \left[0.333 \log_2 \left(\frac{1}{0.333} \right) + 0.333 \log_2 \left(\frac{1}{0.333} \right) \right]$$

$$+ 0.3 \left[0.333 \log_2 \left(\frac{1}{0.333} \right) + 0 \right]$$

$$= 0.4 [0.5] + 0.3 [0.528 + 0.528] + 0.3 [0.528] = 0.675 \text{ bit/symbol}$$

(iii) The source entropy H is:

$$H = \sum_i P(i_{TX}) \log_2 \frac{1}{P(i_{TX})}$$

$$= 0.5 \log_2 \left(\frac{1}{0.5} \right) + 0.2 \log_2 \left(\frac{1}{0.2} \right) + 0.3 \log_2 \left(\frac{1}{0.3} \right)$$

$$= 1.48 \text{ bit/symbol}$$

(iv) The effective entropy, equation (9.14), is therefore:

$$H_{eff} = H - E$$

$$= 1.48 - 0.67 = 0.81 \text{ bit/symbol}$$

Note that the effective entropy could have been found without calculating the equivocation using equations (9.13) and (9.12), i.e.:

$$H_{eff} = P(A_{RX}) \log_2 \frac{P(A_{TX}|A_{RX})}{P(A_{TX})} + P(B_{RX}) \log_2 \frac{P(B_{TX}|B_{RX})}{P(B_{TX})}$$

$$+ P(C_{RX}) \log_2 \frac{P(C_{TX}|C_{RX})}{P(C_{TX})}$$

$$= 0.4 \log_2 \left(\frac{0.75}{0.5} \right) + 0.3 \log_2 \left(\frac{0.333}{0.2} \right) + 0.3 \log_2 \left(\frac{0.667}{0.3} \right)$$

$$= 0.80 \text{ bit/symbol}$$

9.5 Source coding

Source coding does not change or alter the source entropy, i.e. the average number of information bits per source symbol. In this sense source entropy is a fundamental property of the source. Source coding does, however, alter (usually increase) the entropy of the source coded symbols. It may also reduce fluctuations in the information rate from the source and avoid symbol 'surges' which could overload the channel when the message sequence contains many high probability (i.e. frequently occurring, low entropy) symbols.

9.5.1 Code efficiency

Recall the definition of entropy (equation (9.3)) for a source with statistically independent symbols:

$$H = \sum_{m=1}^{M} P(m) \log_2 \frac{1}{P(m)} \quad \text{(bit/symbol)} \tag{9.17}$$

The maximum possible entropy, $H_{\max}$, of this source would be realised if all symbols were equiprobable, $P(m) = 1/M$, i.e.:

$$H_{\max} = \log_2 M \quad \text{(bit/symbol)} \tag{9.18}$$

A code efficiency can therefore be defined as:

$$\eta_{code} = \frac{H}{H_{\max}} \times 100\% \tag{9.19}$$

If source symbols are coded into another symbol set, Figure 9.3, then the new code efficiency is given by equation (9.19) where H and $H_{\max}$ are the entropy and maximum possible entropy of this new symbol set.

If source symbols are coded into binary words then there is a useful alternative interpretation of η_{code}. For a set of symbols represented by binary code words with lengths l_m (binary) digits, an overall code length, L, can be defined as the average codeword length, i.e.:

$$L = \sum_{m=1}^{M} P(m) \, l_m \quad \text{(binary digits/symbol)} \tag{9.20}$$

The code efficiency can then be found from:

$$\eta_{code} = \frac{H}{L} \quad \text{(bit/binary digit)} \times 100\% \tag{9.21}$$

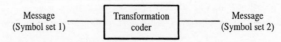

Figure 9.3 *Translating a message between different alphabets.*

Equations (9.21) and (9.19) are seen to be entirely consistent when it is remembered that the maximum information conveyable per L digit binary code word is given by:

$$H_{max} \text{ (bit/symbol)} = L \text{ (bit/codeword)} \qquad (9.22)$$

EXAMPLE 9.3

A scanner converts a black and white document, line-by-line into binary data for transmission. The scanner produces source data comprising symbols representing runs of up to six similar image pixel elements with the probabilities as shown below:

No. of consecutive pixels	1	2	3	4	5	6
Probability of occurrence	0.2	0.4	0.15	0.1	0.06	0.09

Determine the average length of a run (in pixels) and the corresponding effective information rate for this source when the scanner is traversing 1000 pixel/s.

$$H = \sum_{m=1}^{6} P(m) \log_2 \frac{1}{P(m)}$$

$$= 0.2 \times 2.32 + 0.4 \times 1.32 + 0.15 \times 2.74 + 0.1 \times 3.32 + 0.06 \times 4.06 + 0.09 \times 3.47$$

$$= 2.29 \text{ bit/symbol}$$

Average length:

$$L = \sum_{m=1}^{6} P(m) \, l_m = 0.2 + 0.8 + 0.45 + 0.4 + 0.3 + 0.54 = 2.69 \text{ pixels}$$

At 1000 pixel/s scan rate we generate 1000/2.69 = 372 symbol/s. Thus the source information rate is $2.29 \times 372 = 852$ bit/s.

We are generally interested in finding a more efficient code which represents the same information using fewer digits on average. This results in different lengths of codeword being used for different symbols. The problem with such variable length codes is in recognising the start and end of the symbols.

9.5.2 Decoding variable length codewords

The following properties need to be considered when attempting to decode variable length codewords:

(1) Unique decoding.

This is essential if the received message is to have only a single possible meaning. Consider an $M = 4$ symbol alphabet with symbols represented by binary digits as follows:

$$A = 0$$
$$B = 01$$
$$C = 11$$
$$D = 00$$

If we receive the code word 0011 it is not known whether the transmission was D, C or A, A, C. This example is not, therefore, uniquely decodable.

(2) Instantaneous decoding.

Consider now an $M = 4$ symbol alphabet, with the following binary representation:

$$A = 0$$
$$B = 10$$
$$C = 110$$
$$D = 111$$

This code can be instantaneously decoded using the decision tree shown in Figure 9.4 since no complete codeword is a prefix of a larger codeword. This is in contrast to the previous example where A is a prefix of both B and D. The latter example is also a 'comma code' as the symbol zero indicates the end of a codeword except for the all ones word whose length is known. Note that we are restricted in the number of available codewords with small numbers of bits to ensure we achieve the desired decoding properties.

Using the representation:

$$A = 0$$
$$B = 01$$
$$C = 011$$
$$D = 111$$

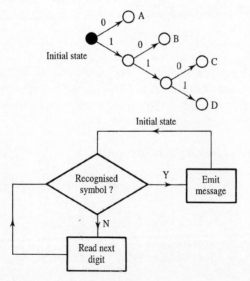

Figure 9.4 *Algorithm for decision tree decoding and example of practical code tree.*

the code is identical to the example just given but the bits are time reversed. It is thus still uniquely decodable but no longer instantaneous, since early codewords are now prefixes of later ones.

9.6 Variable length coding

Assume an $M = 8$ symbol source $A, \cdots, H$ having probabilities of symbol occurrence:

m	A	B	C	D	E	F	G	H
$P(m)$	0.1	0.18	0.4	0.05	0.06	0.1	0.07	0.04

The source entropy is given by:

$$H = \sum_m P(m) \log_2 \frac{1}{P(m)} = 2.55 \text{ bit/symbol} \tag{9.23}$$

and, at a symbol rate of 1 symbol/s, the information rate is 2.55 bit/s. The maximum entropy of an 8 symbol source is $\log_2 8 = 3$ bit/symbol and the source efficiency is therefore given by:

$$\eta_{source} = \frac{2.55}{3} \times 100\% = 85\% \tag{9.24}$$

If the symbols are each allocated 3 bits, comprising all the binary patterns between 000 and 111, the coding efficiency will remain unchanged at 85%.

Shannon-Fano coding [Blahut, 1987], in which we allocate the regularly used or highly probable messages fewer bits, as these are transmitted more often, is more efficient. The less probable messages can then be given the longer, less efficient bit patterns. This yields an improvement in efficiency compared with that before source coding was applied. The improvement is not as great, however, as that obtainable with another variable length coding scheme, namely Huffman coding, which is now described.

9.6.1 Huffman coding

The Huffman coding algorithm comprises two steps – reduction and splitting. These steps can be summarised by the following instructions:

(1) Reduction: List the symbols in descending order of probability. Reduce the two least probable symbols to one symbol with probability equal to their combined probability. Reorder in descending order of probability at each stage, Figure 9.5. Repeat the reduction step until only two symbols remain.

(2) Splitting: Assign 0 and 1 to the two final symbols and work backwards, Figure 9.6. Expand or lengthen the code to cope with each successive split and, at each stage, distinguish between the two split symbols by adding another 0 and 1 respectively to the codeword.

The result of Huffman encoding the symbols $A, \cdots, H$ in the previous example (Figures 9.5 and 9.6) is to allocate the symbols codewords as follows:

Symbol	C	B	A	F	G	E	D	H
Probability	0.40	0.18	0.10	0.10	0.07	0.06	0.05	0.04
Codeword	1	001	011	0000	0100	0101	00010	00011

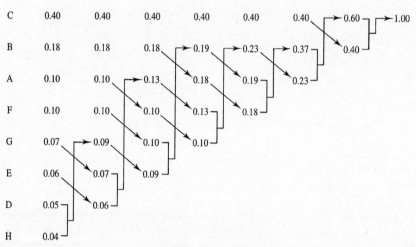

Figure 9.5 *Huffman coding of an 8-symbol alphabet – reduction step.*

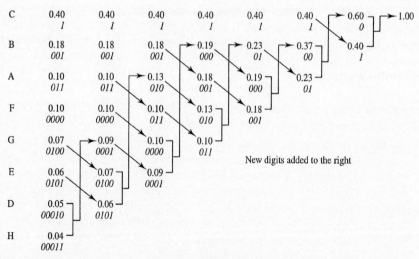

Figure 9.6 *Huffman coding – allocation of codewords to the 8 symbols.*

The code length is now given by equation (9.20) as:

$$L = 1(0.4) + 3(0.18 + 0.10) + 4(0.10 + 0.07 + 0.06) + 5(0.05 + 0.04) = 2.61 \quad (9.25)$$

and the code efficiency, given by equation (9.21), is:

$$\eta_{code} = \frac{H}{L} \times 100\% = \frac{2.55}{2.61} \times 100\% = 97.7\% \quad (9.26)$$

The 85% efficiency without coding would have been improved to 96.6% using Shannon-Fano coding but Huffman coding at 97.7% is even better. (The maximum efficiency is obtained when symbol probabilities are all negative, integer, powers of two, i.e. $1/2^n$.) Note that the Huffman codes are formulated to minimise the average code word length. They do not necessarily possess error detection properties but are uniquely, and instantaneously, decodable, as defined in section 9.5.2.

9.7 Source coding examples

An early example of source coding occurs in Morse where the most commonly occurring letter 'e' (dot) is allocated the shortest code (one bit) and less used consonants, such as 'y' (dash dot dash dash), are allocated 4 bits. A more recent example is facsimile (fax) transmission where an A4 page is scanned at 3.85 scan lines/mm in the vertical dimension with 1728 pixels across each scan line. If each pixel is then binary quantised into black or white this produces 2 Mbit of data per A4 page. If transmitted at 4.8 kbit/s over a telephone modem (see Chapter 11), this takes approximately 7 min/page for transmission.

In Group 3 fax runs of pixels of the same polarity are examined. (Certain run lengths are more common than others, Figure 9.7.) The black letters, or drawn lines, are generally not more than 10 pixels wide while large white areas are much more common. Huffman coding is employed in the ITU-T standard for Group 3 fax transmission, Figure 9.8, to allocate the shortest codes to the most common run lengths. The basic scheme is modified to include a unique end of line code for resynchronisation which determines

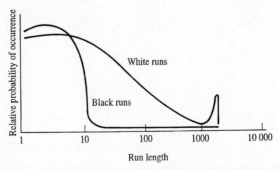

Figure 9.7 *Relative probability of occurrence of run lengths in scanned monochrome text.*

whether lines are received in error, as these should be always 1728 pixels apart.

In this application, Huffman coding improves the efficiency and reduces the time to transmit the page to less than 1 minute. Facsimile is now very significant with 25% of international telephone traffic representing fax transmissions between the 15 million terminals in use worldwide. Continued improvements in the Group 3 fax have made the transition to (the more advanced) Group 4 digital system unattractive. There is now an ITU-T V.17 data modem standard (see Section 11.6) which transmits fax signals at 14.4 kbit/s and which is very similar to the V.32 9.6 kbit/s standard used to achieve the Figure 11.55(c) result.

There is current interest in run length codes aimed at altering their properties to control the spectral shape of the coded signal. In particular it is desirable to introduce a null at DC, as discussed in section 6.4, and/or place nulls or minima at those locations where there is expected to be a null in the channel's frequency response. Such techniques are widely used to optimise the performance of optical and magnetic recording systems [Schouhammer-Immink]. Run length codes can also be modified to achieve some error control properties, as will be described in the next chapter.

9.7.1 Source coding for speech signals

Vocoders (voice coders) are simplified coding devices which extract, in an efficient way, the significant components in a speech waveform, exploiting speech redundancy, to achieve low bit rate (< 2.4 kbit/s) transmission. Speech basically comprises four *formants*, (Figure 9.9), and the vocoder analyses the input signal to find how the position, F_1, F_2, F_3 and F_4, and magnitudes, A_1, A_2, A_3 and A_4, of these formants vary with time. In this instance we fit a model to the input spectrum and transmit the model parameters rather than, for example, the ADPCM quantised error samples, $\varepsilon_q(kT_s)$, of

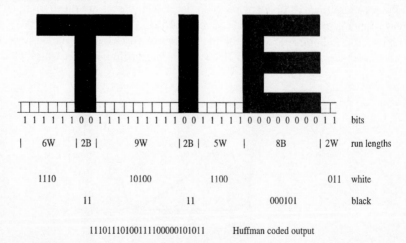

Figure 9.8 *Redundancy removal by source coding of printed character data (source: Pugh, 1991, reproduced with the permission of the IEE).*

section 5.8.3.

Two major vocoder designs exist at present, the channel vocoder and the linear predictive coder (LPC). The channel vocoder is basically a non-linearly spaced filterbank spectrum analyser with 19 distinct filters. The LPC vocoder is usually implemented as a cascade of linear prediction error filters which remove from the speech signal the components that can be predicted [Jayant and Noll, Gray and Markel] from its previous history by modelling the vocal tract as an all-pole filter. (See standard signals and systems texts such as [Jackson, Mulgrew and Grant] for a discussion of the properties of poles and zeros and their effect on system, and filter, frequency responses.) The LPC vocoder extends the DPCM system described in Chapter 5 to perform a full modelling of the speech production mechanism and removes the necessity for any quantised error sample transmissions.

In the LPC vocoder, Figure 9.10, the analyser and encoder normally process the signal in 20-ms frames and subsequently transmit the coarse spectral information via the filter coefficients. The residual error output from the parameter estimation operation is not transmitted. The error signal is used to provide an estimate of the input power level which is sent, along with the pitch information and a binary decision as to whether the input is voiced or unvoiced (Figure 9.10). The pitch information can be ascertained by using autocorrelation (see sections 2.6 and 3.3.3) or zero crossing techniques on either the input signal or the residual error signal, depending on the sophistication and cost of the vocoder implementation. The decoder and synthesiser apply the received filter coefficients to a synthesising filter which is excited with impulses at the pitch frequency if voiced, or by white noise if unvoiced. The excitation amplitude is controlled by the input power estimate information. This excitation with a synthetic signal reduces the transmission bit rate requirement but also reduces the speech quality. (Speech quality is assessed using the mean opinion score (MOS) scale (see section 5.7.3).)

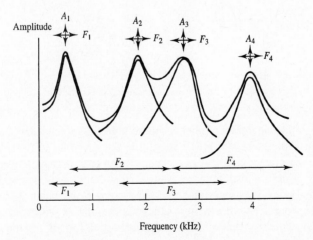

Figure 9.9 *Instantaneous spectral representation of speech as a set of four formant frequencies.*

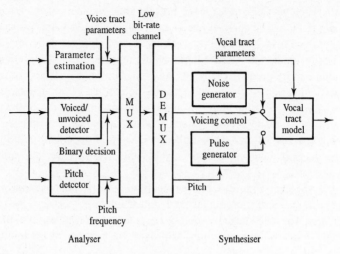

Figure 9.10 *Block diagram of the linear predictive vocoder.*

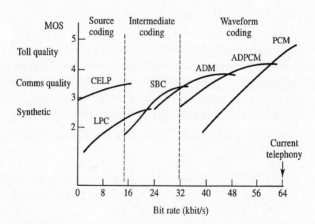

Figure 9.11 *Comparison of quality and transmission rate for various speech coding systems, using the mean opinion score (MOS) rating.*

With delays in the vocal tract of about 1 ms and typical speech sample rates of 8 to 10 kHz the number of predictor stages is normally in the range 8 to 12. 10 is the number adopted in the integrated NATO LPC vocoder standard (LPC-10, also US Federal Standard 1015) which transmits at a rate of 2.4 kbit/s and achieves a MOS of approximately 2, Figure 9.11 and Table 9.3, for a complexity which is 40 times that of PCM. The MOS score is low due to the synthesiser's excitation with noise or regenerated pitch information. This causes LPC to lose all background sounds and the speech to have a characteristic metallic sound. The basic spoken words, however, are still quite intelligible. These vocoders can be implemented as single chip DSP designs. In addition to vocoder applications, LPC synthesisers are used in several commercial speech systems, including the Texas Instruments speak and spell children's toy.

Much current research is in progress to improve the quality of vocoders. This includes studies of hybrid coder systems such as LPC excited from a codebook of possible signal vectors (CELP) or excited with multiple pulses (MPE) rather than simple pitch information as in Figure 9.10. US Federal Standard 1016 defines a 4.8 kbit/s CELP implementation for secure systems. In developing these systems low coder delay (not greater than 2 ms) is an important design goal.

CELP systems achieve a MOS of 3.2 at 4 kbit/s but they are 50 to 100 times as complex as the basic PCM coder. For higher quality transmission ITU-T G.728 defines a 16 kbit/s low delay CELP implementation. These two approaches, which also involve feedback to compare the synthesised speech with the original speech and minimise the difference, are being developed for mobile systems such as GSM (see section 15.4).

Table 9.3 *Rate, performance and complexity comparison for various speech coder designs.*

Class	Technique	Bit rate (kbit/s)	MOS quality	Relative complexity
Waveform coders	PCM G.711	64	4.3	1
	ADPCM G.721	32	4.1	10
	DM	16	3	0.3
Intermediate coder	SBC (2) G.722 (7 kHz)	64 – 48	4.3	30
Enhanced source coder	CELP/MPE for ½ rate GSM	4.8	3.2	50 – 100
Source coder	LPC-10 vocoder	2.4	2	40

9.7.2 High quality audio coders

Other audio coding techniques are based on frequency domain approaches where the 0.2 to 3.2 kHz speech bandwidth is split into 2 to 16 individual sub-bands by a filterbank or a discrete Fourier transform (DFT) (see section 13.5), Figure 9.12, to form a sub-band coder (SBC). Band splitting is used to exploit the fact that the individual bands do not all contain signals with the same energy. This permits the accuracy of the quantiser in the encoder to be reduced, in bands with low energy signals, saving on the coder transmission rate. While this was initially applied to achieve low bit rate intermediate quality speech at 16 kbit/s, the same technique is now used in the ITU-T G.722 0 to 7 kHz high quality audio coder which employs only two subbands to code this wideband signal at 64/56/48 kbit/s with a MOS between 3.7 and 4.3. This is a low complexity coder, implemented in one DSP microprocessor, which can operate at a P_e of 10^{-4} for ISDN teleconferencing and telephone loudspeaker applications. The MOS drops to 3.0 at a P_e of 10^{-3}. Compact disc player manufacturers are also investigating an extension of this technique with 32 channels within the 20 kHz hi-fi bandwidth. Simple 16-bit PCM at 44.1 ksample/s requires 700 kbit/s but the DFT based coder offers indistinguishable music quality at 88 kbit/s or only 2 bit/sample. This development of the ISO/MPEG standard is very important for the digital storage of broadcast quality signals, for example HDTV and digital audio (see Chapter 16).

Finally it has been shown that variants of these techniques can achieve high quality transmission at 64 kbit/s, which is also very significant for reducing the storage requirements of digital memory systems.

Similar coding techniques are applied to encode video signals but due to the large information content the bit rates are much higher. Video is usually encoded at 140 or 34 Mbit/s with the lower rate involving some of the prediction techniques discussed here. In video there is also redundancy in the vertical and horizontal dimensions, and hence spatial Fourier, and other, transformation techniques [Mulgrew and Grant] are attractive for reducing the bit rate requirements in, for example, video telephony applications (see Chapter 16).

9.7.3 String coding

Lempel-Ziv, or its common implementation, the LZW algorithm [Welch, 1984], trains on data to identify commonly occurring strings of input characters and builds up a table of these strings. Each string in the table is then allocated a unique 12 bit code. Commonly encountered word strings, e.g. 000, 0000, or 'High Street' experience compression when they are allocated their 12-bit codes, especially for long (> 10 character) strings. Thus the scheme relies on redundancy in the character occurrence, i.e. individual character string repetitions, but it does not exploit, in any way, positional redundancy.

LZW compression operates typically on blocks of 10,000 to 30,000 symbols and achieves little compression during the adaptive phase when the table is being constructed. It uses simplified logic, which operates at three clock cycles per symbol. After 2,000 to 5,000 words have been processed typical compression is 1.8:1 on text, 1.5:1 on object files and > 2:1 on data or program source code. There is no compression however on

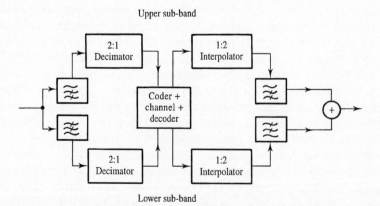

Figure 9.12 *Sub-band (speech) coder with two, equal width, sub-bands and separate encoding of sub-band signals after sample rate reduction.*

floating point arrays which present noise-like inputs.

9.8 Summary

The information content of a symbol is defined as the negative logarithm of the symbol's selection probability. If 2 is chosen as the base of the logarithm then the unit of information is the bit. The entropy of an information source is the average information per selected, or transmitted, symbol. The entropy of a binary source with statistically independent, equiprobable, symbols is 1 bit/binary digit. Statistical dependence of symbols, and unequal selection probabilities, both reduce the entropy of a source below its maximum value. Redundancy is the difference between a source's maximum possible entropy and its actual entropy. Noise at the receiver makes the decision process uncertain and thus reduces the information content of the received symbol stream. The information content of the symbols after detection is called the effective entropy and is equal to the entropy on transmission minus the equivocation introduced by the noise. In this sense noise may be interpreted as negative information added to the transmitted symbol stream.

Source coding removes redundancy, making transmission and/or storage of information more efficient. Source codes must, in general, be uniquely decodable. Source coding of discrete information sources is, normally, also loss-less, i.e. the coding algorithm, in the absence of symbol errors, is precisely reversible. Codes may, in addition, be instantaneously decodable, i.e. contain no code words which form a prefix of any other code words. The Huffman code, used for Group 3 fax transmission, is an example of a unique, loss-less, instantaneous code.

Vocoders use models of the physical mechanism of speech production to transmit intelligible, but low bit rate, speech signals. In essence it is the model parameters, in the form of adaptive filter coefficients, which are transmitted instead of the voice signals themselves. The adaptive filters are excited at the receiver by a pulse generator, of the correct frequency (pitch), for voiced sounds or white noise for unvoiced sounds.

Sub-band coding splits the transmitted signal into two or more frequency sub-bands using hardware or (DFT) software. This allows the quantisation accuracy for each sub-band to reflect the importance of that band as measured by the signal power it contains. Other transform coding techniques can be used to encode speech and images. Transform coding is particularly significant in two-dimensional image transmission as it minimises redundancy which takes the form of correlation in intensity or luminance values between closely spaced pixels. Vocoder and transform coding techniques are lossy in that there is an inevitable (but often small) loss in the information content of the encoded data.

String coding algorithms are particularly efficient at compressing character-based data but coding times are long since commonly occurring strings must be initially searched for in the data. String coding is used for compression of computer files in order to speed up transport between machines and save disc space.

The next chapter shows how the opposite technique to source coding, i.e. increasing redundancy, can be employed to detect, and sometimes correct, errors in data which may occur during its transmission or storage.

9.9 Problems

9.1. (a) Consider a source having an $M = 3$ symbol alphabet where $P(x_1) = \frac{1}{2}$; $P(x_2) = P(x_3) = \frac{1}{4}$ and symbols are statistically independent. Calculate the information conveyed by the receipt of the symbol x_1. Repeat for x_2 and x_3. $[I_{x_1} = 1$ bit, $I_{x_2} = I_{x_3} = 2$ bits$]$

(b) Consider a source whose, statistically independent, symbols consist of all possible binary sequences of length k. Assume all symbols are equiprobable. How much information is conveyed on receipt of any symbol. $[k$ bits$]$

(c) Determine the information conveyed by the specific message $x_1 x_3 x_2 x_1$ when it emanates from each of the following, statistically independent, symbol sources: (i) $M = 4$; $P(x_1) = \frac{1}{2}$, $P(x_2) = \frac{1}{4}$, $P(x_3) = P(x_4) = 1/8$, [7 bits] (ii) $M = 4$; $P(x_1) = P(x_2) = P(x_3) = P(x_4) = \frac{1}{4}$. [8 bits]

9.2. (a) Calculate the entropy of the source in Problem 9.1(a). $[1\frac{1}{2}$ bit/symbol$]$

(b) Calculate the entropy of the sources in Problem 9.1(c). $[1\frac{3}{4}$ bit/symbol, 2 bit/symbol)$]$

(c) What is the maximum entropy of an 8 symbol source and under what conditions is this situation achieved? What are the entropy and redundancy if $P(x_1) = \frac{1}{2}$, $P(x_i) = 1/8$ for $i = 2, 3, 4$ and $P(x_i) = 1/32$ for $i = 5, 6, 7, 8$? [3 bit/symbol, 2.25 bit/symbol]

9.3. Find the entropy, redundancy and code efficiency of a three symbol source A, B, C, if the long term probabilities of symbol selection are $P(A) = 0.4$, $P(B) = 0.3$, $P(C) = 0.3$ and the following statistical dependence exists between symbols. There is a 20% chance of each symbol being succeeded by the next symbol in the *cyclical* sequence $A\ B\ C$ and a 30% chance of each symbol being succeeded by the previous symbol in this sequence? [1.418 bit/symbol; 0.167 bit/symbol; 89.5%]

9.4. Show that the number of redundant symbols per bit of information transmitted by an M-symbol source with code efficiency η_{code} is given by $(1 - \eta_{code})/(\eta_{code} \log_2 M)$ symbol/bit.

9.5. Estimate the maximum information content of a black and white television picture with 625 lines and an aspect ratio of 4/3. Assume that 10 brightness values can be distinguished and that the picture resolution is the same along a horizontal line as along a vertical line. What maximum data rate does a picture rate of 25 picture/s correspond to and what, approximately, must be the bandwidth of the (uncoded and unmodulated) video signal if it is transmitted using binary symbols? (If necessary you should consult Chapter 16 to obtain TV scanning format information.) [3.322 bit/symbol and 1.73 Mbit/picture; 43.25 Mbit/s; 21.62 MHz]

9.6. Calculate the loss in information due to noise, per transmitted digit, if a random binary signal is transmitted through a channel, which adds zero mean Gaussian noise, with an average signal-to-noise ratio of: (a) 0 dB; (b) 5 dB; (c) 10 dB. [0.2493; 0.0565; 0.0012 bit/binit]

9.7. An information source contains 100 different, statistically independent, equiprobable symbols. Find the maximum code efficiency, if, for transmission, all the symbols are represented by binary code words of equal length. [7 bit words and 94.9%]

9.8. (a) Apply Huffman's algorithm to deduce an optimal code for transmitting the source defined in Problem 9.1(a) over a binary channel. Is your code unique?

(b) Define the efficiency of a code and determine the efficiency of the code devised in part(a).

(c) Construct another code for the source of part (a) and assign equal length binary words irrespective of the occurrence probability of the symbols. Calculate the efficiency of this source. [(a) 0, 10, 110, 111, Yes; (b) 100%; (c) 87.5%]

CHAPTER 10

Error control coding

10.1 Introduction

The fundamental resources at the disposal of a communications engineer are signal power, time and bandwidth. For a given communications environment (summarised in this context by an effective noise power spectral density) these three resources can be traded against each other. The basis on which the trade-offs are made will depend on the premium attached to each resource in a given situation. A general objective, however, is often to achieve maximum data transfer, in a minimum bandwidth *while maintaining an acceptable quality of transmission*. The quality of transmission, in the context of digital communications, is essentially concerned with the probability of bit error, P_b, at the receiver. (There are other factors which determine transmission quality, in its widest sense, of course, but focussing on P_b makes the discussion, and more especially the analysis, tractable.)

The Shannon-Hartley law (see section 11.4.1) for the capacity of a communications channel demonstrates two things. Firstly it shows (quantitatively) how bandwidth and signal power may be traded in an ideal system, and secondly it gives a theoretical limit for the transmission rate of (reliable, i.e. error free) data from a transmitter of given power, over a channel with a given bandwidth, operating in a given noise environment. In order to realise this theoretical limit, however, an appropriate coding scheme (which the Shannon-Hartley law assures us exists) must be found. (It should, perhaps, be noted at this point that there is one more quantity which must be traded in return for the advantage which such a coding scheme confers, i.e. time delay which results from the coding process.)

In practice, the objective of the design engineer is to realise the required data rate (often determined by the service being provided) within the bandwidth constraint of the available channel and the power constraint of the particular application. (In a mobile radio application, for example, bandwidth may be determined by channel allocations or frequency coordination considerations, and maximum radiated power may be determined by safety considerations or transceiver battery technology.) Furthermore this data rate must be achieved with an *acceptable* BER and time delay. If an, essentially, uncoded PCM transmission cannot achieve the required BER within these constraints then the application of error control coding may be able to help, providing the constraints are not such as to violate the Shannon-Hartley law.

Error control coding (also referred to as channel coding) is used to detect, and often correct, symbols which are received in error. Error *detection* can be used as the initial step of an error *correction* technique by, for example, triggering a receiving terminal to generate an automatic repeat request (ARQ) signal which is carried, by the return path of a duplex link, to the originating terminal. A successful retransmission of the affected data results in the error being corrected. If ARQ techniques are inconvenient, as is the case, for example, when the propagation delay of the transmission medium is large, then forward error correction coding (FECC) may be appropriate [Blahut 1983, Clark and Cain, MacWilliams and Sloane]. FECC incorporates extra information (i.e. redundancy) into the transmitted data which can then be used not only to detect errors but also to correct them without the need for any retransmissions.

Table 10.1 *A taxonomy of error control codes.*

ARQ			FECC				
Stop & wait	Continuous ARQ (pipelining)		Block codes				Convolutional codes
	Go-back-N	Selective repeat	Others (non-linear)	Group (linear)			
				Others (non-cyclic)	Polynomially generated (cyclic)		
					Golay	BCH	
						Reed-Solomon	Binary BCH
							Hamming ($e = 1$) / $e > 1$

This chapter begins with a general discussion of error rate control, in its widest sense, as may be applied in digital communications systems. Five particular error control methods are identified and briefly described. It is one of these methods, namely FECC, which is then treated in detail during the remainder of the chapter. Some typical applications of FECC are outlined and the threshold phenomenon is highlighted. The Hamming distance between a pair of codewords and the weight of a codeword are defined. The discussion of FEC codes, that follows, is structured loosely around the taxonomy of codes shown in Table 10.1 starting with a description of block codes and including special mention of linear group codes, cyclic codes, the Golay code, BCH codes, Reed-Solomon codes and Hamming codes. The Hamming bound on the performance of a block code is derived and strategies for nearest neighbour, or maximum likelihood, decoding are discussed. Convolution coding is treated towards the end of the chapter. Tree, trellis and state transition diagrams are used to illustrate the encoding process. The significance of constraint length and decoding window length are discussed and Viterbi decoding is illustrated using a trellis diagram. Decoding is accomplished by finding the most likely path through the trellis and relating this back to the transmitted data sequence.

Coding concepts are presented, here, with an essentially non-mathematical treatment, in the context of specific examples of group, cyclic and convolutional codes. The chapter concludes with a brief discussion of practical coders.

10.1.1 Error rate control concepts

The normal measure of error performance is bit-error rate (BER) or the probability of bit error (P_b). P_b is simply the probability of any given transmitted binary digit being in error. The bit error rate is, strictly, the average rate at which errors occur and is given by the product $P_b R_b$, where R_b is the bit transmission rate in the channel. Typical long term P_b for linear PCM systems is 10^{-7} while, for companded PCM, it is 10^{-5} and for ADPCM (Chapter 5) it is 10^{-4}. If the error rate of a particular system is too large then what can be done to make it smaller? The first and most obvious solution is to increase transmitter power, but this may not always be desirable, for example in man-portable systems where the required extra battery weight may be unacceptable.

A second possible solution, which is especially effective against burst errors caused by signal fading, is to use diversity. There are three main types of diversity: space diversity, frequency diversity, and time diversity. All these schemes incorporate redundancy in that data is, effectively, transmitted twice: i.e. via two paths, at two frequencies, or at two different times. In space diversity two or more antennas are used which are sited sufficiently far apart for fading at their outputs to be decorrelated. Frequency diversity employs two different frequencies to transmit the same information. (Frequency diversity can be in-band or out-band depending upon the frequency spacing between the carriers.) In time diversity systems the same message is transmitted more than once at different times.

A third possible solution to the problem of unacceptable BER is to introduce full duplex transmission, implying simultaneous 2-way transmission. Here when a transmitter sends information to a receiver, the information is 'echoed' back to the transmitter on a separate feedback channel. Information echoed back which contains errors can then be retransmitted. This technique requires twice the bandwidth of single direction (simplex) transmission, however, which may be unacceptable in terms of spectrum utilisation.

A fourth method for coping with poor BER is automatic repeat request (ARQ). Here a simple error *detecting* code is used and, if an error is detected in a given data block, then a request is sent via a feedback channel to retransmit that block. There are two major ARQ techniques. These are *stop and wait*, in which each block of data is positively, or negatively, acknowledged by the receiving terminal as being error free before the next data block is transmitted, and *continuous* ARQ, in which blocks of data continue to be transmitted without waiting for each previous block to be acknowledged. (In stop and wait ARQ data blocks are *timed out* if neither a positive nor negative acknowledgement is received within a predetermined time window. After timing out the appropriate data block is retransmitted in the same way as if it had been negatively acknowledged.) Continuous ARQ can, in turn, be divided into two variants. In the *go-*

back-n version, data blocks carry a sequence or reference number, *n*. (This is a *different n* to that used in the (*n, k*) block code notation later.) Each acknowledgement signal contains the reference number of a data block and effectively acknowledges all data blocks up to *n* − 1. When a negative acknowledgement is received (or a data block timed out) all data blocks starting from the reference number in the last acknowledgement signal are retransmitted. In the *selective repeat* version only those data blocks explicitly negatively acknowledged (or timed out) are retransmitted (necessitating data block reordering buffers at the receiver). Go-back-*n* ARQ has a well defined storage requirement whilst selective repeat ARQ, although very efficient, has a less well defined, and potentially much lager, storage requirement, especially when deployed on high speed links. ARQ is very effective, for example in facsimile transmission, Chapter 9. On long links with fast transmission rates, however, such as is typical in satellite communications, ARQ can be very difficult to implement.

The fifth technique for coping with high BER is to employ forward error correction coding (FECC). In common with three of the other four techniques FECC introduces redundancy, this time with data check bits interleaved with the information traffic bits. It relies on the number of errors in a long block of data being close to the statistical average and, being a forward technique, requires no return channel. The widespread adoption of FECC was delayed, historically, because of its complexity and high cost of implementation relative to the other possible solutions. Complexity is now less of a problem following the proliferation of VLSI custom coder/decoder chips.

FECC exploits the difference between the transmission rate or information bit rate R_b and the channel capacity R_{max} as given by the Shannon-Hartley law (see equation (11.38)). P_b can be reduced, at the expense of increasing the transmission delay [Schwartz, 1987], by using FECC with a sufficiently long block or constraint length. The increased transmission delay arises due to the need to assemble the data blocks to be transmitted and the time spent in examining received data blocks to correct errors. The benefits of error control, however, usually outweigh the inherent FECC processor delay disadvantages.

10.1.2 Threshold phenomenon

Figure 10.1 illustrates the error rate for an uncoded system in which P_b increases gradually as SNR decreases, as shown previously in Figure 6.3. Figure 10.1 is plotted as the ratio of bit energy to noise power spectral density (E_b/N_0), which is defined later in Chapter 11. With FECC the P_b versus E_b/N_0 curve is steeper. If the SNR is above a certain value, which here corresponds to an E_b/N_0 of around 6 dB, the error rate will be virtually zero. Below this value system performance degrades rapidly until the coded system is actually poorer than the corresponding uncoded system. (The reason for this is that there is a region of low E_b/N_0 where, in attempting to correct errors, the decoder approximately doubles the number of errors in a decoded codeword.) This behaviour is analogous to the threshold phenomenon in wide band frequency modulation. A coding gain can be defined, for a given P_b, by moving horizontally in Figure 10.1 from the uncoded to the coded curve. The value of the coding gain in dB is relatively constant for

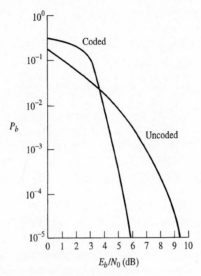

Figure 10.1 *The threshold phenomenon in FECC systems.*

$P_b \leq 10^{-5}$, and is dependant on the precise details of the FECC system deployed.

10.1.3 Applications for error control

Compact disc players provide a growing application area for FECC. In CD applications the powerful Reed-Solomon code is used since it works at a symbol level, rather than at a bit level, and is very effective against burst errors, particularly when combined with interleaving to randomise the bursts. The Reed-Solomon code is also used in computers for data storage and retrieval. Cosmic particles create, on average, one error every two to three days in a 4 Mbyte memory, although small geometry devices are helping to reduce this probability. Digital audio and video systems are also areas in which FECC is applied. Error control coding, generally, is applied widely in control and communications systems for aerospace applications, in mobile (GSM) cellular telephony and for enhancing security in banking and barcode readers.

10.2 Hamming distance and codeword weight

Before embarking on a detailed discussion of code performance the following definitions are required. The Hamming distance between two codewords is defined as the number of places, bits or digits in which they differ. This distance is important since it determines how easy it is to change or alter one valid codeword into another. The weight of a binary codeword is defined as the number of ones which it contains.

EXAMPLE 10.1
Calculate the Hamming distance between the two codewords 11100 and 11011 and find the minimum codeword weight.

The two codewords 11100 and 11011 have a Hamming distance of 3 corresponding to the differences in the 3rd, 4th and 5th digit positions. Thus with three appropriately positioned errors in these locations the codeword 11100 could be altered to 11011.
In this example, 11011 has a weight of 4 due to the four ones and 11100 has a weight of 3. The minimum weight is thus 3. Hamming distance and weight will be used later to bound the error correcting performance of codewords.

10.3 (n, k) Block codes

Figure 10.2 illustrates a block coder with k information digits going into the coder and n digits coming out after the encoding operation. The n-digit codeword is thus made up of k information digits and $(n - k)$ redundant *parity check* digits. The rate, or efficiency, for this code (R) is k/n, representing the ratio of information digits to the total number of digits in the codeword. Rate is normally in the range ½ to unity. (Unlike source coding in which data is compressed, here redundancy is deliberately added, to achieve error detection.) This is an example of a *systematic* code in that the information digits are explicitly transmitted together with the parity check digits, Figure 10.2. In a non-systematic code the n digit codeword may not contain any of the information digits explicitly. There are two definitions of systematic codes in the literature. The stricter of the two definitions assumes that, for the code to be systematic, the k information digits must be transmitted contiguously as a block, with the parity check digits making up the codeword as another contiguous block. The less strict of the two definitions merely stipulates that the information digits must be included in the codeword but not necessarily in a contiguous block. The latter definition is the one which is adopted here.

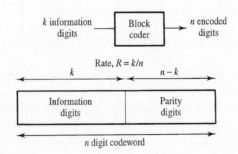

Figure 10.2 *(n, k) systematic block code.*

10.3.1 Single parity check code

The example in Figure 10.3 will be familiar to many as an option in ASCII coded data transmission (see Chapter 8). Consider the data sequence 1101000, to which a single parity check digit (P) is added. For even parity in this example sequence, P will be 1. For odd parity, P will be 0. Since the seven information digits contain three ones, another 1 must be added giving an even number of ones to achieve even parity. Alternatively if a zero is added, ensuring an odd number of ones, i.e. odd parity, the transmission of the all zero codeword is avoided. Even parity is, however, much more common.

Here the rate, $R = k/n$, is seven eighths which represents a very low level of redundancy. This scheme can only identify an odd number of errors because an even number of errors will not violate the chosen parity rule. Single error detection, as illustrated in Figure 10.4, is often used to extend a 7-bit word, with a checksum bit, into an 8-bit codeword. Another example, used in libraries, is the 10-digit ISBN codeword, Figure 10.5. This uses a modulo 11, weighted, checksum in which the weightings are 10 for the first digit, 9 for the second digit, etc. down to 2 for the ninth digit (and 1 for the checksum). The weighting can be applied either left to right or vice versa and a checksum digit of 10 is represented by the symbol C.

Single parity checks are also used on rows and columns of simple two-dimensional data arrays, Figure 10.6. Single errors in the array will be detected *and located* via the corresponding row and column parity bits. Such errors can therefore be corrected. Double errors can be detected but not necessarily corrected as several error patterns can produce the same parity violations. When the data is an array of ASCII characters the row and column check words can also be sent as ASCII characters.

In the English language there is a high level of redundancy. This is why spelling mistakes can be corrected and abbreviations expanded. There is, in fact, an approximate correspondence between the words of a language and code words as being discussed here, although in language contextual information goes beyond isolated words whilst in a block code each codeword is decoded in isolation.

Modulo $2^n - 1$ checksums are in widespread use for performing error detection on byte-serial network connections. They are usually computed by software during the data block (or packet) construction. One such error detection code is the Internet checksum for protocol messages [Comer], Chapter 18. If a message of length w (16-bit) bytes:

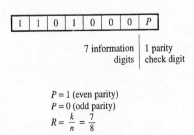

$P = 1$ (even parity)
$P = 0$ (odd parity)
$R = \dfrac{k}{n} = \dfrac{7}{8}$

Figure 10.3 *Example of a single parity check digit codeword.*

Example: Even parity

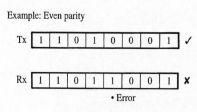

• Error

Detects odd number of errors

Figure 10.4 *Block code with single parity check error detecting capability.*

Checksum, $C = 11 - \sum_{i=1}^{9}(11 - i)k_i \,(\text{mod-}11)$

i	1	2	3	4	5	6	7	8	9	C
ISBN	0	1	9	8	5	3	8	0	4	9

Figure 10.5 *ISBN codeword and checksum calculated to satisfy* $\sum_{i=1}^{10}(11 - i)k_i = 0$ *(mod-11).*

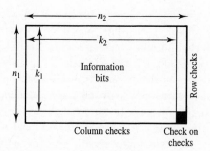

Figure 10.6 *Two-dimensional row–column array code.*

$m_{w-1}, \cdots, m_0$ is to be checksummed, the one-byte checksum is just the complement of:

$$\sum_{i=0}^{w-1} m_i \,(\text{mod} - 65535)$$

The main operation required for this is summation modulo-65535 (i.e. summation using one's complement word addition). The checksum is included with the transmitted message allowing a recipient to check for transmission errors by performing a similar summation over the received data. If this sum is not zero then a channel error has occurred. Another error detection code is the ISO two-byte checksum [Fletcher]. The checksum is again included within the transmitted message and a recipient can perform a summation over the received data to confirm that both checksum bytes are zero; if not, a channel error has occurred.

EXAMPLE 10.2

Figure 10.7 illustrates a seven digit codeword with four information digits (I_1 to I_4) and three parity check digits (P_1 to P_3), commonly referred to as a (7,4) block code. The circles indicate how the information bits contribute to the calculation of each of the parity check bits. Assuming even parity, show the realisation of this encoder using 3-input modulo-2 adders. Calculate the individual parity check bits and encoding of P_1, P_2 and P_3 for the information digits 1011.

Figure 10.7 shows P_1 represented by the modulo-2 sum of I_1, I_3 and I_4. P_2 is the sum of I_1, I_2 and I_4, etc. (Modulo-2 arithmetic was used previously in Table 8.1.) The parity check digits are generated by the circuit in Figure 10.8. For the data sequence 1011, the 3-input modulo-2 adders count the total number of ones which are present at the inputs, and output the least significant bit as the binary coded sum. Thus $P_1 = 1$, $P_2 = 0$ and $P_3 = 0$, giving a coder output of 1011100.

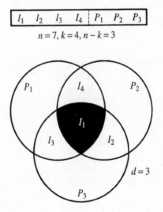

Figure 10.7 *Representation of relationship between parity check and data bits*

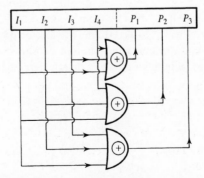

Figure 10.8 *(7,4) block code hardware generation of three parity check digits.*

Figure 10.9 shows how parity check equations for P_1, P_2 and P_3 in the above example may be written using $\oplus$ to represent the modulo-2 or exclusive-or arithmetic operation. Figure 10.9 also shows how these equations can be reduced to matrix form in a *parity check* matrix **H**. The coefficients of the information digits I_1, I_2, I_3 and I_4 are to the left of the dotted partition in the parity check matrix. The top row of the matrix contains the information about parity check P_1, the second row about parity check P_2 and the third row about parity check P_3.

Consider the top left hand part of the matrix (1011). The coefficients 1011 correspond to the information digits I_1, I_3 and I_4 in the equation for P_1. Similarly, the second row is 1101 to the left of the partition because I_3 is not involved in calculating parity check P_2 and the corresponding part of the bottom row is 1110 because I_4 is not involved in calculating the parity check P_3, Figure 10.8. To the right of the dotted partition there is a 3×3 diagonal matrix of ones. Each column in this diagonal matrix corresponds to a particular parity check digit. The first column (100) indicates parity check P_1. The second column indicates parity check P_2 and the third column parity check P_3. Later, in section 10.7, a *generator* matrix will be used to obtain the codeword directly from the information vector.

10.4 Probability of error in n-digit codewords

What is the probability of having more than R' errors in an n-digit codeword? First consider the case of exactly j errors in n digits with a probability of error per digit of P_e. From Chapter 3, equation (3.8):

$$P(j \text{ errors}) = (P_e)^j (1 - P_e)^{n-j} \times {}^nC_j \tag{10.1}$$

The probability of having more than R' errors can be written as:

$$P(> R' \text{ errors}) = 1 - \sum_{j=0}^{R'} P(j) \tag{10.2}$$

Statistical stability controls the usefulness of this equation, statistical convergence occurring for long code words or blocks (see Figure 3.2). A long block effectively embodies a large number of trials, to determine whether or not an error will occur, and the number of errors in such a block will therefore be close to $P_e n$. Furthermore, the fraction of blocks containing a number of errors that deviates significantly from this value

$$P_1 = 1 \times I_1 \oplus 0 \times I_2 \oplus 1 \times I_3 \oplus 1 \times I_4$$
$$P_2 = 1 \times I_1 \oplus 1 \times I_2 \oplus 0 \times I_3 \oplus 1 \times I_4$$
$$P_3 = 1 \times I_1 \oplus 1 \times I_2 \oplus 1 \times I_3 \oplus 0 \times I_4$$

$$\mathbf{H} = \begin{bmatrix} 1 & 0 & 1 & 1 & \vdots & 1 & 0 & 0 \\ 1 & 1 & 0 & 1 & \vdots & 0 & 1 & 0 \\ 1 & 1 & 1 & 0 & \vdots & 0 & 0 & 1 \end{bmatrix}$$

Figure 10.9 *Representation of the code in Figure 10.7 by parity check equations and an* **H** *matrix.*

becomes smaller as the block length, n, becomes larger, Figure 3.2. Choosing a code that can correct $P_e n$ errors in a block will ensure that there are very few cases in which the coding system will fail. This is the rationale for long block codes. The attraction of block codes is that they are amenable to precise performance analysis. By far the most important, and most amenable, set of block codes are the linear group codes.

10.5 Linear group codes

The codewords in a linear group code have a one-to-one correspondence with the elements of a mathematical group. Group codes contain the all-zeros codeword and have the property referred to as closure. That is, taking any two codewords C_i and C_j, then $C_i \oplus C_j = C_k$. (For the all zeros codeword, when $i = 0$, then $k = j$.) Thus adding, modulo-2, corresponding pairs of digits in each of the codewords produces another codeword C_k. The presence of the all-zeros codeword and the closure property together make performance calculations with linear group codes particularly easy, as will be seen later. Figure 10.10 illustrates a simple group code which will be used as an example in this chapter. It is first used to illustrate the property of closure. Figure 10.10 depicts a source alphabet with 4 members: a, b, c and d (i.e. the number of information digits is $k = 2$). Each symbol is coded into an n-digit codeword (where $n = 5$) as shown. This is therefore a (5,2) code. (By the less strict definition, this is also a systematic code where the information digits are in columns 1/2 and 4/5.) Consider the codewords corresponding to c and b. Modulo-2 summing c and b gives the codeword d, illustrating the closure property.

10.5.1 Members of the group code family

Group codes can be divided into two types: those which are 'polynomial generated' in simple feedback shift registers; and others. The simplicity of the former have rendered the rest irrelevant. The polynomial generated codes can be further divided into subgroups, the main ones being the binary Bose–Chaudhuri–Hocquenghem (BCH) codes

Codewords

$$
\begin{array}{ll}
a = 0\,0 & \quad 0\ 0\ 0\ 0\ 0 \\
b = 0\,1 & \quad 0\ 0\ 1\ 1\ 1 \\
c = 1\,0 & \quad 1\ 1\ 1\ 0\ 0 \\
d = 1\,1 & \quad 1\ 1\ 0\ 1\ 1
\end{array}
$$

$$
\underbrace{}_{k = 2} \qquad \underbrace{}_{n = 5}
$$

$$
\begin{array}{l}
c \oplus b = d \\
c = 1\ 1\ 1\ 0\ 0 \\
b = 0\ 0\ 1\ 1\ 1 \\
\hline
d = 1\ 1\ 0\ 1\ 1
\end{array}
$$

Figure 10.10 *Illustration of the closure property of a group code.*

and their important, non-binary counterpart, the Reed-Solomon codes. BCH codes are widely tabulated up to $n = 255$ with an error correcting capability of up to 30 digits [Blahut, 1983]. Generally speaking for the same error correcting capability a larger (e.g. $n = 255$) block size offers a higher rate than a shorter (e.g. $n = 63$) block size. Reed-Solomon, non-binary, byte organised codes are used extensively in compact disc players and computer memories.

10.5.2 Performance prediction

Normally all possible codeword pairs would have to be examined, and their Hamming distances measured, to determine the overall performance of a block code. For the case of group codes, however, consideration of each of the codewords with the all-zeros codeword is sufficient. This is a significant advantage of linear group codes and one reason why these codes are so important in relation to other block codes. (Analysis for large n becomes much simpler for group codes, since the number of combinations of codewords, which would otherwise have to be searched, is very large.)

The important quantity, as far as code performance prediction is concerned, is the minimum Hamming distance between any pair of codewords. For the four five-digit codewords in Figure 10.10 – 0 0 0 0 0, 0 0 1 1 1, 1 1 1 0 0, 1 1 0 1 1, inspection reveals a minimum Hamming distance of 3, i.e. $D_{min} = 3$ for this (5,2) code. Other (n, k) block codes with this minimum distance of 3 are (3,1), (15,11), (31,26), etc.

The weight *structure* of a set of codewords is just a list of the weights of all the codewords. Consider the previous example with 4 codewords: the weights of these are 0, 3, 3 and 4. Ignoring the all-zeros word (as interest is concentrated in the distances from this codeword), the minimum weight in the weight structure (3) is equal to D_{min}, the minimum Hamming distance for the code.

Consider the probability of the ith codeword (C_i) being misinterpreted as the jth codeword (C_j). This probability depends on the distance between these two codewords (D_{ij}). Since this is a linear group code, this distance D_{ij} is equal to the weight of a third codeword C_k which is actually the modulo-2 sum of C_i and C_j. The probability of C_i being mistaken for C_j is therefore equal to the probability of C_k being mistaken for C_0. Furthermore, the probability of C_k being mistaken for the all-zeros codeword (C_0) is equal to the probability of C_0 being misinterpreted as C_k (by symmetry). The probability of C_0 being misinterpreted as C_k, depends only on the weight of C_k.

This reasoning reveals the importance of a linear group code's weight structure since the performance of such a code can be determined completely by consideration of C_0 and the weight structure alone.

10.5.3 Error detection and correction capability

The maximum possible error correcting power, t, of a code is defined by its ability to correct *all* patterns of t or less errors. It is related to the code's minimum Hamming distance by:

$$t = int\left(\frac{D_{min} - 1}{2}\right) \tag{10.3(a)}$$

where:

$$D_{min} - 1 = e + t \tag{10.3(b)}$$

Here *int*( ) indicates 'the integer part of', e is the number of *detectable* errors and $t \le e$. Taking the case where D_{min} is 3, then there are at least two possible binary words which lie between each pair of valid codewords. In Example 10.1 these could be the binary words 11000 and 11001. If any single error occurs in one of the code words it can therefore be corrected. Alternatively, $D_{min} - 1$ errors can be detected (2 in this case as both 11000 and 11001 are detectable as errors). Note that the code *cannot* work in both these detection and correction modes simultaneously (i.e. detect two errors and correct one of them).

Longer codes with larger Hamming distances offer greater detection and correction capability by selecting different t and e values in equation (10.3). $D_{min} = 7$ can offer $t = 1$ bit correction with $e = 5$ bit error detection. If t is increased to 2 then e must decrease to 4. The UK Post Office Code Standards Advisory Group (POCSAG) code with $k = 21$ and $n = 32$ is an $R \simeq 2/3$ code with $D_{min} = 6$. This provides a 3-bit error *detection* or a 2-bit error *correction* capability, for a codeword which is widely used in pager systems, see Chapter 15. The $n = 63$, $k = 57$, BCH code gives $R = 0.9$ with $t = 1$ bit, while reducing k to 45 gives $R = 0.7$ with $t = 3$ bit. Further reducing k to 24 reduces R to below 0.4 but achieves a $t = 7$ bit correction capability. This illustrates the important trade-off between rate and error correction power. BCH codes can correct burst, as well as random, errors.

10.6 Nearest neighbour decoding of block codes

Encoding is achieved by use of a feedback shift register and is relatively simple as will be shown later. The two most important strategies for decoding are nearest neighbour and maximum likelihood decoding. These are equivalent if the probability of t errors is much greater than that of $t + 1$ errors, etc. as in Example 3.4. Using a decoding table based on nearest neighbours, therefore, implies the maximum likelihood decoding strategy, as discussed in the context of decision theory in Chapter 9. This is illustrated with a simple example.

Figure 10.11 is a nearest neighbour decoding table for the previous four-symbol example of Figure 10.10. The codewords are listed along the top of this table starting with the all-zeros codeword in the top left hand corner. Below each codeword all possible received sequences are listed which are at a Hamming distance of 1 from this codeword. (In the case of the all-zeros codeword these are the sequences 10000 to 00001.) If this were a t error correcting code this list would continue with all the patterns of 2 errors, 3 errors, etc. up to all patterns of t errors. Any detected bit pattern appearing in the table is interpreted as representing the codeword at the top of the relevant column, thus allowing the bit errors to be corrected. Below the table in Figure 10.11 there are eight 5-bit words which lie outside the table. These received sequences are equidistant from two possible codewords, so these sequences lie on a decision boundary. It is not possible, therefore, to decide which of the two original codewords they came from, and consequently the errors cannot be corrected. These sequences were referred to previously as *detectable* error sequences.

10.6.1 Hamming bound

Consider the possibility of a code with codewords of length n, comprising k information digits and having error correcting power t. There is an upper bound on the performance of block codes which is given by:

$$2^k \leq \frac{2^n}{1 + n + {}^nC_2 + {}^nC_3 + \cdots + {}^nC_t} \tag{10.4}$$

The simplest way to derive equation (10.4) is to inspect the nearest neighbour decoding table for the (n, k), t-error correcting code. Figure 10.12 develops Figure 10.11 into the general case of a t-error correcting code with 2^k codewords. There are, thus, 2^k columns

Codewords	00000	11100	00111	11011
Single-bit error correctable patterns	10000	01100	10111	01011
	01000	10100	01111	10011
	00100	11000	00011	11111
	00010	11110	00101	11001
	00001	11101	00110	11010
Double-bit error detectable patterns	10001	01101	10110	01010
	10010	01110	10101	01001

Figure 10.11 *Nearest neighbour decoding table for the group code of Figure 10.9.*

2^k columns				
1	0 0 ... 0	C_2	....	C_{2^k}
n	100 ... 0 010 ... 0 ⋮ ⋮ 000 ... 1		Single errors	
${}^nC_2$			Double errors	
⋮	⋮	⋮	⋮	⋮
${}^nC_t$			t errors	

Rows $= [1 + n + {}^nC_2 + ... + {}^nC_t]$

Figure 10.12 *Decoding table for a t-error correcting (n, k) block code.*

in the decoding table. Consider the left hand column. The all-zeros codeword itself is, obviously, one possible correctly received sequence or valid codeword. Also there are n single error patterns associated with that all-zeros codeword. Further, there are ${}^nC_2$ patterns of 2 errors, etc. down to ${}^nC_t$ patterns of t errors. Totalling the number of entries in this column reveals the total number of rows in the table and the value of the denominator in equation (10.4). Taking this number of rows and dividing into 2^n (which is the total number of possible received sequences), as in equation (10.4), gives the maximum possible number of columns and hence the maximum number of codewords in the given code. If the left hand side of equation (10.4) is greater than the right hand side then no such code exists and n must be increased, k decreased, or t decreased, until equation (10.4) is satisfied. For a *perfect* code, equation (10.4) is an equality. This implies that there are no bit patterns which lie outside the decoding table, avoiding the problem of equidistant errors which occurred in the code of Figure 10.11.

10.7 Syndrome decoding

The difficulty with decoding of block codes using the nearest neighbour decoding table of Figure 10.12 is the physical size of the table for large n. The syndrome decoding technique described here provides a solution to this problem.

10.7.1 The generator matrix

The generator matrix is a matrix of basis vectors. The rows of the generator matrix $\mathbf{G}$ are used to derive the actual transmitted codewords. This is in contrast with the $\mathbf{H}$ (or parity check) matrix, Figure 10.9, which does not contain any codewords. The generator matrix $\mathbf{G}$ for an (n, k) block code can be used to generate the appropriate n-digit codeword from any given k-digit data sequence. The $\mathbf{H}$ and corresponding $\mathbf{G}$ matrices for the example (7,4) block code of Figure 10.8 are shown below:

$$\mathbf{H} = \begin{bmatrix} 1\,0\,1\,1:1\,0\,0 \\ 1\,1\,0\,1:0\,1\,0 \\ 1\,1\,1\,0:0\,0\,1 \end{bmatrix} \tag{10.5}$$

$$\mathbf{G} = \begin{bmatrix} 1\,0\,0\,0:1\,1\,1 \\ 0\,1\,0\,0:0\,1\,1 \\ 0\,0\,1\,0:1\,0\,1 \\ 0\,0\,0\,1:1\,1\,0 \end{bmatrix} \tag{10.6}$$

Study of $\mathbf{G}$ shows that on the left of the dotted partition there is a 4×4 unit diagonal matrix and on the right of the partition there is a parity check section. This part of $\mathbf{G}$ is the transpose of the left hand portion of $\mathbf{H}$. As this code has a single error correcting capability then $D_{\min}$, and the weight of the codeword, must be 3. As the identity matrix has a single one in each row then the parity check section must contain at least two ones. In addition to this constraint, rows cannot be identical.

Continuing the (7,4) example, we now show how **G** can be used to construct a codeword, using the matrix equation (4.13) [Spiegel]. Assume the data sequence is 1001. To generate the codeword associated with this data sequence the data vector 1001 is multiplied by **G** using modulo-2 arithmetic:

$$[1\ 0\ 0\ 1]\begin{bmatrix} 1\ 0\ 0\ 0\ 1\ 1\ 1 \\ 0\ 1\ 0\ 0\ 0\ 1\ 1 \\ 0\ 0\ 1\ 0\ 1\ 0\ 1 \\ 0\ 0\ 0\ 1\ 1\ 1\ 0 \end{bmatrix} = [1\ 0\ 0\ 1\ 0\ 0\ 1] \qquad (10.7)$$

The 4×4 unit diagonal matrix in the lefthand portion of **G** results in the data sequence 1001 being repeated as the first four digits of the codeword and the right hand (parity check) portion results in the three parity check digits P_1, P_2 and P_3 (in this case 001) being calculated. (This generator matrix could, therefore, be applied to solve the second part of Example 10.2.)

It is now possible to see why the columns to the right of the partition in **G** are the rows of **H** to the left of its partition. From another standpoint the construction of a codeword is viewed as a weighted sum of the rows of **G**. The digits of the data sequence perform the weighting. With digits 1001 in this example, the top row of **G** is weighted by 1, the second row by 0, the third row by 0 and the fourth row by 1. After weighting the corresponding digits from each row are added modulo-2 to obtain the required codeword.

10.7.2 Syndrome table for error correction

Recall the strong inequality that the probability of t errors is much greater than the probability of $t + 1$ errors. This situation always holds in the P_e regime where FECC systems normally operate. Thus nearest neighbour decoding is equivalent to maximum likelihood decoding. Unfortunately, the nearest neighbour decoding table is normally too large for practical implementation which requires a different technique, involving a smaller table, to be used instead. This table, referred to as the syndrome decoding table, is smaller than the nearest neighbour table by a factor equal to the number of codewords in the code set (2^k). This is because the syndrome is independent of the transmitted codeword and only depends on the error sequence as is demonstrated below.

When **d** is a message vector of k digits, **G** is the $k \times n$ generator matrix and **c** is the n digit codeword corresponding to the message **d**, equation (10.7) can be written as:

$$\mathbf{d\,G} = \mathbf{c} \qquad (10.8)$$

Furthermore:

$$\mathbf{H\,c} = \mathbf{0} \qquad (10.9)$$

where **H** is the (even) parity check matrix corresponding to **G** in equation (10.8). Also:

$$\mathbf{r} = \mathbf{c} \oplus \mathbf{e} \qquad (10.10)$$

where **r** is the sequence received after transmitting **c**, and **e** is an error vector representing the location of the errors which occur in the received sequence **r**. Consider the product **H**

$$\mathbf{H\,e} = \mathbf{s}$$

Error pattern	Syndrome
0 0 0 0 0 0 0	0 0 0
1 0 0 0 0 0 0	1 1 1
0 1 0 0 0 0 0	0 1 1
0 0 1 0 0 0 0	1 0 1
0 0 0 1 0 0 0	1 1 0
0 0 0 0 1 0 0	1 0 0
0 0 0 0 0 1 0	0 1 0
0 0 0 0 0 0 1	0 0 1

Figure 10.13 *The complete syndrome table for all possible single error patterns.*

r which is referred to as the syndrome vector **s**:

$$\mathbf{s} = \mathbf{H\,r} = \mathbf{H}\,(\mathbf{c} \oplus \mathbf{e})$$

$$= \mathbf{H\,c} \oplus \mathbf{H\,e} = \mathbf{0} \oplus \mathbf{H\,e} \tag{10.11}$$

Thus **s** is easily calculated and, if there are no received errors, the syndrome will be the all-zero vector **0**. Calculating the vector **s** provides immediate access to the vector **e** and hence the position of the errors. A syndrome table is constructed by assuming transmission of all the zeros codeword and calculating the syndrome vector associated with each correctable error pattern:

$$\begin{bmatrix} 1\,0\,1\,1\,1\,0\,0 \\ 1\,1\,0\,1\,0\,1\,0 \\ 1\,1\,1\,0\,0\,0\,1 \end{bmatrix} \begin{bmatrix} 0 \\ 0 \\ 0 \\ 0 \\ 0 \\ 0 \\ 0 \end{bmatrix} = \begin{bmatrix} 0 \\ 0 \\ 0 \end{bmatrix} \tag{10.12}$$

Equation 10.12 illustrates the case of no errors in the received sequence leading to the all zeros syndrome for the earlier (7,4) code example. Figure 10.13 shows the full syndrome table for this (7,4) code. In this case only single errors are correctable and the syndrome table closely resembles the transposed matrix $\mathbf{H}^T$. If a double error occurs then it will normally give the same syndrome as some single error and, since single errors are much more likely than double errors, a single error will be assumed and the wrong codeword will be output from the decoder resulting in a 'sequence' error. This syndrome decoding technique is still a nearest neighbour (maximum likelihood) decoding strategy.

EXAMPLE 10.3
As an example of the syndrome decoding technique, assume that the received vector for the (7,4) code is **r** = 1001101 and find the correct transmitted codeword.

$$\begin{bmatrix} 1\,0\,1\,1\,1\,0\,0 \\ 1\,1\,0\,1\,0\,1\,0 \\ 1\,1\,1\,0\,0\,0\,1 \end{bmatrix} \begin{bmatrix} 1 \\ 0 \\ 0 \\ 1 \\ 1 \\ 0 \\ 1 \end{bmatrix} = \begin{bmatrix} 1 \\ 0 \\ 0 \end{bmatrix}$$

The above matrix equation illustrates calculation of the corresponding syndrome (100). Reference to the syndrome table (Figure 10.13) reveals the corresponding error pattern as (0000100). Finally $\mathbf{c} = \mathbf{r} \oplus \mathbf{e}$:

$\mathbf{r} = 1\,0\,0\,1\,1\,0\,1$
$\mathbf{e} = 0\,0\,0\,0\,1\,0\,0$
$\mathbf{c} = 1\,0\,0\,1\,0\,0\,1$

to give the corrected transmitted codeword $\mathbf{c}$ as 1001001.

EXAMPLE 10.4

For a (6, 3) systematic linear block code, the codeword comprises $I_1\,I_2\,I_3\,P_1\,P_2\,P_3$ where the three parity-check bits $P_1\,P_2$ and P_3 are formed from the information bits as follows:

$P_1 = I_1 \oplus I_2$
$P_2 = I_1 \oplus I_3$
$P_3 = I_2 \oplus I_3$

Find: (a) the parity check matrix; (b) the generator matrix; (c) all possible codewords. Determine (d) the minimum weight; (e) the minimum distance; and (f) the error detecting and correcting capability of this code. (g) If the received sequence is 101000, calculate the syndrome and decode the received sequence.

(a)
$$\mathbf{H} = \begin{bmatrix} 1\,1\,0\,1\,0\,0 \\ 1\,0\,1\,0\,1\,0 \\ 0\,1\,1\,0\,0\,1 \end{bmatrix}$$

(b)
$$\mathbf{G} = \begin{bmatrix} 1\,0\,0\,1\,1\,0 \\ 0\,1\,0\,1\,0\,1 \\ 0\,0\,1\,0\,1\,1 \end{bmatrix}$$

(c) and (d)

Message $\times \mathbf{G}$	= Codeword	Weight
$[0\,0\,0] \times \mathbf{G}$	= 000000	0
$[0\,0\,1] \times \mathbf{G}$	= 001011	3
$[0\,1\,0] \times \mathbf{G}$	= 010101	3
$[1\,0\,0] \times \mathbf{G}$	= 100110	3
$[0\,1\,1] \times \mathbf{G}$	= 011110	4

$$
\begin{array}{lll}
[1\,0\,1] \times \mathbf{G} & = 1\,0\,1\,1\,0\,1 & 4 \\
[1\,1\,0] \times \mathbf{G} & = 1\,1\,0\,0\,1\,1 & 4 \\
[1\,1\,1] \times \mathbf{G} & = 1\,1\,1\,0\,0\,0 & 3
\end{array}
$$

Minimum weight = 3.

(e) D_{min} = minimum weight = 3.

(f) The code is thus single error correcting (or double error detecting if no correction is required).

(g) Now using $\mathbf{H}\,\mathbf{r} = \mathbf{s}$, equation (10.11) becomes:

$$
\mathbf{s} = \begin{bmatrix} 1\,1\,0\,1\,0\,0 \\ 1\,0\,1\,0\,1\,0 \\ 0\,1\,1\,0\,0\,1 \end{bmatrix} \begin{bmatrix} 1 \\ 0 \\ 1 \\ 0 \\ 0 \\ 0 \end{bmatrix} = \begin{bmatrix} 1 \\ 0 \\ 1 \end{bmatrix}
$$

The decoded codeword could be found by constructing the syndromes for all possible error patterns and modulo-2 adding the appropriate error pattern to the received bit pattern. It is clear, however, that the decoded codeword is [1 1 1 0 0 0] as this is the closest valid codeword to the received bit pattern.

10.8 Cyclic codes

Cyclic codes are a subclass of group codes which do not possess the all-zeros codeword. The Hamming code is an example of a cyclic code. Their properties and advantages are as follows:

• Their mathematical structure permits higher order correcting codes.
• Their code structures can be easily implemented in hardware by using simple shift registers and exclusive-or gates.
• Cyclic code members are all lateral, or cyclical, shifts of one another.
• Cyclic codes can be represented as, and derived using, polynomials.

The third property, listed above, can be expressed as follows. If:

$$
C = (I_1, I_2, I_3, \cdots, I_n) \tag{10.13}
$$

then:

$$
C_i = (I_{i+1}, \cdots, I_n, I_1, I_2, \cdots, I_i) \tag{10.14}
$$

(C_i is an i bit cyclic shift of C, and $I_1, \cdots, I_n$ now represents both parity and information bits.) The following three codewords provide an example:

        1 0 0 1 0 1 1
        0 1 0 1 1 1 0
        0 0 1 0 1 1 1

Full description of these codes requires a detailed discussion of group and field theory and takes us into a discussion of prime numbers and Galois fields. Non-systematic cyclic codes are obtained by multiplying the data vector by a generator polynomial with modulo arithmetic. In general cyclical codes are generated from parity-check matrices which have cyclically related rows.

Cyclic codes encompass BCH codes and the Reed-Solomon non-binary codes. The Reed-Solomon (RS) code is made up of n symbols where each symbol is m bits long. m can be any length depending on the application, for example if $m = 8$ bits then each symbol would represent a byte. Thus RS codes operate on a multiple bit symbol principle and not a bit principle as other cyclic codes do. An important property of the RS codes is their burst error correction property. Their error correcting power is:

$$t = \frac{n-k}{2}$$ (10.15)

where n and k here relate to encoded symbols, not bits. For example the (31,15) Reed-Solomon code has 31 5-bit encoded symbols which represent 15 symbols of input information or 75 input information bits. This can correct 8 independent bit errors or 4 bursts of length equal to or less than the 5-bit symbol duration.

10.8.1 Polynomial codeword generation

Systematic cyclic redundancy checks (CRC) are in widespread use for performing error detection, but not correction, on bit-serial channels. The operation of CRCs can be considered as an algebraic system in which 0 and 1 are the only values and where addition and subtraction involve no carry operations (i.e. arithmetic is modulo-2). (Addition and subtraction are both, therefore, the same as the logical exclusive-or operation.) If a message of length k bits: $m_{k-1}, \cdots, m_1, m_0$ (from most to least significant bit) is to be transmitted over a channel then, for coding purposes, it may be considered to represent a polynomial of order $k - 1$:

$$M(x) = m_{k-1}x^{k-1} + \cdots + m_1 x + m_0$$ (10.16)

The message $M(x)$ is modified by the generator polynomial $P(x)$ to form the channel coded version of $M(x)$. This is accomplished by multiplying, or bit shifting, $M(x)$ by the order of $P(x)$. $P(x)$ is then divided into the bit shifted or extended version of $M(x)$ and the remainder is then appended to $M(x)$ replacing the zeros which were previously added by the bit shifting operation. Note that the quotient is discarded.

EXAMPLE 10.5

Generate a polynomial codeword from the data sequence 1001 and the generator polynomial $1 + x + x^3$.

$$
\begin{array}{r}
1111 \\
1101\,\overline{)1001000} \\
-1101 \\
\hline
100000 \\
-1101 \\
\hline
10100 \\
-1101 \\
\hline
1110 \\
-1101 \\
\hline
011
\end{array}
$$

Figure 10.14 *Long division calculation of remainder for appending to data to obtain codeword.*

For $M(x) = 1001$ (i.e. $1 + x^3$) and the polynomial $P(x) = 1101$ (i.e. $1 + x + x^3$) the bit shifted sequence $kM(x)$ is 1001000. This is now divided by $P(x)$, Figure 10.14, to obtain the remainder 011. The appended zeros in $kM(x)$ are then replaced by the remainder to realise the transmitted codeword as 1001011.

The required division process can be conveniently implemented in hardware. An encoding circuit, which is equivalent to the long division operation of Figure 10.14, is shown in Figure 10.15. The encoding circuit works as follows. The information bits are transmitted as part of the cyclic code but are also fed back via the feedback loop. During this step, the feedback switch is kept closed. As we clock the shift register and input the message codeword the states $A, \cdots, E$ in the shift register follow this bit pattern:

$$
\begin{aligned}
\text{Input } A &= 1001 \\
B &= 1010 \\
\text{previous } B, \text{ i.e. } C &= 0101 \\
D &= 1111 \\
\text{previous } D, \text{ i.e. } E &= 0111 \\
\text{previous } E, \text{ i.e. } F &= 0011
\end{aligned}
$$

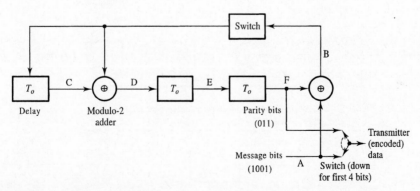

Figure 10.15 *Encoder for a (7,4) cyclic code generated by $P(x) = 1 + x + x^3$.*

When the final bit of information is transmitted, the gate in the feedback loop is opened and the information remaining at the shift register locations E, D, B is then appended to the data for transmission using the output switch. These remaining parity check bits, in this example, would be 110, resulting in the (7,4) Hamming code.

On reception the received data is again divided by $P(x)$ and if the remainder is zero then no errors have been introduced. Further examination of a non-zero remainder in a syndrome table can allow the bit error positions to be determined and the errors corrected by adding in the error pattern in the decoder [Blahut 1983], as previously shown in Example 10.3. The syndrome table can be found either by mathematical manipulation or by successive division for each error location.

The hardware decoding scheme, see Figure 10.16, is basically an inverse version of the encoding scheme of Figure 10.15. If the decoder receives an error it is capable of identifying the position of the error digit via the remainder and the syndrome table.

Thus, the polynomial coded message consists of the original k message bits, followed by $n - k$ additional bits. The generator polynomial is carefully chosen so that almost all errors will be detected. Using a generator of degree k allows the detection of all burst errors affecting up to k consecutive bits. The generator chosen by ITU-T for the V.41 standard, which is the same as that used extensively on wide area networks, is:

$$M(x) = x^{16} + x^{12} + x^5 + 1 \tag{10.17}$$

The generator chosen by IEEE, used extensively on local area and FDDI networks (see Chapter 18), is:

$$M(x) = x^{32} + x^{26} + x^{23} + x^{22} + x^{16} + x^{12} + x^{11} + x^{10} + x^8 + x^7 + x^5 + x^4 + x^2 + x + 1 \tag{10.18}$$

CRC six-bit codewords are transmitted within the plesiochronous multiplex, Chapter 19, to improve the robustness of frame alignment words. The error correcting power of the CRC code is low and it is mainly used when ARQ retransmission is deployed, rather than

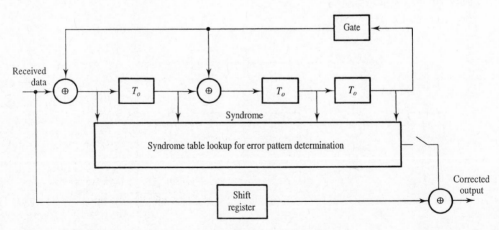

Figure 10.16 *Syndrome calculator and decoder for a (7,4) cyclic code.*

for error correction itself.

10.8.2 Interleaving

The largest application area of block codes is in compact disc players which employ a powerful concatenated and cross-interleaved Reed-Solomon coding scheme to handle random and burst errors. Partitioning data into blocks and then splitting the blocks and interleaving them means that a burst transmission error usually degrades only part of each original block. Thus, using FECC, it is possible to correct for a long error burst, which might have destroyed all the information in the original block, at the expense of the delay required for the interleaver encoder/decoder function. In other applications bit-by-bit interleaving is employed to spread burst errors across a data block prior to decoding. Figure 10.17 shows how an input data stream is read, column by column, into a temporary array and then read out, row by row, to achieve the bit interleaving operation. Now, for example, a burst of three errors in the consecutive transmitted data bits, I_1, I_5, I_9, is converted into isolated single errors in the de-interleaved data.

10.9 Encoding of convolutional codes

Convolutional codes are generated by passing a data sequence through a shift register which has two or more sets of register taps (effectively representing two or more different filters) each set terminating in a modulo-2 adder. The code output is then produced by sampling the output of all the modulo-2 adders once per shift register clock period. The coder output is obtained by the convolution of the input sequence with the impulse response of the coder, hence the name convolutional code. Convolution applies even though there are exclusive or and switch operations rather than multiplies. Figure 10.18

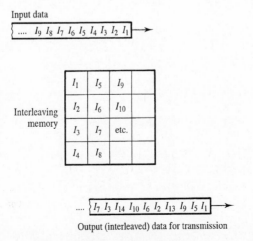

Figure 10.17 *Data block interleaving to overcome burst errors.*

illustrates this with a simple example where the output encoder operation can be defined by two generator polynomials. The first and second encoded outputs, $P_1(x)$, $P_2(x)$, can be defined by $P_1(x) = 1 + x^2$ and $P_2(x) = 1 + x$, as in Example 10.5. This shows a 3-stage encoder giving a constraint length $n = 3$. The error correcting power is related to the constraint length, increasing with longer lengths of shift register.

Assume the data sequence 1101 is input to the three-stage shift register which is initiated or flushed with zeros prior to clocking the sequence through. This example depicts a rate ½ coder ($R = ½$) since there are two output digits for every input information digit. The coder is non-systematic since the data digits are not explicitly present in the transmitted data stream. The first output following a given input is obtained with the switch in its first position and the second output is obtained with the switch in its second position, etc.

This encoder may be regarded as a finite state machine. The first stage of the shift register holds the next input sample and its contents determine the transition to the next state. The final two stages of the shift register hold past inputs and may be regarded as determining the 'memory' of the machine. In this example there are 2 'memory stages'

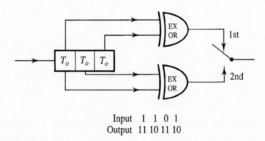

Input   1 1 0 1
Output 11 10 11 10

Figure 10.18 *A simple example of a rate ½ convolutional encoder.*

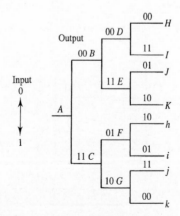

Figure 10.19 *Tree diagram representation of the coder in Figure 10.18.*

and hence four possible states. In general an *n*-stage register would have $2^{(n-1)}$ states. For the $n = 3$ stage coder the four states correspond to the data bit pairs 00, 10, 01, 11 (from prior input data). The convolutional encoder operation may be represented by a tree diagram.

10.9.1 Tree diagram representation

Figure 10.19 depicts the tree diagram corresponding to the example of Figure 10.18. Assume that the encoder is 'flushed' with zeros prior to the first input of data and that it is in an initial state which is labelled *A*. Conventionally the tree diagram is drawn so that inputting a zero results in exiting the present state by the upper path, while inputting a one causes it to exit by the lower path. Assuming a zero is input, the machine will move to state *B* and output 00. Outputs are shown on the corresponding branches of the diagram. Alternatively if the machine is in state *A* and a 1 is input then it proceeds to state *C* via the lower branch and 11 is output. Figure 10.19 depicts the first three stages of the tree diagram, after which there appear to be eight possible states. This is at variance with the previous statement, that the state machine in this example has only four states.

There are, in fact, only four distinct states here, (00, 10, 01, 11), but each state appears twice. Thus *H* is equivalent to *h*, for example. This duplication of states results from identical prior data bits being stored in the shift register of the encoder, The path *B, D* to *H* represents the input of two zeros, as does the path *C, F* to *h*, resulting in identical data being stored in the shift register. After the fourth stage each state would appear four times, etc. Two states are identical if, on receiving the same input, they respond with the same output. Following through input data in Figure 10.19 by the path which this data generates allows the states (00, 10, 01, 11) to be identified and the figure annotated accordingly to identify the redundancy. The apparent exponential growth rate in the number of states can be contained by identifying the identical states and overlaying them. This leads to a trellis diagram.

10.9.2 Trellis diagram

Figure 10.20 shows the trellis diagram corresponding to the tree diagram of Figure 10.19. The horizontal axis represents time while the states are arranged vertically. On the arrival of each new bit the tree diagram is extended to the right. Here five stages are shown with the folding of corresponding tree diagram states being evident at the fourth and fifth stages (states *HIJK*, *LMNO*) by the presence of two entry paths to each state. There are still too many states here and inspection will show that *H* and *L*, for example, are equivalent. Thus four unique states may be identified. These are labelled *a*, *b*, *c* and *d* on this diagram again corresponding to the binary data 00, 10, 01, and 11 being stored in the final two stages of the shift register of Figure 10.18.

The performance of the convolutional coder is basically dependent on the Hamming distance between the valid paths through the trellis, corresponding to all the possible, valid, data bit patterns which can occur. The final step in compacting the graphical

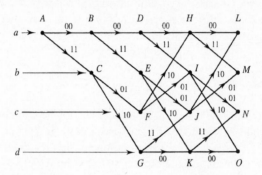

Figure 10.20 *Trellis diagram representation of the coder in Figure 10.18.*

representation of the convolutional encoder is to reduce this trellis diagram to a state transition diagram.

10.9.3 State transition diagram

Here the input to the encoder is shown on the appropriate branch and the corresponding outputs are shown in brackets beside the input, Figure 10.21. For example if the encoder is in state a (the starting state) and a zero is input then the transition is along the self-loop returning to state a. The corresponding output is 00, as shown inside the brackets (and along the top line of Figure 10.20). If, on the other hand, a 1 is input while in state a, then 11 is output and the state transition is along the branch from a to b, etc.

10.10 Viterbi decoding of convolutional codes

There are three main types of decoder. These are based on sequential, threshold (majority logic) and Viterbi decoding techniques. The Viterbi technique is by far the most popular.

Data encoded by modern convolutional coders are usually divided into message blocks for decoding, but unlike the block coded messages, where $n < 255$, the convolutional coded message typically ranges from 500 to >10,000 bits, depending on the application. This makes decoding of convolutional codes potentially onerous. (The decoder memory requirements grow with message length.) The coder operation is

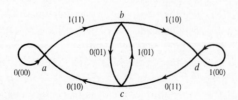

Figure 10.21 *State transition diagram representation of the coder in Figure 10.18.*

illustrated here by the processing of short fixed length blocks, which are fed through the encoder after it has been 'flushed' with zeros to bring it into state a. The block of data is followed with trailing zeros to return the encoder back to state a at the end of the coding cycle. This simplifies decoding and 'flushes' the encoder ready for the next block. The zeros do not, however, carry any information and the efficiency, or rate, of the code is consequently reduced.

Secondly, the Viterbi decoding algorithm is used at each stage of progression through the decoding trellis, retaining only the most likely path to a given node and rejecting all other possible paths on the grounds that their Hamming distance is larger and that they are, thus, less likely events than that represented by the shorter distance path. This leads to a linear increase in storage requirement with block length as opposed to an exponential increase.

The Viterbi decoding algorithm implements a nearest neighbour decoding strategy. It picks the path through the decoding trellis, which assumes the minimum number of errors (the probability of t errors being much greater than that of $t+1$ errors, etc.). Conceptually a decoding trellis, similar to the corresponding encoding trellis, is used for decoding.

EXAMPLE 10.6 – Decoding trellis construction
Assume a received data sequence 1010001010 for the encoder operation of Figure 10.18. Identify the errors and derive the corresponding transmitted data sequence.

The ten transmitted binary digits correspond to five information digits. We assume that the first three of these digits are unknown data and the last two are flushing zeros.

Decoding begins by building a decoding trellis corresponding to the encoding trellis starting at state A as shown in Figure 10.22. We assume that the first input to the encoder is a zero. Reference to the encoding trellis indicates that on entering a zero with the encoder completely flushed 00 would be output, but 10 has been received. This means that the received sequence is a Hamming distance of 1 from the possible transmitted sequence (with an error in the first output bit). This distance metric is noted along the upper branch from A to B.

The possibility that the input data may have been a 1 is now investigated. Again, reference to the encoding trellis indicates that if a 1 is input to the encoder in state A, the encoder will output 11 and follow the lower path to state C. In fact, 10 was received, so again, the actual received sequence is a Hamming distance of 1 from this possible transmitted sequence. (Here the error

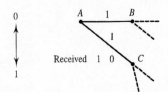

Figure 10.22 *First stage in constructing the decoding trellis for a received sequence from the encoder of Figure 10.18 after receiving two encoded data bits.*

would be in the second bit.) The distance metric is thus noted as 1 along the branch from *A* to *C*.

Now we return to state *B* and assume that the input was zero followed by another zero. If this were the case, the encoder would have gone from state *B* to state *D* and output 00 (Figure 10.23). The third and fourth digits received, however, were 10 and again there is a Hamming distance of 1 between the received sequence and this possible transmitted sequence. This distance metric is noted on the branch *B* to *D* and a similar operation is performed on branch *B* to *E* where the distance is also 1. Next we consider inputting zero while in state *C*. This would create 01 whilst, in fact, 10 was received. The Hamming distance here is 2. This metric is noted on branch *C* to *F* and attention is turned finally to branch *C* to *G*. Starting in state *C* and inputting a 1 would have output 10 and, in fact, 10 was received, so at last there is a received pair of digits which does not imply any errors. The *cumulative* distances along the various paths, i.e. the path metrics, are now entered in square brackets above the final states in Figure 10.23.

Figure 10.24 illustrates a further problem. Decoding is now at stage 3 in the decoding trellis and the possibility of being in state *J* is being considered. From this stage on, each of the four states in this example has two entry, and two exit, paths. Conventionally on reaching a state like *J* with two input paths, the cumulative Hamming distance or path metric of the upper route (*ABEJ*) is shown first and the cumulative Hamming distance for the lower path (*ACGJ*) is shown second in the square brackets adjacent to state *J*. The real power of the Viterbi algorithm lies in its rejection of one of those two paths, retaining one path which is referred to as the 'survivor'. If the two paths have different Hamming distances then, since this is a nearest neighbour (maximum likelihood) decoding strategy, the path with the larger Hamming distance is rejected and the path with the smaller Hamming distance, or path metric, is carried forward as the survivor.

In the case illustrated in Figure 10.24 the distances are identical and the decoder must flag an uncorrectable error sequence, if using incomplete decoding. If using complete decoding a random choice is made between the two paths (bearing in mind there is a 50% probability of being wrong). Fortunately, this situation is rare in practice. (The probability of error has been deliberately increased here for illustrative purposes.)

There are two paths to state *H* with path metrics [2,4], Figure 10.25. The path of distance 4 may thus be rejected as being less likely than the path of distance 2, etc. To state *I* there are also two paths of equal length [4,4]. In the final stage state *P* has been labelled as being the finishing point of the decoding process since in this example only three unknown data digits are being transmitted followed by 00 to flush the encoder and bring the decoder back to state *a*. Only the

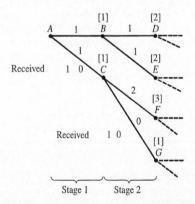

Figure 10.23 *Second stage of trellis of Figure 10.18 after decoding four data bits.*

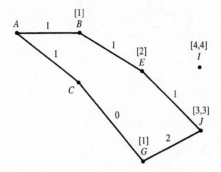

Figure 10.24 *Illustration of a sequence containing a detectable but uncorrectable error pattern.*

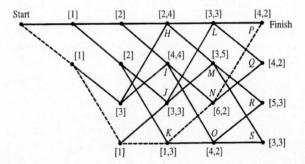

Figure 10.25 *Complete decoding trellis for Example 10.6 with the dashed preferred path and resulting decoded sequence 11100.*

more likely of the two paths to state P is retained. This is the lower path with a Hamming distance of 2. Note that although state O also has distances of 2 we cannot progress from O to P to complete the decoding operation. Figure 10.25 shows the complete decoding trellis. Tracing back along the most likely (dashed) path provides the corresponding decoded sequence as (11100) and the implied correct received data as 1110001110. Although state Q also has a cumulative distance of 2 this cannot terminate the correct path as this decoder must be flushed with zeros ready for the next block of data.

10.10.1 Decoding window

In a practical convolutional decoder the block length would usually be very much larger than in the simple example above. The data from a complete frame of a video coded image, Chapter 16, may be sent as a single message block, for example. In such a case the overhead requirement, for accommodating the flushing zeros, becomes negligible. There is a constraint on the length of data which can be retained in the decoder memory, however, when performing the Viterbi decoding operation. The practical limitation is

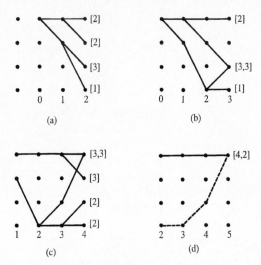

Figure 10.26 *Viterbi decoding within a finite length decoding window.*

known as the decoding window. In a practical decoder the new distance metrics are added to the previous path metrics to obtain updated path metrics. Details of the paths, which correspond to these various distances, are carried forward in the decoding process. In Figure 10.25 the decoding was performed over a window length of 5 input data bits. (With long block lengths the same procedure would be followed.) Figure 10.26 shows an example of decoding with a window length of only 4. This demonstrates how the decoding of Figures 10.22 to 10.25 moves through the window (sequence (a) to (d)) with only the most likely paths being retained.

When the window length is restricted to 4 then, on the arrival of the final received bit pair, it is no longer possible to continue to examine the start bit as this would have propagated out of the restricted length of the decoding window (through which the trellis is viewed). The window length should be long enough to cover all bursts of decoding errors but, since longer lengths involve more computation, a compromise must be made and the decoder's performance verified by computer simulation. (For practical coders a good rule of thumb is for the decoding window to be set at five times the constraint length of the encoder.)

10.10.2 Sequential decoding

Viterbi's algorithm requires all the surviving sequences to be followed throughout the decoding process and leads to excessive memory requirements for long constraint lengths. Complexity can be reduced by sequential decoding, which directly constructs the sequence of states by performing a distance measure at each step. Sequential decoding proceeds forwards until complete decoding is accomplished or the cumulative distance exceeds a preset threshold. When this occurs the algorithm backtracks and selects an alternative path until a satisfactory overall distance is maintained. This works well at low error rates but, when the error rate is high, the number of backward steps can become very large.

The Viterbi decoder has three main components, Figure 10.27. The first is the branch metric value (BMV) calculation unit which finds the Hamming distances for each new branch in the trellis. The add-compare-select unit is the second which calculates and updates the overall path history or path metric values (PMVs) for each path arriving at each node in the trellis. In the third component, the output determination unit, only the surviving paths are retained (i.e. selected) as these have the smallest distances, the paths with higher distances being discarded. When working through the complete trellis, only the most likely overall path is finally retained as the ultimate survivor. Trellis decoding is not restricted to convolutional codes as it is also used for soft decision decoding of block codes [Honary and Markarian]. (A similar decoding trellis is also used later for trellis coded modulation, section 11.4.8 [Biglieri *et al.*, Ungerboeck].)

10.11   Practical coders

Examples of practical block codes are the BCH (127, 64) which has an error correction capability of $t = 10$ bit; the Reed-Solomon (16,8) or (64,32) codes achieve $t = 4$ symbol capability while the shorter block length of the Golay (23,12) code has a $t = 3$ bit capability. The error rate performance of these, and some other, codes is compared in Figure 10.28 [Farrell], all for DPSK modulation, in which the horizontal axis is energy per input *information* data bit divided by one sided noise power spectral density E_b/N_0. Figure 10.28 echoes Figure 10.1, showing clearly the point at which the FECC systems outperform uncoded differential phase shift keyed (DPSK) transmission (see Chapter 11).

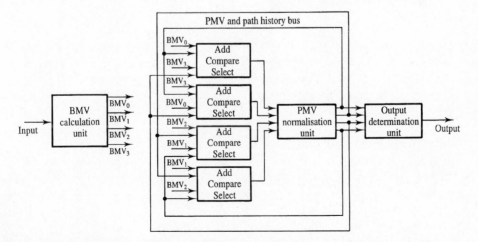

Figure 10.27 *Viterbi decoder circuit for decoding the trellis of Figure 10.20.*

Generally for a $t = 1$ BCH block code, which has a larger block length than the simple Hamming (7,4) example, i.e. a (31,26) or a (63,57) code, then the coding gain over the uncoded system, at a P_b of 10^{-5}, is >2 dB. For the longer $t = 3$ bit BCH (127,106) code the coding gain approaches 4 dB. Linear block codes are usually restricted to $n < 255$ by the decoder complexity. In Figure 10.28 the $t = 4$ symbol Reed-Solomon code is inferior to $t = 3$ bit Golay code at high P_b, but this performance is reversed at lower P_b. The (23,12) Golay code, which corrects triple errors, is a perfect code as equation (10.4) becomes:

$$2^{12} \leq \frac{2^{23}}{{}^{23}C_0 + {}^{23}C_1 + {}^{23}C_2 + {}^{23}C_3} \tag{10.19(a)}$$

i.e.:

$$2^{12} \leq \frac{2^{23}}{1 + 23 + 253 + 1771} \tag{10.19(b)}$$

and hence is satisfied as the equality:

$$2^{12} = \frac{2^{23}}{2048} = \frac{2^{23}}{2^{11}} \tag{10.20}$$

In general Reed-Solomon codes are attractive for bursty error channels where extremely low P_b values (e.g. 10^{-10}) are required. Convolutional codes are favoured for Gaussian noise channels where more moderate P_b values (e.g. 10^{-6}) are required.

Practical convolution codes often have constraint lengths of $n = 7$ with a rate ½ coder which employs 7 delay stages. This requires a decoding window in the trellis of 35 to

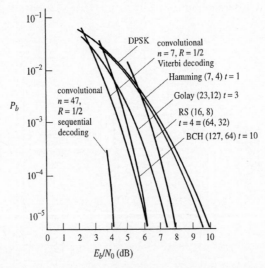

Figure 10.28 *Performance of rate ½ codes for DPSK signals in AWGN with hard decision decoding (source: Farrell, 1989, reproduced with permission of the IEE).*

achieve the theoretical coding gain. With such a window, the encoder message block size can be set appropriate to the specific application. Such a coder then has $D_{min} = 10$ and its performance is equivalent to the BCH (127,64) block code. The convolutional coder is preferred, however, because the encoder is simpler and the decoder can more easily incorporate soft decision techniques (in which a confidence level is retained for the received data). Such soft decisions, which implement maximum likelihood decoding, give approximately 2 dB improvement in coding gain compared to hard decisions. (The Voyager space probes employed soft decision decoders adding 5 to 8 dB coding gain to their link budget, see Chapter 12.)

In general reducing the efficiency, or rate, R, of the coder increases the Hamming distance D_{min} between permissible paths and improves the error correcting power (i.e. the coding gain in dB compared to the uncoded DPSK system). If the performance is normalised by the rate R then there may be little benefit in going to rates of less than ½ as the improvement in error rate is exactly balanced by the reduction in data rate. Whilst additional coding gain improvement is typically 1 dB for an $n = 9$ constraint length, compared to $n = 7$, the improvements beyond this are very small and the decoder complexity is excessively high.

The largest application area for convolutional coders is in rate ¾ compact disc codes which employ a concatenated and cross-interleaved coding scheme. Sequential decoding is a powerful search path technique for use in the decoding trellis, especially with longer constraint length, Figure 10.28. Sophisticated VLSI Viterbi decoders are now available for speeds of 250 kbit/s to 25 Mbit/s with constraint lengths of $n = 7$ at (1996) costs of £10 and above, per FECC decoder. Table 10.2, lists some Qualcom (Q) and Stanford Telecom (ST) convolutional coder chipsets.

Table 10.2 *Examples of commercially available convolutional coder chipsets.*

Data rate (Mbit/s)	25	12	2.5	1	¼
Coder rate, R	$\frac{7}{8}, \frac{3}{4}, \frac{1}{2}, \frac{1}{3}$	$\frac{3}{4}, \frac{2}{3}, \frac{1}{2}$	$\frac{7}{8}, \frac{3}{4}, \frac{1}{2}, \frac{1}{3}$	$\frac{1}{2}, \frac{1}{3}$	$\frac{7}{8}, \frac{3}{4}, \frac{2}{3}, \frac{1}{2}, \frac{1}{3}$
Constraint length	7	7	7	6	6–7
Soft decision capability (bits)	3	3	3	–	3–4
Supplier	Q	Q/ST	Q	ST	Q/ST

10.12 Summary

Error rate control is necessary for many systems to ensure that the probability of bit error is acceptably low. Bit error rates may be reduced by increasing transmitted power, applying various forms of diversity, using echo back and retransmission, employing ARQ, or incorporating FECC. In this chapter the focus has been on channel coding for error detection and correction which are prerequisites, respectively, for ARQ and FECC error control systems.

Channel codes may be systematic or unsystematic. Systematic codes use codewords which contain the information digits, from which the codeword is derived, explicitly. The rate, R, of a code is the ratio of information bits transmitted to total bits transmitted. The Hamming distance between a pair of binary codewords, which is given by their modulo-2 sum, is a measure of how easily one codeword can be transformed into the other. The weight of a binary code word is equal to the number of binary ones which it contains.

Block codes divide the precoded data into k bit lengths and add $(n - k)$ parity check bits to create a post-coded block, with length n bits. An (n, k) block code, therefore, has an efficiency, or rate, of $R = k/n$. Single parity check codes are block codes which append a single digital one, or zero, to each codeword in order to ensure that all codewords have either an even (for even parity) or an odd (for odd parity) weight. This allows single error detection. Data can also be arranged in two dimensional, rectangular, arrays allowing parity check digits to be added to the ends of rows, and the bottoms of columns. This achieves single error correction.

Group codes (also called linear block codes) are block codes which contain the all-zeros codeword and have the closed set, or linear group, property, i.e. the modulo-2 sum of any pair of valid codewords in the set is another valid codeword. The error correcting power, t, of a linear group code is given by $int((D_{min} - 1)/2)$ where D_{min} is the minimum Hamming distance between any pair of codewords. The error detecting power, e, of a linear group code is given by $D_{min} - t - 1$. Group codes are the most important block codes due to the ease with which their performance can be predicted. (D_{min} is given by the weight of the minimum weight codeword, excluding the all-zeros codeword, in the group). Block codes can be generated using a generator matrix. Complete codewords are generated by the product of the precoded data vector with the generator matrix.

Block codes are most easily decoded using a nearest neighbour strategy. This is equivalent to maximum likelihood decoding providing that $P(t \text{ errors}) \gg P(t + 1 \text{ errors})$ which is always the case in practice. Nearest neighbour decoding can be implemented using a nearest neighbour table or a syndrome table. (The syndrome vector for a given received codeword is the product of the parity check matrix and the received codeword vector. It is also the product of the parity check matrix and the error pattern vector of the received codeword allowing error patterns to be determined from a table of syndromes.) Syndrome decoding is advantageous when block size is large since the syndrome table for all (single) error patterns is much smaller than the nearest neighbour decoding table.

Cyclic codes are linear group codes in which the codeword set consists of only, and all, cyclical shifts of any one member of the codeword set. They are particularly easily generated using shift registers with appropriate feedback connections. Their syndromes are also easily calculated using shift register hardware. CRC codes are systematic cyclic codes which are potentially capable of error correction but are often used for error detection only. A polynomial representation of a precoded block of information bits is multiplied (i.e. bit shifted) by the order of a generator polynomial and then divided by the generator polynomial. The remainder of the division process is appended to the block of information bits to form the complete codeword. Division of received codeword polynomials by the generator polynomial leaves zero remainder in the absence of errors.

Convolution codes are unsystematic and operate on long data blocks. The encoding operation can be described (in increasing order of economy) using tree, trellis or state diagrams. The Viterbi algorithm, which implements a nearest neighbour decoding strategy, is usually used to decode convolution codes. The error correction capability of convolutional coders is not inherently in excess of that for block coders but their decoder and, particularly, encoder designs are simpler.

FECC combinations of block and convolutional codes are widely applied to accommodate random and burst errors which may both arise in communication channels.

10.13 Problems

10.1. Assume a binary channel with independent errors and $P_e = 0.05$. Assume k digit symbols from the source alphabet are encoded using an (n, k) block code which can correct all patterns of three or fewer errors. Assume $n = 20$. (a) What is the average number of errors in a block? [1] (b) Assuming binary transmission at 20,000 binary digits per second, derive the symbol error rate at the decoder output. [15.8 symbol/s]

10.2. A binary signal is transmitted through a channel which adds zero mean, white, Gaussian noise. The probability of bit error is 0.001. What is the probability of error in a block of 4 data bits? If the bandwidth is expanded to accommodate a (7,4) block code, what would be the probability of an error in a block of 4 data bits? [0.0040, 0.0019]

10.3. Assume a systematic (n, k) block code where $n = 4$, $k = 2$ and the four codewords are 0000, 0101, 1011, 1110. (a) Construct a maximum likelihood decoding table for this code. (b) How many errors will the code correct? Are there any errors which are detectable but not correctable? [8 corrected and 8 detected but not corrected] (c) Assume this code is used on a channel with a $P_e = 0.01$. What is the probability of having a detectable error sequence? What is the probability of having an undetectable error sequence? [0.0388, 0.0006]

10.4. For a (6,3) systematic linear block code, the three parity check digits are:
$$P_1 = 1 \times I_1 \oplus 1 \times I_2 \oplus 1 \times I_3$$
$$P_2 = 1 \times I_1 \oplus 1 \times I_2 \oplus 0 \times I_3$$
$$P_3 = 0 \times I_1 \oplus 1 \times I_2 \oplus 1 \times I_3$$
(a) Construct the generator matrix $\mathbf{G}$ for this code. (b) Construct all the possible codewords generated by this matrix. (c) Determine the error-correcting capabilities for this code. [single] (d) Prepare a suitable decoding table. (e) Decode the received words 101100, 000110 and 101010. [111100, 100110, 101011]

10.5. Given a code with the parity check matrix:
$$\mathbf{H} = \begin{bmatrix} 1110 \ 100 \\ 1101 \ 010 \\ 1011 \ 001 \end{bmatrix}$$

(a) Write down the generator matrix showing clearly how you derive it from $\mathbf{H}$. (b) Derive the complete weight structure for the above code and find its minimum Hamming distance. How many errors can this code correct? How many errors can this code detect? Can it be used in correction and detection modes simultaneously? [3, 1, no] (c) Write down the syndrome table for this code showing how the table may be derived by consideration of the all-zeros codeword. Also comment on the absence of an all-zeros column from the **H** matrix. (d) Decode the received sequence 1001110, indicate the most likely error pattern associated with this sequence and give the correct codeword. Explain the statement 'most likely error pattern'. [0000010, 1001100]

10.6. When generating a (7,4) cyclic block code using the polynomial $1 + x^2 + x^3$: (a) What would the generated codewords be for the data sequences 1000 and 1010? [1000101, 1010011] (b) Check that these codewords would produce a zero syndrome if received without error. (c) Draw a circuit to generate this code and show how it generates the parity bits for the two data sequences in part (a). (d) If the codeword 1000101 is corrupted to 1001101, i.e. an error occurs in the fourth bit, what is the syndrome at the receiver? Check this is the same as for the codeword 1010011 being corrupted to 1011011. [011]

10.7. Given the ½ rate convolutional encoder defined by $P_1(x) = 1 + x + x^2$ and $P_2(x) = 1 + x^2$, and assuming data is fed into the shift register one bit at a time, draw the encoder: (a) tree diagram; (b) trellis diagram; (c) state transition diagram. (d) State what the rate of the encoder is. (e) Use the Viterbi decoding algorithm to decode the received block of data, 10001000.

Note: there may be errors in this received vector. Assume that the encoder starts in state *a* of the decoding trellis in Figure 10.20 and, after the unknown data digits have been input, the encoder is driven back to state *a* with two 'flushing' zeros. [0000]

CHAPTER 11

Bandpass modulation of a carrier signal

11.1 Introduction

Modulation, at its most general, refers to the modification of one signal's characteristics in sympathy with another signal. The signal being modified is called the carrier and, in the context of communications, the signal doing the modifying is called the information signal. Chapter 5 described pulse modulation in which the carrier is a rectangular pulse train and the characteristics adjusted are, for example, pulse amplitude, or, in PCM, the coded waveform. In this chapter, intermediate or radio frequency (IF or RF) bandpass modulation is described, in which the carrier is a sinusoid and the characteristics adjusted are amplitude, frequency or phase.

The principal reason for employing IF modulation is to transform information signals (which are usually generated at baseband) into signals with more convenient (bandpass) spectra. This allows:

1. Signals to be matched to the characteristics of transmission lines or channels.
2. Signals to be combined using frequency division multiplexing and subsequently transmitted using a common physical transmission medium.
3. Efficient antennas of reasonable physical size to be constructed for radio communication systems.
4. Radio spectrum to be allocated to services on a rational basis and regulated so that interference between systems is kept to acceptable levels.

11.2 Spectral and power efficiency

Different modulation schemes can be compared on the basis of their spectral [1] and power efficiencies. Spectral efficiency is a measure of information transmission rate per Hz of

---

[1] See footnote on first page (pp. 260) of Chapter 8.

bandwidth used. A frequent objective of the communications engineer is to transmit a maximum information rate in a minimum possible bandwidth. This is especially true for radio communications in which radio spectrum is is a scarce, and therefore valuable, resource. The appropriate units for spectral efficiency are clearly bit/s/Hz. It would be elegant if an analogous quantity, i.e. power efficiency, could be defined as the information transmission rate per W of received power. This quantity, however, is not useful since the information received (correctly) depends not only on received signal power but also on noise power.

It is thus carrier to noise ratio (CNR) or equivalently the ratio of symbol energy to noise power spectral density (NPSD), E/N_0, which must be used to compare the power efficiencies of different schemes. It is legitimate, however, to make comparisons between different digital communications systems on the basis of the relative signal power needed to support a given received information rate assuming identical noise environments. In practice this usually means comparing the signal power required by different modulation schemes to sustain identical BERs for identical transmitted information rates.

11.3 Binary IF modulation

Binary IF modulation schemes represent the simplest type of bandpass modulation. They are easy to analyse and occur commonly in practice. Each of the basic binary schemes is therefore examined below in detail. The ideal BER performance of these modulation schemes could be found directly using a general result, equation (8.61), from Chapter 8. It is instructive, however, to obtain the probability of error, P_e, formula for each scheme by considering matched filtering or correlation detection (see section 8.3) as an ideal demodulation process followed by an ideal, baseband, sampling and decision process.

11.3.1 Binary amplitude shift keying (and on-off keying)

In binary amplitude shift keyed (BASK) systems the two digital symbols, zero and one, are represented by pulses of a sinusoidal carrier (frequency, f_c) with two different amplitudes A_0 and A_1. In practice, one of the amplitudes, A_0, is invariably chosen to be zero resulting in on-off keyed (OOK) IF modulation, i.e.:

$$f(t) = \begin{cases} A_1 \, \Pi(t/T_o) \cos 2\pi f_c t, & \text{for a digital 1} \\ 0, & \text{for a digital 0} \end{cases} \tag{11.1}$$

where T_o is the symbol duration (as used in Chapters 6 and 8) and Π is the rectangular pulse function.

An OOK modulator can be implemented either as a simple switch, which keys a carrier on and off, or as a double balanced modulator (or mixer) which is used to multiply the carrier by a baseband unipolar OOK signal. A schematic diagram of the latter type of modulator is shown with rectangular pulse input and output waveforms, spectra and allowed phasor states in Figure 11.1. The modulated signal has a DSB spectrum centred

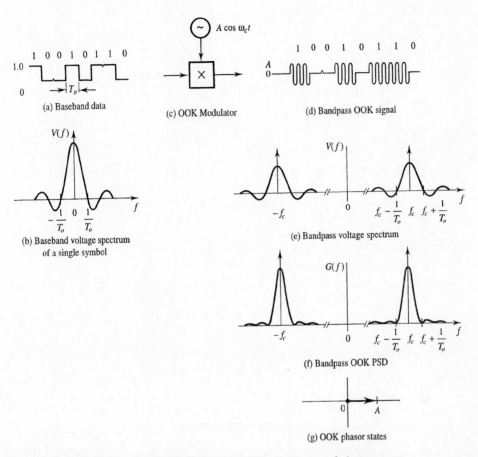

Figure 11.1 *On-off keyed (OOK) modulator, waveforms, spectra and phasor states.*

on $\pm f_c$ (Figure 11.1(e) and (f)) and, since a constant carrier waveform is being keyed, the OOK signal has two phasor states, 0 and $A = A_1$ (Figure 11.1(g)). Detection of IF OOK signals can be coherent or incoherent. In the former case a matched filter or correlator is used prior to sampling and decision thresholding (Figure 11.2(a),(b)). In the latter (more common) case envelope detection is used to recover the baseband digital signal followed by centre point sampling or integrate and dump (I + D) detection (Figure 11.2(d)). (The envelope detector would normally be preceded by a bandpass filter to improve the CNR.) Alternatively, an incoherent detector can be constructed using two correlation channels configured to detect inphase (I) and quadrature (Q) components of the signal followed by I and Q squaring operations and a summing device (Figure 11.2(e)). This arrangement overcomes the requirement for precise carrier phase synchronisation (c.f. Figure 11.2(b)). Whilst such a dual channel incoherent detector may seem unnecessarily complicated compared with the other incoherent detectors, recent advances in VLSI technology mean

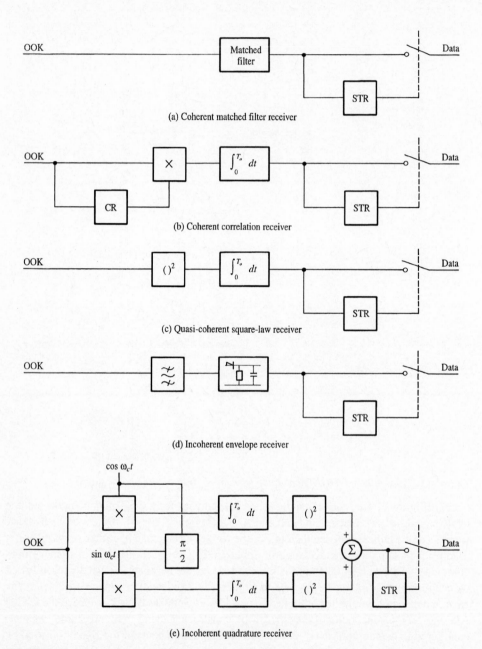

(a) Coherent matched filter receiver

(b) Coherent correlation receiver

(c) Quasi-coherent square-law receiver

(d) Incoherent envelope receiver

(e) Incoherent quadrature receiver

Figure 11.2 *Coherent and incoherent bandpass OOK receivers.*

that they can often be implemented (digitally) as smaller, lighter and cheaper components than the filters and envelope detectors used in more traditional designs. (Symbol timing recovery is discussed in section 6.7 and carrier recovery is discussed later in this chapter.)

The decision instant voltage, $f(nT_o)$, at the output of an OOK matched filter or correlation detector (see equation (8.37)) is:

$$f(nT_o) = \begin{cases} k\,E_1, & \text{digital 1} \\ 0, & \text{digital 0} \end{cases} \tag{11.2}$$

where E_1 (V^{2}s) is the normalised energy contained in symbol 1 and k has units of Hz/V. The normalised noise power, σ^2 (V^2), at the detector output (see equation (8.49)) is:

$$\sigma^2 = k^2 E_1 N_0/2 \tag{11.3}$$

where N_0 (V^2/Hz) is the normalised one-sided noise power spectral density at the matched filter, or correlator, input. (Note that the constant k is not shown in equations (8.37) and (8.49) since it is assumed, there, to be 1.0 Hz/V.) The post filtered decision process is identical to the baseband binary decision process described in Chapter 6. Equation (6.8) can therefore be used with $\Delta V = k(E_1 - 0)$ and $\sigma^2 = k^2 E_1 N_0/2$ to give the probability of symbol error:

$$P_e = \frac{1}{2}\left[1 - \text{erf}\,\frac{1}{2}\left(\frac{E_1}{N_0} \right)^{\!\!1/2} \right] \tag{11.4}$$

Equation (11.4) can be expressed in terms of the time averaged energy per symbol, $\langle E \rangle = \frac{1}{2}(E_1 + E_0)$ where, for OOK, $E_0 = 0$, i.e.:

$$P_e = \frac{1}{2}\left[1 - \text{erf}\,\frac{1}{\sqrt{2}}\left(\frac{\langle E \rangle}{N_0} \right)^{\!\!1/2} \right] \tag{11.5}$$

Finally equations (11.4) and (11.5) can be expressed in terms of received carrier to noise ratios (C/N) using the following relationships:

$$C = \langle E \rangle/T_o \quad \text{(V}^2\text{)} \tag{11.6}$$

$$N = N_0 B \quad \text{(V}^2\text{)} \tag{11.7}$$

$$\langle E \rangle/N_0 = T_o B\, C/N \tag{11.8}$$

where C is the received carrier power averaged over all symbol periods and N is the normalised noise power in a bandwidth B Hz.† This gives:

$$P_e = \frac{1}{2}\left[1 - \text{erf}\,\frac{(T_o B)^{1/2}}{\sqrt{2}}\left(\frac{C}{N} \right)^{\!\!1/2} \right] \tag{11.9}$$

For minimum bandwidth (i.e. Nyquist) pulses $T_o B = 1.0$ and $\langle E \rangle/N_0 = C/N$. Bandlimited signals will result in symbol energy being spread over more than one symbol period, however, resulting (potentially) in ISI. This will degrade P_e with respect to that given in equations (11.4), (11.5) and (11.9) unless proper steps are taken to ensure ISI free sampling instants at the output of the receiver matched filter (as in section 8.4).

Incoherent detection of OOK is, by definition, insensitive to the phase information contained in the received symbols. This lost information degrades the detector's performance over that given above. The degradation incurred is typically equivalent to a 1 dB penalty in receiver CNR at a P_e level of 10^{-4}. The modest size of this CNR penalty means that incoherent detection of OOK signals is almost always used in practice.

EXAMPLE 11.1

An OOK IF modulated signal is detected by an ideal matched filter receiver. The non-zero symbol at the matched filter input is a rectangular pulse with an amplitude 100 mV and a duration of 10 ms. The noise at this point is known to be white and Gaussian, and has an RMS value of 140 mV when measured in a noise bandwidth of 10 kHz. Calculate the probability of bit error.

Energy per non-zero symbol:

$$E_1 = v_{RMS}^2 \, T_o$$

$$= \left(\frac{100 \times 10^{-3}}{\sqrt{2}} \right)^2 10 \times 10^{-3} = 5.0 \times 10^{-5} \quad (\text{V}^2\text{s})$$

Average energy per symbol:

$$\langle E \rangle = \frac{E_1 + 0}{2} = 2.5 \times 10^{-5} \quad (\text{V}^2\text{s})$$

Noise power spectral density (from equation (11.7)):

$$N_0 = \frac{N}{B_N} = \frac{n_{RMS}^2}{B_N}$$

$$= \frac{(140 \times 10^{-3})^2}{10 \times 10^3} = 1.96 \times 10^{-6} \quad (\text{V}^2/\text{Hz})$$

From equation (11.5):

$$P_e = \frac{1}{2} \left[1 - \text{erf} \, \frac{1}{\sqrt{2}} \left(\frac{\langle E \rangle}{N_0} \right)^{1/2} \right]$$

$$= \frac{1}{2} \left[1 - \text{erf} \, \frac{1}{\sqrt{2}} \left(\frac{2.5 \times 10^{-5}}{1.96 \times 10^{-6}} \right)^{1/2} \right]$$

† If pulse shaping, or filtering, has been employed to bandlimit the transmitted signal, the obvious interpretation of B is the predetection bandwidth, equal to the signal bandwidth. Here C/N in equation (11.9) will be the actual CNR measured after a predetection filter with bandwidth just sufficient to pass the signal intact. In the case of rectangular pulse transmission (or any other non-bandlimited transmission scheme), B must be interpreted simply as a convenient bandwidth within which the noise power is measured or specified (typically chosen to be the doublesided Nyquist bandwidth, $B = 1/T_o$). C/N in equation (11.9) does not then correspond to the CNR at the input to the matched filter or correlation receiver since, strictly speaking, this quantity will be zero.

$$= \tfrac{1}{2}\,[1 - \operatorname{erf}\,(2.525)]$$

Using error function tables:

$$P_e \;=\; \tfrac{1}{2}\,[1 - 0.999645] \;=\; 1.778 \times 10^{-4}$$

11.3.2 Binary phase shift keying (and phase reversal keying)

Binary phase shift keying (BPSK) impresses baseband information onto a carrier, by changing the carrier's phase in sympathy with the baseband digital data, i.e.:

$$f(t) \;=\; \begin{cases} A\;\Pi(t/T_o)\,\cos 2\pi f_c t, & \text{for a digital 1} \\ A\;\Pi(t/T_o)\,\cos(2\pi f_c t + \phi), & \text{for a digital 0} \end{cases} \tag{11.10}$$

In principle any two phasor states can be used to represent the binary symbols but usually antipodal states are chosen (i.e. states separated by $\phi = 180°$). For obvious reasons this type of modulation is sometimes referred to as phase reversal keying (PRK). A PRK transmitter with typical baseband and IF waveforms, spectra and allowed phasor states is

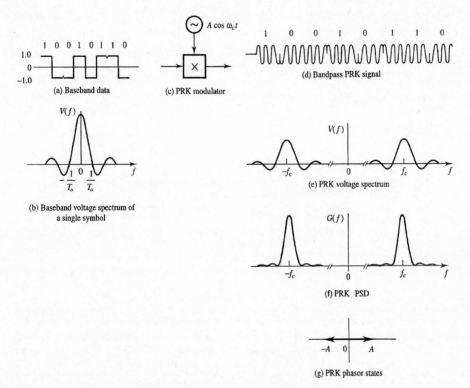

Figure 11.3 *Phase reversal keyed (PRK) modulator waveforms, spectra and phasor states.*

shown in Figure 11.3. PRK systems must obviously employ coherent detectors which can be implemented as either matched filters or correlators. Since the zero and one symbols are antipodal, Figure 11.3(g), only one receiver channel is needed, Figure 11.4. The post-filtered decision instant voltages are $\pm kE$ (V) where E (V^{2}s) is the normalised energy residing in either symbol. The normalised noise power, σ^2 (V^2), at the filter output is the same as in the BASK case. Substituting $\Delta V = 2kE$ and $\sigma^2 = k^2 E N_0/2$ into equation (6.8) gives:

$$P_e = \frac{1}{2}\left[1 - \mathrm{erf}\left(\frac{E}{N_0}\right)^{\frac{1}{2}}\right] \tag{11.11}$$

(Note that in this case $E = \langle E \rangle$.) Using equation (11.8) the PRK probability of symbol error can be expressed as:

$$P_e = \frac{1}{2}\left[1 - \mathrm{erf}\,(T_o B)^{\frac{1}{2}}\left(\frac{C}{N}\right)^{\frac{1}{2}}\right] \tag{11.12}$$

For more general BPSK modulation where the difference between phasor states is less than 180° the P_e performance is most easily derived by resolving the allowed phasor states into a residual carrier and a reduced amplitude PRK signal, Figure 11.5. The residual carrier, which contributes nothing to symbol detection, can be employed as a pilot transmission and used for carrier recovery purposes at the receiver. If the difference between phasor states is $\Delta\theta$ (Figure 11.5) and a BPSK 'modulation index', m, is defined by:

$$m = \sin\left(\frac{\Delta\theta}{2}\right) \tag{11.13}$$

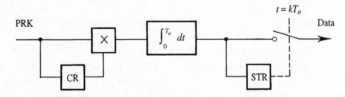

Figure 11.4 *Phase reversal keyed correlation detector.*

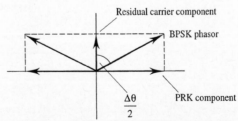

Figure 11.5 *Resolution of BPSK signal into PRK signal plus residual carrier.*

then m is the proportion of the transmitted signal voltage which conveys information and the corresponding proportion of total symbol energy is:

$$m^2 = \sin^2\left(\frac{\Delta\theta}{2}\right) = \frac{1}{2}(1 - \cos\Delta\theta) \tag{11.14}$$

$\cos\Delta\theta$ is the scalar product of the two (unit amplitude) symbol phasors. Denoting this quantity by the normalised correlation coefficient ρ then:

$$m = \sqrt{\tfrac{1}{2}(1 - \rho)} \tag{11.15}$$

The BPSK probability of symbol error is found by replacing E in equation (11.11) with $m^2 E$, i.e.:

$$P_e = \frac{1}{2}\left[1 - \operatorname{erf} m\left(\frac{E}{N_0}\right)^{1/2}\right] \tag{11.16}$$

and equation (11.16) can be rewritten as:

$$P_e = \frac{1}{2}\left[1 - \operatorname{erf}\frac{\sqrt{1-\rho}}{\sqrt{2}}\left(\frac{E}{N_0}\right)^{1/2}\right] \tag{11.17}$$

It is, of course, no coincidence that equation (11.17) is identical to equation (8.61) since, as pointed out both in Chapter 8 and section 11.3, this is a general result for all binary systems having equal energy, equiprobable, symbols. The same caution must be exercised when using equations (11.11), (11.12), (11.16) and (11.17) as when using equations (11.4), (11.5) and (11.9) since all those equations assume ISI free reception. Filtering of PRK signals to limit their bandwidth may, therefore, result in a P_e which is higher than these equations imply.

EXAMPLE 11.2
A 140 Mbit/s ISI free PRK signalling system uses pulse shaping to constrain its transmission to the double sideband Nyquist bandwidth. The received signal power is 10 mW and the one-sided noise power spectral density is 6.0 pW/Hz. Find the BER expected at the output of an ideal matched filter receiver. If the phase angle between symbols is reduced to 165° in order to provide a residual carrier, find the received power in the residual carrier and the new BER.

The double sided Nyquist bandwidth is given by:

$$B = \frac{1}{T_o}$$

i.e.:

$$T_o B = 1.0$$

$$N = N_0 B = N_0/T_o = N_0 R_s$$

$$= 6.0 \times 10^{-12} \times 140 \times 10^6 = 8.4 \times 10^{-4} \quad \text{W}$$

Now using equation (11.12):

$$P_e = \frac{1}{2}\left[1 - \text{erf}\ (T_o B)^{\frac{1}{2}}\left(\frac{C}{N}\right)^{\frac{1}{2}}\right] = \frac{1}{2}\left[1 - \text{erf}\left(\frac{10 \times 10^{-3}}{8.4 \times 10^{-4}}\right)^{\frac{1}{2}}\right]$$

$$= \tfrac{1}{2}\,[1 - \text{erf}\,(3.450)] = \tfrac{1}{2}\,[1 - 0.999\,998\,934] = 5.33 \times 10^{-7}$$

$$\text{BER} = P_e R_s$$

$$= 5.33 \times 10^{-7} \times 140 \times 10^6 = 74.6 \text{ error/s}$$

If $\Delta\theta$ is reduced to $165°$ then:

$$m = \sin\left(\frac{\Delta\theta}{2}\right) = \sin\left(\frac{165}{2}\right) = 0.9914$$

Proportion of signal power in residual carrier is:

$$1 - m^2 = 1 - 0.9914^2 = 0.0171$$

Power received in residual carrier is therefore:

$$C(1 - m^2) = 10 \times 10^{-3} \times 0.0171 = 1.71 \times 10^{-4} \text{ W}$$

Information bearing carrier power is:

$$C\,m^2 = 10 \times 10^{-3}\,(0.9914)^2 = 9.829 \times 10^{-3} \text{ W}$$

Now from equation (11.12):

$$P_e = \frac{1}{2}\left[1 - \text{erf}\ (T_o B)^{\frac{1}{2}}\left(\frac{C\,m^2}{N}\right)^{\frac{1}{2}}\right] = \frac{1}{2}\left[1 - \text{erf}\left(\frac{9.829 \times 10^{-3}}{8.4 \times 10^{-4}}\right)^{\frac{1}{2}}\right]$$

$$= \tfrac{1}{2}\,[1 - \text{erf}\,(3.421)] = \tfrac{1}{2}\,[1 - 0.999\,998\,688] = 6.55 \times 10^{-7}$$

$$\text{BER} = P_e R_s$$

$$= 6.55 \times 10^{-7} \times 140 \times 10^6 = 91.7 \text{ error/s}$$

11.3.3 Binary frequency shift keying

Binary frequency shift keying (BFSK) represents digital ones and zeros by carrier pulses with two distinct frequencies, f_1 and f_2, i.e.:

$$f(t) = \begin{cases} A\ \Pi(t/T_o)\cos 2\pi f_1 t, & \text{for a digital 1} \\ A\ \Pi(t/T_o)\cos 2\pi f_2 t, & \text{for a digital 0} \end{cases} \tag{11.18}$$

Figure 11.6 shows a schematic diagram of a BFSK modulator, signal waveforms and signal spectra. (In practice the BFSK modulator would normally be implemented as a numerically controlled oscillator.) The voltage spectrum of the BFSK signal is the superposition of the two OOK spectra, one representing the baseband data stream modulated onto a carrier with frequency f_1 and one representing the inverse data stream modulated onto a carrier with frequency f_2 (Figure 11.6(e)). It is important to realise, however, that the PSD of a BFSK signal is not the superposition of two OOK PSDs. This is because in the region where the two OOK spectra overlap the spectral lines must be added with due regard to their phases. (In practice, when the separation of f_1 and f_2 is large, and the overlap correspondingly small, then the BFSK PSD is *approximately* the superposition of two OOK PSDs.)

Detection of BFSK can be coherent or incoherent although the latter is more common. Incoherent detection suffers the same CNR penalty, compared with coherent detection, as is the case for OOK systems. Figure 11.7 shows coherent and incoherent BFSK receivers. The receiver in Figure 11.7(c) is an FSK version of the quadrature receiver shown in Figure 11.2. The coherent receiver (Figure 11.7(a)) is an FSK version of the PSK receiver in Figure 11.4.

FSK does not provide the noise reduction of wideband analogue FM transmissions. If, however, a BFSK carrier frequency, f_c Figure 11.6(e), is defined by:

$$f_c = \frac{f_1 + f_2}{2} \quad \text{(Hz)} \tag{11.19}$$

and a BFSK frequency deviation, Δf, is defined (Figure 11.6(e)) as:

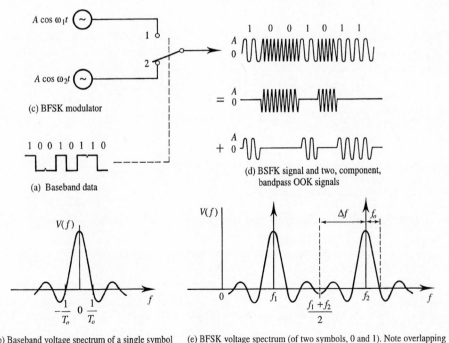

(c) BFSK modulator

(a) Baseband data

(d) BSFK signal and two, component, bandpass OOK signals

(b) Baseband voltage spectrum of a single symbol

(e) BFSK voltage spectrum (of two symbols, 0 and 1). Note overlapping OOK spectra

Figure 11.6 *Binary frequency shift keyed (BFSK) modulators, waveforms and spectra.*

$$\Delta f = \frac{f_2 - f_1}{2} \quad \text{(Hz)} \tag{11.20}$$

then using the first zero crossing points in the BFSK voltage spectrum to define its bandwidth, B, gives:

$$B = 2\Delta f + 2f_o \tag{11.21}$$

Here $f_o = 1/T_o$ is both the nominal bandwidth and the baud rate of the baseband data stream. Equation (11.21) is strongly reminiscent of Carson's rule for the bandwidth of an FM signal [Stremler]. If the binary symbols of a BFSK system are orthogonal (see section 2.5.3), i.e.:

$$\int_0^{T_o} \cos(2\pi f_1 t) \cos(2\pi f_2 t) \, dt = 0 \tag{11.22}$$

then, when the output of one channel of a coherent BFSK receiver is a maximum the output of the other channel will be zero. After subtracting the post-filtered signals arising from each receiver channel the orthogonal BFSK decision instant voltage is:

$$f(nT_o) = \begin{cases} kE, & \text{for a digital 1} \\ -kE, & \text{for a digital 0} \end{cases} \tag{11.23}$$

Table 11.1 *P_e formulae for ideal coherent detection of baseband and IF modulated binary signals.*

		P_e	
Baseband signalling	Unipolar (OOK)	$\frac{1}{2} \operatorname{erfc} \sqrt{\dfrac{1}{2} \dfrac{E}{N_0}}$ (8.61)	$\frac{1}{2} \operatorname{erfc} \sqrt{\dfrac{1}{4} \dfrac{S}{N}}$ (6.10(b))
	Polar	$\frac{1}{2} \operatorname{erfc} \sqrt{\dfrac{E}{N_0}}$ (8.61)	$\frac{1}{2} \operatorname{erfc} \sqrt{\dfrac{1}{2} \dfrac{S}{N}}$ (6.12)
IF/RF signalling	OOK	$\frac{1}{2} \operatorname{erfc} \sqrt{\dfrac{1}{2} \dfrac{E}{N_0}}$	$\frac{1}{2} \operatorname{erfc} \sqrt{\dfrac{T_o B}{2} \dfrac{C}{N}}$
	BFSK (orthogonal)	$\frac{1}{2} \operatorname{erfc} \sqrt{\dfrac{1}{2} \dfrac{E}{N_0}}$	$\frac{1}{2} \operatorname{erfc} \sqrt{\dfrac{T_o B}{2} \dfrac{C}{N}}$
	PRK	$\frac{1}{2} \operatorname{erfc} \sqrt{\dfrac{E}{N_0}}$	$\frac{1}{2} \operatorname{erfc} \sqrt{T_o B \dfrac{C}{N}}$

If the one sided NPSD at the BFSK receiver input is N_0 (V^2Hz^{-1}) then the noise power, $\sigma_1^2 = k^2 E N_0/2$, received via channel 1 and the noise power, $\sigma_2^2 = k^2 E N_0/2$, received via channel 2 will add power-wise (since the noise processes will be uncorrelated). The total noise power at the receiver output will therefore be:

$$\sigma^2 = \sigma_1^2 + \sigma_2^2 = k^2 E N_0 \quad (\text{V}^2) \tag{11.24}$$

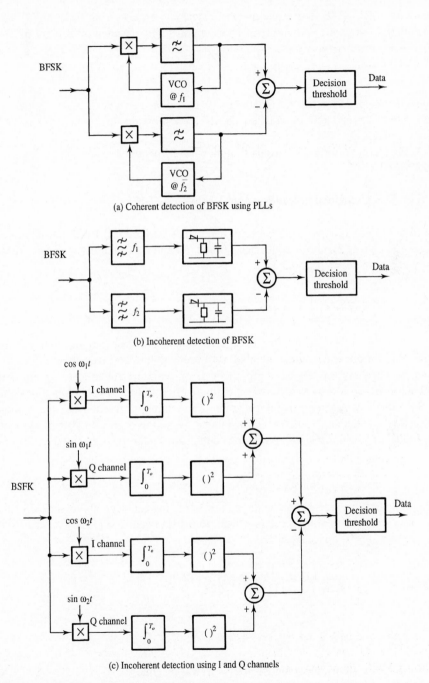

(a) Coherent detection of BFSK using PLLs

(b) Incoherent detection of BFSK

(c) Incoherent detection using I and Q channels

Figure 11.7 *Coherent and incoherent BFSK receivers.*

Substituting $\Delta V = 2kE$ from equation (11.23) and $\sigma = k\sqrt{EN_0}$ from equation (11.24) into equation (6.8) gives the probability of error for coherently detected orthogonal BFSK:

$$P_e = \frac{1}{2}\left[1 - \mathrm{erf}\,\frac{1}{\sqrt{2}}\left(\frac{E}{N_0} \right)^{\frac{1}{2}} \right] \tag{11.25}$$

Notice that here, as for BPSK signalling, $E = \langle E \rangle$. Since equation (11.25) is identical to equation (11.5) for OOK modulation the BFSK expression for P_e in terms of CNR is identical to equation (11.9).

Table 11.1 compares the various formulae for P_e obtained in section 11.3, Chapter 6 and Chapter 8.

11.3.4 BFSK symbol correlation and Sunde's FSK

Equation (11.25) applies to the coherent detection of orthogonal BFSK. It is, of course, possible to choose symbol frequencies (f_1 and f_2) and a symbol duration (T_o) such that the symbols are not orthogonal, i.e.:

$$\int_0^{T_o} \cos(2\pi f_1 t)\cos(2\pi f_2 t) \neq 0 \tag{11.26}$$

If this is the case then there will be non-zero sampling instant outputs from both BFSK receiver channels when either symbol is present at the receiver input. The key to understanding how this affects the probability of symbol error is to recognise that the common fraction of the two symbols (i.e. the fraction of each symbol which is common to the other) can carry no information. The common fraction is the normalised correlation coefficient, ρ, of the two symbols. Denoting a general pair of symbols by $f_1(t)$ and $f_2(t)$ the normalised correlation coefficient is defined by equation (2.89):

$$\rho = \frac{\langle f_1(t) f_2(t) \rangle}{\sqrt{\langle |f_1(t)|^2 \rangle}\,\sqrt{\langle |f_2(t)|^2 \rangle}} \tag{11.27}$$

(The normalisation process can be thought of as scaling each symbol to an RMS value of 1.0 before cross correlating.) For practical calculations equation (11.27) can be rewritten as:

$$\rho = \frac{1}{f_{1RMS}\,f_{2RMS}}\,\frac{1}{T_o}\int_0^{T_o} f_1(t) f_2(t)\, dt$$

$$= \frac{1}{\sqrt{E_1 E_2}}\int_0^{T_o} f_1(t) f_2(t)\, dt \tag{11.28}$$

and for BFSK systems, in which $E_1 = E_2 = E$, this becomes:

$$\rho = \frac{2}{T_o}\int_0^{T_o} \cos(2\pi f_1 t)\cos(2\pi f_2 t)\, dt \tag{11.29}$$

Figure 11.8 shows BFSK symbol correlation, ρ, plotted against the difference in the number of carrier half cycles contained in the symbols. The zero crossing points represent orthogonal signalling systems and $2(f_2 - f_1)T_o = 1.43$ represents a signalling system somewhere between orthogonal and antipodal, see section 6.4.2. (This is the optimum operating point for BFSK in terms of power efficiency and yields a 0.8 dB CNR saving over orthogonal BFSK. It corresponds to symbols which contain a difference of approximately 3/4 of a carrier cycle.)

Whilst any zero crossing point on the $\rho - T_o$ diagram of Figure 11.8 corresponds to orthogonal BFSK it is not possible to use the first zero (i.e. $2(f_2 - f_1)T_o = 1$) if detection is incoherent. (This can be appreciated if it is remembered that for BFSK operated at this point the two symbols are different by only half a carrier cycle. A difference this small can be detected as a change in phase (of 180°) over the symbol duration but not measured reliably in this time as a change in frequency.) The minimum frequency separation for successful incoherent detection of orthogonal BFSK is therefore given by the second zero crossing point (i.e. $2(f_2 - f_1)T_o = 2$) in Figure 11.8 which corresponds to one carrier cycle difference between the two symbols or $\Delta f = f_o/2$ in Figure 11.6. BFSK operated at the second zero of the $\rho - T_o$ diagram is called Sunde's FSK. The voltage and power spectra for this scheme are shown in Figure 11.9. If the first spectral zero definition of bandwidth (or equivalently Carson's rule) is applied to Sunde's FSK the bandwidth would be given by:

$$B = (f_2 - f_1) + \frac{2}{T_o} = \frac{3}{T_o} \quad \text{(Hz)} \tag{11.30}$$

The overlapping spectral lines of the component OOK signals, however, result in cancellation giving a $1/f^4$ roll-off in the envelope of the power spectral density. This rapid roll-off allows the practical bandwidth of Sunde's FSK to be taken, for most applications, to be:

$$B = f_2 - f_1 = \frac{1}{T_o} \quad \text{(Hz)} \tag{11.31}$$

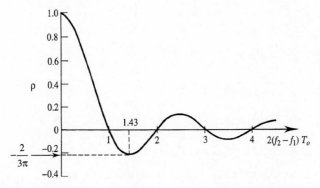

Figure 11.8 *BFSK $\rho - T_o$ diagram, $\rho = (\sin[\pi(f_2 - f_1)T_o]\cos[\pi(f_2 - f_1)T_o])/([\pi(f_2 - f_1)T_o])$*

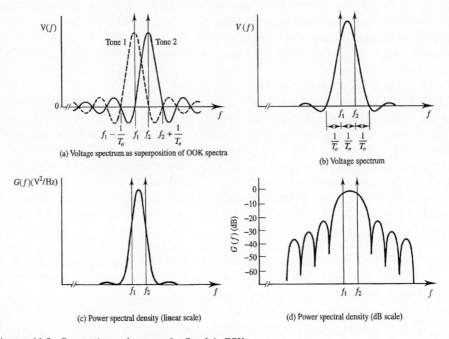

(a) Voltage spectrum as superposition of OOK spectra

(b) Voltage spectrum

(c) Power spectral density (linear scale)

(d) Power spectral density (dB scale)

Figure 11.9 *Comparison of spectra for Sunde's FSK.*

EXAMPLE 11.3

A rectangular pulse BFSK system operates at the 3rd zero crossing point of the $\rho - T_o$ diagram of Figure 11.8. The maximum available transmitter power results in a carrier power at the receiver input of 60 mW. The one-sided NPSD referred to the same point is 0.1 nW/Hz. What is the maximum bit rate which the system can support if the probability of bit error is not to fall below 10^{-6}? What is the nominal bandwidth of the BFSK signal if the lower frequency symbol has a frequency of 80 MHz?

$$P_e = \frac{1}{2}\left[1 - \text{erf}\, \frac{1}{\sqrt{2}}\left(\frac{E}{N_0} \right)^{\frac{1}{2}} \right]$$

i.e.:

$$\frac{E}{N_0} = 2\left[\text{erf}^{-1}(1 - 2P_e) \right]^2 = 2\left[\text{erf}^{-1}\left(1 - 2 \times 10^{-6} \right) \right]^2$$

$$= 2\left[\text{erf}^{-1}(0.999\,998) \right]^2 = 2\,[3.361]^2 = 22.59$$

Therefore:

$$E = 22.59\,N_0 = 22.59 \times 0.1 \times 10^{-9} = 2.259 \times 10^{-9}\ \ \text{J}$$

Now from equation (11.6):

$$T_o = \frac{E}{C} = \frac{2.259 \times 10^{-9}}{60 \times 10^{-3}} = 3.765 \times 10^{-8} \text{ s}$$

$$R_b = R_s = \frac{1}{T_o} = \frac{1}{3.765 \times 10^{-8}} = 2.656 \times 10^7 \text{ bit/s}$$

Since operation on the diagram of Figure 11.8 is at $n = 3$ then:

$$3 = 2 (f_2 - f_1) T_o$$

$$f_2 = \frac{3}{2T_o} + f_1$$

$$= \frac{3}{2 \times 3.765 \times 10^{-8}} + 80 \times 10^6 = 1.198 \times 10^8 \text{ Hz}$$

$$\Delta f = \frac{f_2 - f_1}{2} = \frac{119.8 - 80}{2} = 19.9 \text{ MHz}$$

and from equation (11.21):

$$B = 2 (\Delta f + f_o)$$

$$= 2 (19.9 + 26.6) = 93.0 \text{ MHz}$$

11.3.5 Comparison of binary shift keying techniques

Table 11.2 *Relative power efficiencies of binary bandpass modulation schemes.*

	Bandpass OOK	*Orthogonal BFSK*	*PRK*
$\dfrac{E_1}{N_0}$	4	2	1
$\dfrac{\langle E \rangle}{N_0}$	2	2	1

It is no coincidence that OOK and orthogonal FSK systems are equally power efficient, i.e. have the same probability of symbol error for the same $\langle E \rangle / N_0$, Table 11.1. This is because they are both orthogonal signalling schemes. PRK signalling is antipodal and, therefore, more power efficient, i.e. it has a better P_e performance or, alternatively, a power saving of 3 dB. If comparisons are made on a peak power basis then orthogonal FSK requires 3 dB more power than PRK, as expected, but also requires 3 dB less power than OOK since all the energy of the OOK transmission is squeezed into only one type of symbol. These relative power efficiencies are summarised in Table 11.2.

Although the spectral efficiency for all unfiltered (i.e. rectangular pulse) signalling systems is, strictly, 0 bit/s/Hz (see section 8.2) it is possible to define a nominal spectral efficiency using the signal pulse's main lobe bandwidth, $B = 2/T_o$. (This does not correspond to the theoretical minimum channel bandwidth which is $B = 1/T_o$.) Table 11.3 summarises the nominal spectral efficiencies of unfiltered bandpass BASK, BFSK and BPSK systems and compares them with the nominal efficiency of unfiltered baseband OOK. It is interesting to note that Sunde's FSK could be regarded as being more spectrally efficient than BASK and BPSK due to the more rapid roll-off of its spectral envelope. (If the null-to-null definition of bandwidth is adhered to strictly then Sunde's FSK has a nominal efficiency of 0.33 bit/s/Hz which is less than BASK and BPSK.) The probability of symbol error versus $\langle E \rangle / N_0$ for orthogonal (OOK and orthogonal BFSK) and antipodal (PRK) signalling is shown in Figure 11.10.

It is useful to develop the generalised expression for P_e which can be applied to all binary (coherently detected, ideal,) modulation schemes. For any two-channel coherent receiver the decision instant voltages for equal energy symbols will be:

$$
f(nT_o) = \begin{cases} k\left[E - \displaystyle\int_0^{T_o} f_1(t)f_2(t)\,dt \right], & \text{symbol 0 present} \\[4mm] k\left[-E + \displaystyle\int_0^{T_o} f_1(t)f_2(t)\,dt \right], & \text{symbol 1 present} \end{cases}
$$

$$
= \begin{cases} k\left[E - \rho E \right], & \text{symbol 0 present} \\[2mm] k\left[-E + \rho E \right], & \text{symbol 1 present} \end{cases} \tag{11.32}
$$

The decision instant voltage difference between symbols is:

$$
\Delta V = k\left[2E - 2\rho E \right] = k2E(1 - \rho) \tag{11.33(a)}
$$

Table 11.3 *Relative spectral efficiencies of binary bandpass modulation schemes.*

	Baseband BASK	Bandpass BASK	Orthogonal BFSK $(n \geq 3)$*	Sunde's BFSK $(n = 2)$*	BPSK
Data rate (bit/s)	$1/T_o$	$1/T_o$	$1/T_o$	$1/T_o$	$1/T_o$
Nominal bandwidth (Hz)	$1/T_o$	$2/T_o$	$(n+4)/2T_o$	$1/T_o$† or $3/T_o$	$2/T_o$
Nominal spectral efficiency (bit/s/Hz)	1	1/2	$2/(n+4)$	1† or 1/3	1/2
Minimum ‡ (ISI free) bandwidth	$1/2T_o$	$1/T_o$	$(n+2)/2T_o$	$2/T_o$	$1/T_o$
Maximum ‡ spectral efficiency	2	1	$2/(n+2)$	1/2	1

*n = zero crossing operating point on $\rho - T_o$ diagram of Figure 11.8
† Depends on definition of bandwidth
‡ Based on absolute bandwidth

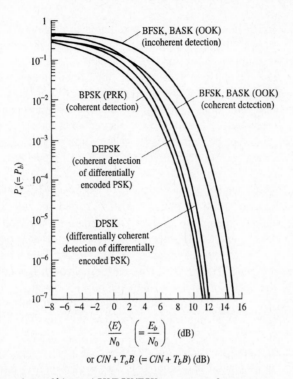

Figure 11.10 *Comparison of binary ASK/PSK/FSK systems performance.*

The two channels result in a total noise power of:

$$\sigma^2 = k^2 E N_0 (1 - \rho) \tag{11.33(b)}$$

which using equation (6.8) gives a probability of error of:

$$P_e = \frac{1}{2} \left[1 - \mathrm{erf} \sqrt{\frac{1 - \rho}{2}} \left(\frac{E}{N_0} \right)^{1/2} \right] \tag{11.34(a)}$$

where the correlation coefficient $\rho = 0$ for OOK, $\rho = 0$ for orthogonal BFSK, $\rho = -2/3\pi$ for optimum BFSK and $\rho = -1$ for PRK. (It can now be seen why ρ was chosen to represent $\cos \Delta\theta$ in equation (11.15).) The equivalent expression in terms of C/N is:

$$P_e = \frac{1}{2} \left[1 - \mathrm{erf} \sqrt{\frac{1 - \rho}{2}} (T_o B)^{1/2} \left(\frac{C}{N} \right)^{1/2} \right] \tag{11.34(b)}$$

11.3.6 Carrier recovery, phase ambiguity and DPSK

Coherent detection of IF modulated signals is required to achieve the lowest error rate, Figure 11.10. This usually requires a reference signal which replicates the phase of the signal carrier. A residual or pilot carrier, if present in the received signal, can be extracted using a filter or phase locked loop or both, then subsequently amplified and employed as a coherent reference. This is possible (though not often implemented) for ASK and FSK signals since both contain discrete lines in their spectra. BPSK signals with phasor states separated by less than 180° also contain a spectral line at the carrier frequency and can therefore be demodulated in the same way. PRK signals, however, contain no such line and carrier recovery must therefore be achieved using alternative methods.

One technique which can be used is to square the received PRK transmission creating a double frequency carrier (with no phase transitions). This occurs because $\sin^2 2\pi f_c t$ and $\sin^2(2\pi f_c t + \pi)$ are both equal to $\sin 2\pi 2 f_c t$. A phase locked loop can then be used as a frequency divider to generate the coherent reference. Such a carrier recovery circuit is shown in Figure 11.11. (Note that the phase locked loop locks to a reference 90° out of phase with the input signal. Thus a 90° phase shifting network is required either within the loop, or between the loop and the demodulator, to obtain the correctly phased reference signal for demodulation.) A second technique is to use a Costas loop [Dunlop and Smith, Lindsey and Simon] in place of a conventional PLL. The Costas loop, Figure 11.12, consists essentially of two PLLs operated in phase quadrature and has the property that it will lock to the suppressed carrier of a PRK signal.

Both the squaring loop and Costas loop solutions to PRK carrier recovery suffer from a 180° phase ambiguity, i.e. the recovered carrier may be either in-phase or in antiphase

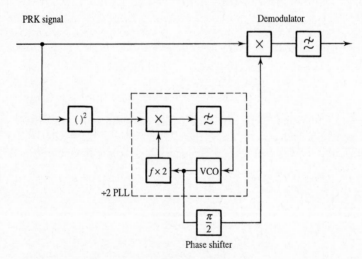

Figure 11.11 *Squaring loop for suppressed carrier recovery.*

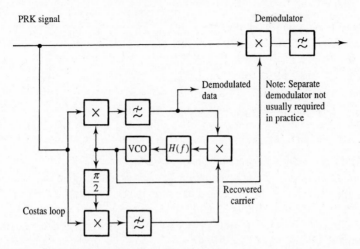

Figure 11.12 *Costas loop for suppressed carrier recovery.*

with the transmitted (suppressed) carrier. This ambiguity can lead to symbol inversion of the demodulated data. Whilst data inversion might be tolerated by some applications, in others it would clearly be disastrous. There are two distinct approaches to resolving phase ambiguities in the recovered carrier.

The first involves periodic transmission of a known data sequence. (This sequence is normally one part of a larger 'preamble' sequence transmitted prior to a block of data, see Chapter 19.) If the received 'training' sequence is inverted this is detected and a second inversion introduced to correct the data. The second approach is to employ differential encoding of the data before PRK modulation (Figure 11.13). This results, for example, in digital ones being represented by a phase transition and digital zeros being represented by no phase transition. The phase ambiguity of the recovered carrier then becomes irrelevant. Demodulation of such encoded PSK signals can be implemented using conventional carrier recovery and detection followed by baseband differential decoding (Figure 11.14). (STR techniques are described in section 6.7.) Systems using this detection scheme are called differentially encoded PSK (DEPSK) and have a probability of symbol error, P'_e, given by:

$$P'_e = P_e(1 - P_e) + P_e(1 - P_e)$$

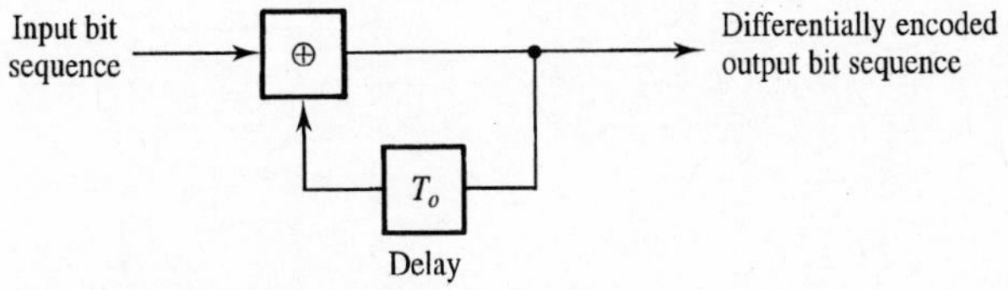

Input bit sequence → ⊕ → Differentially encoded output bit sequence

T_o

Delay

Figure 11.13 *Differential encoding.*

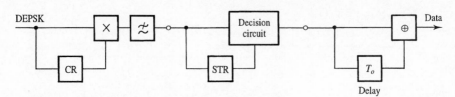

Figure 11.14 *Differentially encoded PSK (DEPSK) detection.*

$$= 2(P_e - P_e^2) \tag{11.35}$$

where P_e is the probability of error for uncoded PSK.

The first term, $P_e(1 - P_e)$, in equation (11.35) is the probability that the current symbol in the decoder is in error and the previous symbol is correct. The second (identical) term is the probability that the current symbol is correct and the previous symbol is in error. (If both current and previous symbols are detected in error the decoded symbol will, of course, be correct.)

An alternative method of demodulating PSK signals with differential coding is to use one symbol as the coherent reference for the next symbol (Figure 11.15). Systems using this detection scheme are called differential PSK (DPSK). They have the advantage of simpler, and therefore cheaper, receivers compared to DEPSK systems. They have the disadvantage, however, of having a noisy reference signal and therefore degraded P_e' performance compared to DEPSK and coherent PSK, Figure 11.10. It might appear that since reference and signal are equally noisy the P_e' degradation in DPSK systems would correspond to a 3 dB penalty in CNR. In practice the degradation is significantly less than this (typically 1 dB) since the noise in the signal and reference channels is correlated. DEPSK and DPSK techniques are easily extended from biphase to multiphase signalling (see section 11.4.2).

11.4 Modulation techniques with increased spectral efficiency

Spectral efficiency, η_s as defined in equation (8.1), depends on symbol (or baud) rate, R_s, signal bandwidth, B, and entropy, H (as defined in equation (9.3)), i.e.:

$$\eta_s = \frac{R_s H}{B} \quad \text{(bit/s/Hz)} \tag{11.36(a)}$$

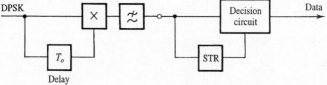

Figure 11.15 *Differential PSK (DPSK) detection.*

Since $R_s = 1/T_o$ and $H = \log_2 M$, for statistically independent, equiprobable symbols (section 9.2.4) then η_s can be expressed as:

$$\eta_s = \frac{\log_2 M}{T_o B} \quad \text{(bit/s/Hz)} \tag{11.36(b)}$$

It is apparent from equation (11.36(b)) that spectral efficiency is maximised by making the symbol alphabet size, M, large and the $T_o B$ product small. This is exactly the strategy employed by spectrally efficient modulation techniques.

Pulse shaping (or filtering) to decrease B for a given baud rate, $R_s = 1/T_o$, has already been discussed in Chapter 8. There, however, the discussion was in the context of baseband signals where the minimum $T_o B$ product (avoiding ISI) was limited by:

$$T_o B \geq 0.5 \tag{11.37(a)}$$

and the minimum bandwidth $B = 1/(2T_o)$ was called the single sided Nyquist bandwidth. In the context of IF modulation (i.e. bandpass signals) the modulation process results in a double sideband (DSB) signal. The minimum ISI free $T_o B$ product is then given by:

$$T_o B \geq 1.0 \tag{11.37(b)}$$

and the minimum bandwidth is now $B = 1/T_o$, sometimes called the double sided Nyquist bandwidth. Dramatic increases in η_s, however, must usually come from increased alphabet size. Operational systems currently exist with M's of 64, 128, 256 and 1024. In principle such multi-symbol signalling can lead to increased spectral efficiency of MASK, MPSK and MFSK systems (the first letter in each case represents M-symbol). Only MASK, MPSK and hybrid combinations of these two are used in practice, however. This is because MFSK signals are normally designed to retain orthogonality between all symbol pairs. In this case increasing M results in an approximately proportional increase in B which actually results in a decrease in spectral efficiency. MASK and MPSK signals, however, are limited to alphabets of 2 (OOK) and 4 (4-PSK) symbols respectively if orthogonality is to be retained. MASK and MPSK therefore must sacrifice orthogonality to achieve values of M greater than four.

Sub-orthogonal signalling requires greater transmitted power than orthogonal signalling for the same P_e. MASK and MPSK systems are therefore spectrally efficient at the expense of increased transmitted power. Conversely, (orthogonal) MFSK systems are power efficient at the expense of increased bandwidth. This is one manifestation of the general trade-off which can be made between power and bandwidth represented at its most fundamental by the Shannon-Hartley law.

11.4.1 Channel capacity

The Shannon-Hartley channel capacity theorem states that the maximum rate of information transmission, $R_{\max}$, over a channel with bandwidth B and signal to noise ratio S/N is given by:

$$R_{\max} = B \log_2 \left(1 + \frac{S}{N}\right) \quad \text{bit/s} \tag{11.38(a)}$$

Thus as B increases S/N can be decreased to compensate [Hartley]. Note that in a 3.2 kHz wide audio channel with a SNR of 1000 (30 dB) the theoretical maximum bit rate, R_{max}, is slightly in excess of 30 kbit/s. The corresponding maximum spectral efficiency, R_{max}/B, for a channel with 30 dB SNR is 10 bit/s/Hz. We can determine the value that the channel capacity approaches as the channel bandwidth B tends to infinity:

$$R_\infty = \lim_{B \to \infty} B \log_2 \left(1 + \frac{S}{N_0 B} \right)$$

$$= \lim_{B \to \infty} \frac{S}{N_0} \log_2 \left(1 + \frac{S}{N_0 B} \right)^{N_0 B/S} \qquad (11.38(b))$$

$$= \frac{S}{N_0} \log_2 e = 1.44 \frac{S}{N_0}$$

This gives the maximum possible channel capacity as a function of signal power and noise power spectral density. In an actual system design, the channel capacity might be compared to this value to decide whether a further increase in bandwidth is worth while. In a practical system it is realistic to attempt to achieve a transmission rate of one half R_∞. Some popular modulation schemes which trade an increase in SNR for a decrease in bandwidth, to achieve improved spectral efficiency, are described below.

11.4.2 *M*-symbol phase shift keying

In the context of PSK signalling M-symbol PSK (i.e. MPSK) implies the extension of the number of allowed phasor states from 2 to 4, 8, 16, ... (i.e. 2^n). The phasor diagram (also called the constellation diagram) for 16-PSK is shown in Figure 11.16, where there are now 16 distinct states. Note that as the signal amplitude is constant these states all lie on a circle in the complex plane. 4-phase modulation (4-PSK) with phase states $\pi/4$, $3\pi/4$, $5\pi/4$ and $7\pi/4$ can be considered to be a superposition of two PRK signals using quadrature ($\cos 2\pi f_c t$ and $\sin 2\pi f_c t$) carriers. This type of modulation (usually called quadrature or quaternary phase shift keying, QPSK) is discussed separately in section 11.4.4.

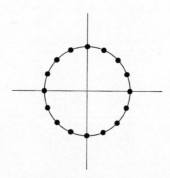

Figure 11.16 *Phasor states (i.e. constellation diagram) for 16-PSK.*

The probability of symbol error for MPSK systems is found by integrating the two-dimensional pdf of the noise centred on the tip of each signal phasor in turn over the corresponding error region and averaging the results. The error region for phasor state 0 is shown as the unhatched region in Figure 11.17. It is possible to derive an exact, but complicated, integral expression for the probability of symbol error in the presence of Gaussian noise. Accurate P_e curves found by numerical evaluation of this expression are shown in Figure 11.18(a). A simple but good approximation [Stein and Jones] which is useful for $M \geq 4$ is:

$$P_e \approx 1 - \mathrm{erf}\left[\sin\left(\frac{\pi}{M}\right)\left(\frac{E}{N_0}\right)^{\frac{1}{2}}\right] \tag{11.39(a)}$$

(Note that when $M = 2$ this expression gives a P_e which is twice the correct result.) The approximation in equation (11.39(a)) becomes better as both M and E/N_0 increase. Equation (11.39(a)) can be rewritten in terms of CNR using $C = E/T_o$ and $N = N_0 B$ from equations (11.6) and (11.7) as usual, i.e.:

$$P_e \approx 1 - \mathrm{erf}\left[(T_o B)^{\frac{1}{2}}\sin\left(\frac{\pi}{M}\right)\left(\frac{C}{N}\right)^{\frac{1}{2}}\right] \tag{11.39(b)}$$

Multisymbol signalling can be thought of as a coding or bit mapping process in which n binary symbols (bits) are mapped into a single M-ary symbol (as discussed in section 6.4.7), except that here each symbol is an IF pulse. A detection error in a single symbol can therefore translate into several errors in the corresponding decoded bit sequence. The probability of bit error, P_b, therefore depends not only on the probability of symbol error, P_e, and the symbol entropy, $H = \log_2 M$, but also on the code or bit mapping used and the types of error which occur. For example, in the 16-PSK scheme shown in Figure 11.17 the most probable type of error involves a given phasor state being detected as an adjacent state. If a Gray code is used to map binary symbols to phasor states this type of error results in only a single decoded bit error. In this case, providing the probability of errors other than this type is negligible, then the bit error probability is:

$$P_b = \frac{P_e}{\log_2 M} \tag{11.40(a)}$$

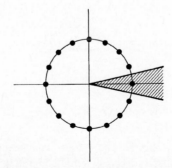

Figure 11.17 *Error region (unhatched) for $\phi = 0°$ state of a 16-PSK signal.*

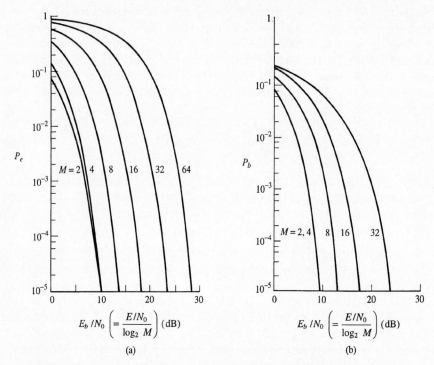

Figure 11.18 *MPSK symbol and bit error probabilities: (a) probability of symbol error, P_e against E_b/N_0; and (b) probability of bit error, P_b against E_b/N_0.*

In order to compare the performance of different modulation schemes on an equitable basis it is useful to express performance in terms of P_b as a function of average energy per bit, E_b. Since the energy, E, of all symbols in an MPSK system are identical the average energy per bit is:

$$E_b = \frac{E}{\log_2 M} \qquad (11.40(b))$$

Substituting equations (11.40) into (11.39):

$$P_b = \frac{1}{\log_2 M}\left\{ 1 - \mathrm{erf}\left[\sin\left(\frac{\pi}{M} \right)\sqrt{\log_2 M}\left(\frac{E_b}{N_0} \right)^{\!\frac{1}{2}} \right] \right\} \qquad (11.41(a))$$

In terms of CNR this becomes:

$$P_b = \frac{1}{\log_2 M}\left\{ 1 - \mathrm{erf}\left[(T_o\, B)^{\frac{1}{2}} \sin\left(\frac{\pi}{M} \right)\left(\frac{C}{N} \right)^{\!\frac{1}{2}} \right] \right\} \qquad (11.41(b))$$

More accurate closed form expressions for Gray coded and natural binary bit mapped MPSK P_b performance have been derived [Irshid and Salous]. Figure 11.18(a) shows the probability of symbol error, P_e, against E_b/N_0, and Figure 11.18(b) shows P_b against E_b/N_0. (Figure 13.41 clarifies the distinction between P_e and P_b).

Differential MPSK (DMPSK) can be used to simplify MPSK receiver design and circumvent the phase ambiguity normally present in the recovered carrier. Binary digits are mapped to phase difference between adjacent symbols and each symbol is detected at the receiver using the previous symbol as a coherent reference. The P_e performance of DMPSK is degraded over that for MPSK since the reference and received symbol are now equally noisy. As for DPSK the degradation is not equivalent to a CNR reduction of 3 dB because the noise present in the signal and reference channels is correlated and therefore, to some extent, cancels. Phase noise correlation makes increasingly little difference, however, to the performance degradation as M increases and the phasor states become crowded together. In the limit of large M the degradation approaches the expected 3 dB limit.

Since each symbol of an MPSK signal has an identical amplitude spectrum to all the other symbols the spectral occupancy of MPSK depends only on baud rate and pulse shaping and is independent of M. For unfiltered MPSK (i.e. MPSK with rectangular pulses) the nominal (i.e. main lobe null-to-null) bandwidth is $2/T_o$ Hz. In this case the spectral efficiency (given by equation (11.36(b))) is:

$$\eta_s = 0.5 \log_2 M \quad \text{(bit/s/Hz)} \tag{11.42(a)}$$

The maximum possible, ISI free, spectral efficiency occurs when pulse shaping is such that signalling takes place in the double sided Nyquist bandwidth $B = 1/T_o$ Hz, i.e.:

$$\eta_s = \log_2 M \quad \text{(bit/s/Hz)} \tag{11.42(b)}$$

Thus we usually say BPSK has an efficiency of 1 bit/s/Hz and 16-PSK 4 bit/s/Hz. Table 11.4 compares the performance of several PSK systems.

Table 11.4 *Comparison of several PSK modulation techniques.*

	Required E_b/N_0 for $P_b = 10^{-6}$	Minimum channel bandwidth for ISI free signalling (R_b = bit rate)	Max spectral efficiency (bit/s/Hz)	Required CNR in minimum channel bandwidth
PRK	10.6 dB	R_b	1	10.6 dB
QPSK	10.6 dB	$0.5R_b$	2	13.6 dB
8-PSK	14.0 dB	$0.33R_b$	3	18.8 dB
16-PSK	18.3 dB	$0.25R_b$	4	24.3 dB

EXAMPLE 11.4

An MSPK, ISI free, system is to operate with 2^N PSK symbols over a 120 kHz channel. The minimum required bit rate is 900 kbit/s. What minimum CNR is required to maintain reception with a P_b no worse than 10^{-6}?

Maximum (ISI free) baud rate:

$$R_s = 1/T_o = B$$

Therefore $R_s \leq 120$ kbaud (or k symbol/s). Minimum required entropy is therefore given by:

$$H \geq \frac{R_b}{R_s} = \frac{900 \times 10^3}{120 \times 10^3} = 7.5 \text{ bit/symbol}$$

Minimum number of symbols required is given by:

$$H \leq \log_2 M$$

$$M \geq 2^H = 2^{7.5}$$

Since M must be an integer power of 2:

$$M = 2^8 = 256 \quad \text{and } H = \log_2 M = 8$$

For Gray coding of bits to PSK symbols:

$$P_e = P_b \log_2 M = 10^{-6} \log_2 256 = 8 \times 10^{-6}$$

$$P_e = 1 - \mathrm{erf}\left[(T_o B)^{\frac{1}{2}} \sin\left(\frac{\pi}{M} \right) \left(\frac{C}{N} \right)^{\frac{1}{2}} \right]$$

Now:

$$R_s = \frac{R_b}{\log_2 M} = \frac{900 \times 10^3}{8} = 112.5 \times 10^3 \text{ baud}$$

$$T_o = \frac{1}{R_s} = 8.889 \times 10^{-6} \text{ s}$$

$$T_o B = 8.889 \times 10^{-6} \times 120 \times 10^3 = 1.067$$

Thus:

$$\frac{C}{N} = \left[\frac{\mathrm{erf}^{-1}(1 - P_e)}{(T_o B)^{\frac{1}{2}} \sin\left(\dfrac{\pi}{M} \right)} \right]^2 = \left[\frac{\mathrm{erf}^{-1}(1 - 8 \times 10^{-6})}{(1.067)^{\frac{1}{2}} \sin\left(\dfrac{\pi}{256} \right)} \right]^2$$

$$= \left[\frac{\mathrm{erf}^{-1}(0.999\,992)}{1.033 \sin\left(\dfrac{\pi}{256} \right)} \right]^2 = \left(\frac{3.157}{0.01268} \right)^2 = 61988 = 47.9 \text{ dB}$$

11.4.3 Amplitude phase keyed and quadrature amplitude modulation

For an unsaturated transmitter operating over a linear channel it is possible to introduce amplitude as well as phase modulation to give an improved distribution of signal states in the signal constellation. The first such proposal (C. 1960) introduced a constellation with two amplitude rings and eight phase states on each ring, Figure 11.19(a). For obvious reasons modulation schemes of this type are called amplitude/phase keying (APK). Subsequently it was observed that with half the number of points on the inner ring, Figure 11.19(b), a 3 dB performance improvement could be gained, as the constellation points are more evenly spaced over 12 distinct phases. The square constellation (Figure 11.19(c)), introduced C. 1962, is easier to implement and has a slightly better P_e performance yet. Since a square constellation APK signal can be interpreted as a pair of multilevel ASK (MASK) signals modulated onto quadrature carriers it is normally called quadrature amplitude modulation (QAM). (The terms APK and QAM are sometimes used interchangeably – here, however, we always use the term QAM to represent the square constellation subset of APK.)

Figure 11.20 shows the time domain waveforms for a 16-state QAM constellation which are obtained by encoding binary data in 4-bit sequences. 2-bit sequences are each encoded in the I and Q channels into 4-level signals, as shown in Figure 11.20(a)/(c) and (b)/(d). These are combined to yield the full 16-state QAM, complex, signal, Figure (11.20(e)). Ideally all of the 16 states are equiprobable and statistically independent. Note that at the symbol sampling times there are 4 possible amplitudes in the inphase and quadrature channels reflecting the constellation's 4×4 structure, Figure 11.19(c). In fact the complex signal has three possible amplitude levels with the intermediate value being the most probable. A simple approximation for the probability of symbol error for MQAM (M even) signalling in Gaussian noise is [Carlson]:

$$P_e = 2 \left\{ \frac{M^{\frac{1}{2}} - 1}{M^{\frac{1}{2}}} \right\} \left[1 - \text{erf} \sqrt{\frac{3}{2(M-1)} \left(\frac{\langle E \rangle}{N_0} \right)^{\frac{1}{2}}} \right] \qquad (11.43)$$

Where $\langle E \rangle$ is the average energy per QAM symbol. For equiprobable rectangular pulse symbols $\langle E \rangle$ is given by:

$$\langle E \rangle = \frac{1}{3} \left(\frac{\Delta V}{2} \right)^2 (M - 1) T_o \qquad (11.44)$$

where ΔV is the voltage separation between adjacent inphase or quadrature MASK levels and T_o is the symbol duration. Using $C = \langle E \rangle / T_o$ and $N = N_0 B$, equation (11.43) can be

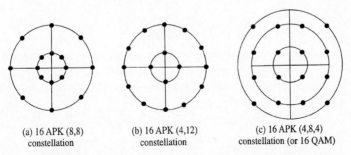

(a) 16 APK (8,8)   (b) 16 APK (4,12)   (c) 16 APK (4,8,4)
constellation   constellation   constellation (or 16 QAM)

Figure 11.19 *16-state quadrature amplitude modulated (QAM) signal constellations.*

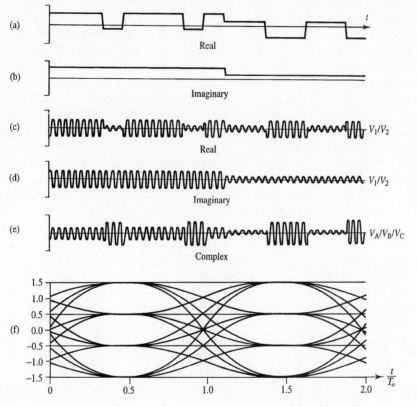

Figure 11.20 *16-state QAM signal: (a) 4-level baseband signals in the inphase and (b) quadrature branches; (c), (d) corresponding 4-level modulated complex signals; (e) resulting combined complex (3-level) QAM signal; (f) demodulated signal eye diagram over 2 symbols (in the I or Q channels).*

rewritten as:

$$P_e = 2\left\{\frac{M^{\frac{1}{2}}-1}{M^{\frac{1}{2}}}\right\}\left[1 - \mathrm{erf}\sqrt{\frac{3\,T_oB}{2(M-1)}}\left(\frac{C}{N}\right)^{\frac{1}{2}}\right] \qquad (11.45)$$

For Gray code mapping of bits along the inphase and quadrature axes of the QAM constellation the probability of bit error P_b is given approximately by equation (11.40(a)). Denoting the average energy per bit by $E_b = \langle E\rangle/\log_2 M$, equation (11.43) can be written as:

$$P_b = \frac{2}{\log_2 M}\left\{\frac{M^{\frac{1}{2}}-1}{M^{\frac{1}{2}}}\right\}\left[1 - \mathrm{erf}\sqrt{\frac{3\log_2 M}{2(M-1)}}\left(\frac{E_b}{N_0}\right)^{\frac{1}{2}}\right] \qquad (11.46(a))$$

and equation (11.45) as:

$$P_b = \frac{2}{\log_2 M} \left\{ \frac{M^{1/2}-1}{M^{1/2}} \right\} \left[1 - \mathrm{erf} \sqrt{\frac{3\,T_o B}{2(M-1)}} \left(\frac{C}{N} \right)^{1/2} \right] \qquad (11.46(\mathrm{b}))$$

Like MPSK all the symbols in a QAM (or APK) signal occupy the same spectral space. The spectral efficiency is therefore identical to MPSK and is given (for statistically independent equiprobable symbols) by equation (11.36b). Unfiltered and Nyquist filtered APK signals therefore have (nominal and maximum) spectral efficiencies given by equations (11.42(a)) and (11.42(b)) respectively. Figure 11.21(a) compares the bit error probability of MPSK and MQAM systems. Note that the superior constellation packing in QAM over MPSK gives a lower required E_b/N_0 for the same P_b value. Figure 11.21(b) shows the spectral efficiencies of Nyquist filtered ($T_o B = 1$) MQAM and MPSK systems plotted against the CNR required for a bit error probability of 10^{-6}. Table 11.5 compares the C/N and E_b/N_0 required for a selection of MPSK and QAM schemes to yield a probability of bit error of 10^{-6}.

Table 11.5 *Comparison of various digital modulation schemes ($P_b = 10^{-6}$, $T_o B = 1.0$).*

Modulation	C/N ratio (dB)	E_b/N_0 (dB)
PRK	10.6	10.6
QPSK	13.6	10.6
4-QAM	13.6	10.6
8-PSK	18.8	14.0
16-PSK	24.3	18.3
16-QAM	20.5	14.5
32-QAM	24.4	17.4
64-QAM	26.6	18.8

EXAMPLE 11.5
Find the maximum spectral efficiency of ISI free 16-QAM. What is the noise induced probability of symbol error in this scheme for a received CNR of 24.0 dB if the maximum spectral efficiency requirement is retained? What is the Gray coded probability of bit error?

$$\eta_s = \frac{R_s\,H}{B} = \frac{H}{T_o B} \quad (\mathrm{bit/s/Hz})$$

For maximum (ISI free) spectral efficiency $T_o B = 1$ and $H = \log_2 M$, i.e.:

$$\eta_s = \log_2 M = \log_2 16 = 4 \quad \mathrm{bit/s/Hz}$$

Probability of symbol error is given by equation (11.45) with $T_o B = 1$ and $C/N = 10^{2.4} = 251.2$:

$$P_e = 2 \left\{ \frac{4-1}{4} \right\} \left[1 - \mathrm{erf} \sqrt{\frac{3}{2(16-1)}}\,(251.2)^{1/2} \right]$$

$$= 1.5[1 - \mathrm{erf}(5.012)]$$

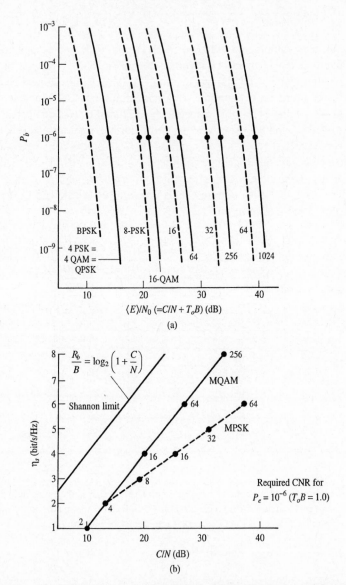

Figure 11.21 P_e *and spectral efficiency for multi-phase PSK and M-QAM modulation: (a) bit error probability against CNR with PSK shown as dashed and QAM as solid curves; (b) comparison of the spectral efficiency of these modulation schemes.*

For $x \geq 4.0$ the error function can be approximated by:

$$\mathrm{erf}(x) = 1 - \frac{e^{-x^2}}{(\sqrt{\pi}\, x)}$$

Therefore:

$$P_e = \frac{1.5\, e^{-5.012^2}}{(\sqrt{\pi}\, 5.012)} = 2.08 \times 10^{-12}$$

For Gray coding:

$$P_b = \frac{P_e}{\log_2 M} = \frac{2.08 \times 10^{-12}}{\log_2 16} = 5.2 \times 10^{-13}$$

11.4.4 Quadrature phase shift keying (QPSK) and offset QPSK

Quadrature phase shift keying (QPSK) [Aghvami] can be interpreted either as 4-PSK with carrier amplitude A (i.e. quaternary PSK) or as a superposition of two (polar) BASK signals with identical 'amplitudes' $\pm A/\sqrt{2}$ and quadrature carriers, $\cos 2\pi f_c t$ and $\sin 2\pi f_c t$, i.e. 4-QAM. The constellation diagram of a QPSK signal is shown in Figure 11.22. The accompanying arrows show that all transitions between the four states are possible. Figure 11.23 shows a schematic diagram of a QPSK transmitter and Figure 11.24 shows the receiver.

The transmitter and receiver are effectively two PRK transmitters and receivers arranged in phase quadrature, the inphase (I) and quadrature (Q) channels each operating at half the bit rate of the QPSK system as a whole. If pulse shaping and filtering are absent the signal is said to be unfiltered, or rectangular pulse, QPSK. Figure 11.25 shows an example sequence of unfiltered QPSK data and Figure 11.26(a) and (b) shows the corresponding power spectral density with $T_o = 2T_b$.

The spectral efficiency of QPSK is twice that for BPSK. This is because the symbols in each quadrature channel occupy the same spectral space and have half the spectral width of a BPSK signal with the same data rate as the QPSK signal. This is illustrated in the form of orthogonal voltage spectra in Figure 11.27.

The P_e performance of QPSK systems will clearly be worse than that of PRK systems, Figure 11.18, since the decision regions on the constellation diagram, Figure

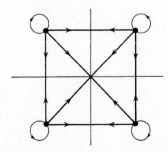

Figure 11.22 *Quadrature phase shift keyed (QPSK) signal constellation showing allowed state transitions.*

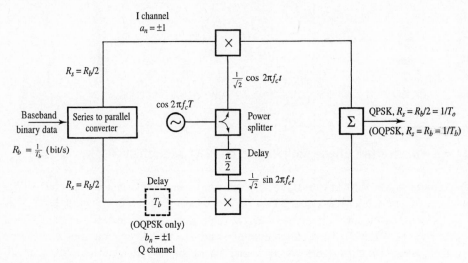

Figure 11.23 *Schematic for QPSK (OQPSK) modulator.*

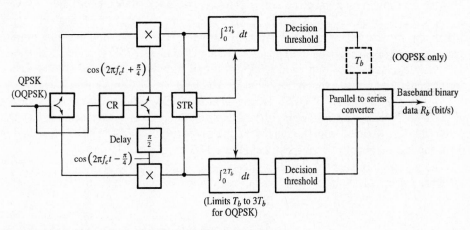

Figure 11.24 *Schematic for QPSK (OQPSK) demodulator.*

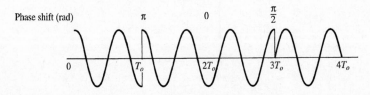

Figure 11.25 *Unfiltered (constant envelope) QPSK signal ($T_o = 2T_b$).*

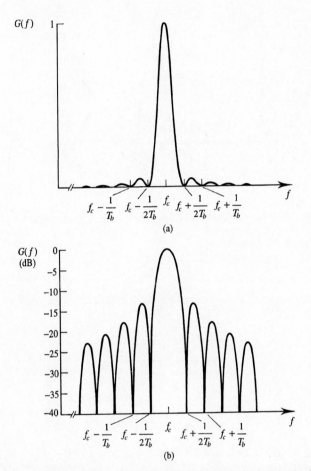

Figure 11.26 *Power spectral density of QPSK/OQPSK signals.* $G(f) \propto [(\sin(4\pi fT_b))/(4\pi fT_b)]^2$
on (a) linear; and (b) dB scales.

11.22, are reduced from half spaces to quadrants. The P_b performances of these modulation schemes are, however, identical. This is easily appreciated by recognising that each channel (I and Q) of the QPSK system is independent of (orthogonal to) the other. In principle, therefore, the I channel (binary) signal could be transmitted first followed by the Q channel signal. The total message would take twice as long to transmit in this form but, because each QPSK I or Q channel symbol is twice the duration and half the power of the equivalent PRK symbols, the total message energy (and therefore the energy per bit, E_b) is the same in both the QPSK and PRK cases, Figure 11.28. The P_b performance of ideal QPSK signalling is therefore given by equation (11.11) with E interpreted as the energy per bit, E_b, i.e.:

$$P_b = \frac{1}{2}\left[1 - \mathrm{erf}\left(\frac{E_b}{N_0} \right)^{\frac{1}{2}} \right] \tag{11.47(a)}$$

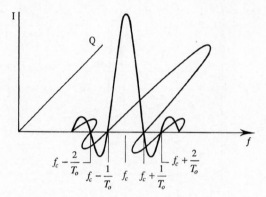

Figure 11.27 *QPSK orthogonal inphase (I) and quadrature (Q) voltage spectra ($T_o = 2T_b$).*

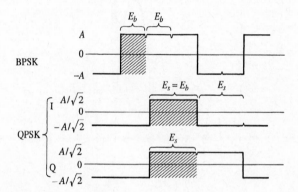

Figure 11.28 *Envelopes of (unfiltered) BPSK and equivalent QPSK signals showing distribution of bit energy in I and Q channels.*

In terms of CNR, therefore:

$$P_b = \frac{1}{2}\left[1 - \text{erf}\,(T_bB)^{1/2}\left(\frac{C}{N}\right)^{1/2}\right] \tag{11.47(b)}$$

where the bit period T_b is half the QPSK symbol period, T_o. This is the origin of the statement that the BER performances of QPSK and PRK systems are identical. (Note that the minimum T_oB product for ISI free QPSK signalling is 1.0, corresponding to $T_bB = 0.5$.) The probability of symbol error, P_e, is given by the probability that the I channel bit is detected in error, or the Q channel bit is detected in error, or both channel bits are detected in error, i.e.:

$$P_e = P_b(1 - P_b) + (1 - P_b)P_b + P_bP_b = 2P_b - P_b^2 \tag{11.48}$$

Offset QPSK (OQPSK), also sometimes called staggered QPSK, is identical to QPSK

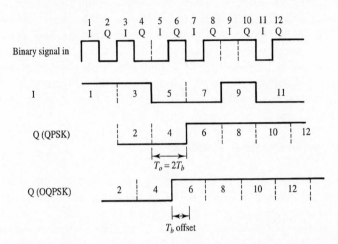

Figure 11.29 *Input bit stream and I and Q channel bit-streams for QPSK and OQPSK systems.*

except that, immediately prior to multiplication by the carrier, the Q channel symbol stream is offset with respect to the I channel symbol stream by half a QPSK symbol period (i.e. one bit period), Figure 11.23. The OQPSK relationships between the input binary data stream and the I and Q channel bit streams are shown in Figure 11.29, and the OQPSK constellation and state transition diagram is shown in Figure 11.30. The lines on Figure 11.30 represent the allowed types of symbol transition and show that transitions across the origin (i.e. phase changes of 180°) are, unlike QPSK, prohibited. The offset between I and Q channels means that OQPSK symbol transitions occur, potentially, every T_b seconds (i.e. $T_o = T_b$).

The spectrum, and therefore spectral efficiency, of OQPSK is thus identical to that of QPSK. (Transitions occur more often but, on average, are less severe.) OQPSK has a potential advantage over QPSK, however, when transmitted over a non-linear channel. This is because envelope variations in the signal at the input of a non-linear device produce distortion of the signal (and therefore the signal spectrum) at the output of the device. This effect leads to spectral spreading which may give rise, ultimately, to adjacent channel interference. To minimise spectral spreading (and some other undesirable effects, see section 14.3.3) in non-linear channels, constant envelope

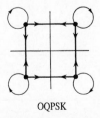

OQPSK

Figure 11.30 *Offset QPSK (OQPSK) constellation diagram showing allowed state transitions.*

modulation schemes are desirable. Since some degree of filtering (i.e. pulse shaping) is always present in a modulated RF signal (even if this is unintentional, and due only to the finite bandwidth of the devices used in the modulator) then the 180° phase transitions present in a QPSK signal will result in severe (∞ dB) envelope fluctuation. In contrast the envelope fluctuation of filtered OQPSK is limited to 3 dB since only one quadrature component of the signal can reverse at any transition instant, Figure 11.31. The spectral spreading suffered by an OQPSK signal will therefore be less than that suffered by the equivalent QPSK signal for the same non-linear channel.

Quadrature partial response systems

Quadrature partial response (QPR) modulation is a quadrature modulated form of partial response signalling (sections 8.2.7 and 8.2.8). A block diagram showing the structure of a duobinary QPR transmitter is shown in Figure 11.32(a) and the resulting duobinary constellation diagram is shown in Figure 11.32(b). The (absolute) bandwidth of the (DSB) QPR signal is $1/T_o$ (Hz) or $1/2T_b$ (Hz) and the spectral efficiency is 2 bit/s/Hz.

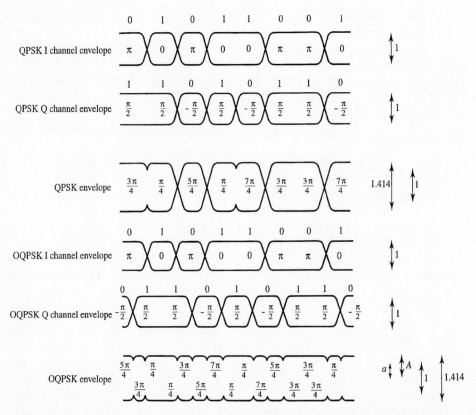

Figure 11.31 *Origin and comparison of QPSK and OQPSK envelope fluctuations (a/A = 1/√2).*

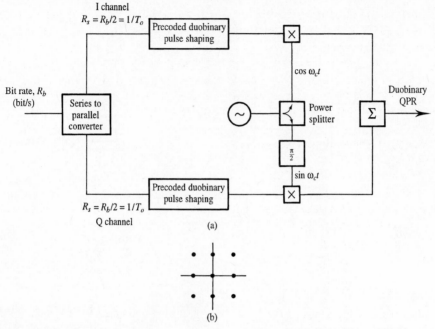

Figure 11.32 *(a) Duobinary QPR transmitter (b) constellation diagram.*

This is the same spectral efficiency as baseband duobinary signalling since a factor of 2 is gained by using quadrature carriers but a factor of 2 is lost by using DSB modulation. The duobinary QPR probability of bit error in the presence of white Gaussian noise [Taub and Schilling] corresponds to a CNR of approximately 3 dB more than that required for the same P_e performance in QPSK and 2 dB less than that required in 8-PSK.

11.4.5 Minimum shift keying

Minimum shift keying (MSK) is a modified form of OQPSK in that I and Q channel sinusoidal pulse shaping is employed prior to multiplication by the carrier, Figure 11.33 [Pasapathy]. The transmitted MSK signal can be represented by:

$$f(t) \;=\; a_n \sin\!\left(\frac{2\pi t}{4T_b}\right)\cos 2\pi f_c t + b_n \cos\!\left(\frac{2\pi t}{4T_b}\right)\sin 2\pi f_c t \qquad (11.49)$$

where a_n and b_n are the nth I and Q channel symbols. Figure 11.34(a) to (c) shows an example sequence of binary data, the corresponding, sinusoidally shaped, I and Q symbols and the resulting MSK signal. MSK signalling is an example from a class of modulation techniques called continuous phase modulation (CPM). It has a symbol constellation which must now be interpreted as a time varying phasor diagram, Figure 11.34(d). The phasor rotates at a constant angular velocity from one constellation point to an adjacent point over the duration of one MSK symbol. (Like OQPSK the MSK

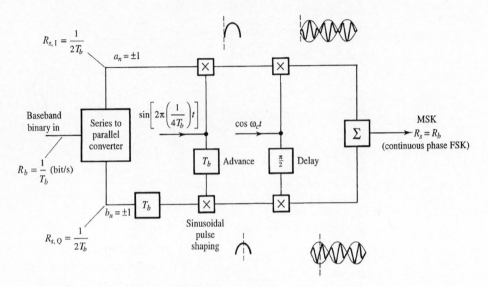

Figure 11.33 *Minimum shift keyed (MSK) transmitter.*

symbol period is the same as the bit period of the unmodulated binary data.) When $a_n = b_n$ the phasor rotates clockwise and when $a_n \neq b_n$ the phasor rotates anticlockwise. A consequence of MSK modulation is that one to one correspondence between constellation points and the original binary data is lost.

An alternative interpretation of MSK signalling is possible in that it can be viewed as a special case of BFSK modulation (and is therefore sometimes called continuous phase FSK). This is obvious, in that when the phasor in Figure 11.34(d) is rotating anticlockwise the MSK symbol has a constant frequency of $f_c + 1/(4T_b)$ Hz and when rotating clockwise it has a frequency of $f_c - 1/(4T_b)$. This corresponds to BFSK operated at the first zero crossing point of the $\rho - T_o$ diagram (see section 11.3.4). When viewed this way MSK can be seen to be identical to BFSK with inherent differential coding (since frequency f_1 is transmitted when $a_n = b_n$ and frequency f_2 is transmitted when $a_n \neq b_n$). A one to one relationship between bits and frequencies can be re-established, however, by differentially precoding the serial input bit stream prior to MSK modulation. The appropriate precoder is identical to that used for differential BPSK, Figure 11.13. When precoding is used the modulation is called fast frequency shift keying (FFSK) although this term is sometimes also used indiscriminately for MSK without precoding.

Since MSK operates at the first zero on the $\rho - T_o$ diagram of Figure 11.8 one of the BFSK signalling frequencies has an integer number of cycles in the symbol period and the other has either one half cycle less or one half cycle more.

The normalised power spectral density of MSK/FFSK is shown in Figure 11.35. The spectra of MSK/FFSK and QPSK/OQPSK are compared, on a dB scale, in Figure 11.36. As can be seen the MSK spectrum has a broader main lobe but more rapidly decaying sidelobes, which is particularly attractive for FDM systems in order to achieve reduced

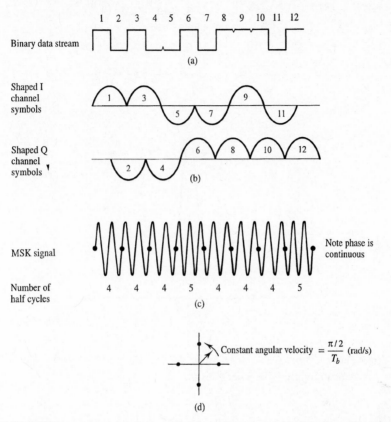

Figure 11.34 *MSK transmitter waveforms and phasor diagrams: (a) binary data; (b) sinusoidally shaped channel symbols; (c) transmitted MSK signal; (d) MSK phasor diagram.*

adjacent channel interference. The probability of bit error for ideal MSK detection, Figure 11.37, is identical to that for QPSK (equations (11.47)) systems since orthogonality between I and Q channels is preserved. (MSK with differentially precoded data, i.e. FFSK, suffers the same degradation in P_b performance as differentially encoded PSK.)

EXAMPLE 11.6
Find the probability of bit error for a 1.0 Mbit/s MSK transmission with a received carrier power of −130 dBW and a NPSD, measured at the same point, of −200 dBW/Hz.

$$C = 10^{\frac{-130}{10}} = 10^{-13} \text{ W}$$

$$E_b = C\,T_b = 10^{-13} \times 10^{-6} = 10^{-19} \text{ J}$$

$$N_0 = 10^{\frac{-200}{10}} = 10^{-20} \text{ W/Hz}$$

$$P_b = \frac{1}{2}\left[1 - \text{erf}\left(\frac{E_b}{N_0}\right)^{\frac{1}{2}}\right] = \frac{1}{2}\left[1 - \text{erf}\left(\frac{10^{-19}}{10^{-20}}\right)^{\frac{1}{2}}\right]$$

$$= \frac{1}{2}[1 - \text{erf}(3.162)] = \frac{1}{2}[1 - 0.999\,999\,24] = 3.88 \times 10^{-6}$$

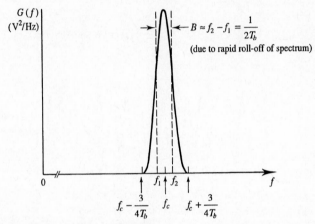

Figure 11.35 *Power spectral density of MSK/FFSK signal.* $G(f) \propto [(\cos(2\pi f T_b))/(1 - 16f^2T_b^2)]^2$. *(Note: no spectral lines are present.)*

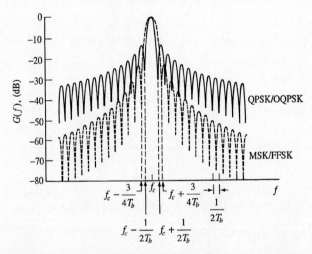

Figure 11.36 *Comparison of QPSK/OQPSK and MSK/FFSK spectra (spectral envelopes roll-off with $(f - f_c)^{-2}$, i.e. −6 dB/octave and $(f - f_c)^{-4}$, i.e. −12 dB/octave).*

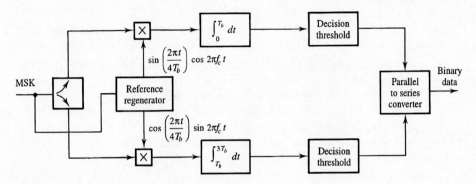

Figure 11.37 *Schematic of MSK demodulator.*

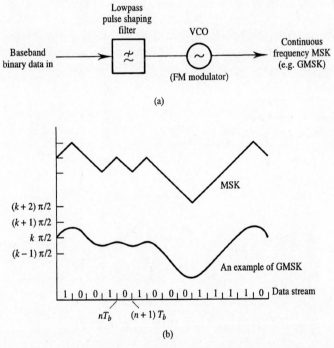

(a)

(b)

Figure 11.38 *(a) Generation of GMSK; (b) MSK and GMSK phase trajectories for typical bit sequences with the MSK trajectory displaced vertically for clarity (source: Hirade and Murota, 1979, reproduced with the permission of the IEEE).*

11.4.6 Gaussian MSK

Although the phase of an MSK signal is continuous, its first derivative (i.e. frequency) is discontinuous. A smoother modulated signal (with correspondingly narrower spectrum) can be generated by reducing this frequency discontinuity. The conceptually simplest way of achieving this is to generate the MSK signal directly as a BFSK signal (with $(f_2 - f_1)T_o = \frac{1}{2}$) using a voltage controlled oscillator, and to employ premodulation pulse shaping of the baseband binary data, Figure 11.38(a). If the pulse shape adopted is Gaussian then the resulting modulation is called Gaussian MSK (GMSK). Figure 11.38(b) compares the phase trajectory of MSK and GMSK signals and Figure 11.39 shows the corresponding power spectra. The parameter B_bT_o in Figure 11.39 is the normalised bandwidth of the premodulation lowpass filter in Figure 11.38(a). (B_b is the baseband bandwidth and T_o is the symbol, or bit, period.) $B_bT_o = \infty$ corresponds to no filtering and therefore MSK signalling. Figure 11.40 shows the impulse response of the premodulation filter for various values of B_bT_o. The baseband Gaussian pulses essentially occupy $1/(B_bT_o)$ bit periods which results in significant sampling instant ISI for $B_bT_o < 0.5$. The improvement in spectral efficiency for $B_bT_o < 0.5$ outweighs the resulting degradation in BER performance, however, in some applications. Figure 11.41 shows measured P_e performance curves for GMSK modulation. GMSK with $B_bT_o = 0.3$ is the modulation scheme adopted by the CEPT's Groupe Spéciale Mobile (Global

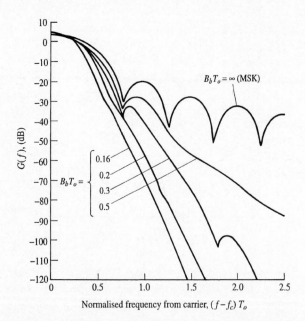

Figure 11.39 *Comparison of GMSK spectra with various values of B_bT_o (B_b is the bandwidth of the premodulation pulse shaping filter, $T_o(= T_b)$ is the symbol period (source: Parsons and Gardiner, 1989, reproduced with the permission of Chapman & Hall).*

System for Mobile), GSM, for the implementation of digital cellular radio systems in Europe, see Chapter 15.

Although Figure 11.38(a) implies a simple implementation for GMSK, in practice the implementation is significantly more complex.

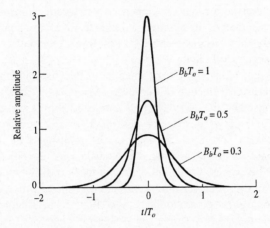

Figure 11.40 *Gaussian filter impulse response for different $B_b T_o$ products.*

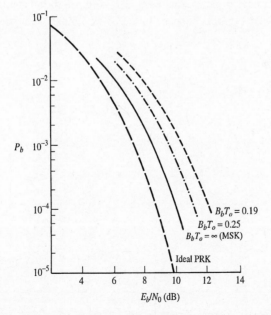

Figure 11.41 *Measured P_b versus E_b/N_0 for MSK with various values of $B_b T_o$ after bandlimiting by an ideal bandpass channel with bandwidth $B = 0.75/T_o$. ($T_o = T_b$) (Source: Hirade and Murota, 1979, reproduced with the permission of the IEEE.)*

11.4.7 Trellis coded modulation

Trellis coded modulation (TCM) [Ungerboeck] is a combined coding and modulation technique for digital transmission over band-limited channels. In TCM there is a restriction on which states may occur so that for a given received symbol sequence the number of possible transmitted symbol sequences is reduced. This increases the decision space across the constellation. Its main attraction is that it allows significant coding gains over the previously described (uncoded) multi-level modulation schemes, without compromising the spectral efficiency.

TCM employs redundant non-binary modulation in combination with a finite-state encoder which governs the selection of transmitted symbols to generate the coded symbol sequences. In the receiver, the noisy symbols are decoded by a soft-decision maximum-likelihood sequence (Viterbi) decoder as described for FEC decoding in Chapter 10. Simple four-state TCM schemes can improve the robustness of digital transmission against additive noise by 3 dB, compared to conventional uncoded modulation. With more complex TCM schemes, the coding gain can reach 6 dB or more. These gains are obtained without bandwidth expansion, or reduction of the effective information rate, as required by traditional error-correction schemes. Figure 11.42 depicts symbol sets and Figure 11.43 the state-transition (trellis) diagrams for uncoded 4-PSK modulation and coded 8-PSK modulation with four trellis states. The trivial one-state diagram in Figure 11.43(a) is shown only to illustrate uncoded 4-PSK from the viewpoint of TCM. Each one of the four connected paths labelled 0, 2, 4, 6 through the trellis in Figure 11.43(a) represents an allowed symbol sequence based on two binary bits (since $2^2 = 4$). In both

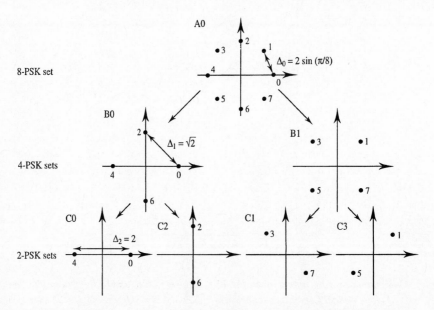

Figure 11.42 *Partitioning of 8-PSK into 4-phase (B0/B1) and 2-phase (C0 $\cdots$ C3) subsets.*

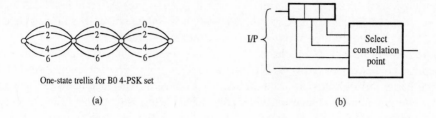

One-state trellis for B0 4-PSK set

(a)

I/P

Select constellation point

(b)

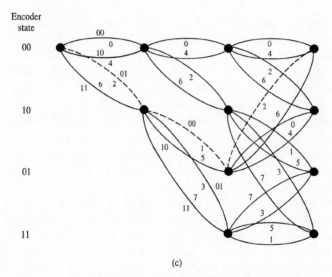

(c)

Figure 11.43 *Trellis coded modulation (TCM): (a) trellis diagram for 4-PSK uncoded transmission where there is no restriction in the choice of constellation point; (b) four-state TCM encoder; (c) resulting four-state TCM 8-PSK trellis diagram.*

systems, starting from any state, four transitions can occur, as required to encode two information bits per modulation interval (and obtain a spectral efficiency of 2 bit/s/Hz).

The four 'parallel' transition paths in the one-state trellis diagram of Figure 11.43(a) for uncoded 4-PSK do not restrict the sequence of 4-PSK symbols that can be transmitted as there is no sequence coding. For unit transmitted amplitude the smallest distance between the 4-PSK signals is $\sqrt{2}$, denoted by Δ_1 in Figure 11.42. 4-PSK at 2 bit/s/Hz has identical spectral efficiency to rate 2/3 FECC data transmitted over an uncoded 8-PSK system.

In order to understand the ideas behind TCM the 8-PSK signal set can be partitioned as shown in Figure 11.42 into B and C subsets with increasing distance or separation between the constellation states [Burr]. The encoder, like the convolutional coder of Chapter 10, consists of a shift register which contains the current data and a number of previous data bits, so that the state of the encoder depends on the previous data samples,

Figure 11.43(b). The nodes in Figure 11.43(c) represent the state of the encoder at a given time, while the branches leaving each node represent the different input data which arrives in that symbol period. Each branch is labelled with the input data bit pairs and the number of the constellation point which is transmitted for that encoder state.

Any pair of paths between the same two nodes represents a pair of codewords which could be confused. In TCM we choose a symbol from the different subsets of the constellation for each branch in the trellis to maximise the distance between such pairs and, therefore, minimise the error probability. This means using points from the same C subset, Figure 11.42, for these parallel branches. Other branches which diverge from, or converge to, the same node require a smaller minimum distance and are chosen from the same B subset.

Consider the trellis diagram of Figure 11.43(c). The closest paths (in terms of Euclidean distance) to the all-zeros symbol path, which is used as a reference, are shown dashed. There are four branches in the trellis from each node, corresponding to two bits per transmitted constellation point (or channel symbol). Thus the scheme can be compared directly with uncoded QPSK, which also transmits two bits per symbol. But in uncoded QPSK the minimum Euclidean distance between transmitted signals is that between two adjacent constellation points, i.e. $\Delta_1 = \sqrt{2}$, Figure 11.42. The TCM scheme achieves $\sqrt{2}$ times this distance, i.e. $\Delta_2 = 2$. As squared distances correspond to signal powers, the squared Euclidean distance between two points determines the noise tolerance between them. The TCM code has twice the square distance of uncoded QPSK, thus it will tolerate twice, or 3 dB, more noise power.

In general TCM offers coding gains of 3 to 6 dB at spectral efficiencies of 2 bit/s/Hz or greater. An 8-state coder (with a 16-state constellation) gives 4 dB of gain while 128 states are required for 6 dB gain. TCM is thus a variation on FECC, Chapter 10, where the redundancy is obtained by increasing the number of vectors in the signal constellation rather than by explicitly adding redundant digits, thus increasing bandwidth. Further design of the TCM signal constellation to cluster the points into *clouds* of closely spaced points with larger gaps between the individual clouds can obtain error control where there are varying degrees of protection on different information bits. Such systems, which offer unequal error protection, are being actively considered for video signal coding where the priority bits reconstruct the basic picture and the less well protected bits add in more detail, to avoid picture loss in poor channels.

Table 11.6 *Commercially available TCM coder chipset examples.*

Data rate (Mbit/s)		75
Rate, R	2/3	3/4
Modulation	8-PSK	16-PSK
Coding gain (dB) at 10^{-5}	3.2	3.0

EXAMPLE 11.7

Compare the carrier to noise ratio required for a P_b of 10^{-6} with: (a) 4-PSK; and (b) TCM derived 8-PSK, using Figure 11.43. How does the performance of the TCM system compare with PRK operation? Is it possible to make any enhancements to the TCM coder to achieve further reductions in the CNR requirement?

For uncoded 4-PSK the distance between states of amplitude 1 is $\sqrt{2}$ and the CNR for $P_b = 10^{-6}$ is given in Figure 11.21 and Table 11.5. The CNR requirement (for $T_oB = 1$) is 13.6 dB.

For TCM based 8-PSK the minimum distance is increased from $\sqrt{2}$ to 2 and the CNR requirement reduces by 3 dB. Thus the minimum CNR will be 10.6 dB.

TCM 8-PSK therefore operates at the same CNR as PRK but achieves twice the spectral efficiency of 2 bit/s/Hz.

The TCM encoder performance can be improved further by extending the constraint length, but this increases the encoder and decoder complexity. (Blahut (1990) shows, in his Table 6.1, that increasing constraint length to 8 increases the coding gain to 5.7 dB.)

11.5 Power efficient modulation techniques

Some communications systems operate in environments where large bandwidths are available but signal power is limited. Such power limited systems rely on power efficient modulation schemes to achieve acceptable bit error, and data, rates. In general data rate can be improved by increasing the number of symbols (i.e. the alphabet size) at the transmitter. If this is to be done without degrading P_e then the enlarged alphabet of symbols must remain at least as widely spaced in the constellation as the original symbol set. This can be achieved without increasing transmitted power by adding orthogonal axes to the constellation space – a technique which results in multidimensional signalling. Power can also be conserved by carefully optimising the arrangement of points in the constellation space. The most significant power saving comes from optimising the lattice pattern of constellation points. This is called symbol packing and results in an increased constellation point density without decreasing point separation. An additional (small) power saving can be obtained by optimising the boundary of the symbol constellation.

11.5.1 Multidimensional signalling and MFSK

Multi-frequency shift keying (MFSK) is a good example of a power efficient, multidimensional, modulation scheme if, as is usually the case, its symbols are designed to be mutually orthogonal, section 2.5.3. Figure 11.44(a), developed from Figure 11.9, shows the voltage spectrum of an orthogonal MFSK signal as a superposition of OOK signals and Figure 11.44(b) shows the power spectrum plotted in dB. The increased data rate realised by MFSK signalling is achieved entirely at the expense of increased bandwidth. Since each symbol (for equiprobable, independent, symbol systems) conveys $H = \log_2 M$ bits of information then the nominal spectral efficiency of orthogonal MFSK is given by:

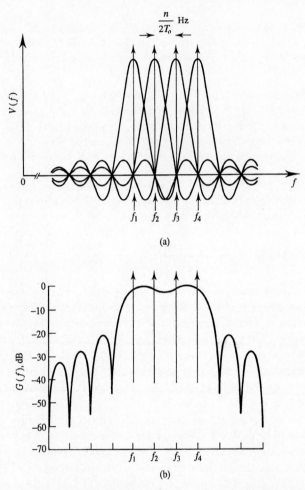

Figure 11.44 *Spectrum of orthogonal MFSK signal ($M = 2$): (a) as a superposition of OOK signal voltage spectra and (b) combined power spectrum on a dB scale. Tone spacing corresponds to second zero crossing point ($n = 2$) on $\rho - T_o$ diagram.*

$$\eta_s = \frac{\log_2 M}{(n/2)(M - 1) + 2} \quad \text{(bit/s/Hz)} \tag{11.50}$$

where n is the selected zero crossing point on the $\rho - T_o$ diagram of Figure 11.8 and the (nominal) signal bandwidth is defined by the first spectral nulls above and below the highest and lowest frequency symbols respectively. For incoherent detection ($n \geq 2$) the maximum spectral efficiency ($n = 2$) in Figure 11.44(b) is given by:

$$\eta_s = \frac{\log_2 M}{M + 1} \quad \text{(bit/s/Hz)} \tag{11.51}$$

EXAMPLE 11.8

Find the maximum spectral efficiency of an 80 kbaud, 8-FSK, orthogonal signalling system which operates with a frequency separation between adjacent tones corresponding to the third zero crossing point of the $\rho - T_o$ diagram of Figure 11.8. Compare this with the nominal spectral efficiency as given by equation (11.50).

The frequency separation between orthogonal tones satisfies:

$$n = 2 (f_i - f_{i-1}) T_o$$

In this case, therefore:

$$(f_i - f_{i-1}) = \frac{3}{2 T_o}$$

$$= \frac{3 \times 80 \times 10^3}{2} = 120 \times 10^3 \quad \text{Hz}$$

The maximum (ISI free) spectral efficiency is realised when $T_o B_{symbol} = 1.0$ and $H = \log_2 M$. The bandwidth of each FSK symbol is therefore $B_{symbol} = 1/T_o$ Hz and for the MFSK signal is:

$$B_{opt\ MFSK} = (M - 1)(f_i - f_{i-1}) + 2 \left(\tfrac{1}{2} \times B_{symbol}\right)$$

$$= (8 - 1)\ 120 \times 10^3 + 80 \times 10^3 = 920 \times 10^3 \quad \text{Hz}$$

The maximum spectral efficiency is therefore given by:

$$\eta_s = \frac{\log_2 M}{T_o B}$$

$$= \frac{(\log_2 8) \times 80 \times 10^3}{920 \times 10^3} = 0.261 \quad \text{bit/s/Hz}$$

The nominal spectral efficiency given by equation (11.50) is:

$$\eta_s = \frac{\log_2 8}{(3/2)\ (8 - 1) + 2} = 0.240 \quad \text{bit/s/Hz}$$

It can be seen that the difference between the nominal and maximum spectral efficiencies (and signal bandwidths) becomes small for MFSK signals as M (and/or tone separation) becomes large.

Multidimensional signalling can also be achieved using sets of orthogonally coded bit patterns, Figure 11.45 [developed from Sklar, p. 253]. (See also equations (15.17) to (15.20).) For an M-symbol alphabet each coded symbol (ideally) carries $H = \log_2 M$ bits of information but, to be mutually orthogonal, symbols must be M binary digits (or 'chips') long, Figure 11.45. The nominal bandwidth of such symbols is M/T_o Hz for baseband signalling and $2M/T_o$ Hz for bandpass signalling where T_o/M is the duration of the binary chips which make up the orthogonal symbols. For equiprobable symbols the nominal spectral efficiency for bandpass orthogonal code signalling is therefore:

$$\eta_s = \frac{\log_2 M}{2M} \quad \text{(bit/s/Hz)} \tag{11.52}$$

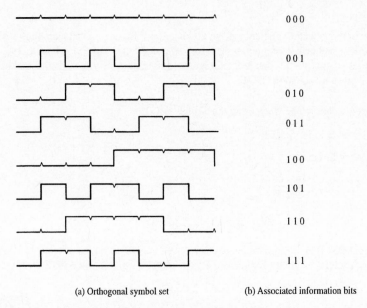

(a) Orthogonal symbol set   (b) Associated information bits

Figure 11.45 *Orthogonal code set comprising eight symbols.*

and the maximum spectral efficiency (with $T_oB = M$) is twice this. For optimum (coherent) detection of any equal energy, equiprobable, M-ary orthogonal symbol set (including MFSK), in the presence of white Gaussian noise, the probability of symbol error is traditionally bounded by the (union bound) formula:

$$P_e \leq \frac{1}{2}(M-1)\left[1 - \mathrm{erf}\sqrt{\frac{E}{2N_0}}\right] \tag{11.53(a)}$$

A tighter bound, however, given by [Hughes] is:

$$P_e \leq 1 - \left\{\frac{1}{2}\left[1 + \mathrm{erf}\sqrt{\frac{E}{2N_0}}\right]\right\}^{M-1} \tag{11.53(b)}$$

where E, as usual, is the symbol energy. The energy per information bit is $E_b = E/\log_2 M$. Figure 11.46 shows P_e against E_b/N_0. The parameter $H = \log_2 M$ is the number of information bits associated with each (equiprobable) symbol. Since the symbols are mutually orthogonal the distance between any pair is a constant. It follows that all types of symbol error are equiprobable, i.e.:

$$P(i_{RX}|j_{TX}) = P_e \quad \text{for all } i \neq j \tag{11.54}$$

From Figure 11.45(b) it can be seen that for any particular information bit associated with any given symbol only $M/2$ symbol errors out of a total of $M-1$ possible symbol errors will result in that bit being in error. The probability of bit error for orthogonal signalling

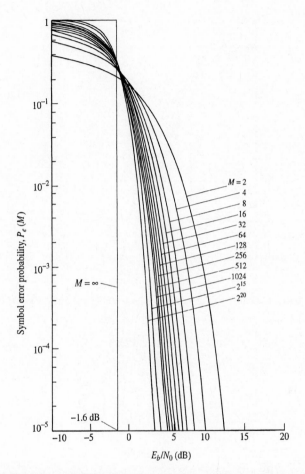

Figure 11.46 *Symbol error probability for coherently detected M-ary orthogonal signalling (source: Lindsey and Simon, 1973, reproduced with permission of Prentice-Hall).*

(including MFSK and orthogonal code signalling) is therefore:

$$P_b = \frac{(M/2)}{M-1} P_e \tag{11.55}$$

Figure 11.47 shows the probability of bit error against E_b/N_0 for orthogonal M-ary systems.

If antipodal symbols are added to an orthogonal symbol set the resulting set is said to be *biorthogonal*. The P_e (and therefore P_b) performance is improved since the average symbol separation in the constellation space is increased for a given average signal power. Figure 11.48 shows the biorthogonal P_b performance. The spectral efficiency of biorthogonal systems is improved by a factor of two over orthogonal systems since pairs of symbols now occupy the same spectral space. Biorthogonal signals must, of course, be

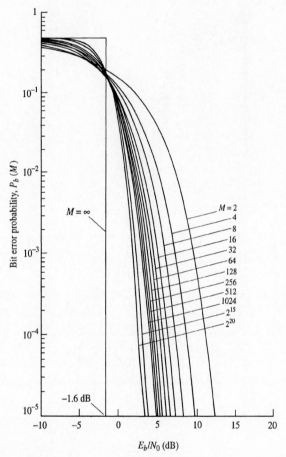

Figure 11.47 *Coherent detection of orthogonally coded transmission (source: Lindsey and Simon, 1973, reproduced with permission of Prentice-Hall).*

detected coherently.

Inspection of Figure 11.45 reveals that for an orthogonal code word set one binary digit (or chip) is identical for all code words. This binary digit therefore carries no information and need not be transmitted. An alphabet of code words formed by deleting the redundant chip from an orthogonal set is said to be *transorthogonal*. It has a correlation between symbol (i.e. code word) pairs given by:

$$\rho = -\frac{1}{M-1} \tag{11.56}$$

and therefore has slightly better P_e (and P_b) performance than orthogonal signalling. Since one binary digit from a total of M has been deleted it has a spectral efficiency which is $M/(M-1)$ times better than the parent orthogonal scheme.

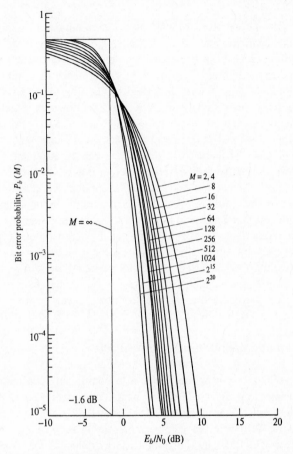

Figure 11.48 *Coherent detection of biorthogonally coded transmission (source: Lindsey and Simon, 1973, reproduced with permission of Prentice-Hall).*

The major application of MFSK type signalling systems is the telegraphy system used for teleprinter messages. This system, called Piccolo [Ralphs], has 32 tones and sends data at only 10 symbol/s with 100 ms duration symbols. (This was chosen to overcome the 0 to 8 ms multipath effects in over-the-horizon HF radio links.)

Another system is currently being developed for multicarrier broadcast to mobiles. It uses *simultaneous* MFSK or orthogonal FDM (OFDM) transmissions [Alard and Lassalle] to reduce a single high data rate signal into many parallel low data rate signals, again with lengthened symbol periods. By suitable FEC coding (Chapter 10) of the 'orthogonal carrier' channels, errors introduced by multipath effects can be tolerated in some of the channels without degrading significantly the overall system error rate. This system is now being developed separately for audio and TV broadcast applications.

11.5.2 Optimum constellation point packing

For a given alphabet of symbols distributed over a fixed number of orthogonal axes there are many possible configurations of constellation points. In general these alternative configurations will each result in a different transmitted power. Power efficient systems may minimise transmitted power for a given P_e performance by optimizing the packing of constellation points. This is well illustrated by the alternative constellation patterns possible for the (two dimensional) 16-APK system shown in Figure 11.19. These patterns are sometimes referred to as (8,8), (4,12) and (4,8,4) APK where each digit represents the number of constellation points on each amplitude ring.

The power efficiency of these constellation patterns increases moving from (8,8) to (4,8,4). The QAM rectangular lattice (represented in Figure 11.19(c) by (4,8,4) APK) is already quite power efficient and as such is a popular practical choice for M-ary signalling. A further saving of 0.6 dB is, however, possible by replacing the rectangular lattice pattern with a hexagonal lattice, Figure 11.49. This represents the densest possible packing of constellation points (in 2 dimensions) for fixed separation of adjacent points.

The power saved by optimising symbol packing increases as the number of constellation dimensions is increased. For a two dimensional constellation the saving gained by replacing a rectangular lattice with a hexagonal lattice is modest. The optimum lattice for a 64-dimensional constellation, however, gives a saving over a rectangular lattice of 8 dB [Forney *et al.*].

11.5.3 Optimum constellation point boundaries

Further modest savings in power can be made by optimising constellation boundaries. Consider, for example, Figure 11.50(a) which represents conventional 64-QAM signalling. If the points at the corners of the constellation are moved to the positions shown in Figure 11.50(b) then the voltage representing each of the moved symbols is reduced by a factor of (9.055/9.899) and their energy reduced by a factor of 0.84. Since the energy residing in each of the other symbols is unaffected the average transmitted power is reduced.

Not surprisingly the most efficient QAM constellation point boundary is that which most closely approximates a circle, and for a three-dimensional signalling scheme the most efficient boundary would be a sphere. It is clear that this idea extends to N-dimensional signalling where the most efficient boundary is an N-dimensional sphere.

Figure 11.49 *Optimum 2-dimensional lattice pattern.*

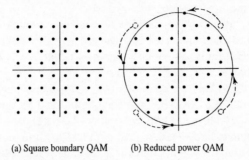

(a) Square boundary QAM   (b) Reduced power QAM

Figure 11.50 *64-QAM constellations.*

The power savings obtained in moving from N-dimensional cubic boundaries to N-dimensional spherical boundaries increase with N. For $N = 2$ (i.e. QAM) the saving is only 0.2 dB whilst for $N = 64$ the saving has increased to 1.3 dB [Forney *et al.*].

If efficient constellation point boundaries are used with efficient constellation point packing then the power savings from both are, of course, realised. Figure 11.51 shows such a (2-dimensional) 64-APK constellation.

11.6 Data modems

Many different techniques used in voiceband (3.2 kHz bandwidth) data modems are covered by the ITU-T V series of recommendations. The QAM constellation of Figure 11.52(a) is used in V.22 data modems. Low bit rate, telephone line based, data modems employ a wide variety of multi-amplitude, multi-phase schemes. The V.32 recommendation covers three different modulation options, 4800 bit/s and 9600 bit/s using conventional modulation techniques and 9600 bit/s TCM. At these low data rates the complexity of TCM is not a problem as current receiver processors have sufficient time to make all the required computations during the symbol period. A standard 9600 bit/s modem (V.29, Figure 11.52(b)) needs a four-wire connection to operate. Leased

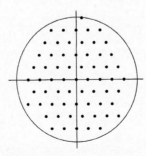

Figure 11.51 *Lattice and boundary efficient 64-APK.*

circuits are available with four-wire connections, but each public switched telephone network (PSTN) connection provides only two wires.

Two PSTN lines are therefore needed to back up a single leased circuit. V.32, modems however, provide major cost savings and other benefits. Firstly, instead of a four-wire leased circuit, only a two-wire circuit is required. In addition only a single PSTN connection per end is needed. Several manufacturers offer products with data rates from 9600 bit/s to 19.2 kbit/s for operation over dial-up networks. With V.33 modems operating at 14.4 kbit/s TCM with full two-way duplex, and the new V.fast standards one now has data rates of 24 kbit/s, and the possibility of 28.8 kbit/s over voiceband circuits. Given that the Shannon-Hartley law, equation (11.38(a)), predicts 32 kbit/s over a 3.2 kHz bandwidth voiceband circuit at 30 dB SNR, then these sophisticated modems are fast approaching the theoretical limits of performance.

11.7 Summary

IF or RF modulation is used principally to shift the spectrum of a digital information signal into a convenient frequency band. This may be to match the spectral band occupied by a signal to the passband of a transmission line, to allow frequency division multiplexing of signals, or to enable signals to be radiated by antennas of practical size.

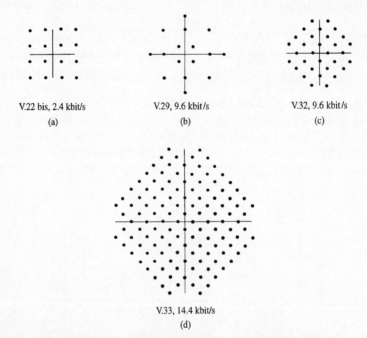

Figure 11.52 *Examples of signal constellations used in speech band data modems: (a) and (c) as used on switched lines; and (b) on leased lines.*

Two performance measures are commonly used to compare different IF modulation techniques – spectral efficiency and power efficiency. The former has units of bit/s/Hz and the latter relates to the required value of E_b/N_0 for a given probability of bit error.

There are three generic IF modulation techniques for digital data. These are ASK, PSK and FSK. Binary ASK and FSK are usually operated as orthogonal signalling schemes. Binary PSK is often operated as an antipodal scheme and therefore requires 3 dB less (average) power. ASK and FSK modulated signals can be detected incoherently which simplifies transmitter and receiver design. A small P_e penalty is incurred in this case over matched filter, or correlation, detection. PSK systems may suffer from phase ambiguity. This can be avoided, however, using either a known phase training sequence, transmitted in a preamble to the information data blocks, or differential coding.

MPSK, APK and QAM are spectrally efficient modulation schemes. QPSK and its variants are also reasonably spectrally efficient. Sophisticated 64-QAM and 256-QAM systems are now being applied in digital microwave long haul radio communications systems (see Chapter 14), but the power amplifier linearity requirements are severe for high level QAM systems. TCM is a technique which increases power efficiency without compromising spectral efficiency by combining modulation with error control coding.

MFSK is usually operated as an orthogonal modulation scheme and is therefore power efficient (as is orthogonal code signalling). Non-orthogonal M-ary schemes can be made more power efficient by distributing the symbols (constellation points) over an increased number of orthogonal axes. Further improvements in power efficiency can be made by choosing the optimum symbol packing arrangement for the number of dimensions used and adopting an (N-dimensional) spherical boundary for the set of constellation points.

Many of these techniques are widely applied in speech band data modems (described in section 11.6) and also in wideband high speed microwave communication systems (described in Chapter 14).

11.8 Problems

11.1. A rectangular pulse OOK signal has an average carrier power, at the input to an ideal correlation receiver, of 8.0 nW. The (one sided) noise power spectral density, measured at the same point, is 2.0×10^{-14} W/Hz. What maximum bit rate can this system support whilst maintaining a P_e of 10^{-6}? [17.7 kbit/s]

11.2. A BPSK, 1.0 Mbaud, communication system is to operate with a P_e of 8×10^{-7}. The signal amplitude at the input to the ideal correlation receiver is 150 mV and the two sided noise power spectral density at the same point is 15 pW/Hz. The impedance level at the correlation receiver input is 50 Ω. What minimum phase shift between phasor states must the system employ and what residual carrier power is therefore available for carrier recovery purposes? [129°, 4.23×10^{-5} W]

11.3. Define on-off keying (OOK), frequency shift keying (FSK), and phase shift keying (PSK) as used in binary signalling. Compare their respective advantages and disadvantages.

In a Datel 600 modem FSK tone frequencies of 1300 Hz and 1700 Hz are used for a signalling rate of 600 bit/s. Comment on the consistency between these tone frequencies, the $\rho - T_o$ diagram, and the signalling rate. What value would the upper tone frequency need to have, to handle a 900 bit/s signalling rate? [1900 Hz]

11.4. A binary, rectangular pulse, BFSK modulation system employs two signalling frequencies f_1 and f_2 and operates at the second zero of the $\rho - T_o$ diagram. The lower frequency f_1 is 1200 Hz and the signalling rate is 500 baud. (a) Calculate f_2. (b) Sketch the PSD of the FSK signal. (c) Calculate the channel bandwidth which would be required to transmit the FSK signal without 'significant' distortion. Indicate in terms of the PSD those frequencies which you assume should be passed to keep the distortion 'insignificant'. [1700 Hz, $B = 1500$ Hz]

11.5. If the CNR of the system described in problem 11.4 is 4 dB at the input to an ideal (matched filter) receiver and the channel has a low-pass rectangular frequency characteristic with a cut-off frequency of 3.6 kHz, estimate the probability of bit error. [1.06×10^{-5}]

11.6. (a) Define the correlation coefficient of two siganls: $s_1(t)$ and $s_2(t)$. (b) Sketch the correlation coefficient of two FSK rectangular RF signalling pulses, such as might be used in an FSK system, as a function of tone spacing. (c) Find the P_e of an orthogonal FSK binary signalling system using ideal matched filter detection if the signalling pulses at the matched filter input have rectangular envelopes of 10 μs duration and 1.0 V peak to peak amplitude. The one sided NPSD at the input to the filter is 5.5 nW/Hz and the impedance level at the filter input is 50 Ω. [6.54×10^{-4}]

11.7. If the spectrum of each signalling pulse described in problem 11.6(c) is assumed to have significant spectral components only within the points defined by its first zeros, what is the minimum bandwidth of the FSK signal? [0.25 MHz]

11.8. A receiver has a mean input power of 25 pW and is used to receive binary FSK data. The carrier frequencies used are 5 MHz and 5.015 MHz. The noise spectral density at the receiver has a value of 2.0×10^{-16} W/Hz. If the error rate is fixed at 2×10^{-4} find the maximum data rate possible, justifying all assumptions. [5550 bits/s] If the transmitter is switched to PRK what new bit rate can be accommodated for the same error rate? [1.11 kbit/s]

11.9. What is meant by orthogonal and antipodal signalling? Can the error performance of an FSK system ever be better than that given by the orthogonal case and, if so, how is the performance improved, and at what cost?

11.10. A DEPSK, rectangular pulse, RZ signal, with 50% duty cycle, has a separation of 165° between phasor states. The baud rate is 50 kHz and the peak received power (excluding noise) at the input of an ideal correlation receiver is 100 μW. The (one sided) noise power spectral density measured at the same point is 160 pW/Hz. What is the probability of symbol error? [4.56×10^{-4}]

11.11. Find the maximum spectral efficiency of ISI free 16-PSK. What is the probability of symbol error of this scheme for a received CNR of 24 dB if the maximum spectral efficiency requirement is retained? What is the Gray coded probability of bit error? [4 bit/s/Hz, 1.227×10^{-5}, 3.068×10^{-6}]

11.12. Sketch the constellation diagram of 64-QAM. What is the spectral efficiency of this scheme if pulse shaping is employed such that $BT_o = 2$? What is the best possible spectral efficency whilst maintaining zero ISI? Why is QAM (with the exception of QPSK) usually restricted to use in linear (or approximately linear) channel applications? [3 bit/s/Hz, 6 bit/s/Hz]

11.13. (a) Draw a block diagram of a QPSK modulator. Show clearly on this diagram or elsewhere the relationship in time between the binary signals in I and Q channels at the points immediately preceeding multiplication by the I and Q carriers. How does this relationship differ in OQPSK systems? What advantage do OQPSK systems have over QPSK systems? Sketch the phasor diagrams for both systems showing which signal state transitions are allowed in each.
(b) Calculate the probability of bit error for an ideal OQPSK system if the single sided NPSD and bit energy are 1.0 pW/Hz and 10.0 pJ respectively. [3.88×10^{-6}]

11.14. Justify the statement: 'The performance of BPSK and QPSK systems are identical'.

11.15. Two engineers are arguing about MSK modulation. One says it is a special case of BFSK and the other asserts it is a form of OQPSK. Resolve their argument. Is incoherent detection of MSK possible? (Explain your answer.) Draw a block diagram of an MSK transmitter and sketch a typical segment of its output.

11.16. A power limited digital communication system is to employ orthogonal MFSK modulation and ideal matched filter detection. P_b is not to exceed 10^{-5} and the available receiver carrier power is limited to 1.6×10^{-8} V^2. Find the minimum required size of the MFSK alphabet which will achieve a bit rate of 1.0 Mbit/s at the specified P_b if the two sided noise power spectral density at the receiver input is 10^{-15} V^2/Hz. (The use of a computer running an interactive maths package such as MATLAB, or a good programmable calculator, will take some of the tedium out of this problem.) [M=6]

11.17. A 3-APK modulation scheme contains the constellation points 0, $e^{j\pi/3}$, $e^{j2\pi/3}$. What is the power saving (in dB) if this scheme is replaced with a minimum power 3-PSK scheme having the same P_e performance? [3 dB]

CHAPTER 12

System noise and communications link budgets

12.1 Introduction

Signal-to-noise ratio (SNR) is a fundamental limiting factor on the performance of communications systems. It can often be improved by a receiving system using appropriate demodulation and signal processing techniques. It is always necessary, however, to know the carrier-to-noise ratio (CNR) present at the input of a communications receiver to enable its performance to be adequately characterised. (The term CNR was used widely in place of SNR in Chapter 11, and is used here, since for most modulation schemes the carrier power is not synonymous with the impressed information signal power.) This chapter reviews some important noise concepts and illustrates how system noise power can be calculated. It also shows how received carrier power can be calculated (at least to first order accuracy) and therefore how a CNR can be estimated.

12.2 Physical aspects of noise

Whilst this text is principally concerned with systems engineering, it is interesting and useful to establish the physical origin of the important noise processes and the physical basis of their observed characteristics (especially noise power spectral density).

Consideration here is restricted to those noise processes arising in the individual components (resistors, transistors, etc.) of a subsystem, often referred to collectively as circuit noise. Noise which arises from external sources and is coupled into a communication system by a receiving antenna is discussed in section 12.4.3.

12.2.1 Thermal noise

Thermal noise is produced by the random motion of free charge carriers (usually electrons) in a resistive medium. These random motions represent small random currents which together produce a random voltage, via Ohm's law, across, for example, the terminals of a resistor. (Such noise does not occur in ideal capacitors as there are no free electrons in a perfect dielectric material.) Thermally excited motion takes place at all temperatures above absolute zero (0 K) since, by definition, temperature is a measure of average kinetic energy per particle. Thermal noise is a limiting factor in the design of many, but not all, radio communications receivers at UHF (0.3 to 3 GHz), Table 1.4, and above.

The power spectral density of thermal noise (at least up to about 10^{12} Hz) can be predicted from classical statistical mechanics. This theory indicates that the average kinetic (i.e. thermal) energy of a molecule in a gas which is in thermal equilibrium and at a temperature T, is $1.5kT$ joules. k is therefore a constant which relates a natural, or atomic, temperature scale to a man-made scale. It is called Boltzmann's constant and has a value of 1.381×10^{-23} J K^{-1}.

The *principle of equipartition* says that a molecule's energy is equally divided between the three dimensions or *degrees of freedom* of space. More generally for a system with any number of degrees of freedom the principle states that:

the average thermal energy per degree of freedom is ½ kT joules.

Equipartition can be applied to the transmission line shown in Figure 12.1. On closing the switches thermal energy will be trapped in the line as standing electromagnetic waves. Since each standing wave mode on the line has two degrees of freedom (its pulsations can exist as sinusoidal or cosinusoidal functions of time) then each standing wave will contain kT joules of energy. The wavelength of the nth mode (Figure 12.2) is given by:

$$\lambda_n = \frac{2l}{n} \quad \text{(m)} \tag{12.1}$$

where l is the length of the transmission line. Therefore the mode number, n, is given by:

$$n = \frac{2lf_n}{c} \tag{12.2}$$

where c is the electromagnetic velocity of propagation on the line. The frequency difference between adjacent modes is:

$$f_n - f_{n-1} = \frac{nc}{2l} - \frac{(n-1)c}{2l}$$

$$= \frac{c}{2l} \quad \text{(Hz)} \tag{12.3}$$

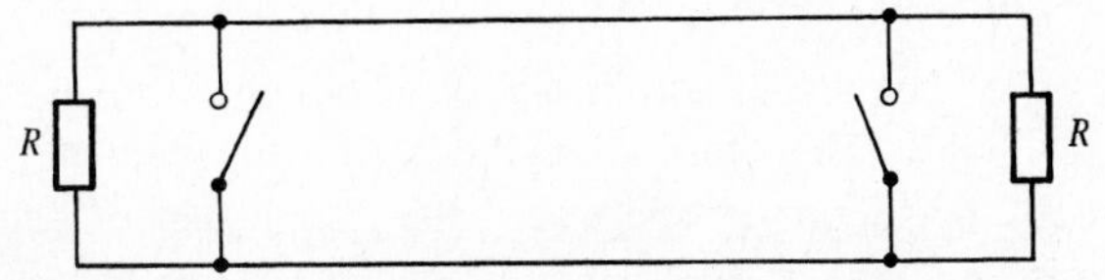

Figure 12.1 *Transmission line for trapping thermal energy.*

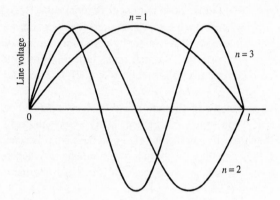

Figure 12.2 *Standing waves of trapped thermal energy on short-circuited transmission line.*

Therefore the number of modes, v, in a bandwidth B is:

$$v = \frac{B}{f_n - f_{n-1}} = \frac{2lB}{c} \tag{12.4}$$

The spatial (line) energy density of each standing wave mode is kT/l (J m^{-1}). Since each standing wave is composed of two travelling waves (with opposite directions of travel) then the spatial energy density in each travelling wave is $\frac{1}{2}kT/l$ (J m^{-1}). The energy per second travelling past a given point in a given direction (i.e. the available power) per oscillating mode is therefore:

$$P_{mode} = \frac{\frac{1}{2} kT}{\tau} \quad \text{(J s}^{-1}\text{)} \tag{12.5}$$

where τ is the energy transit time from one end of the line to the other. Using $\tau = l/c$:

$$P_{mode} = \frac{\frac{1}{2} kTc}{l} \quad \text{(J s}^{-1}\text{)} \tag{12.6}$$

The available power, N, in a bandwidth B (i.e. v modes) is therefore given by:

$$N = P_{mode}v \tag{12.7}$$

$$= \frac{\frac{1}{2} kTc}{l} \frac{2lB}{c} \quad \text{(J s}^{-1}\text{)}$$

i.e.:

$$N = kTB \quad \text{(W)} \tag{12.8}$$

Equation (12.8) is called Nyquist's formula and is accurate so long as classical mechanics can be assumed to hold. The constant behaviour with frequency, however, of the available (one sided) noise power spectral density, $G_n(f) = kT$, incorrectly predicts infinite power when integrated over all frequencies. This paradox, known as the ultraviolet catastrophe, is resolved when quantum mechanical effects are accounted for.

The complete version of equation (12.8) then becomes:

$$N = \int_0^B G_n(f)\, df \quad \text{(W)} \qquad (12.9(a))$$

where:

$$G_n(f) = \frac{hf}{e^{(hf/kT)} - 1} \quad \text{(W Hz}^{-1}) \qquad (12.9(b))$$

h in equation (12.9(b)) is Planck's constant which has a value of 6.626×10^{-34} (J s). For $hf \ll kT$ then:

$$e^{\left(\frac{hf}{kT}\right)} \approx 1 + \frac{hf}{kT} \qquad (12.10)$$

and equation (12.9(b)) reduces to the Nyquist form:

$$G_n(f) \approx kT \quad \text{(W Hz}^{-1}) \qquad (12.11)$$

The classical and quantum mechanical power spectral densities of thermal noise are compared in Figure 12.3.

The Thévenin equivalent circuit of a thermal noise source is shown in Figure 12.4. Since the available noise power for almost all practical temperatures and frequencies (excluding optical applications) is kTB W then the maximum available RMS noise voltage V_{na} across a (conjugately) matched load is:

$$V_{na} = \sqrt{NR} = \sqrt{kTBR} \quad \text{(V)} \qquad (12.12)$$

where R is the load (and source) resistance. Since the same voltage is dropped across the

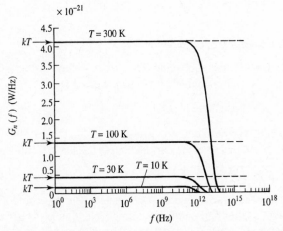

Figure 12.3 *Power spectral densities of thermal noise predicted by classical (- - -) and quantum (---) mechanics at four temperatures.*

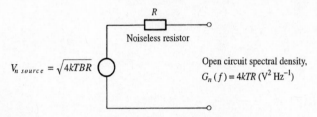

Figure 12.4 *Thévenin equivalent circuit of a thermal noise source.*

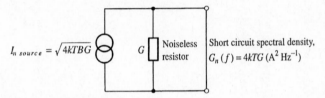

Figure 12.5 *Norton equivalent circuit of a thermal noise source.*

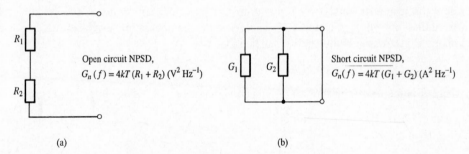

Figure 12.6 *NPSD for resistors in (a) series and (b) parallel.*

source and load resistances the equivalent RMS noise voltage of the source (i.e. the open circuit voltage of the Thévenin equivalent circuit) is given by:

$$V_{n\ source} = 2\sqrt{kTBR} = \sqrt{4kTBR} \quad (V) \tag{12.13}$$

An equivalent random current source can also be defined using the Norton equivalent circuit in Figure 12.5. In this case the current source has a RMS value given by:

$$I_{n\ source} = \sqrt{4kTBG} \quad (A) \tag{12.14}$$

where $G = 1/R$ is the load (and source) conductance. For passive components (such as resistors) connected in series and parallel their equivalent mean square noise voltages and currents add in an obvious way. For example, the series combination of resistances R_1 and R_2 in Figure 12.6(a) results in:

$$V_{n\ source}^2 = V_{n\ source\ 1}^2 + V_{n\ source\ 2}^2 \quad (V^2) \tag{12.15(a)}$$

i.e.:

$$V_{n\,source} = \sqrt{4kTB(R_1 + R_2)} \quad (V) \tag{12.15(b)}$$

and the parallel combination of conductances G_1 and G_2 in Figure 12.6(b) results in:

$$I_{n\,source}^2 = I_{n\,source\,1}^2 + I_{n\,source\,2}^2 \quad (A^2) \tag{12.16(a)}$$

i.e.:

$$I_{n\,source} = \sqrt{4kTB(G_1 + G_2)} \quad (A) \tag{12.16(b)}$$

Alternatively, and entirely unsurprisingly, the equivalent voltage source of the parallel combination is given by:

$$V_{n\,source} = \frac{I_{n\,source}}{G}$$

$$= \sqrt{4kTB\,\frac{R_1\,R_2}{R_1 + R_2}} \quad (V) \tag{12.16(c)}$$

i.e. the two parallel resistors together behave as a single resistor with the appropriate parallel value.

12.2.2 Non-thermal noise

Although the time averaged current flowing in a device may be constant, statistical fluctuations will be present if individual charge carriers have to pass through a potential barrier. The potential barrier may, for example, be the junction of a pn diode, the cathode of a vacuum tube or the emitter-base junction of a bipolar transistor. Such statistical fluctuations constitute shot noise. The traditional device used to illustrate the origin of shot noise is a vacuum diode, Figure 12.7. Electrons are emitted thermally from the cathode. Assuming there is no significant space charge close to the cathode surface due to previously emitted electrons (i.e. assuming that the diode current is temperature limited), then electrons are emitted according to a Poisson statistical process (see Chapter 17).

The spatial (line) charge density, σ, between cathode and anode is given by:

$$\sigma = \frac{nq_e}{l} \quad (C/m) \tag{12.17}$$

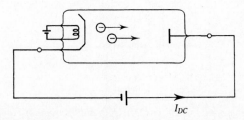

Figure 12.7 *Schematic diagram for a vacuum diode.*

where n is the average number of electrons in flight, l is the distance between cathode and anode and q_e is the charge on an electron. If the average electron velocity is v then the average current, I_{DC}, in (any part of) the circuit is:

$$I_{DC} = \sigma v \quad \text{(C/s)} \tag{12.18}$$

and the average contribution to I_{DC} from any individual electron is:

$$i_e = \frac{I_{DC}}{n} = q_e \frac{v}{l} \quad \text{(C/s)} \tag{12.19}$$

The current associated with any given electron can therefore be considered to flow only for that time the electron is in flight. Since l/v is the average cathode–anode transit time, τ, for the electrons, equation (12.19) can be written as:

$$i_e = \frac{q_e}{\tau} \quad \text{(A)} \tag{12.20}$$

Thus each electron gives rise to a current pulse of duration τ as it moves from cathode to anode. (The precise shape of the current pulse will depend on the way in which the electron's velocity varies during flight. For the purpose of this discussion, however, the pulse can be assumed to be rectangular, Figure 12.8.)

The total current will be a superposition of such current pulses occurring randomly in time with Poisson statistics. Since each pulse has a spectrum with (nominal) bandwidth $1/\tau$ Hz (in the ideal rectangular pulse case the energy spectrum will have a sinc^2 shape, Figure 12.9) then for frequencies very much less than $1/\tau$ (Hz) shot noise has an essentially white spectrum.

The (two sided) current spectral density of the pulse centered on $t = 0$ due to a single electron is:

$$I(f) = \int_{-\infty}^{\infty} \frac{q_e}{\tau} \Pi\left(\frac{t}{\tau}\right) e^{-j2\pi f t} \, dt$$

$$= q_e \, \text{sinc}(\tau f) \quad \text{(A/Hz)} \tag{12.21}$$

The (normalised, two sided) energy spectral density of this current pulse is therefore:

$$E_i(f) = q_e^2 \, \text{sinc}^2(\tau f) \quad \text{(A}^2 \text{ s/Hz)} \tag{12.22}$$

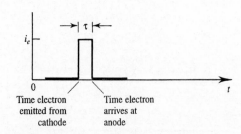

Figure 12.8 *Current pulse due to a single electron in diode circuit of Figure 12.7.*

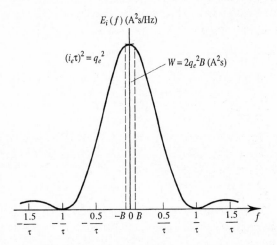

$E_i(f)$ (A^{2}s/Hz)

$(i_e\tau)^2 = q_e^2$

$W = 2q_e^2 B$ (A^{2}s)

1.5 1 0.5 $-B$ 0 B 0.5 1 1.5 f
$-\dfrac{1}{\tau}$ $-\dfrac{1}{\tau}$ $-\dfrac{1}{\tau}$ $\dfrac{1}{\tau}$ $\dfrac{1}{\tau}$ $\dfrac{1}{\tau}$

Figure 12.9 *Energy spectral density of current pulse shown in Figure 12.8.*

For frequencies much less than $1/\tau$ Hz, $E_i(f)$ is a constant, namely q_e^2.

Using this model the energy, W, contained within a bandwidth B ($\ll 1/\tau$) is given by:

$$W = 2q_e^2 B \quad (A^2 \text{ s}) \tag{12.23}$$

For m electrons arriving at the anode in a time T the total shot noise (normalised) power will therefore be:

$$I_n^2 = \frac{2mq_e^2 B}{T} \quad (A^2) \tag{12.24}$$

mq_e/T, however, can be identified as the total DC current flowing in the circuit, i.e.:

$$I_n = \sqrt{2I_{DC}q_e B} \quad (A) \tag{12.25}$$

Equation (12.25) shows that shot noise RMS current is proportional to both the square root of the DC current and the square root of the measurement bandwidth. Although this result has been derived for a vacuum diode it is also correct for a pn junction diode and for the base and collector (with $I_{DC} = I_B$ and $I_{DC} = I_C$ as appropriate) of a bipolar junction transistor. Figure 12.10 shows a bipolar transistor along with its equivalent thermal and shot noise sources. $G_{V_{nB}}$ is the power spectral density of the thermal noise voltage generated in the transistor's base spreading resistance. $G_{I_{nB}}$ is the power spectral density of the shot noise current associated with those carriers flowing across the emitter-base junction which subsequently arrive at the base terminal of the transistor. (The fluctuation in base current caused by $G_{V_{nB}}$ and $G_{I_{nB}}$ will be amplified, like any other base current changes, at the transistor collector.) $G_{I_{nC}}$ is the power spectral density of the shot noise current associated with the carriers flowing across the emitter-base junction which subsequently arrive at the transistor collector. For FETs the shot noise is principally associated with gate current and can be modelled by equation (12.25) with $I_{DC} = I_G$.

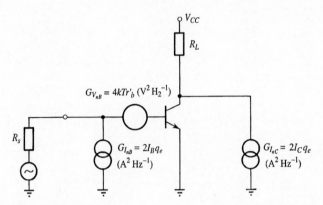

Figure 12.10 *Power spectral densities of thermal (G_{V_n}) and shot (G_{I_n}) noise processes in a bipolar transistor.*

Flicker, or $1/f$, noise also occurs in most active (and some passive) devices. It is rather device specific and is therefore not easily modelled. Its power spectral density has approximately a $1/f$ characteristic for frequencies below a few kilohertz. Above a few kilohertz flicker noise has an essentially flat spectrum but in any case is weak and usually neglected. Because flicker noise power is concentrated at low frequencies it is sometimes called pink noise.

12.2.3 Combining white noise sources

In many applications it is sufficient to account for thermal noise and shot noise only. This also has the practical advantage that circuit noise calculations can be made initially on a power spectral density basis, the total power then being found using the circuit noise bandwidth.

Consider Figure 12.11(a). There are two principal sources of shot noise and, effectively, three sources of thermal noise in this circuit. The former arise from the base-emitter junction of the transistor and the latter arise from the source resistance, R_s (comprising the ouput resistance of the preceding stage in parallel with the biasing resistors R_1 and R_2), the base spreading resistance, $r_{b'}$, and the load resistance, R_L. Figure 12.11(b) shows an equivalent circuit of the transistor, incorporating its noise sources, and Figure 12.11(c) shows an equivalent circuit of Figure 12.11(a) with all noise sources (excluding that due to the load) transferred to the transistor's input and represented as RMS noise voltages. The noise sources in Figure 12.11(c) are given by:

$$V_{nr_{b'}}^2 = 4\,kTr_{b'}B \qquad (12.26(a))$$

$$V_{nB}^2 = I_{nB}^2\,(R_s + r_{b'})^2 = 2q_e\,I_B\,B\,(R_s + r_{b'})^2 \qquad (12.26(b))$$

$$V_{nC}^2 = (I_{nC}/h_{fe})^2\,(R_s + r_{b'} + r_\pi)^2$$

$$= (2q_e\,I_C\,B)\,(R_s + r_{b'} + r_\pi)^2/(g_m\,r_\pi)^2 \qquad (12.26(c))$$

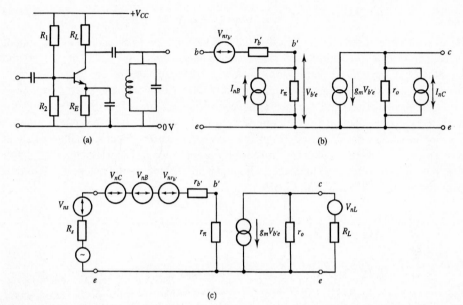

Figure 12.11 *(a) Single stage transistor amplifier; (b) transistor equivalent circuit with noise sources; and (c) equivalent circuit including driving, biasing and load components with most noise sources referred to input.*

$$V_{ns}^2 = 4\,kTR_s B \qquad\qquad (12.26(d))$$

$$V_{nL}^2 = 4\,kTR_L B \qquad\qquad (12.26(e))$$

where B is the noise bandwidth of the circuit, q_e is the electronic charge, I_B and I_C are the quiescent base and collector currents respectively, $h_{fe} = g_m r_\pi$ is the small signal forward current gain of the transistor and g_m is the transistor's transconductance.

The, effective, transistor input noise (which excludes that due to the source resistance) is therefore given by:

$$
\begin{aligned}
V_{n\,tran}^2 &= V_{nC}^2 + V_{nB}^2 + V_{nr_{b'}}^2 \\
&= (2q_e I_C B)(R_s + r_{b'} + r_\pi)^2/(g_m r_\pi)^2 + 2q_e I_B B(R_s + r_{b'})^2 + 4\,kTr_{b'}B \\
&= 4\,kT\left[0.5(R_s + r_{b'} + r_\pi)^2/(g_m r_\pi^2) + 0.5g_m(R_s + r_{b'})^2/h_{FE} + r_{b'}\right]B \\
&= 4\,kTR_{tran}B \qquad\qquad (12.27)
\end{aligned}
$$

where h_{FE} is the transistor's DC forward current gain and use has been made of $g_m = q_e I_C/(kT) \approx 40 I_C$ at room temperature. R_{tran}, defined by the square bracket in equation (12.27), represents a hypothetical resistor located at the (noiseless) transistor's input which would account (thermally) for the actual noise introduced by the real (noisy) transistor. The total noise at the transistor output (including that, V_{nL}^2, generated by the load resistor, R_L) is therefore:

$$V^2_{n\,olp} = (V^2_{ns} + V^2_{n\,tran})\left(\frac{r_\pi}{R_s + r_{b'} + r_\pi}\right)^2 g^2_m \left(\frac{r_o\,R_L}{r_o + R_L}\right)^2 + V^2_{nL} \qquad (12.28)$$

Since the source resistance and load resistance depend, at least partly, on the output impedance of the preceding stage and input impedance of the following stage the importance of impedance optimisation in low noise circuit design can be appreciated. The following example illustrates a circuit noise calculation and also shows, explicitly, how such a calculation can be related to the conventional figure of merit (i.e. noise factor or noise figure, see section 12.3.3) commonly used to summarise the noise performance of a circuit, system or subsystem. (In practice communications circuits usually operate at high frequency where s-parameter transistor descriptions are appropriate. For a full treatment of low noise, high frequency, circuit design techniques the reader is referred to specialist texts, e.g. [Smith, Yip, Liao].)

EXAMPLE 12.1

The transistor in figure 12.11(a)) has the following parameters: $r_{b'} = 150\,\Omega$, $r_o = 40\,k\Omega$, $h_{fe} = 100$, $h_{FE} = 80$. The circuit in which the transistor is embedded has the component values: $R_1 = 10\,k\Omega$, $R_2 = 3.3\,k\Omega$, $R_E = 1\,k\Omega$, $R_L = 3.3\,k\Omega$. The output resistance of the signal source which drives this circuit is 1.7 kΩ. Find: (a) the transistor's equivalent thermal input noise resistance, R_{tran}; (b) the noise power spectral density at the transistor outputs; and (c) the ratio of source plus transistor noise to transistor noise only. (For part (b) assume that, apart from thermal noise arising from the signal source's output resistance, the source is noiseless.) In part (c) the ratio which you have been asked to calculate is a noise factor, see section 12.3.3.

(a) The transistor transconductance, g_m, depends on the collector current which is found in the usual way, i.e.:

$$R_{in} \approx (h_{FE} + 1)\,R_E = (80 + 1)\,1 \times 10^3 = 81\,k\Omega$$

$$V_B = \frac{R_2 R_{in}\,/\,(R_2 + R_{in})}{[R_2 R_{in}\,/\,(R_2 + R_{in})] + R_1}\,V_{CC}$$

$$= \frac{(3.3 \times 81)\,/\,(3.3 + 81)}{[(3.3 \times 81)\,/\,(3.3 + 81)] + 10} \times 10 = 2.4\ V$$

$$V_E = V_B - V_{BE} = 2.4 - 0.6 = 1.8\ V$$

$$I_C \approx I_E = \frac{V_E}{R_E} = \frac{1.8}{1 \times 10^3} = 1.8\ mA$$

The transconductance is the reciprocal of the intrinsic emitter resistance, r_e, i.e.:

$$g_m = \frac{1}{r_e} = \frac{I_C\,(mA)}{25} = \frac{1.8}{25} = 0.072\ S$$

$$r_\pi = \frac{h_{fe}}{g_m} = \frac{100}{0.072} = 1400\ \Omega$$

The source resistance, R_s, is the parallel combination of the signal source's output resistance, R_o, and the transistor's biasing resistors:

$$R_s = \left[\frac{1}{R_o} + \frac{1}{R_1} + \frac{1}{R_2} \right]^{-1}$$

$$= \left[\frac{1}{1700} + \frac{1}{10000} + \frac{1}{3300} \right]^{-1} = 1000 \; \Omega$$

From equation (12.27):

$$R_{tran} = \frac{0.5 \, (R_s + r_{b'} + r_\pi)^2}{g_m \, r_\pi^2} + \frac{0.5 \, g_m \, (R_s + r_{b'})^2}{h_{FE}} + r_{b'}$$

$$= \frac{0.5 \, (1000 + 150 + 1400)^2}{(0.072) \, (1400^2)} + \frac{0.5 \, (0.072) \, (1000 + 150)^2}{80} + 150$$

$$= 23 + 600 + 150 = 770 \; \Omega$$

(b) The total noise power expected at the transistor output is the sum of contributions from the source resistance, the transistor and the load. Thus, using equation (12.28):

$$V_{n\,olp}^2 = (V_{ns}^2 + V_{n\,tran}^2) \left(\frac{r_\pi}{R_s + r_{b'} + r_\pi} \right)^2 g_m^2 \left(\frac{r_o \, R_L}{r_o + R_L} \right)^2 + V_{nL}^2$$

where $V_{ns}^2 = 4 \, kTR_s B$ and $V_{nL}^2 = 4 \, kTR_L B$. Therefore:

$$V_{n\,olp}^2 = 4 \, kT \left[(R_s + R_{tran}) \left(\frac{r_\pi}{R_s + r_{b'} + r_\pi} \right)^2 g_m^2 \left(\frac{r_o \, R_L}{r_o + R_L} \right)^2 + R_L \right] B \; (\mathrm{V}^2)$$

$$= 4 \times 1.38 \times 10^{-23} \times 290 \left[(1000 + 770) \left(\frac{1400}{1150 + 1400} \right)^2 0.072^2 \left(\frac{40000 \times 3300}{40000 + 3300} \right)^2 + 3300 \right] B$$

$$G_{n\,olp} = V_{n\,olp}^2 \, / \, B$$

$$= 4 \times 4 \times 10^{-21} \, [(1000 + 770) \, 15000 + 3300]$$

$$= 16 \times 10^{-21} \left[27 \times 10^6 + 3.3 \times 10^3 \right] = 4.3 \times 10^{-13} \; \mathrm{V^2/Hz}$$

(c) Noise factor, f, (see section 12.3.3) is given by:

$$f = \frac{R_s + R_{tran}}{R_s} = \frac{1000 + 770}{1000} = 1.77$$

12.3 System noise calculations

The gain, G, of a device is often expressed in decibels, as the ratio of the output to input voltages or powers, i.e., $G_{dB} = 20 \log_{10}(V_o/V_i)$ or $G_{dB} = 10 \log_{10}(P_o/P_i)$. Being a ratio gain is not related to any particular power level. There is a logarithmic unit of power, however, which defines power in dB above a specified reference level. If the reference power is 1 mW the units are denoted by dBm (dB with respect to a 1 mW reference) and if the reference level is 1 W the units are denoted by dBW. 10 mW would thus correspond to +10 dBm and 100 mW would be +20 dBm, etc. dBm can be converted to dBW using +30 dBm $\equiv$ 0 dBW. Table 12.1 illustrates these relationships.

The ways in which the noise performance of individual subsystems is specified, and the way these specifications are used to calculate the overall noise performance of a complete system, are now discussed. The overall noise characteristics of individual subsystems are usually either specified by the manufacturer or measured using special instruments (e.g. noise figure meters). If $T = 290$ K (ambient or room temperature) and $B = 1$ Hz, then the available noise power is $1.38 \times 10^{-23} \times 290 \times 1 = 4 \times 10^{-21}$ W/Hz. Expressing this power (in a 1 Hz bandwidth) in dBW gives a noise power spectral density of -204 dBW/Hz or -174 dBm/Hz. Power, in dBm or dBW, can be scaled by simply adding or subtracting the gain of amplifiers or attenuators, measured in dB, to give directly the output power, again in dBm or dBW.

Table 12.1 *Signal power measures.*

dBW	*Power level* (W)	dBm
30 dBW	1,000	60 dBm
20 dBW	100	50 dBm
10 dBW	10	40 dBm
0 dBW	1	30 dBm
−10 dBW	1/10	20 dBm
−20 dBW	1/100	10 dBm
−30 dBW	1/1,000	0 dBm
−40 dBW	1/10,000	−10 dBm

12.3.1 Noise temperature

Consider again equation (12.8) which defines available noise power. The actual noise power delivered is less than this if the source and load impedances are not matched (in the maximum power transfer, i.e. conjugate impedance, sense). A convenient way to specify noise power is via an equivalent thermal noise temperature, T_e, given by:

$$T_e = \frac{N}{kB} \quad \text{(K)} \tag{12.29}$$

The noise temperature of a subsystem (say an amplifier) is *not* the temperature of the room it is in *nor* even the temperature inside its case. It is the (hypothetical) temperature which an ideal resistor matched to the input of the subsystem would need to be at in order to account for the extra available noise observed at the device's output *over and above that which is due to the actual input noise*. This idea is illustrated schematically in Figure 12.12. The total available output noise of a device is therefore given by:

$$N = (kT_sB + kT_eB)G \tag{12.30}$$

where the first term in the brackets is the contribution from the input (or source) noise and the second term is the contribution from the subsystem itself. G is the (power) gain

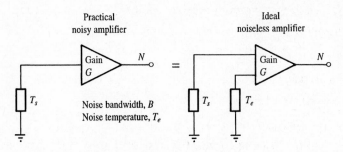

Figure 12.12 *Equivalent noise temperature (T_e) of an amplifier.*

(expressed as a ratio) of the subsystem which can be greater or less than 1.0. For example, an amplifier would have a gain exceeding unity while the mixer in a downconverter, Figure 1.4, would typically have a gain of approximately 0.25. Whilst the contribution of the subsystem to the total output noise, and therefore the noise temperature, will generally be *influenced* by its physical temperature, it will also depend on its design and the quality of its component parts, particularly with respect to the sections at, or close to, its 'front' (i.e. input) end.

Some non-thermal noise processes, such as $1/f$ or flicker noise, have non-white spectra which if included in equation (12.29) would make T_e a function of bandwidth. In practice, however, for most systems work, white noise is assumed to dominate, making T_e bandwidth independent. If thermal noise is dominant (as is often the case in low noise systems) then physical cooling of the device will improve its noise performance.

It has been stated above that the equivalent noise temperature of a device has a dependence on its physical temperature. The strength of this dependence is determined by the relative proportions of thermal to non-thermal noise which the device generates. Strictly speaking, therefore, the noise temperature quoted for a device or subsystem relates to a specific device physical temperature. This temperature is invariably the equilibrium temperature which the device attains (under normal operating conditions which may incorporate heat sinks etc.) when the ambient temperature around it is 290 K. The operating environments of electronic devices normally have ambient temperatures which are sufficiently close to this to make errors in noise calculations based on an assumed temperature of 290 K negligible for most practical purposes.

EXAMPLE 12.2

Calculate the output noise of the amplifier, shown in Figure 12.12, assuming that the amplifier gain is 20 dB, its noise bandwidth is 1 MHz, its equivalent noise temperature is 580 K and the noise temperature of its source is 290 K.

The available NPSD from the matched source at temperature T_0 is given in equation (12.11) by:

$$G_{n,s}(f) = kT_0 \quad \text{(W/Hz)}$$

(T_0 is a widely used symbol denoting 290 K.) The available NPSD from the equivalent resistor at temperature T_e (modelling the internally generated noise) is:

$$G_{n,e}(f) = kT_e \quad (\text{W/Hz})$$

The total equivalent noise power at the input is:

$$N_{in} = k(T_0 + T_e)B \quad (\text{W})$$

The total noise power at the amplifier output is therefore:

$$N_{out} = Gk(T_0 + T_e)B \quad (\text{W})$$
$$= 10^{20/10} \times 1.38 \times 10^{-23} \times (290 + 580) \times 1 \times 10^6$$
$$= 1.2 \times 10^{-12} \quad \text{W} = -119.2 \text{ dBW or } -89.2 \text{ dBm}$$

12.3.2 Noise temperature of cascaded subsystems

The total system noise temperature, at the output of a device or subsystem, $T_{syst\,out}$, can be found by dividing equation (12.30) by kB, i.e.:

$$T_{syst\,out} = (T_s + T_e)G \tag{12.31}$$

If several subsystems are cascaded, as shown in Figure 12.13, the noise temperatures at the output of each subsystem are given by:

$$T_{out\,1} = (T_s + T_{e1})G_1 \tag{12.32(a)}$$

$$T_{out\,2} = (T_{out\,1} + T_{e2})G_2 \tag{12.32(b)}$$

$$T_{out\,3} = (T_{out\,2} + T_{e3})G_3 \tag{12.32(c)}$$

The noise temperature at the output of subsystem 3 is therefore:

$$T_{out\,3} = \{[(T_s + T_{e1})G_1 + T_{e2}]G_2 + T_{e3}\}G_3 \tag{12.33}$$
$$= T_s G_1 G_2 G_3 + T_{e1} G_1 G_2 G_3 + T_{e2} G_2 G_3 + T_{e3} G_3$$

This temperature can be referred to the input of subsystem 1 by dividing equation (12.33)

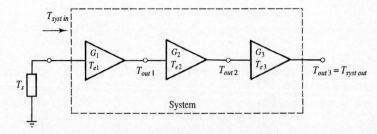

Figure 12.13 *Partitioning of system into cascaded amplifier subsystems.*

by the total gain, $G_1G_2G_3$, i.e.:

$$T_{syst\ in} = T_s + T_{e1} + \frac{T_{e2}}{G_1} + \frac{T_{e3}}{G_1G_2} \tag{12.34}$$

or:

$$T_{syst\ in} = T_s + T_e \tag{12.35}$$

where T_e is the equivalent (input) noise temperature of the system excluding the source and $T_{syst\ in}$ is the overall input noise temperature *including the source*. The total noise temperature output of a set of cascaded subsystems is then simply given by:

$$T_{syst\ out} = T_{syst\ in}G_1G_2G_3 \tag{12.36}$$

It is important to remember that the 'gain' of the individual subsystems can be greater *or less* than 1.0. In the latter case the subsystem is lossy. (A transmission line, for instance, with 10% power loss would have a gain of 0.9 or −0.46 dB.) The equivalent (input) noise temperature, $T_{e,l}$, of a lossy device at a physical temperature T_{ph}, can be found from its 'gain' using:

$$T_{e,l} = \frac{T_{ph}(1 - G_l)}{G_l} \tag{12.37}$$

(The subscript l here reminds us that a lossy device is being considered and that the gain G_l is therefore less than 1.0.) Equation (12.37) is easily verified by considering a transmission line terminated in matched loads as shown in Figure 12.14. Provided the loads, R_s and R_L, and the transmission line are in thermal equilibrium there can be no net flow of noise power across the terminals at either end of the transmission line. From Nyquist's noise formula, equation (12.8), the power available to the transmission line from R_s is:

$$N_s = kT_{ph}B \tag{12.38}$$

If the transmission line has (power) gain, G_l (< 1.0), then the power available to R_L *from* R_s is:

$$N_s' = kT_{ph}BG_l \tag{12.39}$$

Since R_L supplies

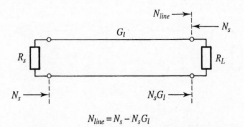

$$N_{line} = N_s - N_s G_l$$

Figure 12.14 *Source, transmission line and load in thermal equilibrium.*

$$N_L = kT_{ph}B \tag{12.40}$$

of power to the transmission line then the transmission line itself must supply the balance of power to R_L, i.e.:

$$N_{line} = kT_{ph}B - kT_{ph}BG_l \tag{12.41}$$

Dividing equation (12.41) by kB gives:

$$T_{line} = T_{ph}(1 - G_l) \tag{12.42}$$

Referring the line temperature to the source terminals then gives equation (12.37).

EXAMPLE 12.3

Figure 12.15 shows a simple superheterodyne receiver consisting of a front end low noise amplifier (LNA), a mixer and two stages of IF amplification. The source temperature of the receiver is 100 K and the characteristics of the individual receiver subsystems are:

Device	Gain (or conversion loss)	T_e
LNA	12 dB	50 K
Mixer	−6 dB	
IF Amp 1	20 dB	1000 K
IF Amp 2	30 dB	1000 K

Calculate the noise power at the output of the second IF amplifier in a 5.0 MHz bandwidth.

Using equation (12.34):

$$T_{syst\ in} = T_s + T_{e1} + \frac{T_{e2}}{G_1} + \frac{T_{e3}}{G_1 G_2} + \frac{T_{e4}}{G_1 G_2 G_3}$$

where:
$G_1 = 12$ dB $= 15.8$, $G_2 = -6$ dB $= 0.25$ $(= G_l)$, $G_3 = 20$ dB $= 100$
$T_s = 100$ K, $T_{e1} = 50$ K, $T_{e2} = T_{ph}(1 - G_l)/G_1 = 290(1 - 0.25)/0.25 = 870$ K, $T_{e3} = 1000$ K, $T_{e4} = 1000$ K, (Notice that the physical temperature of the mixer has been assumed to be 290 K. This is the normal assumption unless information to the contrary is given. Notice also that the gain of the final stage is irrelevant as far as the system equivalent input temperature is concerned.)

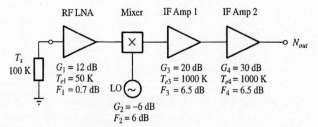

Figure 12.15 *Superheterodyne receiver.*

Therefore:

$$T_{syst\ in} = 100 + 50 + \frac{870}{15.8} + \frac{1000}{15.8 \times 0.25} + \frac{1000}{15.8 \times 0.25 \times 100}$$

$$= 100 + 50 + 55 + 253 + 3 = 461 \ K$$

If the effective noise bandwidth of the system (often determined by the final IF amplifier) is 5.0 MHz then the total noise output power will be:

$$N_{out} = kT_{syst\ in}BG_1G_2G_3G_4$$

$$= 1.38 \times 10^{-23} \times 461 \times 5 \times 10^6 \times 15.8 \times 0.25 \times 100 \times 1000$$

$$= 1.26 \times 10^{-8} \ W = -49.0 \ dBm$$

It is interesting to repeat this calculation with the positions of the LNA and mixer reversed. In this case:

$$T_{syst\ in} = 100 + 870 + \frac{50}{0.25} + \frac{1000}{0.25 \times 15.8} + \frac{1000}{0.25 \times 15.8 \times 100}$$

$$= 100 + 870 + 200 + 253 + 3 = 1426 \ K$$

The equivalent system input noise temperature, and therefore the total system output noise, has been degraded by a factor of 3. The reason for the presence of a low noise amplifier at the front end of a receiver thus becomes obvious.

12.3.3 Noise factor and noise figure

The noise factor, f, of an amplifier is defined as the ratio of SNR at the system input to SNR at the system output *when the input noise corresponds to a temperature of 290 K*, i.e.:

$$f = \frac{(S/N)_i}{(S/N)_o} \Bigg|_{N_i = k\ 290\ B} \tag{12.43}$$

where:

$$(S/N)_i = \frac{\text{signal power at input, } S_i}{k\ 290\ B} \tag{12.44(a)}$$

and:

$$(S/N)_o = \frac{\text{signal power at output, } S_o}{k(290 + T_e)B\ G} \tag{12.44(b)}$$

The specification of the input noise temperature as 290 K allows all devices and systems to be compared fairly, on the basis of their quoted noise factor. f is therefore a figure of merit for comparing the noise performance of different devices and systems. Substituting equations (12.44) into (12.43):

$$f = \frac{S_i\ /\ (k\ 290\ B)}{GS_i\ /\ [G\ k(290 + T_e)B]} = \frac{290 + T_e}{290} \tag{12.45(a)}$$

i.e.:

$$f = 1 + \frac{T_e}{290} \tag{12.45(b)}$$

It is important to remember that, strictly speaking, f is only the ratio of input to output SNR if:

1. The device is operating at its equilibrium temperature in an ambient (290 K) environment. (This is necessary for the quoted T_e to be reliable.)
2. The source temperature at the device input is 290 K.

In practice it is the second condition which is most likely to be unfulfilled. Despite the arbitrary nature of the assumed source temperature in the definition of f, it is still possible to make accurate calculations of overall system noise temperature (and therefore overall SNRs) even when $T_s \neq 290$ K. This is because the noise factor of any device can be converted to an equivalent noise temperature using:

$$T_e = (f - 1)290 \quad (\text{K}) \tag{12.46}$$

and equivalent noise temperature makes no assumption at all about source temperature. If preferred, however, the noise effects due to several subsystems can be cascaded before converting to noise temperatures. Comparing equations (12.34) and (12.35) the equivalent noise temperature of the cascaded subsystems is:

$$T_e = T_{e1} + \frac{T_{e2}}{G_1} + \frac{T_{e3}}{G_1 G_2} \quad (\text{K}) \tag{12.47}$$

This can be rewritten in terms of noise factors as:

$$(f - 1)290 = (f_1 - 1)290 + \frac{(f_2 - 1)290}{G_1} + \frac{(f_3 - 1)290}{G_1 G_2} \tag{12.48}$$

Dividing by 290 and adding 1:

$$f = f_1 + \frac{f_2 - 1}{G_1} + \frac{f_3 - 1}{G_1 G_2} \tag{12.49}$$

This is called the Friis noise formula. The final system noise temperature (referred to the input of device 1) is then calculated from equations (12.46) and (12.35).

Traditionally noise factor is quoted in decibels, i.e.:

$$F = 10 \log_{10} f \quad (\text{dB}) \tag{12.50}$$

and in this form is called the *noise figure*. It is therefore essential to remember to convert F to a ratio before using it in the calculations described above. (Care is required since the terms noise figure and noise factor and the symbols f and F are often used interchangeably in practice.) Table 12.2 gives some noise figures, noise factors and their equivalent noise temperatures.

The noise figure of lossy devices (such as transmission lines or passive mixers) is related to their 'gain', G_l, by:

$$f = 1/G_l \tag{12.51}$$

where f and G_l are expressed as ratios, or alternatively by:

$$F = -G_l \quad (\text{dB}) \qquad (12.52)$$

where F and G_l are expressed in decibels. A transmission line with 10% power loss (i.e. $G_l = 0.9$) therefore has a noise factor given by:

$$f = 1/G_l = 1/0.9 = 1.11 \qquad (12.53)$$

(or 0.46 dB as a noise figure). A mixer with a conversion loss of 6 dB (i.e. a conversion gain of −6 dB) has a noise figure given by:

$$F = -G_l = 6 \text{ dB} \qquad (12.54)$$

(or 4.0 as a noise factor). Strictly speaking a mixer would have slightly greater noise figure than this due to the contribution of non-thermal noise by the diodes in the mixer circuit. For passive mixers such non-thermal effects can usually be neglected. For active mixers this might not be so (although in this case there would probably be a conversion gain rather than loss).

Table 12.2 *Comparison of noise performance measures.*

T_e	f	F	Comments
0 K	1.00	0 dB	Perfect (i.e. noiseless) device
10 K	1.03	0.2 dB	Excellent LNA
100 K	1.34	1.3 dB	Good LNA
290 K	2.00	3.0 dB	Typical LNA
500 K	2.72	4.4 dB	Typical amplifier
1000 K	4.45	6.5 dB	Poor quality amplifier
10,000 K	35.50	15.5 dB	Temperature of a noise source

EXAMPLE 12.4
The output noise of the system shown in Figure 12.15 is now recalculated using noise factors instead of noise temperatures.

The noise factor of the entire system is given in equation (12.49) by:

$$
\begin{aligned}
f &= f_1 + \frac{(f_2 - 1)}{G_1} + \frac{(f_3 - 1)}{G_1 G_2} + \frac{(f_4 - 1)}{G_1 G_2 G_3} \\[2mm]
&= 10^{0.7/10} + \frac{10^{6/10} - 1}{15.8} + \frac{10^{6.5/10} - 1}{15.8 \times 0.25} + \frac{10^{6.5/10} - 1}{15.8 \times 0.25 \times 100} \\[2mm]
&= 1.17 + \frac{2.98}{15.8} + \frac{3.47}{15.8 \times 0.25} + \frac{3.47}{15.8 \times 0.25 \times 100} \\[2mm]
&= 1.17 + 0.19 + 0.88 + 0.01 = 2.25
\end{aligned}
$$

(or $F = 3.5$ dB)

The equivalent system noise temperature at the input to the low noise amplifier (LNA) is:

$$T_e = (f - 1)290$$

$$= (2.25 - 1)290 = 362 \text{ K}$$

and the total noise power at the system output in a bandwidth of 5.0 MHz is therefore:

$$N = k(T_s + T_e)BG_{syst}$$

$$= 1.38 \times 10^{-23} \times (100 + 362) \times 5 \times 10^6 \times 10^{56/10}$$

$$= 1.27 \times 10^{-8} \text{ W} = -49.0 \text{ dBm}$$

12.4 Radio communication link budgets

A communication system link budget refers to the calculation of received signal-to-noise ratio given a specification of transmitted power, transmission medium attenuation and/or gain, and all sources of noise.

The calculation is often set out systematically (in a similar way to a financial budget) accounting explicitly for the various sources of gain, attenuation and noise. For single section line communications the essential elements are transmitted power, cable attenuation, receiver gain and noise figure. The calculation is then simply a matter of applying the Friis formula, or its equivalent, as described in sections 12.3 – 12.3.3. For radio systems the situation is different in that signal energy is lost not only as a result of attenuation (i.e. energy which is dissipated as heat) but also due to its being radiated in directions other than directly towards the receiving antenna. For multi-section communication links the effects of analogue, amplifying, repeaters can be accounted for using their gains and noise figures in the Friis formula, and the effects of digital regenerative repeaters can be accounted for by summing the BERs of each section (providing the BER is small, see section 6.3). Before the details of a radio communications link budget are described some important antenna concepts are reviewed.

12.4.1 Antenna gain, effective area and efficiency

An isotropic antenna (i.e. one which radiates electromagnetic energy equally well in all directions), radiating P_{rad} W of power, supports a power density at a distance R (Figure 12.16) given by:

$$W_{isotrope} = \frac{P_{rad}}{4\pi R^2} \quad (\text{W m}^{-2}) \tag{12.55}$$

The observation point is assumed, here, to be in the far-field of the antenna, i.e. $R \geq 2D^2/\lambda$ where D is the largest dimension (often the diameter) of the antenna. In this region the reactive and radiating antenna near-fields are negligible, and the radiating far-

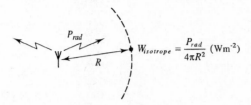

Figure 12.16 *Power density radiated by an isotropic antenna.*

field can be assumed to be a transverse electromagnetic (TEM) wave. The radiating far-field pattern is independent of distance and, in any small region of space, the wavefront can be considered approximately plane. The field strength in the antenna far-field is related to power density by:

$$\frac{E_{RMS}^2}{Z_o} = W \quad (\text{W/m}^2) \tag{12.56}$$

where $Z_o = E/H = 377 \ \Omega$ is the plane wave impedance of free space. Since $1/(4\pi R^2)$ in equation (12.55) represents a purely geometrical dilution of power density as the spherical wave expands, this factor is often referred to as the *spreading loss*. The *radiation intensity*, I, in an isotropic antenna's far-field is given by:

$$I_{isotrope} = \frac{P_{rad}}{4\pi} \quad (\text{W steradian}^{-1} \text{ or W rad}^{-2}) \tag{12.57}$$

The mutually orthogonal requirement on $\mathbf{E}, \mathbf{H}$ and $\mathbf{k}$ in the far-field (where $\mathbf{E}$ is electric field strength, $\mathbf{H}$ is magnetic field strength and $\mathbf{k}$ is vector wave number pointing in the direction of propagation) excludes the possibility of realising isotropic radiators. Thus all practical antennas radiate preferentially in some directions over others. If radiation intensity $I(\theta, \phi)$ is plotted against spherical coordinates θ and ϕ the resulting surface (i.e. radiation pattern) will be spherical for a (hypothetical) isotrope and non-spherical for any realisable antenna, Figure 12.17. The gain, $G(\theta, \phi)$, of a transmitting antenna can be defined as the ratio of radiation intensity in the direction θ, ϕ to the radiation intensity which would be observed if the antenna were replaced with a *lossless* isotrope:

$$G(\theta, \phi) = \frac{I(\theta, \phi)}{I_{lossless\ isotrope}}$$

$$= \frac{I(\theta, \phi)}{P_T/(4\pi)} \tag{12.58(a)}$$

where P_T is the transmitter output power and the antenna is assumed to be well matched to its transmission line feed. A related quantity, antenna directivity, $D(\theta, \phi)$, can be defined by:

$$D(\theta, \phi) = \frac{I(\theta, \phi)}{I_{isotrope}}$$

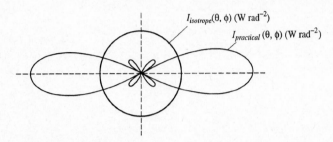

Figure 12.17 *Two-dimensional polar plots of antenna radiation intensity for isotropic and practical antenna.*

$$= \frac{I(\theta, \phi)}{P_{rad}/(4\pi)} \tag{12.58(b)}$$

where the hypothetical isotropic antenna is now assumed to have a loss equal to that of the actual antenna. Directivity and gain therefore have identical shape and are related by:

$$G(\theta, \phi) = \eta_\Omega D(\theta, \phi) \tag{12.58(c)}$$

where $\eta_\Omega = P_{rad}/P_T$ is called the ohmic efficiency of the antenna. The ohmic losses relate physically to the electromagnetic energy dissipated by induced conduction currents flowing in the metallic conductors of the antenna and induced displacement currents flowing in the dielectric of the antenna. (The former usually dominate the latter and are often referred to as I^2R losses.) As an alternative both antenna gain and directivity can be defined as ratios of far-field power densities, i.e.:

$$G(\theta, \phi) = \frac{W(\theta, \phi, R)}{W_{lossless\ isotrope}(R)} \tag{12.59(a)}$$

and

$$D(\theta, \phi) = \frac{W(\theta, \phi, R)}{W_{isotrope}(R)} \tag{12.59(b)}$$

Using equations (12.55), (12.58), (12.59) and $P_{rad} = \eta_\Omega P_T$, the power density at a distance R from a transmitting antenna can be found from either of the following formulas:

$$W(\theta, \phi, R) = \frac{P_{rad}}{4\pi R^2} D(\theta, \phi) \quad (\text{W/m}^2) \tag{12.60(a)}$$

$$W(\theta, \phi, R) = \frac{P_T}{4\pi R^2} G(\theta, \phi) \quad (\text{W/m}^2) \tag{12.60(b)}$$

Figure 12.18 shows a typical, Cartesian coordinate, directivity pattern for a microwave reflector antenna. The effective area, a_e, of a receiving antenna, is defined as the ratio of carrier power C received at the antenna terminals, when the antenna is illuminated by a plane wave from the direction θ, ϕ, to the power density in the plane wave (Figure 12.19),

i.e.:

$$a_e(\theta, \phi) = \frac{C(\theta, \phi)}{W} \quad (m^2) \tag{12.61}$$

(The plane wave is assumed to be polarisation matched to the antenna.) Unsurprisingly, there is an intimate connection between an antenna's gain and its effective area which can be expressed by:

$$G(\theta, \phi) = \frac{4\pi a_e(\theta, \phi)}{\lambda^2} \tag{12.62}$$

Equation (12.62) is a widely used expression of antenna reciprocity which can be derived from the Lorentz reciprocity theorem [Collin]. Usually interest is focused on the gain and effective area corresponding to the direction of maximum radiation intensity. This direction is called antenna boresight and is always implied if θ and ϕ are not specified.

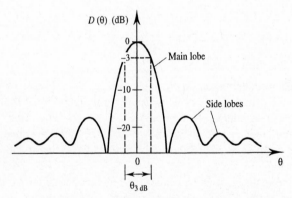

Figure 12.18 *Two-dimensional Cartesian plot of directivity (in dB) for microwave antenna (3 dB beamwidth of axisymmetric reflectors may be estimated using $\theta_{3dB} = 1.2 \, \lambda/D$ rad).*

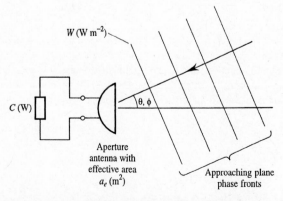

Figure 12.19 *Aperture antenna receiving plane wave from θ, ϕ direction.*

For aperture antennas, such as paraboloidal reflectors, it is intuitively reasonable to associate a_e, at least approximately, with the physical antenna aperture. Unfortunately, they cannot be assumed to be identical since not all parts of the antenna aperture are fully utilised. There are several reasons for this including non-uniform illumination of the aperture by the feed, non-zero illumination of the region outside the aperture by the feed, aperture blockage caused by the presence of the feed and feed struts, and random errors in the definition of the antenna's reflecting surface. These and other effects can be accounted for by reducing the physical area, A_{ph}, of an aperture antenna by an aperture efficiency factor, η_{ap}. This gives the antenna's *effective* aperture, A_e, i.e.:

$$A_e = \eta_{ap} A_{ph} \quad (\text{m}^2) \tag{12.63(a)}$$

These aperture losses are in addition to the ohmic losses referred to earlier. Ohmic loss is accounted for by the ohmic efficiency, which when applied to the effective aperture gives the antenna's effective area as defined in equation (12.61), i.e.:

$$a_e = \eta_\Omega A_e \quad (\text{m}^2) \tag{12.63(b)}$$

The effective area of the antenna is therefore related to its physical area by:

$$a_e = \eta_{ap} \eta_\Omega A_{ph} \quad (\text{m}^2) \tag{12.64}$$

η_{ap} and η_Ω can be further broken down into more specific sources of loss. Any more detailed accounting for losses, however, is usually important only for antenna designers and is of rather academic interest to communications systems engineers. (Care should be taken in interpreting antenna efficiencies, however, since the terms effective aperture and effective area are often used interchangeably, as are the terms aperture efficiency and antenna efficiency.)

It is important to realise that antenna aperture efficiency is certain to be less than unity only for antennas with a well defined aperture of significant size (in terms of wavelengths). Wire antennas, such as dipoles, for example, have aperture efficiencies greater than unity if A_{ph} is taken to be the area presented by the wire to the incident wavefront.

EXAMPLE 12.5

The power density, radiation intensity and electric field strength are now calculated at a distance of 20 km from a microwave antenna having a directivity of 42.0 dB, an ohmic efficiency of 95% and a well matched 4 GHz transmitter with 25 dBm of output power.

The gain is given by:

$$G = \eta_\Omega D$$

$$= 10 \log_{10} \eta_\Omega + D_{dB} \quad (\text{dB})$$

$$= 10 \log_{10} 0.95 + 42.0 = -0.2 + 42.0 = 41.8 \text{ dB}$$

and, from equation (12.60(b)), the received power density is given by:

$$W = \frac{P_T}{4\pi R^2} G_T$$

$$= P_T + G_T - 10 \log_{10} (4\pi R^2) \quad \text{dBm/m}^2$$

$$= 25 + 41.8 - 10 \log_{10} (4\pi \times 20,000^2)$$

$$= -30.2 \quad \text{dBm/m}^2 = -60.2 \quad \text{dBW/m}^2$$

$$= 9.52 \times 10^{-7} \quad \text{W/m}^2$$

The radiation intensity is given by equation (12.58(b)) as:

$$I = \frac{P_{rad}}{4\pi} D = \frac{\eta_\Omega P_T}{4\pi} D$$

$$= \frac{0.95 \times 10^{\frac{25}{10}}}{4\pi} \times 10^{\frac{42}{10}} \quad \text{mW/rad}^2$$

$$= 3.789 \times 10^5 \quad \text{mW/rad}^2$$

and from equation (12.56):

$$E_{RMS} = \sqrt{W Z_o} = \sqrt{9.52 \times 10^{-7} \times 377} \quad \text{V/m} = 18.9 \quad \text{mV/m}$$

If an identical receiving antenna is located 20 km from the first, the available carrier power, C, at its terminals could be calculated as follows:

$$\lambda = \frac{c}{f} = \frac{3 \times 10^8}{4 \times 10^9} = 0.075 \quad \text{m}$$

The antenna effective area, equation (12.62), is:

$$a_e = G \frac{\lambda^2}{4\pi} = 10^{\frac{41.8}{10}} \left[\frac{(0.075)^2}{4\pi} \right]$$

$$= 6.775 \quad \text{m}^2$$

and from equation (12.61):

$$C = W a_e$$

$$= 9.52 \times 10^{-7} \times 6.775 = 6.45 \times 10^{-6} \quad \text{W} = -21.9 \quad \text{dBm}$$

12.4.2 Free space and plane earth signal budgets

Consider the free space radio communication link shown in Figure 12.20. The power density at the receiver radiated by a lossless isotrope would be:

$$W_{lossless\ isotrope} = \frac{P_T}{4\pi R^2} \quad (\text{W m}^{-2}) \tag{12.65}$$

For a practical transmitting antenna with gain G_T the power density at the receiver is actually:

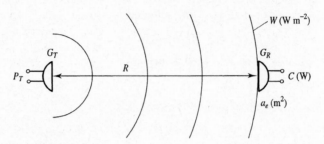

Figure 12.20 *Free space propagation.*

$$W = \frac{P_T}{4\pi R^2} G_T \quad (\text{W m}^{-2})$$

(12.66)

The carrier power available at the receive antenna terminals is given by:

$$C = W a_e \quad (\text{W})$$

(12.67)

which, on substituting equation (12.66) for W, gives:

$$C = \frac{P_T}{4\pi R^2} G_T a_e \quad (\text{W})$$

(12.68)

Using antenna reciprocity (equation (12.62)):

$$C = \frac{P_T}{4\pi R^2} G_T G_R \frac{\lambda^2}{4\pi} \quad (\text{W})$$

(12.69)

where G_R is the gain of the receiving antenna. (Note that $\lambda^2/4\pi$ can be identified as the effective area, a_e, of a lossless isotrope.) Equation (12.69) can be rewritten as:

$$C = P_T G_T \left(\frac{\lambda}{4\pi R} \right)^2 G_R \quad (\text{W})$$

(12.70)

This is the basic free space transmission loss formula for radio systems. The quantity $P_T G_T$ is called the effective isotropic radiated power (EIRP) and the quantity $[\lambda/(4\pi R)]^2$ is called the free space path loss (FSPL). Notice that FSPL is a function of wavelength. This is because it contains a factor to convert the receiving antenna effective area to gain in addition to the (geometrical) spreading loss.

Equation (12.70) is traditionally expressed using decibel quantities:

$$C = \text{EIRP} - \text{FSPL} + G_R \quad (\text{dBW})$$

(12.71(a))

where:

$$\text{EIRP} = 10 \log_{10} P_T + 10 \log_{10} G_T \quad (\text{dBW})$$

(12.71(b))

and:

$$\text{FSPL} = 20 \log_{10} \left(\frac{4\pi R}{\lambda} \right) \quad (\text{dB})$$

(12.71(c))

(The same symbol is used here for powers and gains whether they are measured in natural units or decibels. The context, however, leaves no doubt as to which is intended.)

Figure 12.21 shows a radio link operating above a plane earth. In this case there are two possible propagation paths between the transmitter and receiver, one direct and the other reflected via the ground. Assuming a complex (voltage) reflection coefficient at the ground, $\rho e^{j\phi}$, then the field strength, E, at the receiver will be changed by a (complex) factor, F, given by:

$$F = 1 + \rho e^{j\phi} e^{-j2\pi(d_2 - d_1)/\lambda} \tag{12.72}$$

where d_1 and d_2 are the lengths of the direct and reflected paths respectively and $2\pi(d_2 - d_1)/\lambda = \theta$ is the resulting excess phase shift of the reflected, over the direct, path. Figure 12.22 illustrates the (normalised) phasor addition of the direct and reflected fields. Assuming perfect reflection at the ground (i.e. $\rho e^{j\phi} = -1$) and using $e^{-j\theta} = \cos\theta - j\sin\theta$ the magnitude of the field strength gain factor can be written, for practical multipath geometries (see Problem 12.8), as:

$$|F| = 2\sin\left(\frac{2\pi h_T h_R}{\lambda R}\right) \tag{12.73}$$

where h_T and h_R are transmit and receive antenna heights respectively and R is the horizontal distance between transmitter and receiver, Figure 12.21. The power density at the receiving antenna aperture, and therefore the received power, is increased over that for free space by a factor $|F|^2$. Thus, using natural (not decibel) quantities:

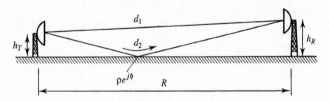

Figure 12.21 *Propagation over a plane earth.*

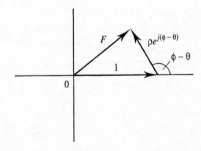

Figure 12.22 *Phasor addition of (normalised) fields from direct and reflected paths in Figure 12.21.*

$$C = \text{EIRP} \times \text{FSPL} \times |F|^2 \times G_R$$

$$= P_T G_T \left(\frac{\lambda}{4\pi R} \right)^2 4 \sin^2 \left(\frac{2\pi h_T h_R}{\lambda R} \right) G_R \quad \text{(W)} \tag{12.74}$$

A typical interference pattern resulting from the two paths is illustrated as a function of horizontal range and height in Figure 12.23(a) and Figure 12.23(b) respectively. Equation (12.74) can be expressed in decibels as:

$$C = P_T + G_T - \text{FSPL} + G_R + 6.0 + 20 \log_{10} \left| \sin \left(\frac{2\pi h_T h_R}{\lambda R} \right) \right| \quad \text{(dBW)} \tag{12.75}$$

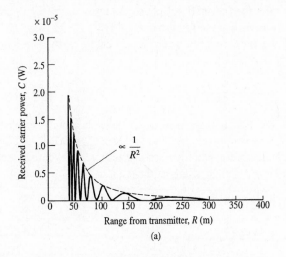

(a)

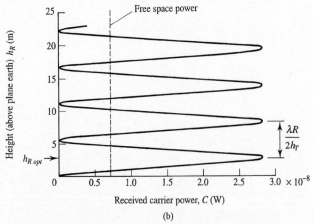

(b)

Figure 12.23 *Variation of received power with: (a) distance at a height of 2 m; (b) height at a range of 1000 m from transmitter. (Parameters as used in equation (12.74):* $P_T G_T = 10$ *W,* $\lambda = 0.333$ *m,* $h_T = 30$ *m,* $G_R = 0$ *dB.)*

Figure 12.24 shows the variation of received power on a dB scale with respect to transmitter–receiver distance. (A plot of power density in dBW m^{-2} or electric field in dBμV m^{-1} has identical shape.) Notice that the peaks, corresponding to points of constructive interference, are 6 dB above the free space value and the troughs, corresponding to points of destructive interference, are, in principle, ∞ dB below the free space value. (In practice the minima do not represent zero power density or field strength because $\rho < 1.0$.) The furthest point of constructive interference occurs for the smallest argument of sin $[(2\pi h_T h_R)/(\lambda R)]$ which gives 1.0, i.e. when:

$$\frac{2\pi h_T h_R}{\lambda R} = \frac{\pi}{2} \tag{12.76(a)}$$

The distance R_{max}, to which this corresponds, is:

$$R_{max} = \frac{4h_T h_R}{\lambda} \quad (m) \tag{12.76(b)}$$

For $R > R_{max}$ the total field decays monotonically. When $2\pi h_T h_R/\lambda R$ is small the approximation sin $x \approx x$ can be used and the received power becomes:

$$C = P_T G_T \left(\frac{h_T h_R}{R^2} \right)^2 G_R \quad (W) \tag{12.77}$$

or, in decibels:

$$C = P_T + G_T - \text{PEPL} + G_R \quad (dBW) \tag{12.78(a)}$$

where PEPL, the plane earth path loss, is given by:

$$\text{PEPL} = 20 \log_{10} \left(\frac{R^2}{h_T h_R} \right) \quad (dB) \tag{12.78(b)}$$

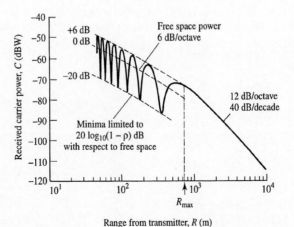

Figure 12.24 *Variation of received power with range for plane earth propagation. (Parameters used are the same as for Figure 12.23.)*

Equation (12.77) shows that over a plane conducting earth the received power, and therefore the power density at the receiving antenna aperture, decays as $1/R^4$. The corresponding field strength therefore decays as $1/R^2$.

The optimum receive antenna height, $h_{R\,opt}$, for fixed λ, R and h_T is found by differentiating $|F|$ in equation (12.73) with respect to h_R and setting the result to zero, i.e.:

$$\frac{d|F|}{dh_R} = 2\cos\left(\frac{2\pi h_T h_R}{\lambda R}\right)\frac{2\pi h_T}{\lambda R} = 0 \tag{12.79}$$

This requires that:

$$\frac{2\pi h_T h_{R\,opt}}{\lambda R} = \frac{n\pi}{2}, \quad n = 1, 3, 5, \cdots \tag{12.80(a)}$$

i.e.:

$$h_{R\,opt} = n\frac{\lambda R}{4h_T} \quad \text{(m)} \tag{12.80(b)}$$

where n is an odd integer. (A second derivative ($d^2|F|/dh_R^2$) confirms that this is a condition for maximum, rather than minimum, received power.) Clearly antenna heights would not normally be chosen to be higher than necessary and so, in the absence of other considerations, n would be chosen to be equal to 1. (It would actually be useful to optimise an antenna height only if the ground reflection was reasonably stable in both magnitude *and phase* – a condition not often encountered.)

EXAMPLE 12.6
A 6 GHz, 40 km, LOS link uses 2.0 m axisymmetric paraboloidal reflectors for both transmitting and receiving antennas. The ohmic and aperture efficiencies of the antennas are 99% and 70% respectively and both antennas are mounted at a height of 25 m. The transmitter power is 0 dBW. Find the received power for both free space and plane earth conditions. Which condition is most likely to prevail if the path is over water?

$$\lambda = \frac{c}{f} = \frac{3 \times 10^8}{6 \times 10^9} = 0.05 \text{ m}$$

From equation (12.64) we have:

$$a_e = \eta_\Omega \eta_{ap} A_{ph}$$

$$= 0.99 \times 0.70 \times \pi \times 1.0^2 = 2.177 \text{ m}^2$$

and using equation (12.62):

$$G = \frac{4\pi a_e}{\lambda^2} = \frac{4\pi \times 2.177}{0.05^2} = 1.094 \times 10^4 = 40.4 \text{ dB}$$

$$\text{FSPL} = 20\log_{10}\left(\frac{4\pi R}{\lambda}\right)$$

$$= 20 \log_{10} \left(\frac{4\pi 40 \times 10^3}{0.05} \right) = 140.0 \text{ dB}$$

For free space conditions equation (12.71(a)) gives:

$$C = P_T + G_T - \text{FSPL} + G_R$$

$$= 0 + 40.4 - 140.0 + 40.4 = -59.2 \text{ dBW}$$

For plane earth conditions (assuming perfect reflection) equation (12.76(b)) gives:

$$R_{\max} = \frac{4h_T h_R}{\lambda}$$

$$= \frac{4 \times 25^2}{0.05} = 50,000 \text{ m}$$

The receive antenna is thus closer to the transmit antenna than the furthest point of constructive interference and equation (12.75) rather than (12.78) must therefore be used.

$$C = C \mid_{\textit{free space}} + 6.0 + 20 \log_{10} \left| \sin \left(\frac{2\pi h_T h_R}{\lambda R} \right) \right|$$

$$= -59.2 + 6.0 + 20 \log_{10} \left| \sin \left(\frac{2\pi \times 25^2}{0.05 \times 40 \times 10^3} \right) \right|$$

$$= -59.2 + 6.0 - 0.7 = -53.9 \text{ dBW}$$

Note that this is only 0.7 dB below the maximum possible received power so the antenna heights must be close to optimum for plane earth propagation. Equation (12.80) shows that optimum (equal) antenna heights would be given by:

$$h_R h_T = \frac{\lambda R}{4} = \frac{0.05 \times 40 \times 10^3}{4} = 500 \text{ m}^2$$

i.e.:

$$h_R = h_T = \sqrt{500} = 22.36 \text{ m}$$

As has been said in the text, however, it is unlikely that such precise positioning of antennas would be of benefit in practice. To assess whether free space or plane earth propagation is likely to occur, a first order calculation based on antenna beamwidth can be carried out as follows:

$$\text{Antenna beamwidth} \approx 1.2 \frac{\lambda}{D} \text{ rad}$$

$$= 1.2 \left(\frac{0.05}{2.0} \right) = 0.03 \text{ rad } (= 1.7 \text{ degrees})$$

Under normal atmospheric conditions for a horizontal path over water, with equal height transmit and receive antennas, the point of specular reflection will be half way along the link. The vertical width of the antenna beam (between −3 dB points) is given by:

$$\text{Vertical width} \approx \text{beamwidth} \times \tfrac{1}{2} \text{ path length}$$

$$= 0.03 \times \tfrac{1}{2} \times 40 \times 10^3 = 600 \text{ m}$$

Since half this is much greater than the path clearance and since water is a good reflector at microwave frequencies then specular reflection (for a calm water surface) is likely to be strong and a plane earth calculation is appropriate.

12.4.3 Antenna temperature and radio noise budgets

The overall noise power in a radio communications receiver depends not only on internally generated receiver noise but also on the electromagnetic noise collected by the receiver's antenna. Just as the equivalent thermal noise of a circuit or receiver subsystem can be represented by a noise temperature, so too can the noise received by an antenna. The equivalent noise temperature of a receiving antenna, T_{ant}, is defined by:

$$T_{ant} = \frac{\text{available NPSD at antenna terminals}}{\text{Boltzmann's constant}} \tag{12.81}$$

$$= \frac{G_N(f)}{k} \quad \text{(K)}$$

where $G_N(f)$ is assumed to be white and the noise power spectral density is one sided. Antenna noise originates from several different sources. Below about 30 MHz it is dominated by the broadband radiation produced in lightning discharges associated with thunderstorms. This radiation is trapped by the ionosphere and so propagates world-wide. Such noise is sometimes referred to as atmospherics. The ionosphere is essentially transparent above about 30 MHz and between this frequency and 1 GHz the dominant noise is galactic. This has a steeply falling spectral density with increasing frequency (the slope is about −25 dBK/decade). It arises principally due to synchrotron radiation produced by fast electrons moving through the galactic magnetic field. Because the galaxy is very oblate in shape, and also because the earth is not located at its centre, galactic noise is markedly anisotropic and is much greater when the receiving antenna is pointed towards the galactic centre than when it is pointed to the galactic pole.

Above 1 GHz galactic noise is relatively weak. This leaves atmospheric thermal radiation and ground noise as the dominant noise processes. Atmospheric and ground noise is approximately flat with frequency up to about 10 GHz, its spectral density depending sensitively on antenna elevation angle. As elevation increases from 0° to 90° the thickness of atmosphere through which the antenna beam passes decreases as does the influence of the ground, both effects leading to a decrease in received thermal noise. In this frequency range a zenith-pointed antenna during clear sky conditions may have a noise temperature close to the cosmic background temperature of 3 K. Above 10 GHz resonance effects (of water vapour molecules at 22 GHz and oxygen molecules at 60 GHz) lead to increasing atmospheric attenuation and, therefore, thermal noise emission. The typical 'clear sky' noise temperature, as would be measured by a lossless narrow-beam antenna, is illustrated over a band of frequencies from HF to SHF in Figure 12.25.

In addition to noise received as electromagnetic radiation by the antenna, thermal noise will also be generated in the antenna itself. A simple equivalent circuit of an antenna is shown in Figure 12.26. The (total) antenna noise temperature, T_{ant}, is the sum

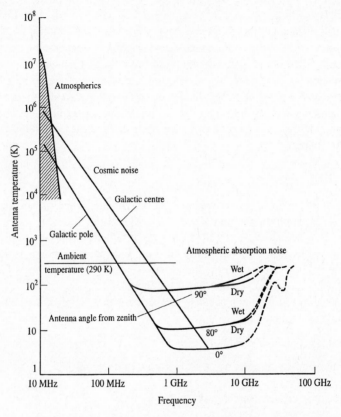

Figure 12.25 *Antenna sky noise temperature as a function of frequency and antenna elevation angle (source: Kraus, 1966, reproduced with his permission).*

of its aperture temperature, T_A (originating from external sources and reduced by ohmic losses) and its equivalent internal, thermal, temperature, i.e.:

$$T_{ant} = T_A \eta_\Omega + T_{ph}(1 - \eta_\Omega) \quad \text{(K)} \tag{12.82}$$

where η_Ω is the antenna ohmic efficiency and T_{ph} is the physical temperature of the antenna. For noise purposes, then, the radiation resistance, R_r, in Figure 12.26 can be assumed to be at the equivalent antenna aperture temperature and the ohmic resistance, R_Ω, can be assumed to be at the antenna physical temperature. The ohmic efficiency of the antenna is related to R_r and R_Ω by:

$$\eta_\Omega = \frac{R_r}{R_r + R_\Omega} \tag{12.83}$$

The calculation of T_A can be complicated but may be estimated using:

$$T_A = \frac{1}{4\pi} \int\int_{4\pi} D(\theta, \phi)\varepsilon(\theta, \phi)T_{ph}(\theta, \phi) \; d\Omega \quad \text{(K)} \tag{12.84}$$

where $\varepsilon(\theta, \phi)$ and $T_{ph}(\theta, \phi)$ are the emissivity and physical temperature, respectively, of the material lying in the direction θ, ϕ, and $d\Omega$ is an element of solid angle. (The emissivity of a surface is related to its voltage reflection coefficient, ρ, by $\varepsilon = 1 - \rho^2$ and the product $\varepsilon(\theta, \phi)T_{ph}(\theta, \phi)$ is sometimes called brightness temperature, $T_B(\theta, \phi)$.)

The quantity $D(\theta, \phi)\varepsilon(\theta, \phi)d\Omega$ in equation (12.84) can be interpreted as the fraction of power radiated by the antenna which is *absorbed* by the material lying in the direction between θ, ϕ and $\theta + d\theta, \phi + d\phi$. If the brightness temperature of the environment around the antenna changes discretely (assuming, for example, different, but individually uniform, temperatures for the ground, clear sky and sun) then equation (12.84) can be written in the simpler form:

$$T_A = \sum_i \alpha_i T_{ph,i} \tag{12.85}$$

where α_i is the fraction of power radiated by the antenna which is absorbed by the *i*th body and $T_{ph,i}$ is the physical temperature of the *i*th body.

For more precise calculations still, scattering of radiation from one source by another (for example scattering of ground noise into the antenna by the atmosphere) must be accounted for. This level of detail, however, is usually the concern of radio remote sensing engineers rather than communications engineers.

Having established an antenna noise temperature the overall noise temperature of a radio receiving system is:

$$T_{syst\ in} = T_{ant} + T_e \tag{12.86}$$

where T_e is the equivalent input noise temperature of the receiver.

12.4.4 Receiver equivalent input CNR

The equivalent input CNR of a radio receiver is given by combining equations (12.70) or (12.74) and (12.86) in the receiver bandwidth, i.e.:

$$\frac{C}{N} = \frac{C}{kT_{syst\ in}B} \tag{12.87}$$

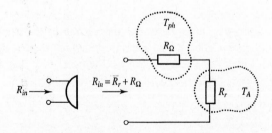

Figure 12.26 *Equivalent circuit of an antenna for noise calculations (R_r is radiation resistance, R_Ω is ohmic resistance and R_{in} is input resistance).*

The baseband BER and/or SNR of the demodulated signal are then found by applying any CNR detection gain provided by the demodulator and/or decoder, and finding the probability of bit error for the particular digital modulation scheme being used (including the mitigating effects of error control coding). If required the PCM decoded SNR of a baseband analogue signal can then be found by applying an equation such as (5.29).

It should be remembered that although the emphasis here has been on naturally occurring noise some systems are interference limited. Such interference may be random, quasi-periodic, intelligible (in which case it is usually called crosstalk) or a combination of these. Cellular radio is an example of a system which usually operates in an interference limited environment.

EXAMPLE 12.7

A 10 GHz terrestrial line of sight link has good clearance over rough terrain such that free space propagation can be assumed. The free space signal power at the receiving antenna terminals is −40. 0 dBm. The overall noise figure of the receiver is 5.0 dB and the noise bandwidth is 20 MHz. Estimate the actual and effective clear sky CNRs at the antenna terminals assuming that the antenna has an ohmic efficiency of 95% and is at a physical temperature of 280 K. Also make a first order estimate of the effective CNR during a rain fade of 2 dB, assuming the rain is localised and occurs close to the receiving antenna.

From Figure 12.25 the clear sky aperture temperature, T_A, at 10 GHz for a horizontal link is about 100 K. The antenna temperature is given in equation (12.82) by:

$$T_{ant} = T_A \eta_\Omega + T_{ph}(1 - \eta_\Omega)$$

$$= 100 \times 0.95 + 280\,(1 - 0.95) = 95 + 14 = 109 \ \text{K}$$

(Notice that the contribution from the physical temperature of the antenna is, in this case, probably within the uncertainty of the estimate of aperture temperature.) Now from equation (12.8):

$$N = kTB = 1.38 \times 10^{-23} \times 109 \times 20 \times 10^6 = 3.01 \times 10^{-14} \ \text{W} = -135.2 \ \text{dBW}$$

The actual clear sky CNR at the antenna terminals with a received power of −70 dBW is given by:

$$\frac{C}{N} = -70.0 - (-135.2) = 65.2 \ \text{dB}$$

The equivalent noise temperature of the receiver is given by equation (12.46) as:

$$T_e = (f - 1)290 = (10^{\frac{5}{10}} - 1)290 = 627 \ \text{K}$$

The system noise temperature (referred to the antenna output) is:

$$T_{syst\ in} = T_{ant} + T_e = 109 + 627 = 736 \ \text{K}$$

The effective system noise power (referred to the antenna output) is:

$$N = kT_{syst\ in}B = 1.38 \times 10^{-23} \times 736 \times 20 \times 10^6$$

$$= 2.03 \times 10^{-13} \ \text{W} = -126.9 \ \text{dBW}$$

The effective carrier to noise ratio is:

$$\left. \frac{C}{N} \right|_{eff} = C - N \quad (dB)$$

$$= -70.0 - (-126.9) = 56.9 \ dB$$

During a 2 dB rain fade carrier power will be reduced by 2 dB to −42 dBm and noise will be increased. A first order estimate of noise power during a fade can be found as follows.

Assuming the physical temperature of the rain (T_{rain}) is the same as that of the antenna then the aperture temperature may be recalculated using the transmission line of equation (12.42) as:

$$T_A = T_{sky} \times \text{fade} + T_{rain}(1 - \text{fade})$$

$$= 100 \times 10^{\frac{-2}{10}} + 280 \ (1 - 10^{\frac{-2}{10}})$$

$$= 63 + 103 = 166 \ K$$

$$T_{ant} = T_A \ \eta_\Omega + T_{ph}(1 - \eta_\Omega)$$

$$= 166 \times 0.95 + 280(1 - 0.95)$$

$$= 158 + 14 = 172 \ K$$

$$N = k(T_{ant} + T_e)B$$

$$= 1.38 \times 10^{-23} \ (172 + 627) \times 20 \times 10^6$$

$$= 2.21 \times 10^{-13} \ W = -126.6 \ dBW$$

The effective CNR during the 2 dB fade is therefore:

$$\left. \frac{C}{N} \right|_{eff} = C - N$$

$$= -72.0 - (-126.6) = 54.6 \ dB$$

The example shown above is intended to illustrate the concepts discussed in the text and probably contains spurious precision. The uncertainties associated with real systems mean that in practice a first order estimate of CNR would probably be based on a worst case antenna noise temperature of, say, 290 K. (The difference between this assumption and the above effective CNR calculation is only $10 \log_{10} ((290 + 627)/(172 + 627)) = 0.6$ dB.)

12.4.5 Multipath fading and diversity reception

Multipath fading occurs to varying extents in many different radio applications [Rummler, 1986]. It is caused whenever radio energy arrives at the receiver by more than one path. Figure 12.27 illustrates how multipath propagation may occur on a point-to-point line-of-sight microwave link. In this case multiple paths may occur due to ground reflections, reflections from stable tropospheric layers (with different refractive index) and refraction by tropospheric layers with extreme refractive index gradients. Other systems suffer multipath propagation due to the presence of scattering obstacles. This is the case for urban cellular radio systems for example.

There are two principal effects of multipath propagation on systems, their relative severity depending essentially on the relative bandwidth of the resulting channel compared with that of the signal being transmitted. For fixed point systems such as the microwave radio relay network the fading process is governed by changes in atmospheric conditions. Often, but not always, the spread of path delays is sufficiently short for the frequency response of the channel to be essentially constant over its operating bandwidth. In this case fading is said to be flat since all frequency components of a signal are subjected to the same fade at any given instant. When several or more propagation paths exist the fading of signal *amplitude* obeys Rayleigh statistics (due to the central limit theorem). If the spread of path delays is longer then the frequency response of the channel may change rapidly on a frequency scale comparable to signal bandwidth, Figure 12.28. In this case the fading is said to be frequency selective and the received signal is subject to severe amplitude and phase distortion. Adaptive equalisers may then be required to flatten and linearise the overall channel characteristics. The effects of flat fading can be combatted by increasing transmitter power whilst the effects of frequency selective fading cannot. For microwave links which are subject to flat fading a fade margin is usually designed into the link budget to offset the expected multipath (and rain induced) fades. The magnitude of this margin depends, of course, on the required availability of the link.

To reduce the necessary fade margin to acceptable levels diversity reception is sometimes employed. Figure 12.29(a) to (c) illustrates the principles of three types of diversity system, namely space (also called height), frequency, and angle diversity. In all cases the essential assumption is that it is unlikely that both main and diversity channel will suffer severe fades at the same instant. Selecting the channel with largest CNR, or combining channels with weightings in proportion to their CNRs, will clearly result in improved overall CNR.

12.5 Summary

Noise is present in all communications systems and, if interpreted to include interference, is always a limiting factor on their performance. Both thermal and shot noise have white

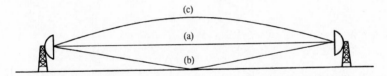

Figure 12.27 *Multipath in line of sight terrestial link due to: (a) direct path plus (b) ground reflection and/or (c) reflection from (or refraction through) a tropospheric layer.*

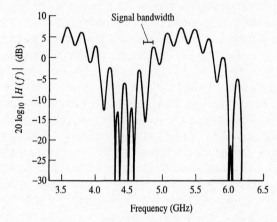

Figure 12.28 *Amplitude response of a frequency selective channel for 3-ray multipath propagation with ray amplitudes and delays of: 1.0, 0 ns; 0.9, 0.56 ns; 0.1, 4.7 ns.*

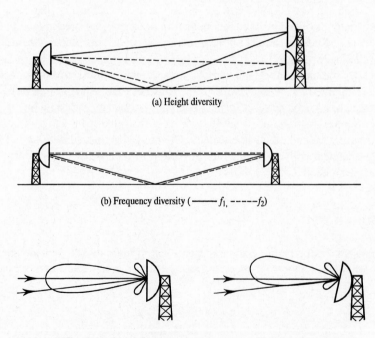

(a) Height diversity

(b) Frequency diversity (—— f_1, -----f_2)

(c) Angle diversity (main and diversity, co-located, antennas unlikely to both respond well to multipath ray(s))

Figure 12.29 *Three types of diversity arrangements to combat multipath fading.*

power spectral densities and are therefore easy to quantify in a specified measurement bandwidth. In many systems calculations, non-white noise is often neglected and white noise is represented by an equivalent thermal spectral density or equivalent thermal noise temperature. The noise factor of a system is the ratio of its input SNR to output SNR when the system's source temperature is 290 K. The total noise available from several cascaded subsystems can be found using the noise factors, and gains, of each subsystem in the Friis formula.

Signal budgets are used to determine the carrier power present at the input to a communications receiver. In a radio system this involves antenna gains and path loss in addition to transmitter power. Two formulas for path loss are commonly used, namely the free space and plane earth formulas. Which is used for a particular application depends on frequency, antenna beamwidths, path clearance and the surface reflection coefficient of the earth beneath the path. Free space propagation results in a power density which decays as $1/R^2$ and a field strength which decays as $1/R$. Plane earth propagation results in a power density which decays as $1/R^4$ and a field strength which decays as $1/R^2$. For paths with particularly stable ground reflections it may be possible to mount antennas at an optimum height in order to take advantage of the 6 dB signal enhancement available at points of constructive interference.

Noise budgets in radio systems must account not only for the noise generated within the receiver itself but also for noise received by the antenna. The latter (assumed white) can be represented by an antenna temperature which is added to the equivalent noise temperature of the receiver to get the effective system input temperature. The effective system input noise power is then found using the Nyquist noise formula.

The link budget includes both signal and noise budgets and establishes the received carrier to noise ratio.

12.6 Problems

12.1. A preamplifier with 20 dB gain has an effective noise temperature of 200 K. What is its noise figure in dB? If this preamplifier is followed by a power amplifier of 30 dB gain with a noise figure F of 4 dB what is the degradation in dB for the overall noise figure of the two amplifier cascade compared to the preamplifier alone? [2.28 dB, 0.04 dB]

12.2. A microwave radio receiver has the following specification:

Antenna noise temperature	= 80K
Antenna feeder physical temperature	= 290 K, Loss = 2.0 dB
Low noise amplifier noise figure	= 2.2 dB, Gain = 12.0 dB
Frequency down converter conversion loss	= 6.0 dB
First IF amplifier noise temperature	= 870 K, Gain 30.0 dB
Second IF amplifier noise figure	= 25.0 dB, Gain 30.0 dB
Receiver bandwidth	=6.0 MHz

Assuming that the bandwidth of the receiver is determined by the second IF amplifier find the noise figure of the receiver, the system noise temperature of the receiver, and the noise power which you would expect to measure at the receiver output. [6.3 dB, 1025 K, –36.7 dBm]

12.3. A digital communications system requires a CNR of 18.0 dB at the receiver's detection circuit input in order to achieve a satisfactory BER. The actual CNR at the receiver input is 23.0 dB and the present (unsatisfactory) CNR at the detection circuit input is 16.0 dB. The source temperature of the receiver is 290 K. What noise figure specification must be met by an additional amplifier (to be placed at the front end of the receiver) for a satisfactory BER to be realised if the gain of the amplifier is: (a) 6.0 dB; (b) 10.0 dB; and (c) 14.0 dB? [3.3 dB, 4.4 dB, 4.8 dB]

12.4. A 36.2 GHz, 15.5 km, single hop microwave link has a transmitter power of 0 dBm. The transmit and receiver antennas are circularly symmetric paraboloidal reflectors with diameters of 0.7 m and 0.5 m, respectively. Assuming both antennas have ohmic efficiencies of 97% and aperture efficiencies of 67% calculate: (a) the transmitted boresight radiation intensity; (b) the spreading loss; (c) the free space path loss; (d) the effective isotropic radiated power; (e) the power density at the receiving antennas aperture; and (f) the received power. [3.64 W/steradian, 94.8 dB m^{-2}, 147.4 dB, 46.6 dBm, −78.2 dBW/m^{-2}, −57.1 dBm]

12.5. If the effective noise bandwidth of the receiver in Problem 12.4 is 100 kHz, the antenna aperture temperature is 200 K, its physical temperature is 290 K and the noise figure of the entire receiver (from antenna output to detector input) is 4.0 dB calculate the CNR at the detector input. [63.4 dB]

12.6. Derive the fundamental free space transmission loss equation using the concepts of spreading loss and antenna reciprocity. Identify the free space path loss in the equation you derive and explain why it is a function of frequency.

12.7. Explain the concept of path gain in the context of plane earth propagation. How does this lead to a $1/R^4$ dependence of power density? What is the equivalent distance dependence of field strength?

12.8. Show that equation (12.73) follows from equation (12.72) when $\rho e^{j\phi} = -1$.

12.9. A VHF communications link operates over a large lake at a carrier frequency of 52 MHz. The path length between transmitter and receiver is 18.6 km and the heights of the sites chosen for the location of the transmitter and receiver towers are 72.6 m and 95.2 m respectively. Assuming that the cost of building a tower increases at a rate greater than linearly with tower height find the most economic tower heights which will take full advantage of the 'ground' reflected ray. Estimate by what height the water level in the lake would have to rise before the 'ground' reflection advantage is lost completely. [80.3 m, 80.3 m, 69.0 m]

12.10. Digital MPSK transmissions carried on a 6 GHz terrestial link require a bandwidth of 24 MHz. The transmitter carrier power level is 10 W and the hop distance is 40 km. The antennas used each have 40 dB gain and filter, isolator and feeder losses of 4 dB. The receiver has a noise figure of 10 dB in the specified frequency band. Calculate the carrier-to-thermal-noise-power ratio at the receiver output. [62 dB]

Comment on any effects which might be expected to seriously degrade this carrier-to-noise ratio in a practical link. If a minimum carrier-to-noise ratio of 30 dB is required for the MPSK modulation what is the fade margin of the above link? [32 dB]

12.11. A QAM 4 GHz link requires a bandwidth of 15.8 MHz. The radiated carrier power is 10 W and the hop distance 80 km. It uses antennas each having 40 dB gain, has filter, isolator and antenna feeder losses totalling 7 dB at each end and a receiver noise figure of 10 dB. If the Rayleigh fade margin is 30 dB, calculate the carrier-to-thermal-noise-power ratio (CNR) at the receiver. What complexity of QAM modulation will this CNR support for an error rate of 10^{-6}? [32.5 dB, 128 state]

12.12. A small unmanned laboratory is established on the moon. It contains robotic equipment which receives instructions from earth via a 4 GHz communications link with a bandwidth of 30 kHz. The EIRP of the earth station is 10 kW and the diameter of the laboratory station's receiving antenna is 3.0 m. If the laboratory's receiving antenna were to transmit, 2% of its radiated power would illuminate the lunar surface and 50% would illuminate the earth. (The rest would be radiated into space which has the cosmic background brightness temperature of 3 K.) The noise figure of the laboratory's receiver is 2.0 dB. If the daytime physical temperature of the lunar surface is 375 K and the brightness temperature of the earth is 280 K estimate the daytime CNR at the laboratory receiver's detection circuit input. (Assume that the laboratory antenna has an ohmic efficiency of 98%, an aperture efficiency of 72%, and a physical temperature equal to that of the lunar surface. Also assume that the earth and moon both behave as black bodies, i.e. have emissivities of 1.0.) The distance from the earth to the moon is 3.844×10^8 m. [23.0 dB]

12.13. If the nighttime temperature of the lunar surface in Problem 12.12 is 125 K find the nighttime improvement in CNR over the daytime CNR. Is this improvement: (a) of real engineering importance; (b) measurable but not significant; or (c) undetectably small? [0.14 dB]

CHAPTER 13

Communication systems simulation

13.1 Introduction

Most of the material in the preceding chapters has been concerned with the development of equations which can be used to predict the performance of digital communications systems. An obvious example is the error function formulae used to find the probability of bit error for ideal systems assuming zero ISI, matched filter receivers and additive, white, Gaussian noise. Such models are important in that they are simple, give instant results, and provide a series of 'reference points' in terms of the relative (or perhaps potential) performance of quite different types of system. In addition they allow engineers to develop a quantitative feel for how the performance of systems will vary as their parameters are changed. They also often provide theoretical limits on performance guarding the design engineer against pursuit of the unobtainable.

The principal limitation of such equations arises from the sometimes unrealistic assumptions on which they are based. Filters, for example, do not have rectangular amplitude responses, oscillators are subject to phase noise and frequency drift, carrier recovery circuits do not operate with zero phase error, sampling circuits are prone to timing jitter, etc.

Hardware prototyping during the design of systems avoids the limitation of idealised models in so far as real-world imperfections are present in the prototype. Designing, implementing and testing hardware, however, is expensive and time consuming, and is becoming more so as communications systems increase in sophistication and complexity. This is true to the extent that it would now usually be impossible to prototype all credible solutions to a given communications problem. Computer simulation of communications systems falls into the middle ground between idealised modelling using simple formulae and hardware prototyping. It occupies an intermediate location along all the following axes:

crude – accurate
simple – complex

cheap – expensive

quick – time consuming

In addition computer simulations are often able to account for system non-linearities which are notoriously difficult to model analytically.

Computer simulations of communications systems usually work as follows. The system is broken down into functional blocks each of which can be modelled mathematically by an equation, a rule, an input/output lookup table or in some other way. These subsystems are connected together such that the outputs of some blocks form the input of others and vice versa. An information source is then modelled as a random or pseudo random sequence of bits (see section 13.4). The signal is sampled and the samples fed into the input of the first subsystem. This subsystem then operates on these samples according to its system model and provides modified samples at its output. These samples then become the inputs for the next subsystem, and so on, typically until the samples represent the received, demodulated, information bit stream. Random samples representing noise and/or interference are usually added at various points in the system. Finally the received information bits are compared with the original information source bits to estimate the BER of the entire communications system. Intermediate results, such as the spectrum of the transmitted signal and the pdf of signal plus noise in the receiver, can also be obtained.

The functional definition of some subsystems, e.g. modulators, is easier in the time domain whilst the definition of others (e.g. filters) is easier in the frequency domain. Conversion between time and frequency domains is an operation which may be performed many times when simulations are run. Convolution, multiplication and discrete Fourier transforms are therefore important operations in communications simulation.

The accuracy of a well designed simulation, in the sense of how closely it matches a hardware prototype, depends essentially of the level of detail at which function blocks are defined. The more detailed the model the more accurate might be expected its results. The penalty paid, of course, is in the effort required to develop the model and the computer power needed to simulate the results in a reasonable time. Only an overview of the central issues involved in simulation is presented here. A detailed and comprehensive exposition of communication system simulation is given in [Jeruchim *et al.*]. In essence, however, the normal simulation process can be summarised as:

1. Derivation of adequate models for all input signals (including noise) and subsystems.
2. Conversion, where possible, of signals and subsystem models to their equivalent baseband form.
3. Sampling of all input signals at an adequately high rate.
4. Running of simulations, converting between time and frequency domains as necessary.
5. Use of Monte Carlo or quasi-analytic methods to estimate bit error rates.
6. Conversion of output signals back to passband form if necessary.
7. Display of intermediate signals, spectra, eye diagrams, etc. as required.

Much of the work in the preceding chapters (especially Chapters 2 to 4) has been directed at modelling the signals, noise and subsystems which commonly occur in digital communications systems. This chapter therefore concentrates on steps 2 to 7 above.

13.2 Equivalent complex baseband representations

Consider a microwave LOS communications system operating with a carrier frequency of 6 GHz and a bandwidth of 100 MHz. To simulate this system it might superficially appear that a sampling rate of $2 \times 6,050$ MHz would be necessary. This conclusion is, of course, incorrect as should be apparent from the bandpass sampling theorem discussed in Chapter 5. In fact, the most convenient way of representing this narrowband system for simulation purposes is to work with equivalent baseband quantities.

13.2.1 Equivalent baseband signals

A (real) passband signal can be expressed in polar (i.e. amplitude and phase) form by:

$$x(t) = a(t) \cos [2\pi f_c t + \phi(t)]$$

$$= \Re \left\{ a(t) \, e^{j2\pi f_c t} \, e^{j\phi(t)} \right\} \tag{13.1(a)}$$

or, alternatively, in Cartesian (i.e. inphase and quadrature) form by:

$$x(t) = x_I(t) \cos 2\pi f_c t - x_Q(t) \sin 2\pi f_c t \tag{13.1(b)}$$

The corresponding complex baseband (or lowpass) signal is defined as:

$$x_{LP}(t) = a(t) \, e^{j\phi(t)} \tag{13.2(a)}$$

or:

$$x_{LP}(t) = x_I(t) + j \, x_Q(t) \tag{13.2(b)}$$

(Multiplying equation (13.2(b)) by $e^{j2\pi f_c t}$ and taking the real part demonstrates the correctness of the $+$ sign here.) $x_{LP}(t)$ is sometimes called the *complex envelope* of $x(t)$. The baseband nature of $x_{LP}(t)$ is now obvious and the modest sampling rate required to satisfy Nyquist's theorem is correspondingly obvious. Notice that the transformation from the real passband signal of equation (13.1(a)) to the complex baseband signal of equation (13.2(a)) can be viewed as a two step process, i.e.:

1. $x(t)$ is made cisoidal (or *analytic*) by adding $j \, a(t) \sin[2\pi f_c t + \phi(t)]$.
2. The carrier is suppressed by dividing by $e^{j2\pi f_c t}$.

Since $a(t) \sin[2\pi f_c t + \phi(t)]$ is derived from $x(t)$ by shifting all positive frequency components by $+90°$ and all negative frequency components $-90°$, step 1 corresponds to adding $j\hat{x}(t)$ to $x(t)$ where ^ denotes the (time domain) Hilbert transform (see section 4.5). The relationship between $x(t)$ and $x_{LP}(t)$ can therefore be summarised as:

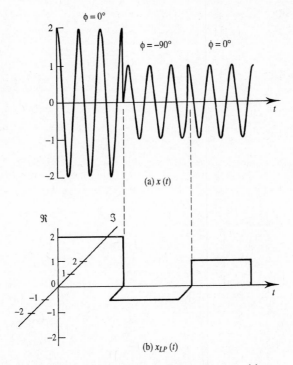

Figure 13.1 *Real passband signal $x(t)$ and its complex envelope $x_{LP}(t)$.*

$$x_{LP}(t) = [x(t) + j\,\hat{x}(t)]\, e^{-j2\pi f_c t} \qquad (13.3)$$

$x(t)$ and $x_{LP}(t)$ are shown schematically, for an APK signal, in Figure 13.1.

The spectrum, $X_{LP}(f)$, of $x_{LP}(t)$ can be found by applying the corresponding frequency domain steps to the spectrum, $X(f)$, of $x(t)$, i.e.:

1. $X(f)$ has its negative frequency components suppressed and its positive frequency components doubled. This is demonstrated using phasor diagrams, for a sinusoidal signal, in Figure 13.2.
2. The (doubled) positive frequency components are shifted to the left by f_c Hz.

Step 1 is more formally expressed as the addition to $X(f)$ of $j\hat{X}(f) = j[-j\mathrm{sgn}(f)X(f)] = \mathrm{sgn}(f)X(f)$. Step 2 follows from the Fourier transform frequency translation theorem. Figure 13.3 shows the relationship between $X(f)$ and $X_{LP}(f)$. Notice that the spectrum of the complex envelope does not have the Hermitian symmetry characteristic of real signals. Steps 1 and 2 together can be summarised by:

$$X_{LP}(f) = 2X(f + f_c)\, u(f + f_c) \qquad (13.4)$$

The first factor on the RHS of equation (13.4) doubles the spectral components, the second moves the entire spectrum to the left by f_c Hz and the third factor suppresses all spectral components to the left of $-f_c$ Hz, Figure 13.4.

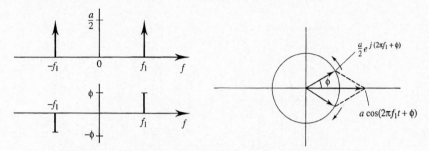

(a) Amplitude/phase spectrum and phasor diagram for a single spectral component

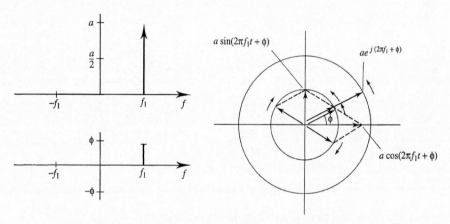

(b) Cancellation of negative frequency phasors, reinforcement of positive frequency phasors

Figure 13.2 *Phasor diagram demonstration of the equivalence between (a) the addition of an imaginary quadrature version of a signal and (b) the suppression of the negative frequency components plus a doubling of the positive frequency components.*

13.2.2 Equivalent baseband systems

Equivalent baseband representations of passband systems can be found in the same way as for signals. A filter with a passband (Hermitian) frequency response $H(f)$ (Figure 13.5(a) to (c)), and (real) impulse response $h(t)$, has an equivalent baseband frequency response (Figure 13.5(d) to (f)):

$$H_{LP}(f) = H(f + f_c)\, u(f + f_c) \tag{13.5(a)}$$

and baseband impulse response:

$$h_{LP}(t) = \tfrac{1}{2}\,[h(t) + j\,\hat{h}(t)]\, e^{-j2\pi f_c t} \tag{13.5(b)}$$

Alternatively the complex baseband impulse response can be expressed in terms of its

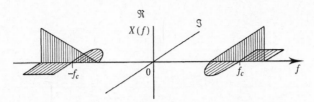

(a) Hermitian spectrum of real passband signal

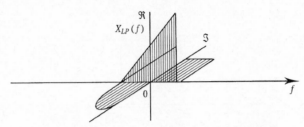

(b) Non-Hermitian spectrum of equivalent complex baseband signal

Figure 13.3 *Spectra of real passband signal and its complex envelope.*

inphase and quadrature components, i.e.:

$$h_{LP}(t) = \tfrac{1}{2}\,[h_I(t) + j\,h_Q(t)] \tag{13.6}$$

where the (real) passband response is:

$$h(t) = h_I(t)\cos 2\pi f_c t - h_Q(t)\sin 2\pi f_c t \tag{13.7}$$

13.2.3 Equivalent baseband system output

The output of an equivalent lowpass linear system when excited by an equivalent lowpass signal is found, in the time domain, using convolution in the usual way (but taking care to convolve both real and imaginary components) and represents the equivalent lowpass output of the system, $y_{LP}(t)$, i.e.:

$$
\begin{aligned}
y_{LP}(t) &= h_{LP}(t) * x_{LP}(t) \\
&= \tfrac{1}{2}\,[h_I(t) + j\,h_Q(t)] * [x_I(t) + j\,x_Q(t)]
\end{aligned}
\tag{13.8}
$$

Notice that when compared with the definition of equivalent baseband signals (equations (13.4), (13.3) and (13.2(b))) the equivalent baseband system definitions (equations (13.5(a)), (13.5(b)) and (13.6)) are smaller by a factor of ½. This is to avoid, for example, the baseband equivalent frequency response of a lossless passband filter having a voltage gain of 2.0 in its passband. (Many authors make no such distinction between the definitions of baseband equivalent signals and systems in which case the factor of ½ usually appears in the *definition* of equivalent baseband convolution.) Recognising that the equivalent baseband system output signal, $y_{LP}(t)$, can be expressed as inphase and

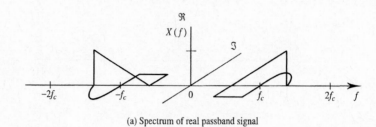

(a) Spectrum of real passband signal

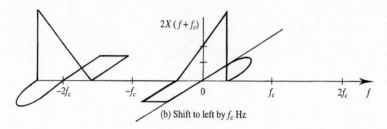

(b) Shift to left by f_c Hz

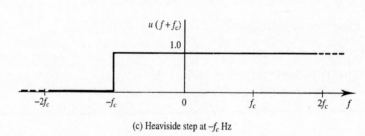

(c) Heaviside step at $-f_c$ Hz

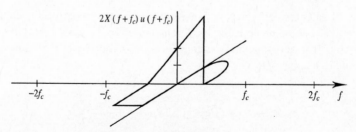

(d) Spectrum of complex equivalent baseband signal

Figure 13.4 *Relationship between passband and equivalent low-pass spectra.*

quadrature components of the passband output signal, $y(t)$, i.e.:

$$y_{LP}(t) = y_I(t) + j\, y_Q(t) \tag{13.9}$$

where:

$$y(t) = y_I(t)\cos 2\pi f_c t - y_Q(t)\sin 2\pi f_c t \tag{13.10}$$

and equating real and imaginary parts in equations (13.8) and (13.9) gives:

$$y_I(t) = \tfrac{1}{2}\,[h_I(t) * x_I(t) - h_Q(t) * x_Q(t)]$$ (13.11(a))

$$y_Q(t) = \tfrac{1}{2}\,[h_Q(t) * x_I(t) + h_I(t) * x_Q(t)]$$ (13.11(b))

These operations are illustrated schematically in Figure 13.6. Equations (13.10) and (13.11) thus give the passband output of a system directly in terms of the inphase and quadrature baseband components of its passband input and impulse response. $y(t)$ can

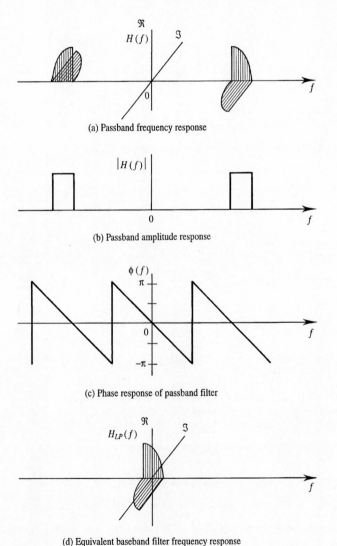

(a) Passband frequency response

(b) Passband amplitude response

(c) Phase response of passband filter

(d) Equivalent baseband filter frequency response

Figure 13.5 *Passband and equivalent lowpass frequency, amplitude and phase responses.*

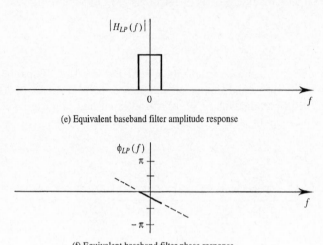

(e) Equivalent baseband filter amplitude response

(f) Equivalent baseband filter phase response

Figure 13.5-ctd. *Equivalent lowpass amplitude and phase responses.*

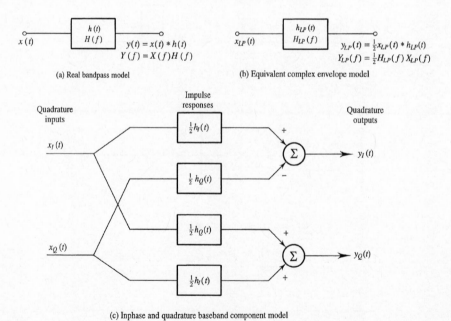

(a) Real bandpass model

(b) Equivalent complex envelope model

(c) Inphase and quadrature baseband component model

Figure 13.6 *Convolution of inputs and impulse responses with equivalent baseband operations.*

also be found from $y_{LP}(t)$ by reversing the steps used in going from equations (13.1(a)) to (13.2(a)), i.e.:

$$y(t) = \Re\left\{y_{LP}(t)\, e^{j2\pi f_c t}\right\}$$

$$= \tfrac{1}{2}\left[y_{LP}(t)\, e^{j2\pi f_c t} + y^*_{LP}(t)\, e^{-j2\pi f_c t} \right] \tag{13.12(a)}$$

Alternatively, if $Y_{LP}(f)$ has been found from:

$$Y_{LP}(f) = H_{LP}(f)\, X_{LP}(f)$$

then $Y(f)$ can be obtained using the equivalent frequency domain quantities and operations of equation (13.12(a)) (see Figure 13.7), i.e.:

$$Y(f) = \text{Hermitian}\{Y_{LP}(f - f_c)\} \tag{13.12(b)}$$

$$= \frac{Y_{LP}(f - f_c) + Y^*_{LP}(-f - f_c)}{2}$$

(That any function, in this case $Y_{LP}(f - f_c)$, can be split into Hermitian and anti-Hermitian parts is easily demonstrated as follows:

$$X(f) = \frac{X(f) + X^*(-f)}{2} + \frac{X(f) - X^*(-f)}{2} \tag{13.13}$$

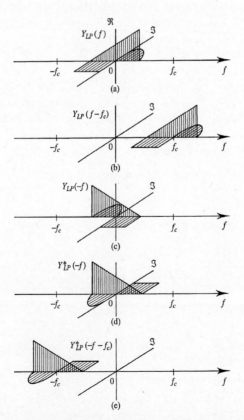

Figure 13.7 *Obtaining the spectrum of real passband signal, (b) plus (e), from the spectrum of an equivalent complex baseband signal, (a).*

The first term has even real part and odd imaginary part and is therefore Hermitian. The second term has odd real part and even imaginary part and is therefore anti-Hermitian.)

13.2.4 Equivalent baseband noise

Equivalent baseband noise, $n_{LP}(t)$, can be modelled in exactly the same way as an equivalent baseband signal, i.e.:

$$n_{LP}(t) = r(t) \, e^{j\theta(t)}$$

$$= n_I(t) + j \, n_Q(t) \qquad (13.14)$$

where the passband noise process is:

$$n(t) = r(t) \, e^{j\theta(t)} \, e^{j2\pi f_c t}$$

$$= n_I(t) \cos 2\pi f_c t - n_Q(t) \sin 2\pi f_c t \qquad (13.15)$$

Figure 13.8 illustrates the relationship between the passband and equivalent baseband processes in time and phase domains. For the special, but very important, case of a strict-sense, zero mean, Gaussian, narrowband noise process the properties of the baseband processes $n_I(t)$ and $n_Q(t)$ are summarised in Table 13.1.

Most of these properties are intuitively reasonable and are, therefore, not proved here. (Proofs can be found in [Taub and Schilling].) One anti-intuitive aspect of the equal variance property, however, is that each of the quadrature baseband processes alone contains the same power (i.e. has the same variance) as the passband noise process. (To the authors, at least, intuition would suggest that each of the baseband process would contain half the power of the passband process.) This problem is easily resolved, however, by considering the power represented by equation (13.15), i.e.:

$$\langle n^2(t) \rangle = \langle [n_I(t) \cos 2\pi f_c t - n_Q(t) \sin 2\pi f_c t]^2 \rangle$$

$$= \tfrac{1}{2}\langle n_I^2(t) \rangle + \tfrac{1}{2}\langle n_Q^2(t) \rangle \qquad (13.16(a))$$

And since $\langle n_I^2(t) \rangle = \langle n_Q^2(t) \rangle$ (acceptable on intuitive grounds) then:

$$\langle n^2(t) \rangle = \langle n_I^2(t) \rangle = \langle n_Q^2(t) \rangle \qquad (13.16(b))$$

Table 13.1 *Properties of equivalent baseband Gaussian noise quadrature processes.*

Property	Definition
Zero mean	$\langle n_I(t) \rangle = \langle n_Q(t) \rangle = 0$
Equal variance	$\langle n_I^2(t) \rangle = \langle n_Q^2(t) \rangle = \langle n^2(t) \rangle = \sigma^2$
Zero correlation	$\langle n_I(t) \, n_Q(t) \rangle = 0$
Gaussian quad. components	$p(n_I) = p(n_Q) = [1/(\sqrt{2\pi} \, \sigma)] \, e^{-\frac{n_x}{2\sigma^2}}$
Rayleigh amplitude	$p(r) = (r/\sigma^2) \, e^{-r^2/2\sigma^2}, \qquad r \geq 0$
Uniform phase	$p(\theta) = 1/(2\pi), \qquad\qquad |\theta| < \pi$

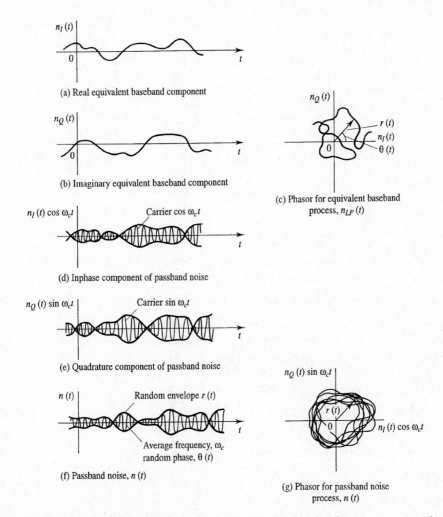

$n_I(t)$

0 t

(a) Real equivalent baseband component

$n_Q(t)$

0 t

(b) Imaginary equivalent baseband component

$n_Q(t)$

$r(t)$
$n_I(t)$
0 $\theta(t)$

(c) Phasor for equivalent baseband
process, $n_{LP}(t)$

$n_I(t)\cos\omega_c t$ Carrier $\cos\omega_c t$

t

(d) Inphase component of passband noise

$n_Q(t)\sin\omega_c t$ Carrier $\sin\omega_c t$

t

(e) Quadrature component of passband noise

$n(t)$ Random envelope $r(t)$

t

Average frequency, ω_c
random phase, $\theta(t)$

(f) Passband noise, $n(t)$

$n_Q(t)\sin\omega_c t$

$r(t)$
0 $n_I(t)\cos\omega_c t$

(g) Phasor for passband noise
process, $n(t)$

Figure 13.8 *Schematic illustration of passband and equivalent baseband noise processes with corresponding phasor trajectories.*

Figure 13.9 shows, in a systems context, how the passband process $n(t)$ could be generated from the baseband processes $n_I(t)$ and $n_Q(t)$. Notice that the power in each quadrature leg is halved after multiplication with the carrier.

Since noise processes do not have a well defined voltage spectrum (preventing equation (13.4) from being used to find an equivalent baseband spectrum) the (power) spectral (density) description of $n_I(t)$ and $n_Q(t)$ is found by translating the positive frequency components of $G_n(f)$ down by f_c Hz, translating the negative frequencies up by f_c Hz, and adding, i.e.:

$$G_{n_I}(f) = G_{n_Q}(f)$$

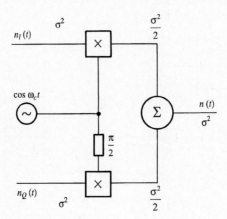

Figure 13.9 *Relationship between baseband quadrature noise components and the passband Gaussian noise process.*

$$= G_n(f + f_c)\, u(f + f_c) + G_n(f - f_c)\, u(-f + f_c) \qquad (13.16(c))$$

The relationship between the PSD of $n(t)$ and that of $n_I(t)$ and $n_Q(t)$ is illustrated in Figure 13.10.

13.3 Sampling and quantisation

Sampling and quantisation, as they affect communications systems generally, have been discussed in Chapter 5. Here these topics are re-examined in the particular context of simulation.

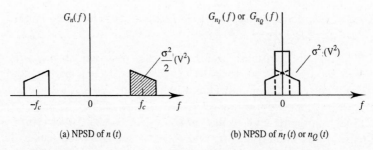

(a) NPSD of $n(t)$ (b) NPSD of $n_I(t)$ or $n_Q(t)$

Figure 13.10 *Relationship between PSD of each equivalent baseband quadrature component of $n(t)$ and PSD of $n(t)$.*

13.3.1 Sampling equivalent baseband signals

Nyquist's sampling theorem, if correctly interpreted, can be applied to any process, including an equivalent, complex, baseband process. In this case the baseband signal has a conventionally defined bandwidth which is only half the bandwidth, B, of the real passband signal, Figure 13.11. Thus a straightforward application of Nyquist's theorem gives a minimum sampling rate:

$$f_s \geq 2\frac{B}{2} = B \text{ Hz} \tag{13.17}$$

Superficially equation (13.17) looks wrong in that it suggests all the information present in a passband signal with bandwidth B Hz is preserved in only half the number of samples expected. This paradox is resolved by remembering that for a complex baseband signal there will be two real sample values for each sampling instant (i.e. an inphase, or real sample, and a quadrature, or imaginary, sample). The total number of real numbers characterising a given passband signal is therefore the same, whether or not an equivalent baseband representation is used.

Sampling at a rate of f_s Hz defines a simulation bandwidth of $f_s/2$ Hz in the sense that any spectral components which lie within this band will be properly simulated whilst spectral components outside this band will be aliased. The selection of f_s is therefore a compromise between the requirement to keep f_s low enough so that simulation can be carried out in a reasonable time with modest computer resources, and the requirement to keep f_s high enough for aliasing errors to be acceptably low. The aliasing errors are quantified in section 5.3.4 by a signal to distortion ratio (SDR) defined as the ratio of unaliased signal power to aliased signal power, Figure 13.12. SDR is clearly a function of the number of samples per symbol (i.e. f_s/R_s). Table 13.2 shows several corresponding pairs of SDR and f_s/R_s for the (worst) case of an unfiltered (i.e. rectangular pulse) symbol stream.

f_s/R_s is typically selected such that SDR is 10 dB greater than the best SNR to be simulated. For signals with significant pulse shaping the SDR for a given value of f_s/R_s is higher than that shown in Table 13.2. Eight samples per symbol, therefore, may often represent sufficiently rapid sampling for the simulation of realistic systems.

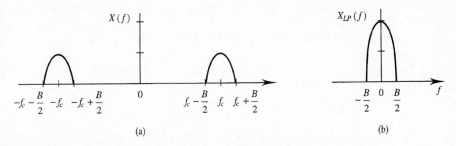

Figure 13.11 *(a) Passband and (b) equivalent baseband frequency spectra.*

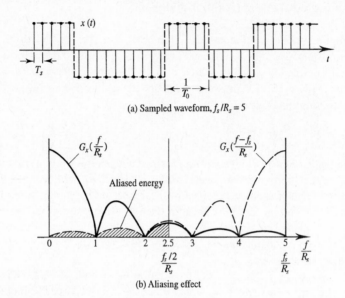

(a) Sampled waveform, $f_s/R_s = 5$

(b) Aliasing effect

Figure 13.12 *Sampling of random binary waveform (source: Jeruchim et al., 1984, reproduced with permission of the IEEE).*

Table 13.2 *Worst case SDR for various numbers of samples per symbol. (After Jeruchim et.al.)*

f_s/R_s (samples/symbol)	SDR (dB)
4	15.8
8	18.7
10	19.8
16	21.9
20	22.9

13.3.2 Quantisation

Simulation quantisation errors can be equated with the limited precision with which a computer can represent numbers. The rounding errors due to this limited precision effectively add noise to the waveform being simulated. In digital communications the system's ADC itself introduces quantisation error (sections 5.5 and 5.6) quantified by a signal to quantisation noise ratio (SN_qR) and given approximately (for linearly quantised voice signals) by $6(n-1)$ dB where n is the number of bits (typically 8) representing each level, equation (5.23). The simulation induced SN_qR will depend in a similar way on the binary word size which the computer uses to represent numbers. This word size would normally be at least 16 bits, typically 32-bits. Whilst this suggests that simulation induced quantisation error will be negligible with respect to system induced quantisation error it may sometimes be the case that simulation quantisation errors accumulate in calculations. For long simulations (perhaps millions of symbols) accumulated errors may become significant. Using double precision arithmetic, where possible, will obviously help in this respect.

13.4 Modelling of signals, noise and systems

One of the strengths of simulation is that it can often be interfaced to real signals and real systems hardware. A real voice signal, for example, might be rapidly sampled and those samples used as the information source in a simulation. Such a simulation might also include measured frequency responses of actual filters and measured input/output characteristics of non-linear amplifiers which are to be incorporated in the final system hardware. (In this respect the distinction between hardware prototyping and software simulation can become blurred.) Nevertheless, it is still a common requirement to model signals, noise and systems using simple (and rather idealised) assumptions. A variety of such models are discussed below.

13.4.1 Random numbers

Statistically independent, random, numbers with a uniform pdf are easily generated by computers. Many algorithms have been proposed, but one [Park and Miller] which appears to have gained wide acceptance for use on machines using 32-bit integer arithmetic is:

$$x(k) = 7^5 x(k-1) (2^{31} - 1) \tag{13.18}$$

$x(k)$ represents a sequence of integer numbers drawn from a uniform pdf with minimum and maximum values of 1 and $2^{31} - 1$ respectively. To generate a sequence of numbers, $x_u(k)$, with uniform pdf between 0 and 1, equation (13.18) is simply divided by $2^{31} - 1$. ($x_u(k)$ never takes on a value of exactly zero since the algorithm would then produce zero values indefinitely.) The initial value of the sequence $x(0) \neq 0$ is called the generator *seed* and is chosen by the user. Strictly speaking only the choice of seed is random since thereafter the sequence is deterministic, repeating periodically. The sequences arising from well designed algorithms such as equation (13.18), however, have many properties in common with truly random sequences and, from an engineering point of view, need not usually be distinguished from them. They are therefore referred to as pseudo-random (or pseudo-noise) sequences. A sequence $y(k)$ with a pdf $p_Y(y)$ can be derived from the sequence $x_u(k)$ using the target cumulative distribution of Y, Figure 13.13. The CD of Y (i.e. $P(Y \leq y)$) is first found (by integrating $p_Y(y)$ if necessary). The values of $x_u(k)$ are then mapped to $y(k)$ according to this curve as shown in Figure 13.13. For simple pdfs this transformation can sometimes be accomplished analytically resulting in a simple formula relating $y(k)$ and $x_u(k)$. Otherwise $P(Y \leq y)$ can be defined by tabulated values and the individual numbers transformed by interpolation.

The method described above is general and could be used to generate numbers with a Gaussian pdf. It is often easier, for this special case, however, to take advantage of the central limit theorem and add several independent, uniformly distributed, random

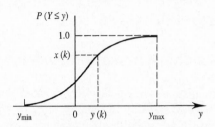

Figure 13.13 *Cumulative distribution used to transform a uniformly distributed random variable to a random variable Y with pdf $p_Y(y)$.*

sequences, i.e.:

$$y(k) = \sum_{i=1}^{N} x_{u,i}(k) - \frac{N}{2} \tag{13.19}$$

Using equation (3.16) we see that $\bar{X}_u$ is 0.5, and subtracting $N/2$ in equation (13.19) ensures that $\bar{Y} = 0$. Furthermore the variance of X_u is 1/12. Choosing $N = 12$ therefore ensures that $\sigma_y^2 = 1.0$ without the need for any additional scaling. (Higher values of N would, of course, improve the accuracy of the resulting Gaussian pdf.)

Correlated random numbers are easily generated from statistically independent random numbers by forming linear combinations of sequence pairs. For example, a random sequence, $y(k)$, with autocorrelation properties, $R_y(\kappa)$, (as defined in section 3.3.3) specified by:

$$R_y(0) = \sigma_Y^2 \tag{13.20(a)}$$

$$R_y(\pm 1) = \alpha \, \sigma_Y^2, \quad \alpha < 1 \tag{13.20(b)}$$

$$R_y(\pm 2) = \beta \, \sigma_Y^2, \quad \beta < 1 \tag{13.20(c)}$$

$$R_y(\pm \kappa) = 0, \quad \kappa > 2 \tag{13.20(d)}$$

$$\overline{y(k)} = 0 \tag{13.20(e)}$$

can be formed from a zero mean, unit variance, statistically independent sequence, $s(k)$, using the linear transform:

$$y(k) = w_1 s(k) + w_2 s(k-1) + w_3 s(k-2) \tag{13.21}$$

Substituting equation (13.21) into equation (13.20(a)) gives:

$$\sigma_Y^2 = \overline{y^2(k)} - [\overline{y(k)}]^2 \tag{13.22(a)}$$

$$= \overline{[w_1 s(k) + w_2 s(k-1) + w_3 s(k-2)]^2} - 0$$

Since $s(k)$ is a sequence of uncorrelated numbers only terms, in the expansion of equation (13.22(a)), having factors with equal arguments of s give non-zero results. Therefore:

$$\sigma_Y^2 = w_1^2 \overline{s^2(k)} + w_2^2 \overline{s^2(k-1)} + w_3^2 \overline{s^2(k-2)} \qquad (13.22(\text{b}))$$

And since the variance of $s(k)$ is 1.0 then:

$$\sigma_Y^2 = w_1^2 + w_2^2 + w_3^2 \qquad (13.22(\text{c}))$$

Similarly:

$$\alpha\,\sigma_Y^2 = \overline{[w_1 s(k) + w_2 s(k-1) + w_3 s(k-2)]\,[w_1 s(k-1) + w_2 s(k-2) + w_3 s(k-3)]}$$

$$= w_2 w_1 \overline{s^2(k-1)} + w_3 w_2 \overline{s^2(k-2)} \qquad (13.23(\text{a}))$$

all other terms being zero. Thus:

$$\alpha\,\sigma_Y^2 = w_2 w_1 + w_3 w_2 \qquad (13.23(\text{b}))$$

and:

$$\beta\,\sigma_Y^2 = \overline{[w_1 s(k) + w_2 s(k-1) + w_3 s(k-2)]\,[w_1 s(k-2) + w_2 s(k-3) + w_3 s(k-4)]}$$

$$= w_3 w_1 \overline{s^2(k-2)} \qquad (13.24(\text{a}))$$

i.e.:

$$\beta\,\sigma_Y^2 = w_3 w_1 \qquad (13.24(\text{b}))$$

Equations (13.22) to (13.24) generalise for a series with non-zero correlation over an N term window to:

$$R_y(\kappa) = \sum_{i=1}^{N-\kappa} w_i w_{\kappa+i} \qquad (13.25)$$

Equation (13.20(e)) is automatically satisfied since $\overline{s(k)} = 0$. Equation (13.25) provides N independent equations which are solved simultaneously to give the appropriate weighing factors, $w_1, w_2, \cdots, w_N$.

The general problem of simultaneously obtaining a specified pdf and specified PSD is a difficult one. This is because the linear system represented by equation (13.21) generally changes the pdf in an unpredictable way. The exception to this, of course, is for random sequences with Gaussian pdf which can be filtered to realise a specified PSD without altering its Gaussian characteristic (see section 4.6.3).

13.4.2 Random digital symbol streams

Random digital symbol streams can be easily generated from a set of random numbers as follows. Consider a sequence of independent random numbers, $y(k)$, with the pdf $p_Y(y)$ shown in Figure 13.14. If three symbols represented by three voltage levels $v = 0, 1, 2$ V are required with probabilities of 0.5, 0.25, 0.25 respectively then $p_Y(y)$ is divided into three areas corresponding to those probabilities. This defines thresholds y_1 and y_2 (Figure 13.14). Each random variable sample can then be mapped to a random symbol using the rule:

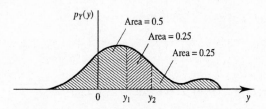

Figure 13.14 *pdf for a random variable used to generate a random digital symbol stream.*

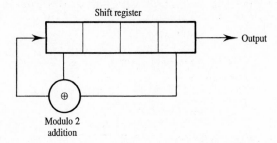

Figure 13.15 *Use of a shift register to generate pseudo-random bit sequences.*

$$
v(k) = \begin{cases} 0, & y(k) \le y_1 \\ 1, & y_1 < y(k) \le y_2 \\ 2, & y(k) > y_2 \end{cases} \tag{13.26}
$$

As an alternative to using a random number algorithm (such as equation (13.18)) followed by equation (13.26), pseudo-random digital symbol streams can be generated using shift registers with appropriate feedback connections. An example generator for a binary sequence is shown in Figure 13.15. Such generators are simple and easily implemented in hardware as well as software. (This makes them useful as signal sources for field measurements of BER when no secure, i.e. errorless, reference channel is available.) The properties of the pseudo-random bit sequences (PRBSs) generated in this way can be summarised as follows:

1. The sequence is periodic.
2. In each period the number of binary ones is one more than the number of binary zeros.
3. Among runs of consecutive ones and zeros one half of the runs of each kind are of length one, one quarter are of length two, one eighth are of length three, etc. as long as these fractions give meaningful numbers.
4. If a PRBS is compared, term by term, with any cyclical shift of itself the number of agreements differs from the number of disagreements by one.

The autocorrelation function of a K-bit, periodic, PRBS, $z(k)$, is defined by:

$$R_z(\tau) = \frac{1}{K} \sum_{k=1}^{K} z(k)z(k-\tau) \tag{13.27}$$

and is shown in Figure 13.16(b) for a polar sequence with amplitude ±1, Figure 13.16(a). Its voltage and power spectra are shown in Figure 13.16(c) and (d).

The general algorithm implemented by an *n*-element shift register with modulo 2 feedback, Figure 13.17, can be expressed mathematically as:

$$v(k) = w_{n-1}v(k-1) \oplus w_{n-2}v(k-2) \oplus \cdots \oplus w_0 v(k-n) \tag{13.28}$$

where $\oplus$ denotes modulo 2 addition. w_i denotes the feedback weighting (1 or 0) associated with the register's $(i+1)$th element and corresponds in hardware to the

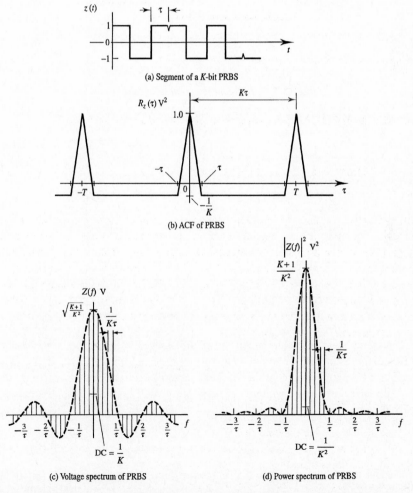

(a) Segment of a *K*-bit PRBS

(b) ACF of PRBS

(c) Voltage spectrum of PRBS

(d) Power spectrum of PRBS

Figure 13.16 *Temporal and spectral properties of a rectangular pulsed polar, NRZ, PRBS.*

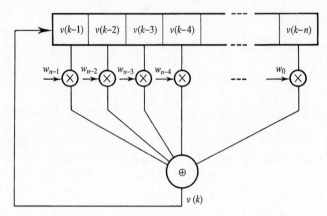

Figure 13.17 *General implementation of PRBS generator using an n-element shift register.*

presence or absence of a connection.

Table 13.3 *PRBS sequence feedback connections.*

Number of shift register elements, n	Non-zero feedback taps in addition to w_0	Sequence length $K = 2^n - 1$
4	w_3	15
5	w_3	31
6	w_5	63
7	w_6	127
8	w_6, w_5, w_4	255
9	w_5	511
10	w_7	1023
12	w_{11}, w_8, w_6	4095
14	w_{13}, w_8, w_4	16383
16	w_{15}, w_{13}, w_4	65535

The all-zeros state is prohibited in the PRBS, as generated above, since this would result in an endless stream of zeros thereafter. A maximal length PRBS is one in which all possible n-bit patterns (except the all-zeros pattern) occur once and once only in each period. The length of such a sequence is therefore $K = 2^n - 1$ bits. Not all arrangements of feedback connection give maximal length sequences. To establish whether a given set of connections will yield a maximal length sequence the polynomial:

$$f(x) = w_0 + w_1 x + w_2 x^2 + \cdots + w_{n-1} x^{n-1} + x^n \qquad (13.29)$$

is formed and then checked to see if it is *irreducible*. (Note that $w_0 = 1$, otherwise the final element of the register is redundant.) An irreducible polynomial of degree n is one which cannot be factored as a product of polynomials with degree lower than n. A test for irreducibility is that such a polynomial will not divide exactly (i.e. leaving zero remainder) into $x^P + 1$ for all $P < 2^n + 1$.

Table 13.3 [adapted from Jeruchim *et al.*] gives a selection of shift register feedback connections which yield maximal length PRBSs.

13.4.3 Noise and interference

Noise is modelled, essentially, in the same way as the signals described in section 13.4.2. The following interesting point arises, however, in the modelling of white Gaussian noise. Because a simulation deals only with noise samples then, providing these noise samples are uncorrelated (and being Gaussian, statistically independent) no distinction can be made between any underlying (continuous) noise processes which satisfy:

$$R_n(kT_s) = 0 \tag{13.30}$$

where $T_s = 1/f_s$ is the simulation sampling period. The simulation is identical, then, whether the underlying noise process is strictly white with an impulsive autocorrelation function, or band limited to $f_s/2$ Hz with the sinc shaped autocorrelation function shown in Figure 13.18. Provided $f_s/2$ (sometimes called the simulation bandwidth) is large with respect to the bandwidth of the system being simulated then the results of the simulation will be unaffected by this ambiguity. White Gaussian noise is therefore, effectively, simulated by generating independent random samples from a Gaussian pdf with variance (i.e. normalised power), σ_n^2, given by:

$$\sigma_n^2 = \frac{N_0 f_s}{2} = \frac{N_0}{2 T_s} \quad (\text{V}^2) \tag{13.31}$$

where N_0 (V^2Hz^{-1}) is the required *one-sided* NPSD.

Impulsive noise is characterised by a transient waveform which may occur with random amplitude at random times, or with fixed amplitude periodically, or with some combination of the two, Figure 13.19. The noise may be generated at baseband or passband. In radio systems, however, the noise will be filtered by the receiver's front end

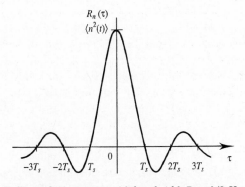

Figure 13.18 *ACF of bandlimited white noise with bandwidth $B = f_s/2$ Hz.*

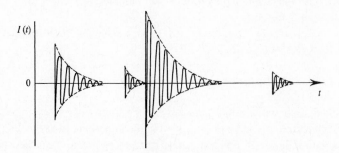

Figure 13.19 *Impulsive noise with random interpulse spacing and random pulse amplitude.*

and IF strip after which even baseband generated impulses will have bandpass characteristics, often resembling the receiver's impulse response. (Impulsive noise does not imply a sequence of impulses in the Dirac delta sense but usually does imply transient pulses with an effective duration which is short compared to the average interpulse spacing.) A useful way of modelling impulsive noise is to separate the statistical aspects from the deterministic pulse shape. This can be done by using a (complex) random number generator to model the amplitude and phase of the pulses, a Poisson counting process to model the arrival times, T_i, of the pulses and a deterministic function, $I(t)$, to model the pulse shape (for example $e^{-at} \cos \omega_c t \, u(t)$). Such a model would be specified at passband by:

$$I(t) = \sum_{i=1}^{N} A_i \, e^{-\frac{t-T_i}{\tau}} \cos(\omega_c t + \theta_i) \, u(t - T_i) \tag{13.32}$$

The equivalent baseband representation would be:

$$I_{LP}(t) = \sum_{i=1}^{N} A_i \, e^{j\theta_i} \, e^{-\frac{t-T_i}{\tau}} \, u(t - T_i) \tag{13.33}$$

Rewriting equation (13.33) as a convolution and expanding the exponential, i.e.:

$$I_{LP}(t) = e^{-\frac{t}{\tau}} u(t) * \sum_{i=1}^{N} A_i \, (\cos \theta_i + j \sin \theta_i) \, \delta(t - T_i) \tag{13.34}$$

emphasises the separation of pulse shape from pulse statistics. θ_i would normally be assumed to have a uniform pdf and a typical pdf for A_i might be log-normal. If the impulse noise is a Poisson process then the pdf of inter-arrival time between pulses is:

$$p_{\Delta T_i} (\Delta T_i) = \lambda \, e^{-\lambda \Delta T_i} \tag{13.35}$$

where $\Delta T_i = T_i - T_{i-1}$ and λ is the average pulse arrival rate (see Chapter 17).

Interference usually implies either an unwanted periodic waveform or an unwanted (information bearing) signal. In the former case pulse trains and sinusoids, for example, are easily generated in both passband and equivalent baseband form. In the latter case the interfering signal(s) can be generated in the same manner as wanted signals.

13.4.4 Time invariant linear systems

Linear subsystems, such as pulse shaping filters in a transmitter and matched filters in a receiver, are usually specified by their frequency response (both amplitude and phase). There is a choice to be made in terms of the appropriate level of idealisation when specifying such subsystems. An IF filter in a receiver may, for example, be represented by a rectangular frequency response at one extreme or a set of tabulated amplitude and phase values, obtained, across the frequency band, from measurements on a hardware prototype, at the other extreme. (In this case, unless the frequency domain points have the same frequency resolution as the simulation Fourier transform algorithm, interpolation and/or resampling with the correct resolution will be required.) In between these extreme cases analytical or tabulated models of classical filter responses (e.g. Butterworth, Chebyshev, Bessel, elliptic) may be used. Digital filter structures, typically implemented using tapped delay lines [Mulgrew and Grant], are especially easy to simulate, at least in principle. The effect of a frequency response on an input signal can be found by convolving the impulse response of the filter (which will be complex if equivalent baseband representations are being used) with its input. (The impulse response is obtained from the frequency response by applying an inverse FFT, see section 13.5.) Alternatively, block filtering can be applied in which the input time series is divided into many equal length segments, Fourier transformed using an FFT algorithm, multiplied by the (discrete) frequency response and then inverse transformed back to a time series. The implementation of block filtering, including important aspects such as appropriate zero padding, is described in [Strum and Kirk].

13.4.5 Non-linear and time varying systems

Amplitude compression, as used in companding (section 5.7.3), is a good example of baseband, memoryless, non-linear, signal processing. This type of non-linearity can be modelled either analytically (using, for example, equation (5.31)) or as a set of tabulated points relating instantaneous values of input and output. (Interpolation may be necessary in the latter case.)

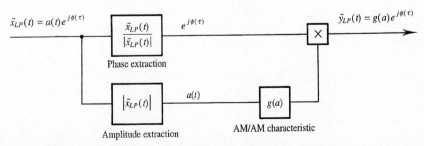

Figure 13.20 *Equivalent baseband model for a non-linear, memoryless, bandpass, system.*

For memoryless, non-linear, systems operating on narrow-band signals the useful equivalent baseband model shown in Figure 13.20 [developed from Tranter and Kosbar] can be used. This model works because, for a sinusoidal input, memoryless non-linearities produce outputs only at the frequency of the input and its harmonics. The harmonics can be filtered out at the non-linearity's output. The overall effect is of a 'linear' system with amplitude dependent gain (AM/AM conversion).

If a non-linearity has memory of intermediate length (i.e. memory which is significant with respect to, and may be many times, the carrier period but which is nevertheless short compared to changes in the carrier's complex envelope) then this too can be simulated as an equivalent baseband system, Figure 13.21. The amplitude of the signal (being related to the envelope) is changed in an essentially memoryless way. The phase of the signal, however, may now be affected by the non-linearity in a way which depends on the signal amplitude (AM/PM conversion). The power amplifier in satellite transponders can often be modelled in this way.

There are other types of non-linear bandpass system which fall into neither of the above special categories. In general it is not possible to obtain equivalent baseband models for these processes. Simulation of these non-linearities must normally, therefore, be executed at passband.

Adaptive equalisers (usually implemented as tapped delay lines with variable tap weights) and adaptive delta modulation (section 5.8.5) are examples of time varying, linear, subsystems. The simulation aspects of such subsystems are essentially straightforward, the design of algorithms to produce the required adaptive behaviour being the principal challenge. It is worth observing, however, that an equaliser operating on the demodulated I and Q components of a quadrature modulated signal generally comprises four separate tapped delay lines. Two lines operate in I and Q channels, individually controlling I and Q channel ISI, whilst two operate across I and Q channels controlling crosstalk, as in Figure 13.6(c). These four, real, tapped lines can be replaced by two lines, in which the weighting factors applied to each tap are complex, and which operate

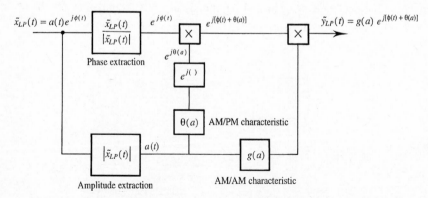

Figure 13.21 *Equivalent baseband model for a non-linear, bandpass system, with memory of intermediate length.*

on the complex envelope of the I and Q channel signals.

13.5 Transformation between time and frequency domains

Both time domain and spectral quantities are of interest in the design and evaluation of communication systems. This alone would be sufficient reason to require transformation between domains when simulating systems. In fact transformation between domains is also desirable since some simulation operations can be implemented more efficiently in the frequency domain than in the time domain.

Computer simulations work with discrete samples of time waveforms (i.e. time series) and discrete samples of frequency spectra. The discrete Fourier transform (DFT) and its relative, the discrete Fourier series (DFS), are the sampled signal equivalent of the Fourier transform and Fourier series described in Chapter 2. From a mathematical point of view the DFT, DFS and their inverses can simply be defined as a set of consistent formulas without any reference to their continuous function counterparts. Almost always, however, in communications engineering the discrete time series and frequency spectra on which these transforms operate represent sampled values of underlying continuous functions. Their intimate and precise connection with continuous Fourier operations is therefore emphasised here.

13.5.1 Discrete Fourier transform

Consider the N-sample time domain signal (or time series), $v_s(t)$, shown in Figure 13.22(a). The sample values can be represented by a series of weighted impulses, Figure 13.22(b), and expressed mathematically by:

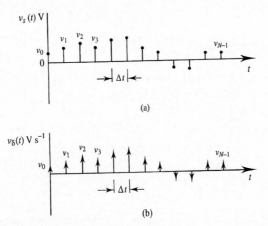

Figure 13.22 *(a) Sampled continuous signal or time series, and (b) its representation as a sum of impulses or delta functions.*

$$v_\delta(t) = v_0\, \delta(t) + v_1\, \delta(t - \Delta t) + v_2\, \delta(t - 2\Delta t) + \cdots \quad \text{(V/s)} \tag{13.36}$$

Notice that $v_\delta(t)$ has units of V/s. This is easily demonstrated by using the sampling property of impulse functions under integration to recover the original time series which has units of V, i.e.:

$$
\begin{aligned}
v_s(t) &= \left\{ \int_{-\infty}^{\infty} v_0\delta(t)\, dt, \ \int_{-\infty}^{\infty} v_1\delta(t - \Delta t)\, dt, \ \int_{-\infty}^{\infty} v_2\delta(t - 2\Delta t)\, dt, \cdots \right\} \\[2mm]
&= \left\{ v_0 \int_{-\infty}^{\infty} \delta(t)\, dt, \ v_1 \int_{-\infty}^{\infty} \delta(t - \Delta t)\, dt, \ v_2 \int_{-\infty}^{\infty} \delta(t - 2\Delta t)\, dt, \cdots \right\} \\[2mm]
&= \{ v_0, \ v_1, \ v_2, \cdots, \ v_{N-1} \} \quad \text{(V)}
\end{aligned}
\tag{13.37}
$$

(Since the sample values, v_i, have units of V this means that the impulses, $\delta(t - \Delta t)$, have units of s^{-1}, i.e. their area, or strength, is dimensionless.) The voltage spectrum of $v_\delta(t)$ is given by:

$$
\begin{aligned}
V_\delta(f) &= \text{FT}\,\{v_\delta(t)\} \\[2mm]
&= v_0 \int_{-\infty}^{\infty} \delta(t)e^{-j2\pi ft}\, dt \ + \ v_1 \int_{-\infty}^{\infty} \delta(t - \Delta t)e^{-j2\pi ft}\, dt + \cdots \quad \text{(V)}
\end{aligned}
\tag{13.38}
$$

(Notice that because equation (13.38) is the transform of a sequence of weighted impulses with units of V/s the voltage spectrum has units of V only, not V/Hz as usual.) Using the sampling property of $\delta(t)$ under integration equation (13.38) becomes:

$$V_\delta(f) = v_0 e^{-j0} + v_1 e^{-j2\pi f\Delta t} + v_2 e^{-j2\pi f2\Delta t} + \cdots \quad \text{(V)} \tag{13.39}$$

$V_\delta(f)$ in equation (13.39) is a *continuous* function, i.e. it is defined for all values of f. Using the summation notation equation (13.39) can be written more succinctly as:

$$V_\delta(f) = \sum_{\tau=0}^{N-1} v_\tau\, e^{-j2\pi f(\tau\Delta t)} \quad \text{(V)} \tag{13.40}$$

The values of the spectrum, $V_\delta(f)$, at the discrete frequencies f_0, f_1, f_2, etc. are given by:

$$V_\delta(f_v) = \sum_{\tau=0}^{N-1} v_\tau\, e^{-j2\pi f_v(\tau\Delta t)} \quad \text{(V)} \tag{13.41}$$

and if the frequencies of interest are equally spaced by Δf Hz then:

$$V_\delta(f_v) = \sum_{\tau=0}^{N-1} v_\tau\, e^{-j2\pi (v\Delta f)(\tau\Delta t)} \quad \text{(V)} \tag{13.42}$$

where $v = 0, 1, 2, 3, \ldots$ etc. Since the time series contains N samples it represents a signal with duration, T, given by:

$$T = N\Delta t \quad \text{(s)} \tag{13.43}$$

Nyquist's sampling theorem (Chapter 5) asserts that the lowest observable frequency (excluding DC) in the time series is:

$$f_1 = \frac{1}{N\Delta t} = \Delta f \quad (\text{Hz}) \tag{13.44}$$

Thus:

$$\Delta f \, \Delta t = \frac{1}{N} \tag{13.45}$$

and equation (13.42) can be written as:

$$V_\delta(f_\nu) = \sum_{\tau=0}^{N-1} v_\tau \, e^{-j2\pi \frac{\nu}{N} \tau} \quad (\text{V}) \tag{13.46}$$

Equation (13.46) is the definition, adopted here, for the *forward* DFT. Comparing this with the conventional (i.e. continuous) FT:

$$V(f) = \int_{-\infty}^{\infty} v(t) \, e^{-j2\pi ft} \, dt \quad (\text{V/Hz}) \tag{13.47}$$

the difference is seen, essentially, to be the absence of a factor corresponding to dt. $V_\delta(f_\nu)$ is therefore related to $V(f)$ by:

$$V(f)|_{f=f_\nu} \approx V_\delta(f_\nu) \, \Delta t \quad (\text{V/Hz}) \tag{13.48}$$

The reason why an approximation sign is used in equation (13.48) will become clear later. A note of caution is appropriate at this point. Equation (13.46) as a definition for the DFT is not universal. Sometimes a factor of $1/N$ and sometimes a factor of $1/\sqrt{N}$ is included in the formula. If the absolute magnitude of a voltage spectrum is important it is essential, therefore, to know the definition being used. Furthermore proprietary DFT software may not include an implementation of equation (13.48). Care is therefore needed in correctly interpreting the results given by DFT software.

13.5.2 Discrete Fourier series

If $v(t)$ is periodic then its FT should represent a *discrete* voltage spectrum (in contrast to discrete values taken from a continuous spectrum). Comparing the DFT (equation (13.46)) with the formula for a set of Fourier series coefficients (Chapter 2):

$$\tilde{C}_\nu = \frac{1}{T} \int_{0}^{T} v(t) \, e^{-j2\pi ft} \, dt \quad (\text{V}) \tag{13.49}$$

and remembering that the length of the time series is given by $T = N\Delta t$ then, to make $V_\delta(f_\nu)$ reflect $\tilde{C}_\nu$ properly, an extra factor $1/N$ is required, i.e.:

$$\tilde{C}_\nu = \frac{1}{N\Delta t} V_\delta(f_\nu) \, \Delta t = \frac{1}{N} V_\delta(f_\nu) \quad (\text{V}) \tag{13.50}$$

The (forward) *discrete Fourier series* (DFS) is therefore defined by:

$$\tilde{C}(f_v) = \frac{1}{N} \sum_{\tau=0}^{N-1} v_\tau \, e^{-j2\pi \frac{v}{N} \tau} \quad (V) \tag{13.51}$$

The need for the factor $1/N$ in equation (13.51) is seen most easily for the DC ($v = 0$) value which is simply the time series average.

13.5.3 DFS spectrum and rearrangement of spectral lines

Providing all the samples in the time series, v_τ, are real then the spectrum defined by equation (13.51) has the following properties:

1. Spectral lines occurring at $v = 0$ and $v = N/2$ are real. All others are potentially complex.
2. $\tilde{C}(f_{N-v}) = \tilde{C}^*(f_v)$, i.e. the DFS amplitude spectrum is even and the DFS phase spectrum is odd.

Figure 13.23 illustrates these properties for a 16-sample time series. The harmonic number, v, along the x-axis of Figure 13.23 is converted to conventional frequency f (in Hz) using:

$$f_v = v f_1 = \frac{v}{N \Delta t} \quad (Hz) \tag{13.52}$$

This leads, superficially, to a paradox in that f_{N-1} appears to correspond to a frequency of (almost) $1/\Delta t$ Hz yet Nyquist's sampling theorem asserts that frequencies no higher than $1/(2\Delta t)$ Hz can be observed. The paradox is resolved by recognising that no additional information is contained in the harmonics $N/2 < v < N$ since these are conjugates of the harmonics $0 < v < N/2$. A satisfying interpretation of the 'redundant' harmonics ($v > N/2$) is as the negative frequency components of a double sided spectrum. The conventional frequency spectrum is therefore constructed from the DFS (or DFT) by shifting all the lines from the top half of the DFS spectrum down in frequency by $N\Delta f$ Hz as shown in Figure 13.24. (Half the component at $v = N/2$ can also be shifted by

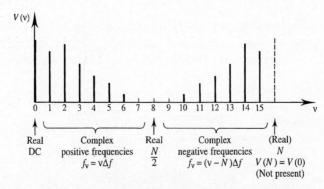

Figure 13.23 *Schematic illustration of discrete Fourier series for a 16-sample time series.*

$-N\Delta f$ Hz to retain overall symmetry, if desired.) The highest observable frequency is then given by:

$$f_{N/2} \; = \; \frac{1}{2\,\Delta t} \quad (\text{Hz}) \tag{13.53}$$

as expected.

13.5.4 Conservation of information

In the time domain an N-sample real time series is represented by N real numbers. In the frequency domain the same signal is represented by $N-2$ complex numbers and 2 real numbers (i.e. $2(N-2)+2$ real numbers). Half of the complex numbers, however, are complex conjugates of the other half. Thus in the frequency domain, as in the time domain, the signal is represented by N *independent* real numbers.

13.5.5 Phasor interpretation of DFS

Consider equation (13.51). Each (complex) spectral line, $\tilde{C}(f_3)$ for example, is a sum of N phasors (one arising from each time sample), all with identical frequency, f_3 in this case, but each with a different phase, $\theta_\tau = -2\pi(3/N)\tau$. Figure 13.25(a) and (b) illustrate the phasor diagrams for $\tilde{C}(f_1)$ and $\tilde{C}(f_2)$ respectively corresponding to an 8-sample time series. In Figure 13.25(a) (where $v = 1$) the phasors advance by $2\pi(1/8)$ rad $= 45°$ each time τ is incremented. In Figure 13.25(b) (where $v = 2$) the phasors advance by $2\pi(2/8)$ rad $= 90°$. (There is an additional π rad phase change when the corresponding time series sample is negative.)

13.5.6 Inverse DFS and DFT

Consider the forward DFS:

$$\tilde{C}(f_v) \; = \; \frac{1}{N} \sum_{\tau=0}^{N-1} v_\tau \, e^{-j2\pi \frac{v}{N}\tau} \quad (\text{V}) \tag{13.54}$$

After rearranging, this represents N cisoids (or phasors) each rotating with a different

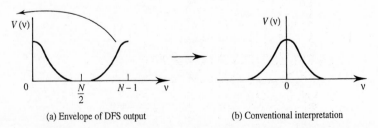

(a) Envelope of DFS output   (b) Conventional interpretation

Figure 13.24 *Interpretation of DFS (or DFT) components $v > N/2$ as negative frequencies.*

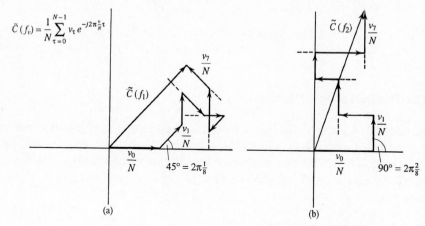

$$\tilde{C}(f_\nu) = \frac{1}{N}\sum_{\tau=0}^{N-1} v_\tau e^{-j2\pi\frac{\nu}{N}\tau}$$

(a)   (b)

Figure 13.25 *Phasor interpretation for DFS showing composition of: (a) fundamental ($\nu = 1$) and (b) second harmonic ($\nu = 2$) components of an 8-sample time series.*

frequency, i.e.:

$$\tfrac{1}{2}\,\tilde{C}(f_{-N/2})\,,\cdots,\,\tilde{C}(f_{-2}),\,\tilde{C}(f_{-1}),\,\tilde{C}(f_0),\,\tilde{C}(f_1),\,\tilde{C}(f_2),\cdots,\,\tfrac{1}{2}\,\tilde{C}(f_{N/2})$$

Each time sample, v_τ, is the sum of these cisoids evaluated at time $t = \tau\Delta t$. Figure 13.26, for example, shows the phasor diagram for $N = 8$ at the instant $t = 3\Delta t$ (i.e. $\tau = 3$). The resultant on this diagram corresponds to the third sample of the time series. (The diagram is referred to as a phasor diagram, here, even though different phasor pairs are rotating at different frequencies.) By inspection of Figure 13.26 the *inverse* DFS can be seen to be:

$$v_\tau = \sum_{\nu=0}^{N-1} \tilde{C}(f_\nu)\, e^{j\frac{2\pi}{N}\nu\tau} \quad \text{(V)} \tag{13.55}$$

Similarly the inverse DFT is:

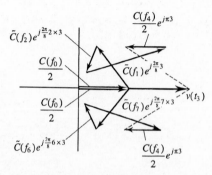

Figure 13.26 *'Phasor' diagram at the instant $t = 3\Delta t$ (i.e. $\tau = 3$) for a real, 8-sample, time series.*

$$v_\tau = \frac{1}{N} \sum_{\nu=0}^{N-1} V_\delta(f_\nu) \, e^{j \frac{2\pi}{N} \nu\tau} \quad \text{(V)} \tag{13.56}$$

(Notice that the factors of $1/N$ are arranged in the DFS, DFT and their inverses, such that applying a given discrete transform, followed immediately by its inverse, results in no change.)

13.5.7 DFT accuracy

Having calculated a spectrum using a DFT, and rearranged the spectral lines as described in section 13.5.3, the question might be asked – how well do the resulting values represent the continuous Fourier transform of the continuous time function underlying the time series? This question is now addressed.

Sampling and truncation errors

There are two problems which have the potential to degrade accuracy. These are:

1. sampling;
2. truncation.

Aliasing, due to sampling, is avoided providing sampling is at the Nyquist rate or higher, i.e.:

$$\frac{1}{\Delta t} \geq 2f_H \quad \text{(Hz)} \tag{13.57}$$

The truncation problem arises because it is only possible to work with time series of finite length. A series with N samples corresponds to an infinite series multiplied by a rectangular window, $\Pi(t/T)$, of width $T = N\Delta t$ s, Figure 13.27. This is equivalent to convolving the FT of $v_\delta(t)$ with the function $T\text{sinc}(Tf)$, Figure 13.28. The convolution smears, or smooths, $V(f)$ on a frequency scale of approximately $2/T$ Hz (i.e. the width of the smoothing function's main lobe). It might be argued that a time series for which the underlying function is strictly time limited to $N\Delta t$ s does not suffer from truncation error. In this case, however, the function cannot be bandlimited and the Nyquist sampling rate would be ∞ Hz making it impossible to avoid aliasing errors. Conversely if the underlying function is strictly bandlimited (allowing the possibility of zero aliasing error)

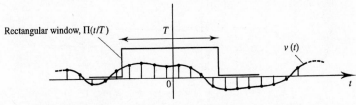

Figure 13.27 *A finite (8-sample) time series interpreted as the product of an infinite series and a rectangular window.*

then it cannot be time limited, making truncation errors unavoidable. (It is, of course, possible to reduce both types of error by sampling at a faster rate and increasing the length of the time series. This means increasing the overall number of samples, however, requiring greater computer resources.)

Frequency sampling, smoothing, leakage and windowing

Discreteness and periodicity are the corresponding properties of a function expressed in frequency and time domains. This is summarised in Table 13.4.

Table 13.4 *Relationship between periodic and discrete signals.*

Time domain	Frequency domain
Periodic	Discrete
Discrete	Periodic

Since the DFT gives discrete samples of $V(f)$ (at least approximately) then the function of which it is the precise FT (or FS) is periodic. Furthermore, since the time series, $v_\delta(t)$, is discrete it should have a periodic spectrum. This implies that a replicated version of a DFT, such as that shown in Figure 13.29(b), is the exact FT (or FS) of a function such as that shown in Figure 13.29(a).

Figure 13.30 illustrates how the implicit periodicity of both the underlying time series and the exact spectrum of this time series impacts on truncation errors. It can be seen that in addition to smoothing of the baseband spectrum, energy leaks into the DFT from the higher frequency spectral replicas via the window spectrum with which the exact spectrum is convolved. This type of error is called leakage and can be reduced by decreasing the sidelobes of the window function's spectrum. Shaping of the time domain window function to realise low sidelobes in its spectrum tends, however, to increase the width of the spectral main lobe. There is, consequently, a trade-off to be made between leakage and smoothing errors. The optimum shape for a time domain window depends on the particular application, the data being transformed and engineering judgement. The commonly encountered window functions are discussed in [Brigham].

The effect of crude windowing, with a rectangular shape for example, does not, *necessarily*, lead to poor accuracy in the calculated spectrum of a time series. If the

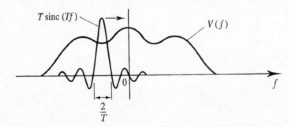

Figure 13.28 *Convolution in frequency domain of FT {v(t)} with FT {Π(t/T)}, corresponding to rectangular windowing in Figure 13.27.*

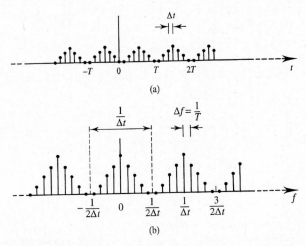

Figure 13.29 *Sampled and replicated versions of a signal in (a) time and (b) frequency domain forming a precise FT or FS pair.*

spectrum of the unwindowed time series is slowly changing on the frequency scale of the oscillations of the window's sidelobes then cancellation tends to occur between the leakage from adjacent sidelobes, Figure 13.31. A corollary of this is a frequency domain manifestation of Gibb's phenomenon, Figure 13.32. If the time series spectrum changes rapidly on the frequency scale of window spectrum oscillations, then as the window spectrum slides across the time series spectrum the convolution process results in window spectrum oscillations being reproduced in the spectrum of the windowed time series. This problem is at its worst in the region of time series spectral discontinuities where it can lead to significant errors. The same effect occurs in the region of any impulses present in the time series spectrum. In particular this means that the DC level (i.e. average value) of a time series should be removed before a DFT is applied. If this is overlooked it is possible that the window spectrum reproduced by convolution with the 0 Hz impulse may obscure, and be mistaken for, the spectrum of the time series data, Figure 13.33. (If impulses in the spectrum of the time series are important they should be identified, and removed, before application of a DFT, and subsequently reinserted into the calculated spectrum.)

Trailing zeros

The frequency spacing of spectral values obtained from a DFT is given by:

$$\Delta f = f_1 = \frac{1}{T} \text{ (Hz)} \tag{13.58}$$

where $T = N\Delta t$ is the (possibly windowed) length of the time series. Δf is sometimes called the resolution of the DFT. (This does not imply that all spectral features on a scale of Δf are necessarily resolved since resolution in this sense may be limited by

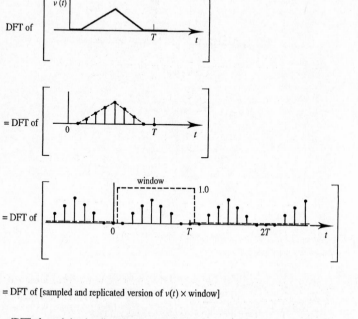

= DFT of [sampled and replicated version of $v(t) \times$ window]

= [DFT of sampled and replicated version of $v(t)$ * [window spectrum]

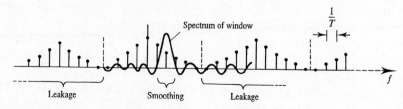

Figure 13.30 *Origin of smoothing and leakage errors in a DFT.*

windowing effects.) Δf can only be made smaller by increasing the length of the time series data record, i.e. increasing T (or N). The alternative is to artificially extend the data record with additional trailing zeros. This is illustrated in Figure 13.34 and is called zero padding. Notice that the highest observable frequency, $f_{N/2}$, determined by the sampling period, Δt, is unaffected by zero padding but, since more samples are included in the DFT operation, there are more samples in the output display, giving finer sampling in the frequency domain. Since the genuine data record has not been extended, however, the underlying resolution is unaltered, the zero padding samples having simply allowed the DFT output to be more finely interpolated [Mulgrew and Grant].

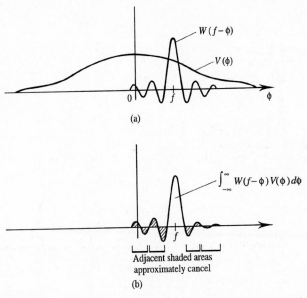

(a)

(b)

Adjacent shaded areas
approximately cancel

Figure 13.31 *Approximate cancellation of leakage errors for a windowed signal with slowly changing spectrum: (a) convolution of underlying spectrum with oscillating window spectrum; (b) result of convolution at frequency shift $\phi = f$.*

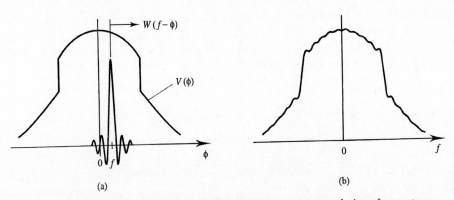

(a)

(b)

Figure 13.32 *Frequency manifestation of Gibb's phenomenon: (a) convolution of a spectrum containing rapidly changing region with oscillating spectrum of a windowing function; (b) resulting oscillations in spectrum of windowed function.*

Random data and spectral estimates

Consider the stationary random process, $v_s(t)$, illustrated in Figure 13.35(a). If several segments of this random time series are windowed and transformed using a DFT then the result is a number of voltage spectra, each with essentially meaningless phase

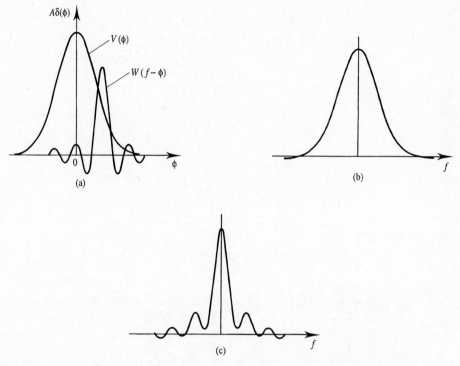

Figure 13.33 *(a) Convolution of a signal spectrum containing an impulse at 0 Hz with a sinc function reflecting rectangular windowing of a signal with a large DC value, (b) result of convolution excluding DC impulse, (c) convolution including DC impulse.*

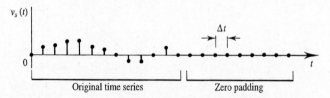

Figure 13.34 *Illustration of zero padding a time series with trailing zeros.*

information. If power spectra, $G_i(f_v)$, are obtained using:

$$G_i(f_v) = \frac{1}{T} \mid \text{DFT} \left[v_\tau \, w_i(\tau) \right] \Delta t \mid^2 \quad (\text{V}^2\text{Hz}^{-1}) \tag{13.59}$$

where $w_i(\tau)$ is the window applied to the ith segment of the data and $G_i(f_v)$ is the power spectrum of the ith segment, then the size of a spectral line at a given frequency will fluctuate randomly from spectrum to spectrum, Figure 13.35(b). Perhaps surprisingly, the random error in the calculated samples of the power spectral density does not decrease if

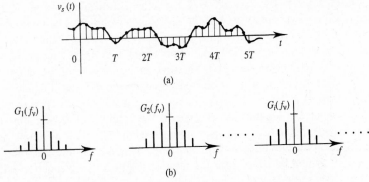

Figure 13.35 *(a) Stationary random process divided into T-second segments, (b) power spectra calculated from each time series segment.*

the length of time series segments is increased. This is because the frequency 'resolution' improves proportionally. An example makes this clear. If the length of a time series segment is doubled the spacing of its spectral lines is decreased by a factor of 2. Twice the number of spectral lines are generated by the DFT and thus the *information per spectral line remains constant*. The random errors in such a spectrum can be decreased by:

1. Averaging the corresponding values (lines) over the spectra found from independent time segments, i.e. using:

$$G(f_v) = \frac{1}{M} \sum_{i=1}^{M} G_i(f_v) \quad (\text{V}^2\text{Hz}^{-1}) \tag{13.60}$$

2. Averaging adjacent values in a single spectra.

An alternative approach is to use the Wiener-Kintchine theorem and transform the autocorrelation function (ACF) of the time series, i.e.:

$$G(f_v) = \text{DFT} \{\text{ACF} [v_\tau \, w(\tau)]\} \, \Delta t \quad (\text{V}^2\text{Hz}^{-1}) \tag{13.61}$$

where the units of the discrete ACF are V^2. The number of spectral lines generated by equation (13.61) now depends on the maximum temporal displacement (or lag) used in the ACF rather than the length of the time series used.

13.6 Discrete and cyclical convolution

For sampled data such as that used in computer simulations the (discrete) convolution between two time series $x_0, x_1, x_2, \cdots, x_{N-1} = \{x_\tau\}$ and $y_0, y_1, y_2, \cdots, y_{M-1} = \{y_\tau\}$ is defined by:

$$z_n = \sum_{\tau=0}^{n} x_\tau \, y_{n-\tau} \quad (\text{V}^2) \tag{13.62}$$

where τ is the (integer) sample number of (both) time series (i.e. $x_\tau = x(\tau \Delta t)$), Δt is the sampling period and $n = 0, 1, 2, \cdots, N + M - 2$ is the shift in sample numbers between x_τ and the time reversed version of y_τ. (If the discrete convolution is being used to evaluate the convolution of a pair of underlying continuous signals, then equation (13.62) must be multiplied by the sampling period, Δt. The result will then have units of V^2s as expected.) The question may be asked, 'does the precise timing of the sampling instants affect the result of the numerical (or discrete) convolution defined in equation (13.62)?'. The slightly surprising answer is generally yes. Figure 13.36 shows a signal sampled at 10 Hz. Dots represent samples which start at the origin (defined by the start of the continuous signal) and crosses represent samples which start half a sample period after the origin. Since convolution involves integration it is usually the case that the set of sampling points which best represent the area of the underlying function will give the best result in the sense that application of equation (13.62) will give an answer closest to the analytical convolution of the underlying functions. This implies that the crosses in Figure 13.36 represent superior sampling instants which is also consistent with an intuitive feeling that a sample should be at the centre of the function segment which it represents.

The crosses in Figure 13.36 also have the advantage that none fall on a point of discontinuity. If sampling at such points cannot be avoided then an improvement in terms of area represented is obtained by assigning a value to that sample equal to the mean of the function value on either side of the discontinuity. In Figure 13.36, therefore, the sample (dot) for the point $t = 1.0$ would be better placed at 0.5 V than 0 V. (This is consistent with the fact that physical signals do not contain discontinuities and that bandlimited signals, corresponding to truncated Fourier series, converge at points of discontinuity, to the mean of the signal value on either side of that discontinuity, see section 2.2.3.)

In practice convolution is often implemented by taking the inverse DFT of the product of the DFTs of the individual time series, i.e.:

$$\{x_\tau\} * \{y_\tau\} \equiv \text{DFT}^{-1}\left[\text{DFT}\{x_\tau\}\,\text{DFT}\{y_\tau\}\right] \quad (V^2) \tag{13.63}$$

For equation (13.63) to yield sensible results the sampling period and length of both time series must be the same. (Zero padding can be used to equalise series lengths if necessary.) Writing out equation (13.63) explicitly (using equations (13.56) and (13.46)), and using primes and double primes to keep track of sample numbers in the different time

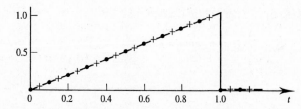

Figure 13.36 *Sampled signal showing alternative sampling instants.*

series:

$$z(\tau) = \{x_{\tau'}\} * \{y_{\tau''}\} = \frac{1}{N} \sum_{v=0}^{N-1} \left[\sum_{\tau'=0}^{N-1} x_{\tau'}\, e^{-j2\pi \frac{v}{N}\tau'} \sum_{\tau''=0}^{N-1} y_{\tau''}\, e^{-j2\pi \frac{v}{N}\tau''} \right] e^{j\frac{2\pi}{N}v\tau}$$

$$= \frac{1}{N} \sum_{v=0}^{N-1} \left[\sum_{\tau'=0}^{N-1} \sum_{\tau''=0}^{N-1} x_{\tau'}\, y_{\tau''}\, e^{-j2\pi \frac{v}{N}(\tau'+\tau'')} \right] e^{j\frac{2\pi}{N}v\tau}$$

$$= \frac{1}{N} \sum_{\tau'=0}^{N-1} \sum_{\tau''=0}^{N-1} x_{\tau'}\, y_{\tau''} \left[\sum_{v=0}^{N-1} e^{j2\pi \frac{v}{N}(\tau-\tau'-\tau'')} \right] \qquad (13.64)$$

The square bracket on the last line of equation (13.64) is zero unless:

$$\tau - \tau' - \tau'' = nN \quad (\text{for } any \ integer \ n) \qquad (13.65)$$

in which case it is equal to N. Equation (13.64) can therefore be rewritten as:

$$z(\tau) = \sum_{\tau'=0}^{N-1} \sum_{\tau''=0}^{N-1} x_{\tau'}\, y_{\tau''} \qquad (\tau'' = \tau - \tau' - nN \ \text{only})$$

$$= \sum_{\tau'=0}^{N-1} x_{\tau'}\, y_{\tau-\tau'-nN} \qquad (13.66)$$

The implication of equation (13.66) is that the τth sample in the convolution result is the same for any n, i.e.:

$$z(\tau) = \ldots$$

$$= \sum_{\tau'=0}^{N-1} x_{\tau'} y_{\tau-\tau'+N}$$

$$= \sum_{\tau'=0}^{N-1} x_{\tau'} y_{\tau-\tau'}$$

$$= \sum_{\tau'=0}^{N-1} x_{\tau'} y_{\tau-\tau'-N}$$

$$= \ldots \qquad (13.67)$$

This means that $y_{\tau-\tau'}$ (i.e. $y_{\tau''}$) is cyclical with period N as shown in Figure 13.37. Alternative interpretations are that as the time shift variable, τ, changes, elements of the time series, $y_{\tau''}$, lying outside the (possibly windowed) time series, $x_{\tau'}$, are recycled as shown in Figure 13.38, or that the time series are arranged in closed loops as shown in Figure 13.39. Figures 13.38 and 13.39 both define the cyclical convolution operation which is the equivalent time series operation to the multiplication of DFTs.

The cyclic (or periodic) convolution of two N-element time series is contained in a series which has, itself, N elements. (For normal, or aperiodic, convolution the series has

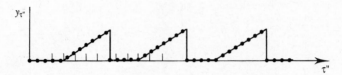

Figure 13.37 *Periodic interpretation of $y_{\tau''}$ resulting in cyclic convolution when using $x_\tau * y_\tau = \mathrm{DFT}^{-1}[\mathrm{DFT}(x_\tau)\mathrm{DFT}(y_\tau)]$ to implement discrete convolution.*

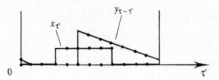

Figure 13.38 *Sample recycling interpretation of y_τ for cyclic convolution.*

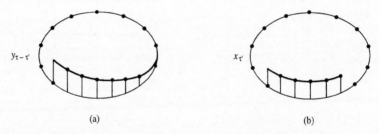

(a)   (b)

Figure 13.39 *Closed loop interpretation of x_τ and y_τ for cyclic convolution.*

$2N-1$ elements.) A consequence is that if two functions are convolved using equation (13.63) (thus implying cyclical convolution) then enough zero padding must be used to ensure that one period of the result is visible, isolated by zeros on either side. If insufficient leading and/or trailing zeros are present in the original time series then the convolved sequences will have overlapping ends, Figure 13.40, making normal interpretation difficult. (In general, to avoid this, the number of leading plus trailing zeros required prior to cyclical convolution is equal to the number of elements in the functions to be convolved from the first non-zero element to the last non-zero element.)

13.7 Estimation of BER

For digital communications the quantity most frequently used as an objective measure of performance is symbol error rate, SER, or equivalently probability of symbol error, P_e. Sometimes more detailed information is desirable, for example it might be important to know whether errors occur independently of each other or whether they tend to occur in

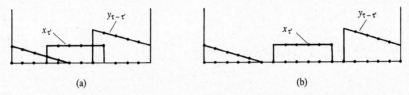

Figure 13.40 *Use of extra zeros to pad x_τ and y_τ in order to avoid overlapping of discrete convolution replicas in cyclical convolution result: (a) insufficient leading/trailing zeros leading to overlapping of replicas; (b) padded functions avoiding overlapping.*

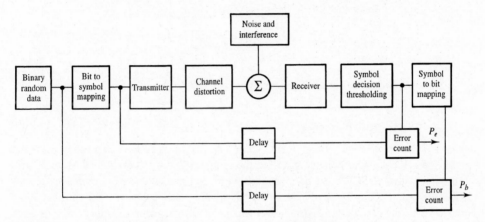

Figure 13.41 *Principles of Monte Carlo evaluation of SER and BER.*

bursts. Here we give a brief outline of two of the methods by which SER (and, if necessary, second order performance information) can be estimated using simulation. These are, the Monte Carlo method and the quasi-analytical method. To some extent these methods represent examples of SER estimation techniques located at opposite ends of a spectrum of techniques. Other methods exist which have potential advantages under particular circumstances. A detailed discussion of many of these methods is given in [Jeruchim *et al.*].

13.7.1 Monte Carlo simulation

This is the conceptually simplest and most general method of estimating SERs. The detected symbol sequence at the receiver is compared symbol by symbol with the (error free) transmitted sequence and the errors are counted, Figure 13.41. The estimated P_e is then given by:

$$P_e = \frac{\text{error count}}{\text{total symbol count}} \tag{13.68}$$

and the SER is given in equation (6.13) as:

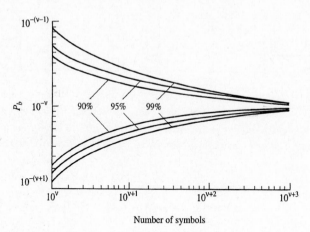

Number of symbols

Figure 13.42 *Confidence bands on P_b when observed value is 10^{-v} for the Monte Carlo technique based on the normal approximation (source: Jeruchim et al., 1984, reproduced with permission of the IEEE).*

$$\text{SER} = P_e R_s \quad \text{(error/s)} \tag{13.69}$$

where R_s is the baud rate. Direct Monte Carlo estimation of BER is effected in a similar way, Figure 13.41. The penalty paid for the simplicity and generality of Monte Carlo simulation is that if SER is to be measured with both precision and confidence for low error rate systems, then large numbers of symbols must be simulated. This implies large computing power and/or long simulation times. If errors occur independently of each other, and a simulation is sufficiently long to count at least 10 errors, then the width of the P_e interval in which we can be 90%, 95% and 99% confident that the true P_e lies may be found from Figure 13.42. If errors are not independent (for example they may occur in bursts due to the presence of impulsive noise) then each error in a given burst clearly yields less information, on average, about the error statistics than in the independent error case. (At its most obvious each subsequent error in a burst is less surprising, and therefore less informative, than the first error.) It follows that a greater number of errors would need to be counted for a given P_e confidence interval in this case than in the independent error case.

13.7.2 Quasi-analytic simulation

Quasi-analytic (QA) simulation can dramatically reduce the required computer power and/or run time compared with Monte Carlo methods. This is because the QA method simulates only the effect of system induced distortion occurring in the signal rather than including the effects of additive noise. It does depend, however, on a knowledge of the total noise pdf at the decision circuit input. The essence of QA simulation is best illustrated by a binary signalling example. Figure 13.43(a) shows an (unfiltered) baseband binary signal, $v_s(t)$, representing the output of a binary information source.

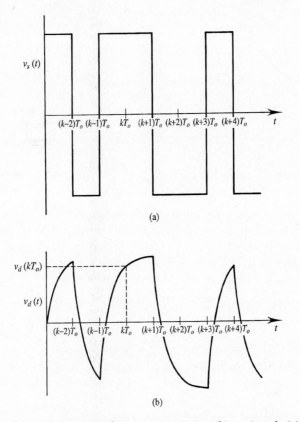

Figure 13.43 *Signals at: (a) transmitter binary source output; (b) receiver decision circuit input after transmission through a distorting but noiseless channel.*

Figure 13.43(b) shows the demodulated, but distorted, signal, $v_d(t)$, at the decision circuit input. If purely additive noise with pdf $p_n(v)$ is present at the decision circuit input the total pdf of signal plus noise at the kth sampling instant is given by:

$$p_{s+n}(v) = p_n[v - v_d(kT_0)] \tag{13.70}$$

where $v_d(kT_0)$ is the decision instant signal. $v_d(kT_0)$ will depend on the history of the bit sequence via the intersymbol interference due to the impulse response (i.e. distorting effect) of the entire system.

The probability, $P_e(k)$, that the noise (if present) would have produced an error at the kth bit sampling instant is:

$$P_e(k) = \int_{-\infty}^{0} p_n[v - v_d(kT_0)] \, dv = \int_{-\infty}^{-v_d(kT_0)} p_n(v) \, dv \tag{13.71(a)}$$

if $v_s(kT_0) > 0$ (i.e. v_s represents a transmitted digital 1), and is:

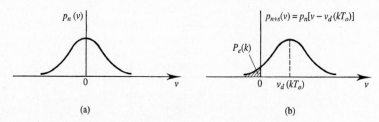

Figure 13.44 *Pdfs of: (a) noise only; (b) signal plus noise at kth sampling instant.*

$$P_e(k) = \int_0^\infty p_n[v - v_d(kT_0)] \, dv = \int_{-v_d(kT_0)}^\infty p_n(v) \, dv \qquad (13.71(b))$$

if $v_s(kT_0) < 0$ (i.e. v_s represents a transmitted digital 0). $p_n(v)$, $p_n[v - v_d(kT_0)]$, and $v_d(kT_0)$ are shown in Figure 13.44. If $p_n(v)$ is Gaussian then the evaluation of the integrals is particularly easy using error function look-up tables or series approximations.

The overall probability of error is found by averaging equations (13.71) over many (N, say) bits, i.e.:

$$P_e = \frac{1}{N} \sum_{k=1}^N P_e(k) \qquad (13.72)$$

N must be sufficiently large to allow essentially all possible combinations of bits, in a time window determined by the memory of the system, to occur. This ensures that all possible distorted signal patterns will be accounted for in the averaging process.

EXAMPLE 13.1
As an example of the power and utility of proprietary simulation packages a simplified model of a satellite communications system is analysed here using Signal Processing WorkSystem (SPW)™ marketed by Comdisco Systems, Inc. SPW is immensely powerful as a simulation tool and only a

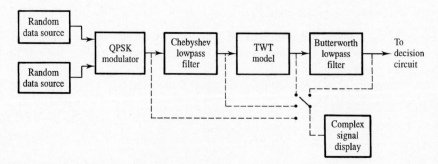

Figure 13.45 *Block diagram for example simulation.*

fraction of its facilities are illustrated in this example. The system is shown in Figure 13.45 which has been redrawn to simplify that produced using SPW's block diagram editor. It consists of a pair of (independent) random binary data sources which provide the input to a QPSK modulator, Figure 11.23. The symbol rate of each source is 1.0 baud.

The output of the modulator is an equivalent baseband signal and is therefore complex. The transmitted, uplink, signal is filtered by a 6-pole Chebyshev filter. The bandwidth of this (equivalent baseband) filter is 1.1 Hz. A satellite channel is modelled analytically using a non-linear input/output characteristic typical of a travelling wave tube (TWT) amplifier operating with 3.0 dB of input back-off which accounts for both AM/AM and AM/PM distortion (see section 14.3.3). A parameter of the TWT model is its average input power which is set to 2.0 W. The received, downlink, signal is filtered by a 6-pole Butterworth filter. Signal sink blocks are used to record the time series data generated as the simulation progresses. It is these signal records which are analysed to produce the required simulation results.

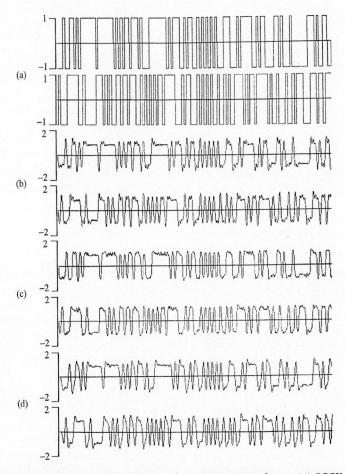

Figure 13.46 *In-phase (upper) and quadrature (lower) time series data at (a) QPSK modulator output and the outputs of: (b) Chebyshev filter; (c) TWT; (d) Butterworth filter.*

The simulation sampling rate used is 16 Hz which, for a baud rate of 1.0, corresponds to 16 sample/symbol. The number of samples simulated (called iterations in SPW) is 3,000. Unrealistic parameters in simulations (e.g. unit baud rate and filter bandwidths of the order of 1 Hz) are typical. It is, usually, necessary only for parameters to be correct relative to each other for the simulation to give useful results. (The interpretation of those results must, of course, be made in the proper context of the parameters used.) Figure 13.46(a), (b), (c) and (d) show the simulated (complex) time series data at the input to the QPSK modulator, and the outputs from the Chebyshev filter, the TWT and the Butterworth filter. Note that the 'random' component of the signals in Figure 13.46(b), (c) and (d) is not due to noise but is due entirely to the distortion introduced by the channel.

Figure 13.47(a) and (b) show the magnitude and phase spectra of the pre- and post-filtered transmitted uplink signals respectively, calculated using a 1024 (complex) point FFT. (The time series were windowed with a Hamming function prior to the FFT in this case.) Figure 13.47(c) shows the spectrum of the TWT output. It is interesting to note the regenerative effect that the TWT non-linearity has on the sidelobes of the QPSK signal, previously suppressed by the Chebyshev filter. (This effect was briefly discussed in section 11.4.4.) The frequency axes of the spectra are easily calibrated remembering that the highest observable frequency (shown as 0.5 in the SPW output) corresponds to half the sampling rate, i.e. $16/2 = 8$ Hz, see section 13.5.3.

A QA simulation routine available in SPW has been run to produce the SER versus E_b/N_0 curve shown in Figure 13.48. QA simulation is applicable if the system is linear between the point at which noise is added and the point at which symbol decisions are made. In this example we assume that noise is added at the satellite (TWT) output. This corresponds to the, often realistic, situation in which the downlink dominates the CNR performance of a satellite communication system, see section 14.3.3.

The way in which SPW implements QA simulation is as follows:

1. The equivalent noise bandwidth, B_N, of the system segment between the noise injection point and the receiver decision circuit is estimated from a (separate) simulation of this segment's impulse response, $h(t)$, i.e.:

$$B_N = \int_0^\infty |h(t)|^2 \, dt \qquad (13.73)$$

2. The signal power at the input to the receiver, C (in this example the input to the Butterworth filter) is estimated from the time series contained in the appropriate signal sink file (i.e. iagsig/twtout.sig, see Figure 13.46(c)).

3. The required range and interval values of E_b/N_0 are specified and, for each value of E_b/N_0, the normalised (Gaussian) noise power, $N (= \sigma^2)$, is found using:

$$\sigma^2 = N_0 B_N \quad (\mathrm{V}^2) \qquad (13.74(a))$$

where:

$$N_0 = \frac{CT_b}{(E_b/N_0)} = \frac{C/(R_s H)}{(E_b/N_0)} \quad (\mathrm{V}^2/\mathrm{Hz}) \qquad (13.74(b))$$

(Here H is the number of binary digits/symbol, i.e. it is the entropy assuming zero redundancy. Thus in this example $H = 2$.)

4. If *centre* point sampling is required then the distances, d_1 and d_2, from the decision thresholds of the (complex) sample at the *centre* of each received symbol is calculated, see Figure 13.49. (For MPSK and MQAM systems SPW assumes constellations which are regular, thus allowing

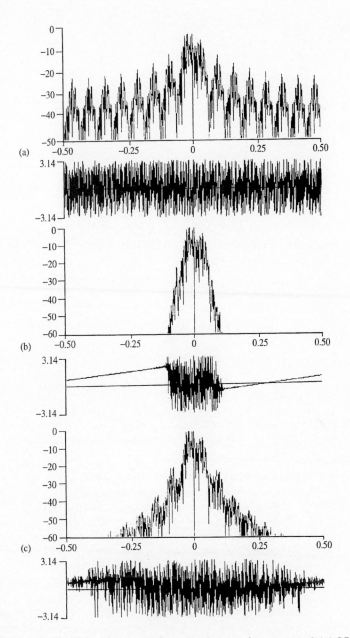

Figure 13.47 *Amplitude (dB) and phase (rad) spectra for signal at outputs of: (a) QPSK modulator; (b) Chebyshev filter; (c) TWT.*

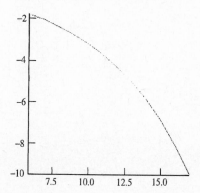

Figure 13.48 *Result of P_e versus E_b/N_0 (dB) for a quasi-analytic (QA) analysis of system shown in Figure 13.45. (P_e is plotted on a logarithmic scale so 10^{-2} is represented by -2.)*

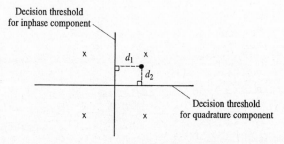

Figure 13.49 *Definition of QPSK received sample distances d_1 and d_2 from the decision thresholds ($\times$ represents transmitted constellation point and $\bullet$ a received sample).*

all constellation points to be folded into the first quadrant before this calculation is carried out.) Although centre point sampling is typical, SPW allows the user to specify which sample within the symbol is to be used as the decision sample.

5. The probability of error for the particular sampling point selected, in the particular symbol being considered, is then calculated using d_1, d_2, σ and the error function.

6. The probability of error is averaged over many (N) symbols. (N is typically a few hundred but should be large enough to allow all sequences of symbols, possible in a time window equal to the duration of the impulse response estimated in 1, to occur at least once.)

Many other analysis routines are provided in SPW, for example the scatter plot of Figure 13.50 which shows the scatter of simulation points, at the Butterworth filter output (decision circuit input), about the nominal points of the QPSK constellation diagram. Figure 13.51 shows the eye diagram (across two symbols) of the received signal's inphase components at the Butterworth filter output, on a much expanded timescale compared to Figure 13.46.

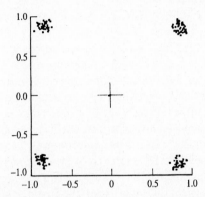

Figure 13.50 *Scatter of simulation QPSK constellation points at Butterworth filter output (decision circuit input) in Figure 13.45.*

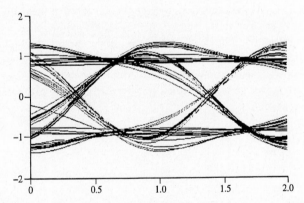

Figure 13.51 *Eye diagram for in-phase component (I) of signal at Butterworth filter output.*

13.8 Summary

Simulation is now a vital part of the design process for all but the simplest of communications systems. It enables performance to be assessed in the presence of noise, interference, and distortion, and allows alternative design approaches to be compared before an expensive construction phase is implemented. Since simulation is based on discrete samples of underlying continuous signals, it is usual for passband systems (with the exception of those which are non-linear and have long memory) to be simulated as equivalent (complex) baseband processes. The distortion introduced by the sampling and quantisation, required for simulation, needs to be carefully considered to ensure it does not significantly alter the simulation results. Noise modelling using random numbers or pseudo-random bit sequences must also be implemented carefully to ensure that simulated noise samples faithfully represent both the spectral and pdf properties of the actual noise present in the system.

Transformation between time and frequency domains during a simulation is frequently required. This is because some simulation processes are more easily (or more efficiently) implemented in one domain rather than the other, and also because some simulation results are more easily interpreted in one domain rather than the other. Digital signal processing algorithms (principally the FFT) are therefore used to translate between domains. An adequate understanding of effects such as smoothing, leakage, windowing and zero padding is required if these algorithms are to be used to best effect. It is also important that the output of DFT/DFS software is properly interpreted and processed if a numerically accurate representation of a power, or energy, spectral density is required.

Finally, in the assessment of most types of digital communications system the principal objective measure of performance is BER. For systems which are linear, between the point at which noise is introduced and the point at which symbol decisions are made, quasi-analytic (QA) simulation is extremely efficient in terms of the computer resources required. QA simulation does require the noise pdf at the decision circuit input to be known, however. The most general method of estimating BER uses Monte-Carlo simulation in which neither a linearity restriction (on the system through which the noise passes prior to detection), nor any apriori knowledge of the noise characteristics at the decision circuit input, is needed. Monte-Carlo simulation can become very expensive in terms of computer power and/or run time, however, if accurate estimates of small error probabilities are required.

Part Three

Applications

Part 3 shows how the principles described in Part Two are applied in a selection of fixed and mobile data applications for voice and video transmission.

The link budget analysis presented in Chapter 12 is extended in Chapter 14 to less idealised fixed service, terrestrial and satellite, microwave communication systems, and includes important propagation effects such as rain fading, multipath fading and signal scintillation. The special problems posed by the exceptionally long range of satellite systems are also discussed as are frequency allocations, multiplexing and multiple accessing schemes.

Mobile, cellular and paging applications are described in Chapter 15 including examples of current systems such as personal cordless, GSM 900, DCS 1800 and DECT. These systems all limit their transmissions to small geographical areas, or cells, permitting frequency reuse in close proximity, without incurring intolerable levels of interference. This maximises the number of active users per cell who can be accommodated per MHz of allocated bandwidth. This chapter also includes evolving standards for CDMA spread spectrum cellular radio, and future satellite-mobile systems.

Finally, Chapter 16 discusses the specific requirements of digitisation, transmission and storage for video applications. It includes examples of HDTV development as well as low bit rate video compression coders using transform and model based coding techniques, as employed in MPEG and other video coding standards.

CHAPTER 14

Fixed point microwave communications

14.1 Introduction

The two most important public service developments in fixed point radio communications over the last 40 years are the installation of national microwave relay networks and international satellite communications. In both cases these systems originally used analogue modulation and frequency division multiplexing and represented simply an increase in capacity, improvement in quality, or increase in convenience, over the traditional wired PSTN. Their subsequent development, however, has been in favour of digital modulation and time division multiplexing. The proliferation of services which can be offered using digital transmission (via ISDN, Chapter 19) has been at least as important in this respect as improvements in digital technology.

14.2 Terrestrial microwave links

Many wideband point-to-point radio communications links employ microwave carriers in the 1 to 20 GHz frequency range. These are used principally by PTTs to carry national telephone and television signals. Their extensive use can be gauged by the number of antenna masts now seen carrying microwave dishes at the summits of hills located between densely populated urban areas. Figure 14.1 shows the main trunk routes of the UK microwave link network. The following points can be made about microwave links:

1. Microwave energy does not follow the curvature of the earth, or diffract easily over mountainous terrain, in the way that MW and LW transmissions do. Microwave transmissions are, therefore, restricted essentially to line of sight (LOS) links.
2. Microwave transmissions are particularly well suited to point-to-point communications since narrow beam, high gain, antennas of reasonable size can be easily designed. (At 2 GHz the wavelength is 0.15 m and a 10-wavelength reflector, i.e. a 1.5 m dish, is still practical. Antenna gains are thus typically 30 to 50 dB.)

Figure 14.1 *The UK microwave communications wideband distribution network.*

3. At about 1 GHz circuit design techniques change from using lumped to distributed elements. Above 20 GHz it becomes difficult and/or expensive to generate reasonable amounts of microwave power.

Antennas are located on high ground to avoid obstacles such as large buildings or hills, and repeaters are used every 40 to 50 km to compensate for path losses. (On a 6 GHz link with a hop distance of 40 km the free space path loss, given by equation (12.71(c)), is 140 dB, see Example 12.6. With transmit and receive antenna gains of 40 dB, however, the basic transmission loss, P_T/C, reduces to 60 dB. This provides sufficient received power for such links to operate effectively.)

Frequencies are allocated in the UK by the Radiocommunication Agency. The principal frequency bands in current use are near 2 GHz, 4 GHz, 6 GHz, 11 GHz and 18 GHz. There are also allocations near 22 GHz and 28 GHz. (The frequency bands listed here are those which are allocated mainly to common carriers. Other allocations for more general use do exist.)

14.2.1 Analogue systems

In the UK microwave links were widely installed during the 1960s for analogue FDM telephony. In these systems each allocated frequency band is subdivided into a number of (approximately) 30 MHz wide radio channels. Figure 14.2 shows how the 500 MHz band, allocated at 4 GHz, is divided into 16 separate channels with 29.65 MHz centre spacings. Each radio channel supports an FDM signal (made up of many individual SSB voice signals) which is frequency modulated onto a carrier. This could be called an SSB/FDM/FM system. Adjacent radio channels use orthogonal antenna polarisations, horizontal (H) and vertical (V), to reduce crosstalk. The allocated band is split into a low and high block containing eight radio channels each, Figure 14.2. One block is used for transmission and the other for reception.

In the superheterodyne receiver a microwave channel filter, centred on the appropriate radio channel, extracts that channel from the block as shown in Figure 14.3. The signal is mixed down to an intermediate frequency (IF) for additional filtering and amplification. The microwave link can accommodate traffic simultaneously in all 16 channels. It is customary, however, not to transmit a given data signal on the same radio channel over

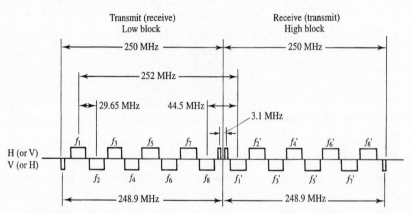

Figure 14.2 *Splitting of a microwave frequency allocation into radio channels.*

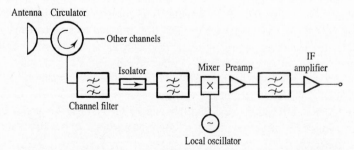

Figure 14.3 *Extraction (dropping) of a single radio channel in a microwave repeater.*

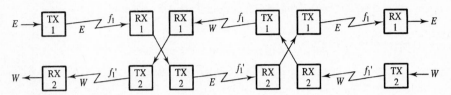

Figure 14.4 *Frequency allocations on adjacent repeaters.*

successive hops. Thus in Figure 14.4, for example, f_1 is used for the first hop but f_1', Figure 14.2, is used on the second hop. Under anomalous propagation conditions, temperature inversions can cause ducting to occur with consequent low loss propagation over long distances, Figure 14.5. This may cause overreaching, with signals being received not only at the next repeater but also at the one after that. Moving the traffic to a different radio channel on the next hop thus ensures that interference arising via this mechanism is essentially uncorrelated with the received signal.

14.2.2 Digital systems

The first digital microwave (PSTN) links were installed in the UK in 1982 [Harrison]. They operated with a bit rate of 140 Mbit/s at a carrier frequency of 11 GHz using QPSK modulation. In more recent systems there has been a move towards 16- and 64-QAM (see Chapter 11). Figure 14.6 shows a block diagram of a typical microwave digital radio terminal. (Signal pre-distortion shown in the transmitter can be used to compensate for distortion introduced by the high power amplifier.)

The practical spectral efficiency of 4 to 5 bit/s/Hz, which 64-QAM systems offer, means that the 30 MHz channel can support a 140 Mbit/s multiplexed telephone traffic signal. Figure 14.7 is a schematic of a digital regenerative repeater, assuming DPSK modulation, for a single 30 MHz channel. Here the circulator and channel filter access the part of the microwave spectrum where the signal is located. With the ever increasing demand for high capacity transmission, digital systems have a major advantage over the older analogue systems in that they can operate satisfactorily at a much lower carrier-to-noise ratio, Figure 11.21. There is also strong interest in even higher level modulation schemes, for example 1024-QAM, to increase the capacity of the traditional 30 MHz channel still further.

Microwave radio links at 2 and 18 GHz are also being applied, at low modulation rates, in place of copper wire connections, in rural communities for implementing the local loop exchange connection [Harrison]. (These are configured as point-to-multipoint

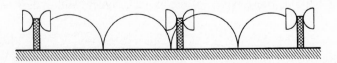

Figure 14.5 *Schematic illustration of ducting causing overreaching.*

links, not unlike the multidrop systems of section 18.2.2. Furthermore the channel access scheme can use derivatives of the digital cellular systems of section 15.4.)

14.2.3 LOS link design

The first-order design problem for a microwave link, whether analogue or digital, is to ensure adequate clearance over the underlying terrain. Path clearance is affected by the following factors:

1. Antenna heights.
2. Terrain cover.
3. Terrain profile.
4. Earth curvature.
5. Tropospheric refraction.

The last of these, tropospheric refraction, occurs because the refractive index, n, of the troposphere depends on its temperature, T, pressure, P, and water vapour partial pressure, e. Since significant changes in P, T and e make only small differences to n, the tropospheric refractive index is usually characterised by the related quantity, refractivity, defined by $N = (n - 1)10^6$.

A good model, [ITU-R, Rec. 453], relating N to P, T and e is:

$$N = 77.6 \frac{P}{T} + 3.73 \times 10^5 \frac{e}{T^2} \tag{14.1}$$

where P and e are in millibars and T is in K, as in Chapter 12. Although refractivity is dimensionless the term 'N units' is usually appended to its numerical value.

Equation (14.1) and the variation of P, T and e with altitude result in an approximately exponential decrease of N with height, h, i.e.:

$$N(h) = N_s e^{-h/h_0} \tag{14.2}$$

where N_s is surface refractivity and h_0 is a scale height. ITU-R Rec. 369 defines a standard reference troposphere, Figure 14.8, as one for which $N_s = 315$ and $h_0 = 7.35$ km. Over the first kilometre of height this is usually approximated as a linear height dependence with refractivity gradient given by:

$$\left. \frac{dN}{dh} \right|_{first\ km} = N(1) - N(0)$$

$$= -40 \ (N \text{ units/km}) \tag{14.3}$$

The vertical gradient in refractive index causes microwave energy to propagate not in straight lines but along approximately circular arcs [Kerr] with radius of curvature, r, given by:

$$\frac{1}{r} = -\frac{1}{n} \frac{dn}{dh} \cos \alpha \tag{14.4}$$

Here α is the grazing angle of the ray to the local horizontal plane, Figure 14.9. For

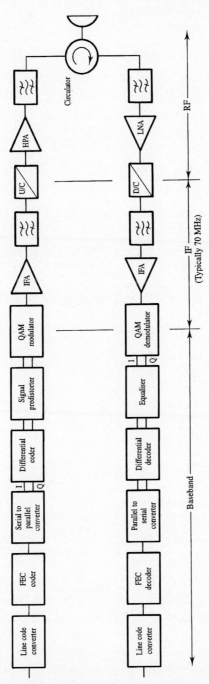

Figure 14.6 *Block diagram of a typical microwave digital radio terminal.*

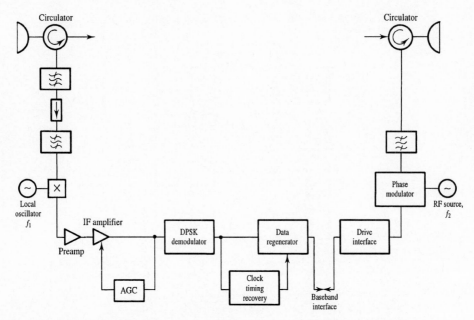

Figure 14.7 *Digital DPSK regenerative repeater for a single 30 MHz radio channel.*

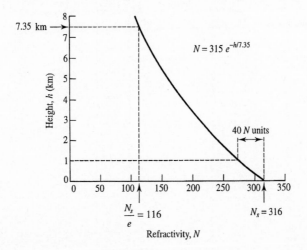

Figure 14.8 *ITU-R standard refractivity profile.*

terrestrial LOS links α is small, i.e. cos $\alpha \approx 1$, and for a standard troposphere with $n \approx 1$ and $dn/dh = -40 \times 10^{-6}$ the radius of curvature is 25,000 km. Microwave energy under these conditions thus bends towards the earth's surface but with a radius of curvature much larger than that of the earth itself, Figure 14.10. (The earth's mean radius, a, may be taken to be 6371 km.) There are two popular ways in which ray curvature is

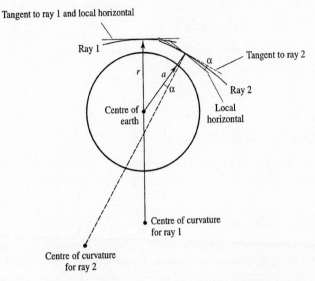

Figure 14.9 *Illustration of circular paths for rays in atmosphere with vertical n-gradient (α = 0 for ray 1, α ≠ 0 for ray 2). Geometry distorted for clarity.*

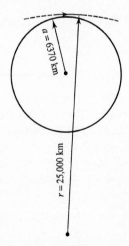

Figure 14.10 *Relative curvatures of earth's surface and ray path in a standard atmosphere.*

accounted for in path profiling. One subtracts the curvature of the ray (equation (14.4)) from that of the earth giving (for $n = 1$ and $\alpha = 0$):

$$\frac{1}{a_e} = \frac{1}{a} + \frac{dn}{dh} \tag{14.5}$$

This effectively decreases the curvature of both ray and earth until the ray is straight,

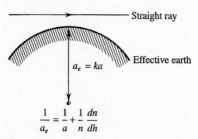

Figure 14.11 *Straight ray model. (Note n ≈ 1.)*

Figure 14.11, resulting in an effective earth radius, a_e. This may be called the straight ray model. Alternatively the negative of this transformation can be applied, i.e.:

$$\frac{1}{r_e} = -\left(\frac{1}{a} + \frac{dn}{dh} \right)$$
(14.6)

which decreases the curvature of both ray and earth until the earth is flat, Figure 14.12, resulting in an effective radius of curvature, r_e, for the ray. This may be called the flat earth model. For the straight ray model the ratio of effective earth radius to actual earth radius is called the k factor, i.e.:

$$k = \frac{a_e}{a}$$
(14.7)

which, using equation (14.5) and the definition of refractivity, is given by:

$$k = \frac{1}{1 + a\, \dfrac{dN}{dh} \times 10^{-6}}$$
(14.8)

k factor thus represents an alternative way of expressing refractivity lapse rate, dN/dh. In the lowest kilometre of the standard ITU-R troposphere $k = 4/3$. Table 14.1 shows the relationship between k and dN/dh and Figure 14.13 shows the characteristic curvatures of rays under standard, sub-refracting, super-refracting and ducting conditions, each drawn (schematically) on a $k = 4/3$ earth profile. For standard meteorological conditions path profiles can be drawn on special $k = 4/3$ earth profile paper.

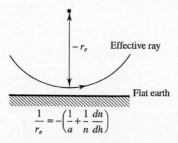

Figure 14.12 *Flat earth model (n≈1 and negative radius indicates ray is concave upwards).*

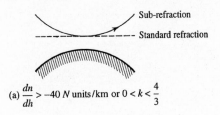

(a) $\dfrac{dn}{dh} > -40\ N$ units/km or $0 < k < \dfrac{4}{3}$

Sub-refraction

Standard refraction

Standard refraction

Super-refraction

Ray for $\dfrac{dn}{dh} = -157\ N$ units/km

(b) $-157 < \dfrac{dN}{dh} < -40$ or $k > \dfrac{4}{3}$

Ducting

(c) $\dfrac{dn}{dh} < -157$ or $k < 0$

Figure 14.13 *Characteristic ray trajectories drawn with respect to a k = 4/3 earth radius.*

Table 14.1 *Equivalent values of refractivity lapse rate and k factor.*

$\dfrac{dN}{dh}$ (N units/km)	157	78	0	−40	−100	−157	−200	−300
k factor	$\dfrac{1}{2}$	$\dfrac{2}{3}$	1	$\dfrac{4}{3}$	2.75	∞	−3.65	−1.09

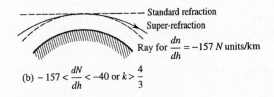

Sub-refraction   Super-refraction   Ducting

<-------------------------->|<----------------->|<-------------------------->

EXAMPLE 14.1

Show that the line-of-sight range over a smooth spherical earth is given by $L = \sqrt{2ka}\,(\sqrt{h_T} + \sqrt{h_R})$ where k is a k-factor, a is earth radius, h_T is transmit antenna height and h_R is receive antenna height.

Consider the geometry shown in Figure 14.14.

$$L_1^2 = (h_T + a_e)^2 - a_e^2$$
$$= h_T^2 + 2\,h_T\,a_e$$

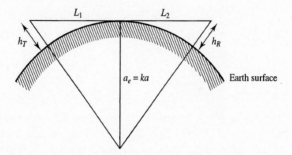

Figure 14.14 *Geometry for maximum range LOS link over a smooth, spherical, earth.*

Similarly:

$$L_2^2 = h_R^2 + 2\,h_R\,a_e$$

Since $h_T \ll a_e$ and $h_R \ll a_e$ then:

$$L_1^2 \approx 2\,h_T\,a_e$$

$$L_2^2 \approx 2\,h_R\,a_e$$

Therefore:

$$L = L_1 + L_2 = \sqrt{2\,a_e}\,\left(\sqrt{h_T} + \sqrt{h_R}\right)$$

Fresnel zones and path profiling

Fresnel zones are defined by the intersection of Fresnel ellipsoids with a plane perpendicular to the LOS path. The ellipsoids in turn are defined by the loci of points, Figure 14.15, which give an excess path length, ΔL, over the direct path of:

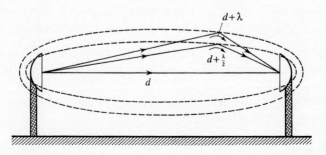

Figure 14.15 *Fresnel ellipsoids.*

$$\Delta L = n\frac{\lambda}{2} \quad (m) \tag{14.9}$$

The first, second, and third Fresnel radii are illustrated in Figure 14.16. For points not too near either end of the link (i.e. for $r_n \ll d_1$ and $r_n \ll d_2$) the nth Fresnel zone radius, r_n, is given by:

$$r_n \approx \sqrt{n\frac{\lambda\, d_1\, d_2}{d_1 + d_2}} \tag{14.10}$$

where d_1 and d_2 are distance from transmitter and receiver respectively. With personal computers it is now practical to plot ray paths for any k-factor on a 4/3 earth radius profile and, making due allowance for terrain cover, find the minimum clearance along the path as a fraction of r_1. (The required clearance of a link is often specified as a particular fraction of the first Fresnel zone radius under meteorological conditions corresponding to a particular k factor.) This design exercise is called path profiling. The principles of path profiling, however, are simple and best illustrated by describing the process as it might be implemented manually:

1. A terrain profile (ignoring earth curvature) is plotted on Cartesian graph paper choosing any suitable vertical and horizontal scales.
2. For all points on the path profile likely to give rise to poor clearance (e.g. local maxima, path midpoint, etc.) the height of the earth's bulge, h_B, is calculated using:

$$h_B = \frac{d_1\, d_2}{2a} \quad (m) \tag{14.11}$$

   (h_B is the height of the, smooth, earth's surface above the straight line connecting the points at sea level below transmit and receive antennas.)
3. The effect of tropospheric refraction on clearance is calculated using h_B and k factor to give:

$$h_{TR} = h_B(k^{-1} - 1) \quad (m) \tag{14.12(a)}$$

4. The required Fresnel zone clearance (in metres) is calculated using:

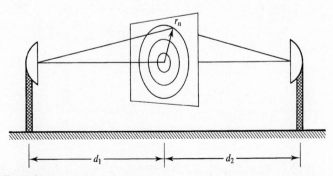

Figure 14.16 *Fresnel zones.*

$$h_{FZC} = f \, r_1 \quad (\text{m}) \tag{14.12(b)}$$

where r_1 is given by equation (14.10) and f is typically 1.0 for $k = 4/3$. (In the UK $f = 0.6$ for $k = 0.7$ is often used.)

5. An estimate is obtained for the terrain cover height, h_{TC}.
6. $h_{TC} + h_{TR} + h_B + h_{FZC}$ is plotted on the terrain path profile.
7. A straight line passing through the highest of the plotted points then allows appropriate antenna heights at each end of the link to be established, Figure 14.17.

There will usually be a trade-off between the height of transmit and receive antennas. Normally these heights would be chosen to be as equal as possible in order to minimise the overall cost of towers. If one or more points of strong reflection occur on the path, however, the antenna heights may be varied to shift these points away from areas of high reflectivity (such as regions of open water).

If adequate Fresnel zone clearance cannot be guaranteed under all conditions then diffraction may occur leading to signal fading. If the diffracting obstacle can be modelled as a knife edge, Figure 14.18, the diffraction loss can be found from Figure 14.19. The parameter v in this figure is called the Fresnel diffraction parameter and is given by:

$$v = \sqrt{\frac{2}{\lambda} \frac{d_1 + d_2}{d_1 \, d_2}} \, h \tag{14.13}$$

where d_1 and d_2 locate the knife edge between transmitter and receiver, and h is the

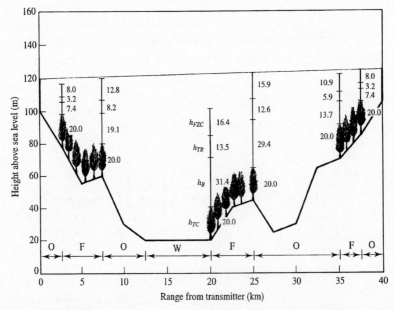

Figure 14.17 *Path profile for hypothetical 4 GHz LOS link designed for 0.6 Fresnel zone (FZ) clearance when k = 0.7 (O = open ground, F = forested region, W = water).*

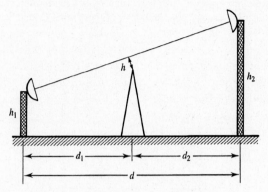

Figure 14.18 *Definition of clearance, h, for knife edge diffraction.*

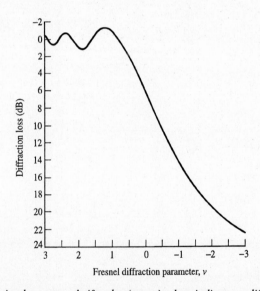

Figure 14.19 *Diffraction loss over a knife-edge (negative loss indicates a diffraction gain).*

minimum clearance measured perpendicularly from the LOS path over the diffracting edge. (If the obstacle blocks the LOS path then h is negative.) If the obstacle cannot be modelled as a knife edge then the diffraction loss will generally be greater. A model for cylindrical edge diffraction, given in [Doughty and Maloney], takes the form of a knife edge loss plus a curvature loss and a correction factor.

EXAMPLE 14.2
A 6 GHz microwave link operating over flat ground is 36 km long. 15 km from the transmitter a

long building with a pitched roof is oriented perpendicularly to the LOS path. The height of the building to the vertex of the roof is 30 m. How high must the transmitter and receiver antenna towers be if, under standard refraction conditions ($k = 4/3$): (i) first Fresnel zone clearance is to be maintained and (ii) diffraction fading is not to exceed 8 dB?

(i)

$$\lambda = \frac{3 \times 10^8}{6 \times 10^9} = 0.05 \text{ m}$$

$$h_B = \frac{d_1 \, d_2}{2a} = \frac{15000 \times 21000}{2 \times 6371000} = 24.72 \text{ m}$$

$$h_{TR} = h_B \, (k^{-1} - 1) = 24.72 \left(\frac{3}{4} - 1\right) = -6.18 \text{ m}$$

$$h_{FZC} = f \, r_1 = 1.0 \sqrt{\frac{\lambda \, d_1 \, d_2}{d_1 + d_2}} = \sqrt{\frac{0.05 \times 15000 \times 21000}{36000}} = 20.92 \text{ m}$$

Required antenna heights are given by:

$$h_T = h_R = h_B + h_{TR} + h_{FZC} = 24.72 - 6.18 + 20.92$$

$$= 39.46 \text{ m}$$

(ii) If diffraction loss is not to exceed 8 dB then, from Figure 14.19, $v = -0.25$. Rearranging equation (14.13):

$$h = \frac{v}{\sqrt{\dfrac{2}{\lambda} \dfrac{d_1 + d_2}{d_1 \, d_2}}} = \frac{-0.25}{\sqrt{\dfrac{2}{0.05} \dfrac{36000}{15000 \times 21000}}} = -3.70 \text{ m}$$

Required antenna heights are now given by:

$$h_T = h_R = h_B + h_{TR} + h = 24.72 - 6.18 - 3.70 = 14.84 \text{ m}$$

14.2.4 Other propagation considerations for terrestrial links

Once a LOS link has been profiled a detailed link budget would normally be prepared. A first-order free space signal budget is described in Chapter 12 (section 12.4.2). The actual transmitter power required, however, will be greater than that implied by this calculation since, even in the absence of diffraction, some allowance must be made for atmospheric absorption, signal fading and noise enhancements. For terrestrial microwave links with good clearance the principal mechanisms causing signal reductions are:

1. Background gaseous absorption.
2. Rain fading.
3. Multipath fading.

   The principal sources of noise enhancement are:

1. Thermal radiation from rain.

2. Interference caused by precipitation scatter and ducting.
3. Crosstalk caused by cross-polarisation.

In addition to reducing the CNR (and/or carrier to interference ratio) signal fading processes also have the potential to cause distortion of the transmitted signal. For narrowband transmissions this will not usually result in a performance degradation beyond that due to the altered CNR. For sufficiently wideband digital transmissions, however, such distortion may result in extra BER degradation.

Background gaseous absorption

Figure 14.20 shows the nominal specific attenuation, $\gamma(f)$ (dB/km), of the atmosphere due to its gaseous constituents for a horizontal path at sea level. (Separate, and more detailed, curves for the dry air and water vapour constituents are given in [ITU-R, Rec. 676].) The rather peaky nature of this curve is the result of molecular resonance effects (principally due to water vapour and oxygen). For normal applications it is adequate simply to multiply $\gamma(f)$ by the path length, L, and include the resulting attenuation in the detailed link budget.

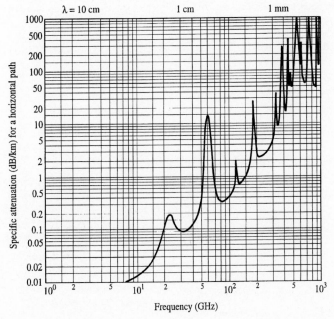

Figure 14.20 *Specific attenuation due to gaseous constituents for transmissions through a standard atmosphere (20°C, pressure one atmosphere, water vapour content 7.5 g/m³). (Source: CCIR Report 719, 1986, reproduced with permission of ITU.)*

Rain fading

Significant rain intensity (measured in mm/h) occurs only for small percentages of time. The specific attenuation which is present during such rain is, however, large for frequencies above a few GHz. When fading of this sort occurs the BER can degrade due to a reduction in signal level and an increase in antenna noise temperature.

For signals with normal bandwidths the fading can be considered to be flat and the distortion negligible. In this case the attenuation exceeded for a particular percentage of time is all that is required. This can be estimated as follows:

1. The point rain rate, R_P (mm/h), exceeded for the required time percentage is found either from local meteorological records or by using an appropriate model.
2. The equivalent uniform (or line) rain rate, R_L, is found using curves such as those in Figure 14.21. This accounts for the non-uniform spatial distribution of rain along the path.
3. The specific attenuation, γ_R, for the line rain rate of interest may be estimated using Figure 14.22, or more accurately, using a formula of the form:

$$\gamma_R = k\, R_L^{\alpha} \ \ (\text{dB/km})\tag{14.14}$$

where the regression coefficients k and α are given in [ITU-R, Rec. 838] for a particular polarisation and frequency of interest.

4. The total path attenuation, A (dB), exceeded for the required percentage of time is then found using:

$$A = \gamma_R L \ \ (\text{dB})\tag{14.15}$$

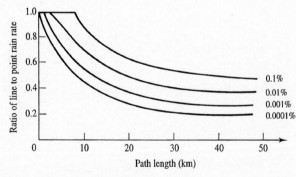

Figure 14.21 *Relationship between point and line rain rates as a function of hop length and percentage time point rain rate is exceeded (source: Hall and Barclay, 1989, reproduced with permission of Peter Peregrinus).*

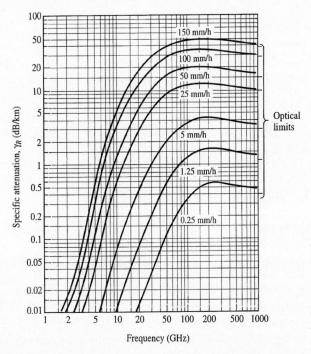

Figure 14.22 *Specific attenuation due to rain (curves derived on the basis of spherical raindrops). (Source: CCIR Report 721, 1990, reproduced with the permission of ITU.)*

Multipath fading on LOS links

Although broadband measurements have shown that several (identifiably discrete) propagation paths can occur on LOS links the consensus at present is that, in practice, the channel can be adequately modelled with only three paths [Rummler, 1979]. The impulse response of such a three-ray model is (to within a multiplicative constant):

$$h(t) = \delta(t) + \alpha\delta(t - T_1) + \beta\delta(t - T_2) \tag{14.16}$$

where T_1 and T_2 define the relative delays between the rays, and α and β define the relative strengths of the rays. The frequency response, $H(f)$, corresponding to equation (14.16) is:

$$H(f) = 1 + \alpha e^{-j2\pi fT_1} + \beta e^{-j2\pi fT_2} \tag{14.17}$$

If $T_1 \ll 1/B$, where B is the bandwidth of the transmitted signal, then fT_1 is approximately constant over the range $f_c \pm B/2$ (where f_c is the carrier frequency). Equation (14.17) can be written as:

$$H(f) = ae^{-j\theta}\left[1 + \frac{\beta}{a}e^{-j(2\pi fT_2 - \theta)}\right] \tag{14.18}$$

where:

$$ae^{-j\theta} = 1 + \alpha e^{-j2\pi fT_1} \tag{14.19(a)}$$

$$a = \sqrt{1 + \alpha^2 + 2\alpha \cos(2\pi fT_1)} \tag{14.19(b)}$$

$$\theta = \tan^{-1}\left(\frac{\alpha \sin 2\pi fT_1}{1 + \alpha \cos 2\pi fT_1} \right) \tag{14.19(c)}$$

The factor $e^{-j\theta}$ in equation (14.18) represents an overall phase shift caused by the first delayed ray which can be ignored.[1] Furthermore, since a notch occurs in the channel's amplitude response at:

$$f(= f_o) = (\theta \pm \pi)/2\pi T_2 \quad \text{(Hz)} \tag{14.20}$$

then equation (14.18) can be rewritten as:

$$H(f) = a\left[1 - be^{-j2\pi(f-f_o)T_2}\right] \tag{14.21}$$

where $b = \beta/a$. If $b < 1$ then equation (14.21) represents a minimum phase frequency response. If unity and b in equation (14.21) are interchanged then the amplitude response, $|H(f)|$, remains unchanged but the phase response becomes non-minimum. Apart from an extra overall phase shift of $-2\pi T_2 f + 2\pi T_2 f_o + \pi$ (the first term representing pure delay, the second term representing intercept distortion and the third term representing signal inversion) the non-minimum phase frequency response can be modelled by changing the sign of the exponent in equation (14.21) which can then be written as:

$$H(f) = a\left[1 - be^{\pm j2\pi(f - f_o)T_2}\right] \tag{14.22}$$

to represent both minimum and non-minimum phase conditions [CCIR, Report 718]. Equation (14.22) has four free parameters (a, b, f_o and T_2) which is excessive, in the sense that if $T_2 < (6B)^{-1}$ then normal channel measurements are not accurate enough to determine all four parameters uniquely. One parameter can therefore be fixed without significantly degrading the formula's capacity to match measured channel frequency responses. The parameter normally fixed is T_2 and the value chosen for it is often set using the rule $T_2 = (6B)^{-1}$. Using this rule, joint statistics of the quantities:

$$A = 10\log_{10}\left[a^2(1 + b^2)\right] \tag{14.23(a)}$$

$$B = 10\log_{10}(2a^2b) \tag{14.23(b)}$$

---

[1] A frequency response with $\phi(f) = \theta$, where θ is a non-zero constant, has zero delay distortion (since $d\phi/df = 0$) but non-zero intercept distortion. Intercept distortion, however, results only in a change in the 'phase' relationship between a wave packet carrier and its envelope, the shape of the envelope remaining unchanged. Providing such distortion changes slowly compared to the time interval between phase training sequences it will have no significant effect on the performance of a PSK communications system.

have been determined experimentally for several specific links [CCIR, Report 338]. *A* and *B* appear to be well described by a bivariate Gaussian random variable. The mean, variance and correlation of this distribution for three different links are shown in Table 14.2.

Table 14.2 *Measured statistics of multipath channel parameters (after CCIR, Report 338).*

Path length	Frequency	Bandwidth	$\bar{A}$	$\bar{B}$	σ_A	σ_B	ρ
37 km	11 GHz	55 MHz	−7.25	−5.5	6.5	6.5	0.45
50 km	11 GHz	55 MHz	−8.25	−3.0	9.0	8.5	0.75
42 km	6 GHz	25 MHz	−24.00	−14.5	7.5	7.5	0

Equations (14.23(a) and (b)) can be inverted to allow the joint distribution of *a* and *b* to be derived.

Multipath fading is a potentially severe problem for wideband digital links. Full transversal filter equalisation, which can compensate for both the amplitude and the phase distortions in the transmission medium, is therefore generally desirable. Current digital repeater and receiver equipment has considerable sophistication, often incorporating equalisers which remove distortion arising not only from propagation effects but also from other sources such as non-ideal filters. ([ITU-R, Rec. 530] details methods for predicting fading due to multipath effects in combination with other clear air mechanisms.)

Mechanisms of noise enhancement

The presence of loss on a propagation path will increase the noise temperature of a receiving antenna due to thermal radiation. Rain is the most variable source of such loss

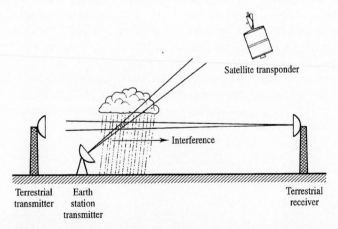

Figure 14.23 *Hydrometeor scatter causing interference between co-frequency systems.*

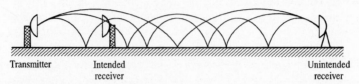

Transmitter   Intended   Unintended
                receiver   receiver

Figure 14.24 *Interference caused by ducting.*

along most microwave paths. During a severe fade the aperture temperature (see Chapter 12) of the antenna, T_A, will approach the physical temperature of the rain producing the loss. A simple model relating T_A to fade depth is:

$$T_A = T_M(1 - \alpha) \tag{14.24}$$

where α is the fade depth expressed as a fraction of unfaded power (i.e. $\alpha = 10^{-\text{atten(dB)}/10}$) and T_M is the absorption temperature of the medium. T_M is not identical to the medium's physical temperature since, in addition to direct thermal radiation, it accounts for scattering and other effects. T_M has been related empirically to surface air temperature, T_S, by [Freeman]:

$$T_M = 1.12\, T_S - 50 \quad \text{(K)} \tag{14.25}$$

where T_S is expressed in K. Equations (14.24) and (14.25) are only appropriate, of course, for values of attenuation which result in aperture temperatures in excess of the clear sky values indicated by Figure 12.25 (i.e. approximately 100 K for terrestrial links between 1 and 10 GHz). Clear sky aperture temperature increases rapidly with frequency above 10 GHz and, for terrestrial links, can be assumed to be 290 K at 20 GHz and above for most purposes (see later, Figure 14.30).

A quite different mechanism of noise enhancement also occurs due to rain, namely precipitation scatter. Here the existence of a common volume between the transmitting antenna of one system and the receiving antenna of a nominally independent system has the potential to couple energy between the two, Figure 14.23. Such coupled energy represents interference but it is thought that its effects are conservatively modelled by an equal amount of thermal noise power. Meteorological ducting conditions can also cause anomalous, long distance, propagation of microwave signals resulting in interference or crosstalk between independent systems, Figure 14.24. Detailed modelling of precipitation scatter, ducting and methods to reduce the interference caused by them has recently been the subject of intensive study [COST 210].

EXAMPLE 14.3
A 7 GHz terrestrial LOS link is 40 km long and operates with good ground clearance (several Fresnel zones) in a location which experiences a rain rate of 25 mm/h or greater for 0.01% of time. The link has a bandwidth of 1.0 MHz, antenna gains of 30.0 dB and an overall receiver noise figure of 5.0 dB. Assuming the ratio of line to point rain rate, R_L/R_p, is well modelled by Figure 14.21,

for this location, estimate the transmitter power required to ensure a CNR of 30.0 dB is achieved or exceeded for 99.99% of time.

From Figure 14.21:

$$R_{L,0.01} = 0.4 \, R_{p,0.01} = 0.4 \times 25 = 10 \text{ mm/h}$$

Using Figure 14.22, specific attenuation, $\gamma_R = 0.2$ dB/km, i.e.:

$$A_{R,0.01} = \gamma_R \, L = 0.2 \times 40 = 8.0 \text{ dB}$$

Assuming a surface temperature of 290 K, equation (14.25) gives:

$$T_M = 1.12 \, T_S - 50 = 1.12 \times 290 - 50 = 275 \text{ K}$$

From equation (14.24):

$$T_A = T_M \left(1 - 10^{-\frac{A}{10}} \right) = 275 \left(1 - 10^{-\frac{8}{10}} \right) = 231 \text{ K}$$

Equivalent noise temperature of receiver:

$$T_e = (f - 1) \, 290 = \left(10^{\frac{5}{10}} - 1 \right) 290 = 627 \text{ K}$$

Total system noise temperature:

$$T_{syst} = T_A + T_e = 231 + 627 = 858 \text{ K}$$

Noise power:

$$N = kTB = 1.38 \times 10^{-23} \times 858 \times 1 \times 10^6 = 1.18 \times 10^{-14} \text{ W} = -139.3 \text{ dBW}$$

Received carrier power:

$$C = P_T + G_T - \text{FSPL} + G_R - A_R$$

$$= P_T + 30.0 - 20 \log_{10} \left(\frac{4\pi \, 40 \times 10^3}{0.0429} \right) + 30.0 - 8.0$$

$$= P_T - 89.4 \text{ dBW}$$

$$\frac{C}{N} = P_T - 89.4 - (-139.3) = P_T + 49.9 \text{ dB}$$

$$P_T = \frac{C}{N} - 49.9 = 30.0 - 49.9 = -19.9 \text{ dBW or 10.1 dBm}$$

(Note that the calculation shown above has been designed to illustrate the application of the preceding material and probably contains spurious precision. In particular a simpler estimate for T_A of 290 K results in an increase in P_T of only 0.3 dB.)

Cross-polarisation and frequency reuse

It is possible to double the capacity of a microwave link by using orthogonal polarisations for independent co-frequency channels. For QPSK systems this combines with the advantage of orthogonal inphase and quadrature signalling to give a four-fold increase in transmission capacity over a simple BPSK, single polarisation, system. Unusual propagation conditions along the radio path can give rise to polarisation changes (called cross-polarisation) which results in potential crosstalk at the receiver. The use of corrugated horns and scalar feeds in the design of microwave antennas reduces antenna induced cross-polarisation during refractive bending or multipath conditions. Similarly the use of vertical and horizontal linear polarisations minimises rain induced cross-polarisation which occurs when the angle between the symmetry axis of falling rain drops and the electric field vector of the signal is other than 0° or 90°.

14.3 Fixed point satellite communications

The use of satellites is one of the three most important developments in telecommunications over the past 40 years. (The other two are cellular radio and the use of optical fibres.) Geostationary satellites, which are essentially motionless with respect to points on the earth's surface and which first made satellite communications commercially feasible, were proposed by the scientist and science fiction writer Arthur C. Clarke. The geostationary orbit lies in the equatorial plane of the earth, is circular and has the same sense of rotation as the earth, Figure 14.25. Its orbital radius is 42,164 km and since the earth's mean equatorial radius is 6,378 km its altitude is 35,786 km. (For simple calculations of satellite range from a given earth station, the earth is assumed to be spherical with radius 6,371 km.) There are other classes of satellite orbit which have advantages over the geostationary orbit for certain applications. These include highly

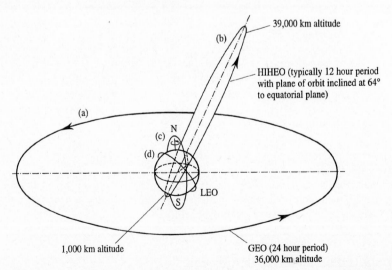

Figure 14.25 *Selection of especially useful satellite orbits: (a) geostationary (GEO); (b) highly inclined highly elliptical (HIHEO); (c) polar orbit; and (d) low earth (LEO).*

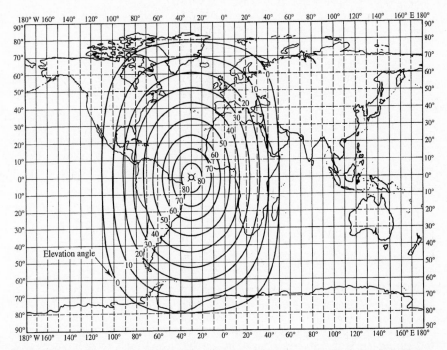

Figure 14.26 *Coverage areas as a function of elevation angle for a satellite with global beam antenna (from CCIR Handbook, 1988, reproduced with the permission of ITU).*

inclined highly elliptical (HIHE) orbits, polar orbits and low earth orbits (LEOs), Figure 14.25. For fixed point communications the geostationary orbit is the most commercially important, for the following reasons:

1. Its high altitude means that a single satellite is visible from a large fraction of the earth's surface (42% for elevation angles > 0° and 38% for elevation angles > 5°). Figure 14.26 shows the coverage area as a function of elevation angle for a geostationary satellite with a global beam antenna. (Elevation angles < 10° are not recommended, and angles < 5° are not used, because of the severe scintillation and fading of the signal, and high antenna noise temperature, which occur due to the large thickness of atmosphere traversed by the propagation path.)

2. No tracking of the satellite by earth station antennas is necessary.

3. No handover from one satellite to another is necessary since the satellite never sets.

4. Three satellites give almost global coverage, Figure 14.27. (The exception is the polar regions with latitudes > 81° for elevation angles > 0° and latitudes > 77° for elevation angles > 5°.)

5. No Doppler shifts occur in the received carrier.

The following advantages apply to geostationary satellites but may also apply, to a greater or lesser extent, to some communication satellites in non-geostationary orbits:

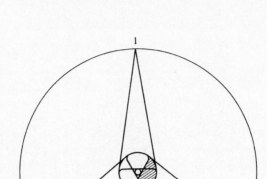

Figure 14.27 *Global coverage (excepting polar regions) from 3 geostationary satellites.*
(Approximately to scale, innermost circle represents 81° parallel.)

1. The communications channel can be either broadcast or point-to-point.
2. New communication network connections can be made simply by pointing an antenna at the satellite. (For non-geostationary satellites this is not entirely trivial since tracking and/or handover are usually necessary.)
3. The cost of transmission is independent of distance.
4. Wide bandwidths are available, limited at present only by the speed of the transponder electronics and receiver noise performance.

Despite their very significant advantages, geostationary satellites do suffer some disadvantages. These include:

1. Polar regions are not covered (i.e. latitudes $> 77°$ for elevation angles $> 5°$).
2. High altitude means large FSPL (typically 200 dB).
3. High altitude results in long propagation delays (approximately 1/8 s for uplink and 1/8 s for downlink).

The latter disadvantage means that inadequately suppressed echoes from subscriber receiving equipment arrive at the transmitting subscriber 0.5 s after transmission. Both the 0.25 s delay between transmission and reception and the 0.5 s delay between transmission and echo can be disturbing to telephone users. This means echo suppression or cancellation equipment, which may use techniques [Mulgrew and Grant] not dissimilar to the equalisers of section 8.5, is almost always required.

14.3.1 Satellite frequency bands and orbital spacing

Figure 14.28 shows the principal European frequency bands allocated to fixed point satellite services. The 6/4 GHz (C-band) allocation is now fairly congested and new systems are being implemented at 14/11 GHz (Ku-band). 30/20 GHz (Ka-band) systems are currently being investigated. The frequency allocation at 12 GHz is mainly for direct broadcast satellites (DBS). Inter-satellite crosslinks use the higher frequencies as here there is no atmospheric attenuation. The higher of the two frequencies allocated for a satellite communications system is invariably the uplink frequency. This is because the satellite has limited antenna size and a high antenna noise temperature (typically 290 K). The gain of the satellite receiving antenna (and therefore the satellite G/T) is maximised by using the higher frequency on the uplink.

The reason why two frequencies are necessary at all (one for the uplink and one for the downlink) is that the isolation between the satellite transmit and receive antennas is finite. Since the satellite transponder has enormous gain there would be the possibility of positive feedback and oscillation if a frequency offset were not introduced.

Although the circumference of a circle of radius 42,000 km is large, the number of satellites which can be accommodated in the geostationary orbit is limited by the need to illuminate only one satellite when transmitting signals from a given earth station. If other satellites are illuminated then interference may result. For practical antenna sizes 4° spacing is required between satellites in the 6/4 GHz bands. Since narrower beamwidths are achievable in the 14/11 GHz band, 3° spacing is permissible here and in the 30/20 GHz band spacing can approach 1°.

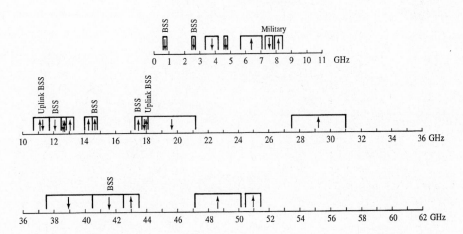

Figure 14.28 *Approximate uplink (↑) and dowlink (↓) allocations for region 1 (Europe, Africa, former USSR, Mongolia) fixed satellite, and broadcast satellite (BSS), services.*

14.3.2 Earth station look angles and satellite range

Figure 14.29 shows the geometry of an earth station (E) and geostationary satellite (S). Some careful trigonometry shows that the earth station antenna elevation angle, α, the azimuth angle, β, and the satellite range from the earth station, R_{ES}, are given by:

$$\alpha = \tan^{-1}\left[(\cos\gamma - 0.15127)/\sin\gamma\right] \tag{14.26}$$

$$\beta = \pm\cos^{-1}\left[-\tan\theta_E/\tan\gamma\right] \tag{14.27}$$

$$R_{ES} = 23.188 \times 10^6 \sqrt{3.381 - \cos\gamma} \quad\text{(m)} \tag{14.28}$$

where γ, the angle subtended by the satellite and earth station at the centre of the earth, is given by:

$$\gamma = \cos^{-1}\left[\cos\theta_E \cos(\phi_E - \phi_S)\right] \tag{14.29}$$

θ_E is the earth station latitude (positive north, negative south), ϕ_E and ϕ_S are the earth station and satellite longitude respectively (positive east, negative west).

Notice that azimuth, β, is defined clockwise (eastwards) from north. Negative β therefore indicates an angle anticlockwise (westwards) from north. The negative sign is taken in equation (14.27) if the earth station is east of the satellite and the positive sign is

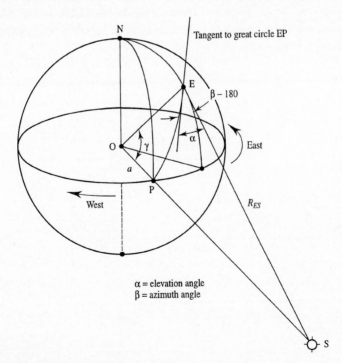

Figure 14.29 *Geostationary satellite geometry.*

taken otherwise.

EXAMPLE 14.4
Find the look angles and range to a geostationary, Indian Ocean, satellite located at 60.0° E as seen from Edinburgh in the UK (55.95° N, 3.20° W).

$$\gamma = \cos^{-1}[\cos\theta_E \cos(\phi_E - \phi_S)]$$

$$= \cos^{-1}[\cos(55.95°)\cos(-3.20° - 60.0°)]$$

$$= 75.38° \ (= 1.316 \text{ rad})$$

$$\alpha = \tan^{-1}[(\cos\gamma - 0.15127)/\sin\gamma]$$

$$= \tan^{-1}[(\cos(1.316) - 0.15127)/\sin(1.316)]$$

$$= 0.1038 \text{ rad} \ (= 5.95°)$$

$$\beta = \pm\cos^{-1}[-\tan\theta_E/\tan\gamma]$$

$$= \pm\cos^{-1}[-\tan(55.95°)/\tan(75.38°)]$$

$$= \pm112.7°$$

Since the earth station is west of the satellite the upper (positive) sign is taken, i.e. the satellite azimuth is +112.7° (eastwards from North). Now from equation (14.28):

$$R_{ES} = 23.188 \times 10^6 \sqrt{3.381 - \cos\gamma}$$

$$= 23.188 \times 10^6 \sqrt{3.381 - \cos(1.316)}$$

$$= 4102 \times 10^4 \text{ m}$$

$$= 41,020 \text{ km}$$

14.3.3 Satellite link budgets

On the uplink the satellite antenna, which looks at the warm earth, has a noise temperature of approximately 290 K. The noise temperature of the earth station antenna, which looks at the sky, is usually in the range 5 to 100 K for frequencies between 1 and 10 GHz, Figure 12.25. For frequencies above 10 GHz resonance effects of water vapour and oxygen molecules, at 22 GHz and 60 GHz respectively, become important, Figure 14.30. For wideband multiplex telephony, the front end of the earth station receiver is also cooled so that its equivalent noise temperature, T_e, section 12.3.1, is very much smaller than 290 K. This therefore achieves a noise floor that is very much lower than −174 dBm/Hz.

The gain, G, of the cooled low noise amplifier boosts the received signal so that the following amplifiers in the receiver can operate at room temperature. Traditional operating frequencies for satellite communication systems were limited at the low end to

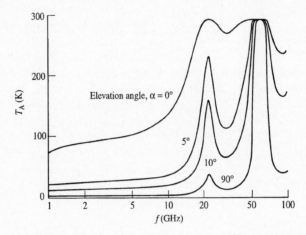

Figure 14.30 *Antenna aperture temperature, T_A, in clear air (pressure one atmosphere, surface temperature 20°C, surface water vapour concentration 10 g/m³). (Source: CCIR Report 720, 1990, reproduced with the permission of the ITU.)*

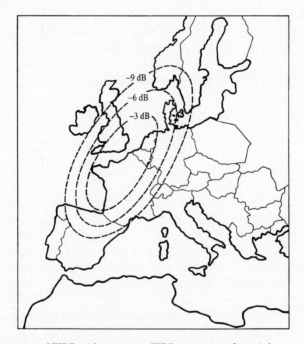

Figure 14.31 *Contours of EIRP with respect to EIRP on antenna boresight.*

>1 GHz by galactic noise and at the high end to <15 GHz by atmospheric thermal noise and rain attenuation, Figure 12.25. However in recent years a more detailed physical and statistical understanding of rain fading, and other hydrometeor effects, has made operation in higher frequency bands practical.

The satellite transponder is the critical component in a satellite link as its transmitter is invariably power limited by the onboard power supply (i.e. solar cell area and battery capacity). The downlink usually has the worst power budget and this often constrains performance. EIRP is also dependent on satellite antenna design and in particular the earth 'footprint'. Spot beam antennas covering only a small part of the earth, e.g. Figure 14.31, have higher gain giving EIRP values of 30 to 40 dBW or greater.

In addition to being power limited the satellite transponder must be as small and light as possible due to the large launching costs, per kg of weight and m^3 of space. This means that the transponder's high power amplifier (HPA), Figure 14.32, must be operated

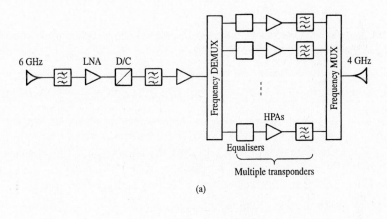

(a)

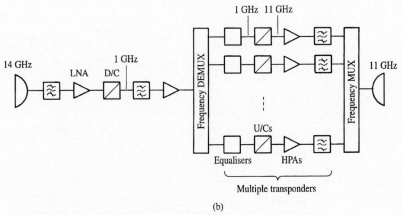

(b)

Figure 14.32 *Simplified block diagram of satellite transponders: (a) single conversion C-band; (b) double conversion Ku-band (redundancy not shown).*

at as high an output power as possible to maintain adequate downlink CNR. As a consequence the transponder is operated in its non-linear region near saturation, resulting in amplitude to amplitude (AM/AM) and amplitude to phase (AM/PM) conversion, Figure 14.33. Intermodulation products (IPs), arising due to mixing of nominally independent signals, simultaneously present in the transponder, can be severe, effectively reducing the overall CNR. Some *back-off* from the transponder saturating input and output power is therefore necessary. (For digital satellite systems the IP problem and resulting need for back-off is usually less serious than for analogue systems, section 14.3.5.) Input and output back-off (BO_i and BO_o respectively) are shown in Figure 14.33. Typical values of back-off are a few dB, BO_i being somewhat greater than BO_o. The transmitted power for the satellite downlink and received power for the satellite uplink during clear sky conditions are therefore given respectively by:

$$P_T = P_{o\,sat} - BO_o \qquad (14.30)$$

$$C = P_{i\,sat} - BO_i \qquad (14.31)$$

where $P_{o\,sat}$ and $P_{i\,sat}$ are the saturated transponder output and input powers. The uplink CNR can therefore be calculated using:

$$\left(\frac{C}{N}\right)_u = W_s - BO_i + 10\log_{10}\left(\frac{\lambda^2}{4\pi}\right) + \left(\frac{G}{T}\right)_s + 228.6 - 10\log_{10} B \qquad (14.32(a))$$

where W_s is the power density (dBW m^{-2}) at the satellite receiving antenna required to saturate the transponder, $10\log_{10}(\lambda^2/4\pi)$ is the effective area of an isotrope (dBm^2), $(G/T)_s$ is the satellite G/T (dB K^{-1}), 228.6 (= $-10\log_{10} k$) is Boltzmann's constant (dBW Hz^{-1}K^{-1}), and $10\log_{10} B$ is bandwidth (dB Hz). Note that the transmitting earth station EIRP, EIRP$_e$, and the received power density at the satellite are related by:

$$W_s - BO_i = EIRP_e - \text{spreading loss} - L_{Au} \quad \text{(dBW m}^{-2}) \qquad (14.32(b))$$

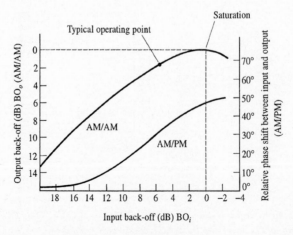

Figure 14.33 *Amplitude and phase characteristic for typical satellite transponder TWT amplifier.*

where L_{Au} is uplink atmospheric attenuation. The equivalent formula, to equation (14.32(a)), for the downlink is:

$$\left(\frac{C}{N}\right)_d = \text{EIRP}_s - \text{BO}_o + \left(\frac{G}{T}\right)_e - \text{FSPL}_d + 228.6 - 10\log_{10} B - L_{Fd} - L_{Ad} \quad (14.33)$$

where EIRP_s is the saturated satellite EIRP, $(G/T)_e$ is the earth station G/T, L_{Fd} represents the downlink fixed losses (such as earth station transmission line loss) and L_{Ad} is the downlink atmospheric attenuation.

Table 14.3 *Typical 4/6 GHz satellite link budget.*

Uplink (6 GHz)		
Saturation flux density		-72.2 dBW m^{-2}
Input back-off, BO$_i$		-5.8 dB
Satellite antenna gain, G_R	23.1 dB	
Satellite system noise temperature, T_{syst}	27.6 dBK	
Satellite G/T	-4.5 dB/K	-4.5 dB/K
Effective area of isotrope		-37.0 dB m^2
Minus Boltzmann's constant		228.6 dBW Hz^{-1}K^{-1}
Minus transponder bandwidth		-75.6 dB Hz
Resulting clear sky CNR$_u$		33.5 dB

Downlink (4 GHz)		
Saturated transponder output power, P_o	7.0 dBW	
Satellite antenna gain, G_T	22.5 dB	
EIRP$_s$	29.5 dBW	29.5 dBW
Output back-off, BO$_o$		-3.2 dB
FSPL, $20\log_{10}(4\pi R/\lambda)$		-196.1 dB
Earth station antenna gain, G_R	58.3 dB	
Clear sky earth station noise temperature, T_{syst}	18.8 dBK	
Clear sky G/T	39.5 dB/K	39.5 dB/K
Minus Boltzmann's constant		228.6 dBW Hz^{-1}K^{-1}
Minus transponder bandwidth		-75.6 dB Hz
Atmospheric attenuation, L_{Ad}		-0.8 dB
Fixed losses, L_{Fd}		-2.0 dB
Resulting clear sky CNR$_d$		19.9 dB

In satellite multiplex telephony systems, large earth station antennas with kW transmitters give very large EIRP_e (typically 90 dBW) and the uplink exhibits good noise, or E_b/N_0, performance. The transmitter power on the satellite is typically restricted by battery and solar cell capacity to 10 to 100 W (and hence EIRP_s is typically restricted to 30 to 50 dBW). The downlink often, therefore, limits overall CNR and BER performance.

A typical, clear sky, link budget for a 6/4 GHz satellite communications system is shown in Table 14.3.

As illustrated schematically in Figure 14.34 the received downlink carrier power, C_d, for a transparent transponder, is given by:

$$C_d = C_u G\, L_d \quad \text{(W)} \tag{14.34}$$

where C_u is the *received* uplink power, G is the operating gain of the transponder and L_d represents *all* downlink losses. Furthermore, the total received downlink noise power, N, can be expressed by:

$$N = N_u G\, L_d + N_d \quad \text{(W)} \tag{14.35}$$

where N_u is the noise power contributed by the uplink and N_d is the noise power contributed by the downlink. Thus:

$$\frac{N}{C_d} = \frac{N_u G\, L_d + N_d}{C_u G\, L_d}$$

$$= \frac{N_u}{C_u} + \frac{N_d}{C_d} \tag{14.36}$$

The overall CNR for the satellite link can therefore be written as:

$$\frac{C}{N} = \frac{1}{\left(\dfrac{C}{N}\right)_u^{-1} + \left(\dfrac{C}{N}\right)_d^{-1}} \tag{14.37}$$

(dropping the subscript d from the received signal power, C, to conform to convention).

More accurate calculations of satellite systems performance must account for the effects of intermodulation products and interference on uplink and/or downlink. Since noise, intermodulation and interference processes are independent, equation (14.37) can be extended as follows:

$$\frac{C}{N} = \frac{1}{\left(\dfrac{C}{N}\right)_u^{-1} + \left(\dfrac{C}{N}\right)_d^{-1} + \left(\dfrac{C}{N}\right)_{IP}^{-1} + \left(\dfrac{C}{I}\right)_u^{-1} + \left(\dfrac{C}{I}\right)_d^{-1}} \tag{14.38}$$

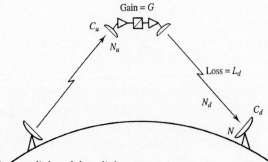

Figure 14.34 *CNRs on uplink and downlink.*

where $(C/N)_{IP}$ is the carrier to intermodulation noise ratio and C/I is the carrier to interference ratio. For digital satellite systems (section 14.3.6) in which only a single carrier is present in the transponder at any given time, then intermodulation noise is absent and $(C/N)_{IP}^{-1}$ is zero. For the example link budget in Table 14.3 the overall clear sky CNR, assuming no intermodulation products and neglecting interference, would be:

$$\frac{C}{N} = \frac{1}{(\text{antilog}_{10} \, 3.35)^{-1} + (\text{antilog}_{10} \, 1.99)^{-1}} \qquad (14.39)$$

$$= \frac{1}{(2239)^{-1} + (97.72)^{-1}} = \frac{1}{0.045 \times 10^{-2} + 1.023 \times 10^{-2}}$$

$$= \frac{1}{1.068 \times 10^{-2}}$$

$$= 93.63 = 19.7 \text{ dB}$$

Notice that here, as in many cases, the CNR performance of the system is dominated by the downlink.

The clear sky carrier to noise ratio calculated above allows only for gaseous background attenuation in the term L_A. In practice a satellite system link budget must also account for fading of the signal and enhancement of the noise (both due, mainly, to the sporadic presence of rain along the propagation path).

Rain induced specific attenuation (in dB/km) varies with frequency, Figure 14.22. The effective length of the propagation path subject to a given rain event depends on elevation angle and climatic factors. *Typical* fade margins included in link budgets to account for 99.99% of meteorological conditions are given in Table 14.4 for the different satellite frequency bands. The detailed calculation of gaseous background attenuation and rain margins is described in section 14.3.4.

Table 14.4 *Typical values for fade margin in different frequency bands.*

Elevation angle	C	Ku	Ka
10°	2.0 dB	8 dB	15 dB
30°	1.0 dB	6 dB	10 dB
90°	0.7 dB	5 dB	8 dB

EXAMPLE 14.5
An 11.7 GHz satellite downlink operates from geosynchronous orbit with 25 W of transmitter output power connected to a 20 dB gain antenna with 2 dB feeder losses. The earth station is at a range of 38,000 km from the satellite and uses a 15 m diameter receive antenna, with 55% efficiency, feeding a low noise (cooled) amplifier which results in a receiver system noise temperature of 100 K. If an E_b/N_0 of 20 dB is required for adequate BER performance what maximum bit rate can be accommodated using BPSK modulation, assuming performance is limited by the downlink and atmospheric attenuation can be neglected?

Transmitter power = 14 dBW

EIRP = 14 + 20 − 2 = 32 dBW = 62 dBm

Free space loss = 20 log10 $(4\pi 38 \times 10^6)/0.0256 = 205.4$ dB

Receiver antenna $G_r = \dfrac{4\pi}{\lambda^2} \dfrac{\pi d^2}{4}\, 0.55 = 62.7$ dB

Received power level = 62 − 205.4 + 62.7 dBm = −80.7 dBm

Receiver noise at 100 K noise temperature = −178.6 dBm/Hz

If $BT_0 = 1.0$ then $E_b/N_0 = C/N$.

Available margin for E_b/N_0 and modulation bandwidth = 178.6 − 80.7 = 97.9 dB Hz.

If $E_b/N_0 = 20$ dB then the margin for modulation = 77.9 dB Hz = 61.6 MHz.

With BPSK at 1 bit/s/Hz then the modulation rate can be 61.6 Mbit/s.

Alternatively:

If $G_r = 62.7$ dB and $T_{syst} = 100$ K then $G/T = 42.7$ dB/K

Then radiated power at receiver antenna = +62 − 205.4 dBm = −143.4 dBm

Power at receiver input = −143.4 + 42.7 dBm/K = −100.7 dBm/K

Boltzmann's constant = −198.6 dBm/Hz/K

Difference = 97.9 dBHz

Allowing 20 dB for acceptable E_b/N_0 leaves 77.9 dBHz which will support a 61.6 Mbit/s BPSK symbol rate.

14.3.4 Slant path propagation considerations

The discussion of satellite link budgets in section 14.3.3 referred to atmospheric effects which must be accounted for to achieve adequate system availability. The principal effects which contribute to changes in signal level on earth–space paths from that expected for free space propagation are:

1. Background atmospheric absorption.
2. Rain fading.
3. Scintillation.

The principal mechanisms of noise and interference enhancement are:

1. Sun transit.
2. Rain enhancement of antenna temperature.
3. Interference caused by precipitation scatter and ducting.
4. Crosstalk caused by cross-polarisation.

Background gaseous absorption

Gaseous absorption on slant path links can be described by $A = \gamma L$ but with L replaced by effective path length in the atmosphere, L_{eff}. L_{eff} is less than the physical path length in the atmosphere due to the decreasing density of the atmosphere with height. In practice the total attenuation, $A(f)$, is usually calculated using curves of zenith attenuation, Figure 14.35, and a simple geometrical dependence on elevation angle, α,

i.e.:

$$A(f) = \frac{A_{zenith}(f)}{\sin \alpha} \tag{14.40}$$

A_{zenith} is the one-way total zenith attenuation and depends on both frequency and surface pressure (reflecting the height of the earth station above sea level). Curve A is for a dry atmosphere and curve B includes the effect of water vapour at a concentration which is typical of temperate climates. (The scale height of the water vapour concentration is 2 km.) Correction factors have been derived which can be used with Figure 14.35 and equation (14.40) to find slant-path gaseous attenuation for other surface pressures and water vapour densities [Freeman].

Rain fading

The same comments can be made for rain fading on slant paths as those which have already been made for terrestrial paths. The slant-path geometry, however, means that the calculation of effective path length depends not only on the horizontal structure of the rain but also on its vertical structure.

One model [ITU-R, Rec. 618] for predicting the rain fading exceeded for a given percentage of time therefore incorporates the following formula which derives an

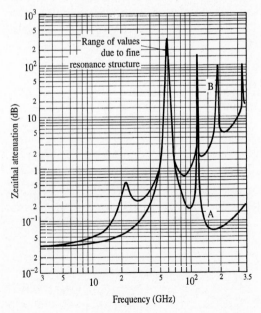

Figure 14.35 *Total ground level zenith attenuation (15°C, 1013 mb) for, A, a dry atmosphere and, B, with a surface water vapour content of 7.5 g/m³ decaying exponentially with height. (Source: ITU-R Rec. 676, 1995, reproduced with the permission of the ITU.)*

effective rain height from earth station latitude, i.e.:

$$h_R \text{ (km)} = \begin{cases} 3.0 + 0.028\phi, & 0 < \phi < 36° \\ 4.0 - 0.075(\phi - 36), & \phi \geq 36° \end{cases} \tag{14.41}$$

The slant path length below the rain height is found from Figure 14.36, i.e.:

$$L_s = \frac{h_R - h_E}{\sin \alpha} \text{ (km)} \tag{14.42}$$

where h_E is the height of the earth station and α is the slant path elevation angle. (A more accurate formula which takes account of earth curvature is used for $\alpha < 5°$.) The ground projection of L_s is calculated using:

$$L_G = L_s \cos \alpha \text{ (km)} \tag{14.43}$$

and a path length reduction formula appropriate for an exceedance value of 0.01% of time is applied, i.e.:

$$r_{0.01} = \frac{1}{1 + L_G(e^{0.015 R_{0.01}})/35} \tag{14.44}$$

(For $R_{0.01} > 100$ mm/h then 100 mm/h is used instead of $R_{0.01}$.) The one-minute rain rate exceeded for 0.01% of time, $R_{0.01}$, is estimated, preferably from local meteorological data, and the corresponding 0.01% specific attenuation is calculated using equation (14.14). (Alternatively a first order estimate of $\gamma_{R\,0.01}$ can be made by interpolating the curves of Figure 14.22.) The total path rain attenuation exceeded for 0.01% of time is then given by:

$$A_{0.01} = \gamma_{R\,0.01} L_s r_{0.01} \text{ (dB)} \tag{14.45}$$

The attenuation exceeded for some other time percentage, p (between 0.001% and 1.0%), can be estimated using the empirical scaling law:

$$A_p = A_{0.01} 0.12 p^{-(0.546 + 0.043 \log_{10} p)} \text{ (dB)} \tag{14.46}$$

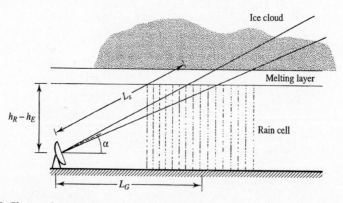

Figure 14.36 *Slant path geometry.*

The slant path attenuation prediction method described here is different from that described in section 14.2.4 (for terrestrial links) in that the former uses an actual rain rate and a path length reduction factor whilst the latter uses an actual path length and a rain rate reduction factor. Clearly both types of model are equivalent in that they take account of the non-uniform distribution of rain along an extended path and either approach can be applied to either type of link. In particular the rain rate prediction method described for satellite links can also be applied to terrestial links [ITU-R, Rec. 530] if L_G and L_s are replaced by the actual terrestial path length, L.

Scintillation

Scintillation refers to the relatively small fluctuations (usually less than, or equal to, a few dB peak to peak) of received signal level due to the inhomogeneous and dynamic nature of the atmosphere. Spatial fluctuations of electron density in the ionosphere and fluctuations of temperature and humidity in the troposphere result in non-uniformities in the atmospheric refractive index. As the refractive index structure changes and/or moves across the slant-path (with, for example, the mean wind velocity) these spatial variations are translated to time variations in received signal level. The fluctuations occur typically on a time scale of a few seconds to several minutes. Scintillation, unlike rain fading, can result in signal enhancements as well as fades. The CNR is degraded, however, during the fading part of the scintillating signal and as such has the potential to degrade system performance. Whilst severe fading is usually dominated by rain and occurs for only small percentages of time the less severe fading due to scintillation occurs for large percentages of time and may be significant in the performance of low-margin, low-availability, systems such as VSATs (see section 14.3.9). At very low elevation angles multipath propagation due to reflection from, and/or refraction through, stable atmospheric layers may occur. Distinguishing between severe scintillation and multipath propagation in this situation may, in practice, be difficult however. Scintillation intensity is sensitively dependent on elevation angle, increasing as elevation angle decreases.

Mechanisms of noise enhancement

Excess thermal noise arising from rain, precipitation scatter, ducting and cross-polarisation may all affect satellite systems in essentially the same way as terrestrial systems (see section 14.2.4). Rain induced cross-polarisation, however, is usually more severe on slant-path links since the system designer is not free to choose the earth station's polarisation. Furthermore, since the propagation path continues above the rain height, tropospheric ice crystals may also contribute to cross-polarisation. Earth–space links employing full frequency reuse (i.e. orthogonal polarisations for independent co-frequency carriers) may therefore require adaptive cross-polar cancellation devices to maintain satisfactory isolation between carriers.

Sun transit refers to the passage of the sun through the beam of a receiving earth station antenna. The enormous noise temperature of the sun effectively makes the system unavailable for the duration of this effect. Geostationary satellite systems suffer sun transit for a short period each day around the spring and vernal equinoxes. (Other celestial noise sources can also cause occasional increases in earth station noise temperature.)

System availability constraints

The propagation effects described above will degrade a system's CNR below its clear sky level for a small, but significant, fraction of time. In order to estimate the constraints which propagation effects put on system availability (i.e. the fraction of time that the CNR exceeds its required minimum value) the clear sky CNR must be modified to account for these propagation effects.

In principle, since received signal levels fluctuate due to variations in gaseous absorption and scintillation, these effects must be combined with the statistics of rain fading to produce an overall fading cumulative distribution, in order to estimate the CNR exceeded for a given percentage of time. Gaseous absorption and scintillation give rise to relatively small fade levels compared to rain fading (at least at the large time percentage end of the fading CD) and it is therefore often adequate, for traditional high availability systems, to treat gaseous absorption as constant (as in the link budget of section 14.3.3) and neglect scintillation altogether. (This approach may not be justified in the case of VSATs (see section 14.3.9), for which transmit power, G/T, and consequent availability are all low.) Once the uplink and downlink fade levels for the required percentage of time have been established then the CNRs can be modified as described below.

The uplink CNR exceeded for $100 - p\%$ of time (where typically $100 - p\% = 99.99\%$, i.e. $p = 0.01\%$), $(C/N)_{u, 100-p}$, is simply the clear sky carrier to noise ratio, $(C/N)_u$, reduced by the fade level exceeded for $p\%$ of time, $F_u(p)$, i.e.:

$$\left(\frac{C}{N}\right)_{u,\,100-p} = \left(\frac{C}{N}\right)_u - F_u(p) \quad \text{(dB)} \tag{14.47}$$

The uplink noise is not increased by the fade since the attenuating event is localised to a small fraction of the receiving satellite antenna's coverage area. (Even if this were not so the temperature of the earth behind the event is essentially the same as the temperature of the event itself.)

If uplink interference arises from outside the fading region then the uplink carrier to interference ratio exceeded for $100 - p\%$ of time will also be reduced by $F_u(p)$, i.e.:

$$\left(\frac{C}{I}\right)_{u,\,100-p} = \left(\frac{C}{I}\right)_u - F_u(p) \quad \text{(dB)} \tag{14.48}$$

In the absence of uplink fading (or the presence of uplink power control to compensate uplink fades) the downlink CNR exceeded for $100 - p\%$ of time is determined by the downlink fade statistics alone. $(C/N)_d$, however, is reduced not only by downlink carrier fading (due to downlink attenuating events) but also by enhanced antenna noise temperature (caused by thermal radiation from the attenuating medium in the earth station's normally cold antenna beam), i.e.:

$$\left(\frac{C}{N}\right)_{d,\,100-p} = \left(\frac{C}{N}\right)_d - F_d(p) - \left(\frac{N_{faded}}{N_{clear\,sky}}\right)_{dB} \quad \text{(dB)} \tag{14.49}$$

where $N_{faded} = k(T_{ant} + T_e)B$, see Chapter 12.

A simple, but conservative, estimate of $(C/N)_{d,\,100-p}$ can be made by assuming, in the 'clear sky' link budget, a (worst case) antenna noise temperature which is equal to the physical temperature of the lossy medium (typically 290 K) and ignoring $(N_{faded}/N_{clear\,sky})_{dB}$ in equation (14.49).

Downlink carrier to interference ratio is usually unaffected by a downlink fade since both the wanted and interfering signals are equally attenuated, i.e.:

$$\left(\frac{C}{I}\right)_{d,\,100-p} = \left(\frac{C}{I}\right)_d \quad \text{(dB)} \tag{14.50}$$

From a system design point of view fade *margins* can be incorporated into the satellite uplink and downlink budgets such that under clear sky conditions the system operates with the correct back-off but with excess uplink and downlink CNR (over those required for adequate overall CNR) of $F_u(p)$ and $F_d(p)$ respectively. Assuming fading does not occur simultaneously on uplink and downlink this ensures that an adequate overall CNR will be available for $100 - 2p\%$ of time. More accurate estimates of the system performance limits imposed by fading would require joint statistics of uplink and downlink attenuation, consideration of changes in back-off produced by uplink fades (including consequent *improvement* in intermodulation noise), allowance for possible cross-polarisation induced crosstalk, hydrometeor scatter and other noise and interference enhancement effects.

EXAMPLE 14.6

A 30 GHz receiving earth station is located near Bradford (54° N, 2° W). It has an overall receiver noise figure of 5.0 dB and a free space link budget which yields a CNR of 35.0 dB. (The free space budget does not account for background gaseous attenuation of the carrier but does allow for normal atmospheric noise.) The station is at a height of 440 m above sea level and the elevation angle to the satellite is 29°. If the one-minute rain rate exceeded for 0.01% of time at the earth station is 28 mm/h estimate the CNR exceeded for 99.9% of time assuming that the CNR is downlink limited and uplink power control is used to compensate all uplink fades.

From Figure 14.35 $A_{zenith} = 0.23$ dB. (This is strictly the value for an earth station at sea level, but since Figure 14.20 shows a specific attenuation *at sea level* of only 0.09 dB/km then the error introduced is less than $0.09 \times 0.44 = 0.04$ dB and can therefore be neglected.)

Clear sky slant path attenuation from equation (14.40):

$$A = \frac{A_{zenith}}{\sin \alpha} = \frac{0.23}{\sin 29°} = 0.5 \text{ dB}$$

$$\left.\frac{C}{N}\right|_{clear\,sky} = \left.\frac{C}{N}\right|_{free\,space} - A$$

$$= 35.0 - 0.5 = 34.5 \text{ dB}$$

From equation (14.41) rain height is:

$$h_R = 4.0 - 0.075 \, (54 - 36) = 2.65 \text{ km}$$

Slant path length in rain (equation (14.42)) is:

$$L_s = \frac{2.65 - 0.44}{\sin 29°} = 4.56 \text{ km}$$

Ground projection of path in rain (equation (14.43)):

$$L_G = 4.56 \cos 29° = 3.99 \text{ km}$$

Path length reduction factor (equation (14.44)):

$$r_{0.01} = \left[1 + \frac{3.99 \, e^{(0.015 \times 28)}}{35} \right]^{-1} = 0.852$$

Using Figure 14.22 specific attenuation, γ_R, for 30 GHz at 28 mm/h can be estimated to be 5.3 dB/km. (A more accurate ITU-R model using the formula aR^b gives 5.6 dB/km for horizontal polarisation and 4.7 dB/km for vertical polarisation.)
   From equation (14.45):

$$A_{0.01} = 5.3 \times 4.56 \times 0.852 = 20.6 \text{ dB}$$

Using equation (14.46):

$$A_{0.1} = 20.6 \times 0.12 \times 0.1^{-(0.546 + 0.043 \log_{10} 0.1)} = 7.9 \text{ dB}$$

Using equations (14.24) and (14.25) and assuming a surface temperature of 290 K:

$$T_A = [(1.12 \times 290) - 50] \, [1 - 10^{-\frac{7.9}{10}}] = 230 \text{ K}$$

Ratio of effective noise powers under faded and clear sky conditions is:

$$
\begin{aligned}
\frac{N_{faded}}{N_{clear\,sky}} &= \frac{T_{syst\,faded}}{T_{syst\,clear\,sky}} \\[2mm]
&= \frac{T_{A\,faded} + (f - 1)\,290}{T_{A\,clear\,sky} + (f - 1)\,290} \\[2mm]
&= \frac{230 + (10^{\frac{5}{10}} - 1)\,290}{50 + (10^{\frac{5}{10}} - 1)\,290} \\[2mm]
&= \frac{230 + 627}{50 + 627} \\[2mm]
&= 1.266 \ = \ 1.0 \text{ dB}
\end{aligned}
$$

(Note that the noise enhancement in this case is within the uncertainty introduced by estimating γ_R from Figure 14.22.) From equation (14.49):

$$\left(\frac{C}{N} \right)_{d,\,99.9} = 34.5 - 7.9 - 1.0 = 25.6 \text{ dB}$$

14.3.5 Analogue FDM/FM/FDMA trunk systems

Figure 14.37 shows a schematic diagram of a large, traditional, earth station. Such an earth station would be used mainly for fixed point-to-point international PSTN communications. The available transponder bandwidth (typically 36 MHz) is subdivided into several transmission bands (typically 3 MHz wide) each allocated to one of the participating earth stations, Figure 14.38. All the signals transmitted by a given earth station, irrespective of their destination, occupy that earth station's allocated transmission band. Individual SSB voice signals arriving from the PSTN at an earth station are frequency division multiplexed (see Figure 5.12) into a position in the earth station's transmission band which depends on the voice signal's destination. Thus all the signals arriving for transmission at earth station 2 and destined for earth station 6 are multiplexed into sub-band 6 of transmission band 2. The FDM signal, consisting of all sub-bands, is then frequency modulated onto the earth station's IF carrier. The FDM/FM signal is subsequently upconverted (U/C) to the 6 GHz RF carrier, amplified (to attain the required EIRP) and transmitted.

A receiving earth station demodulates the carriers from *all* the other earth stations in the network. (Each earth station therefore requires $N - 1$ receivers where N is the number of participating earth stations.) It then filters out the sub-band of each transmission band designated to itself and discards all the other sub-bands. The sub-band signals are then demultiplexed, the resulting SSB voice signals demodulated if necessary (i.e. translated back to baseband) and interfaced once again with the PSTN. This method

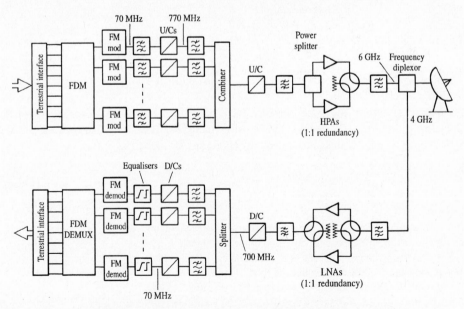

Figure 14.37 *Simplified block diagram of a traditional FDM/FM/FDMA earth station (only HPA/LNA redundancies shown).*

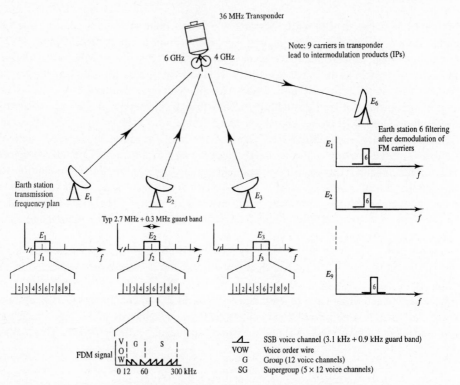

Figure 14.38 *Illustration of MCPC FDM/FM/FDMA single transponder satellite network and frequency plan for the transponder (with nine participating earth stations).*

of transponder resource sharing between earth stations is called frequency division multiple access (FDMA).

When assessing the SNR performance of FDM/FM/FDMA voice systems the detection gain of the FM demodulator must be included. Assuming that operation is at a CNR above threshold, this gain is given by [Pratt and Bostian]:

$$\frac{(S/N_b)_{wc}}{C/N} = \left(\frac{\Delta f_{RMS}}{f_M} \right)^2 \frac{B}{b} \tag{14.51}$$

where:

$(S/N_b)_{wc}$ is the SNR of the worst-case voice channel (see Figure 14.39),

Δf_{RMS} is the RMS frequency deviation of the FM signal,

f_M is the maximum frequency of the modulating (FDM) signal,

B is the bandwidth of the modulated (FM) signal,

b is the bandwidth of a single voice channel (typically 3.1 kHz).

The quantities needed to apply equation (14.51) can be estimated using:

$$f_M = 4.2 \times 10^3 N \quad \text{(Hz)} \tag{14.52}$$

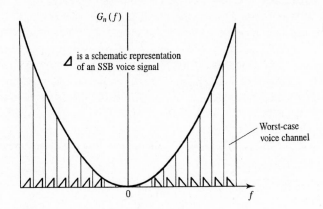

Figure 14.39 *Parabolic noise power spectral density after FM demodulation.*

where N is the number of voice channels in the FDM signal,

$$B = 2(\Delta f + f_M) \quad \text{(Hz)} \tag{14.53}$$

where Δf is the peak frequency deviation of the FM signal, and

$$\Delta f = \begin{cases} 3.16\,\Delta f_{RMS}, & \text{for } N > 24 \\ 6.5\,\Delta f_{RMS}, & \text{for } N \le 24 \end{cases} \tag{14.54}$$

(The FDM signal is usually amplitude limited to ensure the peak to RMS frequency deviation ratio is ≤ 3.16. Since the FDM signal is essentially a sum of many independent voice signals its pdf will be Gaussian, $\Delta f/\Delta f_{RMS} = 3.16$ corresponding to the 0.1% extreme value, Figure 4.22. Clipping will therefore take place for approximately 0.2% of time.) The RMS frequency deviation is set by adjusting the FM modulator constant, K (Hz/V), of the earth station transmitter, i.e.:

$$\Delta f_{RMS} = K\sqrt{\langle g^2(t) \rangle} \tag{14.55}$$

where $g(t)$ is the modulating (FDM) signal. In practice the modulator constant, K, is set using a 1 kHz, 0 dBm test tone as the modulating signal. The test tone RMS frequency deviation, Δf_{RMS}^T, can be related to the required FDM signal frequency deviation, Δf_{RMS}, by:

$$\Delta f_{RMS} = l\,\Delta f_{RMS}^T \tag{14.56}$$

where:

$$20\log_{10} l = \begin{cases} -1 + 4\log_{10} N, & 12 \le N \le 240 \\ -15 + 10\log_{10} N, & N > 240 \end{cases} \tag{14.57}$$

In addition to the FM detection gain given by equation (14.51) an extra SNR gain can be obtained by using a pre-emphasis network prior to modulation and a de-emphasis network after demodulation. Using the ITU pre-emphasis/de-emphasis standards the pre-emphasis SNR gain is 4 dB. Finally, the combined frequency response of a telephone earpiece and the subscriber's ear matches the spectrum of the voice signal better than the spectrum of the noise. This results in a further (if partly subjective) improvement in SNR. This improvement is accounted for by what is called the *psophometric weighting* and has a numerical value of 2.5 dB. Since many voice channels are modulated (as a single FDM signal) onto a single carrier, FDM/FM/FDMA is often referred to as a multiple channel per carrier (MCPC) system. MCPC is efficient providing each earth station is heavily loaded with traffic.

For lightly loaded earth stations MCPC suffers the following disadvantages:

1. Expensive FDM equipment is necessary.
2. Channels cannot be reconfigured easily and must therefore by assigned on essentially a fixed basis.
3. Each earth station carrier is transmitted irrespective of traffic load. This means that full transponder power is consumed even if little or no traffic is present.
4. Even under full traffic load, since an individual user speaks for only about 40% of time, significant transponder resource is wasted.

An alternative to MCPC for lightly loaded earth stations is a single channel per carrier (SCPC) system. In this scheme each voice signal is modulated onto its own individual carrier and each voice carrier is transmitted only as required. This saves on transponder power at the expense of a slightly increased bandwidth requirement. This scheme might be called FM/FDM/FDMA in contrast to the FDM/FM/FDMA process used by MCPC systems. The increased bandwidth per channel requirement over MCPC makes it an uneconomical scheme for traditional point-to-point international trunk applications. The fact that the channels can be demand assigned (DA) as traffic volumes fluctuate, and that the carrier can be switched on (i.e. voice activated) during the 35-40% of active speech time typical of voice signals (thus saving 4 dB of transponder power) makes SCPC superior to MCPC for systems with light, or highly variable, traffic.

Another type of SCPC system dispenses with FM entirely. Compatible single sideband systems simply translate the FDM signal (comprising many SSB voice signals) directly to the RF transmission band (using amplitude modulation). This is the most bandwidth efficient system of all and is not subject to a threshold effect as FM systems are. Compatible single sideband does not, however, have the large SNR detection gain that both FDM/FM/FDMA and FM/FDM/FDMA systems have.

14.3.6 Digital TDM/PSK/TDMA trunk systems

Time division multiplex access (TDMA) is an alternative to FDMA for transponder resource sharing between earth stations. Figure 14.40 illustrates the essential TDMA principle. Each earth station is allocated a time slot (in contrast to an FDMA frequency slot) within which it has sole access to the entire transponder bandwidth. The earth station time slots, or bursts, are interleaved on the uplink, frequency shifted, amplified, and retransmitted by the satellite to all participating earth stations. One earth station

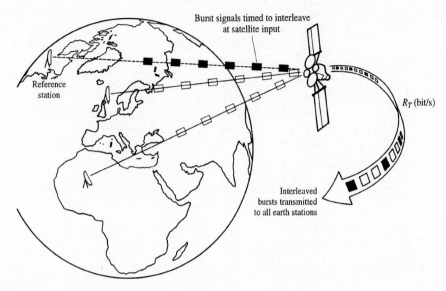

Burst signals timed to interleave
at satellite input

Reference
station

R_T (bit/s)

Interleaved
bursts transmitted
to all earth stations

Figure 14.40 *Principal of time division multiplex accessing (TDMA).*

periodically transmits a reference burst in addition to its information burst in order to synchronise the bursts of all the other earth stations in the TDMA system. Time division multiplexing and digital modulation are obvious techniques to use in conjunction with TDMA. In order to minimise AM/PM conversion in the non-linear transponder, Figure 14.33, constant envelope PM is attractive. MPSK is therefore used in preference to MQAM, for example (see Chapter 11). Since some filtering of the PSK signal prior to transmission is necessary (for spectrum management purposes) even MPSK envelopes are not, in fact, precisely constant. QPSK signals, for instance, have envelopes which fall to zero when both inphase and quadrature symbols change simultaneously, Figure 11.31. Offset QPSK (OQPSK) reduces the maximum envelope fluctuation to 3 dB by offsetting inphase and quadrature symbols by half a symbol period (i.e. one information bit period), Figure 11.31. Chapter 11 discusses bandpass modulation (including OQPSK) in detail. Figure 14.41 shows a schematic diagram of a TDM/PSK/TDMA earth station.

For digital satellite systems having only a single carrier present in the transponder at any one time then intermodulation products are absent and $(C/N)_{IP}^{-1}$ in equation (14.38) is zero. Recall (Chapter 11) that the quantity E_s/N_0 is related to C/N by:

$$\frac{\langle E_s \rangle}{N_0} = \frac{C}{N} BT_o \tag{14.58}$$

where T_o is the symbol period (i.e. the reciprocal of the baud rate, R_s) and BT_o depends on the particular digital modulation scheme and filtering employed. Equation (14.58) can also be expressed in terms of bit energy, E_b, and bit duration, T_b, i.e.:

$$\frac{E_b}{N_0} = \frac{C}{N} BT_b \tag{14.59}$$

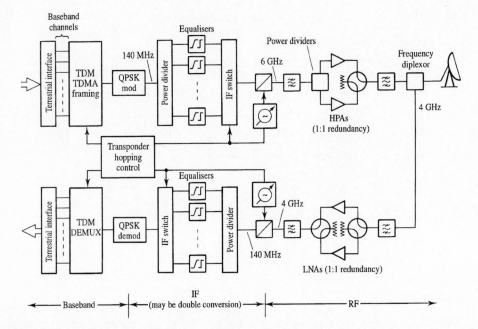

Figure 14.41 *Simplified block diagram of traditional TDM/QPSK/TDMA earth station. (Only HPA/LNA redundancies are shown.)*

For QPSK modulation, which is the primary TDM/PSK/TDMA modulation standard currently used by INTELSAT, the E_b/N_0 required to support a given P_b performance is found using equation (11.47):

$$P_b = \frac{1}{2} \, \text{erfc} \left(\frac{E_b}{N_0} \right)^{1/2} \tag{14.60}$$

and the required CNR is then found using equations (11.6) and (11.7):

$$\frac{C}{N} = \frac{E_b/T_b}{N_0 B} = \frac{E_b R_b}{N_0 B} \tag{14.61}$$

or in decibels:

$$\left(\frac{C}{N} \right)_{dB} = \left(\frac{E_b}{N_0} \right)_{dB} - (BT_b)_{dB} \tag{14.62}$$

(For minimum bandwidth, ISI free, filtering such that the transmitted signal occupies the DSB Nyquist bandwidth then $BT_b = 1.0$ (or 0 dB) and $C/N = E_b/N_0$.) In practice an implementation margin of a few decibels would be added after using equations (14.60) and (14.61) to allow for imperfect modulation, demodulation, etc.

A typical frame structure showing the TDMA slots allocated to different earth stations is shown in Figure 14.42. Two reference bursts are often included (provided by different

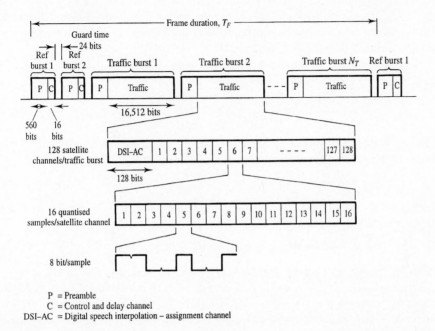

Figure 14.42 *Typical TDMA frame structure. (DSI-AC time slot is discussed in section 14.3.7.)*

earth stations) so that the system can continue to function in the event of losing one reference station due, for example, to equipment failure. Typically the frame period T_F is of the order of 5 ms.

The traffic bursts each consist of a preamble followed by the subscriber traffic. The preamble, which might typically be 280 QPSK symbols long, is used for carrier recovery, symbol timing, frame synchronisation and station identification. In addition it supports voice and data channels (voice order wires) to enable operations and maintenance staff at different earth stations in the network to communicate without using traffic slots. Reference bursts have the same preamble as the traffic bursts followed by control and delay signals (typically 8 symbols in duration) which ensure that the TDMA bursts from participating earth stations are timed to interleave correctly at the transponder input.

Subscriber traffic is subdivided into a number (typically 128) of *satellite channels*, Figure 14.42. In conventional preassigned (PA) systems the satellite channels are pre-divided into groups, each group being assigned to a given destination earth station.

Each satellite channel carries (typically) 128 bits (64-QPSK symbols) representing 16 consecutive 8-bit PCM samples from a single voice channel. For a conventional 8 kHz PCM sampling rate this corresponds to $16 \times 125 \mu s = 2$ ms of voice information. In this case the frame duration would therefore be limited in length to 2 ms so that the next frame could convey the next 2 ms of each voice channel. Channels carrying non-voice, high data rate, information are composed of multiple voice channels. Thus, for example, a 320 kbit/s signal would require five 64 kbit/s voice channels.

Frames may be assembled into master frames as shown in Figure 14.43. The relative lengths of information bursts from each earth station can then be varied from master frame to master frame depending on the relative traffic loads at each station. This would represent a simple demand assigned system.

The frame efficiency, η_F, of a TDM/PSK/TDMA system is equal to the proportion of frame bits which carry revenue earning traffic, i.e.:

$$\eta_F = \frac{b_F - b_o}{b_F} \qquad (14.63)$$

where b_F is the total number of frame bits and b_o is the number of overhead (i.e. non-revenue earning) bits. The total number of frame bits is given by:

$$b_F = R_T T_F \qquad (14.64)$$

where R_T is the TDMA bit rate and T_F is the frame duration. The number of overhead bits per frame can be calculated using:

$$b_o = N_R b_R + N_T(b_p + b_{AC}) + (N_R + N_T)b_G \qquad (14.65)$$

where N_R is the number of participating reference stations, N_T is the number of participating traffic stations, b_R is the number of bits in a reference burst, b_P is the number of preamble bits (i.e. all bits excluding traffic bits) in a traffic burst, b_{AC} is the number of digital speech interpolation–assignment channel bits per traffic burst and b_G is the number of guard bits per reference or traffic burst.

Typically η_F is about 90% for a TDMA system with 15 to 20 participating earth stations and a frame period of several milliseconds.

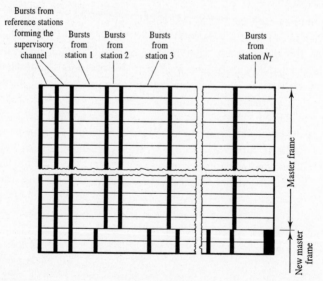

Figure 14.43 *TDMA master frame structure.*

The number of voice channels which a TDMA system can support is called its voice channel capacity, χ. This can be calculated from:

$$\chi = \frac{R_i}{R_v} \tag{14.66}$$

where R_i is the information bit rate and R_v is the bit rate of a single voice channel.

The information bit rate is given by:

$$R_i = \frac{b_F - b_o}{T_F}$$

$$= \eta_F R_T \quad \text{(bit/s)} \tag{14.67}$$

and the bit rate of a single voice channel is given by:

$$R_v = f_s n \quad \text{(bit/s)} \tag{14.68}$$

where f_s is the sampling rate (usually 8 kHz) and n is the number of PCM bits per sample (typically 8), Chapter 6. (For ADPCM, R_v is 32 kbit/s, Table 9.3.) Finally, the average number of voice channels *per earth station access* $\bar{\chi}_A$ is:

$$\bar{\chi}_A = \frac{\chi}{N} \tag{14.69}$$

where N is the number of accesses per frame. (If all earth station only access once per frame then N is also, of course, the number of participating earth stations.)

The primary TDM/PSK/TDMA modulation standard used by INTELSAT is summarised in Table 14.5.

Table 14.5 *INTELSAT TDM/PSK/TDMA modulation standard.*

Modulation	QPSK
Nominal symbol rate	60.416 Mbaud
Nominal bit rate	120.832 Mbit/s
Encoding	Absolute (i.e. not differential phase)
Phase ambiguity resolution	Using unique word in preambles

EXAMPLE 14.7

The frame length of the pure TDMA system illustrated in Figure 14.42 is 2.0 ms. If the QPSK symbol rate is 60.136 Mbaud and all traffic bursts are of equal length determine: (i) the maximum number of earth stations which the system can serve and (ii) the frame efficiency.

$$R_T = 2 \times \text{QPSK baud rate} = 2 \times 60.136 \times 10^6 = 1.20272 \times 10^8 \text{ bit/s}$$

$$b_F = R_T T_F = 1.20272 \times 10^8 \times 2.0 \times 10^{-3} = 240544 \text{ bits}$$

Overhead calculation:

$$b_o = N_R b_R + N_T (b_p + b_{AC}) + (N_R + N_T) b_G$$

Now for a 560 + 16 bit reference, b_R, 560 bit preamble, b_P, 24 bit guard interval, b_G and 128 bit DSI assignment channel:

$$b_o = 2(560 + 16) + (560 + 128) N_T + (2 + N_T) 24$$

$$= 1200 + 712 N_T$$

$$b_F - b_o = 240544 - (1200 + 712 N_T) = 239344 - 712 N_T$$

(i)

$$N_T = \frac{b_F - b_o}{16512 - 128} = \frac{239344 - 712 N_T}{16384}$$

i.e.:

$$N_T (16384 + 712) = 239344$$

Therefore: $N_T = 14$, i.e. a maximum of 14 earth stations may participate.

(ii)

$$\eta_F = \frac{b_F - b_o}{b_F}$$

$$= \frac{240544 - (1200 + 712 \times 14)}{240544} = 0.954$$

i.e. frame efficiency is 95.4%.

14.3.7 DA-TDMA, DSI and random access systems

Preassigned TDMA (PA-TDMA) risks the situation where, at a certain earth station, all the satellite channels assigned to a given destination station are occupied whilst free capacity exists in channels assigned to other destination stations. Demand assigned TDMA (DA-TDMA) allows the reallocation of satellite channels in the traffic burst as the relative demand between earth stations varies. In addition to demand assignment of satellite channels within the earth station's traffic burst DA-TDMA may also allow the number traffic bursts per frame, and/or the duration of the traffic bursts, allocated to a given earth station to be varied.

Digital speech interpolation (DSI) is another technique employed to maximise the use made of available transponder capacity. An average speaker engaged in conversation actually talks for only about 35% of the time. This is because for 50% of time he, or she, is passively listening to the other speaker and for 30% of the remaining 50% of time there is silence due to pauses and gaps between phrases and words. DSI systems automatically detect when speech is present in the channel, and during speech absences reallocate the channel to another user. The inevitable clipping at the beginning of speech which occurs as the channel is being allocated is sufficiently short for it to go unnoticed.

Demand assigned systems require extra overhead in the TDMA frame structure to control the allocation of satellite channels and the relative number per frame, and lengths, of each earth station's traffic bursts. For systems with large numbers of earth stations each contributing short, bursty, traffic at random times then random access (RA) systems may use transponder resources more efficiently than DA systems. The earth stations of RA systems attempt to access the transponder (i.e. in the TDMA context, transmit bursts) essentially at will. There is the possibility of course, that the traffic bursts (usually called packets in RA systems) from more than one earth station will collide in the transponder causing many errors in the received data. Such collisions are easily detected, however, by both transmitting and receiving earth stations. After a collision all the transmitting earth stations wait for a random period of time before retransmitting their packets.

Many variations and hybrids of the multiple access techniques described here have been used, are being used, or have been proposed, for satellite communications systems. A more detailed and quantitative discussion of these techniques and their associated protocols can be found in [Ha].

14.3.8 Economics of satellite communications

The cost of a long distance point-to-point terrestrial voice circuit is about 2000 dollars p.a. Lease of a private, high quality, landline from New York to Los Angeles for FM broadcast use costs 13,000 dollars p.a. Satellite communication systems can provide the equivalent services at lower cost. The monthly lease for a video bandwidth satellite transponder is typically 50,000 to 200,000 dollars whilst the hourly rate for satellite TV programme transmission can be as low as 200 dollars. Access to digital audio programmes can be as low as 2,500 dollars per month for high quality satellite broadcasts. The advent of the space shuttle has greatly reduced satellite launch costs below the 30,000 dollars/kg of conventional rocket techniques making them much more competitive. (The figures given here reflect 1990 costs.)

14.3.9 VSAT systems

Satellite communication is predominantly used for international, point-to-point, multiplex telephony, fax and data traffic, making high EIRP transmissions necessary due to the wideband nature of the signal multiplex. For low data rates (2.4 to 64 kbit/s) with narrow signal bandwidth the reduced noise in the link budget allows the use of satellites with very small aperture (typically 1 m diameter antenna) terminals (VSATs) and modest power (0.1 to 10 W) earth station transmitters [Everett]. This has given rise to the development of VSAT low data rate networks, in which many remote terminals can, for example, access a central computer database. These systems use roof, or garden, located antennas which permanently face a geostationary satellite. VSAT systems are widely deployed in the USA and may have up to 10,000 VSAT earth station terminals in a single network. They are used for retail point-of-sale credit authorisation, cash transactions, reservations, stock control and other data transfer tasks. They are also now being expanded to give worldwide coverage in order to meet the telecommunications requirements of large international companies.

The use of non-geostationary satellites for mobile and personal communications with even lower antenna gains is discussed in Chapter 15.

14.3.10 Satellite-switched-TDMA and on-board signal processing

Satellites operating with small spot beams have high antenna gains. This implies either a low on-board power requirement or a large bandwidth and therefore high potential bit rate. If many spot beams with good mutual isolation are used, frequency bands can be reused thus increasing spectrum utilisation efficiency, Figure 14.44. Connectivity between a system's participating earth stations is potentially decreased, however, since a pair of earth stations in different spot beams can communicate only if their beams are connected. Satellite switched TDMA has the potential to re-establish complete connectivity between earth stations using a switching matrix onboard the satellite, Figure 14.45. The various sub-bursts (destined for different receiving stations) of a transmitting station's traffic burst can be directed by the matrix switch to the correct downlink spot beams. Furthermore, for areas with a sparse population of users, such that many fixed spot beams are uneconomic, the beams may be hopped from area to area and the uplink bursts from each earth station demodulated and stored. On-board signal processing is then used to reconfigure the uplink bursts into appropriately framed downlink bursts before the signals are remodulated and transmitted to the appropriate earth stations as the downlink spot beam is hopped. On-board demodulation and remodulation also has the normal advantage of digital communications, i.e. the uplink and downlink noise is decoupled. The NASA advanced communications satellite was used in the middle 1990s to evaluate these types of system.

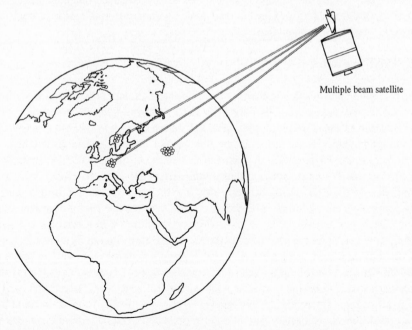

Multiple beam satellite

Figure 14.44 *Satellite switched time division multiplex access (SS-TDMA).*

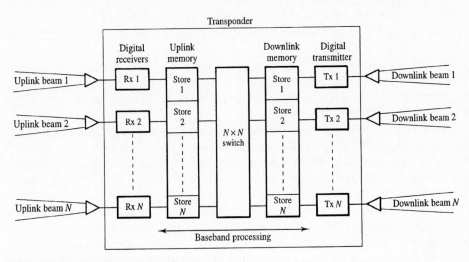

Figure 14.45 *SS-TDMA transponder.*

14.4 Summary

Point-to-point, terrestrial, microwave links now play a major part in the arterial trunk routes of the PSTN. Originally carrying SSB/FDM/FM telephony they now carry mixed services, principally using PCM/TDM/QAM. Typically, repeaters are spaced at 40 to 50 km intervals and the total microwave link bandwidth is split in to high and low frequency blocks, each of which contains eight 30 MHz channels. Adjacent channel isolation is improved by alternating vertical and horizontal polarisations across the transmission band. Correlation between the expected signal and interference due to overreaching caused by ducting, is avoided by alternating transmit and receive channels between high and low channel blocks on adjacent hops.

Proper path profiling of LOS microwave links ensures adequate clearance under specified propagation conditions. If clearance is less than the first Fresnel zone then serious diffraction losses may occur. Propagation paths may be conveniently plotted as straight rays over an earth with appropriately modified curvature, or as rays with modified curvature over a plane earth. Bulk refractive index conditions are described by k factor which is the ratio of effective earth radius to actual earth radius in the straight ray model. Standard propagation conditions correspond to a k factor of 4/3 (which represents a refractivity lapse rate of 40 N units/km). Positive $k < 4/3$ represents sub-refractive conditions and $k > 4/3$ represents super-refraction. Negative k represents ducting.

For accurate clear sky link budgets, background gaseous absorption must be accounted for in addition to free space path loss. Gaseous absorption is small below 10 GHz but rises rapidly with frequency due primarily to water vapour and oxygen resonance lines which occur at 22 GHz and 60 GHz respectively. Rain attenuation also increases rapidly with frequency and must be accounted for if links are to meet a specified availability. Multipath propagation may, like rain, result in deep signal fades but, unlike rain fading which is essentially flat, can also cause amplitude and phase distortion to wide band signals. Microwave paths which suffer from severe frequency selective fading can benefit significantly from the application of diversity techniques and/or the use of adaptive equalisation.

Hydrometeor scatter may result in interference between nominally independent cofrequency systems. Hydrometeor induced cross polarisation may also result in cross talk between the channels of systems using polarisation, in whole or in part, as the channel isolating mechanism.

Fixed service satellite communications makes use of the geostationary orbit. Satellites in this orbit are located above the earth's equator at a height of 36,000 km. They are stationary with respect to a point on the earth's surface and tracking is therefore unnecessary. A single geostationary satellite is visible from 42% of the earth's surface. With the exception of the polar regions three such satellites can give global coverage. The combined propagation delay of uplink and downlink is about 0.25 s making echo cancellation essential. FSPL is of the order of 200 dB making large transmitter powers, large antennas and sensitive receivers necessary for high data rate transmissions. Like terrestrial microwave paths gaseous absorption and rain fading must be considered when engineering an earth-space communications link. Scintillation on earth-space paths might also be important for low margin, low availability systems. Fixed point satellite services currently operate in the C and Ku bands although services using the Ka band will probably follow shortly.

The predominant multiplexing/modulation/multiple accessing technique used currently for international PSTN satellite telephony is FDM/FM/FDMA. Since satellite HPAs are operated near saturation the use of FDMA, which results in multiple FM carriers in the transponder, gives rise to large intermodulation products which must normally be accounted for in the overall CNR calculation. As with terrestrial LOS relay links, the trend in satellite communications is towards digital transmission. FDM/FM/FDMA systems will therefore gradually be replaced by TDM/PSK/TDMA systems. Intermodulation products will be of less concern since in (pure) TDMA systems only one carrier is present in the transponder at any particular time. TDMA systems require less back-off, therefore, than the equivalent FDMA systems. This means that constant envelope modulation is even more desirable to minimise AM/AM and AM/PM distortion. QPSK and OQPSK are attractive modulation techniques in these circumstances.

The bits transmitted by an earth station in a TDMA network are grouped into frames (and frames into master frames). Each frame carries overhead bits which serve various synchronisation, monitoring and control purposes. The frame efficiency is defined as the proportion of frame bits which carry (revenue earning) traffic.

For systems which have fluctuating traffic loads demand assignment may be used to reallocate capacity to those earth stations with highest demand. For systems with large numbers of lightly loaded earth stations, random access techniques may be appropriate.

VSAT systems can use modest transmitter powers and small antennas to transmit low bit rate (and therefore narrow bandwidth) data. Such technology is relatively inexpensive allowing commercial (and other) organisations to operate their own networks of earth stations. Switched spot-beam satellites with on-board signal processing, and satellite based mobile communications systems using both high and low earth orbits (HEOs and LEOs) are currently being developed.

14.5 Problems

14.1. Show that the effective height of the (smooth) earth's surface above a cord connecting two points on the surface, for a straight ray profile, is given (for practical microwave link geometries) by $d_1 d_2/(2ka)$, where k is k-factor and a is the earth radius. Hence confirm the correctness of equations (14.11) and (14.12(a)).

14.2. A 10.3 GHz microwave link with good ground clearance is 60 km long. The link's paraboloidal reflector antennas have a diameter of 2.0 m and are aligned to be correctly pointed during standard ($k = 4/3$) atmospheric conditions. What limiting k-factors can the link tolerate before 6 dB of power is lost from its signal budget due to refractive decoupling of the antennas? To what refractivity lapse rates do these k-factors correspond and in what regime (sub-refraction/super-refraction/ducting) do these lapse rates reside? (Assume that the antennas are mounted at precisely the same heights above sea level and that their 3 dB beamwidths are given by $1.2\lambda/D$ rad where λ is wavelength and D is antenna diameter.)

14.3. If the transmitting antenna in Problem 14.2 is 50 m high and is aligned under standard refractive index conditions such that its boresight is precisely horizontal, what must be the difference between transmit and receive antenna heights for the receive antenna to be located at the centre of the transmitted beam? [158.9 m]

14.4. Show that the radius of the nth Fresnel zone is given by equation (14.10).

14.5. A microwave LOS link operates at 10 GHz over a path length of 32 km. A minimum ground clearance of 50 m occurs at the centre of the path for a k-factor of 0.7. Find the Fresnel zone clearance of the path in terms of (a) the number of Fresnel zones cleared and (b) the fraction (or multiple) of first Fresnel zone radius. [10, 3.2]

14.6. A 6 GHz LOS link has a path length of 13 km. The height of both antennas above sea level at each site is 20 m. The only deviation of the path profile from a smooth earth is an ancient earth works forming a ridge 30 m high running perpendicular to the path axis 3 km from the receiver. If the transmitted power is 0 dBW, the transmit antenna has a gain of 30 dB and the receive antenna has an effective area of 4.0 m^2 estimate the power at the receive antenna terminals assuming knife edge diffraction. By how much would both antennas need to be raised if 0.6 first Fresnel zone clearance were required? (Assume knife edge diffraction and no atmospheric losses other than gaseous background loss.) [−74 dBW]

14.7. A terrestrial 20 GHz microwave link has the following specification:

Path length	35.0 km
Transmitter power	−3.0 dBW
Antenna gains	25.0 dB
Ground clearance	Good (no ground reflections or diffraction)
Receiver noise figure	3.0 dB
Receiver noise bandwidth	100 kHz

Estimate the effective CNR exceeded for 99.9% of time if the 1 minute rain rate exceeded for 0.1% of time is 10 mm/h. (Assume that Figure 14.21 applies to the climate in which the link operates and that the surface temperature is 290 K.) [30.4 dB]

14.8. A geostationary satellite is located at 35° W. What are its look angles and range from: (a) Bradford, UK (54° N, 2° W); Blacksburg, USA (37° N, 80° W); (c) Cape Town, South Africa (34° S, 18° E)? [22°, −140°, 39406 km; 27°, 121°, 38914 km, 22°, 292°, 39365 km]

14.9. A geostationary satellite sits 36,000 km above the earth's surface. The zenith pointing uplink is at 6 GHz and has an earth station power of 2 kW with $G_T = 61$ dB. If satellite antenna gain $G_R = 13$ dB, calculate the power in mW at the satellite receiver. Calculate the effective carrier-to-noise ratio, given that the receiver bandwidth is 36 MHz, its noise figure F is 3 dB and the thermal noise level is −174 dBm/Hz. [6.2×10^{-7} mW, 33.3 dB]

14.10. A receiver for geostationary satellite transmission at 2 GHz has an equivalent noise temperature of 160 K and a bandwidth of 1 MHz. The receiving antenna gain is 35 dB and the antenna noise temperature is 50 K. If the satellite antenna gain is 10 dB and expected total path losses are 195 dB, what is the minimum required satellite transmitter power to achieve a 20 dB CNR at the output of the receiver? [24.6 dBW]

14.11. Ship–shore voice communications are conducted via an INTELSAT IV satellite in geostationary orbit at an equidistant range from ship and shore stations of 38,400 km. One 28 kbit/s voice channel is to be carried in each direction. The shipboard terminal contains a 2.24 m diameter parabolic antenna of 65% efficiency. An uncooled low-noise receiver tuned to 6040 MHz, having a bandwidth of 28 kHz and a system noise temperature of 150 K, is mounted directly behind the antenna. The satellite EIRP in the direction of the ship is 41.8 dBm. The shore based terminal has an antenna of gain 46 dB at the satellite–shore link frequency of 6414.6 MHz. The satellite EIRP in the direction of the shore is 31.8 dBm. The shore's receiver has a bandwidth of 28 kHz and a noise temperature of 142 K. Calculate the carrier-to-noise ratios in dB for reception at the ship and the shore from the two satellite downlinks. [15 dB, 10.1 dB]

14.12. A 14/11 GHz digital satellite, transparent transponder, communications link has the following specification:

Uplink saturating flux density	−83.2 dBW/m^2
Input back-off	8.0 dB
Satellite G/T	1.8 dB/K
Transponder bandwidth	36.0 MHz
Uplink range	41,000 km
Saturated satellite EIRP	45.0 dBW
Output back-off	2.8 dB
Downlink range	39,000 km
Earth station G/T	31.0 dB/K
Downlink earth station fixed losses	3.5 dB

If the uplink and downlink earth station elevation angles are 5.6° and 20.3° respectively estimate the overall link, clear sky, CNR. (Assume pure TDMA operation such that there is no intermodulation noise, and assume that interference is negligible.) What clear sky earth station EIRP is required on the uplink? [15.3 dB, 72.8 dBW]

14.13. If uplink power control is used to precisely compensate uplink fading in the system described in Problem 14.12 and the downlink earth station is located in Bradford (54° N, 2° W) at a height of 440 m, where the 1 minute rain rate exceeded for 0.01% of time is 25 mm/h, estimate the overall CNR exceeded for the following time percentages: (a) 0.1%; (b) 0.01%; (c) 0.001%. (Assume that under clear sky conditions one third of the total system noise can be attributed to the aperture temperature of the antenna and two thirds originates in the receiver.) What would be the maximum possible ISI free bit rate if the modulation is QPSK and what values of BER would you expect to be exceeded for 0.1, 0.01, and 0.001% of time? [12.8 dB; 7.1 dB; −3.9 dB, 459 error/s;

8.5 kerror/s; 35.7 Merror/s]

14.14. A particular satellite communication system has the following TDMA frame structure:

Single reference burst containing 88 bits.

Preamble to each traffic burst containing 144 bits.

Frame duration of 750 μs.

Guard time after each burst of 24 bits duration.

Overall TDMA bit rate of 90.389 Mbit/s.

10 earth stations are each allocated 2 traffic bursts per frame, and one station provides (in addition) the single reference burst. What is the frame efficiency assuming DSI is not employed? If the satellite were used purely for standard (64 kbit/s) PCM voice transmission, what would be the TDMA voice channel capacity of the system? How many consecutive samples from each voice channel must be transmitted per frame? [94.9%, 1340, 6]

CHAPTER 15

Mobile and cellular radio

15.1 Introduction

In comparison to the relative stability and modest technical developments which are occurring in long-haul wideband microwave communication systems there is rapid development and expanding deployment of new mobile personal communication systems. These range from wide coverage area pagers, for simple data message transmission, through to sophisticated cellular systems, which employ common standards and hence achieve contiguous coverage over large geographical areas, such as all the major urban centres and transport routes in Europe or the continental USA. This chapter discusses the specific channel characteristics of mobile systems and examines the typical cellular clusters, adopted to achieve continuous communication with the mobile user. It then highlights the important properties of current, and emerging, TDMA and code division multiple access (CDMA), mobile digital communication systems.

15.1.1 Private mobile radio

Terrestrial mobile radio works best at around 250 MHz as lower frequencies than this suffer from noise and interference while higher frequencies experience multipath propagation from buildings, etc. In practice modest frequency bands are allocated between 60 MHz and 2 GHz. Private mobile radio (PMR) is the system which is used by taxi companies, county councils, health authorities, ambulance services, fire services, the utility industries, etc. for mobile communications.

PMR has three spectral allocations at VHF, one just below the 88 to 108 MHz FM broadcast band and one just above this band with another allocation at approximately 170 MHz. There are also two allocations at UHF around 450 MHz. All these spectral allocations provide a total of just over 1000 radio channels with the channels placed at 12½ kHz channel spacings or centre frequency offsets. Within the 12½ kHz wide channel the analogue modulation in PMR typically allows 7 kHz of bandwidth for the signal transmission. When further allowance is made for the frequency drift in the oscillators of these systems a peak deviation of only 2 to 3 kHz is available for the speech traffic. Traffic is normally impressed on these systems by amplitude modulation or frequency modulation and again the receiver is of the ubiquitous superheterodyne design, Figure 1.4. A double conversion receiver with two separate local oscillator stages is usually required to achieve the required gain and rejection of adjacent channel signals.

One of the problems with PMR receivers is that they are required to detect very small signals, typically -120 dBm at the antenna output, corresponding to 0.2 μV (RMS into 50 Ω), and, after demodulating this signal, produce an output with perhaps 1 W of audio power. It is this stringent gain requirement which demands double conversion. In this type of equipment the first IF is normally at 10.7 MHz and the second IF is very often at 455 kHz. Unfortunately, with just over 1,000 available channels for the whole of the UK and between 20,000 and 30,000 issued licences for these systems, it is inevitable that the average business user will have to share the allocated channel with other companies in his same geographic area.

There are various modes of operation for mobile radio communications networks, the simplest of which is single frequency simplex. In simplex communication, traffic is broadcast, or one way. PMR uses half duplex where, at the end of each transmission period, there is a handover of the single channel to the user previously receiving, in order to permit them to reply over the same channel. This is efficient in that it requires only one frequency allocation for the communications link but it has the disadvantage that all units can hear all transmissions provided they are within range of the mobile and base station. An improvement on half duplex is full duplex operation where two possible frequencies are allocated for the transmissions. One frequency is used for the forward or downlink, that is base-to-mobile communications, while a second frequency is used for the reverse or uplink channel, that is mobile-to-base communications. This permits simultaneous two-way communication and greatly reduces the level of interference, but it halves the overall capacity. One possible disadvantage is that mobiles are now unable to hear each other's transmissions, which can lead to contention with two mobiles attempting to initiate a call, at the same time, on the uplink in a busy system.

Although PMR employs relatively simple techniques with analogue speech transmission there have been many enhancements to these systems over the years. Data transmission is possible in PMR systems using FSK modulation (see section 11.3.3). Data transmission allows the possibility of hard copy graphics output and it also gives direct access to computer services such as databases, etc. Data preambles can also be used, in a *selective calling* mode when initiating a transmission to address a specific receiver and thus obtain more privacy within the system.

The problems in PMR are basically two-fold. One is the very restricted number of channels which are available. The second concerns the fact that mobile equipment will only operate when it is close to the base station transmitter which is owned by the company or organisation using the system. It has been the desire to design a wider coverage system, which also overcomes the restrictions of the limited number of channels, that has given rise to cellular radio. Following the development of national coverage cellular systems, PMR has come to be used mainly by taxi operators and the emergency services.

15.1.2 Radio paging systems

Another simple communications system, which is similar to broadcast, is one way paging. These started as on-site private systems employing 1 W transmitters with, approximately, 1 μV sensitivity superheterodyne receivers. The early systems used sequential tone (FSK) transmission in the UHF band [Macario] and simply alerted a specific receiver that it had been called. Digital pagers were then further developed for wide area public paging, with the POCSAG coded service opening in London in 1981. POCSAG has a capacity of 2 million pager equipments, a message rate of 1 per second and a message length of 40 characters. This is achieved with a data rate of 512 bit/s using NRZ FSK modulation and a tone separation of 9 kHz.

A key feature, necessary to achieve long battery life, is that these simple receivers switch to standby when no message is being transmitted. Paging messages are batched and preceded by a preamble to ensure that all pager receivers are primed to receive the signals. The alphanumeric message, which is displayed on the paged receiver, is sent as sets of (32,21) $R = 2/3$ POCSAG coded data vectors, Chapter 10 and [Macario]. More recently a new European radio message system (ERMES) has been standardised within ETSI for *international* paging. At 9.6 kbit/s ERMES supports four of the current 1.2 kbit/s POCSAG messages in simultaneous transmission. Also, in 1994, INMARSAT launched a worldwide satellite based pocket pager with a £300 receiver cost.

15.2 Mobile radio link budget and channel characteristics

The mobile communications channel suffers from several, potentially serious, disadvantages with respect to static, line of sight (LOS) links. These are:

1. Doppler shifts in the carrier due to relative motion between the terminals.
2. Slow spatial fading due, principally, to topographical shadowing (i.e. diffraction) effects along the propagation path.
3. Rapid spatial fading due to regions of constructive and destructive interference between signals arriving along different propagation paths. (Such multipath fading may also occur on fixed point-to-point systems but is usually less severe and more easily mitigated.)
4. Temporal fading due, principally, to the mobile terminal's motion through the spatially varying field.
5. Frequency selective fading when the signals are broadband.
6. Time dispersion due to multipath propagation.
7. Time variation of channel characteristics due, principally, to movement of the mobile terminal.

Some of these effects are, of course, intimately connected and represent different manifestations of the same physical processes. Figure 15.1 illustrates the origin of these effects, with the signals reflecting off neighbouring buildings before being summed at the mobile terminal receiving antenna.

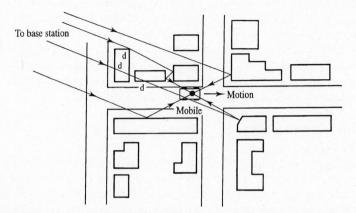

Figure 15.1 *Multipath origin of Doppler shift, fading and dispersion in a mobile radio channel (d indicates possible points of diffraction).*

15.2.1 Prediction of median signal strength

The fading processes referred to above mean that, in general, only statistical statements can be made about the signal strength in a particular place at a particular time [1]. Usually these statements are in the form of cumulative distributions of signal strength. Ideally such distributions could be measured for each possible location of mobile and base station during the design of a mobile communications system. This, however, is usually neither practical nor economic and simplified models are used to predict these cumulative distributions. A common assumption used in these models is that propagation is essentially governed by equation (12.78) with a correction factor, incorporated to account for departures from a perfectly reflecting plane earth and the resulting random fading. A semi-empirical model developed by [Ibrahim and Parsons] uses exactly this approach. The received median carrier strength in the model is derived from equation (12.81) as:

$$C = P_T + G_T - \text{PEPL} + G_R - \beta \quad (\text{dBW}) \tag{15.1(a)}$$

where:

$$\beta = 20 + \frac{f_{MHz}}{40} + 0.18L - 0.34H + K \quad (\text{dB}) \tag{15.1(b)}$$

and:

$$K = \begin{cases} 0.094U - 5.9, & \text{for inner city areas} \\ 0, & \text{elsewhere} \end{cases} \tag{15.1(c)}$$

[1] Recent spectacular improvements in the memory size, and speed, of computers combined with a similar reduction in hardware costs have made possible the simulation of mobile radio channels using detailed topographical databases combined with geometrical optic methods. Much effort is currently being invested in producing practical systems design tools from programs using such 'deterministic' modelling techniques.

β is called a clutter factor, L is called the land usage factor and U the degree of urbanisation. L and U are defined by the percentage of land covered by buildings, and the percentage of land covered by buildings with four or more storeys, respectively, in a given 0.5 km square. These quantities were chosen as parameters for the model because they are collected, and used by, UK local authorities in their databases of land usage. H is the height difference in metres between the 0.5 km squares containing the transmitter and receiver. The RMS error between predicted and measured path loss using this model in London is about 2 dB at 168 MHz and about 6 dB at 900 MHz.

15.2.2 Slow and fast fading

Slow fading, due to topographic diffraction along the propagation path, tends to obey log-normal statistics and occurs, in urban areas, typically on a scale of tens of metres. Its statistics can be explained by the cascading (i.e. multiplicative) effects of independent shadowing processes and the central limit theorem (see section 3.2.9). Fast fading, due to multipath propagation where the receiver experiences several time delayed signal replicas, tends to obey Rayleigh statistics. This is explained by the additive effects of independently faded and phased signals and the central limit theorem (see sections 3.2.9 and 4.7.1). The spatial scale of Rayleigh fading is typically half a wavelength. On a spatial scale of up to a few tens of metres fading can, therefore, usually be assumed to be a purely Rayleigh process given by:

$$p(r) = \begin{cases} (r/\sigma^2)e^{\frac{-r^2}{2\sigma^2}}, & r \geq 0 \\ 0, & r < 0 \end{cases} \tag{15.2}$$

where r is the signal amplitude and σ is the standard deviation of the parent Gaussian distribution. The corresponding exceedance is:

$$P(r > r_{ref}) = e^{\frac{-r_{ref}^2}{2\sigma^2}} \tag{15.3}$$

And since the median value of a Rayleigh distributed quantity is related to the standard deviation of its parent Gaussian distribution by:

$$r_{median} = 1.1774\sigma \tag{15.4}$$

then the exceedence can be rewritten as:

$$P(r > r_{ref}) = e^{-\left(\frac{r_{ref}}{1.2\,r_{median}}\right)^2} \tag{15.5}$$

Equation (15.5) allows the probability that the signal amplitude exceeds any particular reference value for a given median signal level to be found if slow, i.e. log-normal, fading can be neglected. If log-normal fading cannot be neglected (which is usually the case in practice) then the signal level follows a more complicated (Suzuki) distribution [Parsons and Gardiner].

15.2.3 Time dispersion, frequency selective fading, and coherence bandwidth

Multipath propagation results in a received signal that is dispersed in time. For digital signalling, this time dispersion leads to a form of ISI whereby a given received data sample is corrupted by the responses of neighbouring data symbols. The severity of this ISI depends on the degree of multipath induced time dispersion (quantified by the multipath RMS delay spread of the radio channel) relative to the data symbol period. It is generally agreed that if the ratio of the RMS delay spread to symbol period is greater than about 0.3, then multipath-induced ISI must be corrected if the system's performance is to be acceptable.

In the 900 MHz frequency band used for cellular mobile radio, wideband propagation measurements have shown that worst-case RMS delay spreads are usually less than 12 μs. More particularly, urban areas tend to have RMS delay spreads of about 2 to 3 μs, with significant echo power up to about 5 μs, Figure 15.2, while rural and hilly areas have RMS delay spreads of about 5 to 7 μs [Parsons]. Figure 15.2 also illustrates Doppler shift due to relative motion of the transmitter, receiver or reflectors. At 900 MHz a relative velocity of approximately 1.2 km/h produces a 1 Hz Doppler shift.

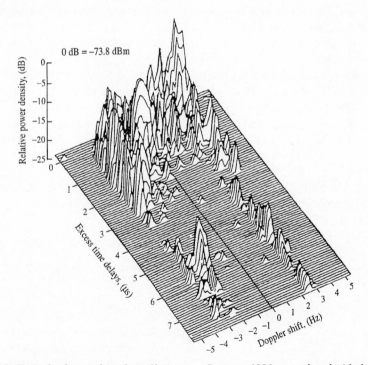

Figure 15.2 *Typical urban multipath profile (source: Parsons, 1991, reproduced with the permission of Peter Peregrinus).*

Multipath delay spread can be, equivalently, quantified in the frequency domain by the channel's coherence bandwidth, which is roughly the reciprocal of the RMS delay spread. The coherence bandwidth gives a measure of spectral flatness in a multipath channel. If two signal components are separated by greater than the coherence bandwidth of a channel, then they tend to fade independently and the overall signal is said to experience frequency-selective fading. On the other hand, if a signal's bandwidth is less than the coherence bandwidth of the channel, then the channel effectively exhibits spectrally flat fading in which all the signal components tend to fade simultaneously.

15.3 Nationwide cellular radio communications

15.3.1 Introduction

The growth in the demand for mobile radio services soon exceeded the capacity of, or possible spectral allocations for, PMR type systems. In the 1970s the concept therefore evolved of using base stations with modest power transmitters, serving all mobile subscribers in a restricted area, or cell, with adjacent cells using different operating frequencies. The key facet in cellular systems, introduced progressively through the 1980s, is that the power level is restricted so that the same frequency allocations can be reused in the adjacent cluster of cells, Figure 15.3. This constitutes a type of FDMA. Note that only cluster sizes of 3, 4, 7, 9, 12, etc. tesselate (i.e. lead to regular repeat patterns without gaps).

The protection ratio for these systems is the ratio of signal power from the desired transmitter (located at the centre of a cell) to the power received from a cochannel cell using the same operating frequency. Small clusters of 3 or 4 cells give 12 to 15 dB protection ratios which will only permit the use of a robust digital modulation method such as BPSK, Figures 11.10 and 11.21. As the cluster size increases so does the protection ratio. This arrangement is complicated as it involves handover as the mobile roams through the cells, but the ability to reuse the frequencies in close geographic proximity is a considerable advantage and this concept has now therefore been adopted worldwide for mobile telephone systems.

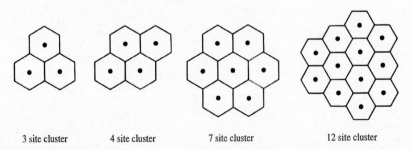

3 site cluster   4 site cluster   7 site cluster   12 site cluster

Figure 15.3 *Examples of possible cluster patterns for cellular systems.*

Cellular radio communication is experiencing rapid growth in North America, Asia and Europe. Hitherto, this unprecedented increase in user demand has been spurred on primarily by the business sector but now cellular equipment is widely used by contractors (such as plumbers and carpenters) as well as business executives. However, with the promise of personal convenience and freedom at relatively low cost, cellular radio is becoming attractive to the general public. In 1994 there were 3 million registered users of the UK cellular systems and in 1996 this grew to almost 7 million, representing 12% of the population. (Scandinavia currently has the highest market penetration worldwide (30%) for cellular telephony.)

It is predicted that, worldwide, user demand will increase ten-fold within the next five years. In fact, mobile cellular telephone systems in some large cities are currently congested to near capacity. Thus, it will be necessary to continuously increase the capacity of mobile cellular systems for the foreseeable future.

15.3.2 Personal cordless communications

The simplest of these systems is the 12 channel domestic, analogue, cordless (CT1) telephone system which operates within 50 to 100 m of the base unit. The second generation 40 channel digital telepoint equipment (CT2) with a 50 to 200 m range allows more mobility but, due to its still restricted range, needs many base stations. The UK Rabbit phone CT2 system, which operated in 1992 and 1993, was a simple system which did not allow the mobile subscriber to be called. This calling facility is, however, available in the French Bibop system. These telepoint FDMA systems, Table 15.1, use simple 32 kbit/s ADPCM speech coders (section 5.8.3), with 10 mW of output power, to achieve two-way communication on one 72 kbit/s circuit with only a 5 ms round trip delay. Channels are allocated at call setup, from the 40 available channels in the public base stations. The base stations are all connected to the PSTN giving telephony coverage similar to wired handsets.

Another system, DECT, was originally conceived as a cordless private branch exchange. However DECT is more advanced than CT2 in that it operates at 1900 MHz with TDMA, Table 15.1, has more advanced signalling and handover, and supports basic rate ISDN access (see Chapter 19). DECT uses a 10 ms frame time which is split into 5 ms for the uplink and 5 ms for the downlink transmissions, using time division duplex. Within this frame it supports 12 separate TDMA timeslots and, with 12 RF channels in the 1880 to 1900 MHz allocation, the base station capacity is approximately 140 simultaneous mobile users.

DECT is designed primarily for indoor operation where the multipath delay spread is less than 50 ns as the bit period is 870 ns. Thus it cannot be extended for use in outdoor cellular systems unless they are of the smaller coverage (microcellular) system design. Due to the simplicity of CT2 and DECT, they cannot handle significant Doppler shift due to handset motion and hence can only be used for mobiles travelling at walking pace. Also, due to the small range and hence cell sizes, DECT systems can only operate within localised areas as it is uneconomic to construct the 25,000 100 m diameter cells, for example, which would be required to achieve full central London coverage. However these microcells are significant in that they use low power transmissions of 1 to 100 mW, which is attractive for battery powered mobiles and, with the base station antenna below roof height, there is very little energy radiated into adjacent cells.

Table 15.1 *Comparison of European digital cordless and cellular telephony systems.*

	CT2	DECT	GSM 900	DCS 1800
Operating band (MHz)	864 – 868	1880 – 1900	890 – 960	1710 – 1880
Bandwidth (MHz)	4	20	2×25	2×75
Access method	FDMA	MF-TDMA	TDMA	TDMA
Peak data rate (kbit/s)	72	1,152	270	270
Carrier separation (kHz)	100	1,728	200	200
Channels per carrier	1	12	8	8
Speech coding	32 kbit/s	32 kbit/s	22.8 kbit/s	22.8 kbit/s
Coding/equalisation	no	no	yes	yes
Modulation	FSK	Gaussian FSK	GMSK	GMSK
Traffic channels/MHz	10	7	19	19
Mobile power output (W)	0.01	0.25	0.8 – 2.0	0.25 – 1
Typical cell size	50 – 200 m	50 – 200 m	0.3 – 35 km	0.02 – 8 km
Operation in motion	walking pace	walking pace	> 250 km/h	> 130 km/h
Capacity (erlangs/km^2)	N/A	10,000+	1,000	2,000

15.3.3 Analogue cellular radio communication

First-generation cellular systems employ analogue narrowband FM techniques [Black, Lee 1993]. For example, the North American advanced mobile phone system (AMPS) provides full duplex voice communications with 30 kHz channel spacing in the 800–900 MHz band, while the UK total access communication system, TACS – and extended TACS (ETACS) – operates with 25 kHz channel spacing in the 900 MHz band.

In a narrowband FM/FDMA cellular system the total available radio spectrum is divided into disjoint frequency channels, each of which is assigned to one user. The total available number of channels are then divided amongst the cells in each cluster and are reused in every cell cluster. Each cell in a given cluster is assigned a different set of frequency channels to minimise the adjacent channel interference while the cell size/cluster spacing is chosen to minimise co-channel interference.

15.3.4 Cell sizes

Cell size is dependent on expected call requirements. From knowledge of the total number of subscribers within an area, the probability of their requiring access and the mean duration of the calls, the traffic intensity in erlangs (see section 17.4.2) can be calculated. Erlang tables can then be used to ascertain the number of required channels for a given blocking, or lost call, probability.

For the cellular geometries of Figure 15.4 it is possible to calculate the carrier to interference ratio (C/I) close to the edge of the cell, when receiving signals from all the adjacent cofrequency cells. In an *n*-cell cluster, using a 4th power propagation law, the carrier to interference ratio is:

$$\frac{C}{I} = \frac{r^{-4}}{(n-1)d^{-4}} \tag{15.6}$$

where *r* is the cell radius and *d* the reuse distance between cell centres operating at the same centre frequency. (This is an approximate formula, arrived at by inspection, which is close, but not equal, to the worst case. A more accurate formula is derived in [Lee 1995].) Figure 15.4 shows the reuse distance between adjacent 7-cell clusters. The cluster size required to achieve a given *d/r* (and therefore the corresponding *C/I*) can be found using:

$$\frac{d}{r} = \sqrt{3n} \tag{15.7}$$

Although essentially general, equation (15.7) is particularly easily proved for the 7-cell cluster (Figure 15.4) as follows:

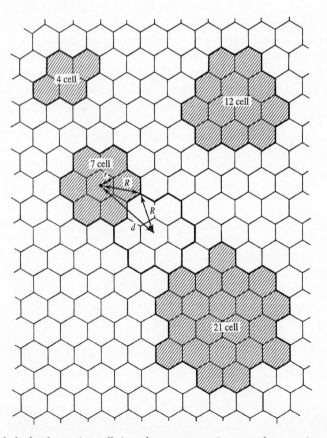

Figure 15.4 *Methods of grouping cells in order to cover a given area by repeating pattern.*

$$n = \frac{\text{cluster area}}{\text{cell area}} \tag{15.8}$$

The ratio of cluster to cell radii is therefore given by:

$$\frac{R}{r} = \frac{\sqrt{\text{cluster area}}}{\sqrt{\text{cell area}}} = \sqrt{n} \tag{15.9}$$

The angle between R and d shown in Figure 15.4 is 30°, therefore:

$$\frac{d}{2} = R\cos 30° \tag{15.10}$$

i.e.:

$$d = \sqrt{3}R \tag{15.11}$$

substituting $R = \sqrt{n}r$ from equation (15.9) into (15.11) gives:

$$\frac{d}{r} = \sqrt{3n} \tag{15.12}$$

It is now possible to refer to Table 15.1 to obtain the available radio channel capacity within each cell. The total number of channels which are required to accommodate the expected traffic within a geographic area defines the required number of cells, or clusters. Division of the area covered, by the total number of cells in use, then provides the required area per individual cell. To accommodate higher traffic capacity, required in dense urban areas such as central London, necessitates smaller cell sizes, Figure 15.5, compared to more rural areas.

At the frequencies presently used for cellular radio, propagation usually follows (approximately) an inverse square law for field strength (i.e. an inverse 4th power law for power density). The median signal strength can be estimated using this law to predict the path loss by subtracting (in dB) a 'clutter factor', as described in section 15.2.1, to

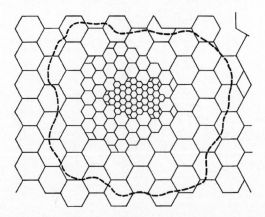

Figure 15.5 *Cell patterns used for London within the M25 orbital motorway, showing smaller cell patterns where most users are expected.*

account for deviations from ideal, plane earth, propagation.

As all cells use the same base station transmitter power, to minimise interference, the power level is set to ensure that there is adequate received signal at the edges of each cell compared to the thermal noise level. Transmission at a higher power level is not desirable as it simply wastes power. In the mobile the transmitter output power is controlled by the base station to minimise both interference and battery drain (and thus to maximise the time between recharges).

15.3.5 System configuration

In the UK cellular telephone network (operated by Vodafone and Cellnet) the base stations are connected to approximately 30 switching centres which hold information on the mobile transceivers which are active, including their location. Thus on calling a mobile transceiver from the wired PSTN one is connected first to the nearest switching centre and then on to the appropriate switching centre to route the call to the current mobile location, Figure 15.6. Within the area covered by a switching centre the mobile experiences handover as it roams from cell to cell. These digital switching centres, which form the interface between the wired PSTN and the mobile network, each have the capacity to handle 20 Merlangs of traffic and up to 100,000 subscribers. There are between six and ten switching centres for the dense London mobile traffic. The location of individual subscribers is tracked and this positional information within the overall mobile network is constantly updated in the home subscriber databases, Figure 15.6.

15.4 Digital TDMA terrestrial cellular systems

Although there are still over one million analogue cellular handsets in use in the UK, the present trend is towards digital cellular technology which has superior speech quality and improved privacy, and offers a natural extension to data transmission with ISDN compatibility. Most importantly, digital cellular systems have the potential to provide significantly higher capacity than analogue cellular systems, while utilising the same available bandwidth, and can achieve interoperability between different countries.

15.4.1 Systems

The digital vehicle mounted cellular systems for Europe, North America and Japan are all based on TDMA. The pan-European Groupe Spéciale Mobile (GSM) or global system for mobile communications [Lee 1995, Black], which was the world's first TDMA cellular system, transmits an overall bit rate of 271 kbit/s in a bandwidth of 200 kHz, with 8 TDMA users per carrier operating at 900 MHz, Figure 15.7. Thus, GSM 900 effectively supports one TDMA user per 25 kHz. On the other hand, the North American narrowband IS-54 TDMA system accommodates three TDMA users per carrier by transmitting an overall bit rate of about 48 kbit/s over the same 30 kHz bandwidth of the existing AMPS system. The Japanese system is similar to the North American system,

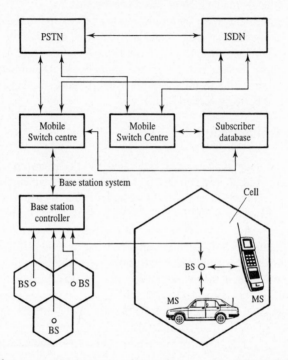

Figure 15.6 *Mobile system operation with mobile and base stations (MS) & (BS).*

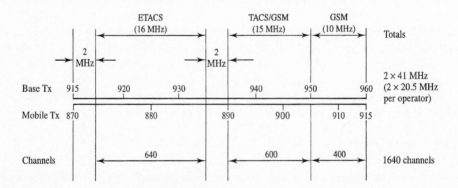

Figure 15.7 *UK allocation of 900 MHz spectrum for cellular communications.*

supporting three TDMA users per carrier with an overall bit rate of about 40 kbit/s transmitted over a 25 kHz bandwidth.

Such narrowband TDMA digital cellular systems improve system capacity by improving frequency reuse through cell size reduction and by accommodating several time-multiplexed users within essentially the same bandwidth that previously supported just one analogue user. Further capacity gains can also be realised by these digital systems through low bit-rate speech coding, higher trunking efficiency and more effective frequency reuse due to increased robustness to co-channel interference. However, TDMA digital cellular systems are susceptible to multipath-induced ISI which must be mitigated with relatively complex adaptive equalisation techniques, Table 15.1.

When it commenced operation in mid-1991, the GSM system provided duplex communications with 45 MHz separation between the frequency bands of the 890 to 915 MHz uplink, and the 935 to 960 MHz downlink. As a vehicle mounted system, the mobile power output ranges from 800 mW to 20 W, permitting up to 30 km cell sizes in low traffic density rural areas. In addition, GSM has been extended, via the DCS 1800 standard, to provide increased capacity for the, potentially larger, hand-held portable market. DCS 1800 operates in the 1800 MHz band, with three times the available bandwidth and capacity of GSM, and accommodates low-power (0.25 to 1 W), small-size portable handsets. At the higher power levels this corresponds to DCS 1800 cell sizes ranging from 0.5 km (urban) to 8 km (rural) with high grade coverage. Table 15.1 provides a comparison of these systems. DCS 1800 is thus optimised more for high density, high capacity situations (not necessarily with the facility to handle fast moving mobile subscribers), to achieve a personal communications network (PCN) meeting the current communication expectations of subscribers. The DCS 1800 system was introduced within the M25 orbital motorway area in London by Mercury in 1993 as their 'one-2-one' PCN service. Handsets are controlled by smart cards to allow several users access to the same handset with separate billing.

15.4.2 Data format and modulation

In GSM 900 and DCS 1800, eight TDMA users communicate over one carrier and share one base station transceiver, thereby reducing equipment costs. At the lowest level, the GSM TDMA format consists of eight user time-slots, each of 0.577 ms duration, within a frame of 4.615 ms duration (see Figure 15.8). This frame duration is similar to DECT

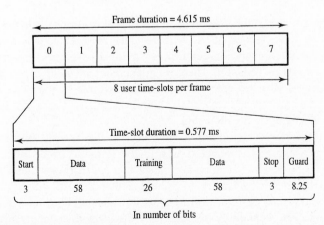

Figure 15.8 *Frame and time-slot details for data bit allocation in GSM.*

but, in GSM, the transmissions are only one way and the uplink and downlink have *separate* frequency allocations as they use frequency division duplex. TDMA users communicate by transmitting a burst of data symbols within their allotted time-slot in each frame. Each TDMA data burst contains 156.25 bits consisting of 116 coded speech bits, 26 training bits and 14.25 start, stop and guard bits (see Figure 15.8). This results in an overall transmitted bit rate of 270.8 kbit/s. GSM is relatively complex with significant overhead for control signalling and channel coding. However, because of the resulting redundancy, signal quality can be maintained even if one out of five frames is badly corrupted due to multipath, interference or noise induced errors.

GSM uses Gaussian minimum shift keying (GMSK) modulation with a bandwidth–time product of 0.3, section 11.4.6. With its Gaussian prefiltering, GMSK generates very low adjacent channel interference (greater than 40 dB below the in-band signal). The GMSK signal is transmitted at the bit rate of 270.8 kbit/s within a bandwidth of 200 kHz, resulting in a spectral (or bandwidth) efficiency of 1.35 bit/s/Hz. With 8 TDMA users sharing a 200 kHz channel, Figure 15.8, GSM has essentially the same spectral capacity (in terms of number of users per unit bandwidth) as the existing analogue TACS system. However, GSM achieves a higher overall capacity (in bit/s/Hz/km^2) by exploiting the digital signal's greater robustness to interference. Specifically, GSM operates at a signal-to-interference ratio of 9 dB while TACS requires about 18 dB. Consequently, a GSM system achieves a higher frequency reuse efficiency by being able to use smaller cells with cell cluster sizes of nine to twelve. GSM in the UK has two operators, Vodafone and Cellnet. The total available radio channels, per base station, is $8 \times 125 = 1,000$ simultaneous mobile users. DCS 1800 which, in the UK, is licensed to Mercury and the Harrison Communications Orange systems, offers three times the number of radio channels and twice the capacity of GSM, Table 15.1. The GSM system was, in 1996, used in 270 networks in 98 countries.

15.4.3 Speech and channel coding

In order to achieve spectrally efficient coding with toll quality speech, GSM employs an enhancement of LPC, section 9.7.1, with residual pulse excitation and a long term prediction [Gray and Markel]. Speech is processed in 20 ms blocks to generate 260 bits, resulting in a bit rate of 13 kbit/s. There is an inherent long delay in the coder, however, of 95 ms which necessitates echo cancellation. There are now provisions in GSM to accommodate half-rate speech coders which will essentially double the system capacity while maintaining similar speech quality.

Because of the low bit rate in this coder, speech quality is quite sensitive to bit errors. GSM overcomes this with channel coding in the form of 3-bit CRC error detection (Chapter 10) on 50 of the 182 most significant bits and half-rate convolutional coding with Viterbi decoding (Chapter 10) to protect further the 182 bits, Figure 15.9. The remaining 78 out of 260 bits are less important in achieving high speech quality and are left unprotected. The redundancy introduced by this channel coding results in an overall bit rate of 22.8 kbit/s (i.e. 456 bits transmitted in 20 ms). Furthermore, these 456 coded bits are interleaved (see Chapter 10) over eight TDMA time-slots to protect against burst

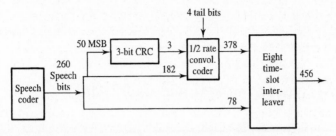

Figure 15.9 *GSM channel coding technique to assemble a 20 ms time-slot.*

errors resulting from Rayleigh fading. In addition to this channel coding, data encryption can also be used to achieve secure communications. It is this coding of digital data that gives GSM increased resilience to interference and hence more intense frequency reuse for high subscriber capacity.

15.4.4 Other operational constraints

Channel equalisation is required to overcome multipath-induced ISI. Moreover, due to the time-varying nature of the multipath fading channel, a user encounters a different channel impulse response at every TDMA burst. Thus, it is necessary to incur the overhead of transmitting a known training sequence in each time-slot, Figure 15.8, to allow the receiver to learn the channel impulse response and to adapt its filter coefficients accordingly. Furthermore, because the fading is potentially rapid in a mobile radio system, the channel response can change appreciably during a single TDMA burst. Consequently, the equaliser must be adaptively updated by tracking the channel variations. The GSM recommendations do not specify a particular equalisation approach, but the signal structure does lend itself to the method of maximum likelihood sequence estimation with Viterbi decoding.

For a given cellular layout and frequency plan, interference is reduced by proper use of adaptive power control, handover, discontinuous transmission and slow frequency hopping. For GSM, adaptive power control is mandatory for the mobile which has a transmit power dynamic range of about 30 dB (i.e. 20 mW to 20 W) adjusted at a rate of 2 dB every 60 ms, to achieve constant received signal power at the base station.

The GSM handover strategy [Lee 1995] is based on finding a base station with equal or higher received signal strength, regardless of the received signal level from the current base station. The use of slow frequency hopping (FSK) amongst carrier channels is a useful diversity method against Rayleigh fading. Since fading is independent for frequency components separated by greater than the coherence bandwidth of the channel, this frequency agility reduces the probability that the received signal will fall into a deep fade for a long duration. This is especially effective for a slowly moving or stationary mobile which might otherwise experience a deep fade for a relatively long time.

Due to the variability in cell sizes and distances, propagation times from transmitter to receiver can range from about 3 to 100 μs. To ensure that adjacent TDMA time-slots from different mobile transmitters do not overlap at the base station receiver, each mobile must transmit its TDMA bursts with timing advances consistent with its distance from the base station. This timing information is intermittently measured at the base station and sent to each mobile to ensure that this is achieved.

15.5 Code division multiple access (CDMA)

One multiple access technique widely adopted in communication systems, frequency division multiplex access (FDMA), allocates to each subscriber a narrow frequency slot within the available channel, Figure 5.12. An alternative TDMA technique allocates the entire channel bandwidth to a subscriber but constrains him to transmit only regular short bursts of wideband signal. Both these accessing techniques are well established for long haul terrestrial, satellite and mobile communications as they offer very good utilisation of the available bandwidth.

15.5.1 The CDMA concept

The inflexibility of these coordinated accessing techniques has resulted in the development of new systems based on the uncoordinated spread spectrum concept [Dixon, Lee 1993]. In these systems the bits of slow speed data traffic from each subscriber are deliberately multiplied by a high chip rate spreading code, $s_k(n)$, forcing the low rate (narrowband data signal) to fill a wide channel bandwidth, Figure 15.10. Spreading ratios, i.e. ratios of the transmitted (chip) bandwidth to data (bit) bandwidth, are typically between 100 and 10,000. (This is in contrast to the longer duration M-ary symbols of Chapter 11. In CDMA, the symbol rate is represented by the chips of spreading code which are of much shorter duration than the information digits.) Many subscribers can then be accessed by allocating a unique, orthogonal, spreading code, $s_k(n)$, to each, Figure 15.10. This constitutes a code division multiple access (CDMA) system. The signals, which are summed in the channel, have a flat, noise-like, spectrum allowing each individual transmission to be effectively hidden within the multiple access interference.

In the receiver, detection of the desired signal is achieved by correlation, section 2.6, against a local reference code, which is identical to the particular spread spectrum code employed prior to transmission, Figure 15.10, (i.e. $s_k(n)$ is used to decode the subscriber k transmissions). The orthogonal property of the allocated spreading codes means that the output of the correlator is essentially zero for all except the desired transmission. Thus correlation detection gives a processing gain or SNR improvement, G_p, equal to the spreading ratio:

$$G_p = 10 \log_{10} \left(\frac{\text{transmitted signal bandwidth}}{\text{original data bandwidth}} \right) \ (\text{dB})$$

$$= 10 \log_{10} \left(\frac{R_c}{R_b} \right) \ (\text{dB}) \tag{15.13(a)}$$

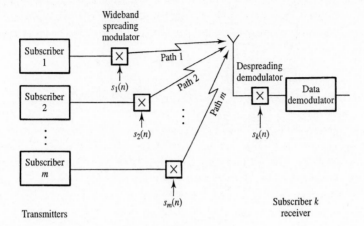

Figure 15.10 *Principle of code division multiple access (CDMA).*

where R_c is the chip rate and R_b the binary digit rate. Equivalently this can be expressed as:

$$G_p = 10 \log_{10} (T_b B) \quad \text{(dB)} \tag{15.13(b)}$$

where B is the bandwidth of the spreading code and T_b is the, relatively long, bit duration of the information signal.

The following (trivial) example shows how a short 3-chip spreading code, $s(n)$, can be added modulo-2 to the slow speed data. Each data bit has a duration equal to the entire 3-chip spreading code, and is synchronised to it, to obtain the transmitted product sequence, $f(n)$.

Data	+1	+1	+1	−1	−1	−1	+1	+1	+1
Spreading code	+1	−1	+1	+1	−1	+1	+1	−1	+1
Product sequence	+1	−1	+1	−1	+1	−1	+1	−1	+1

15.5.2 Receiver design

Both coherent receiver architectures (i.e. the active correlator and matched filter, of Figure 15.11) can be used for decoding or despreading the CDMA signals. One matched filter receiver [Turin 1976] implements convolution (section 4.3.3) using a finite impulse response filter whose coefficients are the time reverse of the expected sequence, to decode the transmitted data. (Note that in the example above, however, the code is symmetric.) Thus the filter coefficients in Figure 15.11(a), which is a repeat of Figure 8.34, would be $h_1 = +1$, $h_2 = -1$ and $h_3 = +1$ for the above 3-chip code. This design of receiver is optimum from an SNR standpoint since the receiver impulse response replicates, or reproduces, the expected transmitted signal. The output is given by the convolution of the received signal with the stored weight values, as shown previously in Chapters 4 and 8.

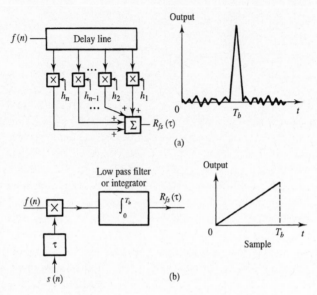

Figure 15.11 *Spread spectrum receiver typical outputs: (a) matched filter; (b) active correlator.*

The signal to interference ratio at the filter input is given by:

$$\left(\frac{S}{I}\right)_{in} = \frac{S}{I_o B} \tag{15.14}$$

where I_o is the power spectral density of all unwanted signals, S is the desired signal power and B is the bandwidth of the spread signal. The output SNR, after adding the processing gain arising from matched filter detection, equation (15.13), is then:

$$\left(\frac{S}{N}\right)_{out} = \frac{S}{I_o f_m} \tag{15.15}$$

where $f_m \ (= R_b)$ is the bandwidth of the desired information signal, i.e. $1/T_b$ in equation (15.13(b)).

If the receiver is not synchronised then the received signal will propagate through the matched filter which outputs the complete correlation function. The large peak confirms that the correct code is indeed being received and provides accurate timing information for the received signal, Figure 15.11(a). Note that with binary stored weight values the filter design is especially simple. Semiconductor suppliers currently market single chip 'correlators' with up to 64 binary-weighted taps at 20 Mchip/s input rates.

When timing information is already available then the simpler active correlator receiver, Figure 15.11(b), can be used (also shown in Figure 15.10). This receiver only operates correctly when the local receiver reference $s_k(t)$ is accurately matched, and correctly timed, with respect to the spreading code within the received signal $f(t)$. Synchronisation can be obtained by sliding the reference signal through the received signal. This can be an extremely slow process, however, for large T_bB spreading waveforms.

The pseudo-random bit sequence (PRBS) or pseudo-noise (PN) sequence, as obtained using a linear feedback shift register (Figure 13.15), can be used as the spreading code. For example one 7-chip PN sequence is $1, 1, 1, -1, -1, 1, -1$ and a 15-chip PN sequence is $1, 1, 1, 1, -1, 1, -1, 1, 1, -1, -1, 1, -1, -1, -1$ [Golomb *et al.*]. As the spreading code is often phase modulated onto the carrier and coherently demodulated $+1/-1$ is a more appropriate representation for the 1/0 binary digits. These codes are repeated cyclically to form continuous spreading codes. For spread spectrum applications the most important properties of PN sequences are that the *cyclic* autocorrelation sidelobe levels are small. For a sequence of K chips the sidelobes are equal to -1, Figure 15.12(a), with a peak amplitude of K. (Figure 15.12 is a repeat of Figure 13.16 without normalisation.) The aperiodic autocorrelation for an isolated PN sequence transmission has peak sidelobes of $< \sqrt{K}$, Figure 15.12(b). The low level of cyclic autocorrelation sidelobes (-1) in Figure 15.12(a) only occurs for the zero Doppler shift case, while the plot with Doppler offset is more like the aperiodic performance of Figure 15.12(b).

The PN sequence produces an approximate balance in the number of 1 and -1 digits and its 2-valued autocorrelation function in Figure 15.12(a) closely resembles that of a white noise waveform, Figure 3.28. The origin of the name pseudo-noise is that the digital signal has an autocorrelation function which is very similar to that of a white noise signal. Note that the rise time and fall time for the peak of this function is one chip period in duration and hence its bandwidth is directly related to the clock, or chip, rate of the spreading sequence.

The key facet of the CDMA system is that there is discrimination against narrowband and wideband interference, such as signals spread with a code sequence which is different

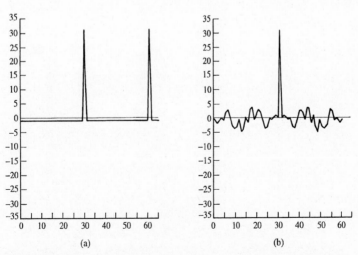

Figure 15.12 *PN code autocorrelation function: (a) for continuous K = 31 chip PN coded transmission; and (b) burst or aperiodic transmission again for K = 31.*

to that used in the receiver, Figure 15.13. The level of the suppression, which applies to CW as well as PN modulated waveforms, is given by the processing gain of equation (15.13).

15.5.3 Spreading sequence design

Maximal length sequences

Maximal length shift register sequences or m-sequences are so called due to their property that all possible shift register states except the all-zero state occur in a single, K length, cycle of the generated sequence [Golomb *et al.*]. Therefore for a shift register with n elements, the longest or maximum (m) length sequence which can be generated is $K = 2^n - 1$, section 13.4.2.

Due to the occurrence of all shift register states (except the all-zero state) each m-sequence will consist of 2^{n-1} ones and $(2^{n-1} - 1)$ zeros. There will be one occurrence of a run of n ones, and one run of $(n - 1)$ zeros. The number of runs of consecutive ones and zeros of length $(n - 2)$ and under will double for each unit reduction in run length and be divided equally between ones and zeros, see section 13.4.2 on PRBS sequences. Typical m-sequence generator feedback connections were shown in Table 13.3 and the overall size of the m-sequence set for different sequence lengths is given in Table 15.2.

The correlation properties of the m-sequence family are interesting due to their flat periodic autocorrelation sidelobes, Figure 15.12(a), which provide a good approximation to an impulsive autocorrelation function. The cross-correlation profile for a pair of m-sequences (forward and time reversed) is more typical of the generalised cross-correlation function.

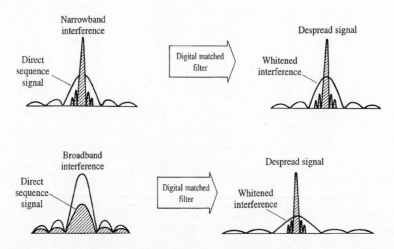

Figure 15.13 *Narrowband and wideband interference reduction in spread spectrum signals.*

Table 15.2 *Set sizes and periodic cross-correlation peak levels for m-sequences and Gold codes.*

		m-sequences		Gold codes	
n	*K*	*Set size*	*Peak level*	*Set size*	*Peak level*
3	7	2	5	9	5
4	15	2	9	17	9
5	31	6	11	33	9
6	63	6	23	65	17
7	127	18	41	129	17
8	255	16	95	257	33
9	511	48	113	513	33
10	1023	60	383	1025	65
11	2047	176	287	2049	65
12	4095	144	1407	4097	129

It can be seen from Table 15.2 that the available set size is very much smaller than the sequence length, K, particularly in the case of the even shift register orders. Therefore, as a multiple access code set, the PN sequence does not provide adequate capability for system subscribers who each require their own, unique, code assignment. An analytical expression for bounds on the maximum cross-correlation level has now become available [Pursley and Roefs] for the aperiodic sequence cross-correlation function. These sequences are not really appropriate for CDMA applications as the cross-correlation levels are too large. m-sequence sets in isolation are not, therefore, favoured for practical, high traffic capacity CDMA systems, hence the development of other sequence sets.

Gold codes

Selected pairs of m-sequences exhibit a three-valued periodic cross-correlation function, with a reduced upper bound on the correlation levels as compared with the rest of the m-sequence set. This m-sequence family subset is referred to as the preferred pair and one such unique subset exists for each sequence length. For the preferred pair of m-sequences of order n, the periodic cross-correlation and autocorrelation sidelobe levels are restricted to the values given by $(-t(n), -1, t(n) - 2)$ where:

$$t(n) = \begin{cases} 2^{(n+1)/2} + 1, & n \text{ odd} \\ 2^{(n+2)/2} + 1, & n \text{ even} \end{cases} \tag{15.16}$$

The enhanced correlation properties of the preferred pair can be passed on to other sequences derived from the original pair. By a process of modulo-2 addition of the preferred pair, Figure 15.14, the resulting derivative sequence shares the same features and can be grouped with the preferred pair as a member of the newly created family. This process can be repeated for all possible cyclically shifted modulo-2 additions of the preferred pair of sequences, producing new family members at each successive shift. A zero-shift 31-chip code can be generated by modulo-2 addition of two parent Gold sequences:

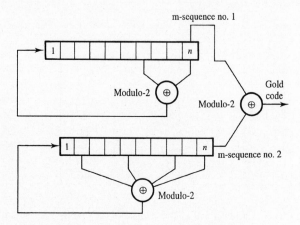

Figure 15.14 *Generation of a Gold code sequence for use in CDMA.*

$+1+1+1+1+1-1-1-1+1+1+1-1+1+1+1-1-1+1-1$ etc $-1+1+1-1-1-1$
$+1+1+1+1+1-1-1+1-1-1-1+1+1-1-1-1-1-1+1$ etc $-1+1+1+1+1-1$

$-1-1-1-1-1-1-1+1+1+1+1+1-1+1+1+1-1+1+1$ etc $-1-1-1+1-1-1$

and the corresponding five-chip shifted Gold code for the second sequence would be:

$+1+1+1+1+1-1-1-1+1+1+1-1+1+1+1-1+1-1$ etc $-1+1+1-1-1-1$
$-1-1+1-1-1+1+1+1-1-1-1-1-1+1-1-1+1+1+1-1-1+1$ etc $+1+1+1+1+1+1$

$+1+1-1+1+1+1+1+1-1+1+1+1-1-1+1+1-1+1+1+1+1$ etc $+1-1-1-1+1+1$

For this case, i.e. with an $n = 5$ generating register, any shift in initial condition from zero to 30 chip can be used (a 31-chip shift is the same as the zero shift). Thus, from this Gold sequence generator, 33 maximal length codes are available when the preferred pair are added as valid Gold codes. Extending this any appropriate two-register generator of length n can generate $2^n + 1$ maximal length sequences (of $2^n - 1$ code length) plus the preferred pair of sequences. The family of sequences derived from and including the original preferred pair are known collectively as Gold codes.

Since there are K possible cyclic shifts between the preferred pairs of m-sequences of length K, the available set size for the Gold code family is $K + 2$. The available sequence numbers for the Gold code set are compared with the available number of m-sequences in Table 15.2. It is apparent that, not only do Gold codes have improved periodic cross-correlation properties, but they are also available in greater abundance than the m-sequence set from which they are derived. However, as with the periodic m-sequence case, this attractive correlation property is lost as soon as the data traffic modulation is added.

Gold code shift register based designs exist, the direct approach consisting of two m-sequence generators employing the necessary preferred pair of feedback polynomials. This type of generator is shown in Figure 15.14 and can be used to synthesise all the preferred pair derived sequences in the Gold code by altering the relative cyclic shift through control of the shift register initial states. Switching control is also required over the output 'exclusive or' operation in order that the pure m-sequences can also be made available.

Gold codes are widely used in spread spectrum systems such as the international Navstar global positioning system (GPS), which uses 1023 chip Gold codes for the civilian clear access (C/A) part of the positioning service.

Walsh sequences

Walsh sequences have the attractive property that all codes in a set are precisely orthogonal. A series of codes $w_k(t)$, for $k = 0, 1, 2, \cdots, K$, are orthogonal with weight K over the interval $0 \leq t \leq T$, section 2.5.3, when:

$$\int_0^T w_n(t)w_m(t)\, dt = \begin{cases} K, & \text{for } n = m \\ 0, & \text{for } n \neq m \end{cases} \tag{15.17}$$

where n and m have integer values and K is a non-negative constant which does not depend on the indices m and n but only on the code length K. This means that, in a *fully synchronised* communication system where each user is uniquely identified by a different Walsh sequence from a set, the different users will not interfere with each other *at the proper correlation instant*, when using the same channel.

Walsh sequence systems are limited to code lengths of $K = 2^n$ where n is an integer. When Walsh sequences are used in communication systems the code length K enables K orthogonal codes to be obtained. This means that the communication system can serve as many users per cell as the length of the Walsh sequence. This is broadly similar to the CDMA capability of Gold codes, Table 15.2.

There are several ways to generate Walsh sequences, but the easiest involves manipulations with Hadamard matrices. The orders of Hadamard matrices are restricted to the powers of two, the lowest order (two) Hadamard matrix being defined by:

$$H_2 = \begin{bmatrix} 1 & 1 \\ 1 & -1 \end{bmatrix} \tag{15.18}$$

Higher-order matrices are generated recursively from the relationship:

$$H_K = H_{K/2} \otimes H_2 \tag{15.19}$$

where $\otimes$ denotes the Kronecker product. The Kronecker product is obtained by multiplying each element in the matrix $H_{K/2}$ by the matrix H_2. This generates K codes of length K, for $K = 2, 4, 8, 16, 32$, etc. Thus for $K = 4$:

$$H_4 = \begin{bmatrix} 1 & 1 & 1 & 1 \\ 1 & -1 & 1 & -1 \\ 1 & 1 & -1 & -1 \\ 1 & -1 & -1 & 1 \end{bmatrix} \tag{15.20(a)}$$

The difference between the Hadamard matrix and the Walsh sequence is only the order in which the codes appear; the codes themselves are the same. The reordering to Walsh sequences is done by numbering the Walsh sequences after the number of zero crossings the individual codes possess, to put them into 'sequency' order. The Walsh sequences thus result when the H_4 rows are reordered:

$$W_4 = \begin{bmatrix} 1 & 1 & 1 & 1 \\ 1 & 1 & -1 & -1 \\ 1 & -1 & -1 & 1 \\ 1 & -1 & 1 & -1 \end{bmatrix} \tag{15.20(b)}$$

Due to orthogonality, when the autocorrelation response peaks the cross-correlations for all other time synchronised Walsh sequenced transmissions are zero. The time sidelobes, however, in autocorrelated and cross-correlated Walsh sequences have considerably *larger* magnitudes than those for Gold or PN sequences. Also the orthogonal performance no longer applies in the imperfect (multipath degraded) channel.

15.5.4 Data modulation

Typically high chip rate short PN, Gold or Walsh coded CDMA spreading sequences are multiplied by the slower data traffic and this signal is used to phase shift key a carrier as described in Chapter 11. When the received signal is fed into a matched filter receiver then the baseband output, as shown in Figure 15.15, results. Figure 15.15 shows a 31-chip spreading code which is keyed by the low rate data traffic sequence $-1, -1, +1, -1, +1, +1, -1, -1$, before reception in the 31-tap matched filter of Figure 15.11(a).

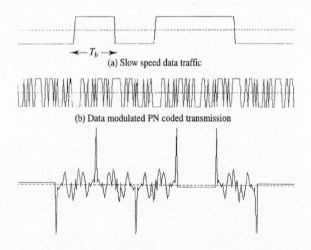

(a) Slow speed data traffic

(b) Data modulated PN coded transmission

(c) Receiver matched filter response to above encoded sequence

Figure 15.15 *Data modulation and demodulation on direct sequence coded signals.*

The matched filter or correlation receiver only provides a sharply peaked output if the correct codeword is received. CDMA systems thus rely on the auto and cross-correlation properties of codes such as those described above to minimise multiple access interference between subscribers.

15.5.5 Multipath responses

Multipath propagation, Figure 15.1, is experienced by nearly all mobile systems resulting in a time varying fading signal as the mobile moves position, Figure 15.16. (Figure 15.16 shows the instantaneous received power for a mobile, with respect to the local spatial average or mean power level, measured when travelling round a path, at approximately fixed range, from the base station.) This signal comprises the sum of two or more delayed path responses, with Doppler frequency offsets. (Note, therefore, that the channel multipath will destroy the orthogonal properties of the spreading codes.) If a CDMA signal has sufficiently wide bandwidth (i.e. greater than the coherence bandwidth of the channel) it is able to resolve the individual multiple paths, producing several peaks at the matched filter output at different time instants.

For efficient operation the receiver must collect and use all the multipath received signals in a coherent manner. It does this by using a channel equaliser which continuously adapts to the time varying mobile channel characteristic resulting, effectively, in a filter which is matched to this characteristic. The RAKE filter [Turin 1980], in which the weights in the sum bus of a tapped delay line are derived by measuring the channel impulse response, is an example. The RAKE filter typically provides 3 dB saving in power for a given error rate in a typical urban channel.

A RAKE receiver can also be configured to talk simultaneously to two separate base stations from different, but closely spaced, cell sites. This provides superior performance at the cell boundary, where the signal strength is weakest, and can also be used to achieve

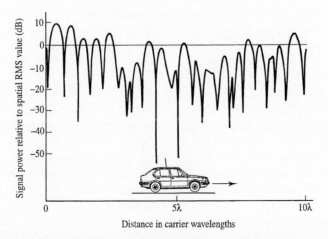

Figure 15.16 *Received power profile for a moving vehicle, as in Figure 12.28.*

a soft handover capability between cells.

15.5.6 The IS-95 system

CDMA cellular systems in which the spreading ratio, or K value, for a cell is approximately 127 are under design. In the US IS-95 CDMA standard, the basic user transmission (speech coder output) rate is 9.6 kbit/s. This is spread with a channel chip rate of 1.2288 Mchip/s (implying a total spreading factor of 128). Detailed analysis has shown that this $T_b B$ product is sufficient to support all the expected users within a typical cell. 1.2288 Mbit/s is also an acceptable speed for the VLSI modem electronics. The spreading process deployed is different on the downlink and uplink paths.

The downlink information, transmitted to the mobile, is split into four channels. These are pilot, synchronisation, paging and traffic channels. The receiver must demodulate these channels, to acquire the necessary synchronisation and systems information etc., before data transmission can start.

The pilot channel, which is transmitted with 20% of the total base station power, contains no data and is coded only with a $2^{15} - 1 = 32,767$ chip (short) PN code and a predetermined (all zeros) Walsh sequence. This signal permits the receiver to measure the strength and delay of all the nearby base station transmissions in order to select the closest ones yielding the greatest received power. (All base stations transmit the same PN code but a different relative delay, or code phase, is allocated to each individual station.) The pilot signal also permits the measurement, using a correlator, of the channel multipath response at the mobile, Figure 15.2, this response being used to continuously adjust the timing delay and gain of the taps in the RAKE receiver. The synchronisation channel, which is convolutionally encoded, interleaved, and then coded with a separate predetermined Walsh sequence, provides the Walsh sequence allocation information for the mobile. The paging channel is convolutionally encoded, interleaved, encrypted with a long PN sequence and then coded with the user specific Walsh sequence. This provides access to the system overhead information and specific messages for the mobile handset. Finally the traffic channel data is encoded using a ½ rate convolutional code (section 10.9), interleaved, Figure 10.23, spread by one of 64 orthogonal Walsh sequences and encripted with the long PN sequence. It contains the embedded power control signals which adjust the transmitted power level every 1.25 ms.

Each mobile in a given cell is assigned a different Walsh spreading sequence, providing perfect separation among the signals from different users, at least for a single-path channel. To reduce interference between mobiles that use the same spreading sequence in different cells, and to provide the desired wideband spectral characteristics (from Walsh functions which have variable power spectral characteristics), all signals in a particular cell are subsequently scrambled using the short PN sequence which is generated at the *same* rate as the Walsh sequence (1.2288 Mchip/s). Orthogonality among users within a cell is preserved because all signals are scrambled identically and the transmissions are synchronous.

On the uplink, the individual transmissions from mobiles are asynchronous and a slightly different spreading strategy is used in which the user data stream is first convolutionally encoded by a rate 1/3 encoder (Chapter 10). Then, after interleaving, each block of six encoded symbols is mapped to one of the 64 encoded orthogonal Walsh sequences (i.e. using a 64-ary orthogonal signalling system). This concept is implied in Figure 11.45 where 3 symbols are effectively mapped into one of 8 orthogonal sequences. A final fourfold spreading, giving the same transmission rate of 1.2288 Mchip/s, is achieved by expanding each data bit in the 307.2 kchip/s data stream with a 4-chip subsequence derived from a long PN spreading code. The rate 1/3 coding and the mapping onto Walsh sequences achieves a greater tolerance to interference than would be realised from traditional PN spreading and offers the possibility of employing more sophisticated receiver (array) processing in the base station.

For CDMA systems to work effectively all signals must be received with comparable power level at the base station. If this is not so then the cross-correlation levels of unwanted stronger, nearby, signals may swamp the weaker wanted, more distant, signal. In addition to the 'near–far' problem, which arises from very different path lengths, different fading and shadowing effects are experienced by different transceivers at the same distance from the base station.

If the receiver input power from one CDMA user is η_1 then for k equal power, active multiple access users, the total receiver input power is $k\eta_1$. The receiver has a processing gain, given by G_p in equation (15.13), to discriminate against other user multiple access interference. Thus the receiver output SNR is given by:

$$\left(\frac{S}{N}\right)_{out} = \frac{\eta_1}{k\eta_1} G_p = \frac{G_p}{k} \tag{15.21}$$

or, for a (given) receiver output SNR, required to reliably detect the data, the multiple access capacity can be estimated from:

$$k = \frac{G_p}{\left(\dfrac{S}{N}\right)_{out}} \tag{15.22}$$

CDMA cellular systems must thus deploy power control on the uplink to adjust mobile transmitter power levels, dependent on their location in the cell, in order to ensure equal received power level at the base station for all users. Fast closed-loop control is used, commands being transmitted at a rate of 800 bit/s, within the speech frames. The reduction in transmitter power away from the cell boundary, i.e. closer to the base station, also saves battery drain further prolonging the mobile's talk time. These CDMA systems reuse identical code sets in each adjacent cell and so form a single cell cluster system.

At both the base station and the mobile, RAKE receivers are used to resolve and combine multipath components, significantly reducing fading. This receiver architecture is also used to provide base station diversity during 'soft' handoffs, whereby a mobile making the transition between cells maintains simultaneous links with both base stations during the transition. Deployment of IS-95 systems in the Los Angeles, California area started in 1995.

Benefits of CDMA

Although CDMA represents a sophisticated system which can only operate with accurate uplink power control, it offers some unique attractions for cellular mobile communications. The voice activity factor in a normal conversation can also be exploited to switch off transmission in the quiet periods further reducing battery power requirement and reducing CDMA interference.

Besides the potential of offering higher capacity and flexibility, there are several attributes of CDMA that are of particular benefit to cellular systems. These are summarised below:

- **No frequency planning needed -** In FDMA and TDMA, the frequency planning for a service region is critical. Because the frequency bands allocated to CDMA can be reused in every CDMA cell, no frequency planning is required. As CDMA deployment grows, there is no need to continually redesign the existing system to coordinate frequency planning.
- **Soft capacity -** For FDMA or TDMA, when a base station's frequency bands or time slots are fully occupied, no additional calls can be accommodated by that cell. Instead of facing a hard limit on the number of available channels a CDMA system can handle more users at the expense of introducing a gradual degradation of C/I, and therefore link quality. CDMA service providers can thus trade off capacity against link quality.
- **Simpler accessing -** Unlike FDMA and TDMA, coordination of signals into prespecified frequency allocations or timeslots is no longer required in CDMA.
- **Micro-diversity -** CDMA mitigates multipath propagation using a RAKE receiver. This combines all the delayed signals within the cell to improve the received signal-to-noise ratio over other cellular systems.
- **Soft handover -** Soft handover refers to a CDMA call being processed simultaneously by two base stations while a mobile is in the boundary area between two cells. CDMA is able to perform soft handover because all users in the system share the same carrier frequency, so no retuning is involved in switching from one base station to another as required by FDMA and TDMA systems.
- **Dynamic power control -** Stringent power control is essential to maximise CDMA system capacity. This is exercised by embedding power control commands into the voice traffic. The average transmitted power of a CDMA mobile is thus significantly less than that of FDMA or TDMA systems.
- **Lower battery dissipation rates -** As the transmitted power in a handset is reduced, so is the battery dissipation, prolonging the available talk time.

All these benefits are incorporated in the IS-95 specification for CDMA mobile cellular systems [IS-95]. It is also proposed to use CDMA at 1.9 GHz for personal communications from 1995 onwards. Other systems which use CDMA, or spread spectrum techniques, are the NAVSTAR-GPS satellite based navigation systems. This is a ranging system in which a correlation peak determines the time of arrival of a signal and hence the range to one of the many satellites orbiting overhead, thus establishing precise timing and position information.

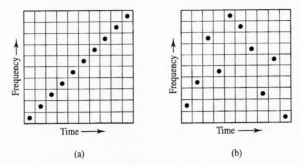

Figure 15.17 *(a) Representation for linear step FH waveform with K = 10; and (b) example of a 10 × 10 random frequency-hop pattern.*

15.5.7 Frequency hopped transmission

An alternative to PSK modulation with the spreading sequence is to alter, hop or shift the transmitter frequency in discrete steps across the spread bandwidth. Figure 15.17 shows in (a) a hopping pattern where the selected frequency increases linearly with time and in (b) a random hopping pattern. With these frequency hopped (FH) techniques a spread spectrum communication system can be designed, in principle, with any desired number of frequency slots, several hundred being typical. Individual frequencies are selected in a frequency synthesiser by detecting several adjacent chips of the spreading code and decoding the word to identify the transmit frequency. Orthogonality between adjacent frequency slots is ensured if the dwell time on each frequency, T_h, equals the reciprocal of the slot separation, see section 11.5.1 on MFSK. FH offers a much flatter transmitted spectrum than spread spectrum PSK, Figure 15.18. Also, for a given RF bandwidth, the FH dwell time, T_h, is much longer than the equivalent chip time, T_c, when direct sequence PSK techniques are used, easing synchronisation acquisition in the FH system. For slow FH systems, such as those sometimes used by the military, coherence between hops is not required as the hopping rate is much less than the bit rate. Here the spread spectrum technique is used primarily to achieve protection against hostile jamming signals and to overcome slow fading.

15.6 Mobile satellite based systems

A current development in satellite communications is to extend communication to the mobile subscriber. This requires the system to handle large numbers of small capacity earth stations which have relatively low gain antennas compared to the stationary antennas which are employed in the INTELSAT and VSAT systems. One mobile communications example is the INMARSAT ship-to-shore telephone and data service which is used on large cruise ships, by construction workers, and by UN peacekeepers. INMARSAT A stations, whose model M (briefcase sized) terminals cost £20,000,

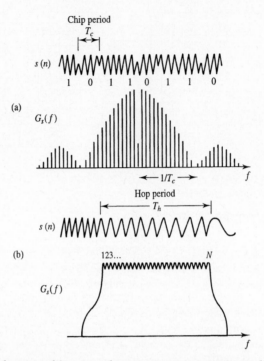

Figure 15.18 *Time domain and frequency domain representations of spread spectrum waveforms: (a) phase modulated transmissions; (b) frequency hopped transmissions.*

provide 64 kbit/s data rate links for speech on which call costs are £3 per minute (1996 prices). The cheaper C terminals have only 600 bit/s capability through the MARISAT geostationary satellites using L-band UHF transmissions. In these systems the mobile has only limited motion and hence a modest gain (10 to 25 dB) steered dish is still employed as the receiving/transmitting antenna.

Another mobile example is the Skyphone system which routes digitally coded speech traffic from commercial aircraft to a satellite ground-station using the INMARSAT satellites as the space-borne repeaters [Schoenenberger]. Here the antenna gain is lower as it has an omnidirectional coverage pattern but with sophisticated speech coding the data rate is reduced to approximately 12 kbit/s to compensate for the low power link budget.

For personal communications, with individual subscribers in mind, the 1992 World Radio Administrative Conference (WRAC) made frequency allocations at L-band (1.6 GHz) for mobile-to-satellite, and S-band (2.5 GHz) for satellite-to-mobile, channels. Here mobile receivers will again use antennas with very modest gain (0 to 6 dB). The satellites will require large antennas to provide the required spot beam coverage, particularly when the link budget is marginal. For this reason alternative systems employing inclined highly elliptical orbit (HEO) or low earth orbit (LEO) satellites are proposed, Figure 14.25.

The most promising inclined HEO is the Molniya orbit, Figure 14.25. It is asynchronous with a period of 12 hours. Its inclination angle (i.e. the angle between the orbital plane and the earth's equatorial plane) is 63.4° and its apogee (orbital point furthest from earth) and perigee (orbital point closest to earth) are about 39,000 km and 1,000 km respectively. The Molniya orbit's essential advantage is that, from densely populated northern latitudes, a satellite near apogee has a high elevation angle and a small transverse velocity relative to an earth station. Its high elevation angle reduces the potential impact on the link budget of shadowing by local terrain, vegetation and tall buildings. (The latter can be especially problematic for mobile terminals in urban areas.) Its small angular velocity near apogee makes a satellite in this orbit appear almost stationary for a significant fraction of the orbital period. Three satellites, suitably phased around a Molniya orbit, can therefore provide continuous coverage at a high elevation angle. Modest gain, zenith pointed, antennas can be used for the mobile terminals, the quasi-stationary nature of the satellite making tracking unnecessary. On either side of apogee satellites in HEOs have the disadvantage of producing large Doppler shifts in the carrier frequency due to their relatively large radial velocity with respect to the earth station. Handover mechanisms must also be employed to allow uninterrupted service as one satellite leaves the apogee region and another enters it.

LEO systems will be implemented using large constellations of satellites in low, circular, earth orbit. Their essential advantage will be the relatively low FSPL compared to GEO and HEO systems (see equation (12.71)). Proposed satellite numbers to implement these systems range from 14 to 77 with spot beams being used to achieve cellular coverage at the earth's surface and to permit the reuse of the same spectral allocation several times through each satellite. It is proposed to operate such systems with < ½ W of mobile transmitter power for only 2 W of battery power drain. One proposal for L-band satellite transmission is the Motorola Iridium system which initially aimed to use 77 LEO satellites, each with 37 spot beams or cells, Figure 14.44. The cells will move rapidly over the earth's surface with the relative satellite motion. This could provide a worldwide space-based personal cellular system for use with simple handsets sometime beyond the year 2000. It is likely that in the future subscriber mobile handsets will be reprogrammable for use with either satellite or terrestial mobile systems.

It will not be possible in the systems described above to use the fixed FDM and TDM access schemes employed in the INTELSAT systems, Chapter 14. Mobile systems may have to employ different modulation and multiple accessing formats on the uplink and on the downlink. In order to allow access from low power narrowband mobile transmissions FDMA will probably be used on the uplink with a single narrowband channel allocated to each carrier or discrete transmission. On the downlink it will be preferable to employ TDMA as this is less affected by non-linearities in the satellite high power amplifiers than FDMA. Such a scheme involves extra complexity, as an on-board processor will be required on the satellite to demodulate the individual FDMA signals, regenerate them and then remodulate them in TDMA format, Figure 15.19. Also with small antenna diameters at mobile (microterminal) ground-stations the bit rates must be low, but commensurate with the bandwidth of encoded speech, Table 9.3, to give sufficient SNR

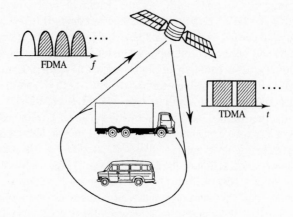

Figure 15.19 *Satellite communication to small mobile transceivers where the uplink and downlink typically use different multiple access techniques.*

for reception at the satellite. Satellite links are widely used in Europe for distributing radio pager messages to VSATs for subsequent terrestial transmission.

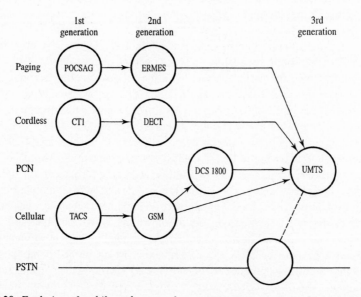

Figure 15.20 *Evolution of mobile and personal communication systems and equipment.*

15.7 The personal communications network

All the systems discussed above are progressing from first generation analogue (TACS) equipment, through second generation digital (GSM) and on to a third generation as represented by the universal mobile telecommunication service (UMTS) and future public land mobile telecommunications system (FPLMTS). These [Dasilva *et al.*] represent the personal communications network (PCN) concept, launched in 1993, for low cost mobile communications for the mass market, Figure 15.20. PCN will be a flexible service with several different price levels to meet individual subscriber needs for speech and data communication. In addressing the mass market it will offer an alternative to the wired PSTN. Figure 15.21 shows the different envisaged levels of radio communication systems in PCN. PCN aims to combine many services such as telephone, fax, teletext, etc. into one system to achieve spectral allocation benefits.

PCN also impacts on the design of short range wireless LANs, which permit interconnected equipment such as computers and printers to be quickly moved around the office. In the USA some of these products use the industrial, scientific and medical (ISM) frequency bands at 902 to 908 MHz, 2400 to 2484 MHz and 5.8 GHz where regulation is less constrained but interference may be severe.

Part of the extra PCN capacity will be achieved by using smaller microcells (picocells) or further splitting the current cellular pattern into cells which have three 120° individual sectors. With base station antennas below rooftop height the RF coverage will be very confined allowing more efficient reuse than in current, larger, cell system designs. PCN will use part of the DCS 1800 allocation, Table 15.1. The band allocations for

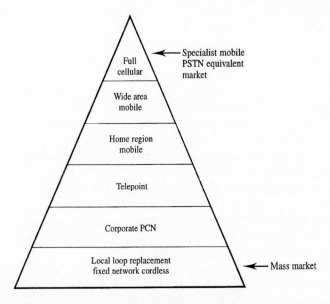

Figure 15.21 *Radio communication systems to be provided within PCN.*

FPLMTS are 1800 to 2025 MHz and 2110 to 2200 MHz which encompasses part of the DECT allocation. These systems will have to use variable rate, low delay, speech coders to carefully match the available radio channel bandwidth. In the USA PCN will be built to the personal communications service standard.

Another feature, which became available in 1997, is personal numbering to avoid the individual subscriber having separate home, office and mobile telephone numbers. This requires reprogramming of the individual handsets and the infrastructure to route incoming call to the correct destination.

15.8 Summary

Private mobile radio utilises VHF and UHF frequencies. It is not usually interfaced with the PSTN and is principally used by the emergency services, the public utilities, the haulage industry and taxi companies. The number of PMR channels available in the UK is a little over 1000. Each channel has an allocated bandwidth of 12½ kHz. Narrowband FM modulation is normally used for speech traffic while FSK modulation can be used for data transmission.

Like satellite communications, first generation cellular radio systems use analogue modulation. In the UK this generation is represented by TACS which operates using narrowband FM/FDMA with 25 kHz channels and a carrier frequency of 900 MHz. The essential feature of cellular systems is the division of the coverage area into cell clusters. The allocated frequencies are divided between the cells of a cluster. The same frequencies are also used (or reused) in the cells of the adjacent clusters. A base station at the (nominal) centre of each cluster uses modest transmit power so that interference in adjacent cells is kept to acceptable levels. Base stations are connected to the wired PSTN via switching centres. The received signal strength is subject to slow (log-normal) fading due to (multiplicative) shadowing effects and fast (Rayleigh) fading due to (additive) multipath effects.

Second generation cellular systems are digital. In Europe this generation is represented by GSM. In this system 8 TDMA users share each 200 kHz wide, 271 kbit/s, channel. The bandwidth per user is therefore 25 kHz as in TACS but overall spectrum is conserved due to the lower C/I tolerated by digital systems allowing the use of smaller cells and, consequently, more intensive frequency reuse. GSM uses GMSK modulation with $BT_b = 0.3$, LPC derivatives for speech bandwidth compression and CRC error correction. For large power delay spread the channel is frequency selective and time dispersion results in serious ISI. Adaptive equalisation is therefore required. Adaptive power control at the mobile is also used to maintain adequate received power at the base station whilst minimising interference. CDMA is more sophisticated than TDMA and it is set for major application in US cellular systems.

Third generation communications systems will integrate further the digital systems described here (and others) to provide an increased range of services to each user. These services will be based on an ISDN infrastructure and will be available to interface to all terminal equipment including those which are mobile and hand-held. This is the medium range objective at the core of the PCN concept.

This chapter has brought together many of the techniques covered in earlier chapters. It has shown how source and channel coding techniques combined with VHF, UHF or microwave transmission frequencies, and advanced receiver processing can combat ISI and multipath effects. This now permits the design and realisation of advanced communications systems which are well matched to the needs of the mobile user in the year 2000 and beyond. Such is the attraction of lightweight handsets that, in 1993, over 2% of the Western European population of 700 million individuals possessed cellular equipment and, in 1996, over 20% of the Danish population had purchased these handsets. Another major user is Asia with 6 million subscribers in 1994 and a 50% p.a. growth rate.

CHAPTER 16

Video transmission and storage

16.1 Introduction

This chapter considers the means of encoding, transmitting, storing and displaying a video picture by both traditional (analogue) and modern (digital) techniques. Here we aim to give only an introduction to the subject; for more advanced and detailed information, the reader is referred to one of the many specialist textbooks, e.g. [Grob, Lenk].

The fundamental requirement of any video display system is the ability to convey, in a usable form, a stream of information relating to the different instances of a picture. This must contain two basic elements, namely some description of the section of the picture being represented, e.g. brightness, and an indication of the *location* (in space and time) of that section. This implies that some *encoding* of the picture is required.

There are many different approaches to the encoding problem. We shall look at some of the common solutions which have been adopted, although other solutions also exist. Initially, we consider analogue based solutions, which transfer and display video information in real time with no direct mechanism for short term storage. Later digitally based solutions, which allow for storage of the video image sequence, are considered. To date, however, most digital video sequences are currently displayed using analogue techniques similar to those employed in current domestic television receivers.

The problem of representing a small section of an image (often called a picture element or pixel) is solved by different encoding mechanisms in different countries. All essentially separate the information contained in a pixel into black and white (intensity) and colour components. The pixel, although normally associated with digital images, is relevant to analogue images as it represents the smallest independent section of an image. The size of the pixel limits *resolution* and, therefore, the quality of the image. It also affects the amount of information needed to represent the image. For real time transmission this in turn determines the bandwidth of the signal conveying the video stream (see Table 1.2 and Chapter 9 on information theory).

The problem of identifying the location of a pixel is addressed by allowing the image to be represented by a series of lines one pixel wide, scanned in a predetermined manner across the image, Figure 16.1. The pixels are then transmitted serially, special pulse sequences being used to indicate the start of both a new line and a new picture or image frame according to the encoding scheme used (see later Figure 16.4). Note that these systems rely upon the transmitter and receiver remaining synchronised. In the UK, the encoding mechanism used for TV broadcasting is known as PAL (phase alternate line), in the USA NTSC (National Television Standards Committee) is used, while in France SECAM (système en couleurs à mémoire) has been adopted.

16.2 Colour representation

A pixel from a colour image may be represented in a number of ways. The usual representations are:

1. An independent intensity (or luminance) signal, and two colour (or chrominance) signals normally known as hue and saturation.
2. Three colour signals, typically the intensity values of red, green and blue, each of which contains part of the luminance information.

In the second technique a white pixel is obtained by mixing the three primary colours in appropriate proportions. The colour triangle of Figure 16.2 shows how the various colours are obtained by mixing. Figure 16.2 also interprets, geometrically, the hue and saturation chrominance information. The hue is an angular measure on the colour triangle whilst the proportion of saturated (pure) colour to white represents radial distance. In practice we also need to take account of the response of the human eye which varies with colour or wavelength as shown in Figure 16.3. Thus for light to be interpreted as white we actually need to add 59% of green light with 30% of red and 11%

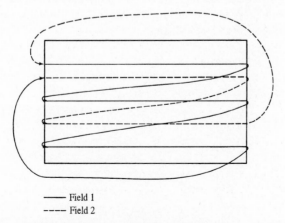

——— Field 1
- - - - Field 2

Figure 16.1 *Line scanning TV format with odd and even fields.*

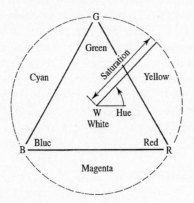

Figure 16.2 *Colour triangle showing hue and saturation.*

of blue light. The luminance component Y is thus related to the intensity values of the red (R), green (G), and blue (B) contributions by the following approximate formula:

$$Y = 0.3R + 0.59G + 0.11B \tag{16.1}$$

In practice, luminance and colour information are mathematically linked by empirical relationships. The principal benefit of separating the luminance and chrominance signals is that the luminance component only may then be used to reproduce a monochrome version of the image. This approach is adopted in colour television transmission for compatibility with the earlier black and white TV transmission system.

The theory governing the production of a range of colours from a combination of three primary colours is known as additive mixing. (This should not be confused with subtractive mixing as used in colour photography.) It is possible for full colour information to be retrieved if the luminance and two colours are transmitted. In practice, colour/luminance difference signals (e.g. R − Y) are transmitted, and these are modified to fit within certain amplitude constraints. The colour difference signals, or colour separated video components, U and V, are:

$$U = 0.88(R - Y) \tag{16.2}$$

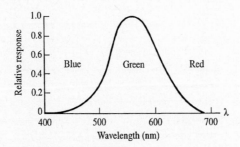

Figure 16.3 *Response of the human eye to colour.*

and

$$V = 0.49(B - Y) \tag{16.3}$$

Most video cameras and cathode-ray tube (CRT) displays produce an image in the so-called RGB format, i.e. using three signals of mixed intensity and colour information. This is transformed into YUV format before transmission and reformed into an RGB representation for colour display.

16.3 Conventional TV transmission systems

16.3.1 PAL encoding

The PAL encoding mechanism, as employed in the UK, provides for both colour information transmission and regeneration of a scanned image after display synchronisation. The latter is achieved by transmitting the image as a sequence of lines (625 in total) which are displayed to form a sequence of frames (25 per second), Figure 16.1. The system is basically analogue in operation, the start of line and start of frame being represented by pulses with predefined amplitudes and durations.

Line scanning to form the image is illustrated in Figure 16.1. In PAL systems, a complete image, or frame, is composed of two fields (even and odd) each of which scans through the whole image area, but includes only alternate lines. The even field contains the even lines of the frame and the odd field contains the odd lines. The field rate is 50 fields/s with 312½ lines per field and the frame rate is 25 frames/s.

The line structure is shown in Figure 16.4. Figure 16.4(a) shows the interval between fields and Figure 16.4(b) shows details of the signal for one scan line. This consists of a line synchronisation pulse preceded by a short duration (1.5 μs) period known as the front porch and followed by a period known as the back porch which contains a 'colour burst' used for chrominance synchronisation, as described shortly. The total duration of this section of the signal (12 μs) corresponds to a period of non-display at the receiver and is known as 'line blanking'. A period of active video information then follows where the signal amplitude is proportional to luminous intensity across the display and represents, in sequence, all the pixel intensities in one scan line. Each displayed line at the receiver is therefore of 52 μs duration and is repeated with a period of 64 μs to form the field display.

625 lines are transmitted over the two fields of one frame. However, only 575 of these contain active video. The non-active lines contain field synchronisation pulses, and also data for teletext type services. The video information carried by a PAL encoded signal is contained within a number of frequency bands, although to save bandwidth, the chrominance information is inserted into a portion of the high frequency section of the luminance signal, Figure 16.4(c). This band is chosen to avoid harmonics of the line scan frequency which contain most of the luminance energy. Fortunately the colour resolution of the human eye is less than its resolution for black and white images and hence we can conveniently transmit chrominance information as a reduced bandwidth signal within the

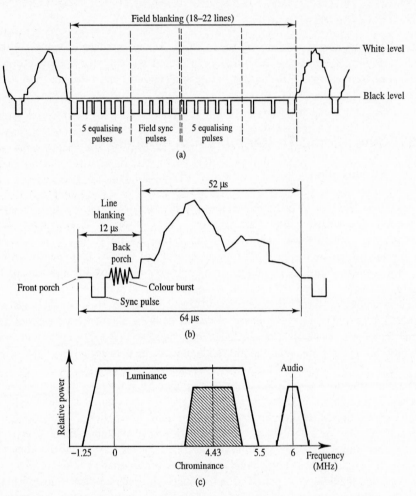

Figure 16.4 *TV waveform details: (a) field blanking information at the end of a frame; (b) detail for one line of video signal; (c) spectrum of the video signal.*

luminance spectrum.

The chrominance information is carried as a quadrature amplitude modulated signal, using two 4.43 MHz (suppressed) carriers separated by a 90° phase shift, to carry the colour difference signals of equations (16.2) and (16.3). This is similar to QAM modulation described in Chapter 11, but with analogue information signals. If the carrier frequency is f_c (referred to as the colour sub-carrier frequency) then the resulting colour signal S_c is:

$$S_c = U\cos(2\pi f_c t) + V\sin(2\pi f_c t) \tag{16.4}$$

The resulting chrominance signal bandwidth is 2 MHz. This is sometimes called the YIQ signal. Figure 16.4 also shows an additional audio signal whose sub-carrier frequency is 6 MHz.

In order to demodulate the QAM chrominance component, a phase locked local oscillator running at the colour sub-carrier frequency f_c is required at the receiver. This is achieved by synchronising the receiver's oscillator to the transmitted colour sub-carrier, using the received 'colour burst' signal, Figure 16.4(b).

Since the chrominance component is phase sensitive, it will be seriously degraded by any relative phase distortion within its frequency band due, for example, to multipath propagation of the transmitted RF signal. This could lead to serious colour errors but the effect is reduced by reversing the phase of one of the colour difference components on alternate line scans. Thus the R − Y channel is reversed in polarity on alternating line transmissions to alleviate the effects of differential phase in the transmission medium.

The PAL signal spectrum shown in Figure 16.4(c) is the baseband TV signal and if RF television transmission is required, this must be modulated on to a suitable carrier and amplified. Typical UHF carrier frequencies extend to hundreds of MHz with power levels up to hundreds of kW (even MW in some cases). Terrestial TV channels 21 to 34 lie between 471.25 and 581.25 MHz and channels 39 to 68 fall between 615.25 and 853.25 MHz, Table 1.4. Satellite TV occupies a band at 11 GHz with 16 MHz interchannel spacings.

16.3.2 PAL television receiver

Figure 16.5 shows a simplified block diagram of the main functional elements of a colour television receiver. The RF signal from the antenna or other source is selected and

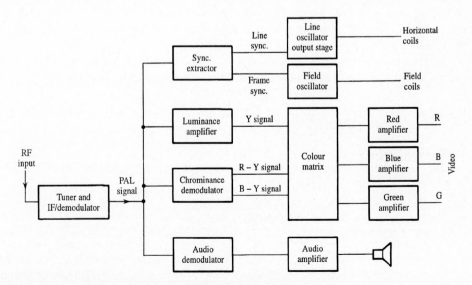

Figure 16.5 *Simplified block diagram of colour TV receiver.*

amplified by the tuner and IF (intermediate frequency) stages and finally demodulated to form a baseband PAL signal. Four information signals are then extracted: the audio, luminance, chrominance, and synchronisation signals.

The chrominance signal is demodulated by mixing with locally generated in phase and quadrature versions of the colour sub-carrier. The resulting colour difference and luminance signals are then summed in appropriate proportions and the R, G and B output signals amplified to sufficient levels to drive the CRT video display.

The x and y deflections of the electron beams are produced by magnetic fields provided by coils situated on the outside of the CRT. These coils are driven from ramp generators, synchronised to the incoming video lines such that the beam is deflected horizontally across the display face of the tube from left to right during the active line period, and vertically from top to bottom during the active field period.

16.3.3 Other encoding schemes

The NTSC system has many fundamental similarities to PAL, but uses different transmission rates. For example, the frame rate is 30 Hz, and the number of lines per frame is 525. PAL is, in fact, an enhancement of the basic principles of the NTSC system. The main difference between the two is in the phase distortion correction provided by PAL. This is not present in the NTSC system, and hence NTSC can be subject to serious colour errors under conditions of poor reception. Any attempted phase correction is controlled entirely by the receiver, and most receivers allow the viewer to adjust the phase via a manual control. The overall spectral width of NTSC is somewhat smaller than PAL.

The SECAM system uses a similar frame and line rate to PAL and, like PAL, was developed in an attempt to reduce the phase distortion sensitivity of NTSC. Essentially, it transmits only one of the two chrominance components per line, switching to the second chrominance component for the next line.

Transmission of digitally encoded video images is currently being developed and implemented. Some of the basic digital techniques are discussed in later sections.

16.4 High definition TV

16.4.1 What is HDTV?

High definition television (HDTV) first came to public attention in 1981, when NHK, the Japanese broadcasting authority, first demonstrated it in the United States. HDTV [Prentiss] is defined by the ITU-R study group as:

'A system designed to allow viewing at about three times the picture height, such that the system is virtually, or nearly, transparent to the quality or portrayal that would have been perceived in the original scene ... by a discerning viewer with normal visual acuity.'

HDTV proposals are for a screen which is wider than the conventional TV image by about 33%. It is generally agreed that the HDTV aspect ratio will be 16:9, as opposed to the 4:3 ratio of conventional TV systems. This ratio has been chosen because psychological tests have shown that it best matches the human visual field. It also enables use of existing cinema film formats as additional source material, since this is the same aspect ratio used in normal 35 mm film. Figure 16.6(a) shows how the aspect ratio of HDTV compares with that of conventional television, using the same resolution, or the same surface area as the comparison metric.

To achieve the improved resolution the video image used in HDTV must contain over 1000 lines, as opposed to the 525 and 625 provided by the existing NTSC and PAL systems. This gives a much improved vertical resolution. The exact value is chosen to be a simple multiple of one or both of the vertical resolutions used in conventional TV. However, due to the higher scan rates the bandwidth requirement for analogue HDTV is approximately 12 MHz, compared to the nominal 6 MHz of conventional TV, Table 1.2.

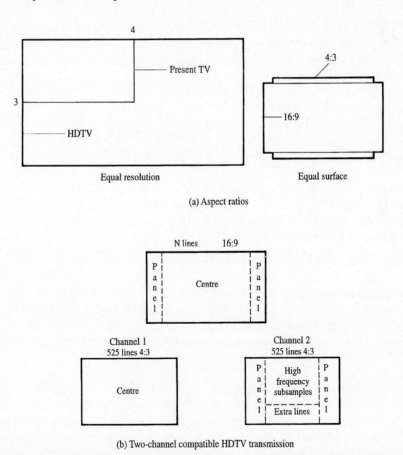

Figure 16.6 *(a) Comparison between conventional TV and HDTV, (b) 2-channel transmission.*

The introduction of a non-compatible TV transmission format for HDTV would require the viewer either to buy a new receiver, or to buy a converter to receive the picture on their old set. The initial thrust in Japan was towards an HDTV format which is compatible with conventional TV standards, and which can be received by conventional receivers, with conventional quality. However, to get the full benefit of HDTV, a new wide screen, high resolution receiver has to be purchased.

One of the principal reasons that HDTV is not already common is that a general standard has not yet been agreed. The XVIth CCIR plenary assembly recommended the adoption of a single, worldwide standard for high definition television. Unfortunately, Japan, Europe and North America are all investing significant time and money in their own systems based on their own, current, conventional TV standards and other national considerations.

16.4.2 Studio standards

Initially there were two main proposals for a worldwide HDTV *studio* system, with characteristics as shown in Table 16.1. The Japanese broadcasting company, NHK, has proposed one of the systems while the joint European project, Eureka, has proposed an alternative standard.

Table 16.1 *Proposed HDTV studio production standards.*

	Europe	North America, Japan
Total lines/picture	1250	1125
Active lines/picture	1192	1035
Scanning method	1:1	2:1 Interlaced
Aspect ratio	16:9	16:9
Field frequency (Hz)	50	60
Line frequency (kHz)	62.5	33.75
Samples/active line	1920	1920

The European standard uses a 50 Hz field rate to provide relatively easy conversion to both 60 and 50 Hz conventional, and HDTV, systems. It is also well suited to transfer from film. 1250 lines were chosen as this is exactly double the number of lines of the European conventional standards. A conversion to the American 525 lines standard is slightly more difficult, involving a ratio of 50/21.

16.4.3 Transmissions

In order to achieve compatibility with conventional TV it is proposed to split the HDTV information and transmit it in two separate channels. When decomposing a 1250 line studio standard, for US transmission, the line interpolator would extract the centre portion of every second line of the top 1050 lines, and send it in channel 1 as the reduced resolution 525 line image, left part of Figure 16.6(b). The panel extractor then removes the side panels, which make up the extra width of the picture, adds in the bottom 200 lines which were not sent with channel 1 plus the missing alternate lines and sends this as channel 2. These two channels are then reconstructed in the TV receiver. An added advantage of a two channel system is that it would allow more than one picture to be displayed simultaneously.

Recent European developments have swung in favour of a digital HDTV transmission standard. This will most likely use the compression schemes described in section 16.7 combined with QAM or QPSK for cable or satellite systems [Forrest] or simultaneous orthogonal frequency division multiplex (OFDM) transmission [Aldard and Lassalle] for the more congested terrestrial systems. This replaces one high bit rate signal with many (1,000 to 8,000) parallel, low bit rate, channels using orthogonal carriers. OFDM can better handle multipath fades as these only degrade a fraction of the parallel channels at any one time. With convolutional coded FECC transmissions, errors can be corrected subsequently in the receiver.

More recently the US grand alliance has pooled all interests in US HDTV into a single consortium to develop the best possible US HDTV system. This will use digital video compression in line with the MPEG standard (see later section 16.7.3) to compress the digital video into a 20 Mbit/s signal which will be transmitted in the conventional 6 MHz wide channel using high level modulation schemes such as those described in Chapter 11. The idea here is to use currently unallocated TV channels with a simultaneously broadcast analogue system rather than augment the existing NTSC system.

16.5 Digital video

Video in either RGB or YUV format can be produced in digital form [Forrest]. In this case discrete samples of the analogue video signal are digitised, to give a series of PCM words which represent the pixels. The words are divided into three fields representing each of the three signals RGB or YUV, section 16.2. Pixel word sizes range from 8 bits to 24 bits. Typical configurations include:

24 bits – where R = G = B = 8 bits and
16 bits – where Y = 8 bits and U = V = 4 bits.

In the latter case the chrominance signals are transmitted with less accuracy than the luminance but the eye is not able to discern the degradation. It is then possible either to store the samples in a memory device (e.g. compact disc (CD)), or to transmit them as a digital signal. Figure 16.7 shows the entropy (see section 9.2.4) of the luminance and chrominance signals in a video sequence comprising a fast moving sports sequence. Cb is the chrominance signal B − Y and Cr is R − Y. Scenes with less movement, such as a head and shoulders newsreader, have correspondingly lower entropy values which can be taken full advantage of if frame to frame differential coding is employed. Note, in Figure 16.7, that the luminance information (Y) requires a higher accuracy in the quantisation operation than does the chrominance information.

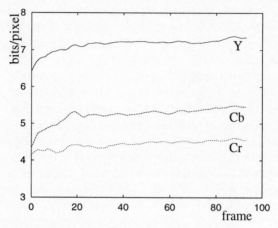

Figure 16.7 *Typical entropy measures in a colour video image sequence.*

It would initially seem ideal to use a computer network, or computer, modem and telephone network for video delivery. Unfortunately, the amount of data which would need to be transmitted is usually excessive. For a conventional TV system, the equivalent digital bit rate is around 140 Mbit/s which is not compatible with a standard modem. Even if the quality of the received image is reduced, data rates are still high. For example, if an image with a reduced resolution of 256×256 pixels is considered where each pixel consists of 16 bits (YUV) and a standard video frame rate of 25 frames per second is used, the bit rate R_b is given by:

$$R_b = 256 \times 256 \times 16 \times 25 = 26.2 \text{ Mbit/s} \tag{16.5}$$

Even a single frame requires 1 Mbit, or 132 kbyte of storage. For full resolution (720 pixel by 480 pixel) ITU-R 601 PAL TV with 8-bit quantisation for each of the three colour components, the rate grows to 207 Mbit/s. A further problem exists in storage and display as the access rate of CD-ROM is 120 kbyte/s while fast hard disc is only 500 kbyte/s. Fortunately, a general solution to the problem of data volume exists. In our sports sequence example, 16 bits were needed to represent each YUV pixel of the image (Figure 16.7). However, the long term average entropy of each pixel is considerably smaller than this and so data compression can be applied.

16.6 Video data compression

Data rate reduction is achieved by exploiting the *redundancy* of natural image sequences which arises from the fact that much of a frame is constant or predictable, because most of the time changes *between* frames are small. This is demonstrated in Figure 16.8 which shows a sequence of frames with a uniform background. The pixels representing the background are identical. The top sequence of frames is very similar in many respects, expect for some movement of the subject. The lower sequence has more movement which increases the entropy. It is thus inefficient to code every image frame into 1 Mbit of data and ignore the predictive properties possessed by the previous frames. Also there are few transmission channels for which we can obtain the 25 to 200 MHz bandwidth needed for uncompressed, digital, TV. For example the ISDN digital telephone channel has 128 kbit/s capacity, while ISDN primary access rate is 2 Mbit/s and the full STM-1 rate is only 45 Mbit/s (see Chapter 19).

Removal of redundancy is achieved by image sequence coding. Compression operations (or algorithms) may work within a single frame (intra-frame), or between frames of a sequence (inter-frame), or a combination of the two. Practical compression systems tend to be hybrid in that they combine a number of different compression mechanisms. For example, the output of an image compression algorithm may be Huffman coded (Chapter 9) to further reduce the final output data rate.

We will look firstly at a few of the basic compression principles, before progressing to review practical systems. General coding techniques such as DPCM discussed in sections 5.8.2 and 5.8.3 are also applicable to image compression. Figure 16.9 shows the interframe difference between two images. DPCM would achieve 2 to 3 times compression compared to conventional PCM quantisation. However, video compression requires 20:1 to 200:1 compression ratios. Video compression and expansion equipment is often referred to as a 'video CODEC' inferring the ability to both transmit and receive images. In practice, not all equipment is able to do this, and the term is sometimes used to refer to the transmitter (coder) or receiver (decoder) only.

16.6.1 Run-length coding

This intra-frame compression algorithm is best suited to graphic images, or video images with large sections made up of identical pixels. The algorithm simply detects the presence of a sequence of identical pixel values (usually operating on the luminance component only), and notes the start point, and number of pixels in the run (the 'run length'). This information is then transmitted in place of the original pixel values. (The

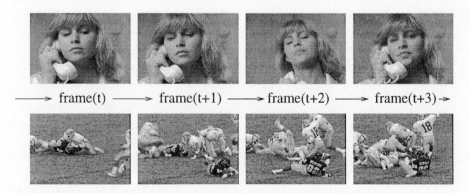

$\longrightarrow$ frame(t) $\longrightarrow$ frame(t+1) $\longrightarrow$ frame(t+2) $\longrightarrow$ frame(t+3) $\rightarrow$

Figure 16.8 *Redundancy in video image sequence.*

Figure 16.9 *Difference image between two frames in a video image sequence.*

technique was described as applied in facsimile transmission in Chapter 9.) To achieve a significant coding gain the run lengths must be large enough to provide a saving when considering the additional overhead of the addressing and control information which is required.

16.6.2 Conditional replenishment

This is an intra-frame coding algorithm which requires a reference frame to be held at the transmitter. The algorithm first divides the frame into small elements, called blocks, although lines can be used. Each pixel element is compared with the same location in the reference frame, and some measure of the difference between the pixel elements is calculated. If this is greater than a decision threshold, the pixel is deemed to have changed and the new value is sent to the receiver and updated in the reference frame. If the difference is not greater than the threshold, no data is sent. The frame displayed is therefore a mixture of old and new pixel element values.

16.6.3 Transform coding

As its name suggests, transform coding attempts to convert or transform the input samples of the image from one domain to another [Clarke 1985]. It is usual to apply a two-dimensional (2-D) transform to the two dimensional image and then quantise the transformed output samples. Note that the transform operation does not in itself provide compression; often one transform coefficient is output for each video amplitude sample input, as in the Fourier transform (Chapters 2 and 13). However, many grey scale patterns in a two dimensional image transform into a much smaller number of output samples which have *significant* magnitude. (Those output samples with insignificant magnitude need not be retained.) Furthermore, the quantisation resolution of a particular output sample which *has* significant magnitude can be chosen to reflect the importance of that sample to the overall image.

To reconstruct the image, the coefficients are input to an inverse quantiser and then inverse transformed such that the original video samples are reconstructed, Figure 16.10. Much work has been undertaken to discover the optimum transforms for these operations, and the Karhunen-Loeve transform [Clarke 1985] has been identified as one which minimises the overall mean square error between the original and reconstructed images. Unfortunately, this is complex to implement in practice and alternative, sub-optimum transforms are normally used.

These alternative transforms, which include sine, cosine and Fourier transforms amongst others, must still possess the property of translating the input sample energy into another domain. The cosine transform has been shown to produce image qualities similar to the Karhunen-Loeve transform for practical images (where there is a high degree of interframe pixel correlation) and is now specified in many standardised compression systems.

The 2-D discrete cosine transform (DCT) [Clarke 1985] can be defined as a matrix with elements:

$$
C_{nk} = \begin{cases} \sqrt{\dfrac{1}{N}} \cos\left[\dfrac{n(2k+1)\pi}{2N} \right], & n = 0, \quad 0 \le k \le N-1 \\[4mm] \sqrt{\dfrac{2}{N}} \cos\left[\dfrac{n(2k+1)\pi}{2N} \right], & 1 \le n \le N-1,\, 0 \le k \le N-1 \end{cases} \tag{16.6}
$$

where N is the number of samples in one of the dimensions of the normally square transform data block. The fact that the video data comprises real (pixel intensity) rather than complex sample values favours the use of the DCT over the DFT.

The next stage in a transform based video compression algorithm is to quantise the output frequency samples. It is at this stage that compression can occur. Recall that the output samples now represent the spatial frequency components of the input 'signal' [Clarke, 1985, Grant *et al.*]. The first output sample represents the DC, or average, value of the 'signal' and is referred to as the DC coefficient. Subsequent samples are AC coefficients of increasing spatial frequency. Low spatial frequency components imply slowly changing features in the image while sharp edges or boundaries demand high spatial frequencies.

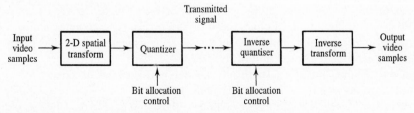

Figure 16.10 *Simplified block diagram illustrating transform coder operation.*

Much of the image information registered by the human visual process is contained in the lower frequency coefficients, with the DC coefficient being the most prominent. The quantisation process therefore involves allocating a different number of bits to each coefficient. The DC coefficient is allocated the largest number of bits (i.e. it is given the highest resolution), with increasingly fewer bits being allocated to the AC coefficients representing increasing frequencies. Indeed, many of the higher frequency coefficients are allocated 0 bits (i.e. they are simply not required to reconstruct an acceptable quality image).

The combined operations of transformation, quantisation and their inverses are shown in Figure 16.10. Note that the quantisation and inverse quantisation operations may be adaptive in the number of bits allocated to each of the transformed values. Modification of the bit allocations allows the coder to transmit over a channel of increased or decreased capacity as required, perhaps due to changes in error rates or a change of application. With the advent of ATM (section 19.7.2), such a variable bit rate (bursty) signal can now be accommodated.

Practical transform coders include other elements in addition to the transformation and quantisation operations described above. Further compression can be obtained by using variable length (e.g. Huffman) type coding operating on the quantiser output. In addition, where coding systems operate on continuous image sequences, the transformation may be preceded by inter-frame redundancy removal (e.g. movement detection) ensuring that only changed areas of the new frame are coded. Some of these supplimentary techniques are utilised in the practical standard compression mechanisms described below.

16.7 Compression standards

16.7.1 COST 211

This CODEC specification was developed in the UK by British Telecom in conjunction with GEC, and the CODEC has been used extensively for professional video conference applications. Its output bit rate is high compared with more recent developments based on transform techniques. A communication channel of at least 340 kbit/s is required for the COST 211 CODECs, the preferred operation being at the 2 Mbit/s ISDN primary access rate (Chapter 19). The COST 211 CODEC provides full frame rate (25 frame/s) video assuming a conference scene without excessive changes between frames. The CODEC has a resolution of 286 lines, each of 255 pixels. The compression uses intra-frame algorithms, initially utilising conditional replenishment where the changed or altered pixel information is transmitted, followed by DPCM coding (section 5.8.2) which is applied to the transmitted pixels. Finally, the output is Huffman encoded.

Further compression modes are available, including sub-sampling (skipping over sections of the frame, or even missing complete frames) to enable the CODEC output to be maintained at a constant rate if replenishment threshold adjustment is not sufficient.

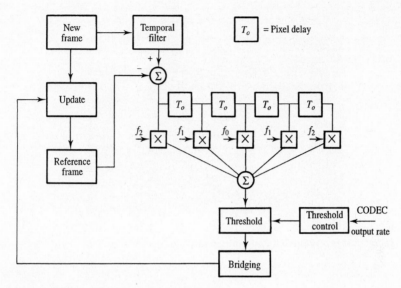

Figure 16.11 *COST 211 codec movement detector.*

The conditional replenishment, or movement detector, part of the CODEC is shown in Figure 16.11. This operates on a pixel-by-pixel basis and calculates the weighted sum of 5 consecutive pixel differences. The difference sum is then compared with a variable threshold value to determine if the pixel has changed. The threshold is adjusted to keep the CODEC output rate constant. Changed pixels are both transmitted, and stored in the local reference frame at the transmitter.

16.7.2 JPEG

JPEG is an international standard for the compression and expansion of single frame monochrome and colour images. It was developed by the Joint Photographic Experts Group (JPEG) and is actually a set of general purpose techniques which can be selected to fulfil a wide range of different requirements. However, a common core to all modes of operation, known as the baseline system, is included within JPEG. JPEG is a transform based coder and some, but not all, operating models use the DCT applied to blocks of 8×8 image pixels yielding 64 output coefficients. Figure 16.12 shows a 352-by-240 pixel image, partitioned into 330 macroblocks. The 16×16 pixel macroblocks are processed slice-by-slice in the DCT coder.

The coefficients are then quantised by a user defined quantisation table which specifies a quantiser step size for each coefficient in the range 1 to 255. The DC coefficient is coded as a difference value from that in the previous block, and the sequence of AC coefficients is reordered as shown in Figure 16.13(a) by zig-zag scanning in ascending order of spatial frequency, and hence decreasing magnitude, progressing from the more significant to the least significant components. Thus the quantiser step

Figure 16.12 *Image partitioning into macroblocks (MB).*

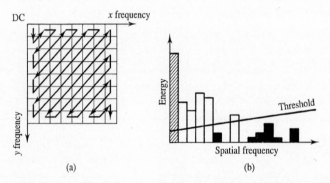

Figure 16.13 *(a) Zig-zag scanning to reorder transformed data into increasing spatial frequency components (b) and permit variable length coding of the scanned components.*

size increases and becomes coarser with increasing spatial frequency in the image to aid the final data compression stage, Figure 16.13(b).

The final stage consists of encoding the quantised coefficients according to their statistical probabilities or entropy (Chapter 9 and Figure 16.7). Huffman encoding is used in the baseline JPEG systems. JPEG also uses a predictor, measuring the values from three adjacent transformed pixels, to estimate the value of the pixel which is about to be encoded. The predicted pixel value is subtracted from the actual value and the difference signal is sent to the Huffman coder as in DPCM (Chapter 5).

JPEG compression ratios range from 2:1 to 20:1. Low compression ratios achieve lossless DPCM coding where the reconstructed image is indistinguishable from the original while high compression ratios can reduce the storage requirements to only ¼ bit per pixel and the transmission rate, for image sequences, to 2 Mbit/s. VLSI JPEG chips existed in 1993 which could process data at 8 Mbyte/s to handle 352×288 pixel (¼ full TV frame) images at 30 frame/s. As JPEG is a single frame coder it is not optimised to exploit the interframe correlation in image sequence coding. The following schemes are therefore preferred for compression of video data.

16.7.3 MPEG

This video coding specification was developed by the Motion Picture Experts Group (MPEG) as a standard for coding image *sequences* to a bit rate of about 1.5 Mbit/s for MPEG1 and 10 Mbit/s for MPEG2. The lower rate was developed, initially, for 352×288 pixel images because it is compatible with digital storage devices such as hard disc drives, compact discs, digital audio tapes, etc. The algorithm is deliberately flexible in operation, allowing different image resolutions, compression ratios and bit rates to be achieved. The basic blocks which can be used include:

- Motion compensation
- DCT
- Variable length coding

A simplified diagram of the encoding operation is shown in Figure 16.14. Because the algorithm is intended primarily for the storage of image *sequences* it incorporates motion compensation which is not included in JPEG. A compromise still exists between the need for high compression, and easy regeneration of randomly selected frame sequences. MPEG therefore allows frames to be coded in one of three ways, Figure 16.15.

Intra-frames (I-frames), which are coded independently of other frames, allow random access, but provide limited compression. They form the start points for replay sequences. Unidirectional predictive coded frames (P-frames) can achieve motion prediction from *previous* reference frames and hence, with the addition of motion compensation, the bit rate can be reduced. Bidirectionally predictive coded frames (B-frames) provide greatest compression but require two reference frames, one previous frame and one future frame in order to be regenerated. Figure 16.15 shows such an I, P, B picture sequence. The precise combination of I, P and B frames which are used depends upon the specific application.

The motion prediction and compensation uses a block based approach on 16×16 pixel block sizes as in Figure 16.12. The concept underlying picture motion

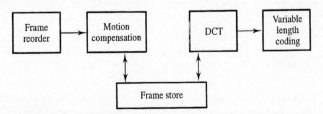

Figure 16.14 *Block diagram for simplified MPEG encoder.*

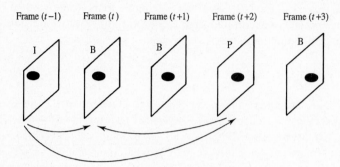

Frame $(t-1)$ Frame (t) Frame $(t+1)$ Frame $(t+2)$ Frame $(t+3)$

I   B   B   P   B

Figure 16.15 *I, P, B image frames, as used in the MPEG coder.*

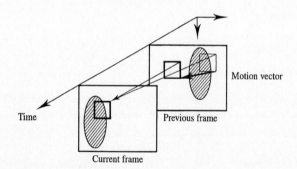

Motion vector

Time   Previous frame

Current frame

Figure 16.16 *Estimation of the motion vector, between consecutive frames, with object movement.*

compensation is to estimate the motion vector, Figure 16.16, and then use this to enable the information in the previous frame to be used in reconstructing the current frame, minimising the need for new picture information. MPEG is a continuously evolving standard. MPEG2 is aimed at HDTV where partial images are combined into the final image. Here the processing is dependent on the image content to enhance the compression capability and permit use with ATM at a P_b of 10^{-4}. MPEG2 data rates vary between VHS video cassette player quality at 1.5 Mbit/s through to 30 Mbit/s for HDTV images. It is generally recognised that fast moving sports scenes require the higher bit rates of 6 to 8 Mbit/s and, by reducing these to 1.5 Mbit/s, then the quality degrades, to much like that of a VHS video cassette player. MPEG3 is included within MPEG2. MPEG4 is now addressing very low bit rates for wireless video on the PSTN and should be standardised around 1998.

16.7.4 H.261 and H.263

The H.261 algorithm was developed for the purpose of image transmission rather than image storage. It is designed to produce a constant output of $p \times 64$ kbit/s, where p is an integer in the range 1 to 30. This allows transmission over a digital network or data link of varying capacity. It also allows transmission over a single 64 kbit/s digital telephone channel for low quality videotelephony, or at higher bit rates for improved picture quality. The basic coding algorithm is similar to that of MPEG in that it is a hybrid of motion compensation, DCT and straightforward DPCM (intra-frame coding mode), Figure 16.17, without the MPEG I, P, B frames. The DCT operation is performed at a low level on 8 × 8 blocks of error samples from the predicted luminance pixel values, with sub-sampled blocks of chrominance data. The motion compensation is performed on the macroblocks of Figure 16.12, comprising four of the previous luminance and two of the previous chrominance blocks.

H.261 is widely used on 176×144 pixel images. The ability to select a range of output rates for the algorithm allows it to be used in different applications. Low output rates ($p = 1$ or 2) are only suitable for face-to-face (videophone) communication. H.261 is thus the standard used in many commercial videophone systems such as the UK BT/Marconi Relate 2000 and the US ATT 2500 products. Video-conferencing would require a greater output data rate ($p > 6$) and might go as high as 2 Mbit/s for high quality transmission with larger image sizes.

A further development of H.261 is H.263 for lower fixed transmission rates. This deploys arithmetic coding in place of the variable length coding in Figure 16.17 and, with other modifications, the data rate is reduced to only 20 kbit/s.

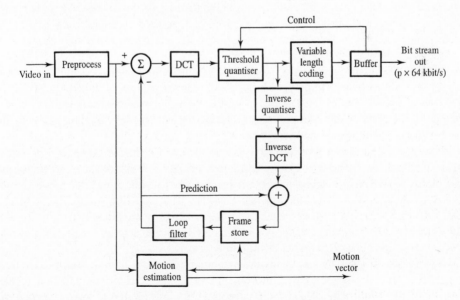

Figure 16.17 *Schematic representation of the H.261 encoder standard.*

16.7.5 Model based coding

At the very low bit rates (20 kbit/s or less) associated with video telephony, the requirements for image transmission stretch the compression techniques described earlier to their limits. In order to achieve the necessary degree of compression they often require reduction in spatial resolution or even the elimination of frames from the sequence. Model based coding (MBC) attempts to exploit a greater degree of redundancy in images than current techniques, in order to achieve significant image compression but without adversely degrading the image content information. It relies upon the fact that the image quality is largely subjective. Providing that the appearance of scenes within an observed image is kept at a visually acceptable level, it may not matter that the observed image is not a precise reproduction of reality.

One MBC method for producing an artificial image of a head sequence utilises a feature codebook where a range of facial expressions, sufficient to create an animation, are generated from sub-images or templates which are joined together to form a complete face. The most important areas of a face, for conveying an expression, are the eyes and mouth, hence the objective is to create an image in which the movement of the eyes and mouth is a convincing approximation to the movements of the original subject. When forming the synthetic image, the feature template vectors which form the closest match to those of the original moving sequence are selected from the codebook and then transmitted as low bit rate coded addresses.

By using only 10 eye and 10 mouth templates, for instance, a total of 100 combinations exists implying that only a 6-bit codebook address need be transmitted. It has been found that there are only 13 visually distinct mouth shapes for vowel and consonant formation during speech. However, the number of mouth sub-images is usually increased, to include intermediate expressions and hence avoid step changes in the image.

Another common way of representing objects in three-dimensional computer graphics is by a net of interconnecting polygons. A model is stored as a set of linked arrays which specify the coordinates of each polygon vertex, with the lines connecting the vertices together forming each side of a polygon. To make realistic models, the polygon net can be shaded to reflect the presence of light sources.

The wire-frame model [Welch 1991] can be modified to fit the shape of a person's head and shoulders. The wire-frame, composed of over 100 interconnecting triangles, can produce subjectively acceptable synthetic images, providing that the frame is not rotated by more than 30° from the full-face position. The model, shown in Figure 16.18, uses smaller triangles in areas associated with high degrees of curvature where significant movement is required. Large flat areas, such as the forehead, contain fewer triangles. A second wire-frame is used to model the mouth interior.

A synthetic image is created by texture mapping detail from an initial full-face source image, over the wire-frame. Facial movement can be achieved by manipulation of the vertices of the wire-frame. Head rotation requires the use of simple matrix operations upon the coordinate array. Facial expression requires the manipulation of the features controlling the vertices.

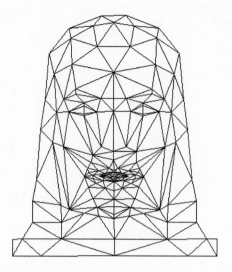

Figure 16.18 *Image representation by wire frame model.*

This model based feature codebook approach suffers from the drawback of codebook formation. This has to be done off-line and, consequently, the image is required to be prerecorded, with a consequent delay. However, the actual image sequence can be sent at a very low data rate. For a codebook with 128 entries where 7 bits are required to code each mouth, a 25 frame/s sequence requires less than 200 bit/s to code the mouth movements. When it is finally implemented, rates as low as 1 kbit/s are confidently expected from MBC systems, but they can only transmit image sequences which match the stored model, e.g. head and shoulders displays.

16.8 Packet video

Packet video is the term used to describe low-bit-rate encoded video, from one of the above CODECs, which is then transmitted over the ISDN as packet data similar to the packet speech example shown later in Figure 17.14. Due to the operation of the CODEC and the nature of the video signal, the resulting data rate varies considerably between low data rates when the image sequence contains similar frames, as in Figure 16.8, to high data rates when there is a sudden change in the sequence and almost the entire image frame must be encoded and transmitted. This is called variable bit rate (VBR) traffic, Figure 16.19, which is accommodated by packet transmission on an ATM network, Chapter 19. Figure 16.19 shows the regular variation in bit rate in the MPEG coder arising from quantising the I, B, P frames, Figure 16.15, plus the major alterations in rate due to motion or dramatic scene changes in the video sequence. At 30 frame/s this represents an overall variation in the required transmission rate between 600 kbit/s and 6 Mbit/s.

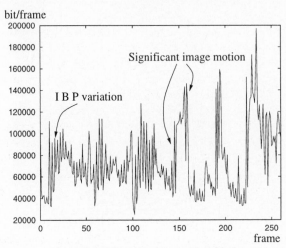

Figure 16.19 *Variable rate bursty traffic from a colour video image codec. Note the regular short-term variation in the MPEG coded data and the major scene changes.*

VBR systems rely on the fact that their average transmission rate has a modest bandwidth requirement. Hence if many VBR sources are statistically multiplexed then their individual bit rate variations will be averaged and they can be carried over a system with a transmission bandwidth which is compatible with the modest *average* bandwidth requirement of each source. Tariffs will then be levied which use this modest data rate requirement, rather than providing each user with access to a peak data rate channel and incurring correspondingly high costs.

The principles underlying packet video are similar to those described in Chapter 17. We must inevitably accept some loss of video data when the overall demand for transmission bandwidth exceeds the available channel allocation since arbitrarily large amounts of data cannot be held within the finite length of the buffer in Figure 16.17 (and Figure 17.1). Much current research is aimed at investigating this problem as the ISDN system expands and ATM techniques start to become more widely used. Some of this work is aimed at the design of the video encoding algorithm to control the statistics of the output VBR traffic; other work is aimed at investigating coding methods which split the video data into high and low priority traffic.

The overall protocol [Falconer and Adams], which corresponds to the lowest two layers of the OSI reference model (see section 18.4.1), represents one potential method of achieving different priority of traffic. By careful control of the video encoding technique we can ensure that high priority bits are allocated to the major video image features while low priority bits add more detail to this basic image. Techniques such as this allow speech and video data to be transmitted, with some packet loss, while still providing a quality of service which is acceptable to the user, for a transmission cost which is much lower than that of providing the full bandwidth requirement for 100% of the time.

16.9 Summary

The smallest elemental area of a two dimensional image which can take on characteristics independent of other areas is called a picture cell or pixel. The fundamental requirement of an image transmission system is the ability to convey a description of each pixel's characteristics: intensity, colour and location in space. For video images, information giving the pixel's location in time is also required.

A pixel from a colour image may be represented by three colour intensity signals (red, green and blue) or by a white (red + green + blue) intensity signal (luminance) and two colour difference (chrominance) signals (white − red and white − blue). The colour difference signals may be transformed geometrically, using the colour triangle, into saturation and hue. The luminance plus chrominance representation makes colour image transmission compatible with monochrome receivers and is therefore used for conventional TV broadcasts.

In the UK the two chrominance signals of a TV broadcast are transmitted using quadrature double sideband suppressed carrier modulation of a sub-carrier. Since this requires a phase coherent reference signal for demodulation, the reconstructed image is potentially susceptible to colour errors caused by phase distortion, within the chrominance signal frequency band, arising from multipath propagation. These errors can be reduced by reversing the carrier phase of one of the quadrature modulated chrominance signals on alternate picture lines which gives the system its name − 'phase alternate line' or PAL. The PAL signal transmits 25 frames per second, each frame being made up of 625 lines. The frame is divided into two fields, one containing even lines only and the other containing odd lines only. The field rate is therefore 50 Hz. The total bandwidth of a PAL picture signal is approximately 6 MHz.

HDTV may be either analogue (for compatibility with traditional broadcast receivers), digital or mixed. The picture resolution of HDTV is about twice that of conventional TV as is the required (analogue) signal bandwidth. Compatibility with conventional TV receivers can be maintained but it is much more likely that HDTV will ultimately adopt an all-digital transmission system.

Digital video can be implemented by sampling and digitising an analogue video signal. Typically, the luminance signal will be subject to a finer quantisation process than the chrominance signals (e.g. 8 bit versus 4 bit quantisation). Unprocessed digital video signals have a bandwidth which is too large for general purpose telecommunications transmissions channels. To utilise digital video for services such as videophone or teleconferencing the digital signal therefore requires data compression. Several coding techniques for redundancy removal, including run-length coding, conditional replenishment, transform coding, Huffman coding and DPCM may be used, either in isolation or in combination, to achieve this.

Several international standards exist for the transmission and/or storage of images. These include JPEG which is principally for single images (i.e. stills) and MPEG which is principally for moving pictures (i.e. video).

Very low bit rate transmission of video pictures can be achieved, using aggressive coding, if only modest image reproduction quality is required. Thus the BT/Marconi standard H.261 CODEC allows videophone services to be provided using a single 64 kbit/s digital telephone channel. Even lower bit rate video services can be provided using model based coding schemes. These are analogous to the low bit rate speech vocoder techniques described in Chapter 9.

Variable bit rate transmissions, which may arise as a result of video coding, can be accommodated by packet transmission over ATM networks. Predicting the overall performance of packet systems requires queuing theory which is discussed in Chapter 17. Networks, and the foundations of ATM data transmission, are discussed in Chapters 18 and 19 respectively.

16.10 Problems

16.1. A 625-line black and white television picture may be considered to be composed of 550 picture elements (pixels) per line. (Assume that each pixel is equiprobable among 64 distinguishable brightness levels.) If this is to be transmitted by raster scanning at a 25 Hz frame rate calculate, using the Shannon-Hartley theorem of Chapter 11, the minimum bandwidth required to transmit the video signal, assuming a 35 dB signal to noise ratio on reception. [4.44 MHz]

16.2. What is the minimum time required to transmit one of the picture frames described in Problem 16.1 over a standard 3.2 kHz bandwidth telephone channel, as defined in Figure 5.12, with 30 dB SNR? [65 s]

16.3. A colour screen for a US computer aided design product may be considered to be composed of a $1,000 \times 1,000$ pixel array. Assume that each pixel is coded with straightforward 24-bit colour information. If this is to be refreshed through a 64 kbit/s ISDN line. Calculate the time required to update the screen with a new picture. [6.25 min]

16.4. Derive an expression for the bandwidth or effective data rate for a television signal for the following scan parameters: frame rate = P frame/s; number of lines per frame = M; horizontal blanking time = B; horizontal resolution = x pixels/line where each pixel comprises one of k discrete levels. Hence calculate the bandwidth of a television system which employs 819 lines, a 50 Hz field rate, a horizontal blanking time of 18% of the line period and 100 levels. The horizontal resolution is 540 pixels/line. [75.5 Mbit/s]

Part Four

Networks

Part Four is devoted to communication networks which now exist on all scales from geographically small LANs to the global ISDN.

It starts with a discussion, in Chapter 17, of queuing theory which may be used to predict the delay suffered by digital information packets as they propagate through a data network.

Chapter 18 describes the topologies and protocols employed by networks to ensure the reliable, accurate and timely, delivery of information packets between network terminals. Rings, buses and their associated medium access protocols are discussed, and international standards such as ISO OSI, X.25 and FDDI are described. The optical transmission medium, which now forms an integral part of many communications networks, is also examined.

Part Four ends, in Chapter 19, by examining public networks. The current plesiochronous digital hierarchy (PDH) is reviewed before introducing a more detailed discussion of the new synchronous digital heirarchy (SDH) which will gradually come to replace it. The Chapter concludes with a brief discussion of PSTN/PDN data access techniques including the ISDN standard, ATM, and the probable future development of the local loop.

CHAPTER 17

Queuing theory for packet networks

17.1 Introduction

In packet switching, secure bundles of information are assembled, addressed and transmitted through a network without the need for dedicated end-to-end connection paths to be established. The packets are individually transported and delivered by the network to the required destination. The network additionally ensures that packets are output in the correct order at the receiver.

Packet switched networks have existed for many years. Early examples were Euronet which links nine EC countries and Switzerland, running the Direct Information Access Network-Europe (DIANE) service. In 1981, British Telecom opened its first national public network, known as the Packet Switched Service (PSS). This system is controlled by a network management centre, based on duplicated minicomputers. PSS uses the X.25 protocol (see Section 18.6.1). The most well known network today is the Internet which supports the information retrieval service known as the World Wide Web (WWW).

The Joint Academic Network (JANET), Figure 18.2, and its successor SuperJANET are funded by the UK research councils. JANET has a node at every university and runs on leased lines. It provides a service for the communication of data, such as computer files and electronic mail, between sites and onward via the Internet (see section 18.7.6). All these networks introduce queuing, with consequent delays and possible loss of traffic data.

Queuing theory [Nussbaumer] can be used to model these or other networks where customers or data packets arrive, wait their turn for handling or service, are subsequently serviced, and are then transmitted through the network. (Supermarket checkouts, ticket booths, and doctors' waiting rooms are all commonly encountered examples of queuing systems.) Queuing theory was developed originally to model analogue teletraffic but is now widely applied to digital packet traffic [Tanenbaum]. A queuing system can be characterised by the following five attributes:

- The interarrival-time probability density function.
- The service-time probability density function.
- The number of servers, or server processes.
- The queuing discipline.
- The amount of buffer, or waiting, space in the queues.

The interarrival-time probability density function (pdf) describes the interval between consecutive arrivals. After a sufficiently long sampling time, the arrival times can be grouped to obtain the pdf which characterises the arrival process.

Each customer requires a certain amount of the server's time which varies from customer to customer. To analyse a queuing system, the service-time pdf, like the interarrival-time pdf, must be known.

The number of servers speaks for itself. Many banks, for example, have one queue for all customers. Whenever a teller is free, the customer at the front of the queue goes directly to that teller. Such a system is a multiserver queuing system. In other banks, each teller has his, or her, own private queue. This corresponds to a collection of independent single-server queues.

The queuing discipline describes the order in which customers are taken from the queue. Supermarkets use first come, first served. Hospital emergency rooms often use sickest attended to first. In friendly office environments, shortest job first often prevails at the photocopy machine. Not all queuing systems have an infinite amount of buffer space. When too many customers are queued up in a finite number of available slots, some customers can get lost or rejected.

This chapter concentrates predominantly on infinite-buffer, single-server systems using a first come first served queuing discipline. The Kendall notation A/B/m [Kleinrock] is widely used in the queuing literature for these systems. A represents the interarrival-time pdf, B the service-time pdf, and m the number of servers employed. The probability densities A and B are usually chosen from the following set:

M – Markov (implying an exponential pdf);

D – deterministic (all customers have the same constant value implying an impulsive pdf);

G – general (i.e. some arbitrary pdf);

E_k – Erlang distributed.

The state of the art ranges from the M/M/1 system, about which everything is known, to the G/G/m system, for which no exact analytical solution is yet available. We will concentrate on the M/M/1 model. Figure 17.1 shows the queue for such a single server process. We now need to develop a mathematical analysis for queuing systems which can be used to show what limits or restricts the practical performance of these systems. This is achieved by first examining arrivals only, before progressing to model the combined arrival and service processes within the queuing buffer memory.

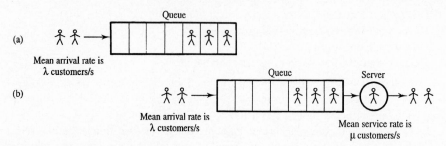

Figure 17.1 *Single server queue model and outcomes for a counting process: (a) arrivals only; (b) arrivals and departures.*

17.2 The arrival process

We make the following assumptions:

- The arrival process is memoryless in the sense that any arrival is statistically independent of all other arrivals;
- The arrival process is statistically stationary. (This implies that the probability of an arrival occuring in any small time interval depends only on the interval's width and not its location in time.)

Queuing systems in which the only transitions are to adjacent states are known as birth–death systems, which is a mathematician's terminology for a counting process with both arrivals and departures. We are considering, at present, *only* arrivals.

We have to model the evolution of the system from one state to another [Gelenbe and Pujolle] where the *state* is synonymous with the number of customers waiting for service. The approach via the Markov probability chain [Chung] is inappropriate, since the probability of transition between any two states at any given point in time, *t*, is zero. While we cannot characterise the *probability* of transition, we *can* characterise the *rate* of transitions between two states. Suppose that for two particular states the rate of transitions between them is a constant λ. What we mean by this is that in a time δt we can expect an average of $\lambda\,\delta t$ transitions. If δt is very small then $\lambda\,\delta t$ is a number much smaller than unity, and the probability of more than one transition in time δt is vanishingly small. Under these conditions, we can think of $\lambda\,\delta t$ as the probability of one transition (P_1) in time δt, and ($1 - \lambda\,\delta t$) as the probability of no transition (P_0) in this time.

This leads us to a transition diagram and associated set of differential equations. The transition diagram, Figure 17.2, associates a node with each state. Within node *j* we denote the probability of being in that state at time *t* as $P_j(t)$.

17.2.1 pdf for *j* arrivals in *t* seconds

Assume arrivals (*A*) are governed by a randomly distributed pure birth process as shown in Figure 17.3 where the arrival rate does not depend on the state of the system. The

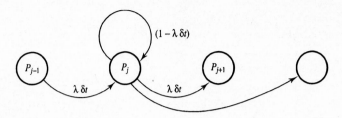

Figure 17.2 *Markov model for queue states corresponding to $j-1$, j, $j+1$ packet arrivals.*

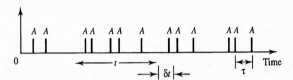

Figure 17.3 *Example of randomly distributed arrivals (A) with interarrival time τ.*

only arrivals then the probability of being in state j after δt seconds is dependent on P_j and P_{j-1}:

$$P_j(t+\delta t) = P_j(t)(1 - \lambda\,\delta t) + P_{j-1}(t)\,\lambda\,\delta t \tag{17.1}$$

hence:

$$\frac{P_j(t+\delta t) - P_j(t)}{\delta t} = \lambda(P_{j-1}(t) - P_j(t))$$

Now let $\delta t \to 0$:

$$\frac{dP_j(t)}{dt} = \lambda(P_{j-1}(t) - P_j(t)) \tag{17.2}$$

For $j = 0$, $P_{j-1} = 0$, as we cannot have less than zero arrivals. Therefore:

$$\frac{dP_0(t)}{dt} = -\lambda P_0(t) \tag{17.3}$$

The solution to equation (17.3) is:

$$P_0(t) = e^{-\lambda t} \tag{17.4}$$

which represents the probability of no arrivals in time t. This is plotted in Figure 17.4, starting with a probability of 1 at $t = 0$, decaying asymptotically to zero with an exponential time constant of λ. Thus at $t = 1/\lambda$, $P_0(1/\lambda) = 0.37$. This is the start of the derivation of the Poisson distribution for the number of arrivals j in t seconds for an exponential interarrival time distribution, with a mean arrival rate of λ. It can further be shown that:

$$P_j(t) = \frac{(\lambda t)^j}{j!}\, e^{-\lambda t} \tag{17.5(a)}$$

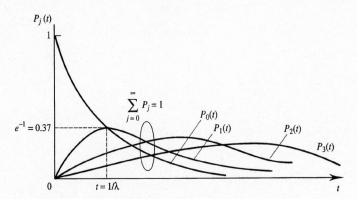

Figure 17.4 *Probability of j arrivals plotted against time for j = 0, 1, 2, 3.*

i.e.:

$$P_1(t) = \lambda t \, e^{-\lambda t} \tag{17.5(b)}$$

$$P_2(t) = \tfrac{1}{2}(\lambda t)^2 \, e^{-\lambda t} \quad etc \tag{17.5(c)}$$

Figure 17.4 shows that the probability of only one arrival, $P_1(t)$, peaks at the time interval $t = 1/\lambda$. If Figure 17.4 is plotted with the horizontal axis as offered traffic, λt, then the peak value of $P_1(t)$ occurs at an offered traffic value of unity. For longer time intervals, it becomes increasingly likely that there will be more than one arrival. The probabilities of there being two or three arrival events, $P_2(t)$ and $P_3(t)$, peak at later time intervals and also have progressively smaller probability peaks, so that, for any specified time interval, all the probabilities sum to unity as required, i.e.:

$$\sum_{j=0}^{\infty} P_j(t) = 1 \tag{17.6}$$

17.2.2 CD and pdf for the time between arrivals

If the time between successive arrivals is τ, as in Figure 17.3, the probability that τ is less than, or equal to, some value of time t, $P(\tau \le t)$, is given by:

$$P(\tau \le t) = 1 - P(\tau > t) \tag{17.7(a)}$$

But $P(\tau > t)$ is the probability of no arrivals in time t. Thus, for the Poisson process:

$$P(\tau > t) = P_0(t) = e^{-\lambda t} \tag{17.7(b)}$$

hence:

$$P(\tau \le t) = 1 - e^{-\lambda t} \tag{17.7(c)}$$

Equation (17.7(c)) is the cumulative distribution (CD) function, equations (3.10) and

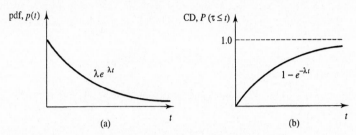

Figure 17.5 *(a) pdf; and (b) CD for time between successive arrivals.*

(3.13(b)), for the time between arrivals, Figure 17.5. This implies an exponential interarrival probability density function (pdf). As shown in section 3.2.4, the pdf is obtained by differentiating the CD, i.e. equation (17.7(c)):

$$P(t \leq \tau < t + dt) \; = \; p(t) \; = \; \frac{dP(\tau \leq t)}{dt} \; = \; \lambda e^{-\lambda t} \qquad (17.7(\text{d}))$$

where the mean or average value of t is $1/\lambda$ and the variance (or second central moment) is $1/\lambda^2$ (see section 3.2.5).

17.2.3 Other arrival patterns

These can be specified to be:

- Unpunctual – which occur at $t = a + E_1, 2a + E_2, \cdots, ma + E_m$, where E_m is a random variable;
- Discrete-time arrivals – these can only occur at a discrete set of allowed instants;
- Non-stationary – where the probabilities vary with time;
- Correlated – where the arrival rate may be affected by the state of the system.

17.3 The service process

The service or transmission mechanism is described by the service time distribution, which defines the capacity, or number of servers, which must be deployed to handle the given traffic.

17.3 1 Service time distributions

As for arrival times there are two extremes. We can have constant (deterministic) service times or alternatively we can have 'completely random' (stastically independent) service times, the latter again leading to an exponential pdf given by:

$$P(t) = \mu \, e^{-\mu t} \qquad (17.8)$$

for a service rate of μ customers/s, Figure 17.1. As before, the mean value, or average service time, is $1/\mu$ and the variance is $1/\mu^2$. The statistical independence of succesive service times (resulting in an exponential distribution of service time pdf) means that service time, like arrival interval, has been modelled as a Poisson process. It should be noted, however, that:

- service times may be discrete (word or packet multiples);
- service time may be non-stationary.

The queuing discipline determines how customers are selected from the queue and allocated to servers.

17.3.2 Single server queues

These typically use one of the following queuing disciplines:

- First-in-first-out (FIFO) – the simple queue;
- Last-in-first-out (LIFO or last come first served, LCFS);
- First-in-random-out (FIRO);
- Priority queuing.

17.3.3 Multiserver queues

Here service is allocated according to rules such as:

- Rotation – customers assigned in strict rotation to each queue;
- Random selection – customers themselves decide which queue to join;
- Single queue – customer at the head of queue goes to the next available server.

17.4 The simple single server queue

It is instructive to find the distribution of queue lengths and waiting times in an M/M/1 system, where total waiting time equals queuing time plus service time.

17.4.1 Simple queue analysis

We define the *state* of the system as the number of customers waiting (i.e. state m implies m customers waiting or a queue of length m), and denote the probability of being in a state m at time t as $P_m(t)$. Assume, when the system is in state m, customers arrive randomly at an average rate λ_m, and are randomly serviced at rate μ_m (i.e. the average service time is $1/\mu_m$).

What is the probability of such a system, Figure 17.6, being in state m at time $t + \delta t$? This is obtained by extending equation (17.1) to include departures as well as arrivals:

$$P_m(t + \delta t) = P_{m-1}(t)\lambda_{m-1}\delta t + P_{m+1}(t)\mu_{m+1}\delta t + P_m(t)(1 - \mu_m\delta t)(1 - \lambda_m\delta t)$$

$$= P_{m-1}(t)\lambda_{m-1}\delta t + P_{m+1}(t)\mu_{m+1}\delta t + P_m(t)[1 - (\mu_m + \lambda_m)\delta t] \qquad (17.9)$$

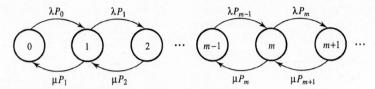

Figure 17.6 *State transition-rate diagram for a simple queue.*

(for small δt). Therefore:

$$\frac{P_m(t + \delta t) - P_m(t)}{\delta t} = \lambda_{m-1} P_{m-1}(t) + \mu_{m+1} P_{m+1}(t) - (\mu_m + \lambda_m) P_m(t) \quad (17.10)$$

In the limit as $\delta t \to 0$:

$$\frac{dP_m(t)}{dt} = \lambda_{m-1} P_{m-1}(t) + \mu_{m+1} P_{m+1}(t) - (\mu_m + \lambda_m)\, P_m(t) \quad (17.11)$$

(for $m \geq 0$ where $P_{-1}(t) = 0$). This is the full Chapman-Kolmogorov equation [Gelenbe and Pujolle] for arrivals and departures. It states that the rate of increase of probability with time for state m is equal to the rate at which transitions into that state, from states $m - 1$ and $m + 1$, are occurring (multiplied by the current probability of those states) minus the rate at which transitions out of state m are occurring (multiplied by the current probability of state m). To solve equation (17.11) we must specify the initial conditions. The process starts in state zero, as there are no arrivals before time t_0, i.e.:

$$P_m(t_0) = \begin{cases} 1, & m = 0 \\ 0, & m > 0 \end{cases}$$

and assuming a stationary solution so that $dP_m(t)/dt = 0$ then:

$$0 = \lambda_{m-1} P_{m-1} + \mu_{m+1} P_{m+1} - (\mu_m + \lambda_m)\, P_m \quad (17.12(a))$$

where:

$$P_{-1} = P_{-2} = \cdots = 0 \quad (17.12(b))$$

$$\lambda_{-1} = \lambda_{-2} = \cdots = 0 \quad (17.12(c))$$

$$\mu_0 = \mu_{-1} = \cdots = 0 \quad (17.12(d))$$

and:

$$P_0 + P_1 + P_2 + \cdots = 1 \quad (17.13)$$

A typical queue result is shown in Figure 17.7. Here the vertical steps in the solid staircase occur with new customers arriving while the vertical steps in the dashed staircase imply service has been completed for these customers. The horizontal displacement measures the queue plus service time for each customer or unique arrival.

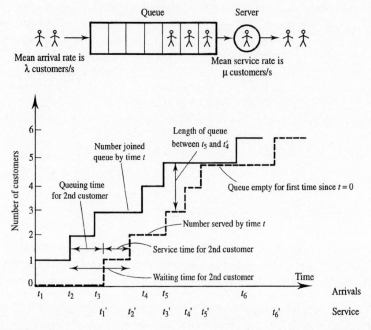

Figure 17.7 *Typical queue performance showing customer (packet) arrivals and departures (after service).*

17.4.2 Queue parameters

The total delay in a time t, $\gamma(t)$, is the sum of the waiting times. If the total number of customers who have arrived in a time t is denoted by $\alpha(t) = \lambda t$ then:

$$\text{average delay,} \qquad T = \frac{\gamma(t)}{\alpha(t)} \qquad\qquad (17.14(a))$$

$$\text{average queue length,} \qquad N = \frac{\gamma(t)}{t} \qquad\qquad (17.14(b))$$

$$\text{average arrival rate,} \qquad \lambda = \frac{\alpha(t)}{t} \qquad\qquad (17.14(c))$$

Now we can rewrite the average queue length as $N = \gamma(t)/t = (\gamma(t)/\alpha(t)) \times (\alpha(t)/t)$, where N is given by the sum, over all m, of the product of queue length, m, and its probability, P_m, to obtain Little's result:

$$N = \sum_{m=0}^{\infty} mP_m = T\lambda \qquad\qquad (17.15)$$

The performance of a queuing system is controlled by its utilisation factor, defined as:

$$\rho = \frac{\text{demand for service}}{\text{maximum rate of supply}} \qquad\qquad (17.16)$$

where the demand for service = arrival rate × mean service time = λ/μ. This is also a measure of the traffic intensity in erlangs [Dunlop and Smith], named after the Danish pioneer of teletraffic theory. (In telephony systems the total applied traffic in erlangs is equal to $\lambda\tau$ where λ is the arrival rate for call connections and τ is the average call duration. A circuit carrying one call continuously then carries one erlang of traffic.)

The service rate is often controlled directly by the output transmission rate and packet length. For example a 2 Mbit/s link (Chapter 19) with 500 byte packets and 8-bit bytes, has a service rate, $\mu = (2 \times 10^6)/(500 \times 8) = 500$ packet/s. For single server queues, maximum capacity or rate of supply is 1 s of service/s, i.e. the maximum rate of supply = 1. In this case:

$$\rho = \lambda/\mu \tag{17.17}$$

and $\rho < 1$ is required to prevent server overload.

17.4.3 Classical queue with single server

Assume a Poisson arrival process and exponentially distributed service times so that the arrival and service rates are independent of the state of the system. Thus λ_m simplifies to λ and μ_m simplifies to μ. Further assume that infinite queuing space is available and that customers are served on FIFO basis.

From equation (17.5) and Figure 17.6, by applying the conditions of equilibrium or detailed balancing, the input and output transitions to and from a state must occur at the same rate. Thus starting with state zero, $\lambda P_0 = \mu P_1$, these balances can be written as:

$$P_1 = \frac{\lambda}{\mu} P_0$$

$$P_2 = \frac{\lambda^2}{\mu^2} P_0$$

$$P_m = \left(\frac{\lambda}{\mu}\right)^m P_0 = \rho^m P_0 \tag{17.18}$$

But $\Sigma_{m=0}^{\infty} P_m = 1$, as shown previously (Figure 17.4 and equation (17.6)). Now $\Sigma_{m=0}^{\infty} P_m = \Sigma_{m=0}^{\infty} \rho^m P_0 = P_0 \Sigma_{m=0}^{\infty} \rho^m = 1$ and the sum of the geometric series $\Sigma_{m=0}^{\infty} \rho^m = 1/(1-\rho)$. Therefore:

$$P_0 = 1 - \rho = 1 - \frac{\lambda}{\mu} \tag{17.19}$$

Figure 17.8 (curve (a)) shows, as ρ increases from 0 to 1, how the probability of an empty queue, P_0, decreases. Also by applying equation (17.18):

$$P_m = \rho^m P_0 = \rho^m(1-\rho) \tag{17.20}$$

implying a geometric distribution for P_m. The probability of the single server being busy is $1 - P_0 = \rho$. $\rho < 1$ ensures that the server has more capacity than is required. Figure 17.9 shows a typical queue length pdf, for $\rho = 0.5$.

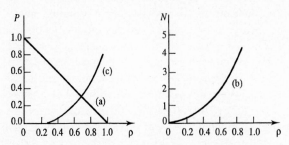

Figure 17.8 *Average queue size or length: (a) probability of an empty queue; (b) mean queue length, N, in packets; (c) probability that queue length exceeds 4.*

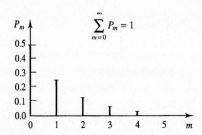

Figure 17.9 *Queue length pdf for $\rho = \frac{1}{2}$.*

EXAMPLE 17.1
For a single server queuing system with Poisson distributed arrivals of average rate 1 message/s and Poisson distributed service of capacity 3 messages/s calculate the probability of receiving no messages in a 5 s period. Also find the probabilities of queue lengths of 0, 1, 2, 3. If the queue length is limited to 4 what percentage of messages will be lost?

From equation (17.5(a)):

$$P_0(t) = e^{-\lambda t}$$

and $\lambda = 1$ and $t = 5$ for a 5 s period. Thus the probability of no arrivals in a 5 s period is given by:

$$P_0(5) = e^{-5} = 0.00674$$

Now from equation (17.18):

$$P_m = \left(\frac{\lambda}{\mu}\right)^m P_0$$

and from equation (17.19):

$$P_0 = 1 - \frac{\lambda}{\mu} = 1 - \frac{1}{3} = \frac{2}{3}$$

Thus the various queue lengths can be calculated as:

$$P_1 = \frac{1}{3} \times \frac{2}{3} = \frac{2}{9}$$

$$P_2 = \left(\frac{1}{3}\right)^2 \times \frac{2}{3} = \frac{2}{27}$$

$$P_3 = \left(\frac{1}{3}\right)^3 \times \frac{2}{3} = \frac{2}{81}$$

This differs slightly from Figure 17.9 in that the P_0 value is larger and, in consequence, the subsequent magnitudes fall off more rapidly. Finally for a queue length limited to 4 the probability of exceeding this restricted length is given by:

$$P(m > 4) = 1 - P(m \leq 4) = 1 - (P_0 + P_1 + P_2 + P_3 + P_4)$$

$$= 1 - \left(\frac{2}{3} + \frac{2}{9} + \frac{2}{27} + \frac{2}{81} + \frac{2}{243}\right) = 0.0041 = 0.41\%$$

17.4.4 Queue length and waiting times

The average queue length, equation (17.15), $N = \Sigma_{m=0}^{\infty} m P_m = (1 - \rho)\Sigma_{m=0}^{\infty} m \rho^m$. Further, as the sum of the geometric series $\Sigma_{m=0}^{\infty} m \rho^m = \rho/(1 - \rho)^2$ the mean queue length is given by:

$$N = \sum_{m=k+1}^{\infty} P_m = \frac{\rho}{1 - \rho} \tag{17.21}$$

Figure 17.8 (curve (b)) shows how N increases with increasing ρ. At $\rho = \frac{1}{2}$, $N = 1$ and for $\rho > \frac{1}{2}$, N increases above unity. As traffic intensity increases and ρ approaches 1 the queue length becomes infinite. If the queue is restricted to some finite value k then the probability of exceeding this value is given by:

$$P(m > k) = \sum_{m=k+1}^{\infty} P_m$$

$$= (1 - \rho) \sum_{m=k+1}^{\infty} \rho^m$$

$$= (1 - \rho) \left[\sum_{m=0}^{\infty} \rho^m - \sum_{m=0}^{k} \rho^m\right]$$

$$= (1 - \rho) \left[\frac{1}{1 - \rho} - \frac{1 - \rho^{k+1}}{1 - \rho}\right]$$

$$= \rho^{k+1} \tag{17.22}$$

This is shown in Figure 17.8 (curve (c)) for various values of ρ. Clearly when ρ exceeds about 0.8 there is a problem. To find average delay or waiting time, T, we use Little's result, equation (17.15) and equation (17.21):

$$T = \frac{\text{average queue length}}{\text{average arrival rate}} = \frac{N}{\lambda} = \frac{\rho}{\lambda(1 - \rho)} \tag{17.23(a)}$$

or, using equation (17.17):

$$T = \frac{1}{\mu(1 - \rho)} = \frac{1}{\mu - \lambda} \tag{17.23(b)}$$

Average delay (normalised by μ) is plotted in Figure 17.10, against ρ in the range $0 \leq \rho < 1$, and it is this key result which forms the basis of network delay analysis. When constrained by a finite queue length of k, then $\Sigma_{m=0}^{k} P_m = 1 = P_0 \Sigma_{m=0}^{k} \rho^m$. For this case $P_0 = (1 - \rho)/(1 - \rho^{k+1})$.

For packets transmitted over a link, at a bit rate of R_b bit/s with packet size K bits, the mean packet delay is, from equation (17.23(b)):

$$T = \frac{1}{R_b/K - \lambda} \tag{17.24}$$

where R_b/K represents the packet transmission or service rate, μ, in packet/s and λ is the packet arrival rate.

EXAMPLE 17.2

Consider a switch at which packets arrive according to a Poisson distribution. The mean arrival rate is 3 packet/s. The service time is exponentially distributed with a mean value of 100 ms. Assume the packet comprises 70 8-bit bytes and the output transmission rate is 5.6 kbit/s. How long does a packet have to wait in the queue?

The mean service rate is $\mu = 5600/(8 \times 70) = 10$ packet/s. From equation (17.23(b)):

$$T = 1/(\mu(1 - \rho)) = 1/(\mu - \lambda)$$

and we find that the mean packet delay is $T = 0.143$ s for queuing plus service. Since the mean service time is 100 ms, the mean queuing time is therefore $143 - 100 = 43$ ms.

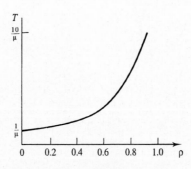

Figure 17.10 *Average time delay (T) for queue against ρ.*

Calculations, such as those shown above, allow us to model the delays on packet data links. The mean overall network delay is given by the sum of the transmitted packets times their delay divided by the total number of transmitted packets. (This type of queue analysis is also used for determining the performance of the Banyan switch networks in Chapter 18 in order to find their blocking performance.)

17.5 Packet speech transmission

Early packet networks for communication (see Chapter 18) were:

- store-and-forward, hard-wired, long-haul networks, e.g. the American ARPANET (with a channel capacity of 50 kbit/s), and the British SuperJANET (Chapter 18);
- satellite networks, e.g. the American Atlantic SATNET (with a channel capacity of 64 kbit/s);
- radio networks, e.g. the American PRNET (with a channel capacity of 100 to 400 kbit/s, for local data distribution).

The most general distinction is that, unlike the circuit switched TDMA telephone system (Figure 17.11), in a packet switched network the individual data (D) and voice (V) packets are switched (Figure 17.12) in accordance with the header information within each packet. The interface between the user and a packet network can be a data terminal, a host computer, or a packet voice terminal which comprises a telephone handset with a full range of control and signalling capabilities.

17.5.1 The components of packet speech

Analogue speech must first be converted into a digital sequence by a coder using waveform processing, e.g. delta modulation, Figure 5.28, or other coding techniques such as LPC, section 9.7. The vocoder is favoured when there is limited channel capacity and speech must be compressed to lower data rates than PCM. LPC produces fewer packet/s (1.7 to 7.4) than delta modulation which generates 9.4 to 38.5 packet/s at a bit rate of 16 kbit/s.

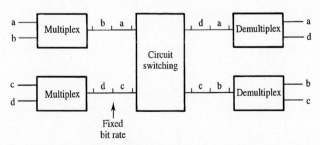

Figure 17.11 *Conventional circuit switched network (e.g. TDMA digital telephony transmission example, as described in section 6.5).*

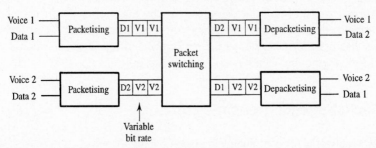

Figure 17.12 *Packet switched network with variable rate traffic where voice rate exceeds the input data rate.*

The digital bit stream is next partitioned into segments and some control (header) information added to each segment to form a packet. The control information consists of a time stamp and a sequence number, Figure 17.13, to assist reconstruction at the receiver. To reduce the bandwidth required for speech transmission, silences between bursts of speech are not packetised or transmitted. The time stamp enables the receiver to generate the appropriate duration of silence before processing the subsequent speech samples. Time stamps also allow reordering of packets which are received out of order. Since the time stamp cannot differentiate silence from packet loss, a sequence number is included to allow detection of lost packets.

It is important that a continuous stream of bits is provided at the receiver in order to produce smooth speech. This is achieved by delaying the arriving packets at the receiver queueing buffer, Figure 17.14. The size of this buffer must be sufficiently long to avoid packet loss due to overflow. Normally the delay will be chosen so that, statistically, a large proportion (e.g. 95%) of packets will be expected to arrive in the time allocated to the buffer delay, which is usually in the range 100 to 170 ms. The delay must not be so long that it dominates the performance of the speech transmission system, however.

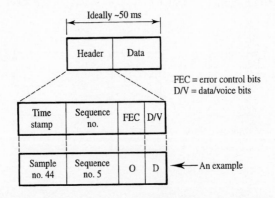

Figure 17.13 *Details of the header information carried within a data packet.*

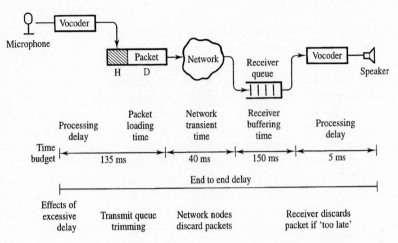

Figure 17.14 *Packet speech transmission system with inherent transmission delays.*

EXAMPLE 17.3

An X.25 packet switch has a single outgoing transmission link at 2 Mbit/s. The average length of each packet is 960 bytes. If the average packet delay through the switch, assuming an M/M/1 queue, is to be less than 15 ms, determine: (i) the maximum gross input packet rate to the switch; (ii) the average length of the queue; (iii) the utilisation factor through the switch if each packet in the input is converted into ATM cells having 48 bytes of data and a 5 byte overhead (see section 19.7.2 for an explanation of ATM).

In the X.25 switch the outgoing link is 2 Mbit/s. An average packet is 960 bytes or 7680 bits. If service time is equal to packet length, then: $\mu = 2 \times 10^6 / 7680 = 260.4$ packet/s
(i) Now if $T \leq 15$ ms, from equation (17.23(b)):

$$T = \frac{1}{\mu(1-\rho)} = \frac{1}{260.4(1-\rho)} \leq 15 \times 10^{-3} \quad \text{and} \quad \rho_{min} = 0.744$$

Maximum gross input rate is $\lambda = \rho\mu = 0.744 \times 260.4 = 193.7$ packet/s.

(ii) Average length of queue from equation (17.21) is:

$$N = \rho/(1-\rho) = 2.91 \text{ packets}$$

(iii) ATM switch

Each packet is 960 bytes, and corresponds to $960/48 = 20$ ATM cells.
The cell input rate is therefore $20 \times 193.7 = 3,874$ cell/s.
Now the output rate is $2 \times 10^6/(48+5) \times 8 = 4,717$ cell/s. Therefore $\rho = 3874/4717 = 0.821$.

17.5.2 Speech service performance

Three key parameters that describe the performance of a packet speech service are end-to-end delay, throughput and reliability. Delay is the time between speech being presented to the system, to the packet carrying that speech being played at the loudspeaker. Experimental tests show that few people notice any quality degradation if delay is kept below 0.3 s while a delay above 1.5 s is intolerable. Throughput is limited by the processing capability of the nodes. Reliability is defined as the proportion of packets that arrive at the destination in time to be used to reconstruct the speech.

Appropriate choice of packet size and rate can minimise delay and allow high throughput. In particular we must control the number of bits in the message header. Since the header is constant for every packet, regardless of size, to maintain high channel utilisation the number of speech bits per packet should be maximised. Large packets are also more desirable from the point of view of network throughput. However, to minimise the effects of lost packets and delay at the transmitter, packets should be short and, ideally, a packet should contain no more than 50 ms of speech. The trade-off is particularly difficult for narrowband speech, e.g. using LPC, because 50 ms of 2.4 kbit/s speech comprises only 120 data bits. Typical packet size for speech transmission across the Internet is about 300 bits comprising 100 to 170 ms speech segments. Channel loading information should be provided to the voice terminal so that packet rate and size can be varied according to the network load. In cases where the network is lightly loaded, it is capable of supporting a higher packet rate, hence smaller packets with less delay are used while, if the network loading is high, packets are made larger and packet rate is reduced.

17.6 Summary

Packets are groups of data bits to which have been added (as headers and/or trailers) addressing and other control information to facilitate routing through a digital data network. Queuing theory can be used to model packet behaviour at the switching nodes of a network and, in particular, can be used to predict average packet delay, average queue length and probability of packet loss at a node, given the packet interarrival-time pdf, the packet service-time pdf, the number of servers, the queuing discipline and the queuing storage space. For real-time applications, such as packet speech, resources can be saved by not transmitting empty or 'silent' packets. This necessitates packets being time stamped, however, for the receiver to regenerate the appropriate speech gaps.

In current Ethernet based networks data rates are in the range 10 to 100 Mbit/s and propagation delays (typically less than 5 μs) are small compared with the time required to transmit a 1 kbit packet. With the trend towards Gbit/s optical fibre links, operating over long distances, propagation delay becomes much longer than the packet duration, which will significantly alter the analysis of these systems.

17.7 Problems

17.1. The number of messages arriving at a particular node in a message switched computer network may be assumed to be Poisson distributed. Given that the average arrival rate is 5 messages per minute, calculate the following:

(a) Probability of receiving no messages in an interval of 2 minutes. [4.5×10^{-5}]

(b) Probability of receiving just 1 message in the next 30 s. [0.2]

(c) Probability of receiving 10 messages in any 30 s period. [2.16×10^{-4}]

17.2. In Problem 17.1: (a) What is the probability of having a gap between two successive messages of greater than 20 s? (b) What is the probability of having gaps between messages in the range 20 to 30 s inclusive? [0.189, 0.106]

17.3. Assuming a classical queue with a single server, determine the probabilities of having: (a) an empty queue, (b) a queue of 4 or more 'customers'. You should assume that the utilisation factor for the service is 0.6. [0.4, 0.13]

17.4. If the queue length in Problem 17.3 is limited to 10, what percentage of customers are lost to the service? [0.36%]

17.5. A packet data network links London, Northampton and Southend for credit card transaction data. It is realised with a 2-way link between London and Northampton and, separately, between London and Southend. Both links operate with primary rate access at 2 Mbit/s in each direction. If the Northampton and Southend nodes send 200 packets/s to each other and the London node sends 50 packets/s to Northampton what is the mean packet delay on the network when the packets have a mean size of 2 kbit? [1.275 ms]

Network topology and protocols

18.1 Introduction

In 1969, the first major packet-switched communications network, the ARPANET, began operation. The network was originally conceived by the Advanced Research Projects Agency (ARPA) of the US Department of Defense for the interconnection of dissimilar computers, each with a specialised capability. Today systems range from small networks interconnecting microcomputers, hard-disks and laser-printers in a single room (e.g. Appletalk), through terminals and computers within a single building or campus (e.g. Ethernet), to large geographically distributed networks spanning the globe, e.g. the Internet. They are often classified as local, metropolitan or wide area networks (LANs, MANs or WANs). Figure 18.1 shows the relationships between LANs, MANs, WANs, the 'plain old telephone system' (POTS) and other more recent types of network. The major features of LANs, MANs and WANs are summarised in Table 18.1, after [Smythe 1991].

UK examples of WANs are the BT packet switched service (PSS) and the new joint academic network (SuperJANET). The original JANET interconnected all UK university

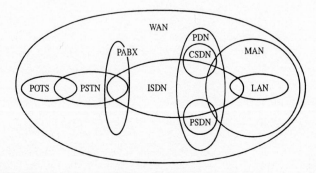

Figure 18.1 *Relationships between network architectures.*

LANs with a 5000 km backbone. With 100 Mbit/s transmission rates available on the individual university LANs the JANET backbone transmission rate has been progressively increased from 9.6 kbit/s to 2 Mbit/s. The main centres (London, Manchester, Edinburgh, etc.) were interconnected at 2 Mbit/s while other sites had 9.6 and 64 kbit/s access rates. The SuperJANET improvement upgraded the backbone bit rate to 34 and 140 Mbit/s using ATM (see section 19.7.2). Figure 18.2 shows the SuperJANET Phase B network, in which access rates are multiples of the 51.8 Mbit/s standard SDH interfaces (see section 19.4), within the PSTN core network. The equivalent US system is NSFnet which operates at 45 Mbit/s.

Table 18.1 *Comparison of LAN/MAN/WAN characteristics.*

Issue	WANs	MANs	LANs
Geographic size	1000s km	1 – 100 km	0 – 5 km
No. of nodes	10,000s	1 – 500	1 – 200
R_b	0.1 – 100 kbit/s	1 – 100 Mbit/s	1 – 100s Mbit/s
P_b	$10^{-3} - 10^{-6}$	$<10^{-9}$	$< 10^{-9}$
Delays	>0.5 s	100 – 100s ms	1 – 100 ms
Routing	sophisticated	simple	none
Linkages	gateways	bridges	bridges

The general distinction between a LAN, MAN and WAN depends on the ratio of the network end-to-end propagation delay to the average packet duration. For a MAN this ratio is close to unity while the more closely coupled LAN has a ratio much smaller than unity. In the loosely coupled WAN, the ratio is much larger than unity [Smythe 1991].

Broadly speaking, networks can be divided into three main categories:

- Circuit-switched: in which a continuous physical link is established between the pair of communicating data terminal equipments (DTEs) for the entire duration of the communications session. Circuit switched networks most commonly utilise portions of the PSTN. Circuit switching is inefficient for variable bit rate transmission since the circuit must always support the highest data rate expected – hence the move towards packet switching and asynchronous transfer mode, ATM. (ATM is discussed in section 19.7.2.)
- Message-switched: in which the complete message (of any reasonable length) is stored and forwarded at each data network node. Physical connections between node pairs are made only for the duration of the message transfer between those node pairs and are broken as soon as the message transfer is complete. No complete physical path need therefore exist between communicating DTEs at any time.
- Packet-switched: in which each message is divided into many standard packets which are then routed individually through the network. Each packet is stored and forwarded at each network node. Messages are reassembled from their constituent packets at the receiving DTE. Two distinct varieties of packet switching exist: virtual circuit packet switching in which all packets follow the same route through the network, and datagram packet switching in which different packets (within the same

Figure 18.2 *SuperJANET wide area network (WAN).*

message) are routed entirely independently. Virtual circuit systems ensure that packets are received in their correct chronological order whilst datagram systems must include packet sequencing information for correct message reassembly.

Hybrids of, and variations on, the above switching philosophies are also sometimes used.

One of the major activities which accelerated the development of packet-switched technology to its present state was the development of the layered communications architecture concept. The proliferation of various architectures, creating possible barriers between different manufacturers' systems, led the International Standards Organisation (ISO) to launch the reference model for Open Systems Interconnection (OSI). This standardised the systems interfaces.

18.2 Network topologies and examples

Any network [Hoiki] must fundamentally be based on some interconnection topology, to link its constituent terminals. The main network topologies are reviewed here.

18.2.1 Point-to-point

This is undoubtedly the simplest wired network type, and is extensively used. It may be transitory and exist only for the duration of the call, as on the circuit-switched PSTN, or may exist permanently, as on a private (leased) line. This configuration is commonly used when a limited number of physically distinct routes are required.

18.2.2 Multidrop

When a large number of locations, which can be partitioned into geographical clusters, are required to be connected the multidrop configuration is often employed, Figure 18.3. All transmissions from node A can be received by nodes B, C and D. However, only node A can receive data from B, C and D, and only one of the latter may transmit at any one time over the network, as there is only one data transmission path. This constraint is enforced by employing a polling protocol at A which addresses B, C and D in turn, permitting only the addressed node to reply.

Multidrop connection provides a way of reducing transmission link costs by utilising a single branched circuit to connect A to B, C and D. The principal current application of this topology is the connection of host computers to terminals – or terminal clusters – at several locations. Multidrop connection is under consideration, however, for optical fibre replacement of the PSTN local loop copper connection (section 19.7.3) with a passive optical network (PON).

18.2.3 Star

Centralised switched-star network configurations have now existed for over a century in the PSTN and, for this reason, represent perhaps the best understood class of network. In the star configuration, the devices comprising the network are connected by point-to-point links, to a central node or computer, Figure 18.4. The star network has two major limitations:

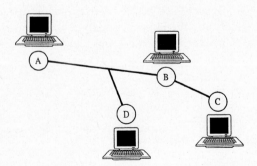

Figure 18.3 *Multidrop configuration.*

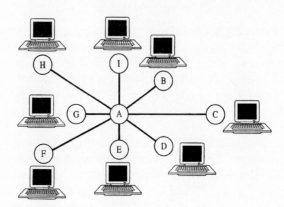

Figure 18.4 *Basic star network.*

- The remote devices are unable to communicate directly and must do so via the central node, which is required to switch these transmissions as well as carrying out its primary processing function.
- Such a network is very vulnerable to failure, either of the central node, causing a complete suspension of operation, or of a transmission link. Therefore, reliability and/or redundancy are particularly important considerations for this topology.

Despite these limitations, the star configuration is important as it has been used extensively for telephone exchange connections and in fibre optic systems. (Until recently it has been difficult to realise cheap, passive, optical couplers for implementation of an optical fibre ring or bus system, hence the attraction of the star network.) This topology is also used for many VSAT networks, section 14.3.9.

18.2.4 Ring

This network consists of a number of devices connected together to form a ring, Figure 18.5. Ring networks employ broadcast transmission, in that messages are passed around the ring from device to device. Each device receives each message, regenerates it, and retransmits it to its neighbour. The message is only retained, however, by the device to which it was addressed. Two variations of the ring network exist. These are:

- Unidirectional: in which messages are passed between the nodes in one direction only. The host, A, controls communication using a mechanism known as 'list polling'. The failure of a single data link will then halt all transmissions.
- Bidirectional: in which the ring is capable of supporting transmission in both directions. In the event of a single data link failing, the host, A, can then maintain contact with the two sectors of the network.

That each network node is involved in the transmission of all data on the network is a potential weakness. The ring topology is simple both in concept and implementation, however, and is popular for fibre optic LANs in which regenerative repeaters are required

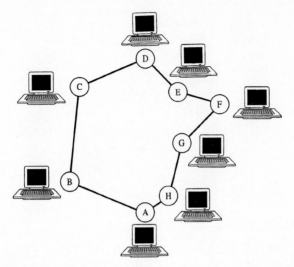

Figure 18.5 *Ring network.*

at each node. Access is via slots or tokens. A simple token ring operates essentially as follows.

A token bit pattern (e.g. 11111111), which is prevented from appearing in genuine data using a technique called bit stuffing, circulates while the ring is idle. A terminal 'captures' the token by removing it, or altering it (e.g. to 11111110). The terminal 'possessing' the token can then transmit one or more packets around the ring, each node in the ring acting as a repeater. Each transmitted packet contains a destination address in its header which is recognised by the destination terminal. The destination node reads the data and signifies receipt by setting response (or acknowledgement) bits in the packet trailer. It then retransmits the packet. When the sending terminal receives the packet with the response bits correctly set it resets the token to the idle pattern and recirculates it around the ring.

Cambridge ring

The Cambridge ring employs the empty slot principle. A constant number of fixed-length data packets – slots – circulate continuously around the ring in one direction only, a full/empty indicator within the slot header being used to signal the state of the slot. To transmit a node occupies the first empty slot by setting the full/empty flag to full and placing its message in the slot. The data packet then completes one revolution of the ring before the sending node 'empties' the slot and resets the indicator to empty. (A minor variation is to allow the receiver to empty the slot.)

The CR82 Cambridge ring, Figure 18.6, is a baseband implementation of a slotted ring using twisted wire pair cable, a bit rate of 10 Mbit/s and, typically, 10 slots. A monitor station checks which stations are active, and fills dummy slots to confirm that the

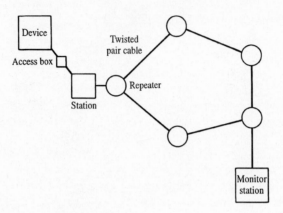

Figure 18.6 *Cambridge backbone ring example.*

ring is unbroken. With optical fibre implementations the data rate can be increased to 600 Mbit/s and beyond.

18.2.5 Mesh

While the star configuration is best suited for host computer/slave terminal connections on a one-to-many basis, mesh networks, Figure 18.7, are primarily used in a many-to-many situation, such as typically exists in WANs. Fully interconnected mesh networks, for more than a small number of nodes, are generally expensive as they require a total of $\frac{1}{2}n(n-1)$ links for n nodes. They are very resilient to failure, however, since alternative routes are available if a link fails. Where link lengths are long or data volumes low, a public packet-switched service may offer a significant cost advantage over a private mesh network. Unlike the ring or star topologies, adding a node to an existing (n node) mesh

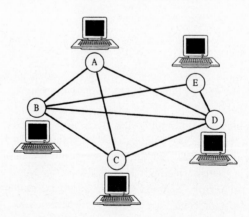

Figure 18.7 *Mesh network.*

network necessitates a further $n - 1$ new connections.

18.2.6 Bus

Bus networks employ a broadcast philosophy. The bus is formed from a length of cable to which devices are attached by cable interfaces, or taps, Figure 18.8. Messages from a device are transmitted (bidirectionally) to all devices on the bus simultaneously; however, devices accept only those messages addressed to themselves. Since devices are not required to retransmit the messages, there is none of the delay (latency) or complexity associated with the ring topology.

The bus configuration is extremely tolerant of terminal failures, since operation of the network will usually continue if one of the active-component devices fails. A further advantage is that bus networks are easily reconfigured and extended. The bus topology is often expanded to form a tree structure, especially appropriate for use in multistorey buildings. (Another broadcast technique, similar in concept to a physical bus, is that traditionally used by satellite linked networks.) Access to the bus transmission medium is discussed in section 18.5. The small computer system interface (SCSI) is a dedicated bus used to connect discs and tape drives directly to the processor of a computer system.

Ethernet bus

Ethernet [Hoiki] is an example of a proprietary bus network, Figure 18.9(a). In this implementation a simple passive medium and transparent high impedance taps are employed. Of particular importance are the (coaxial cable) line terminators, which serve as matched loads to ensure reflections do not corrupt data transmission. Typical data rates are 1 Mbit/s and 10 Mbit/s, with a maximum bus length of 500 m. Figure 18.9(b) shows, for a 10 Mbit/s system, the expected number of attempts required to access the bus for different levels of network load. Ethernet uses 32-bit polynomial codes (section 10.8.1) for error detection.

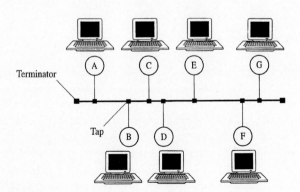

Figure 18.8 *Bus network.*

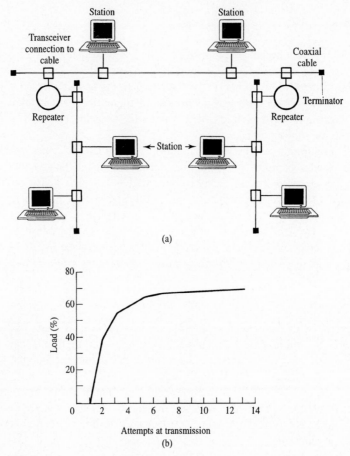

Figure 18.9 *Ethernet: (a) example configuration; (b) typical system response showing how the number of attempts at transmission varies with network load.*

18.2.7 Transmission media

Networks can utilise wire, coaxial cable, fibre optic or wireless links. Table 18.2 compares the different cabled and wireless transmission media. Metallic cable is preferred for many systems as simple, passive, tapped junctions cannot easily be realised for optical fibres. In all systems propagation loss, distortion, delay and noise are potential impairment mechanisms. Many systems use coaxial cable operating with baseband pulse rates up to 10 Mbit/s over 500 m paths. At higher rates 'broadband' data is modulated onto an RF carrier of, typically, 100 MHz, or fibre optic transmission (section 19.5) is used. Recently short runs of simple twisted wire pair have also become attractive for broadband transmission, because optical fibres are 5 to 10 times more expensive than twisted pair cable installations. In the near future broadband wireless LAN technology using carrier frequencies of 5 GHz and 17 GHz will become important.

Table 18.2 *Typical cost per node, data rate and ranges for different transmission media.*

Transmission medium	Twisted pair	Coaxial	Fibre optic	Radio	Infra-red
Range, m	1 – 1,000	10 – 10,000	10 – 10,000	50 – 10,000	0.5 – 30
Data rate, kbit/s	0.3 – 2,000	300 – 10,000	1 – 100,000	1 – 10	0.05 – 20
Cost/node, US dollars	10 – 30	30 – 50	75 – 200	50 – 100	20 – 75

18.3 Network coverage and access

The first computer WAN, ARPANET, spanned the US and was later extended to Europe. With the rise in the number of potential digital communications users, many of the bodies responsible for post, telegraph and telephone (PTT) services have now built such networks. These WANs generally use store and forward systems in which messages are held in a node store before being switched, via an appropriate transmission link, to another node nearer the message's ultimate destination. With 10 to 50 kbit/s data rates the end-to-end delay is of the order of seconds. WANs can be classified as follows:

- Public networks: For example, the British Telecom Gold electronic mail service, the international EURONET (a packet switched service (PSS), which exists primarily for accessing information databases) and the Internet.
- Private networks: For many users, public data networks do not provide the services required. Private telecommunications networks leased on a semi-permanent basis, such as SWIFT employed by the banking fraternity, are one alternative.
- Value added networks: A value added network (VAN) uses conventional PTT facilities combined with specialised message-processing services to add value to the network. The user is offered flexibility and the economies of shared usage. Examples include TRANSPAC and the specialised banking EFTPOS networks.

Local area networks (LANs) are specifically designed for the interconnection of computer systems and peripherals within a geographically small site, such as a single building or university campus, and are generally privately owned. A LAN has many of the features of a WAN, but it also has its own, distinct, characteristics, i.e.:

- Wide bandwidth, of the order of tens of Mbit/s.
- Low (1 to 10 μs) delay due to resource sharing, and absence of buffering.
- Low probabilities of bit error, typically 10^{-9} to 10^{-11}.
- Simple protocols (compared with those necessary for the longer ranges of WANs).
- Low cost and easy installation.
- High degree of connectability and compatibility of physical connections.
- Geographically bounded, with a maximum range of approximately 5 km.

Metropolitan networks or MANs which may be up to 50 km in diameter lie between LANs and WANs. An access method (or protocol) defines the set of procedures for LAN access to a WAN and vice versa. Since LAN transmission rates are much higher than those of interconnecting WANs one LAN network node must normally be dedicated to the WAN interface. With high speed MAN interconnection of LANs it is possible to transfer large files electronically, rather than physically using, for example, discs, tapes or CD-ROMs. Two of the most common network interconnection techniques utilise bridges and gateways.

18.3.1 Bridges and gateways

A bridge is a device that interconnects two networks of the same type (using the same protocol). The bridge utilises a store and forward feature to receive, regenerate and retransmit packets while filtering the addresses between connected segments. A gateway on the other hand [Smythe 1995] connects networks using different protocols, typically LANs and WANs. They can therefore provide transparent access to resources on other, remote, networks. It is a similar device to the bridge but also performs the necessary protocol conversion. The JANET network of Figure 18.2 has transparent gateway connections to other X.25 international networks, which link it to the Internet, section 18.7.6.

18.3.2 Network switches

With the increasing power of VLSI technology, a large switch array can now be implemented on a single chip. National Semiconductor developed a 16×16 switching matrix in 2 μm CMOS gate array technology in the late 1980s. Figure 18.10 shows an 8 $\times$ 8 *Banyan* switch, consisting of twelve 2×2 switching elements. There is only one route through the switch from each input to each output. At each stage in this switch the upper or lower output is chosen depending on whether a specific digit in the route control overhead is 1 or 0. This is one of the simplest switch arrays that can be constructed and illustrates the self-routing capability of a packet navigating a network composed of such switches. When routing and sorting is performed locally at each switch in a network

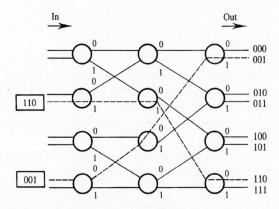

Figure 18.10 *8 × 8 Banyan packet switching network.*

there is no need for recourse to a centrally located processing centre and the throughput is therefore greater than that of a circuit switched network. Dashed lines in Figure 18.10 show two example paths that packets with specific route headers would take through the network.

A problem inherent in all packet switched networks is that of packet collision or blocking (i.e. packets arriving simultaneously on the two inputs of a single switching element). To reduce packet collisions, the network can be operated at higher speed than the inputs or storage buffers can be introduced at the switch sites.

To remove packet collisions completely, the input packets must be sorted in such a way that their paths never cross. The technique normally used to perform this operation is known as Batcher sorting, in which the incoming packets are arranged in ascending or descending order according to route header. (Another non-blocking technique is the crossbar switch which was used in the 1970s for PSTN design.)

The Starlight network switch, which uses both Batcher sorting and Banyan switching, is shown in Figure 18.11. A CMOS 32 × 32 Batcher–Banyan switch chip in the early 1990s operated at 210 Mbit/s. Batcher–Banynan switching is not able to resolve the problem of output blocking, when multiple packets are destined for the same output port, but this can be overcome by the insertion of a trap network between the sorting and switching networks, Figure 18.11. The loss of trapped packets is overcome by recirculating the duplicate address packets into the next sorting cycle to eliminate the need for buffered switch elements.

The network is packet synchronous if it requires all packets to enter the network at the same time. Banyan networks can be operated packet asynchronously because each packet's path through the self-routing network is, at least to first order, unaffected by the presence of other packets. Packet switches such as these, which can operate at input rates of up to 150 Mbit/s, have been realised. (Note that the Banyan name is now often associated with a particular commercial system.)

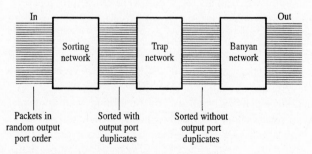

Figure 18.11 *Starlight packet sorting–switching network.*

18.4 Reference model for terminal interfacing

Most network architectures are organised as a series of layers. Previously the number, name and function of each layer differed from network to network. In all cases, however, the purpose of each layer was, and is, to offer certain services to higher layers, while shielding the details of exactly how those services are implemented in the lower layers.

Layer n at one node holds a conversation with layer n at another node. The rules and conventions used in this conversation are collectively known as the layer n protocol. The main functions of such a protocol are:

- Link initiation and termination.
- Synchronisation of data unit boundaries.
- Link control, with reference to polling, contention restriction and/or resolution, time-out, deadlock and restart.
- Error detection and correction.

The messages or blocks of data passed between entities in adjacent layers (i.e. across interfaces) are known as data units. With the exception of layer 1 no data is directly transferred from layer n at one node to layer n at another. Data and control information is passed by each layer to the layer immediately below, until the lowest layer is reached. It is only here that there is physical communication between nodes. The interfaces between layers must be clearly defined so that:

- The amount of data exchanged between adjacent layers is minimised.
- Replacing the implementation of a layer (and possibly its subordinate layers) with an alternative (which provides exactly the same set of services to its upstairs neighbour) is easily achieved.

18.4.1 The ISO model

This is a set of layers and protocols used to model network architecture, Figure 18.12. The overall purpose of the International Standards Organisation (ISO) model is to define standard procedures for the interconnection of network systems, i.e. to achieve open systems interconnection (OSI). ISO OSI processing is normally performed in software but, with the continuous rise in data rates, hardware protocol processors are, increasingly, being deployed. Several major principles were observed in the design of the ISO OSI model. These principles are that:

- A new layer should be created whenever a different level of abstraction is required.
- Each layer should perform a well defined service related to existing protocol standards.
- The layer boundaries should minimise information flow across layer interfaces.
- The number of layers should be sufficient so that distinct functions are not combined in the same layer, but remain small enough to give a compact architecture.

This has resulted in agreement to use seven layers as the OSI standard. The layers of the model are presented from the viewpoint of connection-mode transmission, starting from the interface to the physical medium. A key feature of the ISO OSI model is that it

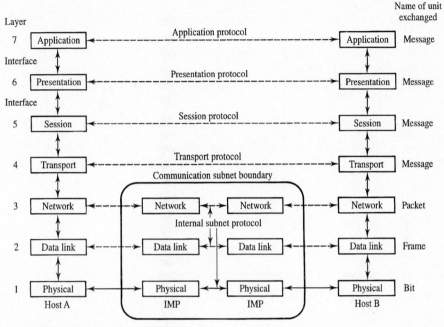

Figure 18.12 *ISO OSI reference model.*

achieves standardisation for data communications combined with error free transmission. A summary of function distribution is provided in Figure 18.12. The physical layer (1) directly interfaces with the transmission medium. This layer therefore includes: mechanical aspects, e.g. specification of cables and connectors; electrical aspects, e.g. specification of voltage levels, current levels and modulation techniques; functions and procedural aspects for the interface to the physical circuit connection.

The task of the data link layer (2) is to take the raw transmission facility and transform it into a link that appears free of transmission errors. It accomplishes this task by breaking the input data into data frames, transmitting the frames sequentially, and processing acknowledgement frames returned by the receiver. This is illustrated in Figure 18.13. Here once the packet is transmitted the transmitter stops and waits for an acknowledgement before the next packet is sent. If the acknowledgement does not arrive from the remote terminal within a prescribed time interval then the original packet is retransmitted, as shown in Figure 18.13 for packet 2.

The data link layer must both create and recognise frame boundaries. It thus defines the protocols for access to the network. Cyclic redundancy or polynomial codes (Chapter 10) are widely used in the data link layer to achieve an error detection capability on the bit-serial data. These processing functions are implemented in hardware. Software packages, such as Kermit, are located here to provide terminal emulation and file transfer facilities. The first two layers together are sometimes called the hardware layer.

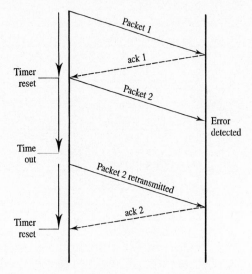

Figure 18.13 *Protocol for acknowledging (ack) successful packet receipt.*

The purpose of the network layer (3) is to provide an end-to-end communications circuit. It has responsibility for tasks such as routing, switching, and interconnection, including the use of multiple transmission resources, to provide a virtual circuit. The bottom three layers, collectively, implement the network function – and can be envisaged as a 'transparent pipe' to the physical medium, Figure 18.12. Data routed between networks, or from node to node within a network, require these functions alone.

The next layer (4) provides a transport service, suitable for the particular terminal equipment, that is independent of the network and the type of service. This includes multiplexing independent message streams over a single connection and segmenting data into appropriately sized units for efficient handling by the lower layers. Modulo $2^n - 1$ checksums (Chapter 10) are frequently computed within the software communications protocols of this layer, and compared with the received value, to determine whether data corruption has occurred. The transport layer can implement up to five different levels of error correction, but in LANs little error correction is required.

The functions of the session layer (5) are to negotiate, initiate, maintain and terminate sessions between application processes. This could for example be a transaction at a bank cash card machine. The layer operates in either half or full duplex, section 15.1.1.

While the session layer (5) selects the type of service, the network layer (3) chooses appropriate facilities, and the data link layer (2) formats the messages. The presentation layer (6) ensures that information is delivered in a form that the receiving system can interpret, understand and use. For example it defines the standard format for date and time information. The services provided by this layer include classification, compression and encryption of the data using codes and ciphers and also conversion, where necessary, of text between different types, e.g. to and from ASCII. The overall aim of layer 6 is to make communication machine independent.

The application layer (7) defines the network applications that support file serving. Hence it provides resource management for file transfer, virtual file and virtual terminal emulation, distributed processing and other functions. Conceptually this is where X.400 electronic mail [Wilkinson] and other network utility software resides. In electronic mail applications, layer 7 contains the memories which store the messages for forwarding when network capacity becomes available.

These ISO layers are equivalent to similar functions in other layered communications systems and hence there is commonality between OSI and other standards which have OSI equivalence. As one moves down the layers overhead bits (added by each layer) come to dominate each data packet making the effective data rate at layer 7 at least an order of magnitude less than the actual bit rate at the data link layer.

18.4.2 Connectionless transmission

The ISO OSI model can accommodate both connection and connectionless transmission modes. In the former an association between two or more peer entities, termed a connection, which has a clearly defined lifetime is established for data transfer purposes. This mechanism is the logical equivalent to a (circuit-switched) PSTN telephone call, and is employed in the virtual circuit strategy of X.25 packet-switched networks which operate at up to 64 kbit/s.

In connectionless-mode transmission data transfer can be achieved by a single access to a service, without the requirement for establishment and release phases. Here each self-contained data unit can be routed independently. This mode is particularly appropriate where interaction between pairs of entities is intermittent and infrequent, as in electronic mail or when bursty, variable bit rate, traffic occurs.

18.4.3 Physical layer protocol

For many years the most common, physical layer, connection standard has been RS232C, which defines the signal functions and their electrical characteristics in the two (forward and reverse) channel directions. RS232C is widely used for interconnecting terminals, PCs, printers, etc., and uses a minimum of three wires for the two channel connection. It can accommodate data rates of a little over 1 kbit/s on paths of up to 20 m. For longer distances (up to 1 km) and higher data rates, balanced transmission systems such as RS423A (10 kbit/s) and RS422A (1 Mbit/s) have been developed which reduce the effects of noise pickup on cables.

18.4.4 Synchronisation and line coding

A key requirement of the protocol processor is that it must be able to handle clock synchronisation, line encoding/decoding and frame synchronisation. A special synchronisation problem in LANs and MANs is that of the clock oscillators in the network's terminal equipment and nodes. For synchronised clock operation all transmissions are synchronised with a common master station clock. For plesiochronous operation (section 19.3) each individual node receives data with a clock signal locked to the clock of the preceding node (or terminal). It then transmits data with its own local clock signal. Due to clock oscillator tolerance, this implies that the number of transmitted bits can differ from the number of information bits, requiring justification.

Line coding usually introduces redundancy into the data stream (see Chapter 6). This redundancy can be utilised for reliable clock recovery and for the coding of special control symbols. Long streams of consecutive identical bits are not desirable if this means that timing information cannot be regenerated easily from the incoming signal. Line codes can be divided into three classes, namely scrambled codes, bit insertion codes, and block codes. The function of a scrambler, in this context, is to generate transitions in the line-encoded data stream when the original data stream contains long sequences of identical bits, as in HDB3, section 6.4.5.

Examples of bit insertion codes are the 5B6B and 8B1C line codes. In the 5B6B-PMSI (periodic mark space insertion) scheme, the encoder alternately inserts a mark (one) and a space (zero) for every five information bits. In the 8B1C (8 binary with 1 complement insertion) scheme, every eight bits are preceded by one extra bit, which is the complement of the first of the eight bits. Block codes allow unique words to be transmitted for synchronisation purposes.

18.5 Medium access control

High-speed channel access schemes are divided into four main classes: random access, demand assignment, fixed assignment, and adaptive assignment protocols [Skov]. For each class, the protocols are further subdivided according to network topology.

18.5.1 Random access and demand assignment protocols

Random access is typified by the ALOHA packet switched systems which simply transmit whenever a message is ready to send and then waits to see if this transmission collides with other data on the system. This is very inefficient giving a maximum channel utilisation (or normalised throughput) of 18% [Kleinrock]. Slotted ALOHA, in which data is constrained to time slots, avoids partial packet overlap and therefore achieves a maximum utilisation of double this.

The best known access scheme for bus systems, because of its low installation cost, is carrier sense multiple access with collision detection (CSMA/CD). This scheme allows nodes to contend for use of the network and, in this sense, is similar to ALOHA. In CSMA/CD, however, any node ready to transmit first listens, sensing the carrier, to check whether another node is transmitting an information frame. If another transmission is already in progress, the node defers operation until the end of the current transmission. Once the network is free, the node transmits its addressed message. However, due to the non-zero propagation delay, two or more nodes can attempt to transmit simultaneously causing contention, which results in a collision. Given this situation, the transmission attempt is aborted and the node waits a random period before retransmission to avoid possible step-lock. This is a more sophisticated access scheme than ALOHA and can realise a channel utilisation of 75%. At this utilisation efficiency, however, collisions introduce unpredictable network delay or latency.

With demand assignment access protocols, message transmissions are governed by serving, or allocating access to, the attached stations in a predetermined (typically cyclic) order. A representative token ring protocol (see section 18.2.4) for packet switched multimode optical fibre systems is the 125 Mbit/s fibre distributed data interface (FDDI). FDDI networks can employ ring lengths of 100 km with 2 km spacings between repeaters. The active optical repeaters receive light from the incoming fibre and regenerate data pulses which are coupled into the outgoing fibre. FDDI (see section 18.7.4) has been developed primarily as a dual broadband backbone ring with a high data rate for image transmission applications. Current developments in FDDI II are aimed at circuit switching for voice and video multimedia transmission.

EXAMPLE 18.1 – Token passing
A token passing ring operates at 2 Mbit/s with 10 stations, each with a latency of 2 bits, and uses a 4-bit token. If the cable delay round the ring is 10 μs what is the maximum waiting time for access?

The total ring latency is $10 + (10 \times 2)/2 = 20$ μs. Maximum waiting time is given by the total ring latency plus the token processing time which is $20 + 4/2 = 22$ μs.

A variation on the conventional token ring is to partition time into slots and let these propagate round the ring. Access to a slot is possible, as it passes, provided it is not already filled with data. Such *slotted rings* can be divided into two groups depending on which station empties a full slot, the source station or the destination station. The fibre optic 600 Mbit/s Cambridge fast ring, section 18.2.4, employs source release.

Unidirectional buses integrate traffic with different priorities by means of rounds. Access to the bus typically follows one of three access schemes. When attempt-and-defer access is employed, a station wishing to transmit waits until the media is idle. It then begins its transmission and continues to listen to the media. If it detects a transmission from any upstream station, it aborts its transmission. The other schemes use polled or reservation access.

18.5.2 Fixed assignment protocols

In fixed assignment protocols the total network resource is divided among the participating stations in time, frequency, or code domain, in a fixed way. Thus, a station always occupies a part of the channel capacity, whether or not it has data to transmit. The protocols are TDMA, Figure 5.2, FDMA, Figure 5.12, and contention free CDMA, see section 15.5. Most of the FDMA and CDMA experimental ring and star networks use optical signalling. CDMA permits uncoordinated accessing but may not offer the capacity of FDMA and TDMA, unless power control is adopted to avoid the near-far problem (see section 15.5.6).

18.5.3 Adaptive assignment protocols

At low loads, a device can usually transmit sucessfully as soon as it senses that the channel is idle. As the load rises, however, packet collisions may occur, thus destroying data. In such cases, adaptive protocols may switch to a more restricted, conflict-free, mode of operation such as token passing, attempt-and-defer, or TDMA access. An example of a high-speed network based on adaptive assignment access is the US 100 Mbit/s fibre optic HYPERchannel-100 network. This is based on a bus topology, and employs an access protocol that starts in CSMA/CD mode and switches to TDMA mode when collisions become too frequent.

18.6 International standards for data transmissions

18.6.1 X.25

This is the PSS standard which effectively replaces the telephone network, using the V-series recommendations (see section 11.6), with a digital system having superior error performance and fast switching. The X.25 recommendation defines the interface between terminal equipments and the public data network for packet-switched communication. A set of associated standards, X.3, X.28, X.29, have been developed to enable simple terminals to access an X.25 network. X.25 is actually a layered network access (interface) protocol that exhibits many of the properties of network architectures. The functionality of the X.25 specification corresponds entirely to the lower three layers in the ISO OSI model, Figure 18.12. In X.25, error checking is conducted at each node in the network. In the new frame relay systems this is only performed at the terminal stations.

X.25 PSS is now available within the UK as a core network for data communications, electronic mail, etc. One use of the network is for credit card verification and connections are available via telephone lines or radio access at 8 kbit/s [Davie and Smith]. Radio coverage in 1991 extended to 75% of the UK population for this data service.

18.6.2 IEEE 802

The IEEE 802.n specifications also map to the bottom three layers of the ISO OSI reference model. The IEEE split the data link layer into sub-layers: logical link control and medium-access control, which are used for bridging between networks. The 802.3 to 802.6 standards describe physical connections and define how access to the physical medium is coordinated for each LAN type. They therefore correspond to layer 1, and a sub-layer of layer 2, in the ISO OSI model.

- 802.2: Defines a logical link control layer.
- 802.3: Defines a CSMA/CD protocol, which is the basis of Ethernet, as implemented on coaxial, and twisted pair, cables.
- 802.4: Defines a token-passing bus protocol.
- 802.5: Defines a token-ring protocol (FDDI).
- 802.6: Is a broadband MAN standard (45 Mbit/s) for voice, data and video.
- 802.7: Specifies a broadband FDMA LAN (with 400 MHz of bandwidth).

These IEEE standards are now internationally accepted as ISO equivalent. 802.4 may eventually be replaced by FDDI-II.

18.7 Network examples

18.7.1 Manufacturers application protocol (MAP)

The manufacturers application protocol is implemented on a 10 Mbit/s broadband multidrop bus LAN using an 802.4 token passing bus channel employing RF carrier modulation. By operating on two separate carriers at frequencies between 59.74 and 264 MHz, simultaneous (FDM) data transmission and reception is accomplished using QAM signalling (Chapter 11). MAP is designed to interconnect plant (such as robots) in a manufacturing environment. Another commercial development is the fieldbus for interconnection of measuring instruments [Jordan]. In one network the bit rate is 1 Mbit/s with a maximum access time of 5 ms for a 32-node system. There are various other similar network designs, many of which share the IEEE 488 interface standard for digitising, encoding and transmitting signal samples from remote locations.

18.7.2 Admiral

The Alvey Admiral multimedia prototype network (Figure 18.14), developed in the mid 1980s, consists of baseband CSMA/CD Ethernet LANs at each of the five project sites, interconnected by a high speed network. This early example of a high speed network is configured as a star, with 2 Mbit/s primary rate access circuits from each user site to a central switch located at the star hub. At the centre of the network is a non-blocking switch providing configurable interconnection of any number of consecutive 64 kbit/s time slots within the 2 Mbit/s bearer to provide point-to-point links from site to site as required. User sites are able to alter the configuration of the network by sending commands to the switch controller, via the public X.25 network. This network was the first practical combination of fast local Cambridge rings and Ethernets with primary rate ISDN interconnections, as described in the following chapter.

18.7.3 Military LAN systems

The US MIL-STD-1553B and its UK equivalent DEF STAN 00-18 or STANAG 3838 applies to military ship, submarine and aircraft LANs. These 1970s systems used

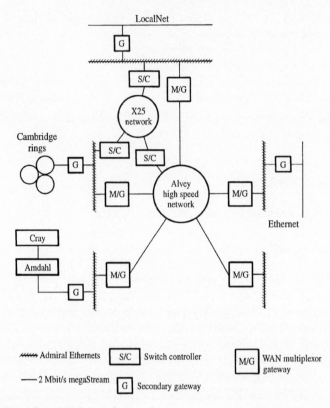

Figure 18.14 *The Admiral high-speed network.*

screened twisted pair cable to achieve 1 Mbit/s transmission rates with up to 31 remote terminals. Further US work has achieved high speed (20 to 40 Mbit/s) fibre optic networks while, in the UK, DEF STAN 00-19 is a 300 m multi-drop bus topology on a screened twin coaxial cable which signals at 3 Mbit/s. The bus for the European fighter aircraft, STANAG 3910, is a dual redundant 20 Mbit/s fibre optic implementation. It defines low speed 1 Mbit/s dual redundant wired channels plus the high speed 20 Mbit/s fibre optic channels. These specialised systems are not generally fully ISO compatible.

In contrast to civilian LANs, military ones require priority accessing schemes with short access delays for threat messages and must also be secure and free from transmission errors. These constraints inevitably result in specialised networks being developed for military applications.

18.7.4 Fibre distributed data interface (FDDI)

FDDI is a backbone broadband ring to which other, slower speed, tree networks and peripherals can be connected, Figure 18.15. (FDDI was conceived as a fast or broadband service to handle multimedia data transfer for applications such as desktop conferencing.)

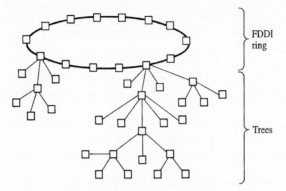

Figure 18.15 *Network comprising FDDI high speed ring with tree connections.*

The overall structure then looks like a ring-of-trees. FDDI operates at 100 Mbit/s as a plug-in multimode fibre ring (section 19.5) with 4B5B coding (which is a variation on the techniques described in section 6.4.7) implying a 125 MHz clock rate for this bit insertion code. Link P_b is 2.5×10^{-10} and packet error probability is 10^{-9}. On the maximum ring length of 200 km, ring transit time is 2.7 ms, corresponding to 60 full length packets. Packet latency for access to a network at 45 Mbit/s load is typically 50 to 200 μs. Currently there is interest in copper distributed data interface (CDDI) using twisted pair cables, but this interest is more intense in North America than Europe. The FDDI structure of Figure 18.15 is not dissimilar to the future SDH inner core proposals employing ring networks (see section 19.4.1).

18.7.5 Wireless networks

There is much current interest in developing wireless systems to obviate the need for Ethernet wired connections between terminals and printers which are located in close proximity to each other. The US NCR WaveLAN is such a system which typically caters for 50 to 300 m link connections. In the indoor environment, if equalisation is not employed to combat multipath responses, then the data rate is typically limited to several hundreds of kbit/s. In WaveLAN systems 2 Mbit/s data is spread by a wideband (spread spectrum) modulator into an 11 MHz bandwidth channel. This wideband transmission permits the multipath signals to be separately resolved and their presence compensated for by equalisation in the receiver. Another system, HYPERLAN, operates using a 5.2 GHz carrier frequency with a 20 Mbit/s BPSK transmission rate [Halls].

EXAMPLE 18.2
A wireless network is to be designed for interfacing several portable, battery powered, laptop computers dispersed around an office environment using the HYPERLAN standard. Due to obstructions in the office path loss follows an inverse cube law with distance (rather than the free space inverse square law of Chapter 12). The portable computer power consumption is 15 W and

the wireless network should not *significantly* degrade the battery operating life. Estimate, using the link budget analysis of Chapter 12, the maximum operating range you might expect to achieve. Assume that the receiver noise figure is 10 dB, the (BPSK) signal requires a receiver CNR of 15 dB, a fade margin of 6 dB is necessary, the total implementation loss is 2 dB and the transmitter efficiency is 40%.

If the transmitter uses 10 to 15% of the battery power then the power available for the radio modem is 1.5 to 2.2 W. At 40% efficiency this gives approximately 0.75 W or 29 dBm of transmitted power.

The thermal noise floor in a 25 MHz bandwidth, from equation (12.8), is −100 dBm. The received carrier level must be larger than this by 10 dB (noise figure), 2 dB (implementation loss), 15 dB (receiver CNR requirement) plus 6 dB (fade margin). Thus, adding a total of 33 dB, the required received carrier level is $C = -67$ dBm.

The maximum tolerable path loss is thus +29 +67 = 96 dB $= 4.0 \times 10^9$. FSPL is defined in Chapter 12. Using an R^{-3} law, equation (12.71) becomes:

$$\text{FSPL} = \left(\frac{4\pi}{\lambda} \right)^2 R^3$$

For isotropic antennas $G_T = G_R = 0$ dB $\equiv 1$. Thus for an operating frequency f and a free space propagation velocity c, the transmission loss, P_T/C, equals the FSPL, i.e.:

$$\frac{P_T}{C} = \text{FSPL} = \left(\frac{4\pi f}{c} \right)^2 R^3$$

Substituting numerical values and solving to find R:

$$4.0 \times 10^9 = \left(\frac{4\pi \times 5.2 \times 10^9}{300 \times 10^6} \right)^2 R^3 = \frac{4.74}{10^{-4}} R^3$$

$$R^3 = \frac{4.00}{4.74} 10^5 \text{ and } R = 44 \text{ m}$$

This range is typical for an office environment.

18.7.6 Internet

The Internet started life over 25 years ago as the ARPANET, section 18.1, which grew, initially, to link many university and industry laboratories in the US defence research community. It then became linked to Europe and gradually spread beyond the defence community, particularly as people realised that electronic mail could assist groups in many different organisations and countries to communicate effectively. The Internet is a communications medium, and a common set of protocols, which have been unified to form a coherent network.

Two separate developments ensured its rapid and widespread application. Firstly, it was demonstrated how a word in one document, located in one computer connected to the Internet, could be linked electronically to a previously unrelated document (a photograph, for example) stored on another computer, often in a different country. Many millions of these *hypertext* links have now been put in place to better organise, and access, a wide range of different kinds of information. This has resulted in the world wide web (WWW) which uses a global web of linkages between an already huge, but ever growing, collection of information items and databases. Secondly, individuals started sharing information, such as pictures, on the Internet. For example, selecting the key word 'Hubble' would readily access not just data about the space telescope, but also the latest colour images originating directly from those computers controling the telescope and analysing the data it generates. Thus, many individuals first saw the Hubble images of the cometary impact on Jupiter via the Internet.

The Internet is a network of networks, hence the name – *inter-net*work network, and statistics about it are legion. In early 1995 it comprised around 50,000 networks, linking more than 2 million computers and perhaps 10 to 15 million users in many countries around the world. (It is not accurately known what the numbers are, because no single individual owns or controls the Internet.) It is self-regulating, capable of a high degree of rapid evolution and growth, and has been operated up to now by loose federations of individuals and organisations. What is really startling is the 20% per month growth of traffic on the Internet.

At its current rate of expansion, everyone on the planet would be connected to the Internet by 2003. A factor which is contributing to the growth of traffic is that commercial use of the Internet is now beginning to become important as the operation and maintenance of the underlying backbone networks in the US moves into the private sector. In 1994 there were over 21,000 commercial 'domains' registered on the Internet through which companies offer facilities to browse electronic catalogues and order goods and services. These commercial developments will change the fundamental character of the Internet. The Internet will evolve as this new technology matures and is used to provide services for home shopping, video entertainment, electronic banking, etc.

18.8 Summary

LANs, MANs and WANs refer to communications networks typically spanning individual buildings, individual towns, and individual countries, respectively. Such networks may be circuit switched, message switched or packet switched. Two forms of packet switching can be distinguished, i.e. virtual circuit and datagram switching – the former requiring all packets in a given message to traverse the same network route and the latter allowing packets to traverse independent routes. Historically the trend in network operation is from circuit switching (typified by the traditional PSTN), through message switching and virtual circuit switching, to datagram switching.

Network topologies include point-to-point, multidrop, star, ring, mesh and bus systems. Transmission media include twisted wire pairs, coaxial cable and optical fibres. Microwave and infrared links are also used to provide wireless connections between network terminals and nodes.

Bridges are used to interconnect networks using the same protocols (e.g. two similar LANs). Gateways are devices used to interconnect networks using different protocols (e.g. a LAN and a WAN). Switches with self-routing properties can be employed at network nodes to improve switching speed over switches which require centralised control. Packet collisions at switch inputs can be avoided using an input sorting network and collisions at switch outputs avoided using a trap network.

The ISO OSI model for network protocol architectures has seven layers. The lowest three layers (physical, data link and network) specify the operation of the communications sub-net. The upper four layers are concerned with terminal equipment/network compatibility, initiation and control of a communication session, data formatting for correct presentation and the particular application being run by the user.

Medium access control (MAC) is one function of the data link layer in the OSI model. MAC protocols govern the allocation of physical medium resources (time, bandwith, orthogonal codes) between network terminal equipments. These resources can be allocated on a fixed, demand or random assignment basis. Fixed assignment protocols are efficient if data terminal equipments (DTEs) have a heavy, and relatively constant, traffic load. Random access protocols are more efficient if traffic loads are highly variable and uncorrelated between terminals. Random access protocols suffer, however, from potential packet collisions if two or more DTEs attempt to access the medium simultaneously. Such contention can be resolved using CSMA/CD protocols. Fixed assignment systems may be adaptive in that the proportions of medium resource allocated to different terminals may track long term variations in terminal traffic loads. Systems may also be adaptive in the sense that an essentially fixed assignment protocol may be substituted for a (normally) random access protocol if collision detection occurs too frequently. Demand assignment systems allow terminals to commandeer medium resources as and when they are required. They may operate using centralised control, in which terminals are polled to ascertain their need for resources, or distributed control, in which token passing is employed. Bus systems typically use CSMA/CD protocols whilst ring systems typically use token passing.

X.25 is the ITU-T standard defining the interface between DTEs and data communication equipment (DCEs) for packet switched networks. The DCE is the interface with a node of the data network and may be located at the same site as the network node or be remote from it (as in the case of a modem used to connect a computer to the packet switched public data network via an analogue local loop of the PSTN). Other X-series standards specify the protocols for connecting low speed character mode terminals to packet networks using packet assemblers and disassemblers (PADs).

FDDI is a 100 Mbit/s, optical fibre, backbone ring network which supports lower speed tree networks at its nodes. Its maximum circumference is 200 km. Wireless networks, using radio transmission between nodes, are susceptible to frequency selective fading due to multipath propagation. Spread spectrum techniques can be used to mitigate the resulting distortion, allowing transmission at data rates many times that which the channel would ordinarily support.

The Internet is, at present, the ultimate WAN in that it gives, potentially, global coverage. In addition to the conventional data communication uses of a WAN, hypertext connections between different documents held on computers connected to the Internet result in the unparalleled information resource called the World Wide Web.

CHAPTER 19

Public networks and the integrated services digital network (ISDN)

19.1 Introduction

This chapter examines how many of the techniques discussed earlier in the book are applied within the international public communications network. It covers the integrated services digital network (ISDN) [Griffiths], and includes discussion of the older plesiochronous, and newer synchronous, multiplexing techniques. The synchronous digital hierarchy (SDH) [Omidyar and Aldridge] is described, its frame structure and payload capacity defined, and its advantages outlined.

As much of this network now employs wideband fibre optic links, the principles and capabilities of fibre transmission, and optical pulse generation, reception and amplification, are summarised and a typical optical link budget presented. Finally this chapter concludes with a brief account of accessing schemes and on-going developments in the local loop network for digitised speech and data connections using primary, and basic, rate ISDN access.

19.2 The telephone network

The internationally agreed European ITU-T standard for PCM, TDM digital telephony multiplexing (Figure 5.27), is shown in Figure 19.1. Although this shows the basic access level at 144 kbit/s, the multiplexing hierarchy provides for the combining of 32 individual 64 kbit/s channels (Chapter 6) into a composite 2.048 Mbit/s signal. (In the USA the first level in the multiplex combines only 24 channels into a 1.544 Mbit/s data stream.) Multiplexing allocates a complete communications channel to each active user for the duration of his call or connection. In principle, as the channel utilisation factor for voice communications is low, we could concentrate the traffic by switching between users as they require transmission capacity. This resource saving strategy is not used in terrestial telephony but digital speech interpolation (DSI) is employed on international satellite circuits to achieve a significant increase in capacity, (see section 14.3.7).

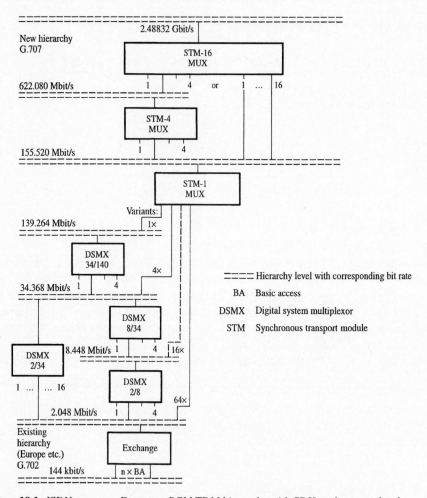

Figure 19.1 *ISDN access to European PCM TDM hierarchy with SDH at the upper levels.*

The ITU-T provides for higher levels of multiplexing, above 2.048 Mbit/s, combining four signals, in the digital system multiplexers (DSMX) 2/8 and 8/34, Figure 19.1, to form the signal at the higher multiplexing level. At each level the bit rate increases by slightly more than a factor of 4 since extra bits are added to provide for frame alignment and to facilitate satisfactory demultiplexing (see section 19.3). The upper levels, beyond 140 Mbit/s, form the synchronous digital hierarchy (SDH), and are only in limited use at present. These will be described later.

The extent of the current digital UK telephone network is shown in Figure 19.2. This is divided into an outer core of main processor exchanges and an inner core of about 55 trunk exchanges (digital main switching centres) which are fully interconnected on 140 Mbit/s, or higher bit rate, transmission links. These are now mainly optical links but also include coaxial, and terrestial microwave relay, paths. On 140 Mbit/s, 1.3 μm, optical fibre links, described later in section 19.5, the repeater spacing is typically 20 km while on microwave radio links, Chapter 14, the spacing is closer to 50 km. International access in Figure 19.2 is provided via satellite links or fibre-optic undersea cables.

Each subscriber telephone has two copper wires that go directly to the local exchange building, Figure 19.3. (The distance is typically 1 to 10 km, being smaller in cities than

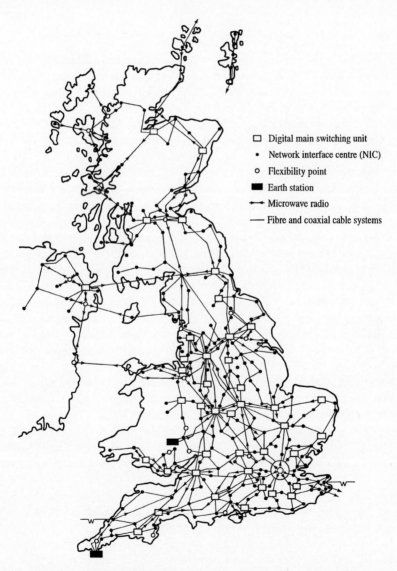

Figure 19.2 *ISDN locations and UK digital trunk telecommunications network (source: Leakey, 1991, reproduced with the permission of British Telecommunications plc.).*

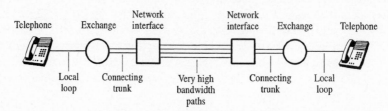

Figure 19.3 *Trunk telecommunications system used for establishing a connection.*

in rural areas where one exchange normally serves 40 km² or 5,000 to 50,000 customers.) The two-wire access connection between each subscriber's telephone and the exchange is known as the *local loop*. Each exchange has a number of outgoing lines to one or more nearby switching centres which provide the network interface. Figure 19.3 shows the typical connection for a long distance (trunk) call involving both inner and outer core connections. In this system there are four wires to the microphone and speaker in the handset, two wires in the local loop and four wires again in the national trunk network. (A four wire circuit implies separate transmit and receive channels as in cellular radio systems, Chapter 15.) In the UK there are approximately 6300 local exchange buildings and the local access network comprises 36 million metallic pair cables.

Traditional control signalling in circuit-switched telephone networks for establishing or initiating call connections may either use a separate communications channel or operate on an in-channel basis. With in-channel signalling, the same channel is used to carry control signals as is used to carry message traffic. Such signalling begins at the originating subscriber and follows the same path as the call itself. This has the merit that no additional transmission facilities are needed for signalling. Two forms of in-channel signalling are in use, in-band and out-of-band. In-band signalling uses not only the same physical path as the call it serves, it also uses the same frequency band as the voice signals that are carried. Out-of-band signalling takes advantage of the fact that voice signals do not use the full 4 kHz of available bandwidth. A separate narrow signalling band, within the 4 kHz, is used to send control signals. A drawback of in-channel signalling schemes is the relatively long delay from the time that a subscriber dials a number to the connection being made.

This problem is addressed by common channel signalling in which control signals are carried by an independent signalling network. Since the control signal bandwidth is small, one separate control signal path can carry the signals for a number of subscriber channels. The common channel uses a signalling protocol, and requires the network architecture to support that protocol, which is more complex than the in-channel signalling case. The control signals are messages that are passed between switches and between a switch and the network management centre.

Over the past decade several different general purpose signalling systems have been developed by ITU and other standards organisations. The most important of these, and the one of major relevance to the ISDN, is the set of procedures known as signalling system No. 7, which is structured in accordance with the ISO OSI model of Figure 18.12. The overall objective is to provide an internationally standardised, general purpose, common channel signalling system which:

- is optimised for use in digital telecommunication networks in conjunction with digital stored program control exchanges utilising 64 kbit/s digital signals;
- is designed to meet present and future information transfer requirements for call control, remote network management, and maintenance;
- provides a reliable means for the transfer of information packets in the correct sequence without loss or duplication;
- is suitable for use on point-to-point terrestrial and satellite links.

19.3 Plesiochronous multiplex

The 2.048 Mbit/s multiplex level (also called the PCM primary multiplex group) in Figures 19.1 and 19.4 comprises frames with 32 8-bit time slots within each frame, Figure 19.5. Time slot zero is reserved for frame alignment and service bits, and time slot 16 is used for multiframe alignment, service bits and signalling. The remaining 30 channels are used for information carrying, or payload capacity, each channel containing one 8-bit voice signal sample.

The system for assembling the TDM telephony data stream assumes that the digital multiplexers (DSMX) in Figure 19.4 are located at physically separate sites which implies separate free running oscillators at each stage in the multiplex hierarchy, hence the name plesiochronous. These oscillators must therefore run at speeds slightly higher than the incoming data, Figure 19.4, to permit local variations to be accommodated. This allows for small errors in the exact data rates in each of the input paths or tributaries but requires some extra bits to be added (i.e. stuffed or justified) to take account of the higher speed oscillator. Elastic stores, Figure 19.6, are used in a typical multiplexer to ensure sufficient bits are always available for transmission or reception. These stores are required because the plesiochronous digital hierarchy (PDH) works by interleaving bytes

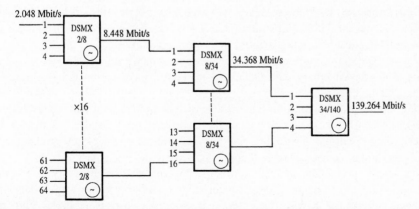

Figure 19.4 *DSMX interconnection to form a plesiochronous multiplex hierarchy.*

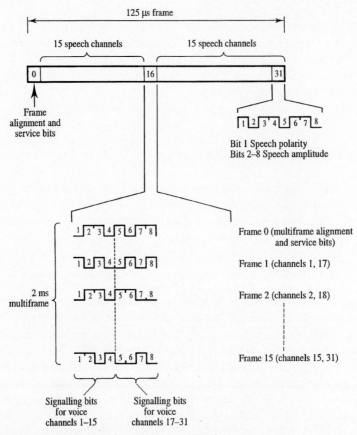

Figure 19.5 *PCM primary multiplex group frame structure.*

or words from each 64 kbit/s tributary, rather than bit interleaving, to form the 2 Mbit/s multiplex. Thus, at the 8 Mbit/s multiplexing level, and above, where bit interleaving is employed, bits must be accumulated for high speed readout.

Figure 19.7 shows more detail of the multiplexer hardware. The code translators in Figure 19.6 convert binary data from, and to, HDB3 (see section 6.4.5). Figure 19.8 shows details of the plesiochronous frame structure at 8 Mbit/s. (Note the bit interleaving, shown explicitly for the justification bits, used at multiplexing levels above 2 Mbit/s.) The ITU-T G series of recommendations (G.702) defines the complete plesiochronous multiplex hierarchy. The frame alignment signal, which is a unique word recognised in the receiver, ensures that the appropriate input tributary is connected to the correct output port. The unique word also permits receiver recovery from loss of synchronisation, if it occurs. The plesiochronous multiplex system was developed at a time when transmission costs were low and switching costs were high. With recent advances in VLSI this premise is now no longer valid, hence the movement to new standards.

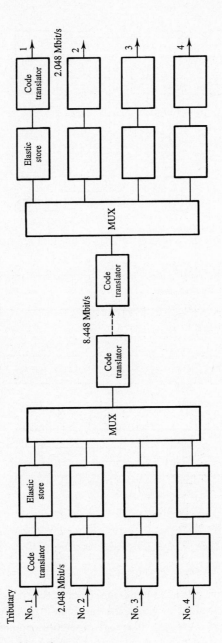

Figure 19.6 *2/8 Multiplexer block diagram.*

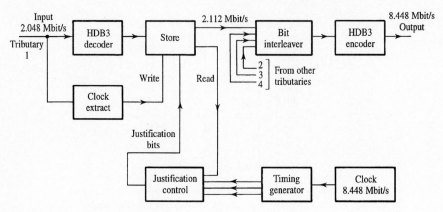

Figure 19.7 *2/8 multiplexer timing details.*

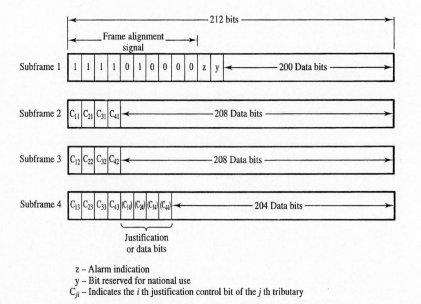

z – Alarm indication
y – Bit reserved for national use
C_{ji} – Indicates the i th justification control bit of the j th tributary

Figure 19.8 *Plesiochronous frame structure.*

A disadvantage of the plesiochronous multiplex is that it was designed for point-to-point transmission applications in which the entire multiplex would be decoded at each end. This is a complicated process since it requires full demultiplexing at each level to recover the bit interleaved data and remove the justification bits. Thus a single 2 Mbit/s channel, for example, cannot be extracted from, or added to, a higher level multiplex signal without demultiplexing down, and then remultiplexing up, through the entire PDH, Figure 19.9. The plesiochronous multiplex does *not* support definitive or clear identification of the various individual channels which are being carried.

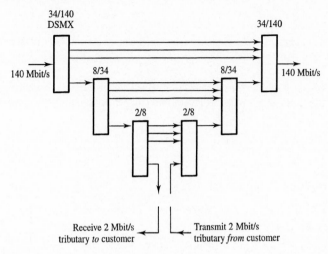

Figure 19.9 *Plesiochronous multiplex add-drop scheme for inserting, and removing, a 2 Mbit/s tributary to, and from, a 140 Mbit/s stream.*

Many high capacity transmission networks are provided by a hierarchy of digital plesiochronous signals. The plesiochronous approach to signal multiplexing is severely limited, however, in its ability to meet the foreseeable requirements of network operators. It does not provide, cost-effectively, the flexible network architecture which is required to respond to the demands of today's evolving telecommunications market. Furthermore, since network management and maintenance strategies were based, historically, on the availability of a manual distribution frame, there was no need to add extra capacity to the plesiochronous frame structure in order to support network operations, administration, maintenance and provisioning (OAM&P) activities. Other PDH drawbacks are the lack of a standard above 140 Mbit/s and the use of different plesiochronous hierarchies in different parts of the world. This leads to problems of interworking between countries whose networks are based on 1.544 Mbit/s (e.g. Japan, North America) and those basing their networks on 2.048 Mbit/s (e.g. Europe, Australia).

The PDH limitations described above, and the desire to move from metallic cable to wideband optical fibres, have been important motivations in the development of the new SONET and synchronous digital hierarchy (SDH) systems which are described in the following sections. These new systems offer improved flexibility and can more readily provide the 2 Mbit/s leased lines which, for example, cellular telephone operators require to connect their base station transmitters to switching centres. (Just prior to the move away from PDH towards SDH systems British Telecom (BT) did develop, in the 1980s, a single 64-channel 2 to 140 Mbit/s multiplexer. This has been included in Figure 19.1 as an input to the new 155.52 Mbit/s standard multiplexer rate.)

EXAMPLE 19.1

Find the number of standard PCM voice signals which can be carried by a PDH level 4 multiplex and estimate the maximum channel utilisation efficiency at this multiplexing level.

PDH Level 1 (primary multiplex group) carries 30 voice signals. Each subsequent level combines four tributaries from previous levels. At level n the number of potential voice signals is therefore given by:

$$\chi = 30 \times 4^{n-1}$$

At level four:

$$\chi = 30 \times 4^{4-1} = 1920 \text{ voice signals}$$

Nominal bit rate at PDH level 4 is 140 Mbit/s. Therefore channel utilisation efficiency is:

$$\eta_{ch} = \frac{\chi}{R_b/R_v} = \frac{1920}{(140 \times 10^6)/(64 \times 10^3)} = 88\%$$

19.4 SONET and SDH

The concept of the digital SONET (synchronous optical network) was initially introduced in the USA in 1986 to establish wideband transmission standards so that international operators could interface using standard frame formats and signalling protocols. The concept also included network flexibility and intelligence, and overhead channels to carry control and performance information between network elements (line systems, multiplexers) and control centres.

In 1988 these concepts were adopted by ITU and ETSI (European Telecommunications Standards Institute) and renamed synchronous digital hierarchy (SDH) with the aim of agreeing worldwide standards for transmission covering optical interfaces, control aspects, equipment, signalling, etc. [Miki and Siller]. Bit transport rates start as low as 52 Mbit/s and go up through 155 Mbit/s with a hierarchy to 622 Mbit/s, 2488 Mbit/s and beyond. The ITU-T G.707/8/9 standards have now reached a mature stage allowing manufacturers to produce common hardware.

19.4.1 Advantages and flexibility

The key advantages of SDH are as follows:
- it is cheaper to add and drop signals to meet customer requirements.
- more bandwidth is available for network management.
- equipment is smaller and cheaper.
- worldwide standards allow a larger manufacturers' marketplace.

- it is easier to introduce new services.
- it is cheaper to achieve remote digital access to services and cross-connections between transmission systems.

Network flexibility implies the ability to rapidly reconfigure networks from a control centre in order to:

- improve capacity utilisation by maximising the number of 2 Mbit/s channels transported in the higher order system;
- improve availability of digital paths by centrally allocating spare capacity and protection schemes to meet service requirements;
- reduce maintenance costs by diverting traffic away from failed network elements;
- provide easier growth with temporary diversion of traffic around areas being upgraded.

Flexibility can be achieved using automatic cross-connect switches between SDH systems or between SDH and plesiochronous systems. Automatic cross-connects will gradually replace existing manual cross-connects and allow remote reconfiguration of capacity within the network at 2 Mbit/s and above. Add-drop multiplexing refers to the ability to extract or insert individual channels without the need to demultiplex the entire high order signal, as required in the plesiochronous system.

19.4.2 Synchronous signal structure

The synchronous signal comprises a set of 8-bit bytes which are carefully interleaved into a frame structure such that the identity of each byte is preserved and known with respect to a framing (or marker) word. The description of the synchronous signal frame structure used here is one in which the bytes of the signal are represented by boxes appearing in rows and columns on a 2-dimensional map, Figure 19.10 [Hawker]. (This is not unlike the masterframe structure of Figure 14.43.)

The signal bits are transmitted in a raster scanned sequence, similar to the lines on a video signal, starting with those in the top left hand byte, followed by those in the 2nd byte in row 1, and so on, until the bits in the Mth (last) byte in row 1 are transmitted. Then the bits in the 1st byte of row 2 are transmitted, followed by the bits in the 2nd byte of row 2, and so on, until the bits in the Mth byte of the Nth (last) row are transmitted.

This transmission sequence repeats, the repetition rate being 8000 frame/s. The duration of each frame is, therefore, 125 μs and the bit rate associated with each byte in the frame is 8×8 kbit/s = 64 kbit/s. Each byte in the frame is thus equivalent in capacity to one PCM voice channel. One or more 8-bit bytes within the synchronous signal structure may be allocated to provide channel capacity for a lower rate (tributary) signal. The fixed sequence frame alignment word allows the positions of all individual data streams within the frame to be identified.

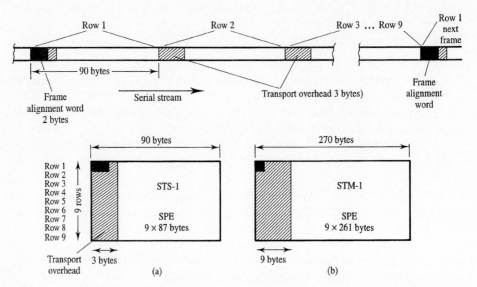

Figure 19.10 *Frame structures showing serial bit stream and the equivalent two dimensional data maps: (a) basic SONET STS-1 frame; and (b) SDH STM-1 frame.*

19.4.3 Frame structure

The lowest level SONET signal is called the synchronous transport signal level 1 (STS-1). It consists of 90 columns and 9 rows of 8-bit bytes giving a total 810 bytes (6480 bits), Figure 19.10(a). With a frame duration of 125 μs, the STS-1 bit rate is 51.84 Mbit/s. The basic SDH frame corresponds to three SONET STS-1 frames and is called a level 1 synchronous transport module (STM-1), Figure 19.10(b). The STM-1 bit rate is therefore 155.52 Mbit/s. Each STS-1 frame can be seen to comprise two parts – an overhead part and a service payload part.

Transport overhead: The first three columns of the STS-1 frame, a total of 27 bytes, Figure 19.10(a), are allocated to overheads that provide operations and maintenance (O&M) facilities. The three bytes in rows 1, 2 and 3 comprise the section overhead while the remaining three bytes in rows 4 to 9 (18 bytes) comprise the line overhead. The section and line terminology is defined in Figure 19.11 for a communications link. (Line terminating equipment might be represented, for example, by an add-drop multiplexer.)

Synchronous payload envelope (SPE): The remaining 87 columns of the STS-1 frame, a total of 783 bytes, provide a channel capacity of 50.112 Mbit/s which supports the transport of the service payload, or traffic data, and also the path overhead, Figure 19.12.

Path overhead: The path overhead supports and maintains the transportation of the SPE between the locations where the SPE is assembled and disassembled. It comprises a total of 9 bytes, Figure 19.12, and is allocated the first column (one byte wide) within the STS-1 SPE.

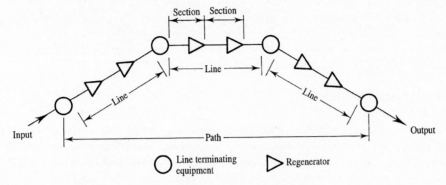

Figure 19.11 *Path, line and section details in a communications link.*

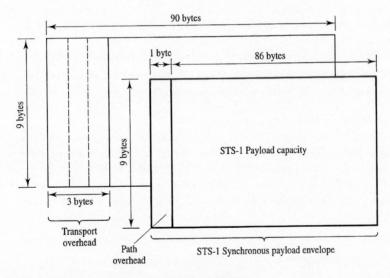

Figure 19.12 *STS-1 synchronous payload envelope (SPE).*

Payload capacity: The STS-1 information capacity comprises the remaining 86 columns of 9 bytes, i.e. a total of 774 bytes providing a 49.536 Mbit/s channel capacity with a frame repetition rate of 8 kHz.

The SONET/SDH frame structure features greatly simplified 'drop and insert' and cross-connect functions by utilising:

- a modular structure made up from SONET/SDH tributaries;
- extensive overheads for monitoring and control;
- byte interleaving with direct visibility of 64 kbit/s channels;
- byte stuffing to improve robustness against loss of synchronisation;

- standard mappings for all common data rates into the SONET/SDH frame format.

The STS-1 SPE represents the unshaded part of Figure 19.10(a). Additional transport capacity is obtained by effectively concatenating SPEs, Figure 19.10(b). Alternatively a reduction in transport capacity may be obtained by partitioning the SPE into smaller segments, called virtual tributaries or VTs (see section 19.4.5).

19.4.4 Payload pointer

To facilitate efficient multiplexing and cross-connection of signals in the synchronous network, the SPE is allowed to 'float' within the payload capacity provided by the STS-1 frames, Figure 19.13. This means that the STS-1 SPE may begin anywhere in the STS-1 frame and is unlikely to be wholly contained in one frame. More likely than not, the STS-1 SPE will begin in one frame and end in the next.

The STS-1 payload pointer, contained in the transport overhead, indicates the location of the first byte of the STS-1 SPE. Byte 1 of the STS-1 SPE is also the first byte of the SPE path overhead. It permits non-synchronous data to be accommodated within the SDH structure, without resorting to justification or bit stuffing.

For synchronous transport, payload pointers provide a means of allowing an SPE to be transferred between network nodes operating plesiochronously. Since the SPE floats freely within the transport frame, with the payload pointer value indicating the location of the first active byte of the SPE, the problem in justified plesiochronous multiplex where the traffic is mixed with stuffing bits is overcome.

Payload pointer processing does, however, introduce a new signal impairment known as 'tributary jitter'. This appears on a received tributary signal after recovery from a synchronous payload envelope which has been subjected to payload pointer movements from frame-to-frame. Excessive tributary jitter will influence the operation of the downstream plesiochronous network equipment processing the tributary signal.

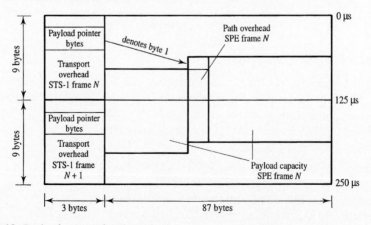

Figure 19.13 *Payload pointer details for locating the start of the SPE.*

19.4.5 Payload capacity partitioning

The virtual tributary (VT) structure (or tributary unit structure in SDH terminology) has been designed to support the transport and switching of payload capacity which is less than that provided by the full STS-1 SPE. There are three sizes of VTs in common use, Figure 19.14. These are:

- VT1.5, consisting of 27 bytes, structured as 3 columns of 9, which, at a frame rate of 8 kHz, provides a transport capacity of 1.728 Mbit/s and will accommodate a US 1.544 Mbit/s DS1 signal.
- VT2, consisting of 36 bytes, structured as 4 columns of 9, which provides a transport capacity of 2.304 Mbit/s and will accommodate a European 2.048 Mbit/s signal.
- VT3, consisting of 54 bytes, structured as 6 columns of 9, to achieve a transport capacity of 3.456 Mbit/s which will accommodate a US DS1C signal.

These and other VTs allow the SDH structure to be electronically reconfigured on demand to handle different customer requirements. It circumvents the fixed nature of the plesiochronous multiplex and, as the VTs are distributed over the entire payload, they can be easily written to, and read from, the system without requiring large data buffer stores.

A 155.52 Mbit/s SDH transmission capability is obtained by combining three STS-1 SPEs into one STM-1 SPE, Figure 19.10(b). Higher order systems are formed by byte interleaving $N \times 155$ Mbit/s channels (e.g. 16×155.5 Mbit/s = 2488 Mbit/s) to form the synchronous transport module level N (e.g. STM-16) rate, Figure 19.1. The SDH frame format allows simultaneous transport of both narrowband (e.g. 2 Mbit/s) and broadband (e.g. 45/140 Mbit/s) services within the 155 Mbit/s capacity system.

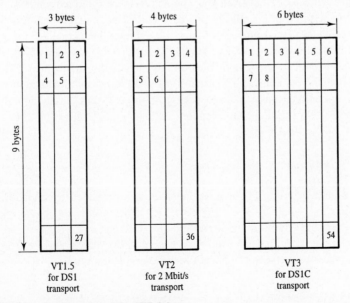

Figure 19.14 *Virtual tributaries of the STS-1 frame.*

EXAMPLE 19.2

Estimate the number of voice channels which can be accommodated by an SDH STM-4 signal assuming that the STM-4 is filled with ITU primary multiplex group signals. Also estimate the channel utilisation efficiency.

Each STS-1 payload envelope has 86 columns (not including the path overhead column). Each primary multiplex group occupies 4 columns. Each STS-1 payload envelope can therefore transport 86/4 = 21 (whole) primary multiplexes.

Each STM-1 payload envelope corresponds to 3 STS-1 payload envelopes and therefore carries $3 \times 21 = 63$ primary multiplexes. STM-4 carries 4 STM-1 signals and therefore carries $4 \times 63 = 252$ primary multiplexes. Each primary multiplex carries 30 voice channels, Figure 19.5. The STM-4 signal can, therefore, carry $30 \times 252 = 7560$ voice channels.

Channel utilisation efficiency, η_{ch}, is thus given by:

$$\eta_{ch} = \frac{7560}{R_b/R_v} = \frac{7560}{(622.08 \times 10^6)/(64 \times 10^3)}$$

$$= \frac{7560}{9720} = 78\%$$

19.5 Fibre optic transmission links

Optical fibres comprise a core, cladding and protective cover and are much lighter than metallic cables [Gowar]. This advantage, coupled with the rapid reduction in propagation loss to its current value of 0.2 dB/km or less, Figure 1.6(e), and the enormous potential bandwidth available, make optical fibre now the only serious contender for the majority of long-haul trunk transmission links. The potential capacity of optical fibres is such that all the radar, navigation and communication signals in the microwave and millimetre wave region, which now exist as free space signals, could be accommodated within 1% of the potential operational bandwidth of a single fibre. Current commercial systems can accommodate 31,000 simultaneous telephone calls in a single fibre and 1M call capacity has been achieved in the laboratory.

19.5.1 Fibre types

In the fibre a circular core of refractive index n_1 is surrounded by a cladding layer of refractive index n_2, where $n_2 < n_1$, Figure 19.15. This results in optical energy being guided along the core with minimal losses.

The size of the core and the nature of the refractive index change from n_1 to n_2 determine the three basic types of optical fibre [Gowar], namely:

- multimode step-index;
- multimode graded-index;

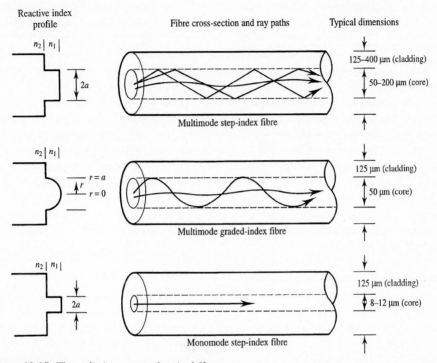

Reactive index profile

Fibre cross-section and ray paths

Typical dimensions

n_2 | n_1 |

$2a$

125–400 μm (cladding)

50–200 μm (core)

Multimode step-index fibre

n_2 | n_1 |

$r = a$
r
$r = 0$

125 μm (cladding)

50 μm (core)

Multimode graded-index fibre

n_2 | n_1 |

$2a$

125 μm (cladding)

8–12 μm (core)

Monomode step-index fibre

Figure 19.15 *Three distinct types of optical fibre.*

- monomode step-index.

The refractive index profiles, typical core and cladding layer diameters and a schematic ray diagram representing the distinct optical modes in these three fibre types are shown in Figure 19.15. The optical wave propagates down the fibre via reflections at the refractive index boundary or refraction in the core. In multimode fibres there was, originally, a 20% difference between the refractive indices of the core and the cladding. In the more recently developed monomode fibres, the difference is much smaller, typically 0.5%. The early multimode step index fibres were cheap to fabricate but they had limited bandwidth and a limited section length between repeaters. In a 1 km length of modern multimode fibre with a 1% difference in refractive index between core and cladding, pulse broadening or dispersion, caused by the difference in propagation velocity between the different electromagnetic modes, limits the maximum data rate to, typically, 10 Mbit/s. Graded index fibres suffer less from mode dispersion because the ray paths representing differing modes encounter material with differing refractive index. Since propagation velocity is higher in material with lower refractive index, the propagation delay for all the modes can, with careful design of the graded-n profile, be made approximately the same. For obvious reasons *modal* dispersion is absent from monomode fibres altogether and in these fibre types *material* dispersion is normally the dominant pulse spreading mechanism. Material dispersion occurs because refractive index is, generally, a function of wavelength, and different frequency components therefore propagate with different velocities. (Material dispersion is exacerbated due to the fact that practical sources often emit light with a narrow, but not monochromatic, spectrum.) The rate of change of propagation velocity with frequency (*dv/df*), and therefore dispersion, in silica fibres changes sign at around 1.3 μm, Figure 19.16, resulting in zero material dispersion at this wavelength. (Fortuitously, this wavelength also corresponds to a local minimum in optical attenuation, see Figure 1.6(e).)

If both modal and material dispersion are zero, or very small, *waveguide* dispersion, which is generally the weakest of the dispersion mechanisms, may become significant. Waveguide dispersion arises because the velocity of a waveguide mode depends on the normalised dimensions (*d/λ*) of the waveguide supporting it. Since the different frequency components in the transmitted pulse have different wavelengths these components will travel at different velocities even though they exist as the same electromagnetic mode. Because both material and waveguide dispersion relate to changes in propagation velocity with wavelength they are sometimes referred to collectively as chromatic dispersion. The various fibre types are further defined in ITU-T recommendation G.652.

19.5.2 Fibre transmission systems

There have been three generations of optical fibre systems operating at 0.85 μm, 1.3 μm and 1.5 μm wavelengths to progressively exploit lower optical attenuation, Figure 1.6(e), and permit longer distances to be achieved between repeaters. Monomode fibres, with core diameters in the range 8 to 12 μm, have been designed at the two longer wavelengths for second and third generation systems. Since material dispersion is zero at around 1.3

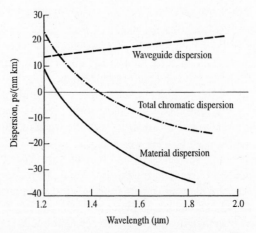

Figure 19.16 *Variation of material dispersion and waveguide dispersion, giving zero total dispersion near λ = 1.5 μm (source, Flood and Cochrane, 1989, reproduced with the permission of Peter Peregrinus).*

μm, where the optical attenuation in silica is also a local minimum, this was the wavelength chosen for second generation systems, which typically operate at 280 Mbit/s.

Third generation systems operate at wavelengths around 1.5 μm and bit rates of 622 Mbit/s to exploit the lowest optical attenuation value of 0.15 dB/km, and tolerate the resulting increased chromatic dispersion which in practice may be 15 to 20 ps nm^{-1}km^{-1}. Thus, the choice at present between 1.3 and 1.5 μm wavelength depends on whether one wants to maximise link repeater spacing or signalling bandwidth.

Figure 19.16 shows that if core dimensions, and core cladding refractive indices, are chosen correctly, however, material dispersion can be cancelled by waveguide dispersion at about 1.5 μm resulting in very low total chromatic dispersion in this lowest attenuation band.

The impact of fibre developments is clearly seen in Figures 1.7 and 19.17. The latter illustrates the evolution of transmission technology for a 100 km wideband link. The coaxial cable used in the 1970s with its associated 50 repeaters had a mean time between failures (MTBF) of 0.4 years, which was much lower than the 2 year MTBF of the plesiochronous multiplex equipment at each terminal station. Multi-mode fibre (MMF) still needed a repeater every 2 km but was a more reliable transmission medium. The real breakthrough came with single mode fibre (SMF) and, with the low optical attenuation at 1.5 μm, there is now no need for any repeaters on a 100 km link, in which the fibre path loss is typically 10 to 28 dB. (This is *very* much lower than the microwave systems of section 12.4.2.) By 1991 there were 1.5 million km of installed optical fibre carrying 80% of the UK telephone traffic. This UK investment represented 20% of the world transmission capability installed in optical fibre at that time. Optical fibre transmission capacity doubles each year with an exponentially reducing cost.

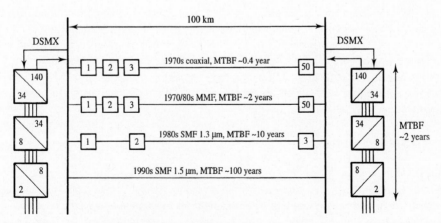

Figure 19.17 *Evolution of a 100 km link from coaxial to optical transmission (source: Cochrane, 1990, reproduced with the permission of British Telecommunications plc.).*

19.5.3 Optical sources

Two devices are commonly used to generate light for fibre optic communications systems: light-emitting diodes (LEDs) and injection laser diodes (ILDs). The edge emitting LED is a PN junction diode made from a semiconductor material such as aluminium–gallium–arsenide (AlGaAs) or gallium–arsenide–phosphide (GaAsP). The wavelength of light emitted is typically 0.94 μm and output power is approximately 3 mW at 100 mA of forward diode current. The primary disadvantage of this type of LED is the non-directionality of its light emission which makes it a poor choice as a light source for fibre optic systems. The planar hetero-junction LED generates a more brilliant light spot which is easier to couple into the fibre. It can also be switched at higher speeds to accommodate wider signal bandwidth.

The injection laser diode (ILD) is similar to the LED but, above the threshold current, an ILD oscillates and lasing occurs. The construction of the ILD is similar to that of an LED, except that the ends are highly polished. The mirror-like ends trap photons in the active region which, as they are reflected back and forth, stimulate free electrons to recombine with holes at a higher-than-normal energy level to achieve the lasing process. ILDs are particularly effective because the optical radiation is easy to couple into the fibre. Also the ILD is powerful, typically giving 10 dB more output power than the equivalent LED, thus permitting operation over longer distances. Finally, ILDs generate close to monochromatic light, which is especially desirable for single mode fibres operating at high bit rates.

19.5.4 Optical detectors

There are two devices that are commonly used to detect light energy in fibre optic systems: PIN (positive–intrinsic–negative) diodes and APDs (avalanche photodiodes). In the PIN diode, the most common device, light falls on the intrinsic material and photons are absorbed by electrons, generating electron-hole pairs which are swept out of the device by the applied electric field. The APD is a positive–intrinsic–positive–negative structure, which operates just below its avalanche breakdown voltage to achieve an internal gain. Consequently, APDs are more sensitive than PIN diodes, each photon typically producing 100 electrons, their outputs therefore requiring less additional amplification.

19.5.5 Optical amplifiers

In many optical systems it is necessary to amplify the light signal to compensate for fibre losses. Light can be detected, converted to an electrical signal and then amplified conventionally before remodulating the semiconductor source for the next stage of the communications link. Optical amplifiers, based on semiconductor or fibre elements employing both linear and non-linear devices, are much more attractive and reliable; they permit a range of optical signals (at different wavelengths) to be amplified simultaneously and are especially significant for sub-marine cable systems.

Basic travelling wave semiconductor laser amplifier (TWSLA) gains are typically in the range 10 to 15 dB, Figure 19.18. These Fabry-Perot lasers are multimode in operation and are used in medium distance systems. Single mode operation is possible with distributed feedback (DFB) lasers for longer distance, high bit rate, systems. In common with all optical amplifiers, the TWSLA generates spontaneous emissions which results in an optical (noise) output in the absence of an input signal. For a system with cascaded TWSLAs these noise terms can accumulate.

Recent research has resulted in fibre amplifiers consisting of 10 m to 50 km of doped or undoped fibre. These amplifiers use either a linear, rare earth (erbium), doping mechanism or the non-linear Raman/Brillouin mechanism [Cochrane *et al.* 1990]. The erbium doped fibre amplifier (EDFA) uses a relatively short section (1 to 100 m) of silica fibre pumped with optical, rather than electrical, energy. Because of the efficient coupling of fibre-to-fibre splices, high gains (20 dB) are achievable, Figure 19.18, over a 30 to 50 nm optical bandwidth. Practical amplifier designs generally have gains of 10 to 15 dB. The key attraction of this amplifier is the excellent end-to-end link SNR which is achievable and the enormous 4 to 7 THz (Hz $\times 10^{12}$) of optical bandwidth. (This far exceeds the 300 GHz of the entire radio, microwave and millimeter-wave spectrum.) Two interesting features of these amplifiers are the precise definition of the operating wavelength via the erbium doping and their relative immunity to signal dependent effects. They can therefore be engineered to maintain wide bandwidths when cascaded, and can operate equally well with OOK, FSK or PSK signals.

Injecting a high power laser beam into an optical fibre results in Raman scattering. Introducing a lower intensity signal-bearing beam into the same fibre with the pump energy, results in its amplification with gains of approximately 15 dB per W of pump power, coupled with bandwidths that are slightly smaller than the erbium amplifiers, Figure 19.18.

Brillouin scattering is a very efficient non-linear amplification mechanism that can realise high gains with modest optical pump powers (approximately 1 mW). However,

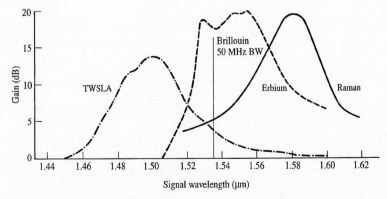

Figure 19.18 *Comparison of the gain of four distinct optical amplifier types (source: Cochrane, 1990, reproduced with the permission of British Telecommunications plc.).*

the bandwidth of only 50 MHz, Figure 19.18, is very limited in pure silica which makes such devices more applicable as narrowband tunable filters. This limited bandwidth fundamentally restricts the brillouin amplifier to relatively low bit rate communication systems.

A comprehensive comparison of the features of optical amplifiers is premature. However, what is clear is that long-haul optical transmission systems (10,000 km) are now feasible, with fibre amplifier based repeaters, and bit rates of 2 to 10 Gbit/s. At the present time TWSLA and erbium amplifiers generally require similar electrical pump power but TWSLAs achieve less gain, Figure 19.18, due to coupling losses. Erbium fibre amplifiers have the lowest noise and WDM channel crosstalk performance of all the amplifier types reported but high splice reflectivities can cause these amplifiers to enter the lasing condition. With the exception of brillouin, these amplifiers can be expected to be used across a broad range of system applications including transmitter power amplifiers, receiver preamplifiers, in-line repeater amplifiers and switches. The key advantage of EDFAs is that the lack of conversion to, and from, electrical signals for amplification gives rise to the 'dark fibre' – a highly reliable data super highway operating at tens of Gbit/s.

19.5.6 Optical repeater and link budgets

The electro-optic repeater is similar to the metallic line regenerator of Figures 6.15 and 6.30. For monomode fibre systems the light emitter is a laser diode and the detector uses an APD. The symbol timing recovery circuit uses zero crossing detection of the equalised received signal followed by pulse regeneration and filtering to generate the necessary sampling signals. Due to the high data rates, the filters often use surface acoustic wave devices [Matthews] exploiting their high frequency operation combined with acceptable Q value. The received SNR is given by:

$$\frac{S}{N} = \frac{I_p^2}{2q_e B(I_p + I_D) + I_n^2} \tag{19.1}$$

where I_p is the photodetector current, I_D is the leakage current, q_e is the charge on an electron, B is bandwidth and I_n is the RMS thermal noise current given by:

$$I_n^2 = \frac{4kTB}{R_L} \tag{19.2}$$

For received power levels of −30 dBm, $I_n^2 \gg 2q_e B(I_p + I_D)$ giving, typically, a 15 to 20 dB SNR and hence an OOK BER in the range 10^{-7} to $< 10^{-10}$ [Alexander].

There are many current developments concerned with realising monolithic integrated electro-optic receivers. Integrated receivers can operate with sensitivities of −20 to −30 dBm and a BER of 10^{-10} at 155, 625 and 2488 Mbit/s, with a 20 dB optical overload capability. They are optimised for low crosstalk with other multiplex channels.

The power budget for a typical link in a fibre transmission system might have a transmitted power of 3 dBm, a 60 km path loss of 28 dB, a 1 dB path dispersion allowance and 4 dB system margin to give a −30 dBm received signal level which is consistent with a low cost 155 Mbit/s transmission rate. The higher rate of 2.5 Gbit/s, with a similar receiver sensitivity of −30 dBm, would necessitate superior optical interfaces on such a 60 km link.

Long haul experiments and trials have achieved bit rates from 140 Mbit/s to 10 Gbit/s using up to 12 cascaded amplifiers spanning approximately 1000 km. BERs of 10^{-4} to 10^{-8} have been measured on a 500 km, five amplifier system, operating at 565 Mbit/s in which the individual amplifier gains were 7 to 12 dB [Cochrane *et al.* 1993].

EXAMPLE 19.3

A monomode, 1.3 μm, optical fibre communications system has the following specification:

Optical output of transmitter, P_T	0.0 dBm
Connector loss at transmitter, L_T	2.0 dB
Fibre specific attenuation, γ	0.6 dB/km
Average fibre splice (joint) loss, L_S	0.2 dB
Fibre lengths, d	2.0 km
Connector loss at receiver, L_R	1.0 dB
Design margin (including dispersion allowance), M	5.0 dB
Required optical carrier power at receiver, C	−30 dBm

Find the maximum loss-limited link length which can be operated without repeaters.

Let estimated loss-limited link length be D' km and assume, initially, that the splice loss is distributed over the entire fibre length.

$$\text{Total loss} = L_T + (D' \times \gamma) + \left(\frac{D'}{d} \times L_S\right) + L_R + M$$

$$= 2.0 + 0.6D' + \left(\frac{0.2}{2.0} \times D'\right) + 1.0 + 5.0$$

$$= 8.0 + 0.7D' \ \text{dB}$$

$$\text{Allowed loss} = P_T - C = 0.0 - (-30) = 30.0 \ \text{dB}$$

Therefore:

$$8.0 + 0.7D' = 30$$

$$D' = \frac{30.0 - 8.0}{0.7} = 31.4 \ \text{km}$$

The assumption of distributed splice loss means that this loss has been over estimated by:

$$\Delta L_S = L_s [\ D' - int\ (D'/d)d\ ]\ /\ 2$$

$$= 0.2\ [\ 31.4 - int\ (31.4/2)2\ ]\ /\ 2$$

$$= 0.14 \ \text{dB}$$

This excess loss can be reallocated to fibre specific attenuation allowing the link length to be extended by:

$$\Delta D = \Delta L_S/\gamma = 0.14/0.6 = 0.2 \text{ km}$$

The maximum link length, D, therefore becomes:

$$D = D' + \Delta D = 31.4 + 0.2 = 31.6 \text{ km}$$

19.5.7 Optical FDM

With the theoretical 50 THz of available bandwidth in an optical fibre transmission system, and with the modest linewidth of modern optical sources, it is now possible to implement optical FDM and transmit multiple optical carriers along a single fibre. The optical carriers might typically be spaced by 1 nm wavelengths. With the aid of optical filters these signals can be separated in the receiver to realise wavelength division multiplex (WDM) communications [Oliphant *et al.*]. It is envisaged that eventually up to 100 separate channels could be accommodated using this technique but the insertion loss of the multiplexers and crosstalk between channels still needs to be assessed. It has been demonstrated that 10 such combined signals, each modulated at 10 Gbit/s, can be transmitted through a practical fibre, and amplified using a single fibre amplifier without having to demultiplex the signals. WDM promises to increase by a hundredfold the information carrying capacity of fibre based systems when the necessary components for modulators and demodulators are fully developed.

Soliton transmission uses pulses that retain their shape for path lengths of thousands of kilometres due to the reciprocal effects of chromatic dispersion and a refractive index which is a function of intensity. Such systems have been constructed for 1,000 km paths with bit rates of 10 to 50 Gbit/s. In the laboratory, 10^6 km recirculating links have been demonstrated, corresponding to many circulations of the earth before the received SNR is unacceptable [Cochrane *et al.* 1993].

19.6 Network advantages of SDH systems

In the current plesiochronous hierarchy, within the transport signal at 140 Mbit/s, we may want to route component signals, for example 2 Mbit/s streams or tributaries, through the network. This requires us to demultiplex the transport signals layer-by-layer through the hierarchy, switch the tributary signals, and then remultiplex them into the next transport signal, Figure 19.9.

In the SDH, individual component signals do not have to be demultiplexed to their original bit rate; instead they are incorporated into a signal called a 'container' which can be handled in a convenient way throughout the network, Figure 19.14. Direct access to these component signals is thus possible, Figure 19.19. The result is a considerable reduction in multiplexer hardware in SDH systems, combined with improved operational flexibility.

With the introduction of SDH the opportunity can now be taken to replace network layers and topologies with those better suited to long haul resilient networks. With the

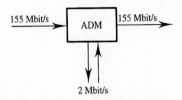

Figure 19.19 *Add-drop multiplexer (ADM) for simplified channel dropping which permits multiplexing into ring networks. (Compare with Figure 19.9.)*

availability of, very high speed, flexible SDH links it now becomes economic to reconsider the structure of the PSTN and replace simple (multiple) two-way transmission paths between the major centres, in which terminal multiplexers are two port or tributary-line systems, by high speed optical rings, as illustrated in Figure 19.20. The three port (input line, output line and tributary) add-drop multiplexers (ADMs), Figure 19.19, provide ring access and egress. This structure (Figure 19.20) will form the heart of the PDN and is fundamentally more reliable and less costly than the previous solution. Figure 19.20 represents an enhanced version of Figure 18.14. When the rings incorporate independent clockwise and anti-clockwise transmission circuits, as in the FDDI example described in Chapter 18, they offer immense flexibility and redundancy allowing information to be transmitted, via the ADM, in either direction around the ring to its intended destination.

Outer-core topologies will be mainly rings of SDH multiplexers linking local exchanges, in contrast to a plesiochronous multiplex where all traffic is routed through a central site. The SDH ring structure with its clockwise and anti-clockwise routing is much more reliable than centre-site routing. Furthermore, the SDH ring based network has less interface, and other, equipment. Access regions will remain, principally, star topologies, but will probably be implemented using optical technology, Figure 19.21.

19.7 Data access

19.7.1 ISDN data access

ISDN digital access was opened in the UK in 1985 and provided basic rate access at 144 kbit/s and primary rate access at 2 Mbit/s to the multiplex hierarchical structure of Figure 19.1. The customer interface for primary rate access provides for up to 30 PCM communications channels (e.g. a PABX) under the control of a common signalling channel.

Basic rate access provides the customer with two independent communications channels at 64 kbit/s together with a common signalling channel at 16 kbit/s. International standards for ISDN access (I.420) have now been agreed within ITU. Communications or bearer channels (B-channels) operate at 64 kbit/s, whilst the signalling or data channel (D-channel) operates at 16 kbit/s giving the basic rate total of

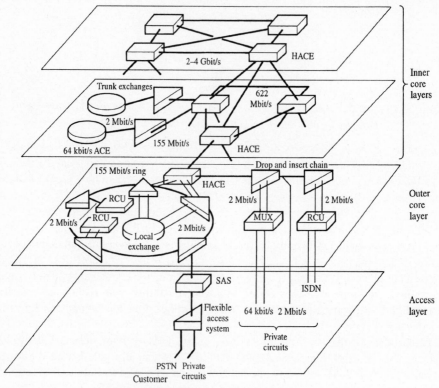

HACE: Higher-order automatic cross-connect equipment
SAS: Service access switch
ACE: Automatic cross-connect equipment
RCU: Remote concentrator unit

Figure 19.20 *Probable future SDH transmission network hierarchy (source: Leakey, 1991, reproduced with the permission of British Telecommunications plc.).*

144 kbit/s. The basic rate voice/data terminal transmits, full duplex, over a two-wire link with a reach of up to 2 km using standard telephone local loop copper cables. The transceiver integrated circuits employ a 256 kbaud, modified DPSK, burst modulation technique, Chapter 11, to minimise RFI/EMI and crosstalk. The D channel is used for signalling to establish (initiate) and disestablish (terminate) calls via standard protocols. During a call there is no signalling information and hence the D channel is available for packet switched data transmission.

Data access at 64 kbit/s is used in low bit rate image coders for videophone applications, Chapter 16. For high quality two way confravision services with full TV (512 × 512 pixel) resolution reduced bit rate coders have been designed to use primary access at 2 Mbit/s. Access at 2 Mbit/s is also required to implement wide area networks, Chapter 18, or to carry cellular telephone traffic between cell sites, Chapter 15.

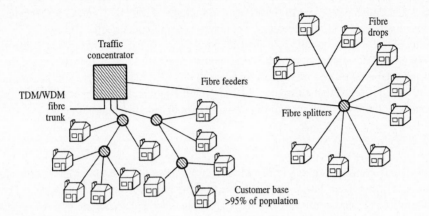

Figure 19.21 *Passive optical network for future local loop implementation.*

The layer 1 specification based on ITU-T I series recommendations [Fogarty] defines the physical characteristics of the user-network interface, Figure 19.22. The NT1 (network termination 1) terminates the transmission system and includes functions which are basically equivalent to layer 1 of the OSI architecture, Figure 18.12. The terminal equipment (TE) includes the ITU-T NT2 (network termination 2) function which

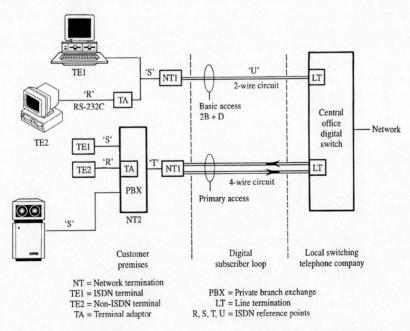

Figure 19.22 *Basic rate (144 kbit/s) and primary rate (2 Mbit/s) access to the ISDN.*

terminates ISO layers 1 to 3 of the interface. Figure 19.22 illustrates how digitised speech and data have access to the ISDN and includes the R, S, T and U ISDN reference points, which can be interconnected by standard interface integrated circuits.

Examples of TE1 equipments are ISDN telephone or fax machines which use the S interface. A TE2 might be a V.24 (RS232) data terminal or computer which requires the terminal adaptor (TA) to interface with the ISDN. The TA performs the processing to establish and disestablish calls over the ISDN and handles the higher level OSI protocol processing. For computer connection the TA is usually incorporated in the PC.

19.7.2 STM and ATM

Synchronous transfer mode (STM) and asynchronous transfer mode (ATM) both refer to techniques which deal with the allocation of usable bandwidth to user services. STM as used here is not to be confused with the synchronous transport module defined in section 19.4.3. In a digital voice network, STM allocates information blocks, at regular intervals (bytes/125 μs). Each STM channel is identified by the position of its time slots within the frame, Figure 19.5, which could easily be extended to a 30 channel, 2 Mbit/s system. STM works best when the network is handling a single service, such as the continuous bit-rate (bandwidth) requirements of voice, or a limited heterogeneous mix of services at fixed channel rates. Now, however, a dynamically changing mix of services requires a much broader range of bandwidths, and a switching capability adequate for both continuous traffic (such as voice and video conferencing) and non-continuous traffic (such as high-speed data and coded video traffic) in which bandwidth may change with time depending on the information rate.

While STM can provide data transfer services, the network operator must supply, and charge for, facilities with the full bandwidth needed by each service for 100% of the time – even if users require the peak bandwidth for only a small fraction of the time. The network therefore operates at low efficiency and the cost to users is prohibitive. This is not an attractive option for variable bit rate (VBR) traffic, such as coded video transmission, in which the data rate is dependent on how fast the image is changing.

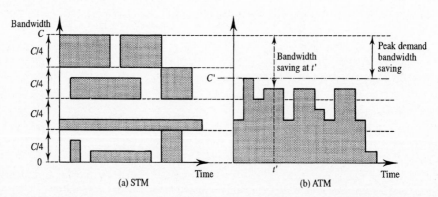

Figure 19.23 *Example of (a) STM and (b) ATM with mix of fixed and variable bit rate traffic.*

Figure 16.19 shows an example of such traffic for which the average data rate is very much smaller than the peak rate. When many VBR sources are averaged then the peak transmission rate requirement comes much closer to the average rate.

The structure of the ATM protocol improves on the limited flexibility of STM by sharing both bandwidth and time [de Prycker]. Instead of breaking down bandwidth into fixed, predetermined, channels to carry information, as shown schematically for STM in Figure 19.23(a), ATM transfers fixed-size blocks of information, called cells, whenever a service requires transmission bandwidth, in a manner analogous to a packet switched system, Chapter 17. Figure 19.23(b) illustrates the (statistical) multiplexing of the ATM traffic. When the transmission bandwidth is dynamically allocated to variable bit rate users, there is a consequent peak bandwidth saving ($C' < C$), provided that the packet overhead is small. ATM uses a 5 byte header combined with a 48 byte data packet representing a compromise between the optimum data packet lengths of 16 bytes, for audio, and 128 bytes for video traffic. (The 53 bit ATM cell also represents, more generally, a compromise between short cell lengths required for real time, low delay, traffic applications such as packet speech, and long cell lengths which are more efficient for data applications due to their reduced proportion of overhead bits.) CRC, section 10.8.1, is implemented on the header information only using the polynomial:

$$M(x) = x^8 + x^2 + x^1 + 1 \qquad\qquad (19.3)$$

to achieve a single error correction, but multiple error detection, capability. In comparison to other packet data networks (e.g. packet speech) ATM achieves high speed (149.76 Mbit/s corresponding to the payload bit rate of the SDH STM-1) by using hardware VLSI chips, rather than software, for the protocol processing and switching at network nodes.

One of the first services to use an ATM network for efficient data transport between LANs and PCs was a high-speed switched data service which carried VBR traffic. The ATM layer allows much higher data transfer rates than is possible with X.25 packet-switched networks (Chapter 18). As a result of ATM's remarkable flexibility and efficiency, end users can enjoy very-high-bandwidth services. These services can be carried over long distances by the PSTN with, potentially, very attractive tariffs. When the power of ATM is joined with the bandwidth and transmission quality of ISDN, a much enhanced service is achieved.

The term asynchronous in ATM refers to the fact that cells allocated to the same connection may exhibit an irregular occurrence pattern, as cells are filled according to the actual demand. This is an unfortunate term as it implies that ATM is an asynchronous transmission technique which is not the case. ATM facilitates the introduction of new services in a gradual and flexible manner, and is now the preferred access technique to the ISDN, for packet speech, video and other variable bit rate traffic. Queuing theory, Chapter 17, allows the analysis of ATM cell throughput rates and losses. The routing switches described in Chapter 18 are used for the ATM interfaces. ATM can fit seamlessly into the SDH frames of Figures 19.12 and 19.13 by accommodating the cells directly into the SDH payload envelope.

19.7.3 The local loop

Half of the investment of a telephone company is in the connections between subscriber handsets and their local exchange. Furthermore, this part of the network generates the least revenue since local calls are often cheap or, as in the USA, free. The length of these connections is 2 km on average and they seldom exceed 7 km. In rural areas the expense of installing copper connections now favours radio access for the local loop implementation. ISDN access demands 144 kbit/s duplex operation over 4 to 5 km which requires sophisticated signal processing if copper pair cables are used.

Assuming that the local loop is to be used for speech telephony and low speed data connections only, its replacement by fibre systems will be very gradual. If, however, we were to combine this requirement with cable TV and many other broadband services such as videophone, Chapter 16, then there would be an immediate demand for wideband local loop connections.

There will thus be a progressive move from copper based conductors to a fibre based passive optical network (PON) for the local connection. This will not be based on the current structure of one dedicated wire pair, or fibre, per household because of the large fibre bandwidth, section 19.5. (Furthermore, with 750 million telephones worldwide it would take more than 300 years for manufacturers to produce all the required cable at current production rates!) The future local network is therefore likely to comprise wideband fibre feeders with splitters and subsequent single fibre drops to each household, Figure 19.21. (One problem with the PON configuration is that there is no longer a copper connection to carry the power required for standby telephone operation.)

19.8 Summary

Multiplexing of PCM–TDM telephone traffic has been traditionally provided using the plesiochronous digital hierarchy. The PDH frame rate is 8000 frame/s and its multiplexing levels, bit rates and constituent signals are as follows:

PDH-1	2.048 Mbit/s	30+2, byte interleaved, 64 kbit/s voice channels – the PCM primary multiplex group
PDH-2	8.448 Mbit/s	4 bit-interleaved PDH-1 signals
PDH-3	34.368 Mbit/s	4 bit-interleaved PDH-2 signals
PDH-4	139.264 Mbit/s	4 bit-interleaved PDH-3 signals

Bit rates increase by a little more than a factor of four at each successive PDH level to allow for small differences in multiplexer clock speeds. Empty slots in a multiplexer output are filled with justification bits as necessary. A serious disadvantage of PDH multiplexing is the multiplex mountain which must be scaled each time a lower level signal is added to, or dropped from, a higher order signal. This is necessary because bit interleaving combined with the presence of justification bits means that complete demultiplexing is required in order to identify the bytes belonging to a given set of voice channels.

The synchronous digital hierarchy (SDH) and its originating North American equivalent, SONET, will eventually replace the PDH. The principal advantage of SDH is that low level signals remain visible in the multiplexing frame structure. This allows lower level multiplexes (down to individual voice channels) to be added or dropped from higher order multiplex signals without demultiplexing the entire frame. This simplifies cross-connection of traffic from one signal multiplex to another. The SDH frame rate is 8000 frame/s and each frame contains one or more synchronous transport modules (STMs). The SONET frame rate is also 8000 frame/s and each contains one or more synchronous transport signals (STSs). The standard SONET/STM payload capacities, bit rates, and constituent signals are as follows:

SONET	STS-1	9×90 bytes	51.84 Mbit/s	
SDH	STM-1	9×270 bytes	155.52 Mbit/s	3 STS-1s
SDH	STM-4	9×1080 bytes	622.08 Mbit/s	4 STM-1s
SDH	STM-16	9×4320 bytes	2.488 Gbit/s	4 STM-4s

SONET signals with capacities based on other multiples of STS-1 are also possible. The capacity of an STM-1 is such that it can carry one PDH-4 signal which will facilitate the operation of PSTNs during the period in which both multiplexing schemes are in use.

SONET and SDH have been designed, primarily, to operate with optical fibre transmission systems. Optical sources may be coherent [Hooijmans] or incoherent. Laser diodes have narrow spectra and are therefore usually the choice for high performance, high bit rate links. LEDs are less spectrally pure, leading to greater dispersion and smaller useful bandwidth, but are cheaper. Optical detectors typically contain PIN diodes or avalanche photodiodes (APDs). APDs are more expensive but more sensitive. Typical optical transmit powers are 0 dBm and typical optical receiver powers are −30 dBm. An optical fibre repeater may be implemented using an optical detector, a conventional electronic repeater and an optical source. Optical amplifiers (e.g. TWSLAs) may also be used. The most recent types of optical amplifier are distributed and consist of doped fibre sections which are optically pumped and lase as the optical signal propagates through them.

Fibres currently operate, in decreasing order of attenuation, at wavelengths of 0.85, 1.3 and 1.5 μm. They can be divided into three types depending on the profile of their refractive index variations. Multimode step-index fibres have relatively large core dimensions and suffer from modal dispersion which limits their useful bandwidth. Multimode graded-index fibres also have large core dimensions but use their variation in refractive index to offset the difference in propagation velocity between modes resulting in larger available bandwidth. Monomode (step-index) fibres suffer only chromatic dispersion which arises partly from the frequency dependence of refractive index (material dispersion) and partly from the frequency dependence of a propagating electromagnetic mode velocity in a waveguide of fixed dimensions. These two contributions can be made to cancel at an operating wavelength around 1.5 μm, however, which also corresponds to a minimum in optical attenuation of about 0.15 dB/km. This wavelength is therefore an excellent choice for long, low dispersion, high bit rate, links. Repeaterless links, hundreds of kilometres long, operating at Gbit/s data rates are now possible. Wavelength division multiplexing (WDM) promises to increase the communications capacity of a single fibre still further.

SDH, combined with optical fibre transmission, has allowed a re-evaluation of the PDN network topology. Future access is likely to be via a passive optical network (PON) at ISDN basic (144 kbit/s) or primary (2 Mbit/s) rates. 155 Mbit/s rings will connect to a network of higher-order automatic cross-connect equipment (HACE) which will themselves be interconnected at bit rates of 155 and 622 Mbit/s and higher. The highest layer of cross-connect equipment will be fully interconnected with 2.4 Gbit/s links.

Asynchronous transfer mode (ATM) will provide efficient use of time and bandwidth resources in the access layers of the network when variable bit rate services are provided. ATM frames consist of 53 byte cells, 48 of which carry traffic and 5 of which carry overhead. At the higher network layers ATM cells will be carried within the payload envelopes of SDH frames.

19.9 Problems

19.1. How does a plesiochronous multiplex function?

19.2. Explain the notion of section, line and path as entities in an SDH transmission network. Taking an STS-1 frame as an example, where is the information concerning these entities carried in the SDH signal? What is the nature of the information?

19.3. What mechanism is used to allow an SDH SPE to pass between SDH networks which are not synchronised? How does this differ from the mechanism used to allow tributaries from the old plesiochronous system (e.g. 2.048 Mbit/s) to be taken in and out of an SDH system?

19.4. Explain the term add-drop in the context of multiplexers. Draw block diagrams to show why this function is simpler to perform in the SDH system than in the older PDH system.

APPENDIX A

Tabulated values of the error function

$$\text{erf}(x) = \frac{2}{\sqrt{\pi}} \int_0^x e^{-y^2}\, dy$$

x	erf x	x	erf x	x	erf x	x	erf x
0.00	0.000000	0.40	0.428392	0.80	0.742101	1.20	0.910314
0.01	0.011283	0.41	0.437969	0.81	0.748003	1.21	0.912956
0.02	0.022565	0.42	0.447468	0.82	0.753811	1.22	0.915534
0.03	0.033841	0.43	0.456887	0.83	0.759524	1.23	0.918050
0.04	0.045111	0.44	0.466225	0.84	0.765143	1.24	0.920505
0.05	0.056372	0.45	0.475482	0.85	0.770668	1.25	0.922900
0.06	0.067622	0.46	0.484655	0.86	0.776100	1.26	0.925236
0.07	0.078858	0.47	0.493745	0.87	0.781440	1.27	0.927514
0.08	0.090078	0.48	0.502750	0.88	0.786687	1.28	0.929734
0.09	0.101282	0.49	0.511668	0.89	0.791843	1.29	0.931899
0.10	0.112463	0.50	0.520500	0.90	0.796908	1.30	0.934008
0.11	0.123623	0.51	0.529244	0.91	0.801883	1.31	0.936063
0.12	0.134758	0.52	0.537899	0.92	0.806768	1.32	0.938065
0.13	0.145867	0.53	0.546464	0.93	0.811564	1.33	0.940015
0.14	0.156947	0.54	0.554939	0.94	0.816271	1.34	0.941914
0.15	0.167996	0.55	0.563323	0.95	0.820891	1.35	0.943762
0.16	0.179012	0.56	0.571616	0.96	0.825424	1.36	0.945561
0.17	0.189992	0.57	0.579816	0.97	0.829870	1.37	0.947312
0.18	0.200936	0.58	0.587923	0.98	0.834232	1.38	0.949016
0.19	0.211840	0.59	0.595936	0.99	0.838508	1.39	0.950673
0.20	0.222703	0.60	0.603856	1.00	0.842701	1.40	0.952285
0.21	0.233522	0.61	0.611681	1.01	0.846810	1.41	0.953852
0.22	0.244296	0.62	0.619411	1.02	0.850838	1.42	0.955376
0.23	0.255023	0.63	0.627046	1.03	0.854784	1.43	0.956857
0.24	0.265700	0.64	0.634586	1.04	0.858650	1.44	0.958297
0.25	0.276326	0.65	0.642029	1.05	0.862436	1.45	0.959695
0.26	0.286900	0.66	0.649377	1.06	0.866144	1.46	0.961054
0.27	0.297418	0.67	0.656628	1.07	0.869773	1.47	0.962373
0.28	0.307880	0.68	0.663782	1.08	0.873326	1.48	0.963654
0.29	0.318283	0.69	0.670840	1.09	0.876803	1.49	0.964898
0.30	0.328627	0.70	0.677801	1.10	0.880205	1.50	0.966105
0.31	0.338908	0.71	0.684666	1.11	0.883533	1.51	0.967277
0.32	0.349126	0.72	0.691433	1.12	0.886788	1.52	0.968413
0.33	0.359279	0.73	0.698104	1.13	0.889971	1.53	0.969516
0.34	0.369365	0.74	0.704678	1.14	0.893082	1.54	0.970586
0.35	0.379382	0.75	0.711156	1.15	0.896124	1.55	0.971623
0.36	0.389330	0.76	0.717537	1.16	0.899096	1.56	0.972628
0.37	0.399206	0.77	0.723822	1.17	0.902000	1.57	0.973603
0.38	0.409009	0.78	0.730010	1.18	0.904837	1.58	0.974547
0.39	0.418739	0.79	0.736103	1.19	0.907608	1.59	0.975462

x	erf x	x	erf x	x	erf x	x	erf x
1.60	0.976348	2.10	0.997021	2.60	0.999764	3.10	0.99998835
1.61	0.977207	2.11	0.997155	2.61	0.999777	3.11	0.99998908
1.62	0.978038	2.12	0.997284	2.62	0.999789	3.12	0.99998977
1.63	0.978843	2.13	0.997407	2.63	0.999800	3.13	0.99999042
1.64	0.979622	2.14	0.997525	2.64	0.999811	3.14	0.99999103
1.65	0.980376	2.15	0.997639	2.65	0.999822	3.15	0.99999160
1.66	0.981105	2.16	0.997747	2.66	0.999831	3.16	0.99999214
1.67	0.981810	2.17	0.997851	2.67	0.999841	3.17	0.99999264
1.68	0.982493	2.18	0.997951	2.68	0.999849	3.18	0.99999311
1.69	0.983153	2.19	0.998046	2.69	0.999858	3.19	0.99999356
1.70	0.983790	2.20	0.998137	2.70	0.999866	3.20	0.99999397
1.71	0.984407	2.21	0.998224	2.71	0.999873	3.21	0.99999436
1.72	0.985003	2.22	0.998308	2.72	0.999880	3.22	0.99999473
1.73	0.985578	2.23	0.998388	2.73	0.999887	3.23	0.99999507
1.74	0.986135	2.24	0.998464	2.74	0.999893	3.24	0.99999540
1.75	0.986672	2.25	0.998537	2.75	0.999899	3.25	0.99999570
1.76	0.987190	2.26	0.998607	2.76	0.999905	3.26	0.99999598
1.77	0.987691	2.27	0.998674	2.77	0.999910	3.27	0.99999624
1.78	0.988174	2.28	0.998738	2.78	0.999916	3.28	0.99999649
1.79	0.988641	2.29	0.998799	2.79	0.999920	3.29	0.99999672
1.80	0.989091	2.30	0.998857	2.80	0.999925	3.30	0.99999694
1.81	0.989525	2.31	0.998912	2.81	0.999929	3.31	0.99999715
1.82	0.989943	2.32	0.998966	2.82	0.999933	3.32	0.99999734
1.83	0.990347	2.33	0.999016	2.83	0.999937	3.33	0.99999751
1.84	0.990736	2.34	0.999065	2.84	0.999941	3.34	0.99999768
1.85	0.991111	2.35	0.999111	2.85	0.999944	3.35	0.999997838
1.86	0.991472	2.36	0.999155	2.86	0.999948	3.36	0.999997983
1.87	0.991821	2.37	0.999197	2.87	0.999951	3.37	0.999998120
1.88	0.992156	2.38	0.999237	2.88	0.999954	3.38	0.999998247
1.89	0.992479	2.39	0.999275	2.89	0.999956	3.39	0.999998367
1.90	0.992790	2.40	0.999311	2.90	0.999959	3.40	0.999998478
1.91	0.993090	2.41	0.999346	2.91	0.999961	3.41	0.999998582
1.92	0.993378	2.42	0.999379	2.92	0.999964	3.42	0.999998679
1.93	0.993656	2.43	0.999411	2.93	0.999966	3.43	0.999998770
1.94	0.993923	2.44	0.999441	2.94	0.999968	3.44	0.999998855
1.95	0.994179	2.45	0.999469	2.95	0.999970	3.45	0.999998934
1.96	0.994426	2.46	0.999497	2.96	0.999972	3.46	0.999999008
1.97	0.994664	2.47	0.999523	2.97	0.999973	3.47	0.999999077
1.98	0.994892	2.48	0.999547	2.98	0.999975	3.48	0.999999141
1.99	0.995111	2.49	0.999571	2.99	0.999977	3.49	0.999999201
2.00	0.995322	2.50	0.999593	3.00	0.99997791	3.50	0.999999257
2.01	0.995525	2.51	0.999614	3.01	0.99997926	3.51	0.999999309
2.02	0.995719	2.52	0.999635	3.02	0.99998053	3.52	0.999999358
2.03	0.995906	2.53	0.999654	3.03	0.99998173	3.53	0.999999403
2.04	0.996086	2.54	0.999672	3.04	0.99998286	3.54	0.999999445
2.05	0.996258	2.55	0.999689	3.05	0.99998392	3.55	0.999999485
2.06	0.996423	2.56	0.999706	3.06	0.99998492	3.56	0.999999521
2.07	0.996582	2.57	0.999722	3.07	0.99998586	3.57	0.999999555
2.08	0.996734	2.58	0.999736	3.08	0.99998674	3.58	0.999999587
2.09	0.996880	2.59	0.999751	3.09	0.99998757	3.59	0.999999617

x	erf x	x	erf x	x	erf x	x	erf x
3.60	0.999999644	3.70	0.999999833	3.80	0.999999923	3.90	0.999999965
3.61	0.999999670	3.71	0.999999845	3.81	0.999999929	3.91	0.999999968
3.62	0.999999694	3.72	0.999999857	3.82	0.999999934	3.92	0.999999970
3.63	0.999999716	3.73	0.999999867	3.83	0.999999939	3.93	0.999999973
3.64	0.999999736	3.74	0.999999877	3.84	0.999999944	3.94	0.999999975
3.65	0.999999756	3.75	0.999999886	3.85	0.999999948	3.95	0.999999977
3.66	0.999999773	3.76	0.999999895	3.86	0.999999952	3.96	0.999999979
3.67	0.999999790	3.77	0.999999903	3.87	0.999999956	3.97	0.999999980
3.68	0.999999805	3.78	0.999999910	3.88	0.999999959	3.98	0.999999982
3.69	0.999999820	3.79	0.999999917	3.89	0.999999962	3.99	0.999999983

For x equal to, or greater than, 4 the following approximation may normally be used:

$$\mathrm{erf}(x) \;\simeq\; 1 - \frac{e^{-x^2}}{\sqrt{\pi}\,x}$$

Some *complementary* error function values for large x are:

x	erfc x
4.0	1.59×10^{-8}
4.1	6.89×10^{-9}
4.2	2.93×10^{-9}
4.3	1.22×10^{-9}
4.4	5.01×10^{-10}
4.5	2.01×10^{-10}
4.6	7.92×10^{-11}
4.7	3.06×10^{-11}
4.8	1.16×10^{-11}
4.9	4.30×10^{-12}

References

Aghvami, H., "Digital Modulation Techniques for Mobile and Personal Communication Systems", *Electronics and Communication Engineering Journal*, Vol. 5, No. 3, pp. 125-132, June 1993.

Alard, M. and Lassalle, R., "Principles of Modulation and Channel Coding for Digital Broadcast for Mobile Receivers", *EBU Collected Papers on Sound Broadcast into the 21st Century*, pp. 47-69, August 1988.

Alexander. S.B., *Optical Communications: Receiver Designs*, IEE, London, 1997.

Arfken, G.B. and Weber, H.J., *Mathematical Methods for Physicists*, (4th edition), Academic Press 1995.

Baker, G., "High-bit-rate Digital Subscriber Lines", *Electronics and Communication Engineering Journal*, Vol. 5, No. 5, pp. 279-283, October 1993.

Beauchamp, K.G., *Walsh Functions and Their Applications*, Academic Press, 1975.

Biglieri, E., Divsalar, D., McLane, P.J. and Simon, M., *Introduction to Trellis Coded Modulation with Applications*, Macmillan, 1991.

Black, U., *Mobile and Wireless Networks*, Prentice Hall, 1996.

Blahut, R.E., *Theory and Practice of Error Control Codes*, Addison-Wesley, 1983.

Blahut, R.E., *Principles and Practice of Information Theory*, Addison-Wesley, 1987.

Blahut, R.E., *Digital Transmission of Information*, Addison-Wesley, 1990.

Bracewell, R., *The Fourier Transform and its Application*, McGraw-Hill, 1975.

Brigham, E.O., *The Fast Fourier Transform and its Applications*, Prentice Hall, 1988.

Burr, A.G., "Trellis Coded Modulation", *Electronics and Communication Engineering Journal*, Vol. 5, No. 4, pp. 240-248, August 1993.

Carlson, A.B., *Communication Systems: An Introduction to Systems and Noise*, (3rd edition), McGraw-Hill, 1986.

CCIR Report 338-5, *Propagation Data and Prediction Methods Required for Terrestial Line-of-Sight Systems*, available from International Telecommunications Union, 1990.

CCIR Report 718-2, *Effects of Tropospheric Refraction on Radio-Wave Propagation*, available from International Telecommunications Union, 1990.

CCIR Report 719-2, *Attenuation by Atmospheric Gasses*, available from International Telecommunications Union, 1986.

CCIR Report 720, *Radio Emission associated with Absorption by Atmospheric Gasses and Precipitation*, available from International Telecommunications Union, 1986.

CCIR Report 721-2, *Attenuation by Hydrometeors, in particular Precipitation, and other Atmospheric Particles*, available from International Telecommunications Union, 1986.

CCIR *Handbook of Satellite Communications (Fixed-Satellite Service)*, available from International Telecommunications Union, 1988.

Cerf, V.G. and Kahn, R., "A Protocol for Packet Network Interconnect", *IEEE Transactions on Communications*, COM-22, No. 5, pp.637-648, May 1974.

Chung, K.L., *Markov Chains with Stationary Transition Probabilities*, Springer Verlag, 1967.

Clark, G.C. and Cain, J.B., *Error-Correction Coding for Digital Communications*, Plenum Press, 1981.

Clarke, A.C. "Extra-terrestial Relays", *Wireless World*, 1945, reprinted in Microwave Systems News, Vol. 15, No. 9, August 1985.

Clarke, R.J., *Transform Coding of Images*, Academic Press, 1985.

Cochrane, P., "Future Directions in Longhaul Fibre Optic Systems", *British Telecom Technology Journal*, Vol. 8, No. 2, pp. 1-17, April 1990.

Cochrane, P., Heatley, D.J.T., Smyth, P.P. and Pearson, I.D., "Optical Communications – Future Prospects", *Electronics and Communication Engineering Journal*, Vol. 5, No. 4, pp. 221-232, August 1993.

Collin, R.E., *Antennas and Radiowave Propagation*, McGraw-Hill, 1985.

Comer, D.E., *Internetworking with TCP/IP*, Volume I, (2nd edition), Prentice Hall, 1991.

COST 210, "Influence of the Atmosphere on Interference between Radiocommunications Systems at Frequencies above 1 GHz", *Commission of the European Communities Report*, EUR 13407 EN, 1991.

Cover, T.M. and Thomas, J.A., *Elements of Information Theory*, Wiley-Interscience, 1991.

Dasilva, J.S., Ikonomou, D. and Erben, H., "European R&D Programs on Third-Generation Mobile Communication Systems", *IEEE Personal Communications*, Vol. 4, No. 1, pp. 46-52, February 1997.

Davie, M.C. and Smith, J.B., "A Cellular Packet Radio Data Network", *Electronics and Communication Engineering Journal*, Vol. 3, No. 3, pp. 137-143, June 1991.

de Prycker, M., *Asynchronous Transfer Mode*, Ellis Horwood, 1991.

Dixon, R.C., *Introduction to Spread Spectrum Systems*, John Wiley, 1993.

Doughty, H. T. and Maloney, L. J., "Application of Diffraction by Convex Surfaces to Irregular Terrain situations", *Radio Science*, Vol. 68D, No. 2, February 1964.

Dunlop, J. and Smith, D.G., *Telecommunications Engineering*, Chapman and Hall, 1989.

Dwight, H.B., *Tables of Integrals and other Mathematical Data*, (4th edition), Macmillan, 1961.

Earnshaw, C.M., Chapter 1 in Flood, J.E. and Cochrane, P. (eds), *Transmission Systems*, Peter Peregrinus, 1991.

Everett, J.L. (ed), *VSATs*, Peter Peregrinus, 1992.

Falconer, R. and Adams, J., "ORWELL: a Protocol for an Integrated Services Local Network", *British Telecom Technology Journal*, Vol. 3, No. 1, pp. 27-35, 1985.

Farrell, P.G., "Coding as a cure for Communication Calamities", *Electronics and Communication Engineering Journal*, Vol. 2, No. 6, pp. 213-220, December 1990.

Feher, K., *Digital Communications – Microwave Applications*, Prentice Hall, 1981.

Feher, K., *Digital Communications – Satellite/Earth Station Engineering*, Prentice Hall, 1983.

Fletcher, J.G., "Arithmetic Checksum for Serial Transmission", *IEEE Transactions on Communications*, COM-30, No. 1, Part 2, pp. 247-252, January 1982.

Flood, J.E. and Cochrane, P., (eds), *Transmission Systems*, Peter Peregrinus, 1991.

Fogarty, K.D., "Introduction to CCITT I. Series Recommendations", *British Telecom Technology Journal*, Vol. 6, No. 1, pp. 5-13, January 1988.

Forney, D.G., Gallacher, R.G., Lang, G.R., Longstaff, F.M. and Qureshi, S.U., "Efficient Modulation for Band-limited Channels", *IEEE Journal on Selected Areas in Communications*, SAC-2, No. 5, pp. 632-647, 1984.

Forrest, J.R., "Special Issue on: Digital Video Broadcasting", *Electronics and Communications Engineering Journal*, Vol. 9, No. 1, February 1997.

Freeman, R.L., *Telecommunication Transmission Handbook*, (3rd edition), John Wiley, 1996.

Furui, S., *Digital Processing, Synthesis, and Recognition*, Marcel Dekker, 1989.

Gelenbe, E. and Pujolle, G., *Introduction to Queueing Networks*, John Wiley, 1987.

Golomb, S.W., *et al., Digital Communications with Space Applications*, Prentice Hall, 1964.

Gower, J., *Optical Communication Systems*, (2nd edition), Prentice Hall, 1993.

Grant, P.M., Cowan, C.F.N., Mulgrew B. and Dripps, J.H., *Analogue and Digital Signal Processing and Coding*, Chartwell Bratt, 1989.

Gray, A.H. and Markel, J.D., *Linear Prediction of Speech*, Springer-Verlag, 1976.

Griffiths, J.M., *ISDN Explained*, John Wiley, 1992.

Grob, B., *Basic TV and Video*, McGraw-Hill, 1984.

Ha, T. T., *Digital Satellite Communications,* (2nd edition), McGraw-Hill, 1990.

Hall, M.P.M. and Barclay, L.W. (eds), *Radiowave propagation*, Peter Peregrinus, on behalf of IEE, 1989.

Halls, G.A., "HIPERLAN: The High Performance Radio Local Area Network Standard", *Electronics and Communication Engineering Journal*, Vol. 6, No. 6, pp. 289-296, December 1994.

Hanks, P. (ed.), *Collins Dictionary of the English Language*, (2nd edition), Collins, 1986.

Harmuth, H.F., *Sequency Theory: Fundamentals and Applications*, Academic Press, 1977.

Harrison, F. G., "Microwave Radio in the BT Access Network", *British Telecommunications Engineering*, Vol. 8, pp. 100-106, July 1989.

Hartley, R.V.L., "Transmission of Information", *Bell System Technical Journal*, Vol. 7, No. 3, pp. 535-563, July 1928.

Hawker, I., "Future Trends in Digital Telecoms Transmission Networks", *Electronics and Communication Engineering Journal*, Vol. 2, No. 6, pp. 251-260, December 1990.

Hioki, W., *Telecommunications*, (2nd edition), Prentice Hall, 1995.

Hirade, K. and Murota, K., "A Study of Modulation for Digital Mobile Telephony", *IEEE 29th Vehicular Technology Conference Proceedings*, pp. 13-19, March 1979.

Honary, B. and Markarian, G., *Trellis decoding of Block Codes*, Kluwer, 1996.

Hooijmans, P.W., *Coherent Optical Systems Design*, John Wiley, 1994.

Hughes, L.W., "A Simple Upper Bound on the Error Probability for Orthogonal Signals in White Noise", *IEEE Transactions on Communications*, COM-40, No. 4, p 670, 1992.

Ibrahim, M.F., and Parsons, J. D., "Signal Strength Prediction in built up Areas – Part 1", *Proceedings IEE*, Vol. 130, Part F, pp. 377–384, August 1983.

Irshid, M.I. and Salous, S., "Bit Error Probability for Coherent M-ary PSK Signals", *IEEE Transactions Communications*, COM-39, No. 3, pp 349-352, March 1991.

IS-95-A, *Mobile Station - Base Station Compatibility Standard for Dual-Mode Wideband Spread Spectrum Cellular Systems*, Electronic Industries Association Publication, May 1995.

ITU-R Recommendation PN.396-6, *Reference Atmosphere for Refraction,* International Telecommunications Union, 1994.

ITU-R Recommendation P.453-5, *The Radio Refractive Index: Its Formula and Refractivity Data,* International Telecommunications Union, 1995.

ITU-R Recommendation P.530-6, *Propagation Data and Prediction Methods Required for the Design of Terrestial Line-of-Sight Systems,* International Telecommunications Union, 1995.

ITU-R Recommendation P.618-4, *Propagation Data and Prediction Methods Required for the Design of Earth-Space Telecommunication Systems,* International Telecommunications Union, 1995.

ITU-R Recommendation P.676-2, *Attenuation by Atmospheric Gasses,* International Telecommunications Union, 1995.

ITU-R Recommendation P.838, *Specific Attenuation Model for Rain for Use in Prediction Methods,* International Telecommunications Union, 1992.

Jackson, L.B., *Signals, Systems and Transforms*, Addison-Wesley, 1988.

Jayant, N.S. and Noll, P., *Digital Coding of Waveforms*, Prentice Hall, 1984.

Jeruchim, H.C. *et al.*, "Techniques for Estimating Bit Error Rate in the Simulation of Digital Communication Systems", *IEEE Journal on Selected Areas in Communications*, SAC-2, No. 1, pp. 153-170, 1984.

Jordan, J.R., *Serial Networked Field Bus Instrumentation*, John Wiley, 1995.

Kao, C.K. and Hockman, G.A., "Dielectric-fiber Surface Waveguides for Optical Frequencies", *Proceedings of the IEEE*, Vol. 113, pp. 1151-1158, 1966.

Kerr, D.E., (ed.), *Propagation of Short-radio Waves*, Peter Peregrinus, 1987, (first published by McGraw-Hill, 1951).

Kleinrock, L., *Queueing Systems*, Vols 1 and 2, John Wiley, 1976.

Kraus, J.D., *Radio Astronomy*, McGraw-Hill, 1966 or Cygnus-Quasar Books, 1986.

Lathi, B.P., *Modern Digital and Analog Communication Systems*, Holt Rinehart and Winston, 1989.

Leakey, D., (ed.), "Special Issue on SDH", *British Telecommunications Engineering*, Vol. 10, Part 2, July 1991.

Lee, W. C. Y., *Mobile Communications Design Fundamentals*, John Wiley, 1993.

Lee, W. C. Y., *Mobile Cellular Telecommunication Systems*, (2nd edition), McGraw-Hill, 1995.

Lender, A., "Correlative Coding for Binary Data Transmission", *IEEE Spectrum*, Vol. 3, No. 2, pp. 104-115, February 1966.

Lenk, J.D., *Video Handbook*, McGraw-Hill, 1991.

Liao, S.Y., *Microwave Circuit Analysis and Design*, Prentice Hall, 1987.

Lindsey, W.C. and Simon, M.K., *Telecommunications Systems Engineering*, Prentice Hall, 1973. Reprinted by Dover Press, New York, 1991.

Macario, R.C.V., *Personal and Mobile Radio Systems*, IEE Telecommunications Series No. 25, Peter Peregrinus, 1991.

MacWilliams, F.J. and Sloane N.J.A., *The Theory of Error Correcting Codes*, North Holland, 1977.

Matthews, H., *Surface Wave Filters: Design, Construction and Use*, John Wiley, New York, 1977.

Miki, T. and Siller, C.A., (eds), "Evolution to a Synchronous Digital Network", *IEEE Communications Magazine*, Vol. 28, No. 8, August 1990.

Mulgrew, B. and Grant, P.M., *Digital Signal Processing*, Prentice Hall, 1997.

North, D.O., *Analysis of Factors which Determine Signal-to-Noise Discrimination in Radar*, RCA Labs Report PTR-6c, Princeton NJ, June 1943, or "Analysis of Factors which Determine Signal-to-Noise Discrimination in Pulsed Carrier Systems", *Proceedings of the IEEE*, Vol. 51, pp. 1016-1027, July 1963.

Nussbaumer, H.J., *Computer Communication Systems*, Vol. 1, John Wiley, 1990.

Nyquist, H., "Certain Topics in Telegraphy Transmission Theory", *Transactions of the AIEE*, Vol. 47, pp. 617-644, April 1928.

Oliphant, A., *et al.*, "RACE 1036 – Broadband CPN Demonstrator using Wavelength and Time Division Multiplex", *Electronics and Communication Engineering Journal*, Vol. 4, No. 4, pp. 252-260, August 1992.

Omidyar, C.G. and Aldridge, A., "Introduction to SDH/SONET", *IEEE Communications Magazine*, Vol. 31, No. 9, pp. 30-33, September 1993.

O'Neal, J.B., "Delta Modulation Quantising Noise Analysis and Computer Simulation Results", *Bell System Technical Journal*, Vol. 45, No. 1, pp. 117-148, January 1966

Papoulis, A., *The Fourier Integral and its Applications*, McGraw-Hill, 1962.

Park, S.K. and Miller, K.W., "Random Number Generators are Hard to Find", *Communications of the ACM*, Vol. 32, No. 10, pp. 1192-1201, October 1988.

Parsons, J.D., "Characterisation of Fading Mobile Radio Channels", Chapter 2 in Macario, R.C.V. (ed.), *Personal and Mobile Radio Systems*, Peter Peregrinus, 1991.

Parsons, J.D. and Gardiner, J., *Mobile Communication Systems*, Blackie, 1989.

Pasapathy, S., "MSK: A Spectrally Efficient Technique", *IEEE Communications Magazine*, Vol. 19, No. 4, p. 18, July 1979.

Pratt, T. and Bostian, C.W., *Satellite Communications*, John Wiley, 1986.

Prentiss, S., *High Definition Television*, McGraw-Hill, 1994.

Pugh, A., "Facsimile Today", *Electronics and Communication Engineering Journal*, Vol. 3, No. 5, pp. 223-231, October 1991.

Pursley, M.B. and Roefs, H.F.A., "Numerical Evaluation of Correlation Parameters for Optimal Phases of Binary Shift Register Sequences", *IEEE Transactions on Communications*, COM-27, No. 10, pp. 1597-1604, October 1979.

Ralphs, J.D., *Principles and Practice of Multi-frequency Telegraphy*, IEE Telecommunications Series No. 11, Peter Peregrinus, 1985.

Reeves, A.H., *Pulse Code Modulation Patent*, 1937, or "The Past, Present and Future of Pulse Code Modulation", *IEEE Spectrum*, Vol. 12, pp. 58-63, May 1975.

Rice, S.O.,"Mathematical Analysis of Random Noise", *Bell System Technical Journal*, Vol. 23, pp. 282-333, July 1944 and Vol. 24, pp. 96-157, January 1945. Reprinted in Wax, N., *Selected Papers on Noise and Stochastic Processes,* Dover, New York, 1954.

Rummler, W.D., "A New Selective Fading Model: Application to Propagation Data", *Bell System Technical Journal*, Vol. 58, pp. 1037-1071, 1979.

Rummler, W.D., "Multipath Fading Channel Models for Microwave Digital Radio", *IEEE Communications Magazine*, Vol. 24, No. 11, pp. 30-42, 1986.

Schoenenberger, J.G., "Telephones in the Sky", *Electronics and Communication Engineering Journal*, Vol. 1, No. 2, pp. 81-89, March 1989.

Schouhammer-Immink, K. A., *Coding Techniques for Optical and Magnetic Recording*, Prentice Hall, 1990.

Schwartz, M., *Information, Transmission, Modulation and Noise,* (3rd edition), McGraw-Hill, 1980.

Schwartz, M., *Telecommunication Networks*, Addison-Wesley, 1987.

Shannon, C.E., "A Mathematical Theory of Communication", *Bell System Technical Journal*, Vol. 27, pp. 379-432 and 623-656, 1948, or "Communications in the Presence of Noise", *IRE*, Vol. 37, No. 10, pp. 10-21 , 1949.

Sklar, B., *Digital Communications: Fundamentals and Applications*, Prentice Hall, 1988.

Skov, M., "Protocols for High-speed Networks", *IEEE Communications Magazine*, Vol. 27, No. 6, pp. 45-53, June 1989.

Smith, J., *Modern Communication Circuits*, McGraw-Hill, 1986

Smythe, C., "Networks and Their Architectures", *Electronics and Communication Engineering Journal*, Vol. 3, No. 1, pp. 18-28, February 1991.

Smythe, C., "Local-area Network Interoperability", *Electronics and Communication Engineering Journal*, Vol. 7, No. 4, pp. 141-153, August 1995.

Spiegel, M.R., *Mathematical Handbook for Formulas and Tables*, Schaum Outline Series, McGraw-Hill, 1968.

Stein, S., and Jones, J.J., *Modern Communications Principles with Application to Digital Signalling*, McGraw-Hill, 1967.

Stremler, F.G., *Introduction to Communication Systems*, (3rd edition), Addison-Wesley, 1990.

Strum, R.D. and Kirk, D.E., *First Principles of Discrete Systems and Digital Signal Processing*, Addison-Wesley, 1988.

Sunde, E.D., "Ideal Binary Pulse Transmission in AM and FM", *Bell System Technical Journal*, Vol. 38, pp. 1357-1426, November 1959.

Tanenbaum, A.S., *Computer Networks*, Prentice Hall, 1981.

Taub, H. and Schilling, D.L., *Principles of Communication Systems*, McGraw-Hill, 1986.

Technical demographics – Article on the technology foresight report, *Electronics and Communication Engineering Journal*, Vol. 7, No. 6, pp. 265-271, December 1995.

Temple, S.R., "The ETSI – Four Years On", *Electronics and Communication Engineering Journal*, Vol. 4, No. 4, pp. 177-181, August 1992.

Tranter, W.H. and Kosbar, K.L., "Simulation of Communication Systems", *IEEE Communications Magazine*, Vol. 32, No. 7, pp. 26-35, July 1994.

Turin, G.L., "An Introduction to Digital Matched Filters", *Proceedings of the IEEE*, Vol. 64, No. 7, pp. 1092-1112, July 1976.

Turin, G.L., "Introduction to Spread Spectrum Antimultipath Techniques", *Proceedings of the IEEE*, Vol. 68, No. 3, pp. 328-353, March 1980.

Ungerboeck, G., "Trellis-Coded Modulation with Redundant Signal Sets: Parts 1 & 2", *IEEE Communications Magazine*, Vol. 25, No. 2, pp. 5-20, February 1987.

Welch, T. A., "A Technique for High-Performance Data Compression", *IEEE Computer*, Vol. 17, No. 6, pp. 8-19, June 1984.

Welch, W.J., "Model-Based Coding of Videophone Images", *Electronics and Communication Engineering Journal*, Vol. 3, No. 1, pp. 29-36, February 1991.

Wilkinson, C.F., "X.400 Electronic Mail", *Electronics and Communication Engineering Journal*, Vol. 3, No. 3, pp. 129-138, June 1991.

Yip, P.C.L., *High-Frequency Circuit Design and Measurements*, Chapman and Hall, 1990.

Young, G., Foster, K.T. and Cook, J.W., "Broadband Multimedia Delivery over Copper", *Electronics and Communication Engineering Journal*, Vol. 8, No. 1, pp. 25-36, February 1996.

Young, P.H., *Electronic Communication Techniques*, (3rd edition), Merrill, 1994.

Index